THE INSTITUTE OF ELECTRICAL AND ELECTRONICS ENGINEERS, INC.

CONFERENCE RECORD OF THE 1996 IEEE INTERNATIONAL SYMPOSIUM ON ELECTRICAL INSULATION

VOLUME 2

HÔTEL DU PARC
MONTRÉAL, QUÉBEC
June 16-19, 1996

Sponsored by the IEEE

DIEELECTRICS AND ELECTRICAL INSULATION SOCIETY

Printed in Canada

IEEE Catalog Number 96CH3597-2
ISBN 0-7803-3531-7 (softbound)
ISBN 0-7803-3532-5 (casebound)
ISBN 0-7803-3533-3 (microfiche)
Library of Congress Number 94-76109
ISSN 1089-084X

Printed in Windsor, Ontario by
Preney Print & Litho, Inc.

PRINTED IN CANADA

TABLE OF CONTENTS

SESSION 2B - DIAGNOSTICS: GIS

SESSION 2C - DIAGNOSTICS: Cables

SESSION 3A - ROTATING MACHINES

SESSION 3B - OUTDOOR INSULATION

SESSION 3C - SPACE AND VACUUM INSULATION

SESSION 4A - PD MEASUREMENTS AND DIAGNOSTICS

SESSION 4B - CAPACITORS AND SURGE ARRESTERS

SESSION 5B - AGING AND INSULATION PROPERTIES: CABLES

SESSION 6A - LIQUID BREAKDOWN

SESSION 6B - FLOW ELECTRIFICATION

SESSION 6C - GAS-INSULATED SUBSTATIONS

SESSION 6D - DIELECTRIC MATERIALS

Conference Record of the 1996 IEEE International Symposium on Electrical Insulation, Montreal, Quebec, Canada, June 16-19, 1996

SURFACE CHARGE GROWTH ON POLYETHYLENE TEREPHTHALATE (PET)

B. G. Scully, H. J. Wintle and A. Younsi
Department of Physics, Queen's University,
Kingston, Canada K7L 3N6

INTRODUCTION

The way in which charge is produced on the surface of an insulator, and the mechanism by which it spreads, are fundamental processes which we need to understand in order to assess the electric field in the neighbourhood of a triple junction prior to flashover. Previous studies of surface charge development have given conflicting results. In view of the importance of these effects for the design of satisfactory insulation systems, we have undertaken an experimental study of the dynamics of charge accumulation under controlled conditions. The present work is a continuation of our earlier study [1], but is deliberately confined to a restricted set of experimental variables.

EXPERIMENTAL

The material tested was 19 μm polyethylene terephthalate (PET). The polymer samples were held under rough vacuum for several weeks to remove any volatiles, and then provided with two evaporated Al electrodes. These electrodes were both on the same side of the sample, and were parallel, 3.5 cm long, 5 mm wide, 30 nm thick, and separated by 9 mm. The samples were mounted in a sample holder, washed with alcohol to remove any surface charges acquired during handling, and dried.

The charge buildup between the electrodes was scanned with a capacitive probe located 120 μm above the sample surface. It had a resolution of about 100 μm. The calibration of the probe presents some problems. Simple capacitive voltage division [2] allows the input capacitance of the electrometer to be taken into account, but our probe and sample cannot be represented by plane parallel capacitors. The λ method [3] accounts for the geometry but not for the instrumental effects, so we have combined both methods to represent our situation. The scans were taken with the electrodes temporarily grounded.

The sample was held under vacuum for 24 h, and a baseline scan was run before any voltages were applied. Typically, the electrodes were set at +250 V and -250 V, and nine scans were run at times between 1 min and 480 min after switching on. The data was collected automatically, and the curves presented here have been filtered to remove some high frequency noise.

The sample environment was varied in order to assess which paramaters affected the surface charging. The sample holder had a number of extra electrodes that allowed us to manipulate the shape of the electric field in the neighbourhood of the electrode edges [1]. Some typical arrangements are shown in Fig. 1. UV radiation was carried out with a 4 W short wavelength tube (254 nm) giving an intensity of about 200 μW/cm^2 at the sample plane. Measurements were made both in air and in vacuum.

RESULTS IN THE DARK

We show in Fig 2 some results taken with no UV illumination and no gap. The charge profile shows a homocharge peak close to each electrode. The effect is rather larger in air than in vacuum, and the charge density is of order 4 x 10^{-6} C/m^2. A plane sheet of charge of this density would generate a normal field of order 10^5 V/m, so it is clear that the charge deposited is rather less than the amount needed to cancel the applied field, which is essentially normal to the surface. When a gap is introduced, we get a slight negative charging of order 10^{-7} C/m^2 over the whole surface. Altering the field by applying various voltages to the grids does not materially change the results. This second result is surprising since in this case there is a tangential field of order 5 x 10^4 V/m, larger near the electrodes, and we might expect the injected charge to be both larger and to spread further.

The featureless overall negative charge seen under the gap field presumably arises from a preferential adsorption of negative ions from the surrounding atmosphere (note that we are working with a diffusion pump vacuum). The other possibility that we are seeing some form of Kelvin probe potential due to work function differences is not viable, because there is no sign of it in the other geometry.

The homocharge has already reached its maximum spread of about 1 mm by the time of our first measurement. Because our readings are taken on a logarithmic time scale, it is easy to see that the magnitude of the deposited charge is tending to saturate.

RESULTS WITH UV

By illuminating the samples with UV light, we produced considerable changes in the charging response for some field geometries. The effects are least when there is a gap underneath the sample, and we describe these situations first. In vacuum, the surface charges positively, suggesting that photoionisation of the polymer is taking place. When the field is shaped to enhance the tangential component, then there is some peaking of the charge density near the cathode. These effects are still at the level of 10^{-7} C/m^2. In air ambient, we see as in the dark case a slight overall negative charging, at about the same order of charge density. There is noticeably no injection or other anomaly near to the electrode edges. The implication of these observations is that the polymer again shows an affinity for negative ions formed in the air, but when this source is severely reduced, then the surface can be ionised and the resulting positive charges can migrate. Whether the UV plays a second role in detrapping these positive carriers is unclear.

Much stronger effects occur when there is no gap beneath the sample. In air, we see a strong surface electrification which extends steadily from the cathode across the interelectrode space (Fig 3). This charge is larger and the spreading is more rapid when the the grids are made negative, thus increasing the tangential field towards the anode. There is also evidence for some positive charge accumulation near the anode, but this shows no tendency to spread. Since the field pattern near the cathode edge is almost semicircular, we assume that negative ions formed near the triple junction fall onto the polymer surface, and can then migrate even under the relatively weak lateral field.

In vacuum, the same negative charge production and migration occurs, but there is in addition a large overall positive response, stronger when the upper grid is made negative. This points to a photoionisation of the polymer surface, acting in addition to the charge migration seen in air. At the same time, there are localised charge concentrations near the electrodes, partially masked by the effects occurring across the main part of the exposed surface.

DISCUSSION

It is clear that there are several different charging mechanisms in play. The charge distributions seem to be quite complicated at first sight, but we believe that in fact the separate mechanisms proceed independently of one another. A general feature seems to be that all effects increase with time but seem to be approaching saturation after several hours.

The easiest feature to understand is the positive charging seen in Fig 4. This appears to be due to straightforward photoionisation. The normal field set up by the grid voltage is about 3 x 10^4 V/m and is cancelled by a surface charge of order 3 x 10^{-7} C/m^2. With an adjacent ground plane, the charge needed to cancel the field above the sample increases, as observed. This positive charge appears to be relatively immobile. In an air ambient, this feature does not occur, presumably because the photoelectrons can be captured by air molecules which then chemisorb on to the surface, thus giving little net effect.

The homocharge that appears close to each electrode is produced in the dark, and does not spread. We conclude that these peaks are due to ions generated at or near to the electrode edge and driven round the almost semicircular field lines until they are deposited on the sample surface and trapped. Direct injection can be ruled out since the effects are reduced when the field is parallel to the surface.

Under UV illumination, the negative charge spreads steadily across the interelectrode region, starting from the cathode. We are unable to say whether the source of charge is a gas ionisation near the triple junction as suggested above, or whether there is now a genuine UV assisted injection from the electrode to the surface. However, it is fairly clear that the UV does serve to detrap the surface charges and so assists their migration. A possible alternative explanation would be to invoke two different negative carrier species, one occurring in the dark (and presumably also under illumination), and the second being excited by the UV. Since even under our vacuum conditions there are considerable numbers of air molecules present, we favour the idea that the negative carriers are simply adsorbed anions in both cases.

The motion of these ions under illumination may be due to the tangential electric field, in which case the mobility is of order 5 x 10^{-12} $m^2/(V\ s)$. Alternatively, the motion may be simply diffusive [4], but in either case it is plainly excited by the illumination, and further experimental and numerical study is required to elucidate the mechanism.

The detailed results for the time dependence have helped to clarify our previous work, while the use of specially shaped fields gives further control over the experimental conditions.

ACKNOWLEDGEMENTS

We wish to thank NSERC (Canada), for financial support, DuPont of Canada for the loan of equipment, and many colleagues in several laboratories for helpful discussions.

REFERENCES

[1] Pepin, M.P., A. Younsi and H.J. Wintle. "Surface Charging: Field and Photo Effects". 1993 Annual Report, Conference on Electrical Insulation and Dielectric Phenomena (IEEE, Piscataway, NJ).

[2] Davies, D.K. "The examination of the electrical properties of insulators by surface charge measurement". J. Sci. Instrum. Vol. 44, 1967, pp. 521-524.

[3] Pedersen, A., G.C. Crichton and I.W. McAllister. "The Functional Relation between Partial Discharges and Induced Charge". IEEE Trans. Dielectrics EI. Vol. 2, 1995, pp. 535-543.

[4] Baum, E.A., T.J. Lewis and R. Toomer. "The lateral motion of charge on thin films of polyethylene terephthalate". J. Phys. D Vol. 11, 1978, pp. 963-977

Model Layouts

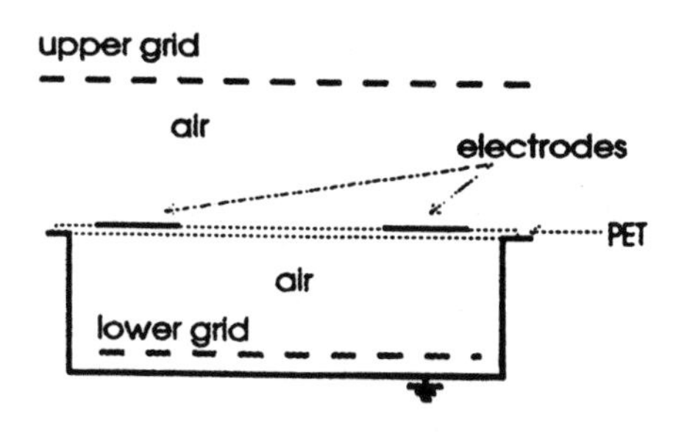

Fig 1

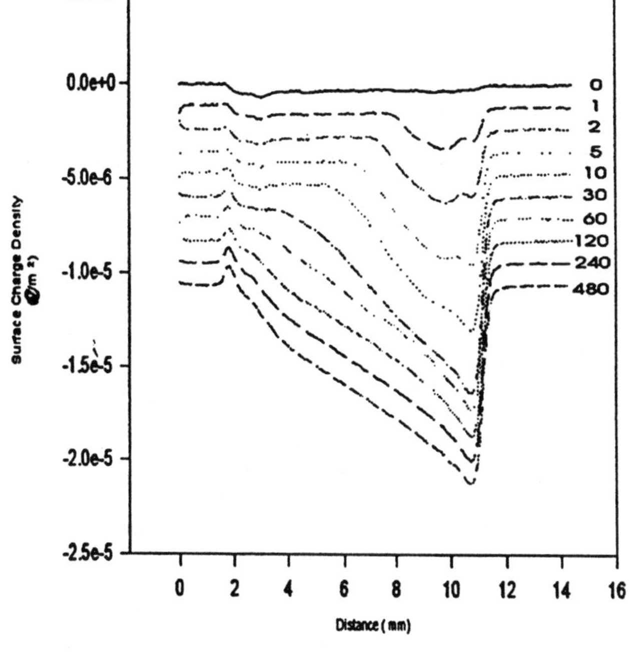

Fig 3

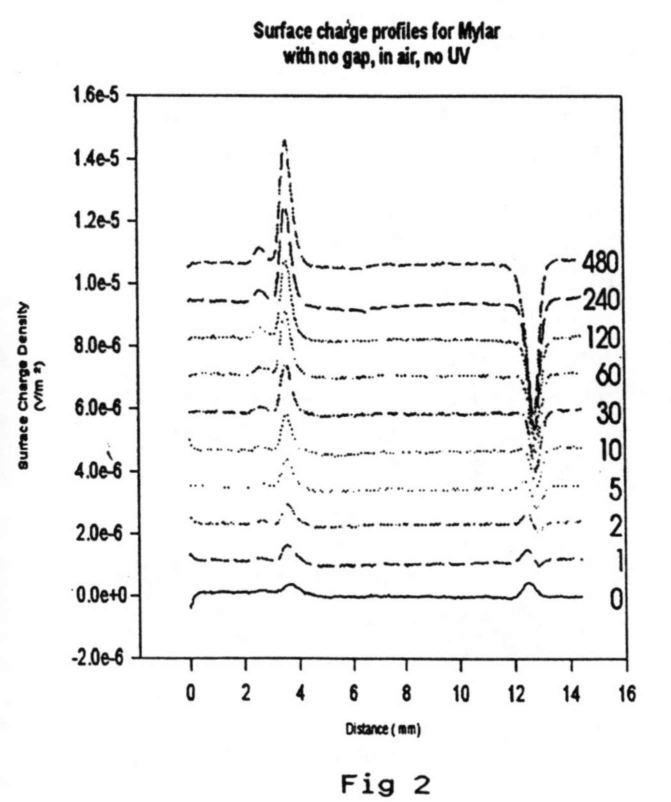

Fig 2

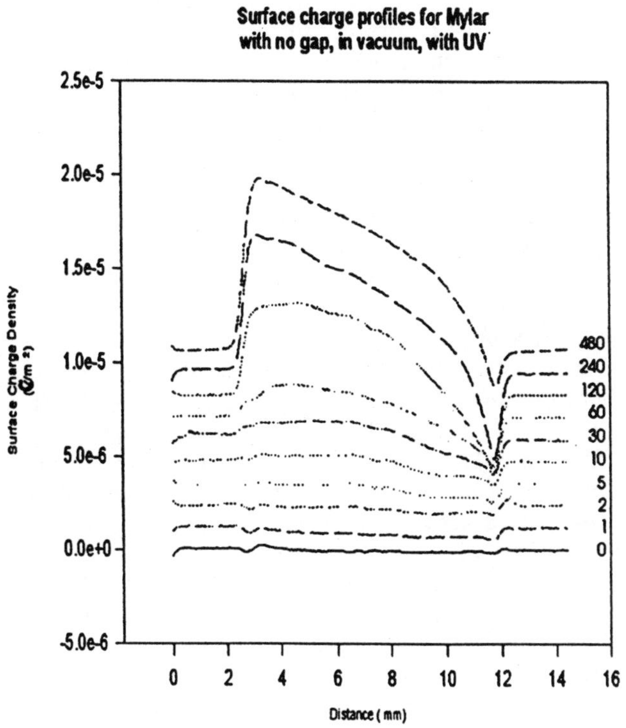

Surface charge profiles for Mylar
with no gap, in vacuum, with UV

2.5e-5
2.0e-5
1.5e-5
1.0e-5
5.0e-6
0.0e+0
-5.0e-6

Surface Charge Density (C/m²)

480
240
120
60
30
10
5
2
1
0

0 2 4 6 8 10 12 14 16

Distance (mm)

Fig 4

Internal Space Charge Distribution Formed by Polymer Solid Electrolytes

K. Fukunaga and T. Maeno
Communications Research Laboratory
4-2-1 Nukui-Kitamachi, Koganei, Tokyo 184, Japan

ABSTRACT: **The internal space charge in solid dielectrics can be observed directly, and is widely discussed dc characteristics of polymer-insulated high-voltage cables. We have experimentally investigated the internal space charge behavior of new anti-electrostatic polymers which include polymer solid electrolytes and additive ion sources and have found that their internal ions form space charge fields that compensate the surface fields of the polymers. The present work focuses on what makes the internal space charges and what determine their distributions. Comparing the space charge distribution of an original anti-electrostatic polymer with that of one without additive ions revealed that the positive charge accumulation near the cathode is caused by the additive anions. Additional experiments using injection-moulded and pressure-moulded specimens indicate that internal space charge behavior is strongly influenced by the polymer structure.**

INTRODUCTION

Improved systems for measuring space charge have been developed, and internal space charge behavior has been widely discussed recently [1-3]. We have developed a high-resolution pulsed-electroacoustic (PEA) method that can observe the space charge distribution in polymer dielectric materials directly during or after the application of a voltage [4], and many researchers have used this method to evaluate the insulating materialsæespecially those used in high-voltage cable systems [5,6]. The results of this research should contribute to the development of high-voltage dc power apparatus and should give information useful in reconsidering the dc test condition.

But because the origin of the internal space charge had not been explained in detail, we have measured the internal space charge behavior of new anti-electrostatic polymers [7, 8]. These materials include polymer solid electrolytes and additive ion sources such as sodium compounds. Although their resistivity is higher than that of other materials using conductive fillers, they prevent electrostatic discharge perfectly. Experimental results indicate that their internal ions forms a space charge field that is orientated in the direction opposite to the applied field and that compensates the surface field of the polymer.

In this paper we discuss two interesting space charge phenomena obtained by using these new polymers. The first is that the occurrence of a space charge distribution depends on the nature of the additive ionic species, and the second is that the space charge behavior is influenced by the polymer structure related to the molding process.

EXPERIMENTS

Table 1 lists the specimens used in this work. Original A and B are practical grade materials used in the manufacturing of electronic devices. These specimens could includes ions in addition to the preselected anions because their polymer solid electrolytes are hydrophilic (so they usually include water in concentrations of ppm order) and because they also contain residues or by-products generated during the reaction process. Polymer solid electrolytes in type-A specimens were blended in polystyrene with miscibility agents, and polymer solid electrolytes in type-B specimens together with some rubber molecules formed a network structure in acrylic resin. Internal space charge distribution was measured by using the high resolution PEA method shown in Fig. 1.

Table I
Specimens

Specimen	Base resin	Polymer solid electrolyte	Additive anion	Moulding process
Original A	Poly-sthyrene	PEG-PA[1]	Na^+	Injection moulding
Anion-free A			—	
Pressure A			Na^+	Presure moulding
Original B	Acrilic resin	PEGMA[2]	K^+	

[1]polyethylene glicole-polyamide copolymer
[2]polyethylene glicole-methacrylate copolymer

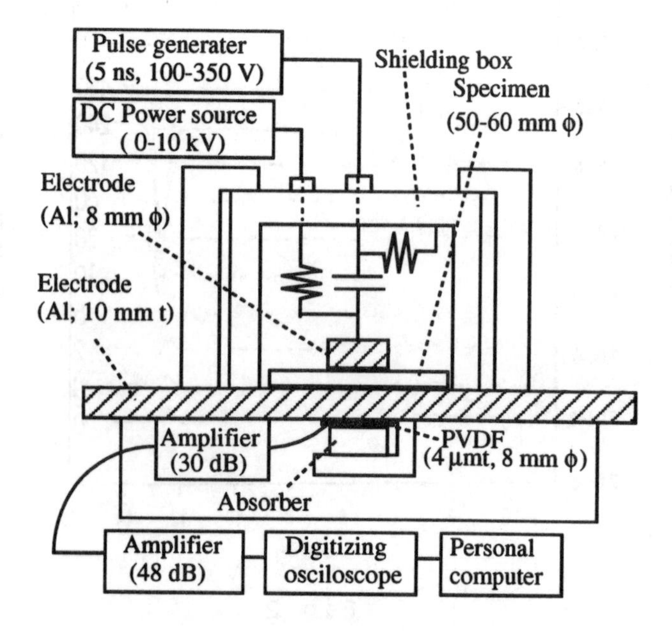

Figure 1. High resolution pulsed-electroacoustic (PEA) measurement system.

SPACE CHARGE ACCUMULATION FORMED BY SODIUM ANIONS

Figure 2 shows the space charge distribution of the specimen Original A under dc electric fields. Internal space charge appears at both of the interfaces between the electrodes and the specimen, resulting in a hetero charge distribution. Since this hetero charge generates a space charge electric field and compensates the applied field, the surface potential can be kept neutral. A space charge distribution also appeared at the center of the specimen. As discussed before in [8], it may have been formed because the specimen was insufficiently uniform.

We also measured the space charge distribution in the anion free-A specimen which was moulded, without the sodium compound, through the same process as used in moulding the Original A specimen. As shown in Fig. 3, the positive charge accumulation at the interface between the cathode and the specimen did not appear. Other internal charges, on the other hand, were observed even in the anion-free A specimen. Thus these charge accumulations were formed by other ion sources, such as water or by-products. These results imply that the positive charge near the cathode (shown in Fig. 2) is produced by the additive sodium anions.

INFLUENCE OF MOULDING PROCESS ON SPACE CHARGE DISTRIBUTION

The anti-ESD polymer of type-A specimens include polymer solid electrolytes blended in polystyrene. Since the polymer solid electrolyte is thought to behave like a solution of ions, additive ions should be able to move easily. In addition, the experimental results shown in the previous section indicate that ions can move by hopping conduction across the interface between the polymer solid electrolyte and polystyrene, or move through micro channels connecting the polymer solid electrolytes. In these cases, the polymer structure is expected to be important. The easiest way to change the polymer structure is by changing the moulding process from injection moulding to pressure moulding, so we compared the results obtained with a pressure-moulded specimen and the results obtained using the Original A specimen.

Figure 4 shows the space charge distribution in the Pressure-A specimen. As shown in this figure, internal space charge did not appear under dc electric field even if the specimen included various ion sources. This charge distribution is the same as that in the base resin polystyrene. The Pressure-A specimen made from the same compound as the original-A did not have anti-ESD performance adequate for practical use.

In general, polymer solid electrolytes exists in thin layers in the injection-moulded polymers æespecially near the surface. Thus the ions in the injection-moulded specimen may be able to move from one layer to another. In the pressure-moulded specimen, on the other hand, the ions can move only in the limited area of the polymer solid electrolyte itself, so that they are difficult to detect by the current measurement method.

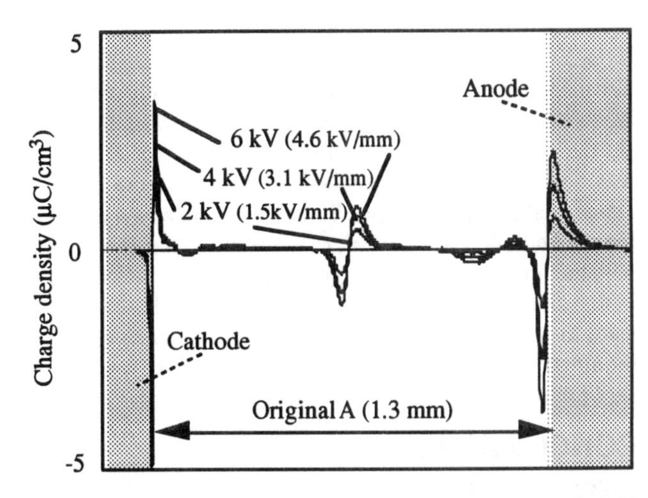

Figure 2 Space charge distribution of Original A under dc electric field.

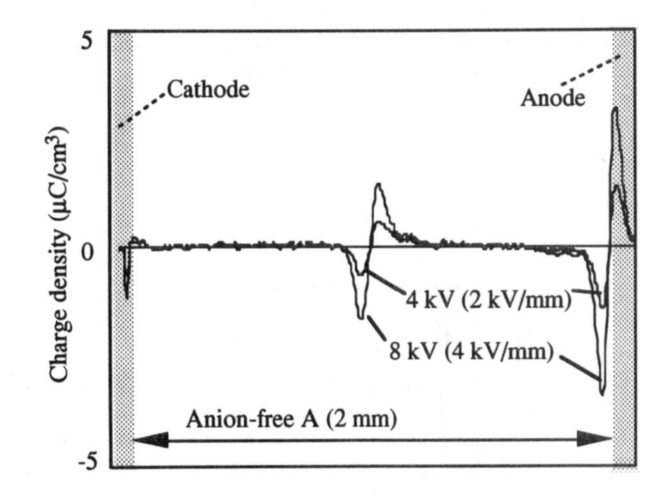

Figure 3 Space charge distribution of Anion-free A under dc electric field.

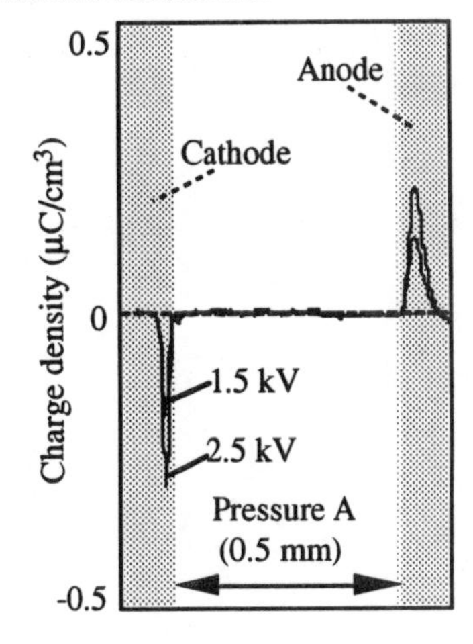

Figure 4 Space charge distribution of Pressure A under dc electric field.

The Original B specimen includes the polymer solid electrolyte formed as a network structure. Figure 5 shows a space charge distribution in the pressure-moulded specimen. Only the positive internal charge accumulates at the interface between the cathode and the specimen. The additive ion source in the type-B specimen was a polymer including potassium ions that can easily be dissociated in the electrolyte. Since the negative ions are part of the polymer, they are difficult to move in the specimen. Thus, only the positive internal charge accumulation can be observed in this case. Here, the most important point in this experiment using the type-B specimen is that ions can move inside a pressure-moulded specimen if the polymer solid electrolyte forms a network structure.

CONCLUSIONS

The internal space charge distribution in an anti-ESD polymer results from the polymer solid electrolyte and the additive ion source. The behavior of preselected additive ions in the polymer can be observed by measuring space charge distribution.

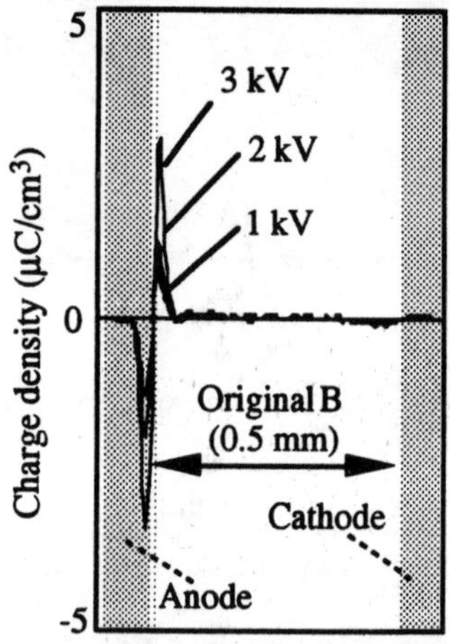

Figure 5 Space charge distribution of Original B under dc electric field.

Internal space charge distribution is strongly influenced by the moulding process because this process determines the polymer structure.

ACKNOWLEDGMENT

We thank Asahi Chemical Industry Co. Ltd. and Kureha Chemical Industry Co. Ltd. for supporting the work described above and for giving permission for the results to be published.

REFERENCES

[1] Li, Y. and T. Takada, "Progress in space charge measurement of solid insulating materials in Japan", IEEE Electrical Insulation Magazine, Vol. 10, no. 5, 1994.

[2] Tornkvist, C., K. Johansson and A. Gustafsson, "Measurement of charge distributions: comparison of the PWP and PEA techniques", Proceedings of 5th ICSD, no. 8. 1, 1995.

[3] Fleming, J., M. Molby Henriksen, M. Henriksen and J. T. Holboll, "LIPP and PEA space charge measurements on LDPE", Proceedings of 5th ICSD, no. 3. 4. 3, 1995.

[4] Maeno, T., K. Fukunaga and T. Takada, "High Resolution PEA Charge Distribution Measurement System", CEIDP Annual Report, no. 2B-7, 1994.

[5] Li, Y., J. Kawai, Y. Ebinuma, Y. Fukiwara, M. Aihara, Y. Yanaka and T. Takada, "Space charge distribution of water tree-degraded polyethylene sheet under ac application", CEIDP Annual Report, no. 2A-1, 1995.

[6] Hozumi, N., H. Suzuki, T. Okamoto, K. Watanabe and A. Watanabe, "Direct observation of time-dependent space charge profiles in XLPE cable under high electric fields", IEEE Trans. on Dielectrics and Electrical Insulation, Vol. 1, no. 6, 1994.

[7] Fukunaga, K. and T. Maeno, "Measurement of the internal space charge distribution of an anti-electrostatic discharge polymer", IEEE Trans. on Dielectrics and Electrical Insulation, Vol. 2, no.1, 1995.

[8] Fukunaga, K. and T. Maeno, "Internal Space Charge Behavior of an Anti-electrostatic Polymer", Proc. of 5th ICSD, no. 12. 5, 1995.

Conference Record of the 1996 IEEE International Symposium on Electrical Insulation, Montreal, Quebec, Canada, June 16-19, 1996

IONIC CHARGE ACCUMULATION AT MICROSCOPIC INTERFACES IN FILLED COMPOSITES

Zhu Yutao Wang Xinheng Xie Hengkun Liu Yaonan
State Key Lab of Electrical Insulation for Power Equipments
Xi'an Jiaotong University, Xi'an, 710049, China

***Abstract*: In this paper the charge accumulation process at microscopic interfaces in insulating materials filled with inorganic fillers is analyzed by using a unit model. Dynamic equations of interfacial ionic charge accumulation are proposed by the authors. The charge accumulation and its regulations are proved by TSC test results obtained on silica filled EPDM samples.**

INTRODUCTION

Composite materials have always been widely used in electrical insulation for HV power equipment, especially polymer composites filled with various inorganic fillers, e.g. composite plastics and rubbers filled with reinforcements, fire retandants and other kinds of fillers applied as cable insulation. Composite insulating materials have many advantages, however problems exists because of the introduction of interfaces between different dielectric phases of the system. Remarkable changes on dielectric and electrical properties after fillers were added have been observed on many polymers, that could include: increased permitivity and loss tangent, decreased volume resistivity and breakdown strength. MWS polarization (or interfacial polarization) models are very commonly used to explain these phenomena[1,2], but it is found not easy for explanation because of a certain lack of connection between traditional models and practical materials. Charges accumulate at microscopic interfaces under electric field when permitivities and volume conductivities of the two kinds of dielectrics differ from each other. Interfacial charge accumulation modifies the absorbing process and influences remarkably the electrical properties of the filled insulating material. In the paper the dynamic process of the interfacial charge accumulation is analyzed by using a kind of unit model from an idealized filler filled system, and TSC measurements are conducted on silica filled EPDM samples, which give proof to interfacial charge accumulation and the regulations achieved by the authors.

In polymer composite filled with inorganic fillers, the polymer and the filler, being as continuous and dispersed phases separatedly, are often insulating materials. There are few electrons on conduction band in insulators under low fields, while ions were introduced into polymer or filler through manufacturing process. Thus in this paper ions are considered as main carriers for conduction in the dielectrics. Inorganic fillers could be crystalline or non-crystalline, whose particless are spheres, discs or needles. But filler particles always aggregate(or agglomerate), as a result they do not appear with uniform or regular shapes when dispersed in polymer matrix, especially at high filling concentration. Figure 1 is the SEM photo of a silica filled EPDM sample made at our lab by the authors.

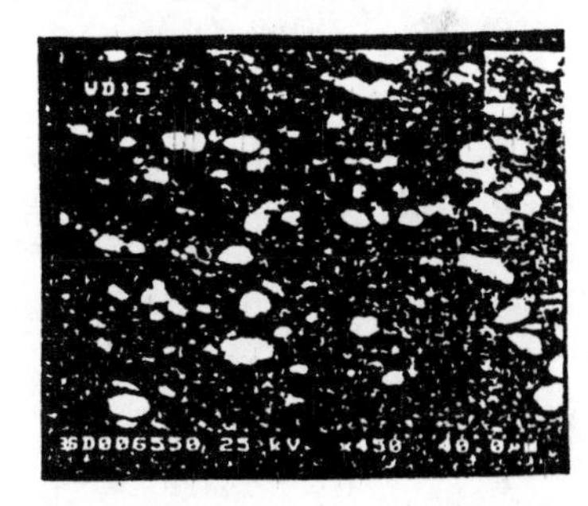

Figure 1 The SEM of silica filled EPDM

An idealized filled system with cubic particles uniformly distributed in matrix is applied for study, as shown in Figure 2.

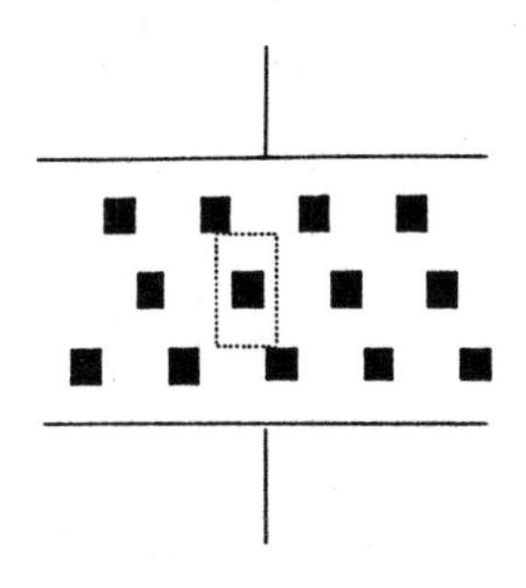

Figure 2 Idealized filled system

DYNAMIC PROCESS

A unit model like that in the frame in Figure 2 can be used for discussion since the total effect of fillers is the result of the field responses of many units like this. When electric field is applied on the unit model in Figure 2 by plain electrodes, different conduction currents are produced in two sides of interfaces due to different values of dielectric constant $(\varepsilon_1, \varepsilon_2)$ and volume conductivity (γ_1, γ_2) of the two kinds of dielectrics. Then ionic charges accumulate at interfaces. In the paper subscripts 1 represent continuous phase—polymer, subscripts 2 represent dispersed phase—filler.

Generally there is $\varepsilon_2 > \varepsilon_1$, but they are in the same order. When $\gamma_2 \gg \gamma_1$, positive and negative ions from dielectric 2(filler) accumulate respectively at interface(A, B) as shown in Figure 3. Suppose the thickness of dielectric 1 is d_1, dielectric 2 d_2, total thickness d, and they satisfy

$$d_2 / d_1 = V_{D2}$$

V_{D2} is the volume density of filler in the composite system.

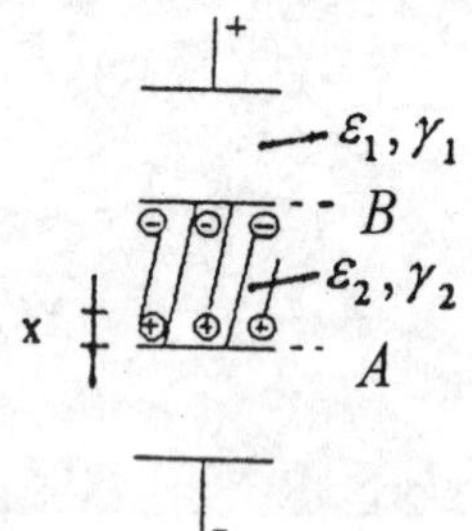

Figure 3 unit Model under field

Assume the positive ions accumulated at interface A near the negative electrode are uniformly distributed in a thickness of x, with a surface density of $\sigma(c/m^2)$ at time t. Then we have:

$$\frac{d\sigma}{dt} = \gamma_2 E_2 - \gamma_1 E_1 \tag{1}$$

where E_1, E_2 are electric strength in the matrix and filler separately。

Equation (1) stands for the accumulating rate of ionic charges at interface A.

The boundary conditions are as follows:

$$\varepsilon_1 E_1 - \varepsilon_2 E_2 = \sigma$$

$$E_1 d_1 + E_2 d_2 = V$$

from which E_1, E_2 can be obtained:

$$E_1 = \frac{\varepsilon_2 V + \sigma d_2}{\varepsilon_1 d_2 + \varepsilon_2 d_1}, \quad E_2 = \frac{\varepsilon_1 V - \sigma d_1}{\varepsilon_1 d_2 + \varepsilon_2 d_1}$$

Considering $\gamma_2 = (N_2 - \frac{\sigma}{x})\mu_2$

where N_2 is the bulk density of ions in dielectric 2 at the beginning moment t=0, μ_2 the movability of ions. Putting E_1, E_2 in (1), we have

$$\frac{d\sigma}{dt} = \frac{V(N_2\mu_2\varepsilon_1 - \gamma_1\varepsilon_2) - (\mu_2\varepsilon_1 V / x + N_2\mu_2 d_1 + \gamma_1 d_2)\sigma + (\mu_2 d_1 / x)\sigma^2}{\varepsilon_1 d_2 + \varepsilon_2 d_1}$$

it can be written as

$$\frac{d\sigma}{dt} = \frac{A\sigma^2 - B\sigma + C}{\varepsilon_1 d_2 + \varepsilon_2 d_1} \tag{2}$$

where $A = \mu_2 d_1 / x$

$$B = \mu_2 \varepsilon_1 V / x + N_2 \mu_2 d_1 + \gamma_1 d_2$$

$$C = V(N_2 \mu_2 \varepsilon_1 - \gamma_1 \varepsilon_2)$$

DISCUSSION

At steady state $(t \to \infty)$

$$\frac{d\sigma}{dt} = 0$$

that is $A\sigma^2 - B\sigma + C = 0$ (3)

It can be proved that $B^2 - 4AC > 0$, so equation (3) has two solutions:

$$\sigma_1 = \frac{B + (B^2 - 4AC)^{1/2}}{2A}$$

$$\sigma_2 = \frac{B - (B^2 - 4AC)^{1/2}}{2A}$$

σ varies with time like this:

$$\sigma(t) = \frac{\sigma_1 \sigma_2 (1 - e^{-R \cdot t})}{\sigma_1 - \sigma_2 e^{-R \cdot t}} \tag{4}$$

where $R = (B^2 - 4AC)^{1/2}$

at time t=0, $\sigma = \sigma_0 = 0$;

when $t \to \infty$, $\sigma = \sigma_\infty$

$$\sigma_\infty = \sigma_2 = \frac{B-(B^2-4AC)^{1/2}}{2A} = \frac{2C}{B+(B^2-4AC)^{1/2}} \quad (5)$$

Detailed discussion on σ are as follows.

(1) Variation of σ with time

$\sigma(t)$ can be written as

$$\sigma(t) = \frac{\sigma_2(1-e^{-R\cdot t})}{1-\frac{\sigma_2}{\sigma_1}e^{-R\cdot t}}$$

generally $\frac{\sigma_2}{\sigma_1} < 1$. The first and second derivatives of σ with respect to time t(σ',σ'')are achieved, $\sigma' > 0, \sigma'' < 0$, that means σ increases with time and at last reaches a saturated value(σ_∞).

(2) The value of σ_∞

Considering the relationships of C, B with applied voltage V , σ_∞ can be written as

$$\sigma_\infty = \frac{XV}{YV+Z} = \frac{X}{Y+Z/V}$$

where X, Y,Z are constants irrelevant to V. The derivative of σ_∞ with t is

$$\sigma_\infty' = \frac{XZ}{(YV+Z)^2}$$

when V increases, σ_∞ also increases but σ_∞' decreases. Relationship between σ_∞ and V are shown in Figure 4.

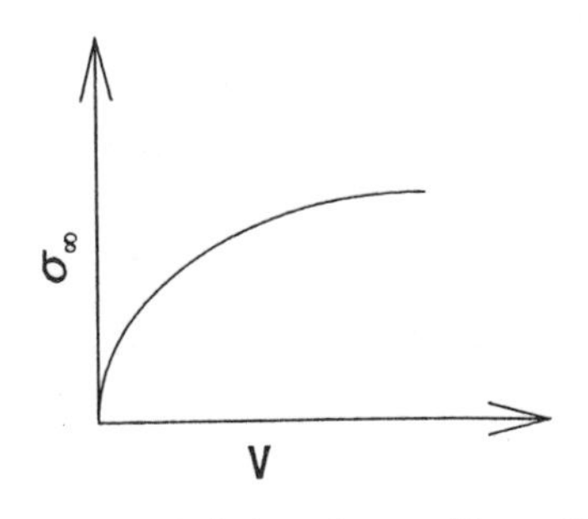

Figure 4 Variation of σ_∞ with applied voltage

TSC TESTS

EPDM used as polymer matrix are composed of molecules with little polarity and stable structure. DSC analyzing results show no remarkable crystalline area. Silica is chosen as filler. It is a kind of reinforcements widely used in rubbers, and can not be replaced by carbon black when applied in insulating composites.

Samples were made as follows: EPDM (Dupont 2722) was mixed with a certain amount of silica on a roll mixer at the temperature 60 ℃ for about 20 minutes, the mixture was then hot pressed into films (about 300μm) at 140 ℃ by a hydraulic press. Silver paste are coated on both sides of the films.

The TSC apparatus was a TOYO SEIKI system, and conventional TSC measurements were conducted in the temperature range from -120 to 120 ℃. The sample was heated at a constant rate of 5 ℃ /min, the current vs. temperature was recorded by a recorder. TSC test results are shown in Figure 5 (polarizing temperature 60℃, 15mins).

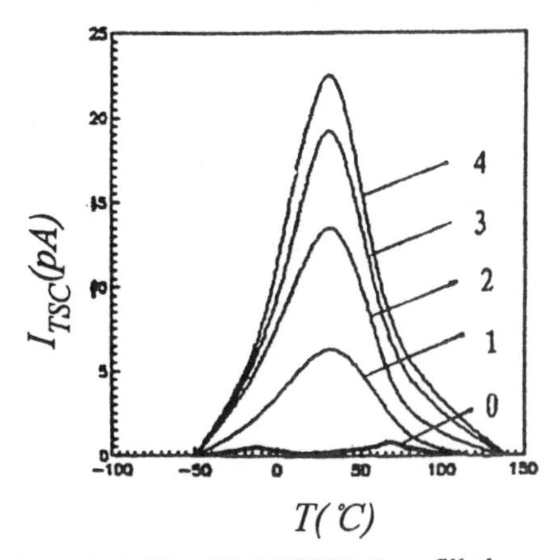

Figure 5 TSC of silica filled EPDM (0: unfilled ssample; 1: filled, polarization voltage 500V; 2: filled, 1000V; 3: filled, 1500V; 4: filled, 2000V)

A large current peak can be seen around 31℃ with filled sample. The peak appeared with the introduction of microscopic interfaces, it was due to the released charges accumulated at interfaces under poling voltage and temperature. When voltage increased TSC values increased too. It was also found through calculation that the augments of the amounts of released charges (equals to the peak area) slowed down with increasing voltage, which obeyed a similar regulation just as that shown in Figure 4 and former discussion. Therefore the theoretical analysis of the dynamic process of interfacial charge accumulation was proved by TSC measurements on filled EPDM samples.

CONCLUSION

In this paper the authors applied a kind of a unit model and analyzed the charge accumulation process at microscopic interfaces in insulating materials filled with inorganic fillers. Dynamic equations of interfacial ionic charge accumulation are proposed , through which it is deduced that the charge density σ increases with time and at last reaches a saturated value(σ_{∞}), and σ_{∞} increases with voltage V. The interfacial charge accumulation will arrive at a steady state at last.

ACKNOWLEDGMENT

The authors wish to thank the Nature Science Foundation of China for the Financial Support of the project "Study on the interfacial dielectric phenomena in insulation system" undertaken at Xi'an Jiaotong University.

REFERENCES

1. Yamanaka, S. and Fukuka, T. "Ultralow-frequency Dielectric Properties of EPR with filler". IEEE Trans. EI. Vol.27, No.6, Dec.1992, pp.1073-1082.
2. Yin, W. Tanaka, J. and Damon, D.H. "A Study of Dielectric relaxation in Aluminosilicate-filled Low-density Polyethylene". IEEE Trans. EI. Vol.1, No.2, April 1994, pp.169-180.

Conference Record of the 1996 IEEE International Symposium on Electrical Insulation, Montreal, Quebec, Canada, June 16-19, 1996

Partial Discharge Measurement for the Investigation of Solid Insulation Ageing

N.Kolev

Faculty of Automation
Technical University of Sofia
1756 Sofia, Bulgaria

Th.Flohr, W. Pfeiffer

Institut für Hochspannungs und Meβtechnik
Technische Hochschule Darmstadt
Landgraf Georg Str. 4
64293 Darmstadt, Germany

Abstract: **The paper deals with the use of partial discharge (PD) measurement for the investigation of solid insulation ageing caused by voltage and temperature stress. Different insulating materials such as insulating foil, insulated wire, insulating tape, and flexible insulating sleeving used in manufacturing of electrical machines were tested. Partial discharge data collected during a certain recording time have been accumulated. Comparisons between degradation caused by voltage stress, degradation caused by thermal stress and degradation due to both factors are given and discussed.**

INTRODUCTION

Solid insulation ageing and failure are initiated by electrical, mechanical, thermal and chemical processes during manufacturing and operation. The existence of local defects within the electrical solid insulation may result in a local overstressing by a non homogeneous electric field causing the occurrence of partial discharges (PD) and leading to a further insulation degradation that reduces the lifetime and reliability of insulating systems. Figure 1 shows the test arrangement which consists of a PD measuring system, HV supply, temperature chamber and computer system. The computerised partial discharge measuring system is presented in [1].

Some experiments were carried out to investigate solid insulation caused by electrical stress, some were carried out separately to investigate ageing caused by thermal stress, and some were performed applying voltage during the thermal treatment of materials.

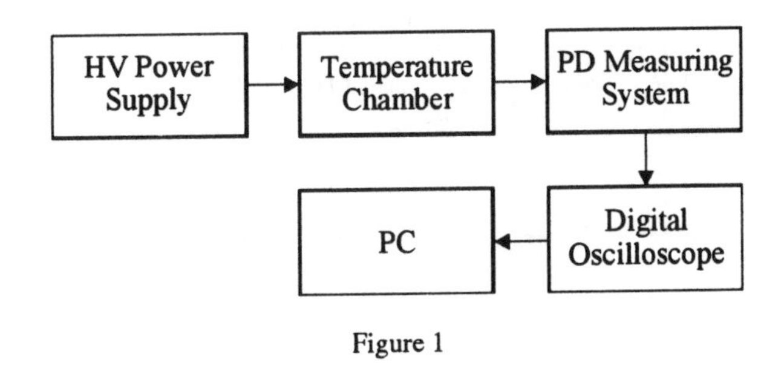

Figure 1

EXPERIMENTS

Electrical Stress of the Insulation

During the electrical ageing tests conventional AC voltage with frequency 50Hz and range from 350V to 2500V (rms) was applied.

Tested materials are showing different discharge activities with the time under electrical stress [2]. Periods of high discharge activities are followed by discharge free intervals. The length and the sequence of the occurrence of the PD pulse packets seems to be chaotic. PD frequency of occurrence and amplitudes are increasing with the increase of voltage but decreasing with the time. Figures 2, 3, 4 and 5 show PD magnitude at certain applied voltages for the different insulating materials. As it indicates amplitudes of the PD are decreasing and extinction voltages are increasing during the electrical stress under standard laboratory environmental conditions.

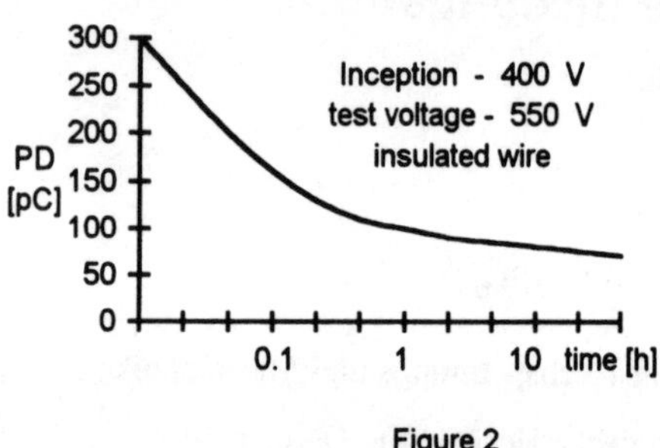

Figure 2

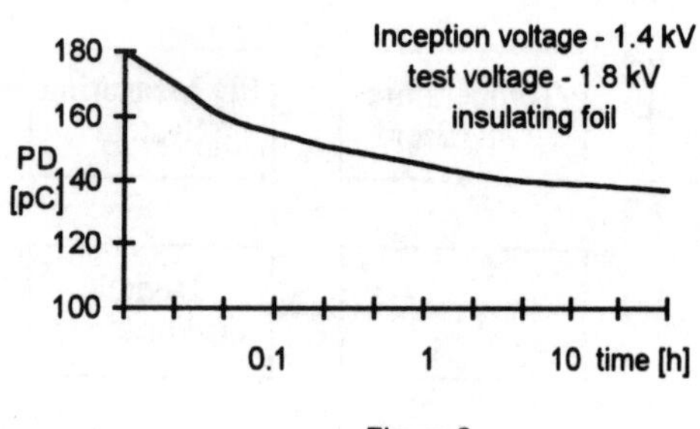

Figure 3

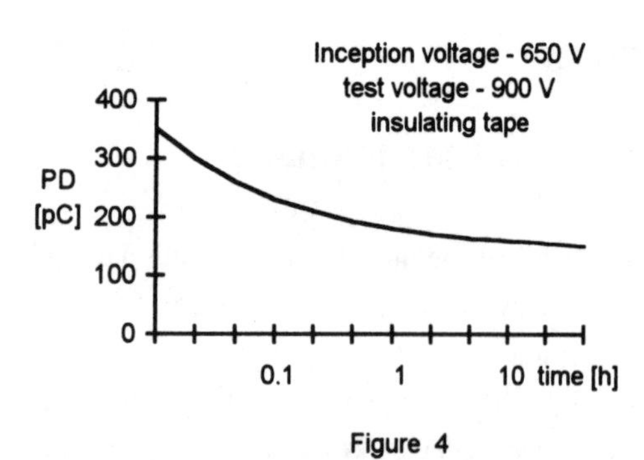

Figure 4

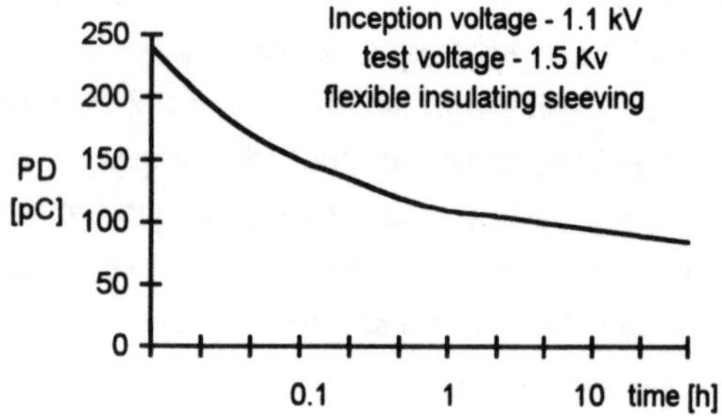

Figure 5

The phase of occurrence of PD tends to 90 and 270 degrees in two to four hours depending on the type of samples after electrical stress was applied. If the sample is disconnected from the HV and the stress is applied again then insulating materials begin to perform as if they have never been under stress. We are assuming that all these effects are due to accumulation of space charge and that charge prevents to some extent material degradation [3].

Thermal Stress of the Insulation

Samples were installed in the temperature chamber and the temperature was adjusted to the desired level. When that level was reached the voltage was increased until the PD started. After recording the PD inception voltage, the test voltage was decreased and the PD extinction voltage was recorded. Thermal treatment was performed within the range of 80 to 120°C and for maximum duration of 100h. At the beginning of the test the inception voltages were similar to those described above (see electrical stress). During the thermal stress the samples are losing their flexibility and became brittle. The review of literature suggests that increasing temperature results in an increase in geometry of voids and the pressure within them [4, 5]. In 24h at the temperature stress at 100°C extinction voltage was twice as high as at the beginning of the test. For the next 24h the extinction voltages were increased by 10 to 20% as shown in Figure 6. These characteristics for the all tested materials are quite similar.

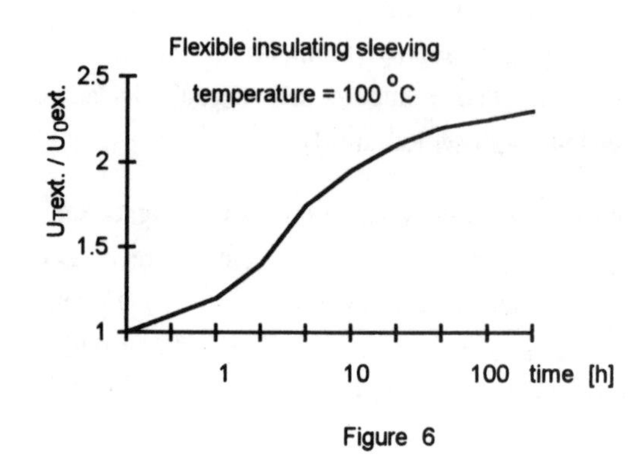

Figure 6

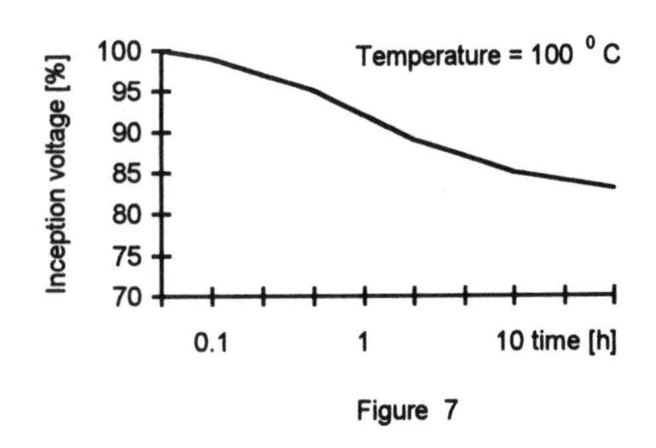

Figure 7

Electrical and Thermal Stress of the Insulation

In practice thermal and voltage stress usually are applied simultaneously. In that case the magnitude of PD signals and their frequency of occurrence is comparatively high. During the 24h test the magnitude of PD signals have increased twice but inception and extinction voltages have decreased by approximately 20% as shown in Figure 5. Obviously this test is the most difficult for the insulating materials. This was proofed by several samples which were completely break down after 24h test.

CONCLUSIONS

Digital signal processing of PD signals registered during the tests of insulating materials indicates that there is no noticeable difference in the signal spectrum. This observation suggests that the test time is probably not sufficient. The type of insulating materials considered seems to require a longer span of time for the ageing processes to be become manifest.

The voltage stress itself does not affect significantly the characteristics and particularly the ageing of insulating materials if the applied voltage is not in excess of the range of PD process.

The most significant changes are observed during the thermal and the voltage - thermal stress as described previously.

ACKNOWLEDGEMENT

This paper is based on the joint research project funded by Volkswagen foundation, Germany.

REFERENCES

1. Buchalla, H., Th. Flohr, W. Pfeiffer, N. Kolev and E. Manov. "Computer Aided Partial Discharge Testing of Electrical Motors and Capacitors". Proceedings: Electrical Electronics Insulation Conference, Chicago'95, September 18-21, 1995, pp. 613 - 617.

2. Patsch, R. and M. Hoof. "Electrical Treeing - Physical Details Obtained by Pulse-Sequence-Analysis". International Conference on Conduction and Breakdown in Solid Dielectrics ICSD'95, Leicester, England, July 10-13 1995.

3. Hikita, M., K. Yamada, A. Nakamura, T. Mizutani, A. Oohasi and M. Ieda. "Measurements of Partial Discharges by Computer and Analysis of Partial Discharge Distribution by the Monte Carlo Method". IEEE Transactions on Electrical Insulation, Vol. 25, No. 3, June 1990, pp. 453-468.

4. Karady, G.G., R.R.Roy. "Effect of Temperature on the Partial Discharge Initiation Voltage of Capacitors". IEEE Transactions on Dielectrics and Electrical Insulation, Vol. 2, No. 3, June 1995, pp. 499-502.

5. Devins, J. C.. "The Physics of Partial Discharges in Solid Dielectrics". IEEE Transactions on Electrical Insulation, Vol. 19, pp. 475-495, 1984.

Conference Record of the 1996 IEEE International Symposium on Electrical Insulation, Montreal, Quebec, Canada, June 16-19, 1996

Estimating Off-line High Voltage Apparatus Dielectric Loss and DC Contact Resistance with On-line Infrared Measurements

James M. Bodah
J. M. Bodah Company
Wilmington, MA 01887

***Abstract*: Simple thermodynamic models can be used to estimate the watt loss of in-service high voltage apparatus. The operating watt loss of many electrical system components can then be analyzed to estimate the dielectric loss or contact resistance. The technique requires temperature data that is normally obtained with an infrared camera, basic apparatus dimensional data, and system conditions at the time of the infrared scan. This paper demonstrates how practical heat transfer calculations can be used to translate on-line thermal test data into estimates of off-line dielectric loss and contact resistance.**

INTRODUCTION

Infrared cameras are used in maintenance programs throughout the utility industry; however, the thermal data provided by this technology has not been fully exploited. Currently, the thermal data is not evaluated to determine the dielectric loss of the insulation or the contact resistance of a suspicious device. Furthermore, temperature limits used to evaluate the serviceability of on-line high voltage apparatus vary widely. Infrared scans can be used in conjunction with practical engineering heat transfer calculations to estimate the on-line watt loss of high voltage apparatus. After calculating the on-line watt loss, a user can estimate the off-line dielectric loss or contact resistance of a device. The estimate can be made by comparison with sister apparatus, previous off-line test results, apparatus nameplate, or manufacturer's recommendations. The calculations provide a scientific interpretation of thermal test data in terms of watts or ohms. In addition, one can use the same technique to establish scientifically based temperature limits for on-line apparatus.

First, a brief discussion of the relevant heat transfer equations is given. Next, the results of dielectric loss and contact resistance predictions for three high voltage components are presented. Last, the author's research is summarized.

HEAT TRANSFER EQUATIONS

The Laws of Thermodynamics, Newton's Law of Cooling, the Stefan-Boltzman Law, and typical heat transfer correlation equations form the basis of this technique. It is necessary to underscore the fact that the following calculations are estimates. At present, electrical energy is the only energy source considered. Additionally, conducted heat is ignored in the given examples. The rationale for ignoring conducted heat is four fold. First, the conducting surface area of most high voltage apparatus is small, as compared to the total surface area. Second, the scatter of experimental data used to develop convection heat transfer coefficient correlation equations is +/-15% or more [1]. Third, the given examples are normally considered serious when the dielectric loss or contact resistance increases by 100% to 300%. Fourth, the calculations are practical, and provide an acceptable degree of accuracy for decision making. Basically, the errors introduced by this simplified method are swamped by the large heat transfer increases experienced in apparatus that are jeopardized by the types of problems addressed.

The First Law of Thermodynamics is incontrovertible. The energy input equals the energy output for a steady state system. Electrical energy that is conducted through a steady state system can be cancelled from both sides of the energy balance equation; therefore, the equation can be written in terms of the electrical energy that is converted to heat by the conductors, contacts and insulation, as stated in (1).

Heat can be transferred across a system boundary in three ways: conduction, convection, and thermal radiation. This discussion ignores conducted heat. Convection, for the purpose of this paper, occurs at a thin film existing at the solid-air interface. The convected heat rate is governed by Newton's Law of Cooling. Thermal radiation is an electromagnetic phenomenon. The net radiated heat transfer rate is determined by the Stefan-Boltzman Law. The Boundary Condition is a combined natural convection-radiation process [2], (2). The energy balance equation can be rewritten, as shown in (3).

$$Q_{IN} = Q_{OUT} \tag{1}$$

Q_{IN} rate of internal heat generation, (W)
Q_{OUT} heat transfer rate to surroundings, (W)

$$Q_{OUT} = Q_{CV} + Q_R \tag{2}$$

Q_{CV} heat transfer rate due to convection, (W)
Q_R heat transfer rate due to thermal radiation, (W)

$$Q_{OUT} = \bar{h}_L A(T_S - T_\infty) + A\epsilon\sigma(T_S^4 - T_\infty^4) \tag{3}$$

$\bar{h}_L$ average convection heat transfer coefficient ($W\,m^{-2}K^{-1}$) subscript L refers to characteristic length
A area (m^2)
T_S surface temperature (K)
T_∞ air temperature (K)
ϵ emissivity
σ Stefan-Boltzman constant ($5.66961 * 10^{-8}\ W\,m^{-2}K^{-4}$)

Calculating the average heat transfer coefficients is key to solving a combined natural convection-radiation problem. It is a three step process. First, the Rayleigh numbers are calculated for the top, side, and bottom of the apparatus (4). The Rayleigh number magnitudes determine whether the film flows are laminar or turbulent. The Rayleigh number is the product of the dimensionless Grashof and Prandtl numbers. The Grashof number[1] has been expanded in (4). Second, appropriate average Nusselt number correlation equations are selected and solved. Selection of an average Nusselt number correlation equation depends upon the magnitude of the Rayleigh number, the surface orientation, and the surface geometry. Equations (5) through (12) are typical average Nusselt number correlation equations. Many Nusselt number correlation equations exist. The average Nusselt number equations presented here were selected for simplicity and consistency in form. Third, the average heat transfer coefficient is calculated for each surface orientation (13).

After calculating the average heat transfer coefficients, the heat transfer rates for the top, side(s), and bottom of the apparatus is estimated using (3). The sum of the heat transfer rates is the estimated on-line watt loss for the apparatus in question.

It is important to note that the film temperature should be used when referring to a table of the thermophysical properties of air. The film temperature is the average of the air and surface temperatures. The thermophysical properties of air used in this paper are calculated from third order interpolation equations [2].

$$Ra_L = g\beta\Delta T L^3 Pr/\upsilon^2 \quad (4)$$

Ra_L	Rayleigh number
g	gravitational acceleration (9.807 m/s^2)
β	coefficient of thermal expansion of air (K^{-1}) using the film temperature
ΔT	temperature difference (K), $T_s - T_\infty$
L	characteristic length (m)
Pr	Prandtl number of air using the film temperature
υ	kinematic viscosity of air (m^2/s) using the film temperature

Vertical plate/cylinder correlation equations:

Laminar flow, $Ra_L \leq 10^8$

$$\overline{Nu_L} = [(0.908\, Pr\, Ra_L)/(0.952 + Pr)]^{1/4} \quad (5)$$

Turbulent flow, $Gr_L \geq 10^9$

$$\overline{Nu_L} = 0.13\, Ra_L^{1/3} \quad (6)$$

[1]Grashof number, $Gr = g\beta\Delta T L^3/\upsilon^2$

Horizontal plate correlation equations:

Top surface laminar flow, $7 \times 10^6 \leq Ra_L \leq 2 \times 10^8$

$$\overline{Nu_L} = 0.16\, Ra_L^{1/3} \quad (7)$$

Top surface turbulent flow, $Ra_L \leq 5 \times 10^8$

$$\overline{Nu_L} = 0.13\, Ra_L^{1/3} \quad (8)$$

Bottom surface laminar and turbulent flows, $10^6 \leq Ra_L \leq 10^{11}$

$$\overline{Nu_L} = 0.58\, Ra_L^{1/5} \quad (9)$$

Horizontal disk correlation equations[2]:

Top surface laminar flow, $7 \times 10^6 \leq Ra_D \leq 2 \times 10^8$

$$\overline{Nu_D} = 0.16\, Ra_D^{1/3} \quad (10)$$

Top surface turbulent flow, $Ra_D \leq 5 \times 10^8$

$$\overline{Nu_D} = 0.13\, Ra_D^{1/3} \quad (11)$$

Bottom surface laminar and turbulent flows, no restrictions

$$\overline{Nu_D} = 0.82\, Ra_D^{1/5}\, Pr^{0.034} \quad (12)$$

$$\overline{h_L} = \overline{Nu_L}\, k/L \quad (13)$$

APPARATUS EXAMPLES

Shunt Bank Capacitor - High Dielectric Loss

This unit is a 14.4 kV 100 kVAR shunt bank capacitor that was reported in [3]. It is an high temperature example that is used to demonstrate the analysis of a box shaped device.

length	0.324 m	top temperature	72°C
width	0.165 m	side temperature	64°C
height	0.648 m	bottom temperature	64°C
ϵ	0.96	air temperature	18°C

The typical dielectric loss for this capacitor is 100 W. Using a dissipation factor test set, the user measured 390 W of loss at operating voltage. The following heat transfer calculations demonstrate how the watt loss is estimated from the thermal data. Equation (4) yields the following Rayleigh numbers:

$Ra_{L\,TOP}$	1.28×10^8
$Ra_{L\,SIDE}$	8.97×10^8
$Ra_{L\,BOTTOM}$	1.16×10^8

[2]The subscript D refers to characteristic diameter.

The Rayleigh numbers suggest that (7), (6), and (9) should be used to determine the average Nusselt numbers, respectively:

$\overline{Nu}_{L\ TOP}$	80.6
$\overline{Nu}_{L\ SIDE}$	125.3
$\overline{Nu}_{L\ BOTTOM}$	23.8

The average heat transfer coefficients are calculated via (13):

$\overline{h}_{L\ TOP}$	6.90
$\overline{h}_{L\ SIDE}$	5.36
$\overline{h}_{L\ BOTTOM}$	2.01

Applying (3) gives the heat transfer rates:

$Q_{OUT\ TOP}$	40.4 W
$Q_{OUT\ 4\ SIDES}$	351.0 W
$Q_{OUT\ BOTTOM}$	21.7 W

The total calculated heat transfer rate is 413 W. This predicted value is 6% above the watt loss measured with instrumentation. The 6% error would not influence the decision to remove the capacitor from service.

Using typical values of capacitance for this device, it is an academic exercise to estimate the values of power factor. Additionally, unlike traditional test methods, the on-line apparatus can be graphically dissected so that the watt loss or power factor can be determined for different parts of the apparatus. That is, for the case of localized heating, the dielectric watt loss can be estimated for the top half of the capacitor, for example. Additionally, it is important to note that the thermal data is obtained at operating voltage, unlike the test data usually obtained by off-line test methods; therefore, voltage sensitive problems may manifest themselves more aggressively.

Oil Circuit Breaker - High Contact Resistance

Tank #1 of a 115 kV 1200 A oil circuit breaker was operating 5°C above tanks #2 and #3. The results of two heat transfer analyses are presented. The tank #1 data is:

diameter	1.22 m	top temperature	21.1°C
height	2.39 m	side temperature	21.1°C
current	490 A	bottom temperature	21.1°C
ϵ	0.96	air temperature	12.0°C

The first method analyzes tank #1 with respect to the ambient air temperature. It yields a total heat transfer of 876 W. The heat is generated by the contacts, conductors, and insulation. In this case it is stated without proof that the predominant heat source is the contacts. It is beyond the scope of this paper to show that this is not an insulation problem. A first approximation of the contact resistance is obtained from Ohm's Law. The estimated contact resistance is 3650 μohm. A second approximation is obtained by subtracting the dielectric loss of the lower portion of the bushings. As a rough guess, half of the bushing losses can be attributed below the bushing flange. Using previous power factor tests results, and extrapolating to the operating voltage, it is estimated that 3.5 W of the tank heat transfer is due to the bushing insulation. The second approximation contact resistance is 3630 μohm.

The second method compares tank #1 to its sister tanks. The assumption is that the sister tanks are normal; therefore, the additional heat transfer experienced by tank #1 must be due to a problem. In this case, the additional heat generated by tank #1 is 521 W. The corresponding additional contact resistance is 2170 μohm. This value is extra contact resistance, as compared to the sister tanks. The circuit breaker was tested off-line. Tank #2 and #3 had a contact resistance of 1000 μohm; therefore, the predicted value of contact resistance for tank #1 is 2170 μohm plus 1000 μohm, or 3170 μohm.

The off-line 100 A contact resistance measurement for tank #1 was 3260 μohm. The first heat transfer method predicted a contact resistance of 3630 μohm, 11% above the off-line test. The second heat transfer method predicted a contact resistance of 3170 μohm, 3% below the off-line test. A practical estimation is achieved with the heat transfer calculations.

Bushing-Overheated Dome Assembly

A 196 kV 2000 A oil circuit breaker bushing was observed with an operating temperature of 19.9°C. The ambient air temperature was -8.9°C. Sister bushings had an operating temperature of 1°C. The purpose for mentioning this item is not to demonstrate the fact that an obvious conductor problem existed in the dome assembly, but to underscore the risk associated with keeping the bushing in service by applying heat transfer calculations..

A first approximation heat transfer analysis of the bushing dome assembly indicates a dome assembly heat transfer rate of 122 W, which corresponds to 473 μohm resistance in the dome. This is approximately 20 times the typical off-line value for this bushing type. Additionally, the circuit breaker manufacturer specifies the maximum pole resistance at 450 μohm. The allowable resistance is exceeded by the single bushing dome assembly.

Since it was strongly desired to delay off-line tests and repairs on the bushing, a recommendation was made to closely monitor the bushing. If the bushing dome assembly temperature approached the manufacturers maximum limit of 70°C then the circuit breaker would be removed from service. Recalling that the ambient temperature at the time was -9.9°C, it later became apparent that heat transfer calculations could also be used to estimate the maximum ambient temperature that could be tolerated given a dome assembly resistance of 450 μohm.

After removing the bushing, the overall off-line resistance was measured; 6300 μohm. It was observed that the resistance decreased non-linearly as test current was increased. This effect may account for the much lower predicted, yet still very high, on-line bushing conductor resistance.

OBSERVATIONS

Twenty eight heat transfer analyses of oil circuit breakers, capacitors, arresters and bushings have been conducted. In every case where overheated conductors/contacts or high dielectric loss occurred the problem was confirmed by a heat transfer analysis of the on-line thermal data. Generally, the experiments predicted the off-line test results within 6% . Two categories of analyses have deviated outside the +/- 6% prediction range. The first category is defects which are non-linear, such as the bushing conductor problem noted earlier. The second category is the insulation of very low power factor devices, for example, condenser type bushings in excellent condition.

Five possible sources of error should be considered. In the author's opinion, one should be primarily concerned with maintaining a normal infrared camera viewing angle and targeting identical spots on sister units for comparison. Next, infrared testing should be conducted at favorable times of the day in order to reduce the influence of the sun. Another unknown is forced convection under windy conditions. Additionally, simplifying the geometry of power apparatus, such as treating an arrester as a cylinder, may be a liability. Finally, although conducted heat can often be ignored, it too is unaccounted. When one is convinced that the calculations and the logic for treating all the contributing sources of heat are sound then consideration should be given to the items mentioned above.

It is instructive to perform the calculations by hand, but only once. The heat transfer, correlation, and interpolation equations can be easily entered into a computer spreadsheet or program. Not only is this efficient, it also provides a means for performing "what if" scenarios and compact data storage.

CONCLUSION

Infrared data collected on operating high voltage apparatus can be used to obtain a practical estimate of the off-line contact/conductor resistance and dielectric watt loss of devices with incipient problems. The analysis is based upon the Laws of Thermodynamics, the combined natural convection-radiation equation, and heat transfer correlation equations. Capacitors, oil circuit breakers, bushings and arresters appear to be good candidates for the technique. It is likely that other apparatus types may also yield acceptable predictions.

The line by line mathematics are simple; however, the number of calculations and tabulated look-up values are tedious. A computer program is essential for conducting regular analyses. A user is not required to have an understanding of thermodynamics when a computer solution is sought. The user only needs to enter the temperature data and apparatus geometry into the program. The necessary computer output is the heat transfer rate, or watt loss. The watt loss is then interpreted by the user from his knowledge of the apparatus and load conditions.

The technique addressed by this paper is not intended to produce precise results; however, it does produce practical results that can help identify apparatus that require further attention by off-line test methods and adds a scientific dimension to the analysis of infrared scans.

REFERENCES

1. Kreith, F. and M.S. Bohn. *Principles of Heat Transfer*. Harper & Row, Publishers, 1986, pp. 245-279

2. Schmidt, F.W., R.E. Henderson, and C.H. Wolgemuth. *Introduction to Thermal Sciences*. John Wiley & Sons, Inc., 1984, pp. 235-309

3. Denis, R.J. and J.M. Stalp. "Expanded Use of Infrared Imaging as a Diagnostic Tool". 1993 Doble Client Conference Minutes. Doble Engineering Company Publication 60AIC93, pp. 2-6.1 - 2-6.6.

4. Incropera, F.P. and D.P. DeWitt. *Introduction to Heat Transfer*. John Wiley & Sons, Inc., 1985, pp. 379-408

5. Bejan, A. *Convection Heat Transfer*. John Wiley & Sons, Inc., 1995, pp. 156-205

6. Faires, V.M. and C.M. Simmang. *Thermodynamics*. Macmillan Publishing Co., Inc., 1978, pp. 516-559

7. Burmeister, L.C. Convective Heat Transfer. John Wiley & Sons, Inc., 1993, pp. 382-444

Conference Record of the 1996 IEEE International Symposium on Electrical Insulation, Montreal, Quebec, Canada, June 16-19, 1996

Optical Tomography of Kerr Electro-Optic Measurements With Axisymmetric Electric Field

A. Üstündag and M. Zahn
Massachusetts Institute of Technology
Department of Electrical Engineering and Computer Science
Laboratory for Electromagnetic and Electronic Systems
Cambridge, MA 02139

Abstract: **Dielectrics become birefringent (Kerr effect) when stressed by high electric fields so that incident linearly or circularly polarized light propagating through the medium becomes elliptically polarized. Most past experimental work has been limited to cases where the electric field magnitude and direction have been constant along the light path, while recent analysis and point/plane electrode measurements have developed the Abel transformation which describes Kerr effect measurements when an axisymmetric electric field has magnitude but not direction varying along the light path.**

The present work develops the governing Kerr effect differential equations for an axisymmetric electric field for the case when both magnitude and direction vary along the light path. The specific case of point/plane electrodes are studied where analytical electric field solutions are used for the space charge free case and finite element computer analyses are used to calculate the electric field distribution for postulated space charge injection from the point electrode. We then calculate the Kerr electro optic fringe patterns that would result. We use the "onion peeling" method previously used for photoelastic analysis to calculate the electric field magnitude and direction from computer-simulated optical measurements.

INTRODUCTION

Kerr effect measurements provide a powerful method to investigate space-charge effects on dielectric liquids. The method has mostly been limited to cases where the applied electric field direction is constant along the light path. In this paper we extend the method for general axisymmetric electric field distributions.

OPTICAL PATTERN PREDICTION

The Kerr effect introduces a refractive index difference between the light electric field components polarized parallel and perpendicular to the applied electric field component, $\mathbf{E}_T$, in the plane perpendicular to the propagation direction of the light (see Figure 1) [1]

$$\Delta n = n_{\parallel} - n_{\perp} = \lambda B E_T^2 \tag{1}$$

Here λ is the free-space light wavelength, B is the Kerr constant and E_T is the magnitude of $\mathbf{E}_T$. We assume that the light electric field is a plane wave that propagates in a straight line in the $+z$ direction through a lossless medium. Because $|\Delta n| << n_{\parallel}$ and $n_{\perp}$, we assume that reflections are negligible. Then source free Maxwell's equations for time harmonic fields ($e^{i\omega t}$ dependence) exactly reduce to [2]

$$\frac{d^2 e_x(z)}{dz^2} = -\mu_0\omega^2\left[\varepsilon_{xx}(z)e_x(z) + \varepsilon_{xy}(z)e_y(z)\right] \tag{2a}$$

$$\frac{d^2 e_y(z)}{dz^2} = -\mu_0\omega^2\left[\varepsilon_{yy}(z)e_y(z) + \varepsilon_{xy}(z)e_x(z)\right] \tag{2b}$$

Here $e_j(z)$ are the complex light electric field components, $\varepsilon_{ij}(z)$ are elements of the dielectric tensor which are real, and $\mu_0=4\pi\times10^{-7}$ henries/meter is the magnetic permeability of free space. Equation (2) governs light propagation in inhomogeneous anisotropic lossless media.

Because the Kerr effect changes the dielectric tensor only slightly

$$\varepsilon_{xy}(z), \varepsilon_{xx}(z)-\varepsilon, \varepsilon_{yy}(z)-\varepsilon \quad << \quad \varepsilon \tag{3}$$

where ε is the isotropic dielectric constant of the medium, we can assume solutions of the form

$$e_j(z) = a_j(z)\exp(-ikz) \qquad j = x,y \tag{4}$$

Here $k=\omega[\varepsilon\mu_0]^{1/2}$ is the isotropic wave number and $a_j(z)$ is a slowly varying amplitude function of z

$$\left|\frac{da_j(z)}{dz}\right| << \left|ka_j(z)\right| \qquad j = x,y \tag{5}$$

which modulates the isotropic solution. Substituting (4) into (2) and using (5) results in the approximate equations for weakly anisotropic media

$$\frac{db_x(z)}{dz} = -i\frac{\omega^2\mu_0}{4k}\left[\varepsilon_d(z)b_x(z) + 2\varepsilon_{xy}(z)b_y(z)\right] \tag{6a}$$

$$\frac{db_y(z)}{dz} = -i\frac{\omega^2\mu_0}{4k}\left[-\varepsilon_d(z)b_y(z) + 2\varepsilon_{xy}(z)b_x(z)\right] \tag{6b}$$

where $\varepsilon_d(z)=\varepsilon_{xx}(z)-\varepsilon_{yy}(z)$ and

$$b_j(z) = a_j(z)\exp\left(i\frac{\omega^2\mu_0}{4k}\int_0^z\left[\varepsilon_{xx}(z') + \varepsilon_{yy}(z') - 2\varepsilon\right]dz'\right) \tag{7}$$

for $j=x,y$. To incorporate the Kerr effect, (6) should be expressed in the $\mathbf{E}_T$ frame (see Figure 1). Substituting the components of $\mathbf{b}$ and the dielectric tensor in the $\mathbf{E}_T$ frame into (6), together with (1) yield

$$\frac{db_{\parallel}(z)}{dz} = -i\pi BE_T^2(z)b_{\parallel}(z) + \frac{d\varphi(z)}{dz}b_{\perp}(z) \tag{8a}$$

$$\frac{db_{\perp}(z)}{dz} = i\pi BE_T^2(z)b_{\perp}(z) - \frac{d\varphi(z)}{dz}b_{\parallel}(z) \tag{8b}$$

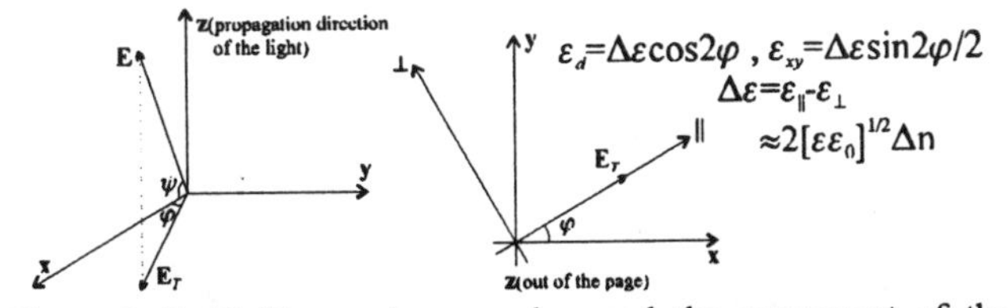

Figure 1 (Left) The angles φ and ψ and the component of the electric field, $\mathbf{E}_T$, in the xy plane perpendicular to the +z direction of light propagation. (Right) $\mathbf{E}_T$ frame of directions parallel and perpendicular to $\mathbf{E}_T$ in the xy plane.

where $b_\parallel = b_x \cos\varphi + b_y \sin\varphi$ and $b_\perp = b_y \cos\varphi - b_x \sin\varphi$ are the vector components of **b** which are respectively parallel and perpendicular to $\mathbf{E}_T$ and we assume that $|\Delta n| << n_\parallel, n_\perp$.

Given the electric field distribution, $\mathbf{E}_T$, and the input light polarization, (8) may be integrated to find the output polarization and consequently the output intensity when various optical polariscope systems are employed. In Figures 2 and 3 we demonstrate the computer generated optical patterns for the point-plane geometry [3] with 1 mm needle to plane distance and needle radius of curvature of 40 μm. For this case study the medium is assumed to be nitrobenzene with Kerr constant $B \approx 3\times10^{-12}$ m/V^2 and the applied voltage is 15 kV. Figures 2 and 3 show the computed optical fringe patterns in the vicinity of the needle when there is no space-charge and when we postulate a step-wise uniform space-charge distribution respectively. A linear polariscope with crossed polarizers with light transmission axes at -45° and 45° with respect to the needle axis, and circular polariscope with crossed polarizers are employed for the patterns. When there is no space-charge the electric field is analytically available [3,4]. The postulated space charge distribution is a strip of revolution on the needle axis with a constant charge density over a radius of 90 *μm* from needle to ground plane. For this case the electric field distribution is found using the finite-element computer package from Ansoft. The electric field is axisymmetric for either case. To interpret the patterns, study of the so called characteristic parameters is necessary.

THE CHARACTERISTIC PARAMETERS

From the theory of differential equations it is well known that given an initial point, z_0 and $\mathbf{b}_0 = \mathbf{b}(z_0)$, the solution to (6) is

$$\mathbf{b}(z) = \Omega_{z_0}^{z}\,\mathbf{b}_0 \tag{9}$$

where $\Omega_{z_0}^{z}$ is often known as the *matricant* of the system [5]. From (6) it can directly be shown that

$$\frac{d\left(|\mathbf{b}(z)|^2\right)}{dz} = 0 \Rightarrow |\mathbf{b}(z)|^2 = |\mathbf{b}_0|^2 \tag{10}$$

Consequently $\Omega_{z_0}^{z}$ is a unitary matrix [5]

$$[\Omega]^{-1} = [\Omega]^{\dagger} = [\Omega^*]^T \tag{11}$$

where † denotes the complex conjugate transpose. Furthermore, by the Jacobi identity [5] the determinant of the matricant is unity. A general unitary matrix with unit determinant may be written in terms of three independent parameters as [2]

$$\Omega_{z_0}^{z} = \Lambda(\alpha_f, \gamma, \alpha_i) \equiv \mathbf{S}(-\alpha_f)\mathbf{G}(\gamma)\mathbf{S}(\alpha_0) \tag{12}$$

Here

$$\mathbf{S}(\theta) = \begin{bmatrix} \cos\theta & \sin\theta \\ -\sin\theta & \cos\theta \end{bmatrix}, \mathbf{G}(\theta) = \begin{bmatrix} \exp(-i\theta) & 0 \\ 0 & \exp(i\theta) \end{bmatrix} \tag{13}$$

α_f, α_0 and γ are well known in photoelasticity and are known as the characteristic parameters [2].

By identical reasoning, the matricant for (8), which we denote by $\Gamma_{z_0}^{z}$, is also a unitary matrix with unity determinant. $\Gamma_{z_0}^{z}$ and $\Omega_{z_0}^{z}$ relate the same vectors **b** and $\mathbf{b}_0$ in different coordinate systems (see Figure 1). They are related by

$$\Omega_{z_0}^{z} = \mathbf{S}(-\varphi(z))\Gamma_{z_0}^{z}\mathbf{S}(\varphi(z_0)) \tag{14}$$

In particular, if the applied electric field magnitude and direction are constants (8) is easily integrated to get

$$\Omega_{z_0}^{z} = \mathbf{S}(-\varphi_e)\mathbf{G}\left(\pi B E_{Te}^2 (z - z_0)\right)\mathbf{S}(\varphi_e) \tag{15}$$

where subscript *e* refers to the applied electric field.

The first two of three properties of the matricant that we make use of are rather intuitive and may be easily proven mathematically [5]. For two points z_a and z_b

$$\Omega_{z_b}^{z_a} = \left[\Omega_{z_a}^{z_b}\right]^{-1} \quad , \quad \Omega_{z_a}^{z_c} = \Omega_{z_b}^{z_c}\Omega_{z_a}^{z_b} \tag{16}$$

The third property follows from the form of (6). If *z* is replaced by *-z* (light propagation in *-z*-direction), the matricant of the new system, which we denote by $\Phi_{z_0}^{z}$, is the complex conjugate of $\Omega_{z_0}^{z}$. Then from (11) and (16)

$$\Phi_{z_b}^{z_a} = \left[\Omega_{z_a}^{z_b}\right]^T \tag{17}$$

In Figure 4 (left) we show a slice of an example axisymmetric electric field distribution. Notice from this distribution that *z*=0 is a symmetry plane for $\mathbf{E}_T$ (for example $\mathbf{E}_T$(o'')=$\mathbf{E}_T$(o''')). In fact this is true for any axisymmetric electric field distribution. Then the matricant from a point to the symmetry plane for *z*-directed light must be identical to the matricant from the symmetric point to the symmetry plane (e.g. $\Phi_v^{o'} = \Omega_w^{o'}$). Using (16) and (17) we conclude that for any symmetric points *z'* and *z''* (e.g. *w* and *v*)

$$\Omega_{z'}^{z''} = \Omega_0^{z''}\Omega_{z'}^{0} = \Omega_0^{z''}\Phi_{z''}^{0} = \Omega_0^{z''}\left[\Omega_0^{z''}\right]^T \tag{18}$$

For an axisymmetric applied electric distribution the matricant between symmetric points is symmetric.

For usual cases of practical Kerr-effect measurements the applied electric field becomes small far from the electrodes. Then in a typical Kerr effect measurement the action of the Kerr medium may be conveniently represented as $\Omega_{-\infty}^{\infty}$ (e.g. in the example geometry for the first ray $\Omega_{-\infty}^{\infty} = \Omega_q^p$). For axisymmetric applied electric field distributions $\Omega_{-\infty}^{\infty}$ is symmetric. Referring to (12) we conclude that only two characteristic parameters are needed to express the action of Kerr media on the light polarization

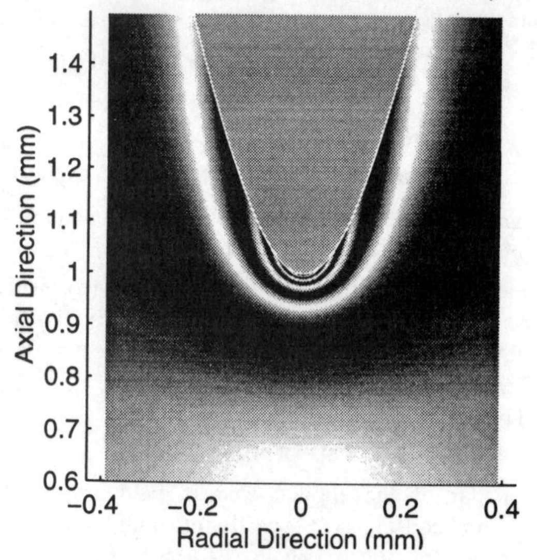

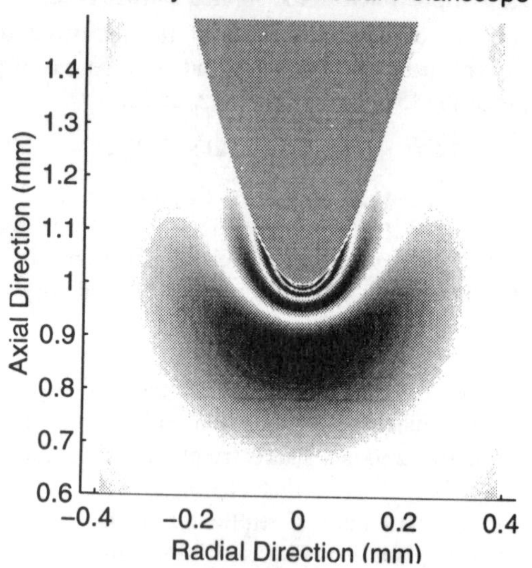

Figure 2 The optical patterns when there is no space charge.

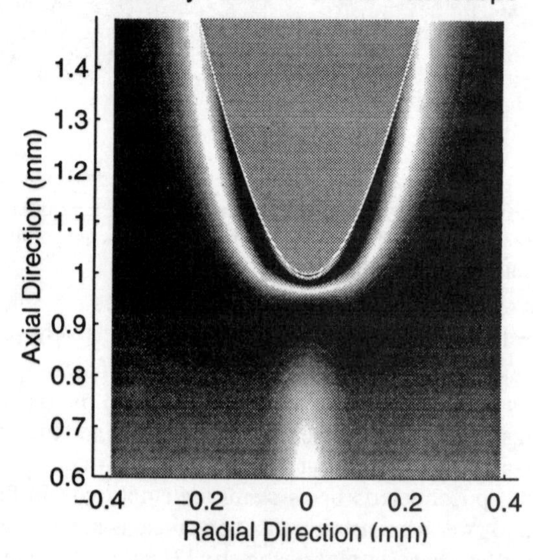

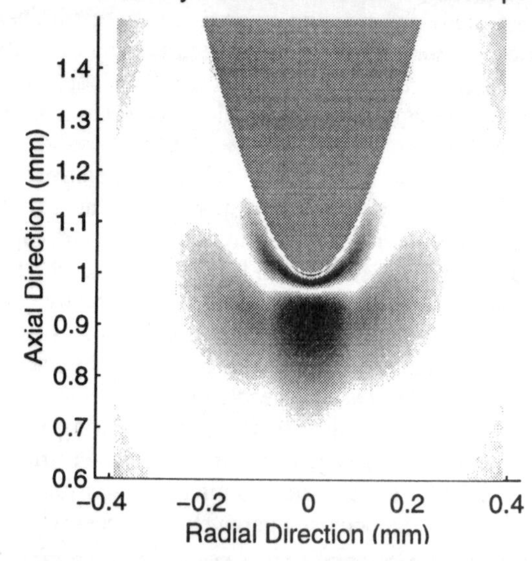

Figure 3 The optical patterns when a strip of space charge with a width of 90 μm is specified on the needle axis.

$$\Omega_{-\infty}^{\infty} = \Lambda(\alpha,\gamma,\alpha) \equiv \mathbf{S}(-\alpha)\mathbf{G}(\gamma)\mathbf{S}(\alpha) \qquad (19)$$

Then, input (I_0) and output (I) light intensity relationships of a linear polariscope (LP) with crossed polarizers with analyzer transmission axis at angle θ_a to the x axis and for a circular polariscope (CP) with crossed polarizers (see for example [1]) are

$$\frac{I}{I_0} = \sin^2\gamma\sin^2(2\alpha - 2\theta_a)\ \text{(LP)}\ , \frac{I}{I_0} = \sin^2\gamma\ \text{(CP)} \qquad (20)$$

respectively which can be used to determine the characteristic parameters, α and γ, from the fringe patterns. In past Kerr measurements where the electric field magnitude and direction between electrodes of length l did not vary along the light path, $\alpha = \varphi_e$ and $\gamma = \pi B E_{Te}^2 l$.

THE ONION PEELING METHOD

The onion peeling method may be used for the recovery of the axisymmetric electric field magnitude and angle as a function of r from the characteristic parameters. The method has been used in photoelasticity and developed for Kerr effect measurements by Aben [6]. However we have been able to simplify the results. We illustrate the method on the slice of the example axisymmetric electric field (Figure 4). The method recovers the electric field on each slice independently.

The measurements from two light rays yield α_1, α_2, γ_1, and γ_2. The onion peeling method systematically solves the unknowns (E_1, E_2, ψ_1 and ψ_2) in terms of these measured quantities. From (19) the experimental matricant for the first and second rays are ${}^{e_1}\Omega_{-\infty}^{\infty} = \Lambda(\alpha_1,\gamma_1,\alpha_1)$ and ${}^{e_2}\Omega_{-\infty}^{\infty} = \Lambda(\alpha_2,\gamma_2,\alpha_2)$ respectively. We use e_1 and e_2 to indicate the experimental matricants for respective rays. The first ray passes through the first layer only. Even though E_1 and ψ_1 are constants along the path, E_T and φ are not. Further approximation is necessary. We assume that E_T and φ are constants along the ray and approximately equal to their values at the middle of the path (point o). Since $E_T(o)=E_1$ and $\varphi(o)=\psi_1$, (15) yields the approximate matricant, indicated by a_1

$$ {}^{a_1}\Omega_{-\infty}^{\infty} = {}^{a_1}\Omega_q^p = \Lambda\left(\psi_1, \pi B E_1^2 l_{11}, \psi_1\right) \tag{21} $$

Comparing (21) with ${}^{e_1}\Omega_{-\infty}^{\infty}$ yields E_1 and ψ_1

$$ \psi_1 = \alpha_1 \quad , \quad E_1 = \sqrt{\frac{\gamma_1}{\pi B l_{11}}} \tag{22} $$

For the second ray there are three regions. For each region E_T and φ are approximated by their values at the middle point of the path inside that region. At the point o', $E_T(o')=E_2$ and $\varphi(o')=\psi_2$. Hence, (15) yields the approximate matricant

$$ {}^{a_2}\Omega_s^t = \Lambda\left(\psi_2, \pi B E_2^2 l_{22}, \psi_2\right) \tag{23} $$

Similarly

$$ {}^{a_2}\Omega_t^v = \Lambda\left(\varphi(o''), \pi B E_T^2(o'') l_{21}, \varphi(o'')\right) \tag{24} $$

where

$$ E_T^2(o'') = E_1^2\left(\cos^2\psi_1 + \sin^2\psi_1 \cos^2\theta\right) \tag{25} $$

$$ \varphi(o'') = \arctan\left(\tan\psi_1 \cos\theta\right) \tag{26} $$

${}^{a_2}\Omega_w^s$ is given by the transpose of ${}^{a_2}\Omega_t^v$ by (17). Using (16) the matricant in each region ${}^{a_2}\Omega_{-\infty}^{\infty}$ is given by

$$ {}^{a_2}\Omega_{-\infty}^{\infty} = {}^{a_2}\Omega_w^v = {}^{a_2}\Omega_t^v \, {}^{a_2}\Omega_s^t \left[{}^{a_2}\Omega_t^v\right]^T \tag{27} $$

Equating experimental and approximate matricants yields

$$ {}^{a_2}\Omega_s^t = \Lambda\left(\psi_2, \pi B E_2^2 l_{22}, \psi_2\right) = \left[{}^{a_2}\Omega_t^v\right]^{\dagger} \, {}^{e_2}\Omega_{-\infty}^{\infty} \left[{}^{a_2}\Omega_t^v\right]^{*} \tag{28} $$

Since the right hand side is known, ψ_2 and E_2 can be found.

The method is easily generalized to arbitrary axisymmetric electric field distributions. The distribution is first discretized into slices that are perpendicular to x and each slice is discretized into the n layer discretization shown in Figure 4. There are 2n unknowns (E_i, ψ_i, $i=1,2...n$) and 2n measured values (α_i, γ_i, $i=1,2...n$) for each slice. Like the two layer geometry the method begins with the outermost layer. In the i'th step E and ψ of all the outer layers are known and E_i and ψ_i are found from

$$ \Lambda\left(\psi_i, \pi B E_i^2 l_{ii}, \psi_i\right) = \left\{\prod_{j=i-1}^{1}\left[\Theta_i^j\right]^{\dagger}\right\} \Lambda\left(\alpha_i, \gamma_i, \alpha_i\right) \left\{\prod_{j=1}^{i-1}\left[\Theta_i^j\right]^{*}\right\} $$

$$ \Theta_i^j = \Lambda\left(\chi_{ij}, \kappa_{ij}, \chi_{ij}\right) \quad , \quad \chi_{ij} = \arctan\left(\tan\psi_j \cos\theta_{ij}\right) $$

$$ \kappa_{ij} = \pi B E_j^2 l_{ij}\left(\cos^2\psi_j + \sin^2\psi_j \cos^2\theta_{ij}\right) \tag{29} $$

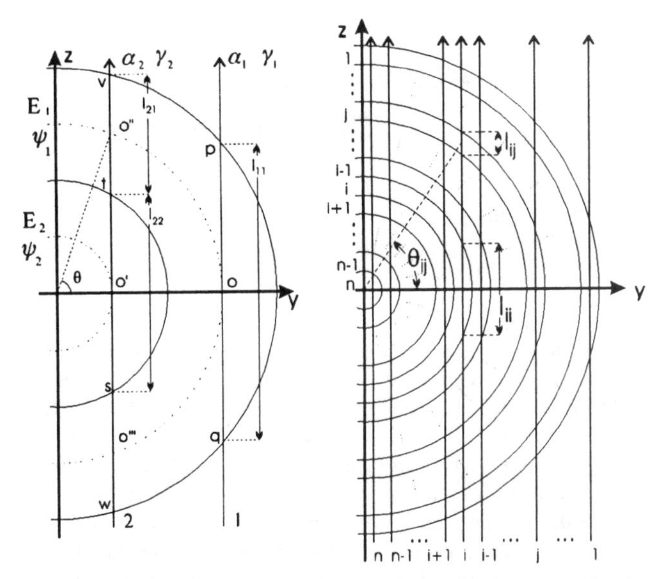

Figure 4 (left) The example two layer geometry with the axisymmetric-axis in the x direction. In the outer region the electric field vanishes. In the inner two regions the magnitude of the electric field (E_1 and E_2,) and ψ_1 and ψ_2 (see Figure 1) are constants. (Right) A n-layer discretization of a general axisymmetric electric field distribution to be used with the onion peeling method.

CONCLUSION

Using the governing differential equations of light propagation in weakly anisotropic media, the optical patterns of Kerr effect measurements may be predicted with knowledge of the electric field distribution. In an actual experiment the measured optical patterns recover the axisymmetric electric field distributions using the onion peeling method.

REFERENCES

1. Zahn, M. "Space Charge Effects in Dielectric Liquids". The Liquid State and Its Electrical Properties. Edited by E. E. Kunhardt, L. G. Christophorou, and L. H. Luessen. Plenum Publishing Corporation , 1988, pp. 367-430.
2. Aben, H. Integrated Photoelasticity. McGraw-Hill, 1979, p. 2 and pp.15-23.
3. Zahn, M. and R. Hanaoka. "Kerr Electro-optic Field Mapping Measurements Using Point-Plane Electrodes". Proceedings of the 2nd International Conference on Space Charge in Solid Dielectrics, April 2-7, 1995, Antibes-Juan-Les-Pins, France, pp 360-372.
4. Coelho R. and J. Debeau. "Properties of the tip-plane configuration". J. Phys. D: Appl. Phys. Vol. 4., 1971, pp. 1266-1280.
5. Gentmacher, F. R. The Theory of Matrices. Chelsea Publishing Company, 1959, p. 114 and p. 127.
6. Aben, H. K. "Kerr effect tomography for general axisymmetric field". Applied Optics. Vol.26. No.14, July 1987, pp. 2921-2924.

Conference Record of the 1996 IEEE International Symposium on Electrical Insulation, Montreal, Quebec, Canada, June 16-19, 1996

Applying Simulation Model to Uniform Field Space Charge Distribution Measurements by the PEA Method

Y. Liu
Department of Electrical and Computer Engineering
University of Waterloo
Waterloo, Canada N2L 3G1

M.M.A. Salama
Department of Electrical and Computer Engineering
University of Waterloo
Waterloo, Canada N2L 3G1

***Abstract*: Signals measured under uniform fields by the PEA method have been processed by the deconvolution procedure to obtain space charge distributions since 1988[1]. To simplify data processing, a direct method has been proposed recently in which the deconvolution is eliminated[2]. However, the surface charge cannot be represented well by the method because the surface charge has a bandwidth being from zero to infinity. The bandwidth of the charge distribution must be much narrower than the bandwidths of the PEA system transfer function in order to apply the direct method properly. When surface charges can not be distinguished from space charge distributions, the accuracy and the resolution of the obtained space charge distributions decrease. To overcome this difficulty a simulation model is therefore proposed. This paper shows our attempts to apply the simulation model to obtain space charge distributions under plane-plane electrode configurations. Due to the page limitation for the paper, the charge distribution originated by the simulation model is compared to that obtained by the direct method with a set of simulated signals.**

INTRODUCTION

It is very important to investigate space charge accumulation inside solid dielectrics under high electric field. This has stimulated the development of space charge distribution measuring techniques in solids in the high voltage insulation area. At present the research goal is not just to obtain the charge distribution profile but to represent the space charge distribution with higher accuracy and higher resolution. For all the distribution measuring techniques, it has been a difficulty to distinguish surface charges from the space charge distribution measurements. Under high electric fields space charge distributions are mostly homocharges or hetrocharges distributed near the electrode sample interfaces where surface charges are usually present. These surface charges may be caused by the applied voltages and/or be induced by the inside space charges. Unless specially compensated, they cannot be treated as zeros even if the electrodes are short-circuited. Therefore, if surface charges can not be distinguished from the space charge distributions, treating the mixed distributions as the space charge distributions would definitely lead to the inaccurate representation of the space charge distributions.

Among the available measuring techniques, the Pressure Wave Propagation (PWP) method and the Pulsed Electro-acoustic (PEA) method are being considered most promising. It is known that the PWP has higher resolution than the PEA, but its experimental setup is more sophisticated[3]. For the PEA method, the deconvolution procedure was first used to originate charge distributions. As it is well know in the signal processing theory, the practical implementation of this procedure is tricky[4], since the direct inverse is not necessarily stable and may strongly amplify all the additional disturbances in the frequency ranges around the zeros of the system transfer function. The direct method was thereafter proposed. It attempted to apply a wide-bandwidth transducer and short duration pulse such that the bandwidth of the system transfer function is much wider than the bandwidth of the space charge distribution. The system transfer function could be then considered a constant and the measured voltage signal could be considered directly proportional to the space charge distribution. However, the surface charge could not be represented well and the resolution would be related to the width of the voltage response of the system transfer function.

The PEA method is chosen as base for the proposed method due to its simple experimental setup. As shown in the later sections, the proposed simulation model enables surface charges to be distinguished from space charge distributions and the resolution to be related to the rise time of the voltage response of the system transfer function. The improvement to the accuracy and the resolution is discussed in this paper.

UNIFORM FIELD SIMULATION MODEL

Figure 1 shows a typical uniform field PEA setup. Under this electrode configuration, the electric pulse field inside the sample is supposed to be uniform, $e_p(t)$ equals to $v_p(t)/d$. Assume that surface charge σ_1 is distributed on the lower electrode sample interface, surface charge σ_2 is on the upper electrode sample interface, and space charges $\rho(x)$ are distributed inside the sample. When the electric pulse is applied, the pressure waves are generated at these charge locations. Part of the pressure waves, $p(t)$, can be transmitted through lower electrode and can be detected by the transducer. The transducer then transfer the detected pressure waves into a voltage signal which can be amplified by an amplifier and then recorded as $v_s(t)$ by an oscilloscope.

Suppose the whole detecting system(transducer, amplifier & oscilloscope) is linear time-invariant, and $h(t)$ accounts for its system transfer function,

$$v_s(t) = h(t) * p(t), \quad (1)$$

where $p(t)$ is the total detected pressure waves.

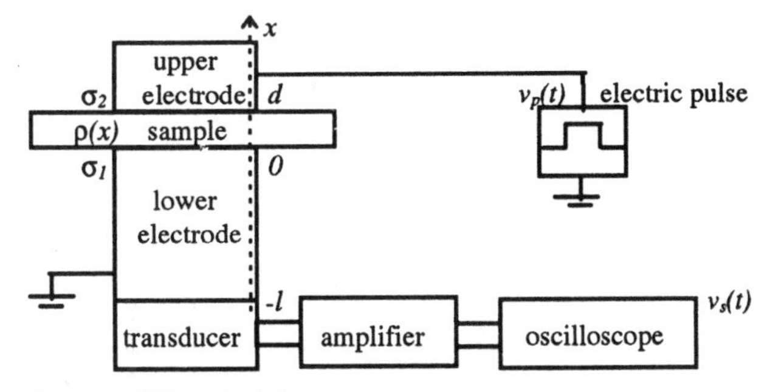

Figure 1. PEA principle setup

$p(t)$ includes the detected pressure wave $p_0(t)$, $p_\rho(t)$ and $p_d(t)$, generated by σ_1, $\rho(x)$, and σ_2 respectively[1-3]:

$$p(t) = p_0(t) + p_\rho(t) + p_d(t), \quad (2)$$

$$p_0(t) = k_1[\sigma_1 e_p(t - t_{delay}) - \frac{1}{2}\varepsilon_0\varepsilon_r e_p{}^2(t - t_{delay})], \quad (3)$$

$$p_\rho(t) = k_2 \int_0^d \rho(x) e_p(t - \frac{x}{c_2} - t_{delay})dx, \quad (4)$$

$$p_d(t) = k_3[\sigma_2 e_p(t - \frac{d}{c_2} - t_{delay}) + \frac{1}{2}\varepsilon_0\varepsilon_r e_p{}^2(t - \frac{d}{c_2} - t_{delay})]. \quad (5)$$

In (3)-(5), $t_{delay}=l/c_1$, where l is the length of the delay line (lower electrode), c_1 is the acoustic velocity of its material, and c_2 is the acoustic velocity of the sample. k_1, k_2, and k_3 are constants related to the acoustic impedances of the lower electrode, the sample, the upper electrode, and the transducer, $k_2= k_1$, and $k_3=k_1 k'$[2].

If we apply a positive pulse first to obtain $v_{s+}(t)$, and then apply a negative pulse to obtain $v_{s-}(t)$, it can be shown that:

$$v_s'(t) = v_{s+}(t) - v_{s-}(t) = h(t) * p'(t) \quad (6)$$

where,

$$p'(t) = p_0'(t) + p_\rho'(t) + p_d'(t) \quad (7)$$

$$p_0'(t) = 2k_1\sigma_1 e_p(t - t_{delay}) \quad (8)$$

$$p_\rho'(t) = 2k_1 \int_0^d \rho(x) e_p(t - \frac{x}{c_2} - t_{delay})dx \quad (9)$$

$$p_d'(t) = 2k_1 k'\sigma_2 e_p(t - \frac{d}{c_2} - t_{delay}) \quad (10)$$

For a same thickness sample contains no space charges, *i.e.*, for $\rho(x)=0$, surface charges σ_{10} and σ_{20} are distributed on the lower and upper electrode sample interfaces when the sample is subjected to a dc voltage. Under the same setup and experimental conditions, the voltage signal response $v_{s10}'(t)$ to the surface charge σ_{10} has the relationship:

$$v_{s10}'(t) = 2k_1 h(t) * \sigma_{10} e_p(t - t_{delay}). \quad (11)$$

Let $g(t)$ be the voltage response to a unit surface charge,

$$g(t) = \frac{v_{s10}'(t)}{\sigma_{10}} = 2k_1 h(t) * e_p(t - t_{delay}), \quad (12)$$

(6)-(10) can then be combined and simplified as:

$$v_s'(t) = \sigma_1 g(t) + \int_0^d \rho(x) g(t - \frac{x}{c_2})dx + k'\sigma_2 g(t - \frac{d}{c_2}), \quad (13)$$

The basic idea of the simulation model is to obtain the charge distribution in the discrete-time domain through digital simulation. Many algorithms in discrete-time signal processing, system identification and control theory can be applied[5]. This paper shows one way that we used to obtain the charge distributions. It includes the following three steps.

Step 1: Discreting $\rho(x)$

Consider N points equally spaced within the sample. Replace space charge distribution $\rho(x)$ with (N+2) pseudo surface charges $\sigma_\rho(x_i)$ to simulate the actual space charge distribution, $\sigma_\rho(x_i)=\rho(x_i)\Delta x$, $(i=0\text{-}N+1)$, $x_0=0$, $x_{N+1}=d$ and $\Delta x= x_{i+1} - x_i = d/(N+1)$. (13) can be written in the discrete form as:

$$v_s'(t) = \sigma_1 g(t) + \sum_{i=0}^{N+1} \sigma_\rho(x_i) g(t - \frac{x_i}{c_2}) + k'\sigma_2 g(t - \frac{d}{c_2}). \quad (14)$$

If we define $\Delta\tau=\Delta x/c_2$, then $x_i=ic_2\Delta\tau$. (14) can be further simplified and written as:

$$v_s'(t) = \sum_{i=0}^{N+1} \sigma(i\Delta\tau) g(t - i\Delta\tau), \quad (15)$$

where $\sigma(0)=\sigma_1+\sigma_\rho(0)$, $\sigma(i\Delta\tau)=\sigma_\rho(x_i)$, $(i=1\text{-}N)$ and $\sigma((N+1)\tau) =\sigma_2+ \sigma_\rho(d)$.

Step 2: Obtaining σ Impulse Train

Discrete the $v_s'(t)$ and $g(t)$ by the period $\Delta\tau$, (15) becomes:

$$v_s'(k\Delta\tau) = \sum_{i=0}^{N+1} \sigma(i\Delta\tau) g(k\Delta\tau - i\Delta\tau), \quad (16)$$

or, $$v_s'(k) = \sum_{i=0}^{N+1} \sigma(i) g(k - i). \quad (17)$$

According to discrete signal processing, the Z transform can be applied to (17):

$$Vs'(z) = G(z)\sigma(z) \quad (18)$$

where $V_s'(z)$, $G(z)$ and $\sigma(z)$ are the Z transforms of $v_s'(t)$, $g(t)$ and σ train respectively. $g(t)$ is the system response to a surface charge. Let ΔT stand for the width of $g(t)$, $\Delta T=m\Delta\tau$, then, $g(t)=0$ when $t\le 0$ or $t\ge m\Delta\tau$.

$$G(z) = Z(g(t)) = \sum_{n=0}^{\infty} g(n\Delta\tau) z^{-n}$$

$$= \sum_{n=0}^{\infty} g(n) z^{-n} = 0 + \frac{g(1)}{z} + \frac{g(2)}{z^2} + \ldots\ldots + \frac{g(m)}{z^m}. \quad (19)$$

Therefore,

$$Vs'(z) = [0 + \frac{g(1)}{z} + \frac{g(2)}{z^2} + \ldots\ldots + \frac{g(m)}{z^m}]\sigma(z). \quad (20)$$

The difference equation that represents the system is:

$$v_s'(k) = g(1)\sigma(k-1) + g(2)\sigma(k-2) + \ldots\ldots + g(m)\sigma(k-m) \quad (21)$$

From this difference equation, we can get the relationship:

$$\begin{cases} v_s'(1) = g(1)\sigma(0) \\ v_s'(2) = g(1)\sigma(1) + g(2)\sigma(0) \\ v_s'(3) = g(1)\sigma(2) + g(2)\sigma(1) + g(3)\sigma(0) \\ \ldots\ldots \\ v_s'(N+2) = g(1)\sigma(N+1) + g(2)\sigma(N) + \ldots + g(N+2)\sigma(0) \end{cases} \quad (22)$$

Therefore,

$$\begin{cases} \sigma(0) = \dfrac{v_s'(1)}{g(1)} \\ \sigma(1) = \dfrac{v_s'(2) - \sigma(0)g(2)}{g(1)} \\ \sigma(2) = \dfrac{v_s'(3) - \sigma(1)g(2) - \sigma(0)g(3)}{g(1)} \\ \ldots\ldots \\ \sigma(N+1) = \dfrac{v_s'(N+2) - \sigma(N)g(2) - \sigma(N-1)g(3) \ldots - \sigma(0)g(N+2)}{g(1)} \end{cases} \quad (23)$$

Step 3. Reconstructing $\rho(x)$

Since the signal outputs of $v_s'(t)$ and $g(t)$ from the digital oscilloscope are also in discrete forms, $\Delta\tau$ is equal to an integer number times Δt where Δt stands for the signal sample duration. This integer number is actually the step used in the simulation. We can use Δ to stand for the step. When the impulse train of $\sigma(j)$(j=0-N+1) under step Δ is obtained, we change the simulation step to another step, for example Δ', to get another impulse train of $\sigma'(j')$(j'=0-N'+1). From the definition of the pseudo surface charges, their densities $\sigma_\rho(x_i)$ are almost proportional to the step, while the real surface charges σ_1 and σ_2 will not change their values when the step is changed. The present of surface charges can then be extracted from the two impulse trains.

The space charge densities within the sample at each step point can be determined from:

$$\rho(x_i) = \frac{(N+1)\sigma_\rho(x_i)}{d} = \frac{\sigma(i)}{c_2\Delta\tau} = \frac{1}{\Delta}\cdot\frac{\sigma(i)}{c_2\Delta t}. \quad (24)$$

The surface charges σ_1,,σ_2 and the space charges $\rho(0)$, $\rho(d)$ can be determined by solving:

$$\begin{cases} \sigma_1 + \rho(0)\cdot\Delta\cdot c_2\Delta t = \sigma(0) \\ \sigma_1 + \rho(0)\cdot\Delta'\cdot c_2\Delta t = \sigma'(N+1) \\ \sigma_2 + \rho(d)\cdot\Delta\cdot c_2\Delta t = \sigma(d) \\ \sigma_2 + \rho(d)\cdot\Delta'\cdot c_2\Delta t = \sigma'(N'+1) \end{cases} \quad (25)$$

A data reconstructor such as the zero-order hold is then applied to obtain the space charge density between x_i to x_{i+1}, where i=0-N+1. Thus, the continuous space charge distribution in the sample is obtained.

RESULTS

This section compares the results obtained from the direct method and the simulation model for a set of simulated signals. The charge distribution is assumed as shown in Figure 2 with $\sigma(0)=\sigma_1+\sigma_\rho(0)=\sigma_1=2\text{C/m}^2$ distributed at the lower electrode sample interface, space charges described by $\sigma(i)=\sigma_\rho(x_i)=\rho(x_i)c_2\Delta t$, (i=1-299) distributed inside the sample and $\sigma(300)=\sigma_2+\sigma_\rho(d)=\sigma_2+\rho(d)c_2\Delta t=1\text{C/m}^2$ at the upper electrode sample interface($\sigma_2\approx0.6\text{C/m}^2$).

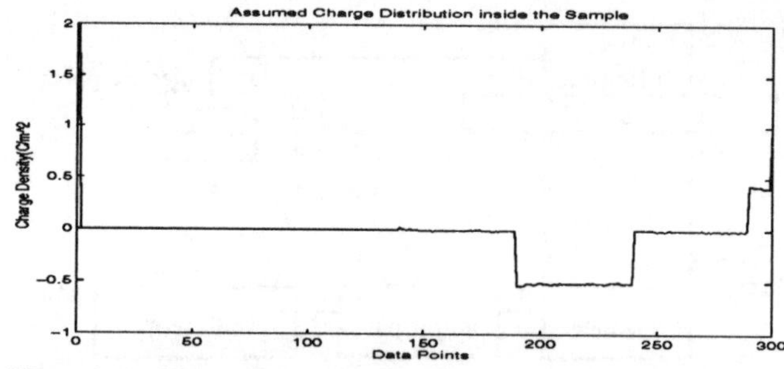

Figure 2. An assumed charge distribution

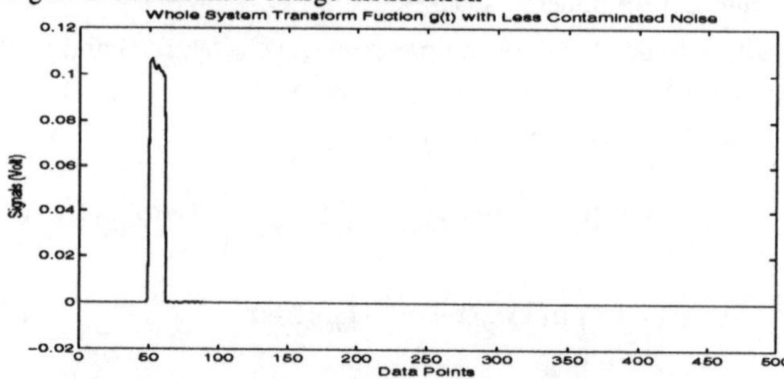

Figure 3. An assumed whole system transfer function $g(t)$

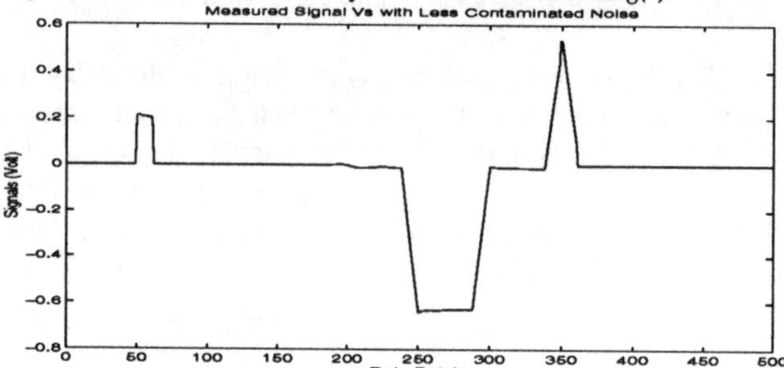

Figure 4. Signal response $v_s'(t)$

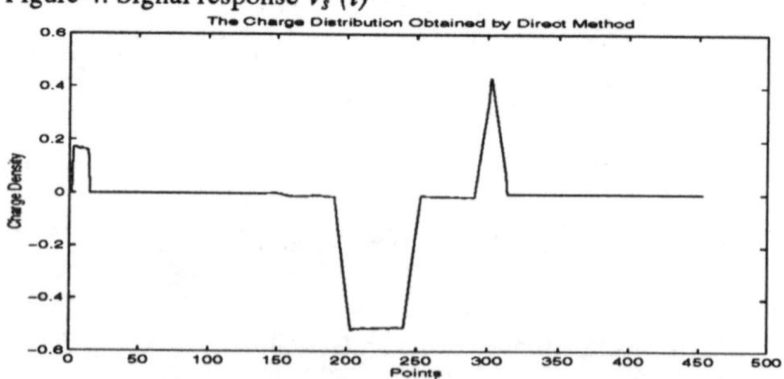

Figure 5. Charge distribution obtained by the direct method

Figure 3 is an assumed whole system transfer function $g(t)$ which is the system response to a surface charge with surface charge density $\sigma_{10}=1\text{C/m}^2$ at the lower electrode sample interface. Figure 4 is the signal response $v_s'(t)$ of Figure 2 charge distribution with Figure 3 g(t) applied.

Figure 5 shows the result of charge distribution obtained by the direct method[2]. The difference between Figure 4 and 5 is only a linear calibration. Comparing Figure 5 with Figure 2(the assumed charge distribution), surface charges, the rapid change of space charges and the space charges distributed near the surface charges are obviously not presented well.

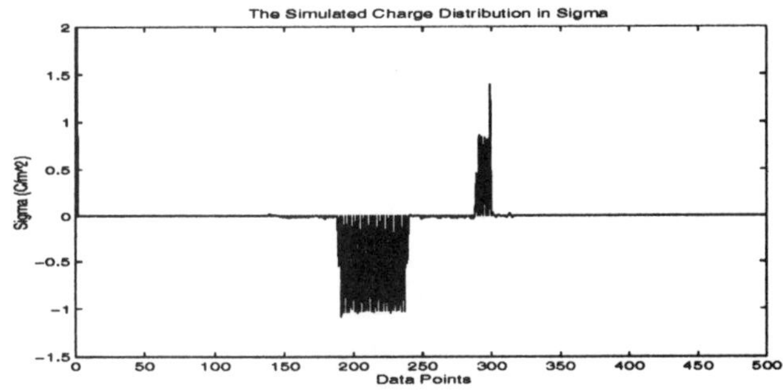

Figure 6. Impulse σ results for Δ=2

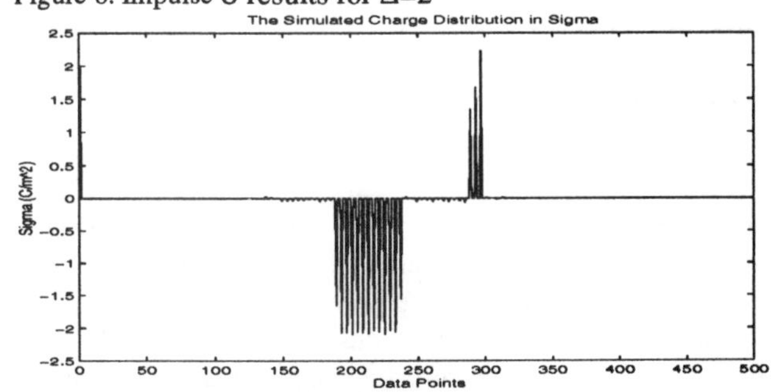

Figure 7. Impulse σ results for Δ'=4

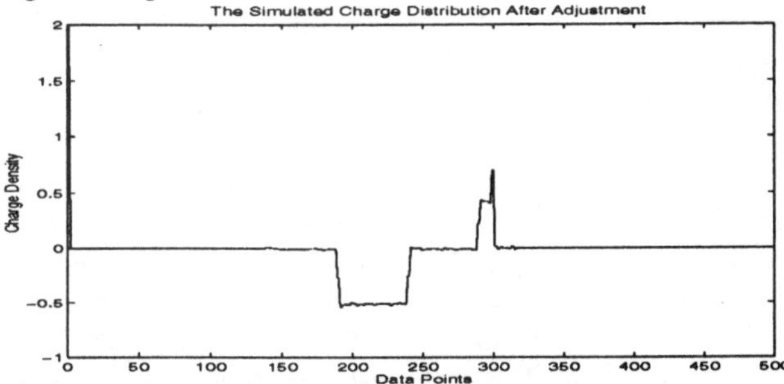

Figure 8 Charge distribution mainly based on Δ=2

Figures 6 and 7 show the discrete σ and σ' using the simulation model for step Δ=2 and Δ'=4 ($g(0)$ sets at point 49 which is part of the t_{delay}). Comparing these two figures, we found that: 1) at $i=0$, $\sigma(0)=\sigma'(0)=2C/m^2$ which indicates that only surface charge is present at this point; 2) at i=1-299, $\sigma(i) \approx \sigma'(i)/2$, which indicates that space charges are distributed within the sample, and 3) at i=300, $\sigma(300) \approx 1.4C/m^2$, $\sigma'(300) \approx 2.3C/m^2$ which indicates that surface charge and space charge may be present. Solving (25), we found that $\sigma_2 \approx 0.5C/m^2$ and $\sigma_\rho(d)=\rho(d)c_2\Delta t \approx 0.45C/m^2$.

Figure 8 shows the charge distribution obtained from the results of Figures 6 and 7. Surface charges and volume charges are differently treated. The distributions inside the sample are based on the Δ=2 results. The distribution obtained is closed to the distribution in Figure 2.

DISCUSSIONS

Equation (13) indicates the PEA system can be represented by a linear time-invariant system with charge distribution as the input and $v_s'(t)$ as the output. Surface charges are represented as δ functions in the charge distributions. $g(t)$ is then the system transfer function corresponding to a unit δ impulse. If we apply integration to both sides of (13) in respect to t, the integration of $g(t)$ is the system transfer function corresponding to a step function and the integration of $v_s'(t)$ is then the step response of the system to the charge distribution. The three simulation model steps can be applied similarly to get charge distributions from the step response.

The resolution and the accuracy of the simulation model depend upon the simulation step. Generally, the smaller step is used, the higher the resolution and the accuracy. However, this will lead to more computation time. Usually in the simulation a larger step is used first, then the step is decreased such as halved. If the two results are about the same, this means the larger step can be used in the simulation and the simulation results are correct. Apart from that, changing the simulation steps here enables surface charges to be distinguished from space charge distributions. This can improve the accuracy of the obtained charge distribution.

The highest resolution happens when the step $\Delta=1$, i.e., $\Delta\tau=\Delta t$. However, as shown in (23), the accuracy of σ depends upon the accuracy and stability of $g(1)$, i.e., $g(\Delta\tau)$. Practically, $\Delta\tau$ should be around the rise time of the $g(t)$. Compared to the direct method, the wave shape of system transfer function is considered. The resolution is improved.

CONCLUSIONS

The application of the simulation model to the uniform field space charge distribution measurements improves the resolution and the accuracy. It utilizes the rise time of the system transfer function $g(t)$ instead of its width as proposed in the direct method. Surfaces charges may be distinguished from the impulse results when two simulation steps are used. Under the simulation model, various algorithms in system identification, modern control, and signal processing can be borrowed to maximize using the measured signals for more charge distribution information.

REFERENCES

1. Maeno T., T. Futami, H. Kushibe, T. Takada and C.M.Cooke, " Measurement of Spatial Charge Distribution in Thick Dielectrics Using the Pulsed Electro-acoustic Method". IEEE Trans EI. Vol.23 No.3, June 1988, pp.433-439
2. Li Y., M. Yasuda and T. Takada, "Pulsed Electroacoustic Method for Measurement of Charge Accumulations in Solid Dielectrics". IEEE Trans. DEI. Vol.1 No.2, April 1994, pp.188-195
3. Bernstein J.B.,"Analysis of the Electrically Stimulated Acoustic-Wave Method for Observing Space charge in Semi-insulating Films", Phys. Rev. B. Vol.44 No.19, November 1991, pp.10804-10814
4. F. de Coulon, *Signal Theory and Processing*. Artech House, Inc., 1986, p262
5. Hartley T.T., G.O. Beale and S.P. Chicatelli, *Digital Simulation of Dynamic Systems: a Control Theory Approach*. PTR Prentice Hall, 1994

Conference Record of the 1996 IEEE International Symposium on Electrical Insulation, Montreal, Quebec, Canada, June 16-19, 1996

On The Measurement of The Resistivity of Insulating Solids

J. C. Filippini, C. Marteau and R. Tobazéon
Laboratoire d'Electrostatique et de Matériaux Diélectriques
CNRS, BP 166, 38 042 Grenoble, France

Abstract: **The question of the relevance of the methods presently used to characterize the resistivity of insulating solids is discussed from a critical examination of the standards. A brief review of the involved phenomena is given, which, if necessary, emphasizes the complexity of the subject. The interest of a method of characterization derived from the alternate square wave method, perfectly suited to liquids, is pointed out. Results of measurements on polymers are presented and discussed.**

INTRODUCTION

The objective of this paper is not penetrate the physics of conduction of solid materials but rather to present some considerations on the methods to be followed for a convenient characterization of the volume conduction in insulators, and on the means to reach this goal.

The measurement of the resistivity of insulating materials is of prime importance in the industrial practice for their characterization after processing, or for reception tests and monitoring of high voltage apparatus. The procedure of measurement of the "resistivity under dc voltage" was described several years ago, as well for liquids as for solids, in american and european standards. Briefly, for so widely different materials, the standards curiously recommend to measure the current under a constant voltage (and not mean field) after an arbitrary period of time of 1 minute. Such a procedure is questionable since it does not lie in any well-established physical basis.

In this paper, we will first summarize the content of the standards, then we will comment on the present status of knowledge about liquids and the future possible evolution of the liquid standards. Afterwards, we will give an overview of the present knowledge concerning the conduction in solids. By comparison with liquids, we will discuss the problems raised by the attempts to characterize the resistivity of these materials. Then we will present some results of measurements at low field stress under sinusoidal ac voltage and under alternate square wave voltage performed on solid materials. In the light of our results we will make some new proposals for a better characterization of conduction in insulators.

GENERAL

Standard Methods for Resistivity Measurement

Insulating Solids: According to ASTM [1] and IEC [2] standards the insulating quality of solid insulating materials can be characterized by two parameters called surface and volume resistivity derived from d.c. voltage and current measurements. The standards state that, when a d.c. voltage is applied between the electrodes in contact with the tested material, the current through it generally decreases asymptotically toward a limiting value, and the decrease to within 1 % of this minimum value can be reached after a time between a few seconds and many hours [1] and even several weeks [2]. A conventional arbitrary time duration of voltage application has been chosen and the standards recommend the current measurement to be made after 60 s. However, the standards remark that, for some materials, misleading conclusions many be drawn from the test results obtained at this arbitrary time [1] and suggest a resistance - time curve to be given to characterize the material [2]. The other "electrical" factors which can influence the measurements are the thickness of the specimen between the electrodes and the voltage. None of the two standards specifies any value of the thickness. Both standards state that commonly specified test voltages to be applied to the specimen are between 100 V and 15 000 V. Generally recommended values are 500 V for [1] and 100 V, 500 V and 1 000 V for [2].

Insulating liquids: According to ASTM [3], IEC [4] and JIS [5] the volume resistivity of insulating liquids should also be determined from current measurements under dc voltage. ASTM [3] recommends the dc electric stress in the liquid to lie between 200 V/mm and 1 200 V/mm with liquids gaps of 1 to 2.5 mm.

The specifications of the IEC [4] are the same except that the dc electric stress is 250 V/mm. For JIS [5], the voltage should be 250 V and the electrode gap 1 mm. For all the standards the current measurement should be made after 1 minute of electrification.

Characterization of Conduction in Liquids by Resistivity Measurements

The question of the characterization of conduction in highly insulating liquids has been thoroughly studied at the CNRS these last ten years.

It was shown that the "resistivity" values derived from the recommendations of the standards were generally not representative of the ionic purity of the liquid - which determines the "genuine resistivity" - because the thermodynamic equilibrium was disturbed during the measurements [6]. A new procedure using a low voltage alternate square wave (ASW) was proposed, in which the current is measured after a time of the electrification short enough for the thermodynamic equilibrium to be followed and for the Ohm's law to be obeyed. This method has been tested since 1992 by the Task Force "Characterization of conduction losses in dielectric liquids", of the WG 15-02 of the CIGRE. It was checked that for insulating liquids with no dipolar losses, i.e. those used in electrical engineering - the resistivity ρ obtained from the initial current was the same as that derived from measurements of the dissipation factor:

$$\tan \delta = 1/\varepsilon\rho\omega \qquad (1)$$

($\omega = 2\pi f$, f frequency of the ac voltage; ε permittivity of the liquid). For highly insulating liquids, usual bridges are not sensitive enough to measure $\tan \delta$, but the ASW method can measure the current.

Thus, the evaluation of the resistivity ρ of insulating liquids is unambiguously performed either by measurement of the initial value of the current following the application of a voltage step, or by measurement of the dissipation factor $\tan \delta$. A project of new standard recommending the ASW method has been submitted this year by the CIGRE to the International Electrotechnical Commission.

Overview of The Present Knowledge of Conduction in Insulating Solids

Unfortunately the available knowledge of the electrical conduction of insulating solids has not allowed to propose theories successfully tested on well-chosen "model" materials, especially at low - field, whatever the shape of the applied voltage. It has been known for a long time that after application of a constant voltage, the charging current decays according to a power law of the type:

$$I(t) = A\, t^{-n} \qquad (2)$$

over several decades of the time (with $0.6 \leq n \leq 1.1$) and eventually levels off in all types of solids (mineral, organic; polar, non polar; impregnated by a liquid or not). Unlike a pure capacitance - using a vacuum or gas as dielectric - where no transient currents take place, solids exhibit a time-dependent impedance. The interpretation proposed to this behaviour can be divided into two groups:

- those which favour polarization phenomena due to the motion of dipoles;
- those which favour the motion of charge carriers, more or less impeded in the bulk of the material or at the electrodes.

We could be tempted in order to calculate the resistivity of solids, following the way successfully used for liquids [6], to use the initial value of the current, i.e. the value taken as soon as possible following the application of the step of voltage.

However:

- the absorption current I_a (charging current) and resorption current I_r (after short - circuit of the sample) are, as a general rule, quasi identical over decades of time (they follow the superposition principle), which suggests that polarization mechanisms are dominant;
- it has presently never been possible to measure a constant initial current, even at very short times (a few microseconds) after the voltage application [7].

On the contrary, at elapsed times long enough, a more slowly decaying time dependence of I_a is observed while I_r is still sharply decreasing. Then I_a becomes much larger than I_r, which seems to indicate that conduction mechanism are prevailing. I_a reaches eventually a quasi constant value after waiting times t_w ranging from say a fraction of second to several days, which indeed depends on the nature of the material, but also of many other factors: temperature T, amplitude voltage V, conditioning of the sample before the measurement.

A compilation of the results published on different kinds of insulating solids indicates that the ohmic behaviour of these materials has not been demonstrated: the effect of difference in sample thickness has been scarcely investigated, and sometimes reported to produce no significant change in apparent conductivity. As concerns the field-strength dependence (at a given thickness) of the currents, it appears from the published results that both I_a and I_r vary linearly with the field E below around 10^4 V/cm, but can increase more rapidly at higher stresses.

Indeed, except the case of highly conducting glasses or of certain polymers plasticized with conducting polar liquids, it is not possible to reach the resistivity of solids from measurements with ac voltage (as it is for liquids): the dissipation factor $\tan \delta$ measured in the usual frequency range (f = 10 Hz to 10^4 Hz) of commercial bridges does not follow the $1/f$ variation characteristic of an ohmic behaviour, but is rather constant (as expected from the variation of I_a) and very often impossible to measure because of the lack of sensitivity of the bridges.

Conclusion

The above overview of the present knowledge of conduction in insulating solids confirms the great difficulty in characterizing the conduction properties of solids by their "resistivity". The arbitrary decision of deriving the "resistivity" from current measurement after 60 seconds of dc voltage application is a source of error, especially when comparing materials with very different *I(t)* laws. Indeed, this had been pointed out by the authors of the standards who suggested in some cases to characterize the material by its *I(t)* law. In fact, in actual practice, this suggestion seems to be forgotten, and this can be understood because it is not easy to use the information contained in the response of the material to a unique step of voltage. In the following we examine if we could get a better characterization using the ASW method as we do for liquids.

EXPERIMENTAL RESULTS AND COMMENTS

From the above analysis we can infer that the following requirements should be fulfilled in order to elaborate a correct method to evaluate the "resitivity" of solids:

1 - employ low voltages and rather thick samples with the view of subjecting the samples to low field. So, we used voltages $\leq$ 10 V and samples 0.1 to 1 mm thick. In these conditions we expect not to disturb the thermodynamic equilibrium and to minimize the space charge influence.

2 - measure and compare the conductance *G* obtained with different methods. We used: (1) a transformer ratio bridge (GENRAD, 1621 A) with ac sinusoidal voltage in the range 10^2 Hz to 10 Hz; (2) an alternate square wave apparatus (IRLAB, LDTRP-2) operating at 1 s half period; (3) a generator specially built for this study capable of generating highly stable ASW voltages of half-period 1 s, 10 s, 100 s, the transient currents being recorded from 0.2 s to the end of each step.

3 - use a test cell with a convenient system of guard electrodes to avoid surface leakage. The active sample area in our experiments was 28 cm^2.

4 - use as far as possible "model" materials: we tested two materials of the same family (olefinic elastomers made with metallocene catalyst referred to as A and B, B containing a filler).

Frequency and Time Dependence of The Conductance *G*

Figure 1 shows the "conductance" *G* at temperature 22 °C given by the bridge used in "conductance" mode as a function of the reciprocal of the frequency. Figure 2 shows the conductance *G* at temperature 22 °C given by the ASW device as a function of time. We can see a considerable difference between materials A and B. The filler contained in material B modifies the reponse at the largest used frequencies (Figure 1), which corresponds to short times, and also at longer times (Figure 2). In material A a limiting value of *G* was reached from about 1 s whereas, in material B, G decreased in the whole explored range (Figure 2). We noticed that the measured values of the current were identical for each period, even for the longest times, and that they were reproducible whatever the value of the half-period in the explored range of 1 s to 100 s. Thus, in these measurements, the conditioning action of usual dc voltage measurements is avoided. This effect had already been observed at higher electric field for periods of 1 000 s [8]. Measurements made using the IRLAB apparatus gave, as expected, the same values of *G* as in Figure 2 for $t = 0.7$ s.

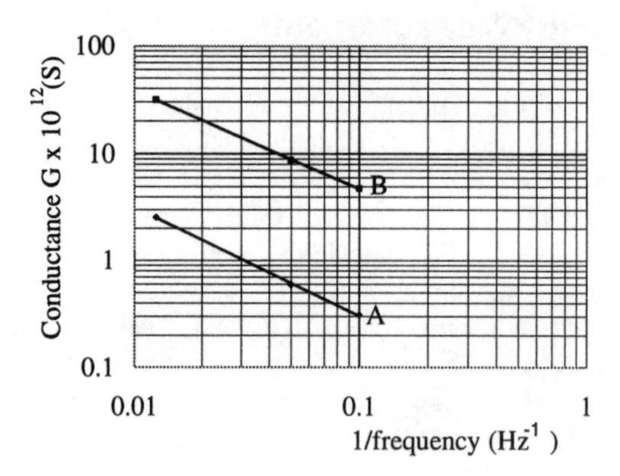

Figure 1. The frequency dependence of the conductance *G* of polymers *A* and *B* from measurements under sinusoidal ac voltage at 22°C.

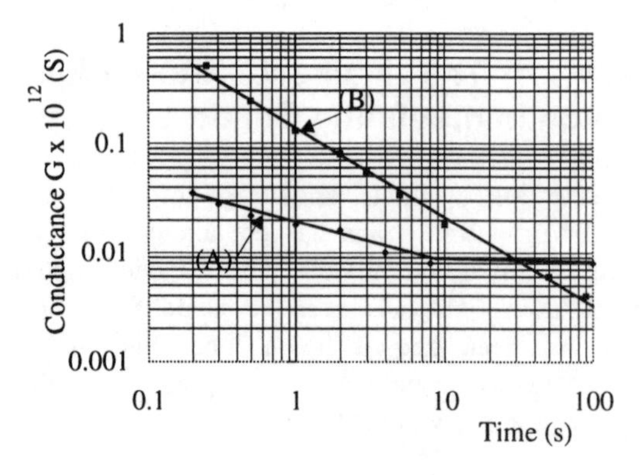

Figure 2. The conductance *G* of polymers *A* and *B* at 22 °C as a function of time elapsed after each step of voltage during the application of an alternate square wave of voltage.

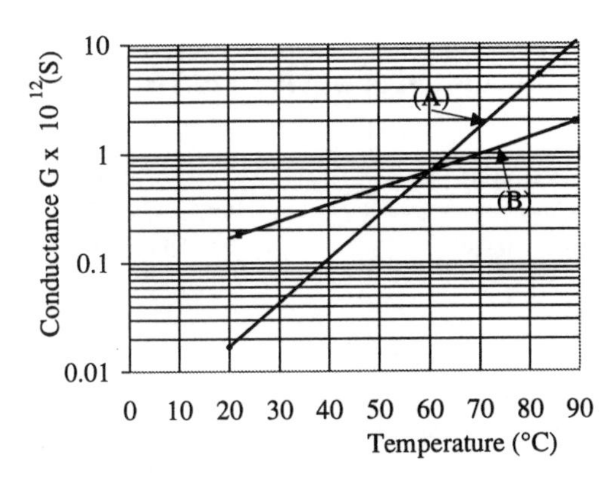

Figure 3. The temperature dependence of the conductance of polymers A and B from measurements under an alternate square wave voltage.

Temperature Effect on The Conductance G

We measured the temperature dependence of *G* in materials A and B using the ASW method and the IRLAB apparatus was used because of its simplicity and rapidity of use. Figure 3 shows the results. We notice that the conductance of specimen A is much more sensitive to temperature than that of specimen B. Such a result is consistent with the fact that, for material A at 22 °C, the time at which the measurement is performed (≈ 0.7 s) lies in the range where the current has reached a nearly constant value (Figure 2). It has been often reported that, in the temperature range 20 °C - 100 °C, the electrical conduction could varies by 3 orders of magnitude, I_a being more temperature dependent than I_r [7,8,9].

CONCLUSION

The methods recommended by the present standards to determine the "resistivity " of insulating solids can lead to errors because the experimental conditions are at once arbitrary - e.g. the electrification time - and permissive - no conditions about the electric field in the specimen. The reason is that, unlike the insulating liquids, our knowledge of the conduction phenomena is still insufficient to determine precise conditions in which the "resistivity " of the insulating solids should be characterized.

However, benefiting from the experience from liquids, we can approach a better characterization using a low electric field and a method in which the dc voltage is periodically reversed, which limits space charge influence and suppresses the errors due to successive applications of the voltage - for example when testing the effect of temperature.

The ASW method allows us to approach conditions in which the measurements are performed at the thermodynamic equilibrium.

The available low voltage equipments based on this method are sensitive enough to give reproducible measurements at low electric field and, following this way, we can hope to be able to better characterize the "resistivity" of insulating solids.

REFERENCES

[1] ASTM Standard: "Standard Test Methods for DC Resistance or Conductance of Insulating Materials". ASTM D257-93, August 1993.

[2] IEC Standard: "Method of test for volume resistivity and surface resistivity of solid electrical insulating materials". IEC 93, 1980.

[3] "Standard Method of Test for Specific Resistance (Resistivity) of Electrical Insulating Liquids". American National Standard and ANSI/ASTM D, 1979, pp. 1169-1179.

[4] IEC Standard: "Measurement of Relative permittivity, Dielectric Dissipation Factor and dc Resistivity of Insulating Liquids". International Electrotechnical Commission, Publication 247, 1978.

[5] "Testing Methods of Electrical Insulating Oils". Japanese Industrial Standard, JIS C 2101, 1988.

[6] Tobazéon, R. Filippini, J.C. and Marteau, C. "On the measurement of the conductivity of highly insulating liquids". IEEE Trans. Dielectrics and Electr. Insul. Vol. 1, 1994, pp. 1000 - 1004.

[7] Tavakoli, M. and Hirsch, J. "Absorption current and depolarisation measurements on polypropylene film", J. Phys. D: Appl. Phys., 21, 1988, pp. 454-462.

[8] Adamec, V. and Calderwood, J.H., "On the determination of electrical conductivity in polyethylene". J. Phys. D: Appl. Phys., 14, 1981, pp. 1487-1494.

[9] Munick, R.J., "Transient electric currents from plastic insulators", J. Appl. Phys. Vol. 27, 1956, pp. 1114-1118.

Conference Record of the 1996 IEEE International Symposium on Electrical Insulation, Montreal, Quebec, Canada, June 16-19, 1996

Some Remarks Regarding the Test Cells Used for Electric Strength Measurements

G. Mazzanti
Electrical Engineering Department, University of Bologna, Italy

G.C.Montanari
Electrical Engineering Department, University of Bologna, Italy

F. Peruzzotti
Pirelli Cavi S.p.A., Milano, Italy

A. Zaopo
Pirelli Cavi S.p.A., Milano, Italy

Abstract: This paper deals with the influence of the test cell on the results of electric strengh measurements performed on polymeric materials. Two different types of electrodes are considered, i.e. rods and Pseudo-Rogosky configurations, surrounded by different media, that is, oil, SF_6, and epoxy resin. An Ethylene-Propylene-Rubber (EPR)-based insulation was tested, in the form of flat specimens having different thickness. The results show that the scale and shape parameters of the two-parameter Weibull function employed to fit the breakdown data significantly changes when different specimen thickness, surrounding media and types of electrodes are used.

INTRODUCTION

Measurements of breakdown voltage or electric strength are often performed on electrical insulating materials in order to evaluate their short-term performance. Moreover, electric strength is a diagnostic property for thermal, electrical and mechanical aging of insulation, as well as a criterion for the quality control for insulating materials used in electrical (high-field) applications. Thus, it can be certainly stated that electric stength is a fundamental property for insulating material and system characterization.

For this reason, several national and international standards take care of aspects of electric-strength measurement techniques. In particular, IEC 243 is now under revision.

Limiting the scope of this paper to tests at power frequency, covered by IEC 243-1 [1], it can be pointed out that standards generally leave a quite large amount of freedom degrees as regards test apparatus and procedures. For example, electrode geometry changes with material type and shape, and the choice of the surrounding medium in which the tests are performed is left to the operator, being aware that it can significantly affect the results of the measurements (due to specimen impregnation, as well as to the presence of contaminants and surface discharges). IEC 243-1 states, however, that the above factors (plus the area between electrodes, the ambient temperature, pressure, humidity, the material by which the electrodes are made, the rate of rise of applied voltage, the thickness of specimen and its pre-conditioning) should be considered when investigating materials for which no experience exists. Moreover, the results given by different measurement methods are not directly comparable, but may provide information on relative electric strengths of materials.

Actually, when electric strength is used for aging diagnosis, the problem comes out of setting a system which is able to provide measurements sensitive to insulation degradation. This means that the measurement system must not be so affected by the above-mentioned factors to hide the real changes of the detected property. The same holds when electric strength is the property considered for quality control of materials, so that even small differences between batches can be detected (this is the case, for example, of quality control of the base material, either EPR or PE, used for cable insulation).

The results presented in this paper come from experiments made with the purpose to assess a system for electric strength measurements which fits the requirements of correlation with aging tests, significance and reproducibility. Different electrode systems and specimen surrounding media are considered, and, moreover, the effect of specimen thickness is pointed out.

TEST RESULTS AND DISCUSSION

The material used for this research is EPR. Flat specimens of different thickness, from 0.1 to 0.65 mm, were subjected to electric strength measurements, changing electrodes and specimen surrounding medium. The measurement results where statistically processed according to the two-parameter Weibull function, that is

$$F(E) = 1-\exp[-(E/\alpha)^{\beta}] \quad (1)$$

where E is electric strength, α and β are scale and shape parameters. These parameters were estimated by different methods [2-5]; the values reported in the following are those obtained by the Bain Engelhardt method. The confidence intervals of parameters and failure percentiles were calculated by the Monte Carlo method.

Test procedures

Two different electrode geometries were considered. One, called ASTM cell, consists of two metal rods, each 6 ± 0.2 mm in diameter, coaxial, with rounded edges (radius of approximately 1 mm), arranged according to ASTM D149, [6], and IEC 243-1 (configuration suggested for tapes, films and narrow strips). The other, called Pseudo-Rogowski cell, is basically a uniform-field cell. It has a profile which should maintain the electric field along the curved contour (see Fig. 1) at the same value as in the central area between the electrodes [4]. The diameter of the cell is 57 mm. Both cells were made by stainless steel electrodes. The two test assemblies are sketched in Fig. 1.

Different specimen surrounding media were used for each electrode geometry. The data reported below are relevant to the following configurations:

A: ASTM electrodes immersed in silicone oil;

B: ASTM cell immersed in SF_6 at pressure of 2 bar;

C: Pseudo-Rogowski cell immersed in silicone oil;

D: Pseudo-Rogowski cell immersed in SF_6 at pressure of 2 bar;

E: Pseudo-Rogowski cell surrouded by casted epoxy bisphenolic resin.

50 Hz ac linear voltage ramps at rates of rise of 12 kV/min (with the exception of the thicknest specimens, subjected to 50 kV/min) were employed for breakdown measurements of all the specimens. Breakdown sites were examined with an optical microscope, in order to observe possible surface tracking. Afterwards, the thickness was measured at the breakdown site to calculate the breakdown electric field. After each test, the electrodes were carefully cleaned with CCl_4 and periodically changed or cleaned by mechanical machining. Samples of 15 to 5 specimens were tested (specifically, 10 or 15 specimens for the ASTM configurations, A and B, 5 to 7 specimens for C, D and E). Before each measurement, specimens were thermally pre-treated for 48 hours at 80°C, in order to expel cross-linking by-products.

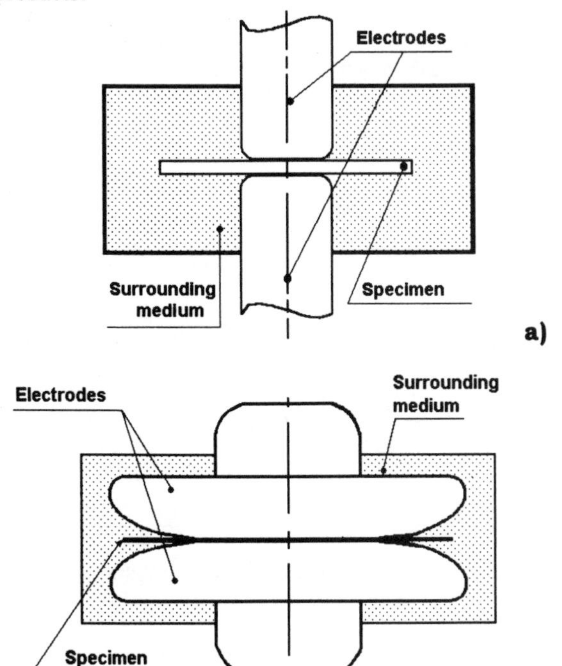

Fig. 1. Test assemblies considered: a) ASTM cell; b) Pseudo-Rogowsky cell (the real dimensions are not respected in the figure).

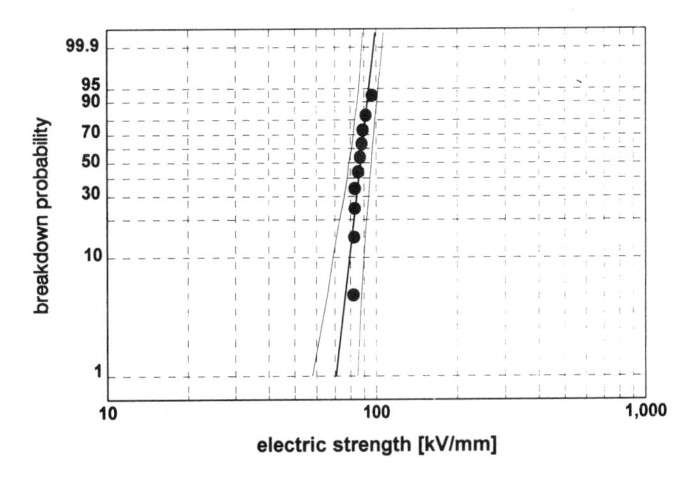

Fig. 2. Weibull plot obtained from measurements performed by the ASTM cell in oil, specimen thickness 0.10 mm.

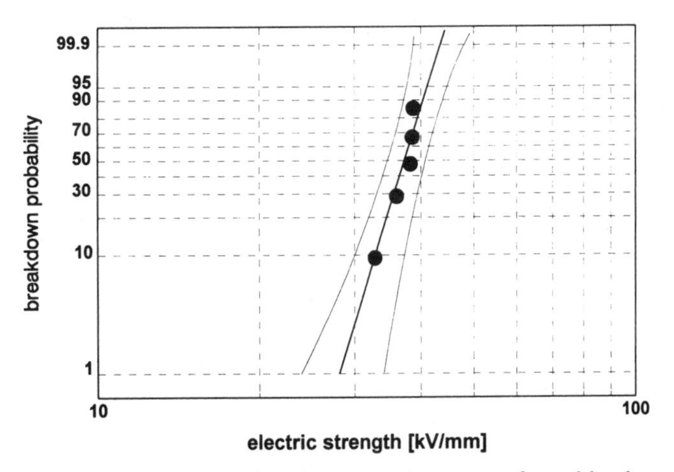

Fig. 3. Weibull plot obtained from measurements performed by the Pseudo-Rogowsky cell in oil, specimen thickness 0.67 mm.

Results and discussion

Figures 2, 3 show typical Weibull plots obtained from measurements performed by the ASTM and the Pseudo-Rogowsky cells. The quality of fitting the two parameter Weibull function was generally quite good (as expected with these kinds of cell, [6]).

It must be pointed out that the electric strength data reported in Figs. 2 and 3, as well as in the following figures, are obtained (from [1]) dividing the measured values of breakdown voltage by the specimen thickness, even if the electric field generated by the ASTM cell is not uniform. This was, indeed, one of the reasons of investigating the behavior of the Pseudo-Rogowsky cell (providing uniform field), since the knowledge of the real stress is fundamental in aging tests for electrical endurance characterization (i.e. the ultimate purpose of the research).

A summary of the behavior of the different cells, in relation to specimen thickness, is shown by Figs. 4-7. On the basis of the results obtained for the thickest specimens, cell B and, further on, cell D, were not considered for the thinner specimens.

The first consideration coming from the observation of Figs. 4-7 is that the electric strength data change significantly with type of cell and specimen thickness. This result is pointed out by Fig. 8, where the scale parameter of the Weibull function, eq. (1), is plotted as function of specimen thickness. Such a finding underlines again a known concept: the value of electric stength cannot be generalized as an intrinsic property of a material, but must be closely associated to test set and procedures.

In particular, for the thickest specimens (Fig. 4) we note a considerable difference between the values of α and ß obtained by the ASTM cells (both in oil and in SF_6) and the Pseudo-Rogowsky cells. The latter have smaller values of α when the specimen surrounding medium is oil or SF_6, while we get the largest values of electric strength when specimen and Pseudo-Rogowsky electrodes are embedded in epoxy resin. Even if a possible explanation for this finding is that the epoxy resin may penetrate between electrodes and specimen, forming a thin layer which increases the electric strength, visual inspection would exclude this hypotesis. On the other hand, the test length is not long enough to allow diffusion of the resin into the specimen (casting is made at 20°C) and, anyway, oil and SF_6 would be favoured in this respect. Thus, the significant difference among electric strength values might be explained by the large values of breakdown voltage required by specimen thickness, which gives rise to high field in the surrounding medium, thus increased surface discharges on the specimens. This favours specimen breakdown (only the punctures occurred between the electrodes, or at the edges for the ASTM cells, were considered valid).

The difference among the values obtained by various cells could be also attributed, in part, to the different specimen area stressed by the electric field, i.e. to the dimensional effect. Actually, the area of specimen free of epoxy in cell E is a bit larger than that of the ASTM cell, while the evaluation is difficult for cells C and D. Moreover, the difference in the values of shape paramater ß (pointed out in Fig. 9), which is the slope of the lines of Figs. 4-7 (eq. (1)), influences in turn the dimensional effect.

The previous explanation relevant to the thickest sample seems to be supported by Figs. 5-7, since the differences between the Weibull graphs becomes less and less significant as specimen thickness, thus breakdown voltage, decreases. Figure 8 shows, in fact, that the scale parameter values tend to converge for specimens 0.14 mm thick, while some difference is again detected at 0.10 mm (which might be explained by the dimensional effect).

As regards the behavior of ß, shown in Fig. 9, it is considerably affected by the kind of test cell, displaying the lowest values for cell E. β is also markedly influenced by thickness, increasing as thickness value diminishes, with the exception of the lowest values. This points out again the influence of the surrounding medium on measured results.

Finally, it must be noted from Fig. 8 that the choice of the cell can even affect, for the reasons raised above, the physical dependence of electric strength, ES, on thickness, t, which is expressed by the relationship $ES=K/t^{1/2}$. Depending on cell and specimen surrounding medium, in fact, the quality of fitting the above law considerably changes.

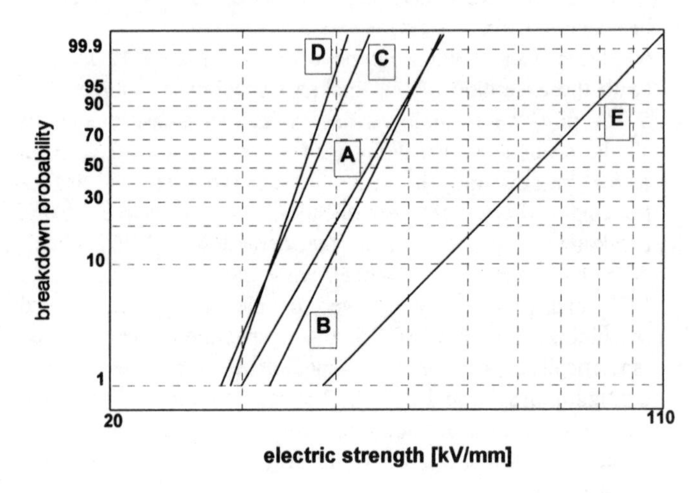

Fig. 4. Weibull plots obtained by different cells for specimens having mean thickness 0.67 mm. The experimental points are omitted for the sake of graphical clearness.

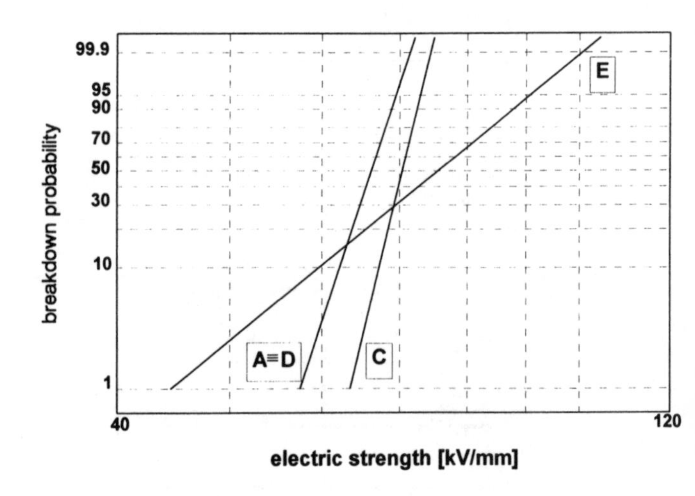

Fig. 5. Weibull plots obtained by different cells for specimens having mean thickness 0.20 mm. The experimental points are omitted for the sake of graphical clearness.

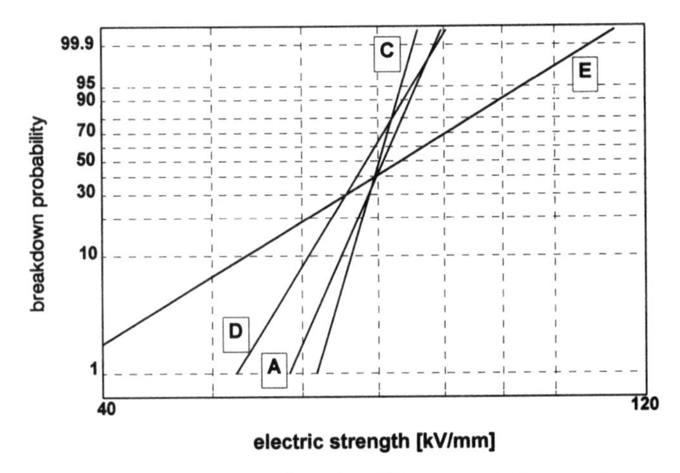

Fig. 6. Weibull plots obtained by different cells for specimens having mean thickness 0.14 mm. The experimental points are omitted for the sake of graphical clearness.

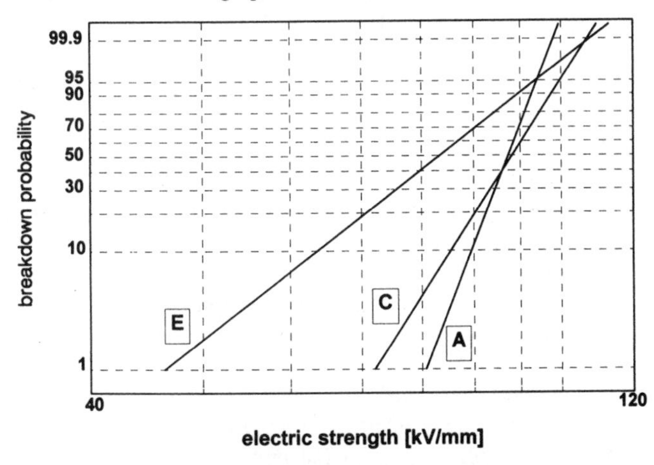

Fig. 7. Weibull plots obtained by different cells for specimens having mean thickness 0.10 mm. The experimental points are omitted for the sake of graphical clearness.

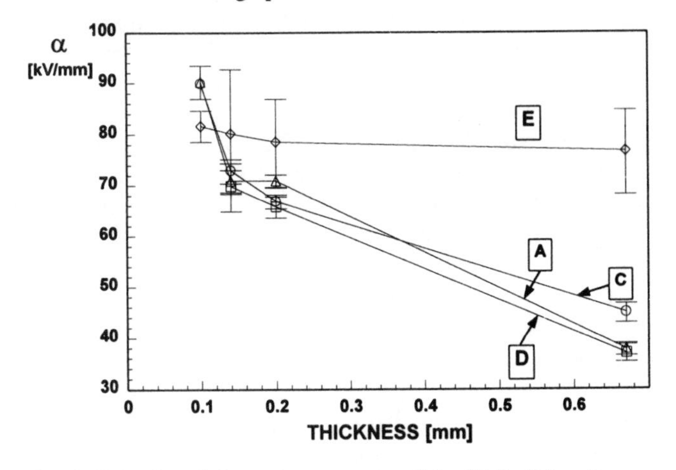

Fig. 8. Behavior of the scale parameter of the Weibull function, α, as function of mean specimen thickness, for cells A, C, D and E. The confidence intervals are calculated at probability 95%.

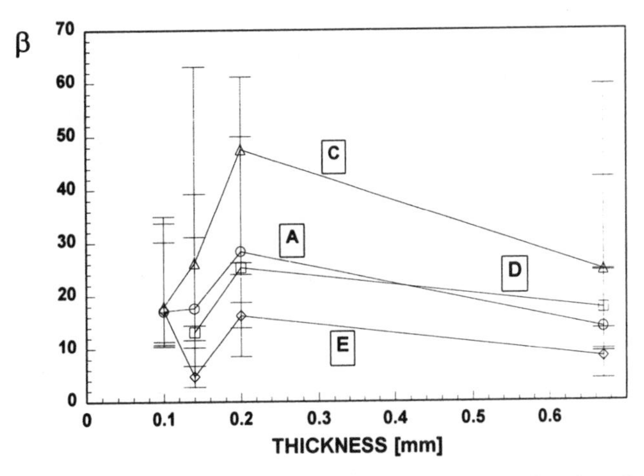

Fig. 9. Behavior of the scale parameter of the Weibull function, β, as function of mean specimen thickness, for cells A, C, D and E. The confidence intervals are calculated at probability 95%.

CONCLUSIONS

The dependence of the electric strength data on the cell chosen for the measurement of breakdown voltage may be so significant to affect material evaluation. If the purpose of electric strength measurements is aging diagnosis or, e.g. quality control, it must be ascertained if the chosen test cell is really able to measure a material property, not hidden by the experimental features underlined in this paper. Filled this requirement, the same cell must be used for the whole measurement set, even if one must be careful, in the case of aging, that modifications of material structure (e.g. strong bulk oxidation), will not affect test cell efficiency.

REFERENCES

1. IEC 15A(Sec)94, "Revision of IEC 243-1: Methods of test for electric strength of solid insulating materials. Part 1: Tests at power frequencies", November 1994.
2. W. Nelson, Accelerated Testing, John Wiley & Sons, New York, 1990.
3. "Procedures for Goodness-of-fit Tests, confidence Intervals and Lower Confidence Limits for Weibull Distributed Data", IEC/TC 56, January 1991.
4. M. Cacciari, G. Mazzanti, G.C. Montanari, "Electric Strength Measurements and Weibull Statistics on Thin EPR Films", IEEE Trans. on Diel. and El. Ins., Vol. 1, n. 1, pp. 153-159, February 1994.
5. Cacciari, G. Mazzanti, G.C. Montanari, "Comparison of Maximum Likelihood Unbiasing Methods for the Estimation of the Weibull Function Parameters", IEEE Trans. on Diel. and El. Ins., Vol. 3, n. 1, February 1996.
6. ANSI/ASTM D 149-75, "Standard Test Methods for Dielectric Breakdown Voltage and Dielectric Strength of Electrical Insulating Materials at Commercial Power Frequencies", July 1975.

Conference Record of the 1996 IEEE International Symposium on Electrical Insulation, Montreal, Quebec, Canada, June 16-19, 1996

The Effect of Thickness and Area on the Electric Strength of Thin Dielectric Films

Güneş Yılmaz
Türk Siemens Kablo ve Elektrik Sanayii
R&D Department
16941 Mudanya - Bursa - TURKEY

Özcan Kalenderli
Istanbul Technical University
Electric - Electronics Faculty
80626 Maslak - Istanbul - TURKEY

Abstract: **When the short-time electric strength (E_d) of polymers is determined in accordance with IEC 243, breakdown is usually caused by surface discharges. Unless there are gross isolated defects, E_d is more dependent on the thickness d than on the area of the samples. E_d is affected by the permittivity and electric strength of the surrounding medium or the immersion medium. In this paper, the ac electric breakdown strength of thin polyester films as a function of sample thickness in air and oil was investigated. When d < 0.2 mm, E_d is usually higher when samples are tested in oil than in air.**

INTRODUCTION

Electric breakdown in polymers, which is an important limiting phenomenon in most electric devices and components, has been subject to numerous investigations during the years. The mechanisms for electric breakdown in solids have been extensively studied and reviewed. However, few studies of the effect of the thickness on the electric strength of thin polyester films have appeared.

Polyesters have excellent dielectric properties and superior surface hardness, and are highly resistant to most chemicals. They represent a whole family of thermosetting plastics produced by the condensation of dicarboxylic acids and dihydric alcohols, and are classified as either saturated or unsaturated types. Unsaturated polyesters are used in glass laminates and glass fibre reinforced mouldings, both of which are widely used for making small electrical components to very large structures. Saturated polyesters are used in producing fibres and film. Polyester fibre is used to make paper, mat and cloth for electrical applications. The film is used for insulating wires and cables in motors, capacitors and transformers. Mylor polyesters film is being largely used in preference to paper insulation. At power frequencies, its dissipation factor is very low, and it decreases as the temperature increase.

IEC 243-1 gives methods of test for the determination of the short-time electric strength of solid insulating materials at power frequency.

Cygan and Laghari have found that the electric strength of polypropylene (PP) film decreases more rapidly if the sample thickness is increased than with a proportionate, increase in electrode area [1]. In contrast, Morton and Stannett found that the impulse strength of polyethylene terephthalate (PET), polystyrene (PS) and polyethylene (PE) films depended on the volume of sample, irrespective of its thickness [2].

Helgee and Bjellheim have found an exponential relationship between the short-time electric strength E_{st} at 50 Hz and the thickness d of films of five aromatic polymers, when 0.013 to 0.270 mm films were tested between 5 mm diameter cylindrical electrodes immersed in clean mineral oil.

Tests with other polymers have also that $E_{st}=kd^{-n}$, but for any material the value of n varies with the electrode configuration and with the permittivity of the immersion medium (n=0.55 for 25 mm dia. cylinders).

As the PET and PE films were immersed in deionized water (ε_r=80), the stress at the periphery of the electrodes would have been much lower than in the films. This would not have been the case with the PP films, which were immersed in transformer oil (ε_r=2.2); unless the electric strength of the oil in gaps comparable with the thickness of film, exceeded that of the film discharges would have occured in the oil, causing local stress concentration and heating in the film and reducing its electric strength.

If the PP films had been immersed in an oil of higher permittivity, their electric strength might have been considerably higher.

The occurrence of discharges around the electrodes can be readily observed during ac step tests on sheet insulations.
Initially the discharges caused gassing in the oil, but did not appear to damage the polyethylene. When the voltage was raised to a critical value, the discharges penetrated the surface of the polyethylene and the thinner samples suffered immediate breakdown. With the 3 and 6 mm sheets, the discharges propagated in channels just below the surface, for

some seconds, until a branch caused breakdown. It appeared that the discharges acted as hot conducting extensions of the electrodes.

Cygan and Laghari found that the electric strength of 0.008 mm films was usually higher when tested in oil, compared with tests in air. The different effect of the immersion medium found with very thin films, may arise because the breakdown voltage in air at atmospheric pressure.

Even when tests are made under conditions which prevent surface discharges, the ac strength of polymers may show a greater dependence on sample thickness than on electrode area, for samples with recessed electrodes. In these tests internal discharges were detected at 10 to 20% of the electric strength. Alternatively, the discharges might have occured if tree initiation occured at microscopic conducting inclusions.

AC electric strength tests are primarily useful for assessing composite materials. These tests can also gross defects (e.g., pinholes, or cracks) in low loss polymers, but they are insensitive to small defects which give rise to internal discharges, which can cause progressive degration and eventual breakdown at much lower stresses, than in short-time tests. Even when surface discharges are eliminated, the short-time electric strength of small laboratory molding may be of little value for assessing the relative service performance of materials.

When samples are tested in oil at 50 Hz, V_i is much higher than with similar samples tested in air.

The tests on polymer films were made using sphere and plane electrodes, immersed in liquids having a higher permittivity than that of the films, to reduce stress concentration in the liquid and raise V_i. Tests on 0.2 mm films of polyester at 50 Hz showed that E_{st} increased from 180 kV/mm, when tested in oil with $\varepsilon_r=2.24$ to 195 kV/mm in oil with $\varepsilon_r=2.65$. Nevertheless the variation of E_{st} in oils with ε_r varying 2.24 to 2.65 indicates that breakdown at 50 Hz would have been caused by discharges in the oil.

When samples with permittivity ε_r are tested in liquids with a permittivity ε_r higher than that of mineral oil to reduce stress concentration in the oil and raise the inception stress for surface discharges in the oil, thicker film show a greater increase in E_{st} than thinner films.

The main purpose of this work is therefore to study the breakdown strength of N_2, SF_6 and a mixture of N_2 + 1% SF_6 and to compare the results with previous investigations.

Experiment

Polyester was used as a polymer in the experiment. The polymer films studied were polyester from Hoechst AG. Film thicknesses ranged from 0.012 to 0.200 mm. Film thickness were measured with a film thickness meter with resolution of 0.001 mm. A minimum of five voltage breakdown measurements were conducted on different portions of each sample in order to obtain proper Weibull statistics.

AC breakdown measurements were performed with a Hipotronics OC 60 (60 kV, 2 kVA). For the measurements of electric breakdown fields of films, a Hipotronics OC 60 (60 kV, 2 kVA) liquid dielectric tester was used. Instead of the original electrode setup, test electrodes were mounted. The entire electrode setup was immersed in a vessel containing transformer oil.

The immersion mediums were air, mineral oil and silicon transformer oil. Films under test were held between 20 mm diameter sphere electrode and 50 mm diameter disk (plane) electrode while immersed in silicon transformer oil at room temperature.

The diameters of the upper electrode and the lower electrode were 25 mm and 50 mm, respectively. The sample was sandwiched between brass electrodes. It was immersed in silicon oil maintained at a given temperature.

Experiments were carried out first using 20 mm diameter sphere - 50 mm diameter plane (cylinder) electrodes and 25 mm diameter cylinder - 50 mm diameter plane (cylinder) electrodes.

Relative permittivity and dissipation factor of silicon oil and tranformer oil at 50 Hz were 2.65 and 1×10^{-5} and 2.24 and 1×10^{-5}, respectively.

Applied voltages across the films were ramped at 500 V/s from 0 V until breakdown occured, and did not hold off additional voltage.

All measurements were performed using a continuous increase in voltage at a rate of 500 V/s. All voltage values are rms. Each film thickness was subjected to more than 20 breakdown experiments.

In this case gap spacing was kept at 30 mm in order to obtain a highly divergent field. The rod electrode was made of stainless steel. Electrodes were mounted in an aluminium tank of 550 mm diameter and 883 mm length. In sphere-sphere arrangement one of the spheres was connected to the high-voltage supply while the second was earthed, but in rod-sphere arrangement the rod was connected to the high-voltage supply while the sphere was earthed. Before each series of tests, the electrodes were polished and cleaned thoroughly.

Experimental Results

Typical results of the ac electric breakdown strength measurements of the polyester films are given in Fig.1, Fig. 2 and Fig. 3.

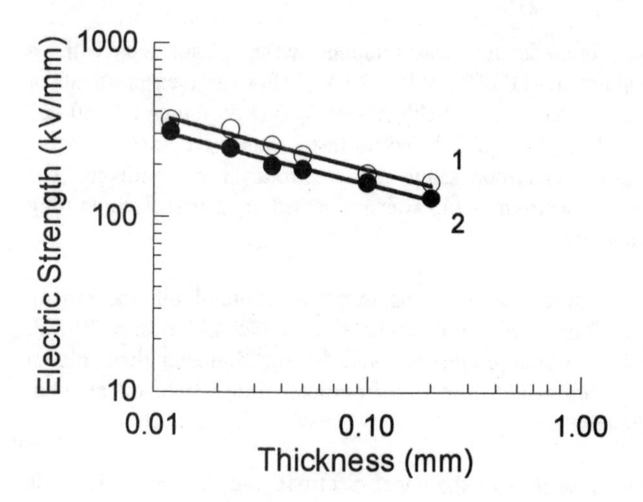

Figure 1. The effect of sample thickness on the ac electric strength of 0.012 to 0.2 mm polyester films in air.

1: Tests using 20 mm dia. sphere and plane electrodes.
2: Tests using 25 mm dia. cylinder and plane electrodes.

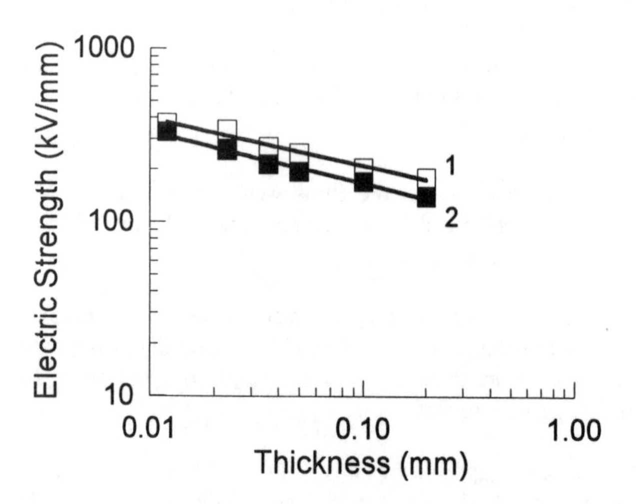

Figure 2. The effect of sample thickness on the ac electric strength of 0.012 to 0.2 mm polyester films immersed in transformer oil.
1: Tests using 20 mm dia. sphere and plane electrodes.
2: Tests using 25 mm dia. cylinder and plane electrodes.

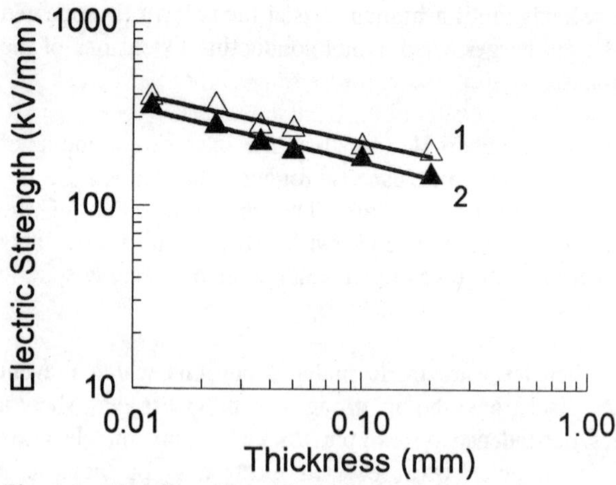

Figure 3. The effect of sample thickness on the ac electric strength of 0.012 to 0.2mm polyester films immersed in silicon oil.

1: Tests using 20 mm dia. sphere and plane electrodes.
2: Tests using 25 mm dia. cylinder and plane electrodes.

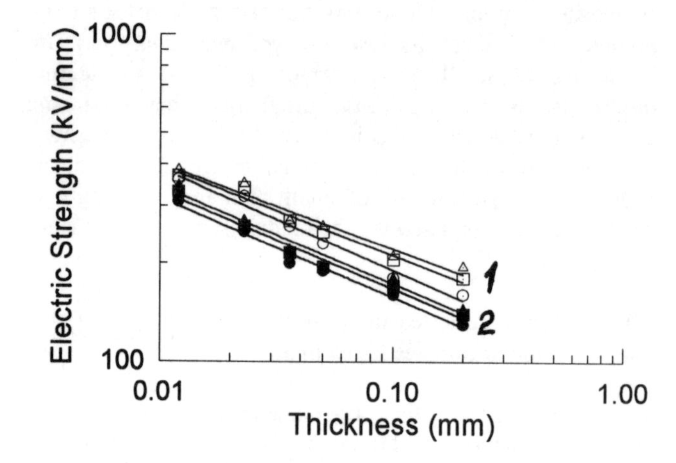

Figure 4. The effect of sample thickness on the ac electric strength of 0.012 to 0.2 mm polyester films.

1: Tests using 20 mm dia. sphere and plane electrodes.
2: Tests using 25 mm dia. cylinder and plane electrodes.

DISCUSSION

By plotting the ac breakdown strength values vs. the film thickness in a log - log plot (Fig. 1 - 3), good linear correlation was achieved for all the studied cases.

Generally speaking, electric strength reduced with increasing thickness of the sample due to so-called volume effect.
In oil, the electric strength of the polymer increases rapidly as the sample thickness is reduced.

Because treeing in polymers occurs at lower stresses when the temperature is raised, it is probable that an increase in the energy dissipated at the point of impact the discharges will reduce the stress necessary for breakdown.

Because the magnitude and energy of surface discharges increase greatly when the stress is raised, samples which are exposed to surface discharges may be expected to suffer cumulative heating and breakdown at lower stresses than samples of similar thickness which are exposed to internal discharges.

CONCLUSION

When low-loss polymers are exposed to internal discharges, or to surface discharges in air, the discharge inception stress, at 50 Hz, is much lower than the short-time electric strength E_{st}.

REFERENCES

[1] Cygan, S. and Laghari, J. R. "Dependence of the Electric Strength on the Thickness, Area and Volume of Polypropylene". IEEE Trans. on Electrical Insulation, Vol.22, No. 4, 1987, pp.835-837.

[2] Morton, V. M. and Stannett, A. W. "Volume Dependence of Electric Strength of Polymers". Proc. IEE, Vol. 115, 1968, pp. 185-187.

[3] Mason, J. H. "Effects of Thickness and Area on the Electric Strength of Polymers". IEEE Trans. on Electrical Insulation, Vol.26, No. 2, April 1991, pp.318-322.

[4] IEC 674. "Specification for Plastic Films for Electrical Purpose". 1974.

[5] IEC 243-1. "Methods of Test for Electric Strength of solid insulating Materials, Part 1: Tests at Power Frequency". 1988.

[6] IEC 212. "Standard Conditions for Use Prior to and During the Testing of Solid Electrical Insulating Materials". 1971.

[7] Mason, J. H. "Effects of Frequency on Electric Strength of Polymers". IEEE Trans. on Electrical Insulation, Vol.27, No. 6, April 1992, pp.1213-1216.

[8] Mason, J. H., Helgee, B. "Electric Breakdown Strength of Aromatic Polymers, Dependence on Film Thickness an Chemical Structure". IEEE Trans. on Electrical Insulation, Vol.27, No. 5, October 1991, pp.1061-1062.

Conference Record of the 1996 IEEE International Symposium on Electrical Insulation, Montreal, Quebec, Canada, June 16-19, 1996

Transient Errors in a Precision Resistive Divider

Steven A. Boggs
Electrical Insulation Research Center
University of Connecticut
Storrs, CT 06269-3136 and
Department of Electrical Engineering
University of Toronto

Gerald J. FitzPatrick
National Institute of Standards
and Technology
Building 220, Room B344
Gaithersburg, MD 20899

Jinbo Kuang
Department of Electrical Engineering
University of Toronto
10 Kings College Road
Toronto, Canada M5S 1A4

Abstract: **Resistive dividers have the advantages of dc response and stability. However unlike capacitive dividers, they inevitably involve power dissipation and also generally involve an appreciable inductance. These aspects of a resistive divider result in transient errors, i.e., errors which are a function of the applied waveform. This paper discusses transient measurement errors of precision high voltage resistive dividers such as the one recently developed by NIST.**

INTRODUCTION

High voltage dividers are essential to electrotechnology as many commercial transactions require tests to be performed at specified transient voltage levels and waveshapes. Impulse voltage dividers have typically been calibrated either (1) through measurements at low voltage together with some verification of linearity such as comparison of the divider output with impulse generator charging voltage or (2) through comparison with another divider calibrated in the manner of (1). Until recently, international standards required such devices to measure impulse voltage peaks with uncertainties of ±3%. New versions of these standards [1,2] require peak voltage measurements with ±1% relative standard uncertainties for reference dividers used to check other dividers. NIST continues to work at reducing measurement uncertainties in high voltage impulse measurements. Kerr cells are useful laboratory standards that complement voltage dividers, but they are limited at low frequencies and are too complex for routine industrial laboratory usage. Precision resistive voltage dividers offer the possibility of easily manufactured laboratory standards.

Resistive high voltage dividers typically have high voltage arms that are wound from uniformly resistive wire having a very low temperature coefficient of resistance. The high voltage arm generally comprises two concentric coils, counter-wound to minimize the residual inductance. Such a divider can have transient errors produced by residual inductance in both the high and low voltage arms, thermally-induced errors caused by unequal changes in overall resistance ratio due to heating of the divider, and by stray capacitances. The focus of this paper is an analysis of possible thermal effects and the effects of residual inductances on the accurate measurement of high voltage impulses.

THE NIST HIGH VOLTAGE DIVIDER

Table I gives basic divider characteristics [3]. The divider consists of high voltage and low voltage arms, each formed of two, counter-wound coils of the same turns density wound with wire of the same diameter but with differing resistance per unit length on mandrels of differing diameters. The high voltage arm is about 26 cm long and is rated at 300 kV for a standard lightning impulse waveform (1.2 x 50 μs).

THERMAL MODELING

The divider was modeled thermally using a finite element program which solves Poisson's equation and the (thermal) diffusion equation with time-dependent boundary conditions and field (electric or thermal) dependent material properties. In the present application, short sections of the top and bottom arms of the divider were each modeled as a layer of homogeneous resistive metal over a cylindrical substrate with the thermal properties of Macor* on one side of the resistance layer and with air on the other side. This divider in question normally operates in oil; however, many such dividers operate in air, and

Table I
NIST High Voltage Divider Construction

High Voltage Arm	
Wire Resistance	128.9 Ω/m (#37 Evanohm*)
Wire heat capacity, density	455 J/(kg K), 6562 kg/m^3
Turns density	0.1344 mm/turn
Total turns per winding	1944
Mandrel diameter	25.4 mm, Macor* ceramic
Macor* heat capacity, density	755 J/(kg K), 2520 kg/m^3
Total resistance of HV arm	10,000 Ω (20 kΩ/winding)
Wire conductor diameter	114.3 μm
Wire insulation thickness	6.4 μm polyesteramide
Insulation heat cap., density	2170 J/(kg K), 920 kg/m3 (est)
Wire temperature coefficient	-11 ppm/^{o}C
Low Voltage Arm (where different from HV arm)	
Wire Resistance	25.82 Ω/m (#10 Evanohm*)
Mandrel diameter	6.35 mm Macor* ceramic
Total turns per winding	7
Total resistance of LV arm	1.80 Ω

*Certain commercial materials are identified for completeness. In no case does this identification imply a recommendation by the National Institute of Standards and Technology, nor does it imply that the materials are the best available.

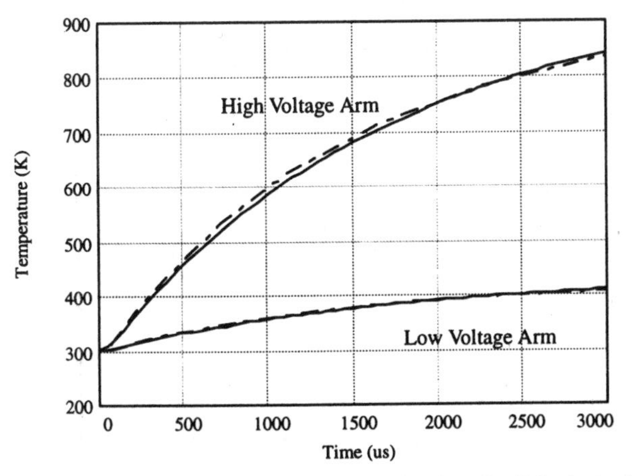

Figure 1. Finite element (FEM) (dashed lines) and adiabatic (solid lines) computations of the temperature vs. time during application of a 200 kV switching impulse. The temperature rise is nearly adiabatic; however, the effect of thermal diffusion can be seen in the crossing of the FEM and adiabatic temperatures at 2300 μs. The adiabatic and finite element computations were undertaken using identical material parameters and are compared with no adjustable parameters.

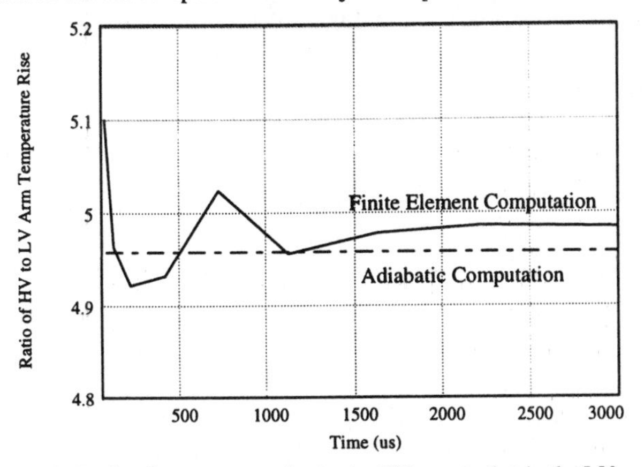

Figure 2. Ratio of temperature rise in the HV arm to that in the LV arm for finite element (solid line) and adiabatic (dashed line) computations. Since the power dissipation goes as I^2R with I common to the two arms, in the adiabatic approximation, we expect the ratio of temperature rise to be given by ratio of the wire resistance per unit length, which is 4.99. The roughness at short times is caused by the small temperature rise.

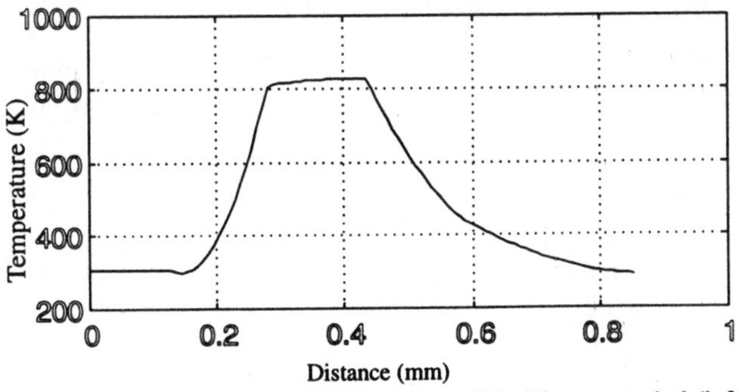

Figure 3. Temperature vs. position from within Macor mandrel (left) across the metal resistive element (flat peak) and into the air (right). Although the air temperature has increased appreciably, little heat has been transferred as the air has a very low heat capacity.

air is the worst case condition. The thickness of the metal layer was adjusted to give the same volume, mass, etc. per unit length of divider arm as provided by the two parallel, concentric windings of the high voltage divider being modeled. Short sections of the high voltage and low voltage windings were so modeled in series, connected by a material of high electrical conductivity but very low thermal conductivity. For comparison, the temperature rise of the winding was computed analytically in an adiabatic approximation, i.e., assuming that all power dissipated in the winding heats the wire. Figure 1 shows the temperature rise of the upper and lower arms for a switching surge (50 x 2500 μs) with an amplitude of 200 kV, as the switching surge rating of power apparatus is generally about two-thirds of the lightning impulse rating.

THERMALLY-INDUCED ERROR

The thermally-induced transient error is caused by the unequal change in temperature of the two arms of the divider during application of the voltage waveform, as shown in Figure 2. Figure 3 shows the radial temperature distribution in the high voltage arm of the divider. Assuming that the temperature coefficient of resistance remains constant at its room temperature value of -11 ppm/K, Figure 4 shows the error in the divider ratio as a function of time during the switching surge. As the high voltage arm heats up more than the low voltage arm and the arms have a negative temperature coefficient of resistance, the division ratio decreases as a function of time and the divider reads high. The error is small at the peak of the switching surge (about 100 μs) and increases during the tail of the waveform as the divider heats up. Similar computations have been undertaken for a 300 kV lightning impulse, for which the temperature rise in the high voltage arm is less than 30 K, which makes the divider error roughly a factor of 20 less than for the switching surge. The divider will take substantial time to return to thermal equilibrium. Although the NIST divider was designed for lightning impulse measurement and is never used with switching surge waveforms, this analysis demonstrates the errors that can arise when a precision resistive divider is used for switching surge measurements. Obviously care must be exercised in using such a divider for repeated lightning impulse measurements, as high repetition rates could cause substantial heating.

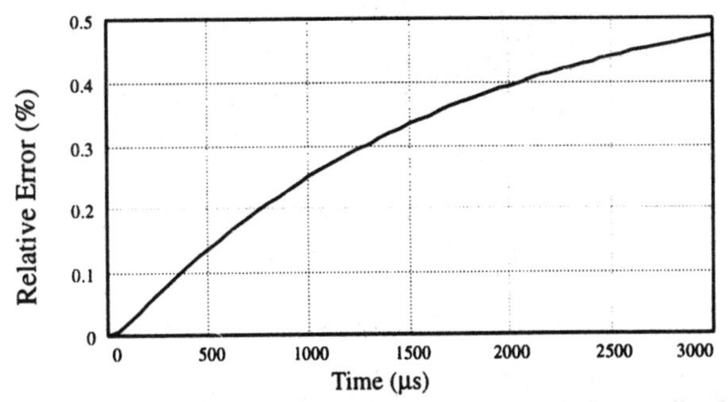

Figure 4. Thermally-induced divider error vs. time during application of a 200 kV switching surge. Positive error indicates that the divider reads high.

INDUCTIVE ERROR

Inductance is caused by energy stored in the magnetic field by current flow through a circuit. Inductance can be computed from the energy condition that

$$\frac{1}{2}LI^2 = E = \frac{\mu_o}{2}\int H^2\,dV \text{ or } L = \frac{2E}{I^2} = \frac{\mu_o}{I^2}\int H^2\,dV \quad (1)$$

where L is the inductance, E is the energy stored in the magnetic field, I is the current through the circuit, H is the magnetic field intensity and μ_o is the magnetic permeability of free space which is assumed constant throughout the volume. The volume integral covers all regions with current-induced magnetic field.

For the divider in question, a magnetic field exists in the region between and within the wires of the two windings of each coil. The inductive voltage drop in each arm of the divider is proportional to the inductance of that arm and the rate of change of the current through the arm. The high voltage arm is much longer, has a much larger number of turns, and has a larger diameter than the low voltage arm, so that the volume integral of magnetic field is much greater with the result that the inductance is much greater. Thus the inductive voltage drop across the high voltage arm is much greater than that across the low voltage arm, which results in an error proportional to dI/dt through the divider.

We compute the inductance of the high voltage arm by assuming that the current is always distributed evenly through the cross section of the wire. As this is resistance wire, skin effect should be negligible. The resistivity of the high voltage arm wire is 1.32×10^{-6} Ω m which results in a skin depth of about 0.5 mm at 1 MHz. The skin depth becomes equal to the wire radius at about 100 MHz, which justifies our assumption. Since the two windings are counter-wound, the outer winding lies on top of the inner winding. Going in from the outside surface of the outer winding, the magnetic field increases with the cross section of the outer wire until the magnetic field generated by the outer coil is achieved in the insulation and air space between the two wires. The magnetic field drops as the cross section of the inner wire is crossed and is zero inside the inner winding.

The cross section, A(x) of the wire as a function of distance, x, into the wire is given by (a = the wire diameter)

$$A(x) = \frac{a^2}{8}\left[2\,\mathrm{acos}\left(\frac{a-2x}{a}\right) - 4\left(\frac{ax-x^2}{a^2}\right)\frac{a-2x}{a}\right] \quad (2)$$

The current vs. position is therefore given by

$$A(r) = \frac{a^2}{8}\left[2\,\mathrm{acos}\left(\frac{a-2R_o-2r}{a}\right) - 4\frac{a(R_o-r)-(R_o-r)^2}{a^3}(a-2R_o+2r)\right] \quad (3)$$

$$I(r) = \frac{4\,A(r)}{\pi\,a^2}\,0.5\ \text{amp} \quad (4)$$

$$E_1 = 2\frac{1}{2}\int_{R_1}^{R_o}\mu_o\,(\,I(r)\,n)^2\,(2\cdot\pi\,r)\,dr \quad (5)$$

where A(r) is the area of the wire as a function of distance r inward from the outer coil surface at R_o, I(r) is the current outside radius R_o-r, and the integral is taken from the inner radius of the outer coil to the outer radius of the outer coil. As we are computing the inductance of the high voltage arm and not the coil, we assume that 1 A flows through the high voltage arm, which means that only 0.5 A runs through the outer coil. The total energy stored in the magnetic field is simply twice this integral, since the same magnetic field exists within the wire of each coil. Thus the total energy stored in the magnetic field of the wire is found to be 50.6 μJ/m for a high voltage arm current of 1 A. This implies an inductance of 101 μH/m length of high voltage arm or about 26.5 μH for the 26.2 cm length of the high voltage arm. In addition, we have the magnetic field within the insulation space between the windings which results in a magnetic field energy of 8.9×10^{-6} J/m which results in an inductance of 4.66 μH so that the total high voltage arm inductance is 31.1 μH. Similar computations for the low voltage arm result in an inductance of 25 nH, which is negligible.

We are now in a position to predict the transient error caused by the residual inductance of the high voltage arm of the divider. As skin effect should not cause significant errors in the resistance of the divider below 100 MHz, we might reasonably expect the divider to be good for risetimes as short as 3.5 ns and certainly for 10 ns risetimes. The current through the divider caused by an applied step voltage is

$$I(t) = \frac{V_P}{2\,R_d}\left[1 + \tanh\left(\frac{t\cdot10^9}{t_r\,0.45\cdot10^9} - 3\right)\right] \quad (6)$$

where t_r is the 10% to 90% risetime in seconds, V_p is the peak voltage, and R_d is the divider resistance. The derivative of this waveform is

$$\frac{dI(t)}{dt} = 1.111\frac{V_P}{R_d}\frac{1-\tanh\left(\frac{2.222\,t}{t_r}-3\right)^2}{t_r} \quad (7)$$

We can take the derivative of (7), set it equal to zero, and find the time at which the maximum in the derivative takes place. We can substitute this back into (7) to find the maximum dI/dt as a function of the waveform risetime, which is given by (8). The peak inductive voltage drop across the high voltage arm is then simply $L_{arm}\,(dI/dt)_{max}$ or

$$V_L(t_r) = 31.1\cdot10^{-6}\,\text{henry}\left[1.111\frac{V_P}{R_d\,t_r}\right] \quad (8)$$

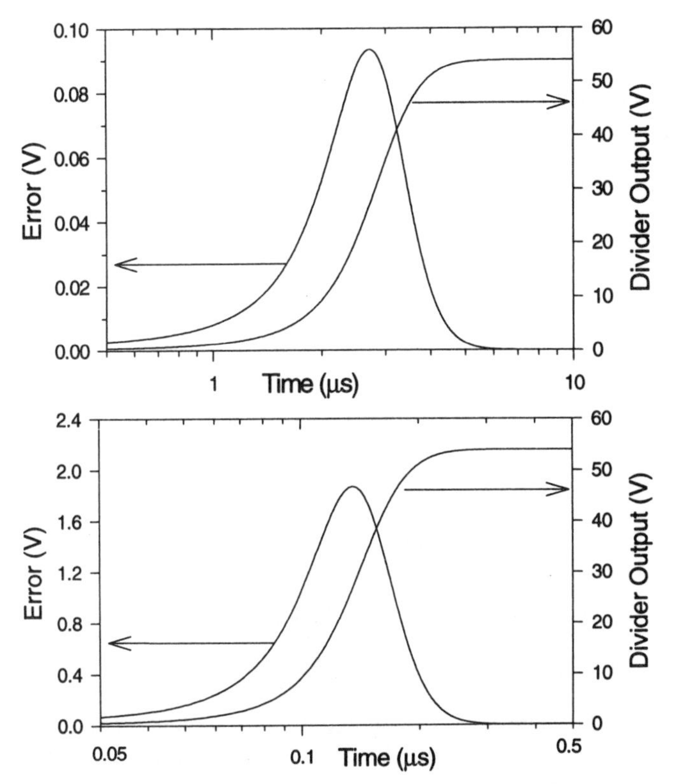

Figure 5. Top, divider output and inductive error in divider output for a 300 kV, 2 μs (10 to 90%) step wave. Below, similar data for a 100 ns risetime (10 to 90%) waveform of 300 kV amplitude. In both cases, the error peaks at the inflection point of the waveform, as it should.

Figure 5 plots divider output and the inductive error in the divider output for 300 kV, 2 μs and 100 ns (10 to 90%) risetime waveforms. For the latter waveform, the relative error (error voltage over output voltage) is roughly constant at 12% early in the waveform, as dI/dt and the voltage are increasing in proportion. The percent error drops later in the waveform and is about 6% at the waveform inflection point, where the error voltage is maximum. In principle, the inductive error should be zero at the peak of a surge waveform, as dV/dt and, therefore dI/dt should be zero. The substantial resistance of this divider should damp any tendency toward oscillation. For example, a 100 ns risetime corresponds to a -3 dB high frequency cutoff of about 3.5 MHz. At that frequency, the inductive impedance of the divider is about 600 Ω (resulting in a 6% error). Given the resistive impedance of 10,000 Ω, the Q is about 0.06, so that the divider should be very well damped. A Q of 1 will be reached at a risetime of about 6 ns corresponding to a bandwidth of 60 MHz. For the 2 μs risetime waveform, which is more typical of a lightning impulse, the relative error is about 20 times less at about 0.6% early in the waveform dropping to about 0.3% at the inflection point

ERROR CORRECTION

Given that transient errors are inevitable in a resistive divider, the obvious question is whether such errors can be corrected easily with digital post processing. In the case of inductive errors, this would be relatively simple as long as the divider remains well damped, as it should. The high voltage arm inductance simply causes a high frequency rolloff and phase shift which can be inverted with an appropriate inverse digital or analog filter.

Correction of the thermally-induced error would be much more difficult, as the thermal effects are long-term and to some degree cumulative. One method of correction would be to insert a miniature thermocouple or thermistor in contact with the winding at the bottom end of the high voltage arm. Such a temperature measurement would allow correction for average, long-term shifts in temperature resulting from application of multiple impulses. Given knowledge of the starting temperatures of the top and bottom windings, a measured waveform could be corrected to first order by assuming an adiabatic temperature rise of the winding during the impulse waveform. As seen above, the adiabatic assumption appears to be good even for switching impulse (50 x 2500 μs) waveforms.

CONCLUSION

We have identified two sources of transient error in a high voltage resistive divider. Thermal errors become significant for long-duration waveforms such as switching surges and affect the tail of the waveform. Inductive errors affect the rise of a waveform (or any region with high dV/dt) and become significant for risetimes less than about 1 μs. Inductive errors will affect measurement of waveform risetime but have little effect on measurement of the peak voltage amplitude.

The present analysis indicates that the range and magnitude of transient errors that can occur in even very well constructed resistive dividers. Interestingly, measurement of peak amplitudes for standard power engineering waveforms are not likely to be much affected as thermal errors early in the waveform are small and inductive errors are theoretically zero at the peak, where dI/dt is zero.

REFERENCES

1. IEEE Standard 4-1995, IEEE Standard Techniques for High Voltage Testing (1995).
2. IEC International Standard 60-2:1994, High Voltage Test Techniques, Part 2: Measuring Systems (1994).
3. Anderson, W.E., ed. "Research for Electric Energy Systems - An Annual Report". NIST Interagency Report, NISTIR 4691, June 1991, pp. 73-88.

ACKNOWLEDGMENT

Two of the authors (JK and SAB) are pleased to acknowledge support from the National Science and Engineering Research Council of Canada. Part of this work was performed by the Electricity Division, Electronics and Electrical Engineering Laboratory, National Institute of Standards and Technology, Technology Administration, U.S. Department of Commerce.

Conference Record of the 1996 IEEE International Symposium on Electrical Insulation, Montreal, Quebec, Canada, June 16-19, 1996

Techniques for Semi-Empirical Characterization of Material and Sensor Properties in Interdigital Dielectrometry

A. Mamishev and M. Zahn
Massachusetts Institute of Technology
Department of Electrical Engineering and Computer Science
Laboratory for Electromagnetic and Electronic Systems
Cambridge, MA 02139

Abstract: **Interdigital frequency-wavenumber dielectrometry can be used for measurement of dielectric properties of insulating materials. It is not always possible to adequately model distributed circuit parameters of the sensor structure. Even in cases when the forward problem can be solved, that is, the distribution of electric field can be calculated using numerical techniques, it may not be suitable for the solution of the inverse problem due to the high sensitivity of the parameter estimation algorithms to variations in circuit and material parameters as well as non-ideality in the models due to finite size of the sensor, fringing field effects, and lead effects. This paper particularly shows the advantage of changing the function of the conducting plane beneath the interdigital electrode structure from a ground plane to a guard electrode that follows the sensing electrode voltage.**

The techniques presented in this paper build a bridge between idealized models of interdigital structures and real measurements. Employment of such techniques allows experimental calibration of the sensor, verification of the intermediate steps in the process of development of parameter estimation algorithms, and provide data necessary for improvement of experimental arrangements.

INTRODUCTION

Several theoretical studies for ω-k (frequency-wavenumber) dielectrometry technology have been conducted over the last decade [1]. Efforts have been made to utilize developed circuit models and electronic circuitry for various practical applications, mainly in the field of dielectric materials characterization [2]. The repeatability and precision of measurements based on this approach varies depending on a particular application. Recent experiments revealed noticeable differences between the theoretically predicted response of an interdigital sensor and measurements. The search and analysis of possible sources of such discrepancies as well as improved characterization methods are the subject of this paper.

Specifically, calculation and measurement of interelectrode capacitance and conductance is discussed. As is commonly done for capacitive sensors, the leads of the interdigital sensor are actively electrically guarded. As a result, it is not possible to measure values of interelectrode capacitance and conductance with a conventional two terminal connection. Alternatives to this approach are discussed.

A direct calibration of the sensor and reduction of the number of unknowns in the governing equation provides a dependable way of determining the dielectric properties of a homogeneous material pressed against the sensor's working surface. The characterization of non-homogeneous materials requires a much more complex approach.

DESCRIPTION OF THE SENSOR

The structure of the three-wavelength interdigital sensor used is shown in Figure 1. The sensor consists of three sets of electrodes deposited on a flexible polyimide substrate (Kapton). The sensing electrodes are shielded by guard electrodes driven by the buffer stage of the interface circuitry, and the guard electrodes are separated by ground electrodes. The sensor is connected to a microprocessor-controlled voltage source through electronic interface circuitry, which creates a resistive-capacitive divider, the output of which is recorded and analyzed.

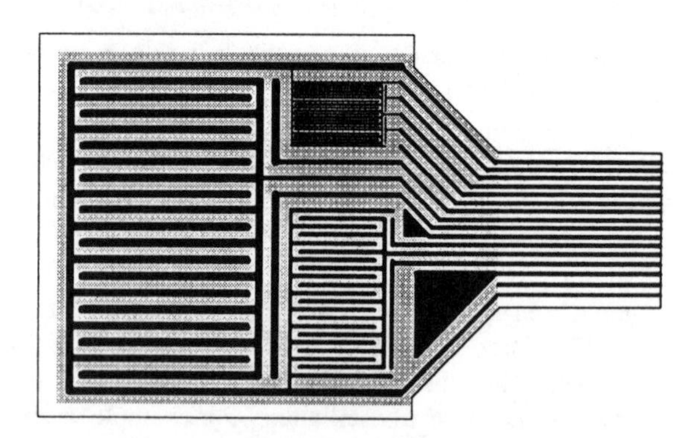

Figure 1. The three-wavelength interdigital sensor [2,3].

The cross-section of a half-wavelength of the sensor is shown in Figure 2. The parylene coating of 5 μm thickness is necessary to prevent moisture absorption into the Kapton.

Using the lumped parameter representation of the sensor electrodes, the test response is equivalent to voltage gain and phase of an RC circuit shown in Figure 3. The sensor can potentially operate in two modes - with the ground plane being held at zero potential (V_T=0), or guarded at the sensing electrode potential (V_T=V_S). The voltage gain as a function of angular frequency ω measured with V_T=0 can be expressed as

$$G = \frac{V_S}{V_D} = \frac{G_{12} + j\omega C_{12}}{(G_{12} + j\omega C_{12})(1+\alpha) + G_{11} + j\omega(C_{11} + C_L)}, \quad (1)$$

where

$$\alpha = \frac{2[G_{11} + j\omega(C_{11} + C_L)]}{j\omega C_P}, \qquad (2)$$

V_S is a measured voltage on a sensing electrode, V_D is the imposed voltage on a driven electrode, and C_P is the parylene coating capacitance on each electrode. When $V_T = V_S$, (1) and (2) are valid with G_{11} and C_{11} set to zero.

When operating in this guarded mode, the circuit can be solved directly for G_{12} and C_{12}, which can be mapped to permittivity ε and conductivity σ of the test dielectric. In the case when the ground plane is held at zero potential, the equipotential lines are like those in Figure 4. In the case when the ground plane is held at the potential of a guard, there are no equipotential lines between the sensing electrode and the ground (now guard) plane.

When V_T is held at zero, the test dielectric material adjacent to the sensor primarily affects $Y_{12}=G_{12}+j\omega C_{12}$, while $Y_{11}=G_{11}+j\omega C_{11}$ primarily depends on the dielectric properties of the Kapton substrate. However, Y_{11} is also affected by the unknown dielectric property of the test dielectric due to fringing field effects. A prime advantage of the guarded mode is that the complicating effects of Y_{11} are removed.

DIRECT MEASUREMENTS

With the wavelength of the interdigital structure being on the order of millimeters, the capacitance C_{12} between the driven and sensing electrodes is around one pF, while the capacitance C_{11} between the sensing electrode and the ground plane is on the order of tens of pF. Often, the capacitance C_P of the parylene coating is so high compared with C_{12}, typically from 10 to 100 times C_{12}, that the parylene capacitance can be assumed to be a short circuit over most of the operating frequency range of .005 to 10,000 Hz ($\alpha \approx 0$). Capacitance C_P can be estimated for each wavelength at very low frequency. Then (1) in the guarded mode reduces to $G \approx 1/(1+2C_L/C_P)$.

One advantage of the guarded mode can be easily seen if the material above the interelectrode structure is highly insulating, so that $G_{11} \approx 0$ and $G_{12} \approx 0$. Then for the grounded mode the gain G in (1) is real, and with $\alpha \approx 0$ it provides only one equation for two unknowns (C_{11} and C_{12}). In the guarded mode with G_{11} and C_{11} effectively set to zero, (1) reduces to one equation in the unknown C_{12}. Similarly, if the test material is sufficiently conductive to have non-zero values of G_{11} and G_{12}, then the complex form of (1) in the grounded mode results in two equations in four unknowns (G_{11}, C_{11}, G_{12}, and C_{12}). In the guarded mode, (1) reduces to two equations in two unknowns (G_{12} and C_{12}).

For the grounded mode, one may try to add a resistor or a capacitor of known value to either branch in order to provide additional equations with the same unknowns. However, the precision of such an approach proved to be low because of the extremely high precision required in knowing element values.

Since the capacitance C_{11} is relatively high, it can also be measured using a conventional impedance meter in a standard three-terminal connection scheme where drive and sense electrodes are driven at the same potential with respect to the ground plane electrode.

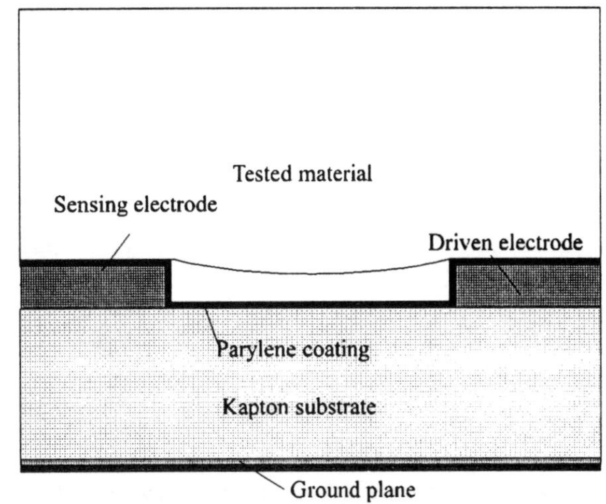

Figure 2. The cross-section of the interdigital sensor with a slightly flexible test material above it.

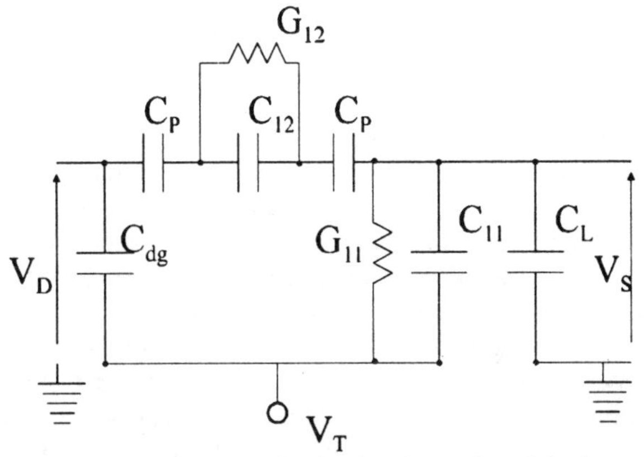

Figure 3. The equivalent circuit of each wavelength in the sensor of Figure 1.

Since the value of C_{12} is too low to accurately measure, it can be determined using (1) in the grounded mode after C_{11} has been found.

COMPARISON OF CALCULATIONS AND MEASUREMENTS

The values of the interelectrode capacitances can also be found by modeling the structure of the interdigital sensor with finite element electric field software (FEM, Ansoft).

Distribution of equipotential field lines for the case when two layers of dielectric materials are placed above the sensor is shown in Figure 4. The distribution shown of the electric field

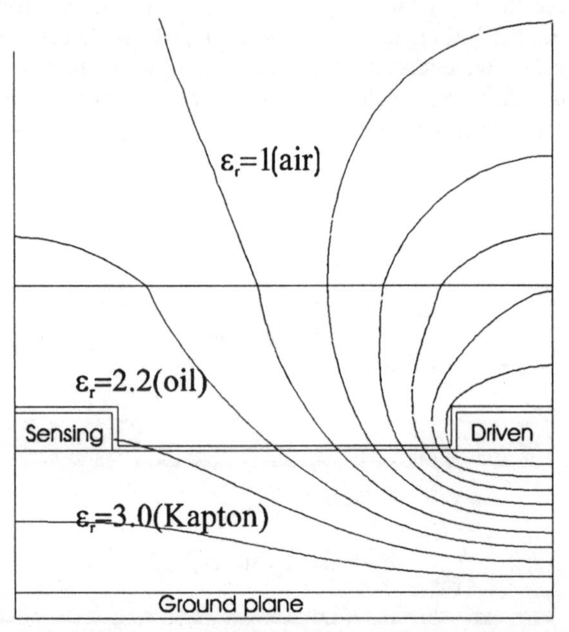

Figure 4. Equipotential field lines in the half-cell geometry of an interdigital sensor. Relative permittivity of parylene ε_r=3.05. The driven electrode (right) is at 1 V potential, the sensing electrode (left) is at 0.2 V potential, and the ground plane is at 0 V.

is for the sensor working in the high frequency mode when conductance effects through the material are relatively unimportant.

A continuum model of an interdigital sensor [1] provides a closed-form solution for a set of idealized interdigitated electrodes (an infinite array of infinitely long microstrip conductors of zero thickness). A 2D finite-element model allows assumption of finite thickness of the electrodes as in Figure 4. Comparisons given in Table 1 demonstrates that although the continuum model of the interdigital sensor is theoretically adequate, the results of actual measurements may differ from predicted values by more than 30 percent. Several factors contribute to this, such as finite size of the interdigitated electrodes, non-zero metallization thickness of the electrodes, and, sometimes, poor contact between the materials on the sensor-tested material boundary.

Table 1. Comparison of calculated and measured values of the interelectrode capacitances in air.

	λ=1.0 mm		λ=2.5 mm		λ=5.0 mm	
(pF)	C_{11}	C_{12}	C_{11}	C_{12}	C_{11}	C_{12}
Continuum Model	11.8	1.19	24.6	0.54	89.1	0.80
FEM (Ansoft)	10.7	1.01	23.4	0.56	88.8	0.80
Direct Measurement	17.1	1.24	32.8	0.67	112	0.82

The measured values in Table 1 were obtained by combinations of guarded mode and grounded mode measurements together with capacitance meter measurements of C_{11}. The values given in Table 1 were self-consistent between different measurement methods.

EMPIRICAL CALIBRATION OF SENSOR

Calibration of C_{12} versus the permittivity of the test dielectric ε is performed using materials with known permittivity verified by capacitive measurements with parallel plate electrodes.

Since ε of air has a most reliable value (ε_0), the calibration is performed according to the formula:

$$C_{12} = C_0 + (\varepsilon / \varepsilon_0 - 1)S, \quad (3)$$

where C_0 is the measured capacitance at $\varepsilon=\varepsilon_0$, and the slope S = $dC_{12}/d(\varepsilon/\varepsilon_0)$ is the slope of the fitted line that goes through the origin. Consequently, ε is found as:

$$\varepsilon / \varepsilon_0 = 1 + (C_{12} - C_0) / S. \quad (4)$$

An example of such a calibration is shown in Table 2. The values of C_0 in Table 2 for air are slightly different from corresponding values in Table 1, because of small changes in configuration for guarding the previous ground plane.

Table 2. High frequency calibration in the guarded mode of the three-wavelength sensor.

		λ=1.0 mm S≈0.61		λ=2.5 mm S≈0.53		λ=5.0 mm S≈0.77	
Material	$\varepsilon/\varepsilon_0$	G, dB	C_{12}, pF	G, dB	C_{12}, pF	G, dB	C_{12}, pF
Air (C_0)	1.0	-33.1	1.34	-29.6	0.80	-28.7	0.89
Teflon	2.1	-29.7	2.02	-24.9	1.41	-23.2	1.74
Polyethylene	2.3	-28.9	2.22	-24.4	1.49	-22.7	1.85
Lexan	3.2	-27.5	2.62	-22.4	1.92	-20.0	2.58
Delrin	3.6	-26.3	2.93	-18.4	2.18	-19.4	2.83

Results of experiments with oil-free transformer pressboard in air are shown in Figure 5. The shape of the phase curve in Figure 5 differs from an ideal one-pole system primarily due to the dependence of the complex permittivity of the dispersive pressboard on the frequency of applied electric field. Estimation of the numerical values of C_P and evaluation of it's importance can be done based on the leftmost part of the gain curve in Figure 5. At the lowest frequency all three wavelengths have approximately zero phase, so that the gain is real in agreement with (1) in the guarded mode at very low frequency. From the gain value at low frequency the value of C_P for the 1, 2.5, and 5 mm wavelengths are calculated to be about 10, 20, and 80 pF, with C_L respectively 60, 24, and 23 pF. The C_P values are in the appropriate ratio to electrode areas as expected and are due to parylene capacitance in series with a small air gap due to the surface texture of pressboard. Solution of the inverse problem for a complex dielectric permittivity ($\varepsilon=\varepsilon'-j\varepsilon''$) for this material are shown in Figures 6 and 7. The permittivity measurements (ε') are most accurate at high frequencies while the conductivity ($\varepsilon''=\sigma/\omega$) is most accurate in the mid-frequency range. For a homogenous material, all three wavelengths should indicate approximately equal values of ε' and ε'' while G_{12} and C_{12} will be different for each wavelength as the electrode lengths and spacings are different.

However, the calibration of ε" versus G_{12} has been done with pressboard in a parallel plate sensor, similarly to the Table 2 capacitive measurements. However, the results are not very reliable, partially due to low repeatability of the measurement. Only smooth, rigid, slightly compressible materials can be used for precise calibration for σ, whereas pressboard has a surface texture pattern. In general, intimacy of contact, and, consequently, applied pressure, plays an important role and may potentially introduce very large errors in the parameter estimation process. In addition, slight surface contamination of the sensor itself may affect calibration of highly insulating materials.

The slope of ε" is not strongly affected by calibration. In our case, it is equal to ≈-0.6, obtained both with the three-wavelength sensor as well as with parallel plate capacitor measurements.

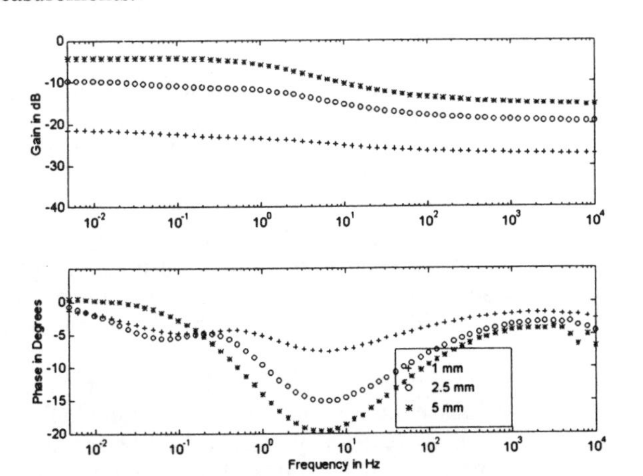

Figure 5. Gain and phase for the three-wavelength sensor with guarded sensing electrode, with a thick layer (4mm) of oil-free transformer pressboard pressed against the surface of the sensor.

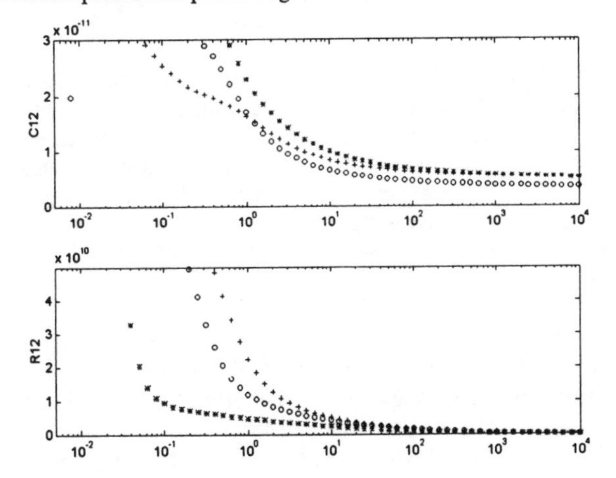

Figure 6. Pressboard in air - solution of the inverse problem for a homogeneous material. The values of $R_{12}=1/G_{12}$ and C_{12} are obtained from gain/phase measurements of Figure 5.

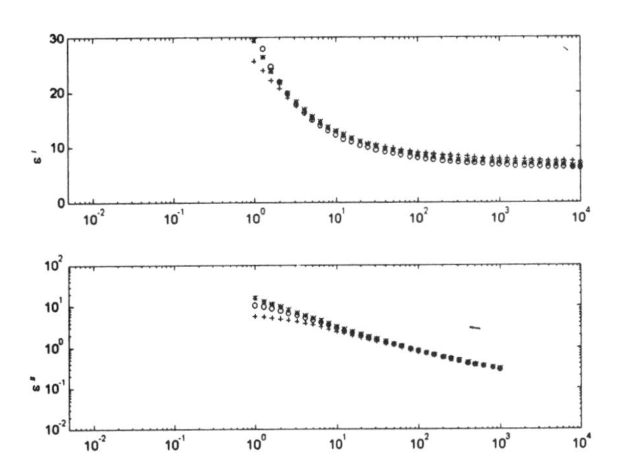

Figure 7. Pressboard in air - values of ε′ and ε″ obtained from Figure 6 using the calibration data of Table 2.

CONCLUSIONS

Application of ω-k interdigital dielectrometry for characterization of insulation materials requires refinement and verification of idealized theoretical models.

Feasibility of the direct calibration approach for the purposes of homogeneous material characterization, and, more importantly, verification of more complete algorithms for parameter estimation, is demonstrated. Comparison between theoretical predictions and actual measurements is made. An example inverse problem of parameter estimation is solved for transformer pressboard. The advantages of guarded mode measurements are demonstrated.

ACKNOWLEDGMENTS

The authors would like to acknowledge the support of the Electric Power Research Institute, under grant WO 3334-1, managed by Mr. S. Lindgren, and the National Science Foundation under grant No. ECS-9523128. The donation of Maxwell software by Ansoft Corp. is gratefully appreciated. The authors would like to thank Mr. Darrell Schlicker, Ms. Yanqing Du, and Prof. B. Lesieutre, all from the MIT Laboratory for Electromagnetic and Electronic Systems, for valuable discussions and assistance with the experiments.

REFERENCES

1. Zaretsky, M.C., L. Mouayad, and J.R. Melcher, "Continuum Properties from Interdigital Electrode Dielectrometry," IEEE Transactions on Electrical Insulation, Vol. 23, No.6, December 1988, pp. 897-917.

2. von Guggenberg, P.A., and J.R. Melcher, "A Three-wavelength Flexible Sensor for Monitoring the Moisture Content of Transformer Pressboard'', Proc. of the 3rd International Conference on Properties and Applications of Dielectric Materials, Vol. 2, pp. 1262-5.

3. Mamishev, A.V., Y. Du, and M. Zahn, "Measurement of Dielectric Property Distributions Using Interdigital Dielectrometry Sensors," IEEE Conference on Electrical Insulation and Dielectric Phenomena, Virginia Beach, VA, October 1995, pp. 309-311.

Conference Record of the 1996 IEEE International Symposium on Electrical Insulation, Montreal, Quebec, Canada, June 16-19, 1996

Field distortion effects by an electro-optical field sensor and the resulting influence on the sensor output

Jörg Oesterheld
Department Electro-Technologies
ABB Corporate Research Baden
CH-5405 Baden-Dättwil (Switzerland)

Abstract:

The use of electro-optical sensors to measure the electrical field in or around technical devices becomes more and more common. The known advantages of those sensors are that the measurement can be performed completely potential free and the sensor characteristic is quite linear within a wide frequency range from DC up to GHz. It is also often claimed that only a small field distortion is caused by the sensor. But on a quantitative level similar questions as known from classical, metal containing field sensors are still waiting for an answer:

1. How and how much does the sensor deform the field by its presence in the field region?
2. Does the field deformation around the sensor have an influence on the validity of the calibration?
3. Which differences according to the field deformation during calibration and measurement have to be expected between the "real" field vectors of the undistorted field distribution and the vectors derived from the measured signals?

The paper deals with these problems. The investigations are performed using a common type of today's field-sensors. At first some possible sources of error due to the change of field vectors inside the given inner structure of the sensor are mentioned. Then the usual calibration procedure of the sensor inside a plane-plane capacitor in air will be discussed on the basis of field simulation results. Particular emphasis is on proximity effects near the electrodes. In a third part of the paper a field simulation of a complete measuring session will be offered. This simulated session includes on the one hand the above mentioned calibration procedure. On the other hand the measurement at several field positions around a cable termination was simulated. Besides the "measured" field vectors at these positions also the undistorted field vectors at the same positions were determined. In order to evaluate the field distortion effect by the sensor and the resulting inaccuracy, each couple of vectors were compared by absolute value and angle.
This comparison shows that despite of the field distortion by the sensor and the change of field vectors inside the sensor an astonishing small difference exists between real field vectors and the vectors reconstructed by the results of the described measuring procedure. Hence, for the given sensor type, the theoretically expectable uncertainty is quantified, which increases the confidence in this measuring tool.

1. Introduction

The development of electro-optical field sensors is on a very promising stage today [1, 2]. Especially the possible potential on miniaturization of sensors, which are capable to measure fields within a wide frequency range - very important for EMC problems - is of high interest. Hence, in high voltage and insulation technology this technique becomes more and more common.
On the other hand some problems remain similar to those known from classical field sensors like spherical electric field sensors. Those are generally made of metal with isolated sphere segments to measure particular field components. Conductive bodies force the electric field vectors to touch them perpendicular to their surface. Consequently, any field distribution is changed or distorted by inserting a conductive body or sensor, respectively. The same behavior could be found if one would insert a body with a permittivity near infinity.
On the other hand a "gas permittivity body" (permittivity 1) would not influence the electric field distribution at all. Solid insulation materials (used as packaging for sensor crystals) have permittivities usually higher than 2 and the crystals themselves mainly have permittivities above 15. That means, field vectors do not touch sensor surfaces perpendicularly, but the higher the sensors permittivity the more the field distribution is changed by the sensor.

The described problem leads to the following questions:
- What effects do exist which might influence the field distribution and consequently the accuracy of the measurement?
- What quantitative inaccuracy has to be expected considering only the electrical field around and inside the sensor during calibration and measuring procedure?
A basic contribution considering these problems is presented in [3].

The present paper will address these questions by a theoretical approach. The first question will be discussed from the simple point of view of field distortion and refraction at interfaces between different dielectrics.
As the discussion will show, the treatment of the second question is not possible on a theoretical or analytical way. The only opportunity to get an appropriate answer is to do careful field simulations. Such simulations were performed using the FEM-solver of the universal field calculation program ACE (from ABB Corporate Research Västeras, Sweden).

2. Sensor structure

The sensor under consideration is a type using the well known Pockels effect. The principle is described in several references [1, 2]. The sensor is able to detect only one field component. To determine complete field vectors it is necessary to rotate the sensor head corresponding to the used co-ordinate system. The active part of the sensor is a $Bi_4Ge_3O_{12}$ (BGO) crystal. As already described in [4], the crystal in this particular sensor is placed between crossed polarizers. Optical fibres connect the electronic devices and the sensor head via graded index lenses. A quarter wave plate is placed between output polarizer and crystal to change the polarization from a linear to circular one. The main components of the sensor head and their permittivities are given in Fig. 1. For simplicity the optical fibres and lenses are omitted. The field component which is measured and considered to be proportional to the sensor output is the component that passes the crystal following the axis polarizer-crystal-polarizer (see arrows).

3. Distortion of the electric field

As mentioned above the presence of any solid body in an electric field changes the field distribution compared to the field without that body, because of several charge separation effects in the bodies. Absolute value and direction of the field vectors around that body are different from the original ones. If this body is a sensor this means it can never detect the real undistorted field.
An additional problem arises if one is using a sensor like in Fig. 1. The measurement of several field components requires a rotation (by 90° in Cartesian co-ordinate system). But, the sensor is not center symmetrical. Therefore, the sensor causes different field distortions dependent on its orientation (according to the measured field component). Hence, apart from the field distortion in general, it has to be expected that this effect causes an additional error. This influence can be seen in Fig. 2, considering the calculated field vectors in the same area around a sensor head at its two measuring orientations. To simplify the comparison, to chosen vectors of the upper picture were copied to the same position in the bottom one in Fig. 2. It is obvious that the absolute value and the direction of the vectors differ significantly.

4. Proximity effect

The nearer two bodies are situated to each other the stronger the interaction between free charges (conductors) and / or bound charges like in dipoles (dielectrics) is. Consequently, the nearer any sensorhead is placed to an electrode the more the field distribution around the sensorhead is changed, compared to the field at the same position without any sensor. On the one hand this problem is relevant for the calibration procedure of the measurement system. Usually the calibration is performed with an air capacitor, which is able to produce a homogeneous field in a sufficient large volume. The sensor is placed inside this volume of the capacitor. For a known voltage the homogeneous field strength is known considering the electrode distance. The comparison between sensor output and homogeneous field strength gives the scaling factor which is later used during measurement. But of course, placing the sensor inside the capacitor changes the field distribution inevitably.
As calculation and measurement show, the sensor output remains rather constant as long as the sensor crystal is not closer to the electrode than about 30 mm (Fig. 3). At smaller distances the proximity effect causes an increase of the field strength inside the crystal and consequently an increase of the sensor output. For the given sensor the maximum difference due to the proximity effect (sensor case and electrode in direct contact) was about 5 %.
Thus, a valid calibration can be expected if a minimal distance between crystal and electrode is realized. To meet the requirement, a sufficiently large capacitor is necessary and the sensor should be positioned near the center between both electrodes.
Furthermore, these considerations touch the question if the proximity effect is a serious influence at practical measurements. One has to take into account that field measurements at technical devices quite seldom take place directly at "electrodes" and that those "electrodes" are mostly covered by an insulating material. Hence, the critical distances are usually not reached. Thus, the proximity effect seems to have a rather small influence on the measurement. But quantification needs a detailed field calculation of a practical situation.

5. Field refraction at sensor interfaces

Besides the mentioned distortion and proximity effects it has to be taken into consideration that at interfaces between dielectrics of different permittivities, the electric field vectors are subjected to a refraction. The conditions at interfaces are given by two equations:

$$\frac{\tan \alpha_2}{\tan \alpha_1} = \frac{\varepsilon_{r2}}{\varepsilon_{r1}} = \frac{E_{1normal}}{E_{2normal}} \qquad (1)$$

$$E_{2tangential} = E_{1tangential} \qquad (2)$$

Here, α is the angle between the field vector and a line perpendicular to the interface. Index 1, 2 stand for the involved materials, E_{normal} is the field component perpendicular to the interface, $E_{tangential}$ the component parallel to the interface.
The main problem of refraction is connected to the form of the sensor. Different side lengths cause the following effects:

1. The distance a field vector has to pass through inside the sensor is different dependent on the orientation of the sensor. Hence, the crystal "sees" different vectors according to its orientation (Fig. 4).
2. The number and kind of interfaces a vector has to pass are different according to the sensor orientation.

3. Vectors which do not penetrate polarizer or crystal along the axis polarizer - crystal - polarizer (i.e. homogeneous field axis during calibration) are not exactly evaluated by the scaling factor determined from the calibration.

These effects could especially gain importance if a quite inhomogeneous field distribution has to be measured.
The point is that the quantitative influence of these mentioned effects can hardly be estimated, not at least because of the wide variety of interactions between all these effects. Analytical methods do not appear to be adequate to solve the problem. Thus, the decision was made to use numerical field simulation to get an idea for the complex effect of all these described influences.

6. Field simulation of a measuring session

For the numerical simulation a typical measuring problem was selected. Test object was a model cable termination installed on a model cable.
At first the calibration procedure was simulated. The geometry of calibration capacitor and sensor (in the center of the capacitor) were prepared in the FEM program. Then, for a given voltage the field strength inside the sensor crystal was calculated. This field strength is considered to be a measure for the sensor output.
With that the scaling factor was calculated from sensor output and homogeneous field strength inside the calibration capacitor (voltage over gap length).
Now it was possible to simulate a measurement. This means to place the sensor geometry in any field geometry, to calculate the field inside the sensor crystal (i.e. sensor output) using the FEM program and to calculate the absolute value of the "measured" field component using the scaling factor.

This procedure was done with the above mentioned cable-termination-geometry. This axis-symmetrical arrangement is presented in Fig. 5. In order to determine absolute value and direction (angle) of the field at the selected positions two components which stand rectangular to each other had to be calculated. Several positions around the termination were selected for calculation to include all possible field conditions (angles relatively to the sensor). This way all "measured" field strengths and directions were determined. There is no difference to real measurement procedures.
However, the advantage of the field simulation is the following: At all these sensor positions the undistorted field (means without a sensor) can be calculated with equal accuracy. This was done for the given arrangement.

The comparison between simulated measurement and real field relations is plotted in Fig. 6. Taking into consideration the variety of possible error sources, the result of the comparison is quite astonishing.

The maximum difference between "measured" and calculated field directions of only 0.4° (i.e. 0.11% of 360°) is the most surprising outcome. But also a maximum error of the absolute field strength, which is smaller than 4% is an unexpected result.
Probably the quantitative influences of the above described single effects are rather weak and possibly the effects are able to compensate each other in their influence on the field inside the sensor crystal.
Due to the selection of the simulated sensor positions all main cases of relative orientation between field vectors and sensor were included. Assuming usual field configurations, which are similar to the investigated field, one can regard the determined uncertainty as generally valid. Thus, the <u>accuracy</u> one can expect from the sensor system <u>is definitely sufficient for qualitative as well as for quantitative investigations</u>.

7. Summary

Exemplary for a common type of electro-optical field sensor possible error sources which could effect the accuracy of a field measurement were discussed. Included were field situations occurring during calibration and measurement procedure. In order to estimate the influence of the potential error effects it was necessary to take into account the whole complexity of their action. Therefore a numerical FEM simulation of a complete measuring session was performed. The outcomes show that the expected differences between measurable field, using the present sensor, and the real field are sufficiently small for typical measuring problems.

8. References

[1] *Matsumoto, T.; Kato, Y.:*
Digital Impulse Voltage Measurement Using Pockels' Cell. 7th ISH, Dresden, Germany, 1991, paper 63.02, Vol. 6, pp. 143-146.

[2] *Hidaka, K.; Kouno, T.; Takahashi, T.:*
New optical-waveguide Pockels sensor for measuring electric fields. 9th ISH, Graz, Austria, 1995, paper 8356, Vol. 8.

[3] *Ye, F.; Zhang, H.:*
The influence of probe on the measuring accuracy of electric field. 9th ISH, Graz, Austria, 1995, paper 8326, Vol. 8.

[4] *Mailand, M.; Aumeier, W.; Zaengl, S.:*
Electric field measurements on MV and HV components with an electrooptic sensor. 4th Int. Conf. on Insulated Power Cables, Paris, France, 1995, HV Technology 15.

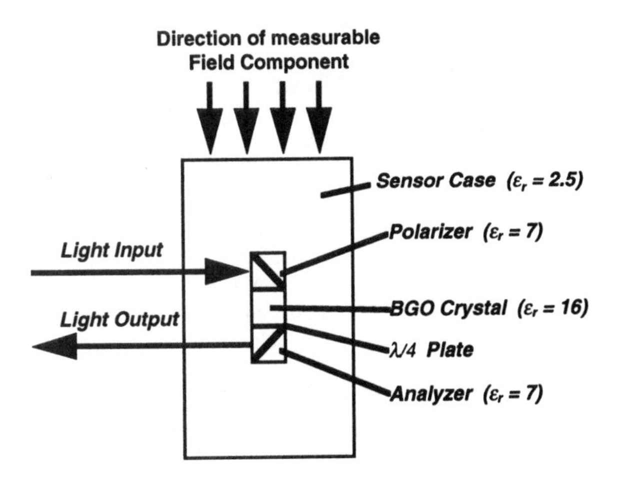

Fig. 1: Structure of the Pockels sensor

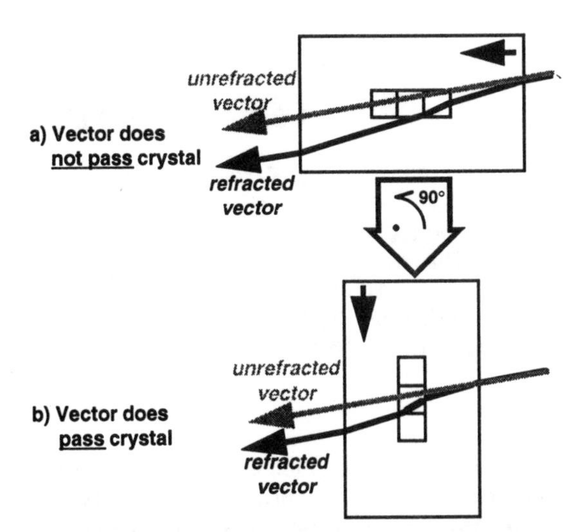

Fig. 4: Example for refraction effects inside the sensor

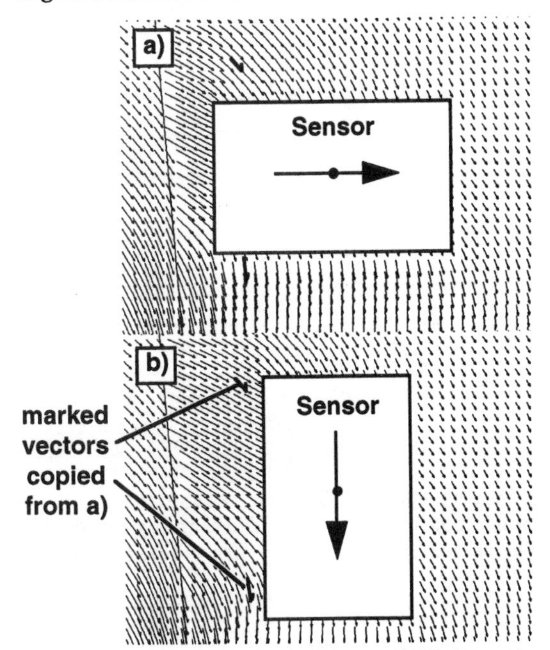

Fig. 2: Sensor orientation effect on field distortion

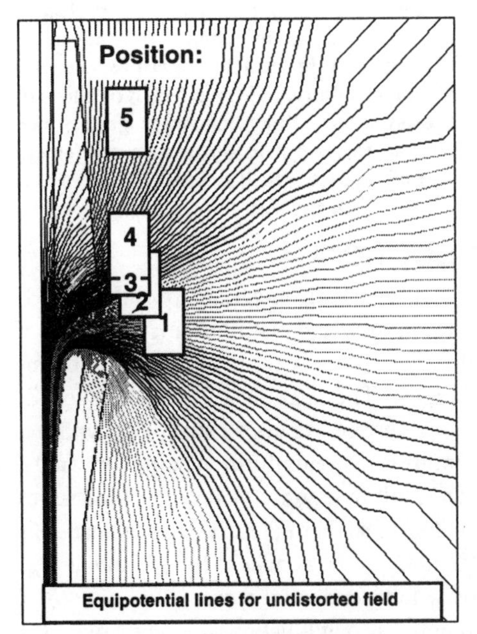

Fig. 5: Arrangment including undistorted equipotential lines and simulated sensor positions

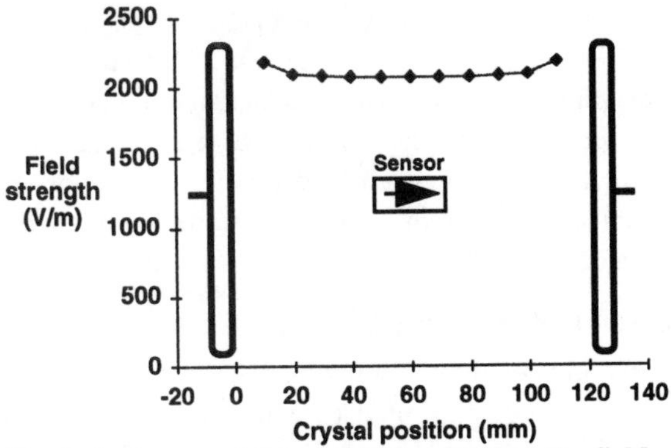

Fig. 3: Influence of the proximity effect on the field strength inside the sensor crystal

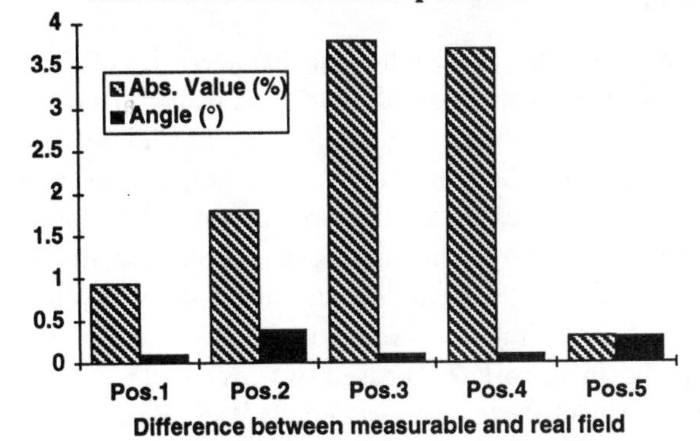

Fig. 6: Comparison between real field and the field "measured" by the sensor

Conference Record of the 1996 IEEE International Symposium on Electrical Insulation, Montreal, Quebec, Canada, June 16-19, 1996

Temperature Distribution in a ZnO Arrester Subjected to Multiple Current Impulses

Jinbo Kuang
Department of Electrical Engineering
University of Toronto
10 Kings College Road
Toronto, Canada M5S 1A4

Jeffery A. Bennett and Ram G. Bommakanti
Surge Arrester Systems
Raychem Corporation
300 Constitution Drive
Menlo Park, CA 94025

Steven A. Boggs
Electrical Insulation Research Center
University of Connecticut
Storrs, CT 06269-3136 and
Department of Electrical Engineering
University of Toronto

Abstract: **A recently developed finite element program for solution of coupled, transient, nonlinear field problems has been used to solve for the temperature distribution in ZnO arrester elements subjected to various current waveforms. Since the low field conductivity of ZnO is a strong function of both electric field and temperature, rapid changes in voltage can result in non-uniform heating which can cause local temperature rise, thermal runaway, and element failure. These effects are being investigated using a 2-D axisymmetric finite element program which solves for electric field and thermal field simultaneously for systems with temperature and/or electric field-dependent conductivity, thermal conductivity, heat capacity, etc.**

INTRODUCTION

The non-linear I-V characteristic of ZnO is used to protect power transmission and distribution systems from transient over-voltages, especially those caused by lightning-induced current surges. The large magnitude and short duration of the lightning current impulses causes substantial and, under some conditions, nonuniform heating of ZnO elements, which can result in damage or failure. Mechanical stresses caused by localized thermal expansion can cause mechanical failure. Although the conductivity of ZnO at high currents is relatively independent of temperature, the low field leakage current is a strong function of temperature. If the steady state leakage current becomes too large, thermal runaway can result.

Analysis of the electro-thermal characteristics of ZnO under impulse conditions is difficult as a result of the double non-linearity in the problem, i.e., the electrical properties of ZnO depend on both the electric and thermal fields, and the thermal properties are temperature-dependent. The problem must be solved by coupling the thermal and electric fields within the ZnO element. Experimental study of ZnO for large, 8/20 μs impulse currents is expensive and time consuming. In the past, ZnO-based systems have been modeled by equivalent electro-thermal circuits [1,2]. Here, we model a ZnO distribution arrester using the finite element method (FEM) with simultaneous solution for the electric and thermal fields and field-dependent material properties.

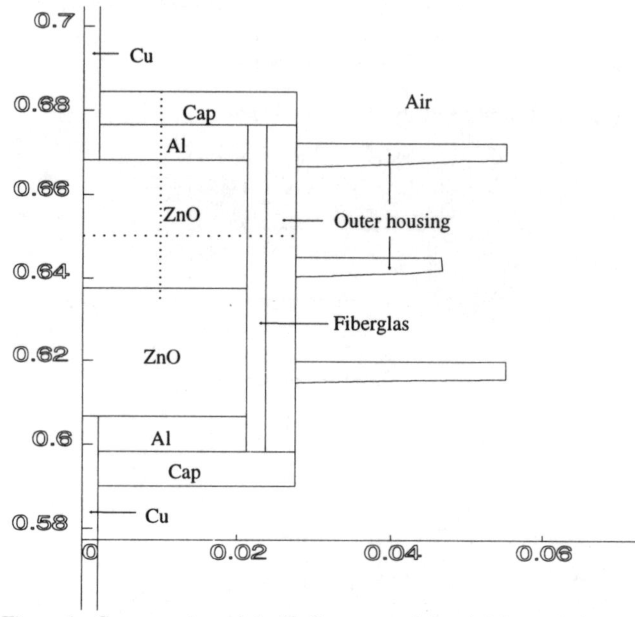

Figure 1. Cross-section of the ZnO arrester. Material thermal characteristics are provided in Table 1. The dotted lines show the positions of the temperature profiles of Figures 7 and 8. The vertical and horizontal scales are in metres.

MODEL CONFIGURATION

The model we studied is based on tests carried out by Raychem. The structure of the model ZnO surge arrester is shown in Figure 1. The two, 3 kV ZnO elements are enclosed by a Fiberglas tube within a shedded housing. The arrester is connected to the current source using heavy Cu braid, which is here modeled by a 60 cm long copper rod with a diameter of 4 mm. Table 1 lists material thermal characteristics.

LABORATORY TEST DATA

The ZnO arrester shown in Figure 1 was subjected to six, 8/20 μs, 10 kA current impulses while energized at a AC voltage of 7.2 kV (3.6 kV per ZnO element). The impulses are spaced 1 minute apart, and the AC voltage source was removed after the 6th current impulse. Data are available for 24 minutes from the beginning of the first current impulse.

Table 1. Material Thermal Parameters

Material	Thermal Conductivity (W/m-K)	Heat Capacity (J/kg-K)	Density (kg/m^3)
Zinc oxide	26	490	5500
Fiberglas	1.5	1600	1720
Outer housing	0.33	2300	1300
Aluminum	180	960	2800
Cap	0.33	2300	1300
Copper	230	384	8900
Air	0.1	1003	1.2

The negative temperature coefficient of low field ZnO conductivity is derived from data for Watts loss vs. temperature at an applied AC voltage of 3.6 kV per element. This is used to compute the power dissipation during the AC voltage application. The I-V characteristic (conductivity vs electric field) of the ZnO element is obtained from the test.

COMPUTATIONAL METHOD

A 2-D, axis-symmetric finite element program developed by one of us (JK) has been used to simulate the laboratory test described in the previous section. The software solves transient non-linear coupled field problems in time domain. In the present case, we are concerned only with the temperature distribution during the test, so that the computation is concentrated on the transient thermal field. We have used circuit theory to represent the ZnO electrically, i.e., we treat the ZnO as a nonlinear resistor subjected to a 8/20 μs impulse current waveform and compute the resistive power dissipation as a function of time based on the I-V curve of the ZnO. The power dissipation during application of the 3.6 kV AC between impulse currents is a function of ZnO temperature and is available from the test data. For the thermal boundary condition, we assume the far ends of the Cu rods are at room temperature, which is 19.4 °C according to the test data. We can assume that the upper and lower boundaries of the air are at room temperature, and for the right side boundary, we assume $\partial T/\partial n = 0$, i.e., the normal component of the temperature gradient along this boundary is zero.

The thermal field is governed by the diffusion equation,

$$-\nabla \cdot k \nabla T + \rho c \frac{\partial T}{\partial t} = Q \tag{1}$$

where k is the thermal conductivity, ρ is the mass density, c is the heat capacity, and Q is the power dissipation per unit volume, which in this study is the resistive power loss. Although the program is capable of accommodating temperature-dependence in k, ρ and c, these parameters are taken as constant with values as given in Table 1.

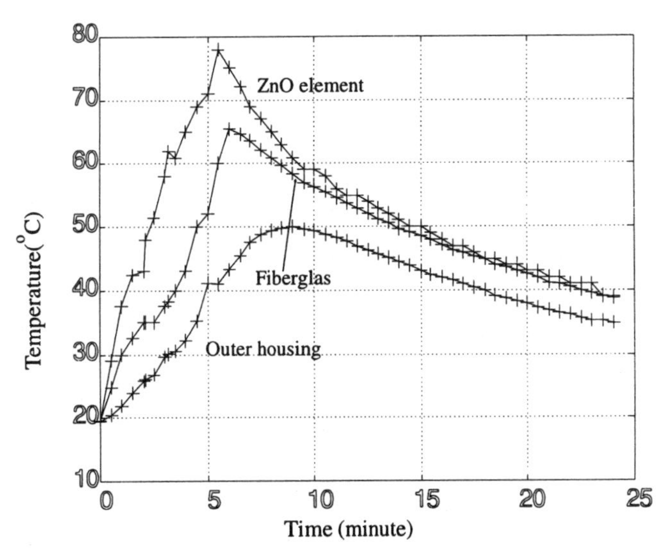

Figure 2. Laboratory test data for temperature as a function of time during a combined AC and current impulse test sequence for the ZnO element, Fiberglas shell, and outer housing of a model arrester.

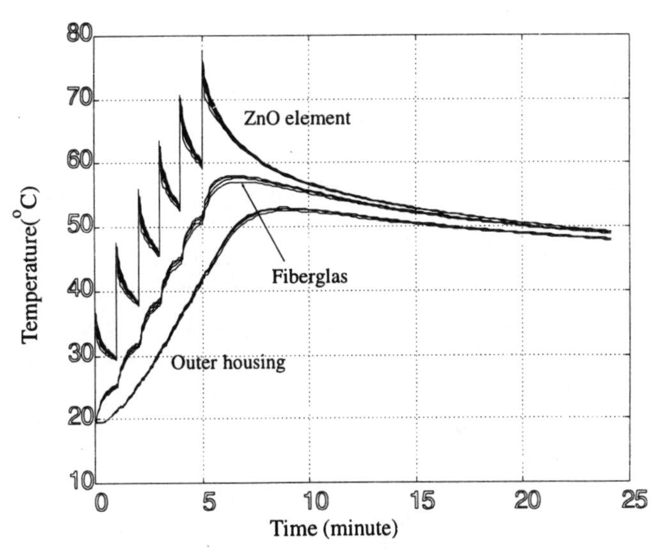

Figure 3. Computed temperature variation as a function of time for various locations on the ZnO element, the Fiberglas shell and outer housing.

COMPUTATIONAL RESULTS

Figure 2 shows the measured time variation of the temperature for some points on the outside surface of the ZnO elements, Fiberglas, and shedded housing. The temporal resolution of the measured thermal data was several seconds. The temperature was measured with thermocouples located at the perimeter of the substrate equidistant between the electrodes. These experimental data can be compared with

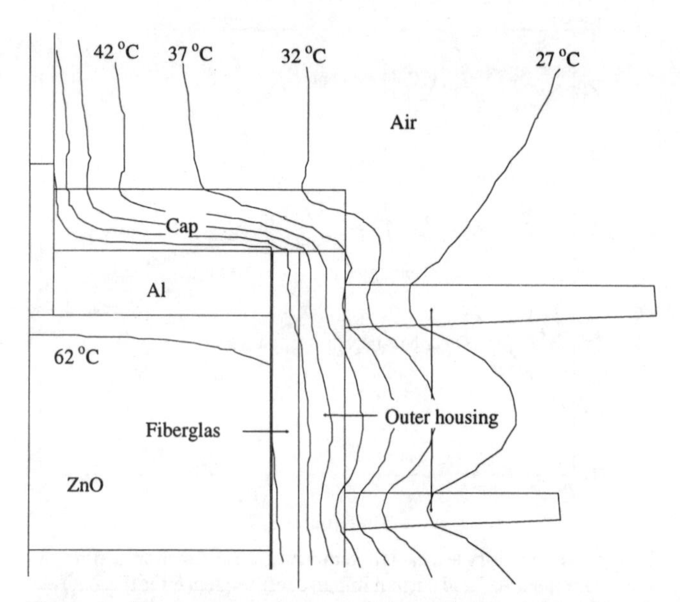

Figure 4. The temperature contour lines at 5 minutes into the test, which is a minute after completion of the 5th current pulse. The temperature difference between contour lines is 5 °C.

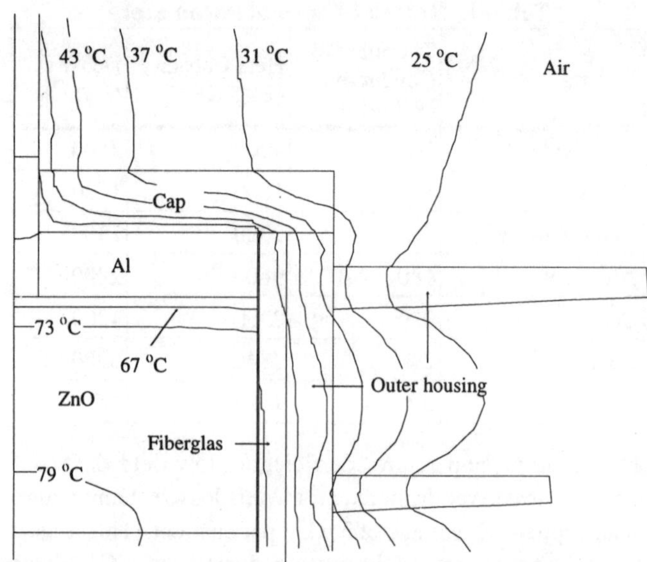

Figure 5. The temperature contour lines at 5 minutes plus 0.93 second into the test, which is 0.93 s into the 6th current pulse. The temperature difference between contour lines is 6 °C.

computed data of Figure 3. The time variation of temperature obtained from computation and test data are similar. The principal discrepancy between the model and measured data is the cooling rate after the peak temperature. This could be attributed to factors such as model assumptions made for convective heat flow and temperature dependence of material thermal parameters. From Figures 2 and 3, we can see the maximum temperature increase is 60 °C. Figure 4 shows thermal contour lines around the upper half of the arrester at 5 minutes into the test, i.e., immediately prior to application of the 6th current impulse. At this time, the ZnO element and the aluminum electrode have almost same temperature as a result of thermal diffusion during the minute since the end of the 5th current impulse. Figure 5 shows the temperature contour lines 0.93 second into the 6th current pulse. A large temperature gradient is apparent at the boundary of ZnO and aluminum, as confirmed by the temperature profiles of Figure 7. Figure 6 shows temperature contour lines around the ZnO elements at 24 minutes into the test, i.e., at the end of the test, by which time the temperature distribution is relatively stable. Figure 7 shows the temperature variation along the horizontal line indicated in Figure 1 at four times before, during, and after the 6th current impulse. Figure 8 shows the temperature variation along the vertical line indicated in Figure 1 for the same four times. The 6th impulse is applied at 300 s with a duration of 20 μs. The duration of the impulse is so short that measuring the temperature change during the impulse would be very difficult. The computation during the current impulse is made with a time step of 2

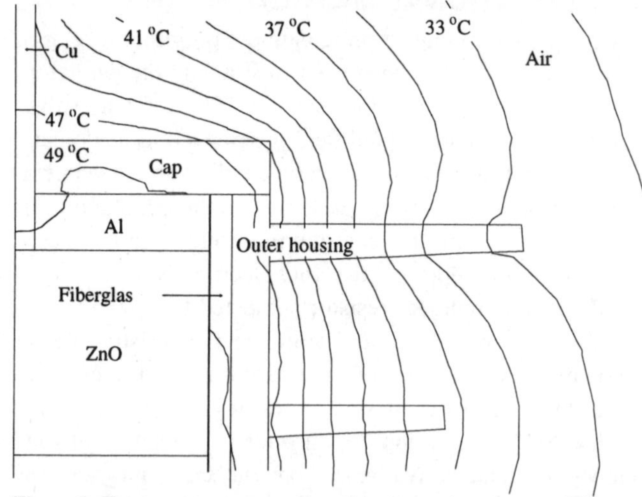

Figure 6. Temperature contour lines 24 minutes into the test. The temperature difference between contour lines is 2 °C.

μs, which allows determination of the temperature within the ZnO elements and arrester components during the current impulse. The spatial resolution of the computation depends on the mesh size. For regions of large temperature gradient, a fine mesh is employed. For the computations presented in this paper, the smallest mesh element is about 3 mm on a side.

CONCLUSIONS

Laboratory testing is time consuming and expensive. In all areas of design, numerically-based design technology is replacing laboratory testing. The ability to undertake such

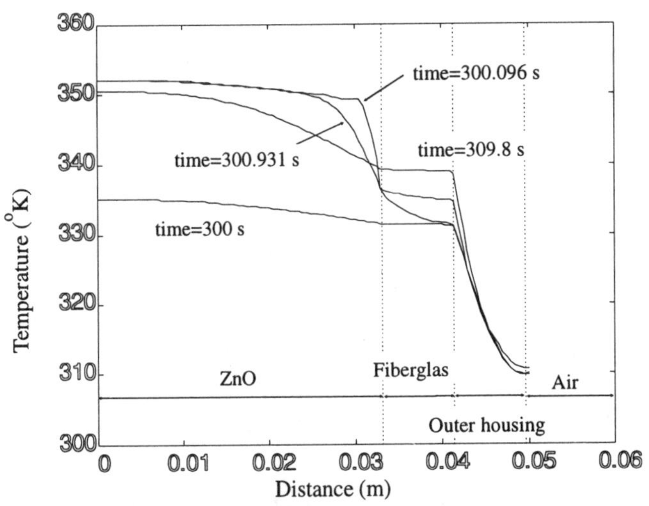

Figure 7. Temperature profile at various times along a horizontal line across the arrester as shown by a dotted horizontal line in figure 1.

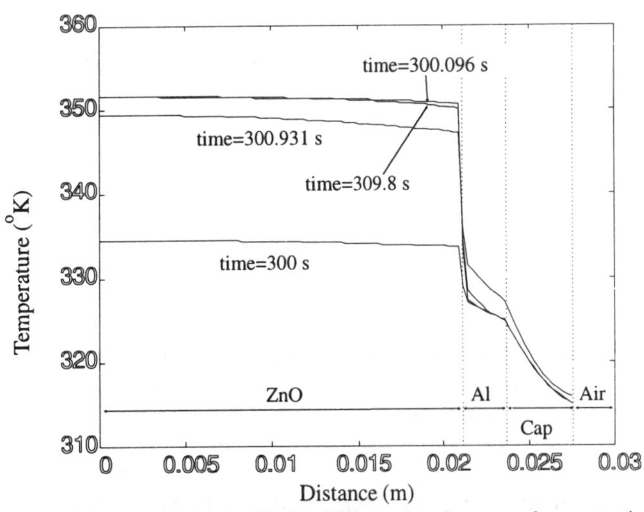

Figure 8. Temperature profile for different time instance along a vertical line across the arrester as shown by a dotted vertical line in figure 1.

computations would both reduce the need for expensive laboratory testing and provide the ability to investigate the effect of various design modifications relating to the metal oxide V-I characteristics.

We have demonstrated reasonable agreement between measured and computed thermal data for a model lightning arrester subjected to combined AC and current impulses over an extended time period. With improved material parameters and further calibration of model parameters, improved agreement can undoubtedly be achieved. The study included the electric field and the temperature dependence of the electric conductivity of the ZnO elements. In principle other nonlinearities can be included as well (e.g., temperature dependence of the thermal conductivity and/or heat capacity). These computational techniques allow temperature gradients and/or conductivity gradients within the ZnO to be determined, from which mechanical stresses and nonuniform element heating can be computed. The data can be displayed in many forms, including profiles along a line, equi-field contour plots, or colorized field magnitude plots.

On the basis of the reasonable agreement demonstrated between measured and computed temperatures, we believe that the temperature distribution of ZnO arresters can be computed accurately given improved material and model parameters.

ACKNOWLEDGMENT

The authors are pleased to acknowledge the financial support to this work from the National Science and Engineering Research Council of Canada.

REFERENCES

1 Zahedi, Ahmad. "Effect of Dry Band on Performance of UHV Surge Arrester and Leakage Current Monitoring, using Developed Model", Proceeding of the 4th International Conference on Properties and Applications of Dielectric Materials, July 3-8, 1994, Brisbane, Australia, pp. 880-883.

2. Darveniza, M., D. Roby and L. R. Tumma. "Laboratory and Analytical Studies of the Effects of Multipulse Lightning Current on Metal Oxide Arresters", IEEE Trans. on Power Delivery, Vol. 9, No. 2, April 1994, pp. 764-771.

3. Dortolina, Carlos A. and Rafael A. Rios. "Surge Arresters, Protection Equipment from Heatstroke", IEEE Potentials, February/March 1996, pp. 34-37.

Conference Record of the 1996 IEEE International Symposium on Electrical Insulation, Montreal, Quebec, Canada, June 16-19, 1996

Comments on Metal Oxide Surge Arresters Surges Energy Absorption Capacity

Manuel L.B. Martinez
Escola Federal de Engenharia de Itajubá - EFEI
P.O. Box 50 - CEP 37.500-000
Itajubá - Minas Gerais - Brazil

Luiz Cera Zanetta Jr.
E.Politécnica Universidade de São Paulo - EPUSP
Av. L. Gualberto, Trav. 3 - 158 - CEP 05.655-010
São Paulo - São Paulo - Brazil

Abstract: **This paper presents an approach to determine the energy absorption capacity of metal oxide surge arrester resistors. The proposed approach deals with the discharge current peak versus discharge current time relation. A testing method and a statistical evaluation are proposed. After determining the discharge current withstanding limit of the tested metal oxide resistors, the prospective energy absorption capacity limit is computed. Finally, comments on the obtained results are presented.**

INTRODUCTION

This paper discusses a non-standard approach to demonstrate the energy withstanding capacity of metal oxide arrester resistors. The developed approach is based in the duty factor method [1]. As it is a general method to check the discharge current withstanding capacity of metal oxide resistors this approach can be also used to define a model to compute margins for the energy absorption capacity. The duty factor approach deals with the discharge current - the stress - discretization and with the relation between the discharge current peak and the discharge current time - the withstanding capacity - of the metal oxide resistors, as stated by equations 1 and 2, respectively .

$$S = \sum_{i=1}^{n} \frac{\Delta T_i}{T_{Max-i}} \qquad (1)$$

where: S - Discharge current duty factor, ΔT_i - i^{th} Rectangular pulse discharge time and T_{Max-i} - i^{th} Maximum rectangular pulse discharge time, computed by equation 2.

$$T_{Max} = \alpha \, I^{-\beta} \qquad (2)$$

where: α and β - are generic constants and I - Current pulse peak.

The duty factor approach takes into account the possibility of current concentration on small conducting channel. This results in a differential heating and, therefore, in a series of mechanical stresses that can destroy the resistor. Then, once reached a local critical temperature gradient, and a defined destroyer energy, the resistor collapsed, normally, by cracking, as observed during all carried out tests.

The development of a localized type of failure is independent of the discharge current wave shape and amplitude. Therefore, in theory, it is possible to use the rectangular pulse discretization to carry out a general discharge current evaluation.

To carry out the testing evaluation it is proposed to apply the *up and down test method* and the *likelihood method*, that are non-standardized techniques. An approach to compute the metal oxide resistors discharge current probability of failure, is presented. Finally, the application of this approach to the determination of the prospective energy absorption capacity of metal oxide resistors is commented.

BASIC DATA DETERMINATION

The discharge current and energy absorption capacity of metal oxide resistors are usually obtained by the transmission line discharge test. This test was designed to provide withstanding-failure data on silicon carbide arresters. However, as verified by field practice, it can be successfully applied to metal oxide arresters.

In order to check a non-standard performance of the metal oxide arresters it is possible to use a suitable special testing set up. As the duty factor approach and model deal with a discharge current peak versus discharge current time relation they can be alternatively applied.

The simplified duty factor laboratory work deals with a test set up able to produce a semi-sinus discharge current wave shape. This set up must generate current impulses covering three current peak and three discharge time decades. The *up and down testing method*, one sample per level, in order to obtain $I_{50\%}$, is presently adopted. As proposed the laboratory work requests sample batches, containing 20-25 metal oxide resistors per current discharge time. It is used a visual inspection to verify the sample withstanding or failure. For each discharge current time, data point characteristics, i.e., mean current value, standard deviations and confidence limits are computed by the likelihood method. To consider the statistical current and discharge time dependence it is proposed the use of the Symmetric Weibull Distribution.

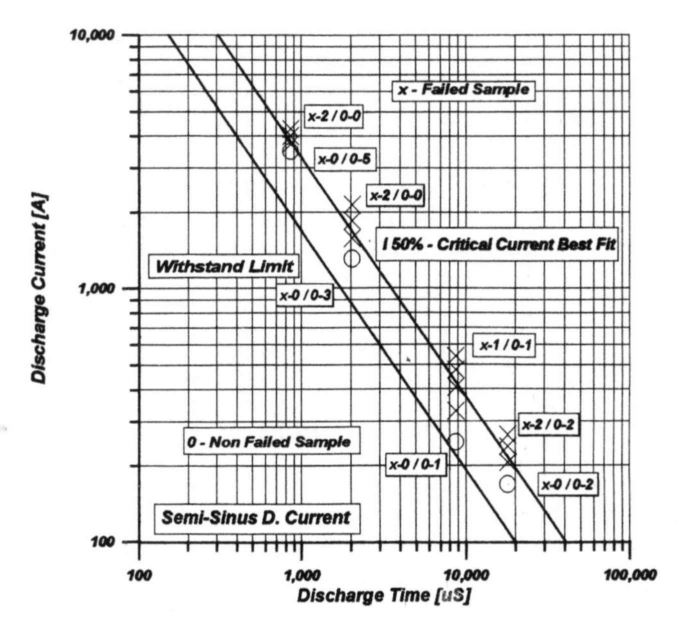

Figure 1 - Samples Discharge Current and Time Data Summary

The results presented in Figure 1, part of a wide study on 350 samples, are related to 82 metal oxide resistors, divided in 4 sets, submitted to semi-sinus discharge current applications. The following rule was applied, high current-short duration and low current-long duration impulses, that is in agreement with equation 2. The obtained data ranged from 850 μs to 18,000 μs. This figure shows a test result summary and also the computed critical current - $I_{50\%}$ and withstanding limit current - $I_{0\%}$ versus time relations. For each discharge current time, for the superior and for the inferior discharge current level, the number of non-failed and the number of failed samples are presented.

Values for the minimum, the mean and the maximum current peak, associated to the figure 1 testing results, are obtained with the likelihood method. The results, for each discharge current time, are presented on table-I.

Table I- Semi-Sinus Discharge Current Data Results

Semi-Sinus Discharge Current			
T_D[μS]	$I_{Med.}$[A]	$I_{Min.}$[A]	$I_{Max.}$[A]
860	3,795	3,400	4,170
2,050	1,730	1,550	1,945
8,700	400	305	470
18,000	230	210	245

The equation 3, obtained from equations 1 and 2, basing on the semi-sinus testing results computes the rectangular discharge current versus time relation. In this case it is considered, for both wave shapes, the same discharge current stress, i.e., a duty factor equals one.

$$T_R = \frac{\pi \alpha}{\int \sin^{\beta} x \, dx} I^{-\beta} \tag{3}$$

where: T_R - Rectangular discharge current duration.

From figure 1 and from equation 3 it is possible to obtain the data issued on figure 2 where, they are plotted, for the semi-sinus wave shape, the critical discharge current $I_{50\%}$ versus time characteristic and the computed rectangular wave shape series, basis for the duty factor approach.

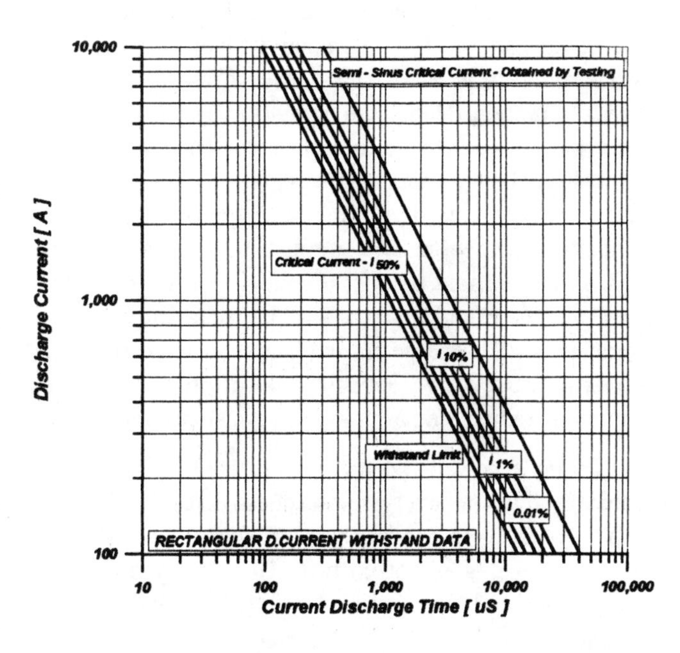

Figure 2 - Semi-Sinus Base Data and Generated Rectangular Data

COMMENTS ON THE OBTAINED RESULTS

Regarding the duty factor approach it is necessary to state two key points. The first is: *As proposed, the obtained data intend to represent the differential heating transient process of metal oxide resistors submitted to a single discharge current impulse. This model does not cover any surge ageing process.* The second is: *The global surge arrester risk of failure is associated to the surge arrester design. It is interesting to observe that to get the same risk of failure, a metal oxide distribution surge arrester can deal with a higher metal oxide resistor probability of failure.*

After computing the complete discharge current failure-withstanding behaviour, as issued in figure 2, it is necessary to define the working limit for the probability of failure of the metal oxide resistors. This depends on the final arrester design and system application. The withstanding limit characteristic, i.e., the 0% - probability of failure characteristic is the logical choice, However, it is sometimes necessary to work with a "non-zero" probability of failure. Therefore, for practical purposes the 0.01% - probability of failure characteristic at which a "duty factor - S=1" is assigned seems to be a reasonable work limit for the probability of failure of the metal oxide resistors. For a 144 kV surge arrester, this results in a withstanding probability of 99.5%.

After defining the resistor probability of failure, according equation 1, it is computed, for any non-standard discharge current wave shape, the current peak versus current duration relation, as issued in figures 1 and 2 for the semi-sinus discharge current shape. Once obtained the basic discharge current parameters, peak and duration, the metal oxide resistor prospective energy absorption capacity is computed by a simple time integration of the voltage and current product. Figure 3, shows the current peak versus specific energy - kJ/kV data for the tested samples. The discharge current peak and the metal oxide arrester prospective energy absorption capacity are common results of a power system transients study.

During a power system transient study it is usual to consider that the energy absorption capacity does not depend on the current peak and wave shape. As can be seen on figure 3 this is not correct. The energy absorption capacity depends on equation 1 or 3 β parameter, as shown by equation 4, that deals with rectangular and semi-sinus discharge current shapes. However, for the tested samples, i.e., for $\beta = 1$, according to equation 4 the energy absorption capacity does not depend on the current shape. Finally, as can be seen in figure 3, a current peak increase results in an increase on the energy absorption capacity limit.

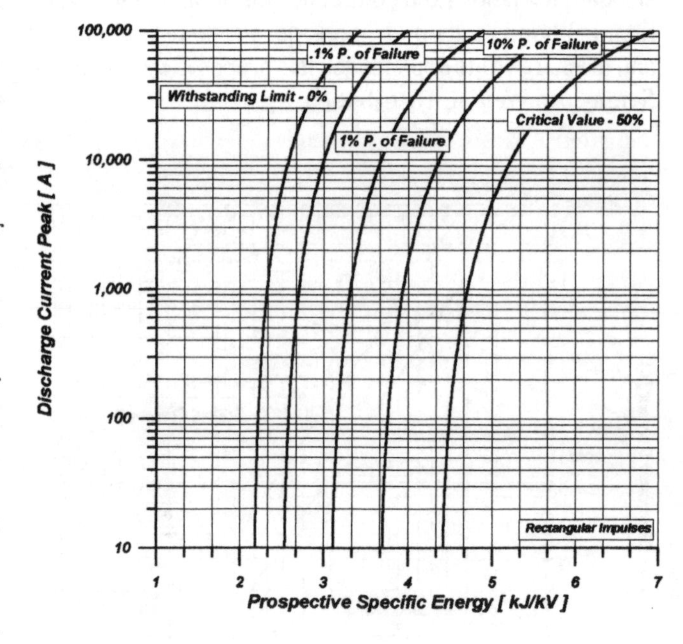

Figure 3 - Discharge Current versus Prospective Specific Energy and Probability of Failure

$$\frac{E_R}{E_S} = \frac{\int_0^{\pi} \sin^{\beta} x \, dx}{(2-\frac{\pi}{2})\frac{k_2 - Ln(I) - 1}{k_2 - Ln(I)} + \frac{\pi}{2}} \tag{4}$$

where: E_R - Energy absorbed by the resistors submitted to a rectangular discharge current, E_s - Energy absorbed by the resistors submitted to a semi-sinus discharge current and I - Discharge Current peak.

To the development of equation 4 it was considered the metal oxide resistors residual voltage model stated on equation 5 .

$$U = \frac{k_1}{k_2 - Ln(I)} + k_3 I \tag{5}$$

where: U - Metal oxide residual voltage, I - Discharge Current peak and $k_1 - k_2 - k_3$ - generic constants.

For the tested resistors, according to equation 4, for a discharge current peak inferior to 1,000 A, the energy absorbed by the resistors submitted to a rectangular current is equal to the energy absorbed by the resistors submitted to a semi-sinus current. Considering a β value of 1.60[1] and the equation 5 residual voltage model, for a discharge current peak inferior to 1,000 A, the ratio of the energy absorbed by the resistors submitted to a rectangular discharge current by the energy absorbed by the resistors submitted to a semi-sinus discharge current is approximately equal to 0.80. The manufacturer published information states for this ratio a value of 0.57 [1]. The equation 4 is obtained considering a linearization of the residual voltage versus discharge current characteristic, therefore, it is suitable for a residual voltage - reference voltage ratio inferior to 1.50, as for the case of the tested samples. In any case, a better correct ratio value can be obtained by numerical integration.

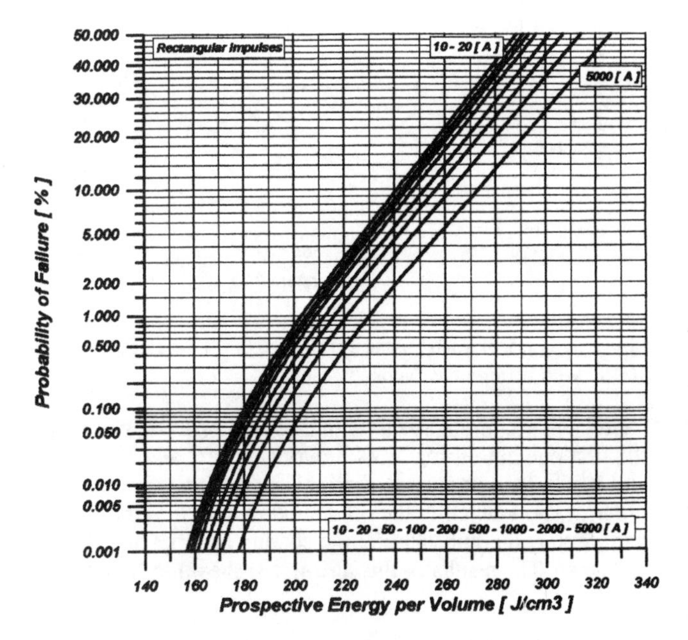

Figure 4 - Prospective Energy and Probability of Failure

Recently published results are in a good agreement with the presented testing results, i.e., with a β value of 1.00 [3]. However, for any new resistor design this must be confirmed otherwise, it will be difficult, or even doubtful, to assume that the energy absorption capacity of the metal oxide resistors does not depend on the current wave shape.

As shown by figure 4, for the presented tested samples, the energy absorption capacity is dependent of the discharge current peak. According to figures 3 and 4 the lowest value of the energy by unit of volume must be considered as the metal oxide resistor design limit. In the present case, depending on the probability of failure, this amounts to a value between $145 J/cm^3$ and $160 J/cm^3$.

FINAL CONCLUSIONS

As commented the energy absorption capability of a metal oxide arrester can be modelled by the duty factor approach.

Depending on the value of the coefficient β - discharge current versus current discharge time model parameter - it is impossible to assume, that in the limit, the metal oxide arrester energy absorption capacity does not depend on the current shape. Even in the case of a suitable coefficient β value the metal oxide arrester energy absorption capacity depends on the discharge current peak.

As shown in figure 4 for a β value of 1.00 the metal oxide resistor energy absorption capacity increases with a current peak increase. For a β value of 1.60 [1] there is a decrease in the metal oxide resistor energy absorption capacity with a current peak increase. This means that it is always necessary to carry out a careful evaluation before to state a limiting value for the metal oxide arrester energy absorption capacity.

An analysis of the transmission line discharge test suggests that, as stated, it can be safely used to define a reliable limit for the metal oxide resistors energy absorption capacity. However, normally, this limit is far bellow the manufacturer published limiting data.

Finally, it is necessary to consider that: *"A standard test method to demonstrate discharge current limits and associated energy absorption capability must be developed and included in the appropriated standards"*[3].

REFERENCES

1. Neugebauer,W.;"Application of Arresters Near Shunt Capacitors Banks"; Application Eng. Conference Roundtable, Schenectady, 1983;

2. Martinez,M.L.B.; Zanetta Jr., L.C.; "An Approach to Metal Oxide Arresters Switching Surges Energy Absorption Capacity Evaluation'; IPST 95 - International Power System Transients; 1995; and

3. Kirkby, P. et alli; "The Energy Absorption Capability and Time to Failure of Varistors used in Station Class Metal Oxide Surge Arresters"; 95 - SM - 364-0 PWRD.

Conference Record of the 1996 IEEE International Symposium on Electrical Insulation, Montreal, Quebec, Canada, June 16-19, 1996

Ageing study of the mineral oil in an oil-immersed ZnO-based surge arrester

Mohammad R. Meshkatoddini, André Loubiere, Ai Bui
Laboratoire de Génie Electrique
Université Paul Sabatier
118, Route de Narbonne
31062, Toulouse Cedex, France

***Abstract*:** **In this work the degradation of transformer mineral oil in oil-immersed ZnO-based surge arresters has been studied. Three different kinds of ZnO varistors: 1-non-coated, 2-resin-coated and 3-glass-coated, have been used in this study. It can be seen that the ZnO varistor and its coating have an influence on degradation of the oil.**

INTRODUCTION

Recently, there has been an interest for installing the ZnO-based varistor elements inside the oil-immersed transformers [1,2,3]. We think that it might also be interesting to realize oil-immersed metal-oxide surge arresters. We have already studied the long-term behavior and the advantages of installing the varistors in oil [4,5,6,7]. But we have to verify that the varistors and their coating do not accelerate the degradation of transformer oil, otherwise installing them inside the transformer during a long period is not reasonable.

In this work the degradation of the oil specimens containing different kinds of varistors, subjected to thermic and electric constraints in three ageing processes has been studied. Firstly, we have accomplished an ageing of the oil with different varistors at 100°C during 6,000 hours. Secondly we have realized another ageing during 5,300 hours at 100°C plus 264 hours at 140°C; and the third ageing has been done by applying the high-amplitude electric impulses to the varistors immersed in oil. The electrical characteristics of the oil before and after the ageing have been measured.

Long-period ageing of the oil with varistors

In this comparative research we wanted to know which type of varistor coating has an accelerating role in the oil degradation.

In order to study the interaction between oil and varistor, we have put differently coated varistors in oil specimens at 100°C without applying any electric constraint to the varistors. In figure 1, we see the dissipation factor (tan δ) of the oils aged with different kinds of varistors during 6,000 hours at 100°C.

We observe a degradation in oil, which is more in the oil samples aged with resin-coated varistors and less for the oil samples aged with glass-coated or non-coated ones.

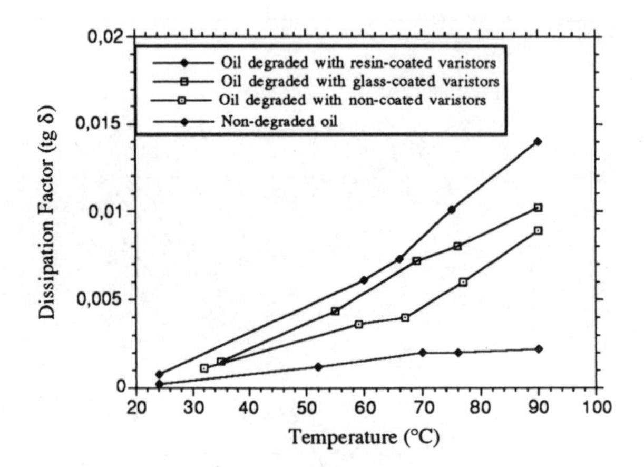

Figure 1: tan δ of transformer oil after 6,000 hours of ageing at 100°C with different varistors.

We have accomplished a second ageing process, similar to the first one but during 5,300 hours at 100°C and 264 hours at 140°C. This important thermal constraint was added in order to know whether the oil degradation in the first ageing had a purely thermic cause or the coating material had also an influence. The results of this ageing are shown in figure 2.

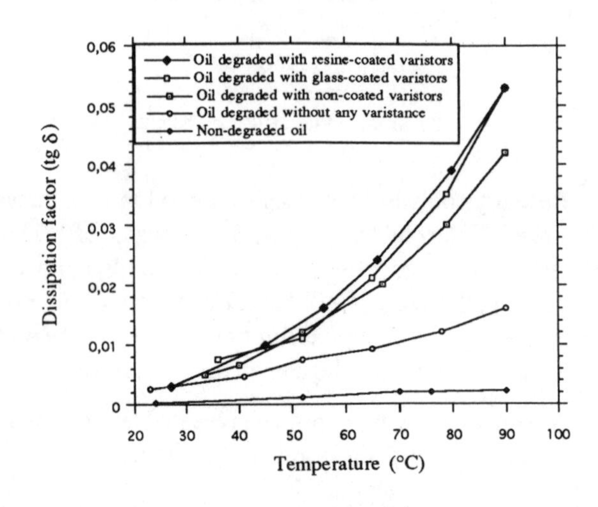

Figure 2: Tan δ of the oils degraded with different varistors during 5,300 hours at 100°C plus 264 hours at 140°C.

It is observed that the degradation of the oil samples is in the same order as the first experiment, but the oil is much more degraded, which is quitely normal because of the important thermal constraint. This shows that there is an acceleration in the oil degradation because of the coating materials.

We have also measured the breakdown voltage of the oil samples by applying an increasing power frequency voltage in a cell with two electrodes having a distance of 2.5 mm from each other, according to the IEC-156 standard. Table 1 shows the measured breakdown voltages.

Table 1: Breakdown voltage (in kV) of the oil samples aged with different varistors during 5,300 hours at 100°C plus 264 hours at 140°C

oil \ number	1 (removed)	2	3	4	5	6	average
non-degraded oil	44	35	42.5	41	44	43.5	**41.2**
oil degraded without any varistor	39.5	39	36.5	42	37.5	34	**37.8**
oil degraded with glass-coated varistors	49	40.5	41	34	35	38	**37.7**
oil degraded with resin-coated varistors	40.5	34	31	34	31.5	28.8	**31.8**
oil degraded with non-coated varistors	47.5	27	31	29	28.5	34.5	**30**

After removing the first measurement, according to the standard, we take the average of the five other ones. The average breakdown voltages are shown in the last column of the table. We observe that the breakdown voltage of the oil samples aged with the non-coated and resin-coated varistors has decreased significantly, while that of the oils with the glass-coated ones has changed very little because of the applied thermal constraint.

For understanding the cause of degradation of the oil specimens aged with the non-coated varistors we say that in the long-period ageing processes without applying any voltage to the varistors, we have found that the non-coated ZnO varistors are degraded in the same manner as the varistors aged in oil but with applying an ac voltage, while the varistors coated by resin or glass are not at all degraded [4]. This shows that there is a chemical interaction between oil and naked ZnO varistors. We have tried to study and analyse the nature of this interaction [7]. The degradation of the naked ZnO varistors in transformer oil is attributed to the reducing character of the oil. According to most of models on varistor action, electron traps in the surface region of the ZnO grains make the normally n-type material to have Schottky-like back-to-back barriers at each grain boundary, which is responsible for the varistor effect. Destruction of the non-linear varistor action by an ambient reducing atmosphere like the mineral oil involves the modification of these potential barriers. The importance of oxygen in barrier formation has been reported in the litterature [8]. Absorption of oxygen from the varistor by the oil creates a surface layer which demonstrates a low resistance parallel path, which gradually brings the varistor to a short circuit. We realized some experiments in order to verify and explain this hypothesis [7]. This phenomenon can help to degradation of the oil.

Ageing the oil under high-amplitude current impulses:

We applied the 65kA, 5/15µs current impulses to different varistors immersed in oil. The experimental set-up is shown in figure 3.

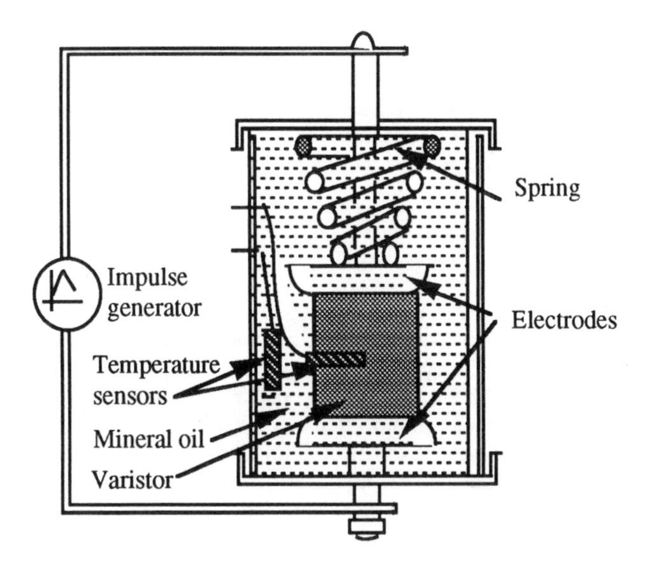

Figure 3: Experimental set-up for applying impulses to the varistors in oil.

Study of the oil characteristics after application of the impulses shows a degradation in the oil supporting the impulses on the naked or resin-coated varistors, while the oil supporting the same impulses on glass-coated varistors is not degraded, as observed in figure 4.

Summary

Experiments show that in an oil-immersed surge arrester, some kinds of varistors have an accelerating influence on degradation of mineral oil. We observe that in a long-period ageing, the resin-coated varistors have a degrading influence on dissipation factor and breakdown voltage of the oil. As well, the non-coated varistors have a degrading influence on breakdown voltage of the oil. Likewise under high-amplitude current impulses the oils with resin-coated and non-coated varistors demonstrate a degradation. On the other hand

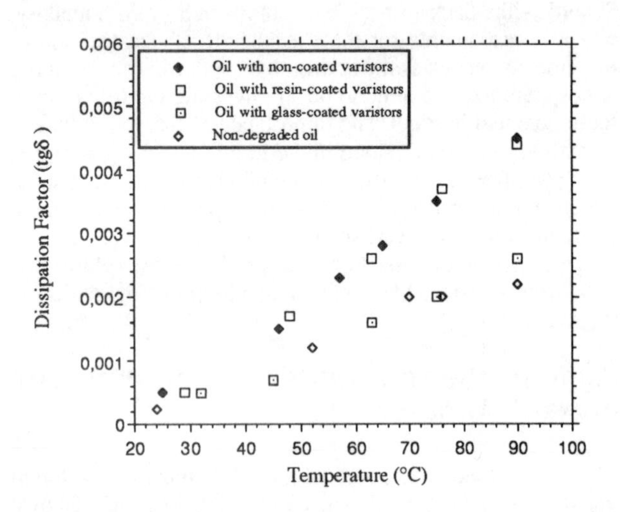

Figure 4: Tan δ of the oils containing different varistors subjected to high-amplitude impulses.

applying high-amplitude current impulses to glass-coated varistors in oil has no influence on its dissipation factor. Furthermore long-period ageing of the oil with glass-coated varistors doesn't provoke any degradation in its breakdown voltage; Nevertheless there is a degrading influence on its dissipation factor.

Conclusion

We can conclude that the non-coated and resin-coated ZnO-based varistors can help to degradation of the mineral oil in an oil-immersed surge arrester, while the glass-coated ones exhibit a better behaviour. It has been shown that the glass-coated ZnO-based varistors have an excellent thermal and electrical performance in oil [5,6]. Nevertheless, in order to avoid the probable problems of the arrester for transformer, the surge arrester should be installed in an oil tank isolated from the transformer oil.

References

[1.] C.J.Mc Millen et al. "The development of an oil-immersed surge arrester for distribution transformers". IEEE Transactions on Power Apparatus and Systems. Vol. PAS-104, No.9,pp.2482-2491, September 1985.

[2] Y.Harumoto et al "Evaluation for application of built-in type zinc oxide gapless surge arresters for power system equipments". IEEE Transactions on Power Delivery, Vol. PWRD-2, No.3, pp. 750-756, July 1987.

[3] W.R.Henning et al."Fault current capability of distribution transformers with under-oil arresters". IEEE Transactions on Power Delivery, Vol.4, No.1, pp.405-412, January 1989.

[4] M.R.Meshkatoddini, A.Loubiere and A.Bui. "Ageing study of ZnO-based varistors in transformer oil". ICSD'95, 5th International Conference on Conduction and Breakdown in Solid Dielectrics,Leicester, U.K., Proceeding pp. 498-502, July 1995.

[5] M.R.Meshkatoddini, A.Loubiere and A.Bui. "Long-term behaviour of ZnO-Based varistors in mineral oil". Proceedings of ICECM-ICSA'95, Xian, China, October 1995.

[6] M.R.Meshkatoddini, A.Loubiere and A.Bui. "Study of the ZnO-based varistors thermal behaviour in mineral oil for realization of an oil-immersed ZnO surge-arrester". PSC95, 10th International Power System Conference, Tehran, Iran, Proceeding pp.375-383, November 1995.

[7] M.R.Meshkatoddini, A.Loubiere and A.Bui. "Degradation study of oil-immersed ZnO-based varistors". MEPCON-96, Fourth Middle East Power Systems Conference, Assiut, Egypt, Proceeding pp.372-376, January 1996.

[8] E.Sonder et al. "Effect of oxidizing and reducing atmospheres at elevated temperatures on the electrical propeties of zinc-oxide varistors". Journal of applied physics, 54(6), pp.49-58, April 1986.

Conference Record of the 1996 IEEE International Symposium on Electrical Insulation, Montreal, Quebec, Canada, June 16-19, 1996

Partial dicharges as a diagnostic tool for full size power capacitors

Charles Hantouche
Electricité De France
Direction des Etudes et Recherches
Les Renardières B.P. 1
77250 Morêt sur Loing - FRANCE

Abstract : In this paper Partial Discharges (PD) measurements on industrial power capacitors have been carried out. Results of some physical parameters are presented after different accelerated aging times and during Transient Overvoltages (TOV) tests on all-film power capacitors with high capacitance values. Many observations emerge from PD measurements and some parameters could be the indicators of dielectric capacitor states.

INTRODUCTION

To know the reliability of power capacitors on electrical network, different procedures of type tests are introduced in international, national and utility standards. In the aim to use PD measurements as a control tool of capacitor manufacture and as a diagnostic tool of dielectric states and also to introduce it in standard, this paper presents the investigations into PD activity in industrial power capacitors. It suggests some parameters to determine the performance of capacitor manufacture and dielectric states.

The bridge method of PD detection is used due to the highest capacitance value, more than many µF, [1,2]. Multi-terminal industrial capacitors were used. This type of capacitor allowed the PD measurements to be carried out on a single element, on parallel elements, on series groups of elements or on the full capacitor (see figure 1). The capacitors had no known defects. Tey had a capacitance of 3.8 µF and rated at a voltage of 6 kV. Therefore, a single element had the same capacitance (3.8 µF) with a rated voltage at 2 kV. The dielectric consisted of polypropylene films impregnated with a nonchlorinated dielectric liquid. The electrodes were made with extended aluminium foils. The first multi-terminal capacitor was submitted to accelerated aging test and PD measurements after each 500 h of aging time. Test results are presented in section 2. The second multi-terminal capacitor was submitted to transient overvoltage (TOV) test at 20°C. The PD spectrum of some TOV cycles are shown in section 3. Finally, many observations emerge from PD measurements and some parameters are selected.

ACCELERATED AGING TESTS AND PD MEASUREMENTS

Test process

The capacitor is placed in a ventilated chamber. The room temperature is defined in order to maintain the dielectric of the capacitor at 75°C. A voltage of 1.4 time rated voltage (1.4 U_N) was applied continuously on a full capacitor for 500 hours. Each accelerated aging hour is equivalent to 70 hours of normal service on the network. Afterwards, the capacitor was placed in a screened cage (Faraday cage) at 20°C where the voltage was increased by step of 0.2 U_N from U_N up to PD Inception Voltage (PDIV). PDs were recorded at each step. PD measurements were carried out on different devices of connections (on a single element, on two elements in parallel, on two series groups of three elements in parallel and on a full size capacitor). Afterwards, the capacitor was put back into the ventilated chamber for a second cycle of 500 hours of aging. These measurements were repeated up to 5500 hours of accelerated aging.

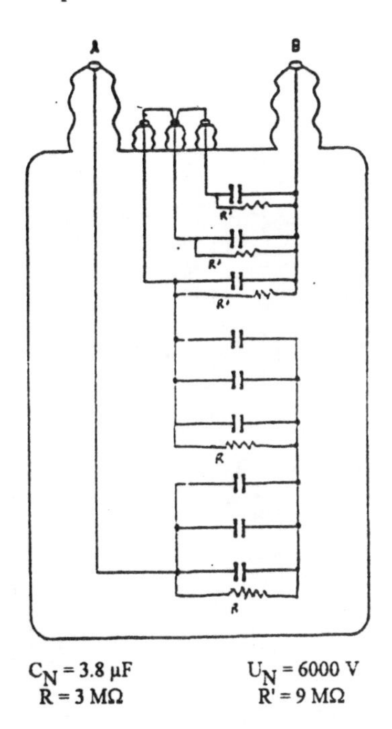

Figure 1 : Device of multi-terminal capacitor

Results

During PD measurements several physical and stastistical parameters were recorded or calculated. Among these parameters : voltage, PD energy released during 10 s, magnitude of pulses and its phase, average magnitude of PD, number of pulses, deviation, kurtosis, skewness, phase of first PD pulses for positive half cycle and for negative half cycle of 50 Hz sine wave, phase of last PD pulses for positive and negative half cycle and width of

PD spectrum also for positive and negative half cycle.

Study and comparaison between acquisitions from different devices and as a function of aging time, found that some parameters were more reproductible and indicate the state and the stress of the dielectric than other parameters.

Aging time	PDIV*	Φ_-	Φ'_-	$\Delta\Phi_-$	Φ_+	Φ'_+	$\Delta\Phi_+$
0 h	2.4 U_N	309°	101°	152°	114°	288°	174°
500 h	2.4 U_N	46°	106°	60°	233°	284°	51°
1000 h	2.2 U_N	38°	101°	60°	212°	283°	71°
1500 h	2.3 U_N	309°	110°	161°	136°	288°	152°
2000 h	2.4 U_N	322°	114°	152°	136°	288°	152°
2500 h	2.5 U_N	300°	101°	161°	112°	288°	176°
3000 h	2.5 U_N	309°	110°	161°	128°	298°	170°
3500 h	2.5 U_N	309°	106°	165°	119°	288°	169°
4000 h	2.5 U_N	308°	107°	159°	129°	290°	161°
4500 h	2.5 U_N	300°	91°	151°	120°	276°	156°
5000 h	2.2 U_N	6°	111°	105°	187°	290°	104°
5500 h	2.5 U_N	301°	106°	165°	119°	275°	156°

* U_N = 2000 V

Table 1 : PDIV acquisitions of a single element as a function of aging time

Aging time	PDIV*	Φ_-	Φ'_-	$\Delta\Phi_-$	Φ_+	Φ'_+	$\Delta\Phi_+$
0 h	2.2 U_N	10°	106°	96°	196°	295°	99°
2500 h	2.2 U_N	338°	102°	124°	160°	285°	125°
3000 h	2.2 U_N	343°	101°	118°	166°	287°	121°
3500 h	2.1 U_N	327°	40°	73°	142°	225°	83°
4000 h	1.9 U_N	100°	184°	84°	284°	2°	78°
4000 h	2.3 U_N	310°	111°	161°	124°	295°	171°
4500 h	1.9 U_N	93°	174°	81°	270°	339°	69°
4500 h	2.4 U_N	315°	103°	148°	131°	288°	157°
5000 h	1.9 U_N	105°	155°	50°	292°	337°	45°
5000 h	2.2 U_N	9°	99°	90°	190°	286°	96°
5500 h	1.9 U_N	100°	183°	83°	283°	20°	97°
5500 h	2.4 U_N	307°	111°	164°	124°	297°	173°

* U_N = 4000 V

Table 2 : PDIV acquisitions of two series groups as a function of aging time

Aging time	PDIV*	Φ_-	Φ'_-	$\Delta\Phi_-$	Φ_+	Φ'_+	$\Delta\Phi_+$
0 h	1.9 U_N	5°	123°	118°	186°	330°	144°
500 h	1.8 U_N	313°	123°	170°	136°	292°	151°
1000 h	1.8 U_N	0°	110°	110°	186°	288°	102°
1500 h	2.3 U_N	305°	110°	165°	123°	284°	161°
2000 h	1.9 U_N	347°	106°	119°	161°	288°	127°
2500 h	1.8 U_N	17°	106°	89°	199°	284°	85°
3000 h	1.8 U_N	11°	106°	95°	191°	284°	93°
3500 h	1.8 U_N	12°	102°	90°	195°	288°	93°
4000 h	1.9 U_N	354°	104°	110°	183°	281°	98°
4500 h	1.9 U_N	9°	102°	93°	190°	289°	99°
5000 h	2.0 U_N	4°	102°	98°	178°	292°	114°
5500 h	1.9 U_N	10°	102°	92°	176°	282°	106°

* U_N = 6000 V

Table 3 : PDIV acquisitions of a full size capacitor as a function of aging time

Notes :

Φ_- = Phase of first PD pulses for negative half cycle

Φ'_- = Phase of last PD pulses for negative half cycle

$\Delta\Phi_-$ = Width of negative half cycle PD spectrum

Φ_+ = Phase of first PD pulses for positive half cycle

Φ'_+ = Phase of last PD pulses for positive half cycle

$\Delta\Phi_+$ = Width of positive half cycle PD spectrum

Tables 1,2 and 3 represent values of these parameters as a function of accelerated aging time from three different devices of connections. Figure 2 presents typical profiles of number-phase distribution at PDIV.

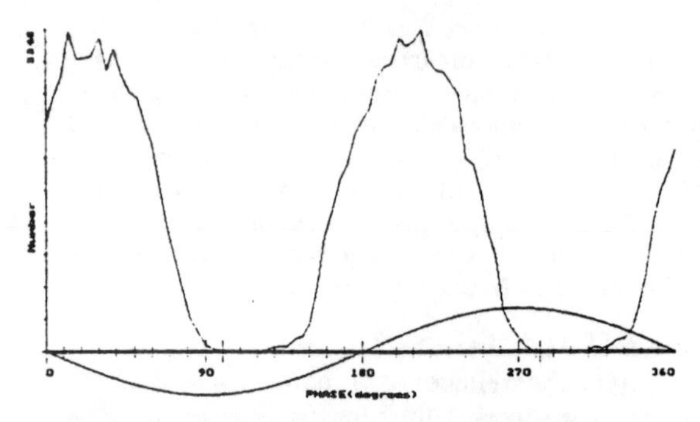

Figure 2 : typical number-phase distribution at PDIV from single element connection after 5500 h of aging.

After 5500 hours of aging, this capacitor was opened and the 9 aged elements were broken down individually on a dc voltage ramp. Table 4 represents values of dielectric rigidity obtained and compares them to values of other 9 unaged elements made at the same time. Also, average dielectric rigidities for aged and unaged elements, were calculated.

Aged elements	Breakdown voltage kV	unaged elements	Breakdown voltage kV
first element	10.4	N°1	10.4
second element	10.1	N°2	11.5
third element	9.3	N°3	11.2
2d series group			
first element	10.3	N°4	9.9
second element	10.2	N°5	11.1
third element	10.5	N°6	11.3
3rd series group			
first element	10.5	N°7	10.8
second element	9.8	N°8	11.4
third element	10.1	N°9	11.1
Average voltage	10.1	Average voltage	10.9

Table 4 : Breakdown voltage of aged and unaged elements

Discussions

Many observations can be made from the above tables and figures.

- PDIV of a single element is higher than PDIV of the other devices.
- PDIV of each device as a function of accelerated aging time is constant or slightly increases when the capacitor has been made well.
- After 5500 hours of aging which was equivalent to 44 years of normal service on the network, the dielectric rigidity measured remains the same as an

unaged elements. This result proves previous observation about results of PDIV values.

- Table 4 shows, hachured lines, that PD measurements of two series groups device recorded after 4 000 h of aging, two distinct PD spectra in each half cycle of voltage. The first one between 90° and 180° has low magnitude of PD pulses (about 106 pC) and appeared at less than 2 U_N. The second spectrum between 300° and 110° has a high magnitude of PD pulses (about 8800 pC) and appeared at close 2.2 U_N. After the opening of the capacitor, traces of discharges were found on the container close to one of terminals. This means the presence of an insulation problem between terminal and container which was detected by PD measurements.
- The following parameters were representative of inside capacitor state :
  - Partial Discharges Inception Voltage (PDIV)
  - Phases of first positive and negative PD pulses
  - width of PD spectrum in each half cycle of 50 Hz sine wave

The state of the capacitor could be assessed by a combination of different previous parameters.

For example, if capacitor PD measurements give a PDIV less than 1.2 U_N with a width of PD spectrum of each half cycle greater than 100° and phases of first positive and negative PD pulses are very far below zero degree (see figure 2), than the capacitor dielectric state could be degraded and nearer failure state. On the other hand, if PD measurements record a PDIV less than 1.5 U_N with a width of PD spectrum is greater than 50° and phases of first positive and negative pulses are higher than 90° then capacitor connections under test could be supposed badly welded or have problems of insulation between the metallic parts of capacitor.

TRANSCIENT OVERVOLTAGE TEST AND PD MEASUREMENTS

Test process

The transient overvoltage test conditions were to subject a capacitor placed in a screen chamber at 20°C, to its rated voltage (U_N) to which was added for 1s, every 5 min, an overvoltage at power frequency of 2.25 U_N (see figure 3).

A second multi-terminal capacitor (see figure 1) was submited to this type test. The magnitude of PD pulses as a function of time were recorded during TOV on different devices of connections.

Results

In order to investigate on PD activity, a number of PD measurements during TOV test was carried out. Figure 4 shows a typical dicharge magnitude distribution as a function of time recorded. Figure 5 shows a zoom curve of magnitude-phase of the first discharge appearing at the 20th cycle. Figures 6 and 7 show a zoom curve of discharge magnitude distributions at the 40th and 61th cycles of the same TOV.

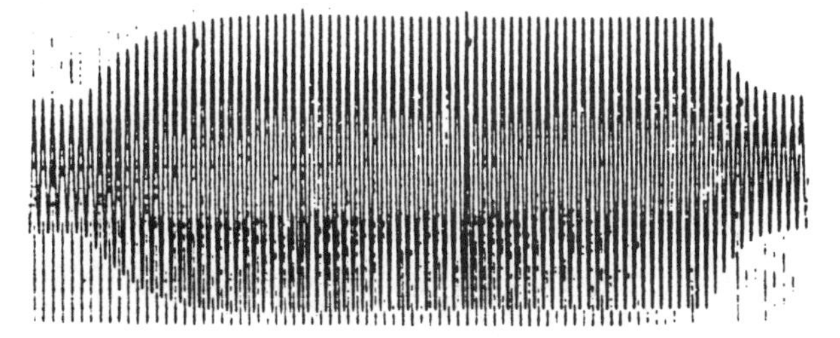

Figure 3 : shape of TOV

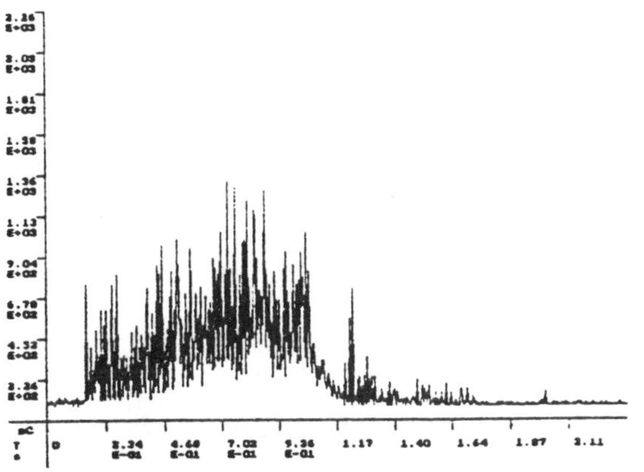

Figure 4 : Discharge magnitude distribution of the 54th TOV applied on 2 series groups.

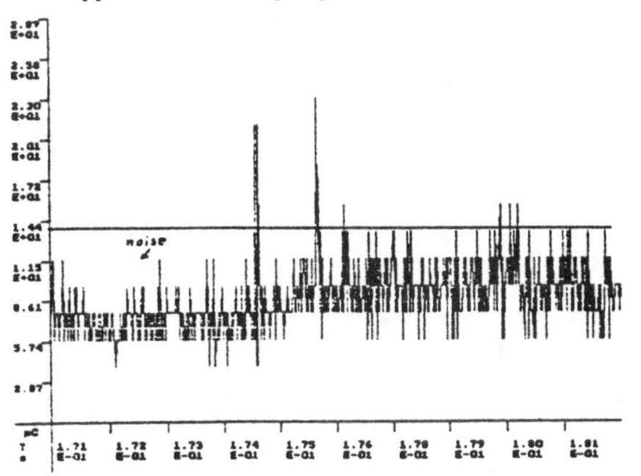

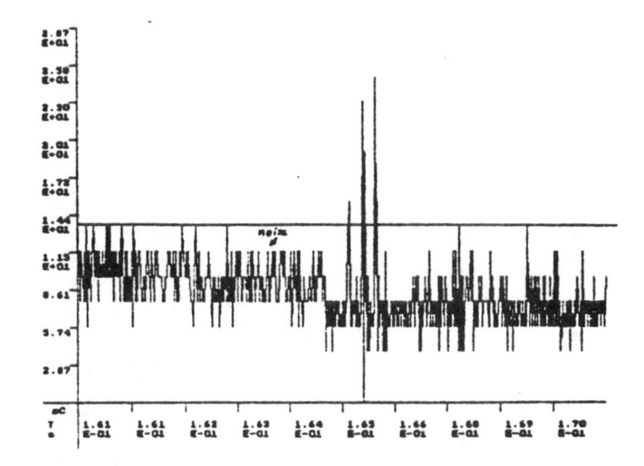

Figure 5 : First PD pulses at 9th cycle (negative and positive half cycles) of the 56th TOV applied on 2 series groups

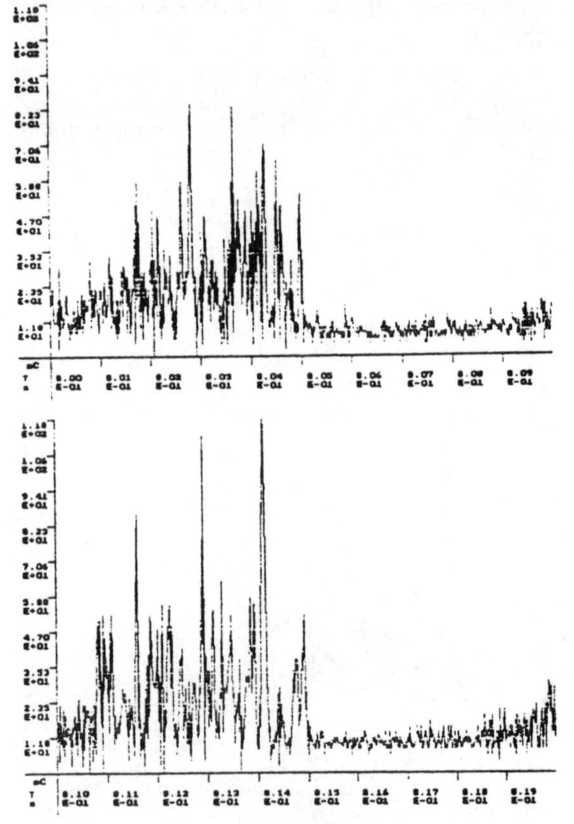

Figure 6 : Discharge magnitude distribution at the 40th cycle of the 56th TOV applied on 2 series groups.

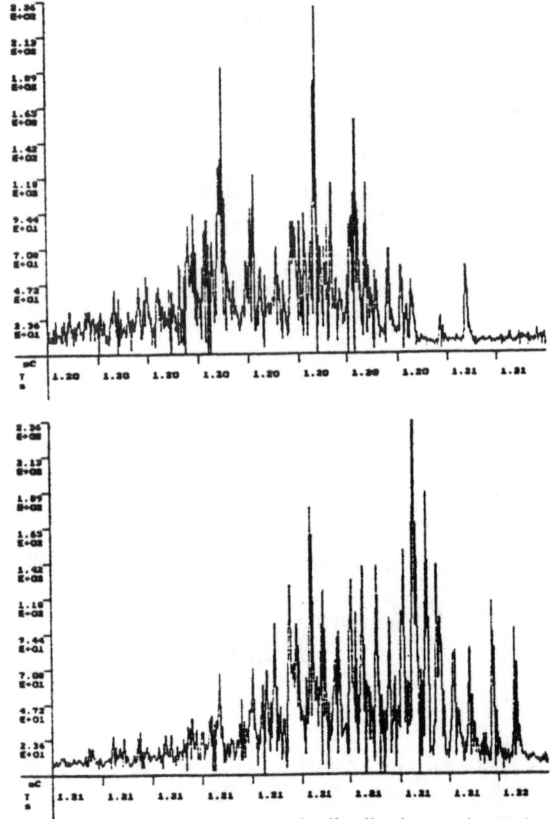

Figure 7 : Discharge magnitude distribution at the 61th cycle of the 56th TOV applied on 2 series groups.

Discussions

Observations and comments can be pointed out from the above measurements.

- PD activity persists about 600 ms to 800 ms only after TOV at 20°C.
- PD number and magnitude increase with the cycle number of the same TOV. 6 PD pulses are recorded at the 20th cycle (2.25 U_N) of the 56th TOV (see figure 5). But, about 250 PD pulses at the 40th cycle and 325 PD pulses at the 61th of the same TOV are recorded (see figures 6 and 7).
- According to figures 6 and 7, PDs occured by package. For instance, we suppose that many cavities of dicharges with different dimensions appear progressively during the first and third quarter cycle. Cavities could be created in the impregnated liquid or between liquid and solid dielectrics. Afterwards, they desappear during the second and forth quarter.
- Figure 7 shows that first PD pluses in this cycle start below zero degree. It is the same observation made in the aging test. This means dielectric stress is higher and it is nearer failure point.

CONCLUSIONS

These PD measurements on industrial capacitors and in an industrial laboratory shown promising results. The combination of the three previous parameters (PDIV, phase of first PD pulses and the width of PD spectrum in half cycle) could assess the dielectric state and the reliability of the manufacturing process. PD measurements as a function of time give the mecanism of discharges during each cycle of 50 Hz sine wave.

These capacitors had no defects. Tests on known defects industrial capacitors are outstanding to improve the investigations and to set up diagnostic and control tools for capacitors used on the electrical network.

REFERENCES

1. Hantouche, C. "Partial Dicharges on power capacitors". Conference record of the 1992 IEEE International Symposium on Electrical Insulation. IEEE Publication 92CH3150-0, pp. 401-406.

2. Hantouche, C. and Fortuné, D. "Digital measurements of Partial Discharges in full-sized power capacitors". IEE Trans EI. Vol. 28. No. 6, December 1993, pp. 1025-1032.

Conference Record of the 1996 IEEE International Symposium on Electrical Insulation, Montreal, Quebec, Canada, June 16-19, 1996

Measuring the Degradation Level of Polymer Films Subjected to Partial Discharges

R. Bozzo, C. Gemme, F. Guastavino
Electrical Eng. Dept., Univ. of Genova
via all'Opera Pia 11A, 16145 Genova, Italy

P. Tiemblo
Instituto de Ciencia y Tecnologia de Polimeros
Juan de la Cierva 3, 28006 Madrid, Espana

Abstract. Polymer films have been subjected to partial discharge aging. It is shown that statistical quantities derived from partial discharges patterns can be related to test conditions, film characteristics and degradation level.

INTRODUCTION

To measure the endurance to Partial Discharges (PD) of polymeric films and laminates the most commonly adopted test electrode geometry is based on the one recommended by IEC [1]. Nevertheless, when films are particularly thin, the time to breakdown can be strongly affected by electrical field aging, if the upper electrode is located onto the film [2]. Furthermore, when coated films have to be tested, the upper electrode should not touch the specimen, both to avoid mechanical strain on the coating layer and to allow the coating layer possible separation from the attached film because of the PDs action (in case of insufficient adhesion between the two elements). Therefore past research [3] allowed to optimize a test assembly and a relevant test procedure useful to discriminate the PD endurance of thin films by measuring the time to breakdown. The procedure appeared easy to use, of low cost and it gave good and reproducible results, if the environment was controlled.

By adopting such a test assembly and relevant procedure, PDs have been measured by means of a digital system. Several resulting PD patterns have been elaborated and about 50 derived and statistical quantities have been obtained for each pattern. The effects of the test conditions on the derived quantities has been studied with relevance to the following items.

a) To recognize the kind of film under test

b) To correlate the value of quantities with the degradation level of the film (i.e. to focus at the quantities which change with time)

c) To find a link between the quantities values and the test ambient conditions (i.e. relative humidity)

d) To determine the influence of the film thickness

e) To evidence the effect of the voltage level

EXPERIMENTAL SETUP AND SPECIMENS

A sphere-plane electrode configuration is obtained by means of the cell described in [3]. The upper hemispherical electrode is made of stainless steel and it is housed in a polymethylmetacrylate (PMMA) support which is positioned onto the film. The upper electrode vertical position can be controlled by means of a precision screw system, allowing to obtain the desired 0.1 mm electrode-film gap with a precision of ±10 μm. The electrode radius of curvature is 3 mm and the electrode surface is optically finished.

The lower plane electrode is made by brass. The adhesion of the film to this electrode is obtained by applying a moderate vacuum (about 100 Pascal) to a series of 1 mm diameter holes present in the plane electrode. The position of the holes is designed to avoid any deformation on the film area where PDs are present. Besides, a graphite paint is applied to the back of the film to avoid occurrence of PDs between film and plane electrode.

The electrode assembly can also be positioned inside a PMMA box (about 4 litres volume) in order to control the atmosphere during tests. Further details of the test cell have been presented in [3].

An AC 50 Hz 2400 V voltage has been applied to the test cell. This voltage level is about three times the inception voltage. To check the effect of voltage on results a few experiments have also been performed at 1800 V.

Two different relative humidity conditions have been adopted:

a) Close configuration with controlled dry air circulation. The test cell is set inside the above box , in order to control the relative humidity level. The adopted air flow of 0.5 l/min allowed a 23 % RH at 21 ± 2 °C.

b) Test cell in open air subjected to ambient conditions (i.e. RH higher than 50%).

The analog signal, relevant to PD phenomen, available across the measuring impedance Zm (RLC), feeds an amplifier. The maximum level of the resulting output signal is sent to the A/D converter of the PRPDA (Phase Reference Partial Discharges Analyser) system. For each partial discharge, the relevant amplitude (apparent charge) and the phase of the supply voltage at the instant of the partial discharge occurrence are acquired.

Thus, during an acquisition time t_{acq} (generally a multiple of the AC supply voltage period) the three basic measured quantities are:

N_{ij} = *number of discharges stored in the* -ij *cell*

Q_i = *apparent charge of the discharge occurred in the* i-th *amplitude window*

φ_j = *phase of occurrence of* Q_i

where i and j values identify one of the available h (64) amplitude and k (128) phase window of PRPDA. The previous measured quantities allow to obtain the so-called "PD pattern".

Several quantities can be directly calculated from the basic ones, to form a set of derived quantities that have been called "fingerprints" by Gulski and others [4,5]. Among them, the following quantities are here considered:

$$Q^+ = \sum_{i=1}^{\frac{h}{2}} \sum_{j=1}^{k} N_{ij} \cdot Q_i \; ; Q^- = \sum_{i=\frac{h}{2}+1}^{h} \sum_{j=1}^{k} N_{ij} \cdot Q_i$$

: total charge, positive and negative

$$I^+ = \frac{Q^+}{t_{acq}} \; ; \; I^- = \frac{Q^-}{t_{acq}}$$

: average positive and negative charge or PD current

Besides, from each PD pattern, the following frequency histograms can be computed together with the relevant statistic parameters such as Maximum, Mean, Median, Mode, Variance, Skewness, Kurtosis, Cross Correlation Factor et al.

$H_q(\varphi)$: frequency histogram of the charge Q (positive or negative) in the j-th phase cell

$H_{qn}(\varphi)$: frequency histogram of the mean charge Q/Nj (positive or negative) in the j-th phase cell

$H_n(Q)$: frequency histogram of the number of discharges $N_i(Q)$ (positive or negative) as function of the amplitude Q_i

$H_n(\varphi)$: frequency histogram of the number of discharges $N_j(\varphi)$ (positive or negative) as function of the phase φ_j

Monitoring the above quantities can allow to identify and to characterise several defects in insulation systems, as well as to characterise insulating materials. Besides, when the discharge sites morphology changes with time, the temporal evolution of the above parameters can be related to such changes and to the degradation level.

The experiments have been carried out on Polyethylenterephalate films (PET), having a 23 or 36 or 50 μm thickness, and on Polyimide films (PI) 25 μm thick.

RESULTS AND COMMENTS

Due to the great number of derived and statistical quantities hereafter only some of the main results are shown.

Relative Humidity

Relative humidity has a strong effect on the time to breakdown. As concerns PI films differences related to two RH levels of 23% and 52% are well evidenced in the PD patterns shown in fig. 1. As consequence, a lot of derived and statistical quantities show differences in their values during all the test duration. For example the Skewness relevant to Phase-Mean Charge distribution of the positive discharges and the Mean value relevant to the Phase-number distribution of positive discharges are showed in fig. 2. The times to breakdown range from 34700 s for RH 23% to 14700 s for RH 52%. Different results have been found when PET films have been considered. Even though time to breakdown still changes depending on the RH value (17900 s at RH 23%, 25000 s at RH 57%, 15400 s at RH 65%) PD patterns, derived and statistical quantities showed smaller differences. The highest differences in the data due to RH influence have been obtained in the case of skewness of the Amplitud-Number distribution, negative discharges, reported in fig. 3. Where it is evidenced that curves relevant to 23 and 57 % RH are similar while a 65 % RH is needed to find different values.

Thickness

To study the effect of thickness on the derived and statistical quantities, PET films having 23, 36 and 50 μm thickness have been tested at 23 % RH. Under those conditions times to breakdown varied respectively from 17900 to 56000 to 78800 s. Nevertheless none of the monitored quantities showed any appreciable variation in value.

Degradation level

As already showed in the previous figures some of the derived quantities assume values that are quite constant with test time, even though high scatter can be present. On the other hand there are quantities whose change significantly with time. Thus a correlation between the value and the degradation level can be proposed. As example the maximum of the amplitude-number histograms or the I+ and I-, average positive and negative charge, can be considered. The trend versus time of the last quantities is shown in fig. 4.

Voltage

The test voltage variation has a noticeable influence on the statistical parameters and on the derived quantities. In fact electric field variations lead to different discharge conditions as concerns both repetition rate and max. amplitude; therefore the times to breakdown depend on them. The least sensitive parameters , as regards such variations, are the fourth and third order moment of the Hn(Q) distribution, which identify the shape of such frequency histogram and which are better linkable to the material type and to the ambient conditions rather than to the geometric electric field variations. For instance the skewness values, for negative discharges, relevant to the distribution Hn(Q) and Hn(φ), obtained from tests performed on 25 μm thick PI films both at 1800 V and 2400 V voltage demonstrate this point, as shown in fig. 5.

Kind of film

The performed experiments showed slight differences in the quantity values recorded at low RH level. In fact the low

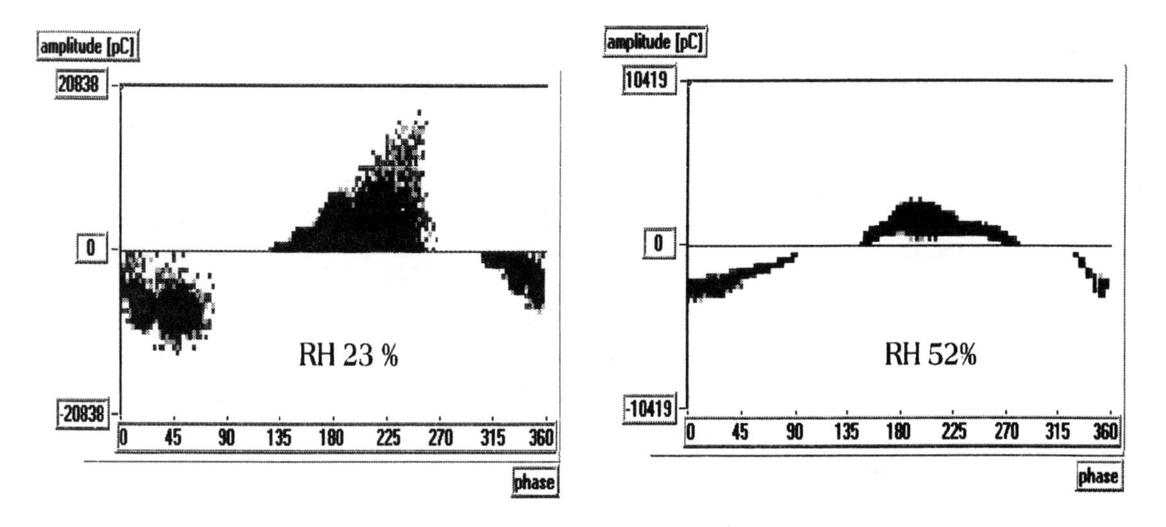

Figure1. Patter relevant to PI 25 μm thickness at two RH levels

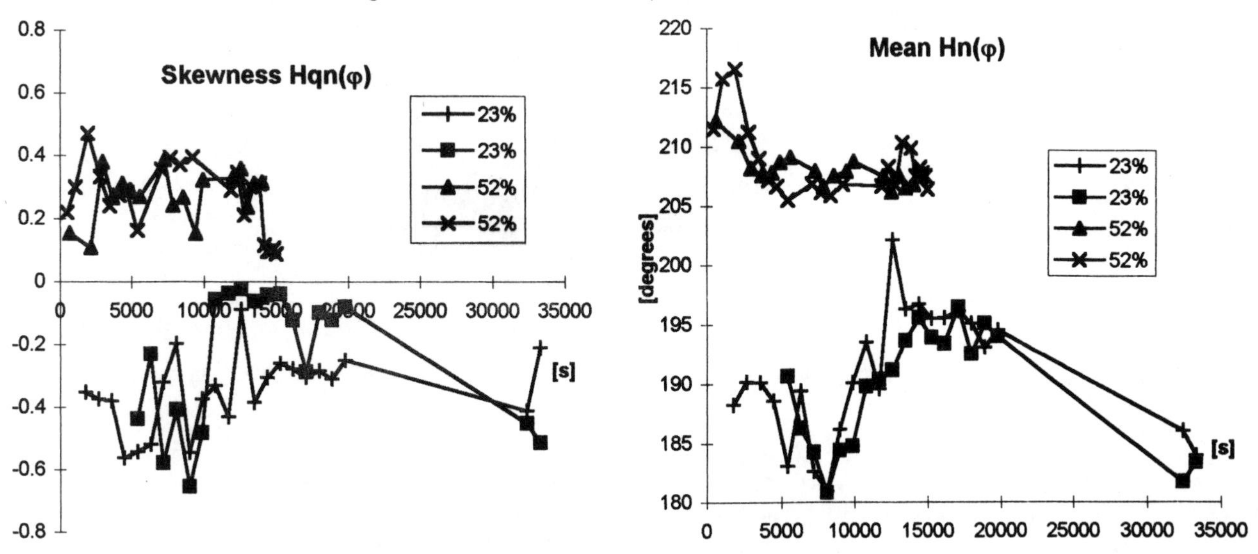

Figure 2. Influence of RH % level on PI 25 μm specimens

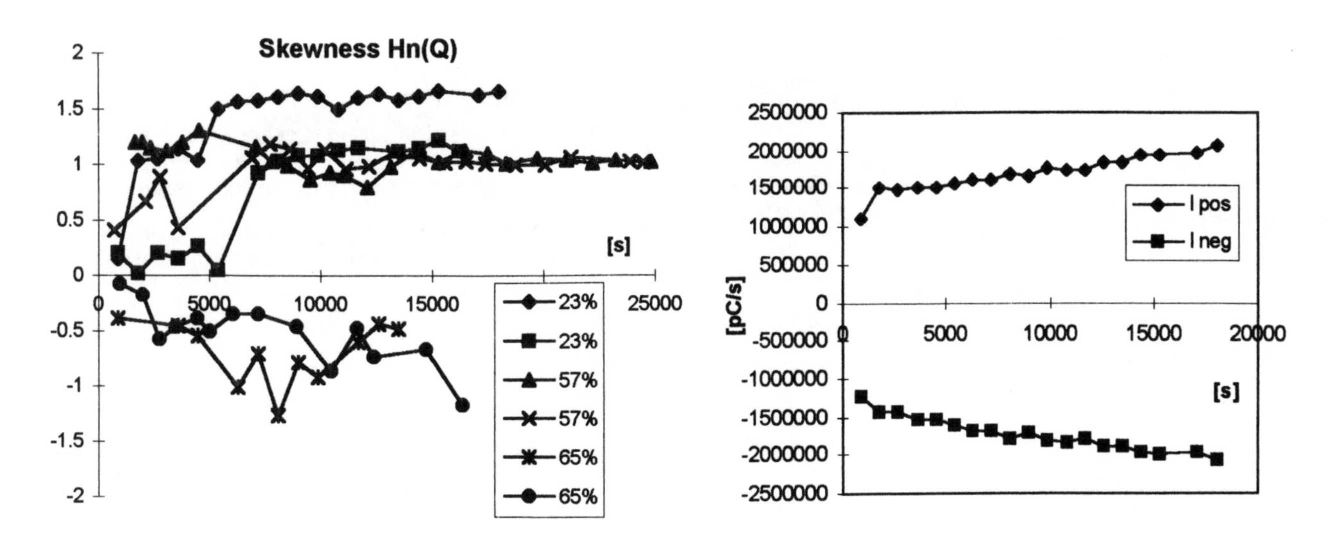

Figure 3. Influence of RH % level on PET 23 μm specimens

Figure 4. Variation with test time of the PD currents

humidity PD patterns of PI and PET are relatively similar. Nevertheless when higher RH value is adopted the differences in the quantities appear evident. An example of the inception phase relevant to negative discharges is showed in fig.6.

CONCLUSIONS

The experiments evidenced that values of studied quantities from PD patterns can be dependent on a few of the test conditions and characteristics of tested specimens, at least for the considered test set-up. Among the adopted distributions, parameters relevant to Phase-Mean Charge and Phase-Number distributions would seem the most sensitive to a modification of the test conditions. Nevertheless the research well demonstrated the existence of interactions among the factors taken into consideration which characterize the test. For example many of the analysed quantities change by changing the tested film, but only at high RH. Again same of them vary with RH when PI films are tested but differences versus RH are not so evident if PET films are tested.

Thus, when derived quantities are considered for diagnostic purposes the influence of tests conditions and insulation systems should be checked with attention.

ACKNOWLEDGEMENTS

The authors acknowledge the financial support of italian MURST 60 % funds.

REFERENCES

1. IEC Publ. 343. "Recommended Test Methods for Determining the Relative Resistance of Insulating Materials to Breakdown by Surface Discharges", 1970

2. Guastavino F. and G. Guerra. "Optimisation of Partial Discharge testing techniques", Proc. of 4th Int. Conf. on Conduction and Breakdown in Solid Dielectrics, Sestri Levante, Italy, June 1992, pp. 193-199

3. Bozzo R., L. Centurioni, F. Guastavino. "Measuring the Endurance of Films in Partial Discharges", IEEE Trans. on EI, Vol 28, No. 6, December 1993, pp. 1050-1056

4. Fruth B. and L. Niemeyer. "The importance of statistics of partial discharge data", IEEE Trans. on Electric. Insulation, Vol. 27, No. 1, February 1992, pp. 60-69

6. Gulski E. and F.H.Kreuger. "Computer aided recognition of discharge sources" IEEE Trans. on Electr. Insul., EI-27, 1992, pp. 82-92

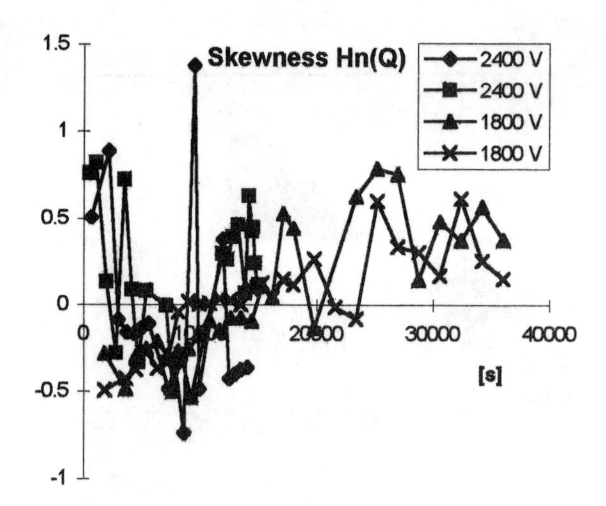

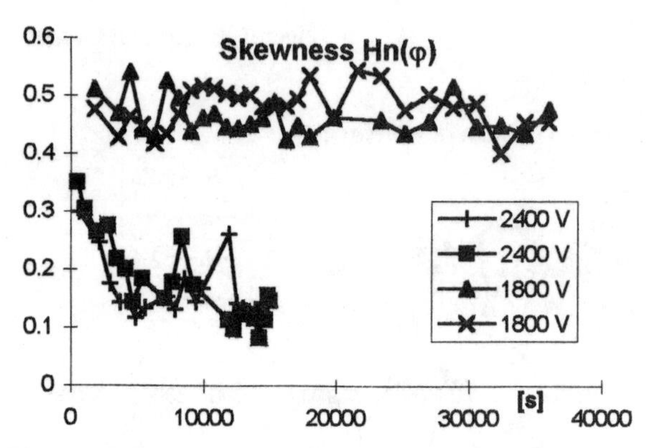

Figure 5. Influence of test voltage on PI 25 mm specimens, negative discharges

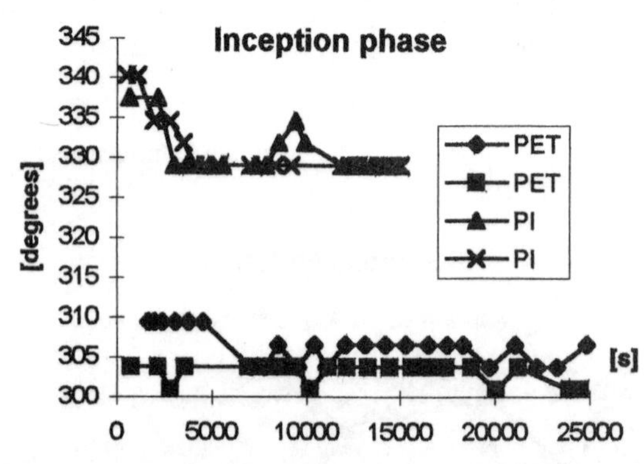

Figure 6. Variation of inception phase, negative discharges, for PET and PI films of the similar thickness tested at high RH level.

Conference Record of the 1996 IEEE International Symposium on Electrical Insulation, Montreal, Quebec, Canada, June 16-19, 1996

BIAXIALLY ORIENTED POLY(ETHYLENE - 2,6 NAPHTHALATE) UNDER DISCHARGE ACTIVITY AT VARIOUS TEMPERATURES

Michael Danikas
Democritus University of Thrace
Department of Electrical and Computer Engineering
67100 Xanthi,Greece
Jean Guastavino,Eric Krause and Christian Mayoux
Laboratoire de Génie Electrique
Université Paul Sabatier,Toulouse France

***Abstractt*: This paper deals with a polymer quite new in the domain of electrical insulation and its behaviour under corona discharge activity. Results on the stability of discharge regime, all along of two hours of experiments, under three different temperatures are presented. A FTIR analysis gives the evolution of the polymer structure and the bulk conductivity measurements reveal the threshold of electrical properties modifications, likely related to trapped charges. It is found that PEN is more sensible to temperature than to chemical modification due to discharge species.**

INTRODUCTION

In many cases partial dicharges are considered to occur in voids created during the manufacturing process of insulating constructions. They might also occur in trapped gases by delamination between two dielectrics or between a dielectric and a conductor or semi-conductor, due to a bad adhesion. The latter situation is considered in the present work since bi-axially oriented films may be used for capacitors or winding insulation of motors. Both mechanisms of partial discharges and material ageing are complicated because they interact, leading to difficulties of detection [1,2], and interpretation [3,4,5]

Because of its high temperature properties, poly(ethylene-2,6 naphthalate dicarboxylate) (PEN) can be used for a variety of applications in electronics and electrical power engineering. The naphthalene group provides rigidity to the polymer backbone, elevating the glass transition (117°C) and enhancing mechanical properties.

Bell et al. [6], investigating the influence of chemical structure on the response of aromatic polyesters to high-energy ionizing radiation (γ), found marked changes in response of the base polymer, poly(ethylene terephthalate) (PET) with both chain scission and crosslinking. They found that, in comparison to other linear polyesters, poly(ethylene-2,6 naphthalate dicarboxylate) is extremely resistant to radiation degradation. However, in terms of the utility of this polymer material like electrical properties after irradiation, few informations are available in literature.

The work was underataken to follow the physical and chemical properties of PEN when submitted to the irradiation of corona discharges in air at different temperatures. In the first part of the present paper we give the results of the discharges characteristics initiated near the inception voltage, in a homogeneous field. However, the long term ageing of different specimens is ordinary carried out for higher gap and voltage. The corresponding results are presented in the second part.

EXPERIMENTAL PROCEDURE

The design is based on a two electrode capacitor of a parallel plate configuration. Discharges are initiated in air in a gap of 0.5 or 1 mm between brass electrodes, the lower being covered with a mica sheet onto which is placed a 12 µm film of PEN. The ageing cell placed in a oven where the temperature may reach 300°C (±0.5°C), is supplied by a transformer at 50 Hz. The type of discharge is recorded with the aid of a 100 MHz oscilloscope and a measuring resistor of 1kΩ. A 1.5MΩ resistor is placed between the transformer and the test cell in order to limit the current in the high voltage circuit in case of breakdown. A preliminary test shows that this resistor does not influence the shape of the detected impulses.

The chemical and electrical characterizations were undertaken on films aged for two hours in air at atmospheric pressure with a gap of 1 mm and a voltage of about 6 kV.

RESULTS AND DISCUSSION

Discharge regime

The aim of this experience is to observe the influence of the working temperature and of the insulating material on the discharge mechanism. Three temperatures are choosen: 30°C, 80°C and 120°C, the last being near the glass transition of the polymer. A gap of 0.5 mm allows to work at low voltage which is applied when the temperature in the oven is reached. During the ageing test there is no automatic measurement procedure but a manual detection every 15 minutes. In the mean time, the full scale sensitivity and trigger level are controlled. The rise time of the oscilloscope (3.5 ns) limits the detection of streamer-like pulses. It is shown in figure 1, the general shape of the positive discharge pulse (a), the rise time and the amplitude of the pulses as a function of temperature (b) (c) (d). The figure corresponds to 75 minutes of ageing, but if we have observed a rise time practically constant the pulse amplitude varied in a large range as it is shown in figure 2 and 3. The voltage was fixed when the first discharges were detected so that it was pointed out a lowering of the inception voltage when the working temperature was 120°C.

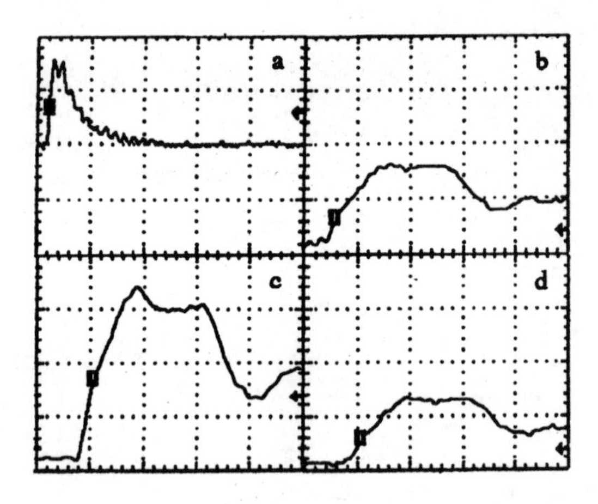

Figure 1.Rise time and amplitude of positive discharge impulses recorded at the same time of PEN ageing, i.e.75 min, at different temperatures.

a - typical shape of a positive impulse.
t = 250 ns / div ; i = 20 V / div.
b - ageing temperature : 30°C
t = 25 ns / div ; i = 20 V / div.
c - ageing temperature : 80°C
t = 25 ns / div ; i = 20 V / div.
d - ageing temperature : 120°C
t = 25 ns / div ; i = 20 V / div.

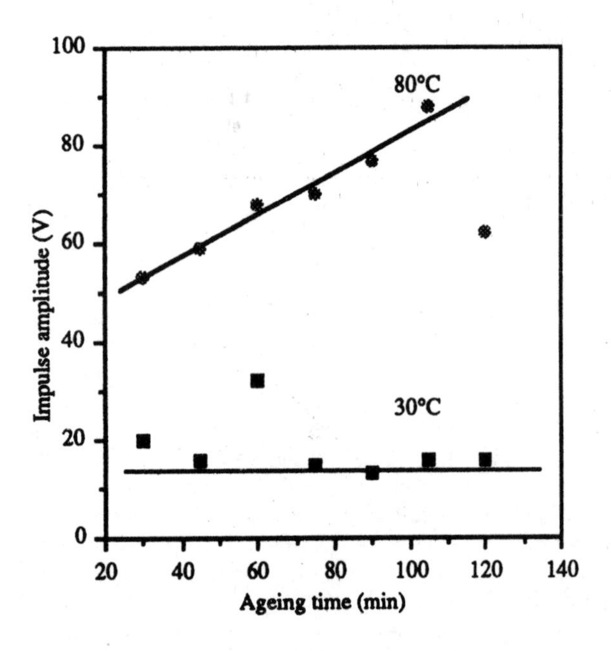

Figure 2 Temperature dependence of discharge amplitude after voltage application, 2 kV rms /0.5 mm. A measuring resistor of 1 kΩ is placed in series with the testing cell.

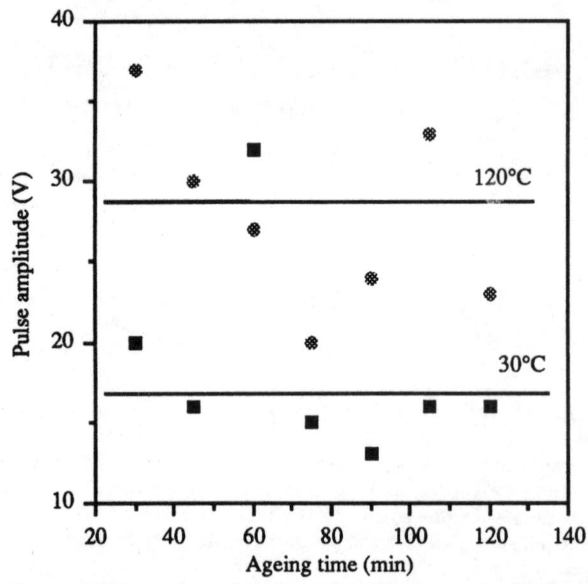

Figure 3 Temperature dependence of discharge amplitude after voltage application, (1kV and 2kV/0.5 mm for 120°Cand 30°C respectively). The threshold voltage is lowered when the temperature is 120°C.

Indeed the inception voltage was 2 kV (rms) when the temperature was 30 or 80 °C and 1 kV at 120°C. The influence of temperature on the discharge type is not obvious. Even if the rise time of the oscilloscope is a limiting factor, the recorded pulses may be related to streamerlike discharges. The thermo-electronic emission of the brass upper electrode is favoured at zero time, but there is no influence of temperature on PEN in the conditions of experience and when the voltage is applied at the beginning of test. The temperature plays a role on the atmosphere modifying the mean free path of gas molecules. Then, after the first events, the temperature variation of ions mobility may be considered. According to Loeb [6], studies on various gases have shown an increase of this parameter from 2.67 (at 292 K) to 3 (at 450 K) when nitrogen ions (N_2^+) in N_2. In their study on breakdown of air at high temperature and 50 Hz, Powell *et al.* [8], have shown that Ei, the corona gradient onset, is significantly lowered when temperature is increased, on the other hand Ei was observed largely independent of gap.

The application of a voltage near the inception voltave leads to a large variation in amplitude of pulses, particularly at 120 °C and 1 kV, as shown in figure 3. We suppose the growth of discharge at different places in the gap, impinging on different areas of the PEN film. When the ionization process is well established, e.g. as observed at 80°C the amplitude of discharges may increase up to a significant modification of surface conductivity of the polymer. Before the latter situation is reached we think that the light emission due to discharge activity allows the injection of a lot of photoelectrons into the plasma by secondary emission. This assumption is based on the chemical structure of the polymer with two aromatic rings. With the aim to look for a difference between the value of the secondary emission coefficient (γ) of virgin and aged insulating materials Morshuis [9] found that the photo-current of polycarbonate, before ageing, is high in comparison to other studied polymers like polyethylene. The presence of aromatic rings could be the reason of such a behaviour. This can support the hypothesis of a streamerlike discharge both near the inception voltage and during a short time of ageing.

Discharge Activity

It is reported in the following the observed consequences of the discharge activity in a well established regime, the applied voltage being 6 kV in a air gap of 1 mm, for two hours at room temperature

Bulk conductivity.

All the specimens were coated with gold electrodes. A dc step voltage was applied to the specimens placed in a vacuum chamber, at room temperature, and the transient current was measured with a vibrating reed electrometer. The steady state was considered to be established after 150 min of the decreasing of absorption current. Examples of current-voltage characteristics are shown in figure 4. The increase of bulk conductivity of the insulating material reveals an ohmic conductivity up to 80 MV/m. The current in this field region, with Au electrodes, was ascribed by Kojima *et al.* [10], to the displacement current of dipole orientation. The injection and creation of charges during discharge irradiation seems to modify this behaviour in a short time after the ionization is initiated. This is in accordance with the chemical changes which were found through ESCA measurements. This analysis has shown a threshold of PEN modifications after 15 min, with aromatic increase and carbonyl decrease, likely due to the bridging by naphthalene open parts [11]. It is the reason why, in the following, we focus our attention to the beginning of the polymer transformation.

The current shows a strong dependence on the electric field in the region of high fields, this is shown in a Poole-Frenkel plot as presented in figure 5. The density of current (J) is given by equation (1):

$$J(E) = K(E)\exp -(\phi_0 - \beta E^{1/2}) \qquad (1)$$

where E is electric field (V/m), K is a preexponential factor, ϕ_0 is the barrier height and β is the Poole-Frenkel coefficient. If we suppose a linear dependence between carriers velocity and the applied field according to $v = \mu E$, μ being the carriers mobility, (1) may be written:

$$J(E) = K_0 E\exp -(\phi_0 - \beta E^{1/2}) \qquad (2)$$

where K_0 is a constant. The Poole-Frenkel plot gives a ß value of 1.04×10^{-5} leading to a dielectric constant of 3.

FTIR analysis

During this study The Fourier Transform InfraRed (FTIR) spectroscopy is used to follow the chemical evolution of the polymer after discharge irradiation at different temperatures.

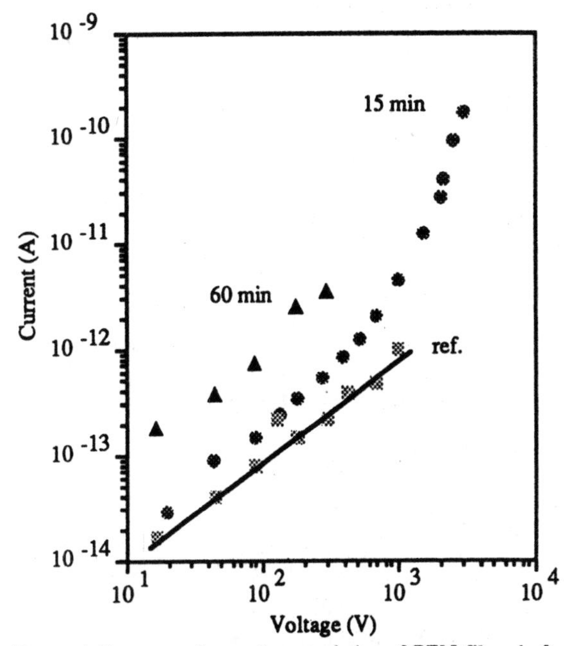

Figure 4 Current-voltage characteristics of PEN films before and after discharge activity. A 60 min aged specimen example is presented to show the rapid increase in bulk current. In many cases the superimposed polarization voltage has led to the breakdown of the film. 30V/µm is the breakdown field in the present case. This field represents more than the tenth of the reference material one. It is supposed that the trapped charges are responsable of this behaviour.

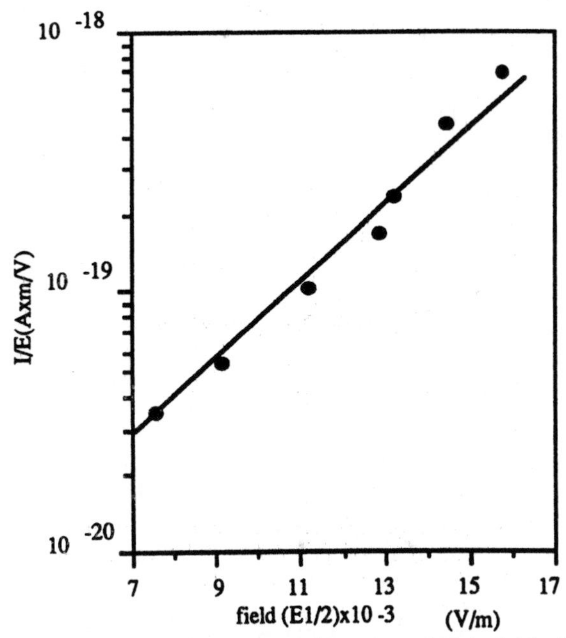

Figure 5 Poole-frenkel plot, in the region of high fields, for a 15 min discharges aged specimen.

Infrared spectra in the range between 4000 and 400 cm^{-1} were recorded by means of a Perkin Elmer 1760 spectrometer. The first observation, made more especially by the transmission technique, shows a better transmission of the aged specimens which is interpreted as due to the etching of the polymer submitted to discharges. It is well known that the activity of discharges in air on polyolefins, leads to an oxidation of the insulating material. This oxidation is characterized by the rate of carbonyl groups formation in the range between 1750-1700 cm^{-1}. But, a strong absorption, due to the streching vibration of C=O (1715 cm^{-1}) exists in the reference PEN, so that other windows in the spectra are to be explored. Some differences with the virgin specimens are observed, for example, near 1040 cm^{-1} which absorption was attributed by Ouchi *et al.* [12] to C-O group. This band is weaker after discharge irradiation. The intermolecular interactions giving birth to 813, 832 and 1002 cm^{-1} bands, are considered by Ouchi to be the close packing of molecules, in the unit cells of the cristalline structure. These bands change after irradiation, and differently when the bulk is compared to the surface of the film. The latter region was investigated by multi-internal reflectance technique (MIR). Two other bands at 529 and 480 cm^{-1}, representing the polymer cristallinity decrease in intensity after discharge irradiation at temperature higher than room temperature. This confirms the evolution to an amorphous structure into which carriers may be trapped. However, when the effect of temperature is compared to the one of discharges for the same duration of stress, the infrared analysis shows that discharges are predominant in the degradation process, particularly below the secondary cristallization temperature [195°C]. This will be published in a next future.

CONCLUSION

The introduction of Poly(ethylene Naphthalene 2,6 Dicarboxylate) (PEN) semicristalline films with thicknesses of 0.9, 1.5, 4 and 6 µm and larger thicknesses in the domains of capacitors and electrical power engineering respectively, brought a great interest. A study is conducted in our laboratory, to have a better knowledge of the electrical properties of this dielectric on one hand, of its behaviour under thermal and electrical discharges stresses on the other hand. In the present paper we report on some observations made on the discharge regime when these discharges are initiated near the threshold voltage and when the working temperature is increased from 30 up to 120 °C. In these conditions a streamerlike discharge was characterized. The ionization process is certainly influenced by temperature and by the insulating material structure as well. It was found a quite fast modification of the bulk conductivity since, in accordance with a chemical analysis (ESCA), after 15 min of discharge irradiation the bulk conductivity increases significantly. This kind of characterization is now in progress, with the aid of thermo-stimulated current and photo-conductivity techniques to distiguish the conductivity processes. However, with film of 12 µm, as previously said, many of the aged specimens were broken under a low polarization voltage. This seems to show the influence of trapped charges and the chemical changes during discharge irradiation. The amount of charges would be underevaluate since PEN has a photoconductive character. The effect of discharges is prevailing on the one of temperature below the secondary cristallization temperature. This is confirmed by the infrared spectroscopy analysis, and a competition of both effects is possible.

ACKNOWLEDGMENT

We gratefully acknowledge Dupont de Nemours Luxembourg for providing polymer films.

REFERENCES

1. Boggs, S. A. "Fundamental Limits in the Measurement and Detection of Corona and Partial discharge", IEEE Trans. EI Vol. 17,pp. 143-146, 1982.
2. Boggs, S. A. "Partial Discharge. Part II, Detection Sensitivity, IEEE Electrical Insulation Magazine, Vol. 26, pp. 35-42, 1990.
3. Morshuis, P. H. R. and F. H.Kreuger. "Transition from Streamer to Townsend Mechanisms in Dielectric Voids", J. Phys.D: Appl. Phys. Vol. 23 N° 12, pp. 1562-1568, 1990.
4. Bartnikas, R. and J.Novak. " On the Spark to Pseudoglow and Glow Transition Mechanism and Discharge Detectability". IEEE Trans. EI, Vol.27 N° 1, pp.3-14, 1992.
5. Danikas, M. G., R.Bartnikas, and J. Novak. " On the Spark to Pseudoglow and Glow Transition Mechanism and Discharge Detectability". IEEE Trans. EI, Vol.28 N° 3, pp.429-431, 1993.
7. Loeb, L. B. *Basic Procees of Gaseous Electronics.* University of California Press,1961, pp.61-63.
8. Powell, C. W., H. M. Ryan and A. Reyrolle. "Breakdown Characteristics of Air at High Temperature", 2nd Intern. Confer. on Gas Discharges, London 1972, IEE N° 90, pp. 285-287.
9. Morshuis, P. H. R. *"Partial Discharge Mechanisms"*. PhD Thesis, Delft University Press, 1993, pp. 76-80.
10. Kojima, K., Y. Takai, and M.Ieda. "Electronic Conduction in Polyethylene Naphthalate at High Electric Fields", J. Appl. Phys., Vol. 59 (8), pp. 2655-2659, 1986.
11. Krause, E., J. Guastavino, J. Rault, R. Grob and C. Mayoux. "Poly(ethylene Naphthalene 2,6 Dicarboxylate) Ageing under Electrical Discharges", 4th ICPADM, Brisbane1994, IEEE Publication 94CH3311-8, Vol.2, pp.788-791.
12. Ouchi, I., M. Osoi and S. Shimotsuma, "Infrared Spectra of Poly(ethylene 2,6-Naphthalate) and Some Related Polyesters". J. Appl. Polym. Sc., Vol.21, pp. 3445-3456, 1977.

Conference Record of the 1996 IEEE International Symposium on Electrical Insulation, Montreal, Quebec, Canada, June 16-19, 1996

CHARACTERISATION OF PARTIAL DISCHARGE PULSES IN ARTIFICIAL VOIDS IN POLYPROPYLENE FILMS USED IN CAPACITORS

B.Ramachandra R.S. Nema

Department of High Voltage Engineering
Indian Institute of Science
Bangalore 560 012
India

Abstract : Partial discharges in voids may cause deterioration of solid insulating materials. They often start in voids enclosed in insulation and or at the interface defects. A method of measuring fast discharge pulses with rise times below 1 ns is reported. Characterisation of partial discharge pulses in artificial voids in polypropylene films at atmospheric pressure is analysed that incorporates inception voltage , apparent and real charge , drift velocity and mobility of electrons.

1 Introduction

Polypropylene film is being widely used as dielectric in high voltage capacitors. One of the prime cause of insulation degradation is partial discharges in gas pockets or voids created between two adjacent films. Interpretation of physical processes in voids is complicated due to lack of knowledge of void shape, location and gas pressure . The purpose of present work is to investigate the mechanisms of discharges inside the void for various depths. To study pre-breakdown events in dielectrics, a time-dependent partial discharge model is presented. Ramo - Shockley theory is considered to study the relation between the motion of charges in an electrode gap and current induced in the external circuit. Very fast partial discharge current pulses in artificial voids in polypropylene are reported. The present model incorporates physics of partial discharge like drift velocity and mobility, real and apparent charge for different void depths. There are indication of streamer type of discharge in voids enclosed in solid insulation.

2 Experimental Setup

The experimental setup is shown in Figure 1. For high frequency response, subdivided electrode arrangement [1] was used. The radius of the measuring electrode should be small to reduce stray capacitances but not too small since otherwise the moving charge carriers also induce a current in the ground connection of the surrounding ring, thus reducing the current through the measuring resistor [2]. A 1 MΩ damping resistor close to the top electrode was provided for protection and to prevent reflections on high frequency currents on the high voltage lead. The diameter of subdivided electrode was 2 cm and 6 cm for measuring disk and guard electrode with 0.1 cm annular gap spacing between them. The upper electrode, encapsulated in epoxy, had a $2\Pi/3$ Rogowski profile and an effective uniform field diameter of 4.8 cm.

The electrodes were made of high quality stainless steel and were cleaned with emery paper, benzene and acetone. The 50 Ω measuring resistor consisting of six small 300 Ω metal film resistors connected in star configuration was placed below measuring electrode. The signal was fed by a 50 Ω coaxial transmission line of one metre length to a 50 Ω input of 1 GHz Tektronix TDS 684A digital storage oscilloscope with a rise time of 350 ps. Recorded data on floppy was analysed by 386 PC.

Electrical grade bi-axially oriented 20 μm thick and 11 cm diameter polypropylene films were used with an artificial cylindrical void of 0.2 cm diameter made at the centre as shown in Figure 2. The AC analogue circuit of Figure 2 is shown in Figure 3. Void depths of 80, 120 and 200 μm were studied. Additionally, samples without void for these thicknesses were also studied to check discharge free levels. Care was taken so that no air gap either between films or between film and electrode existed. The air pressure in the void was the local atmospheric pressure, 680 Torr.

The test voltage from a 50 kV, 2 KVA testing transformer was discharge free upto 22 kV. The pulse measurements described in this paper were performed at inception voltage. The TDS 684A oscilloscope was used in a single shot bandwidth of 1 GHz with 5 G samples/sec sampling rate.

3 Experimental Results

3.1 Inception Voltage

Figure 4 shows experimental and theoretical discharge inception voltages with varying void depth. Theoretical inception Voltage [3] is calculated by

$$V_{inc} = \frac{E_g}{\epsilon_r}[t + t'(\epsilon_r - 1)] \ , \tag{1}$$

where Eg is the breakdown strength of the void and relative permitivity ϵ_r for pp is 2.2. Void inception Voltage is calculated by

$$V_{void} = \frac{V_{app}}{[1 + \frac{(t-t')}{\epsilon_r t'}]} \ , \tag{2}$$

where V_{app} is the applied voltage across the sample. Figure 5 shows void inception voltage together with Paschen's value for different cavity depths.

3.2 Apparent and Real charge

The behavior of internal discharges in voids can be described conveniently with the analogue circuit of dielectric and void as shown in Figure 3. In the analogue circuit, c represents the capacitance of the void, b is the total capacitance in series with the void and a is the remaining capacitance of the dielectric shunting the series combination of b and c. The measurable charge is the apparent charge which is defined as the charge lost by the capacitor c when discharge occurs in void and theoretical apparent charge is given by

$$q_{app}^{theo} = \frac{\epsilon_0 \epsilon_r A}{(t - t')}(V_{inc} - V_{ext}) \ . \tag{3}$$

Theoretical real charge of the void is expressed by

$$q_{real}^{theo} = q_{app}^{theo}\left[\frac{b+c}{b}\right] = q_{app}^{theo}\left[1 + \frac{(t-t')}{\epsilon_r t'}\right] \ . \tag{4}$$

Figures 6 and 7 offer an illustration of the results obtained in Table I. The quoted experimental values are the average of ten measurements.

3.3 Drift Velocity and Mobility

Transit time of electrons can be determined theoretically from the drift velocity in air [4],

$$v_d = 1.334X10^6 + 4.222X10^5\left[\frac{E_{inc}^{void}}{p}\right] \ , \tag{5}$$

with v_d in cm/sec and (E/p) in kV/cm-bar. Extremely fast discharge pulses with rise time of the order of nanosecond and sub-nanosecond range have been recorded. Figure 8 shows the fastest discharge pulse recorded with a rise time of 454 ps for 80 μm void depth. The results of rise time are shown in Table I. From partial discharge pulse obtained, drift velocity of electrons is calculated by

$$v_d^{cal} = \frac{t'}{t_e} \ . \tag{6}$$

Rise time of the pulse is considered for transit time of electrons t_e. Mobility of electrons is calculated by

$$\mu_e = \frac{v_e^{cal}}{E_{inc}^{void}} \ . \tag{7}$$

Figures 9 and 10 offer an illustration of the results obtained in Table II.

4 Discussion and Conclusion

Theoretical analysis of the discharge current in voids shows that pulse rise times of few hundreds of ps are to be expected. In the literature measured rise times [5,6] ranging from 400 to 800 ps are reported. The present data shows correlation between recorded rise times and the void depth; discharge current rise times are faster for smaller void depths. The minimum rise time recorded in these studies was 454 ps as shown in Figure 8.
This work shows that the apparent charge, at inception, calculated with the classical capacitance model is closer with measured values. The measured apparent charge has been calculated from the area of the pulse. It is seen that apparent discharge magnitude increases with void volume, from 50 pC to 180 pC for 80 μm to 200 μm void depth with 0.2 cm void diameter. The present data agrees with measured average discharge magnitude [7] by pulse distribution analysis. The theoretical estimation of apparent charge by Pedersen's model [6] shows decreasing trend with increasing void depth for our model. However for varying void depths with constant insulation thickness, Pedersen's model shows increasing trend of apparent charge with increase in void depth. Real charge of void has been estimated from the calculated and measured apparent charge.

This work reports extremely fast rise time pulses which requires several generation of electron production. The measured waveforms and drift velocity indicate streamer mechanism [4,6] in the present work. Reported drift velocity is based on rise time of discharge pulse. The drift velocity of minimum rise time pulse is 2x10^5 m/sec which is in reasonable agreement with values given [6] for streamer velocity. The electron mobility for 200 μm void depth is 3.7x10^{-2} (m^2/volt-sec) being close to the literature value [6] of 4x10^{-2} (m^2/volt -sec).

5 References

1 H.F.A Verhart and P.C.T Van der lan. " Fast current measurements for avalanche studies ". J. Appl. Phys, Vol 53 (3), March 1982, PP 1430 - 1436.
2 M.G.Danikas. " Discharge studies in solid insulation in voids ". IEEE Conf. on Electrical Insulation and Dielectric Phenomena, 1990, PP 249 - 254.
3 E.Hussain and R.S. Nema. " Analysis of Paschen curves for Air , N2 and SF6 using the Townsend breakdown equation ". Proc. IEEE Trans. on Electrical Insulation, Vol EI - 17, No 4, Aug 1982, PP 350 -353.
4 J.M.Wetzer et al . " Experimental study of the mechanism of partial discharges in voids in polyethylene ". 7 th Int. Symposium on High Voltage Engg, No 71.02, Aug 1991, PP 13 -16.
5 P.H.F Morshuis. " Time - Resolved discharge measurements ". Intl. Conf. on partial discharge, No 378, sept 1993 (IEE).
6 M.G.Danikas and A.M. Bruning. " Comparison of several theoretical sub-corona to corona transition relations with recent experimental results ". IEEE Intl. Symp. on Electrical Insulation, 1992, PP 383 - 388.
7 R. Shobha. " Internal partial discharge and breakdown characteristics of thin polypropylene films ". M.Sc Thesis, 1992, Dept of HVE, IISc, Bangalore.

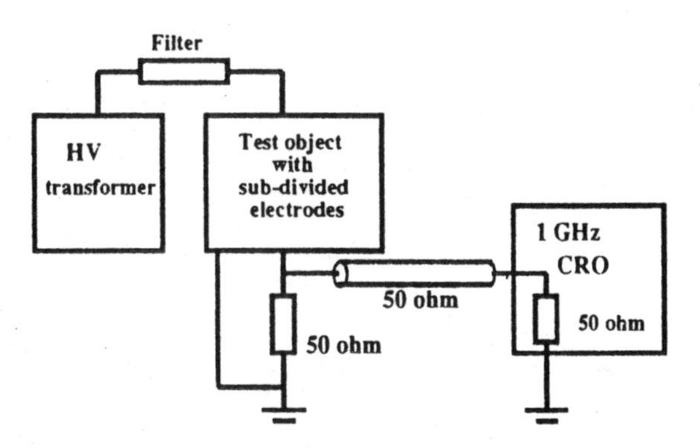

Figure 1 Experimental setup for pd measurements.

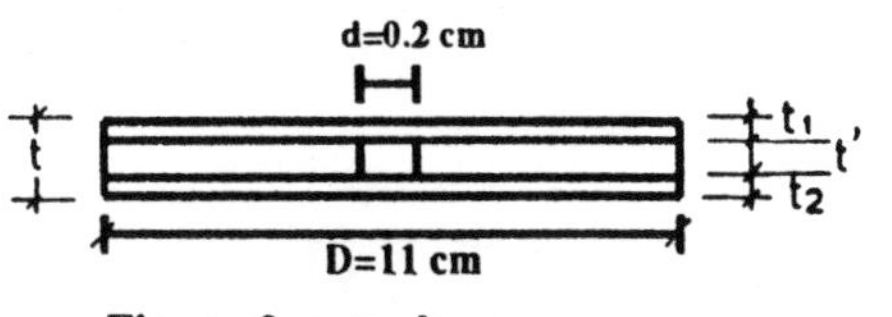

Figure 2 sample.
$t_1 = t_2$ = thickness of pp film = 20 μm
t = thickness of the films containing the void

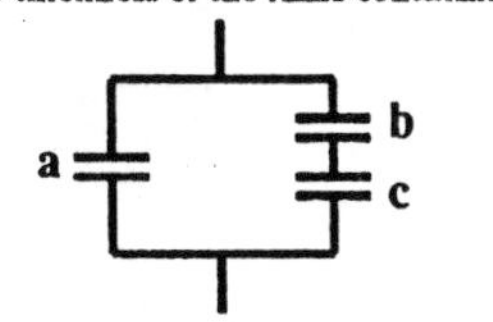

Figure 3 A.C Analogue for void .

Table 1

Void depth (mm)	Apparent charge (pC)			Real charge (pC)			Measured Rise Time (ns)	
	Theo	Measured +ve pulse	Measured -ve pulse	Theo	Measured +ve pulse	Measured -ve pulse	+ve pulse	-ve pulse
0.08	81.6	53.2	52.9	100.1	65.3	64.9	0.754	0.8
0.12	141.4	117.5	108.8	162.8	135.3	125.3	0.991	0.98
0.20	180.0	165.5	176.7	196.4	180.6	192.8	1.43	1.41

Table II

Void depth (mm)	Measured Drift Velocity X 10 ^{5} (m/sec)		Measured Mobility (m^2 /Volt-sec)	
	+ve pulse	-ve pulse	+ve pulse	-ve pulse
0.08	1.06	1.00	1.62	1.53
0.12	1.21	1.22	2.32	2.35
0.20	1.39	1.42	3.64	3.71

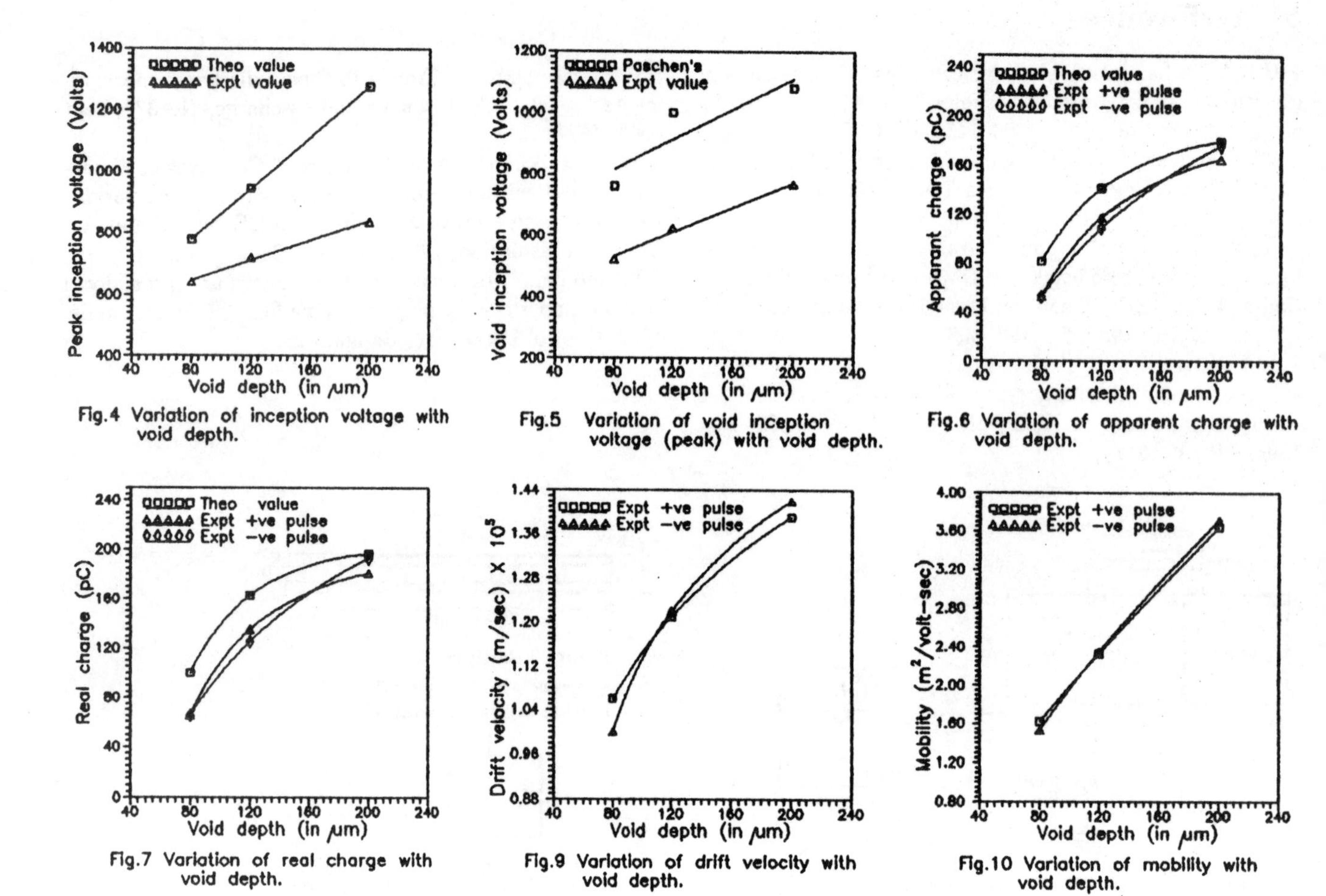

Fig.4 Variation of inception voltage with void depth.

Fig.5 Variation of void inception voltage (peak) with void depth.

Fig.6 Variation of apparent charge with void depth.

Fig.7 Variation of real charge with void depth.

Fig.9 Variation of drift velocity with void depth.

Fig.10 Variation of mobility with void depth.

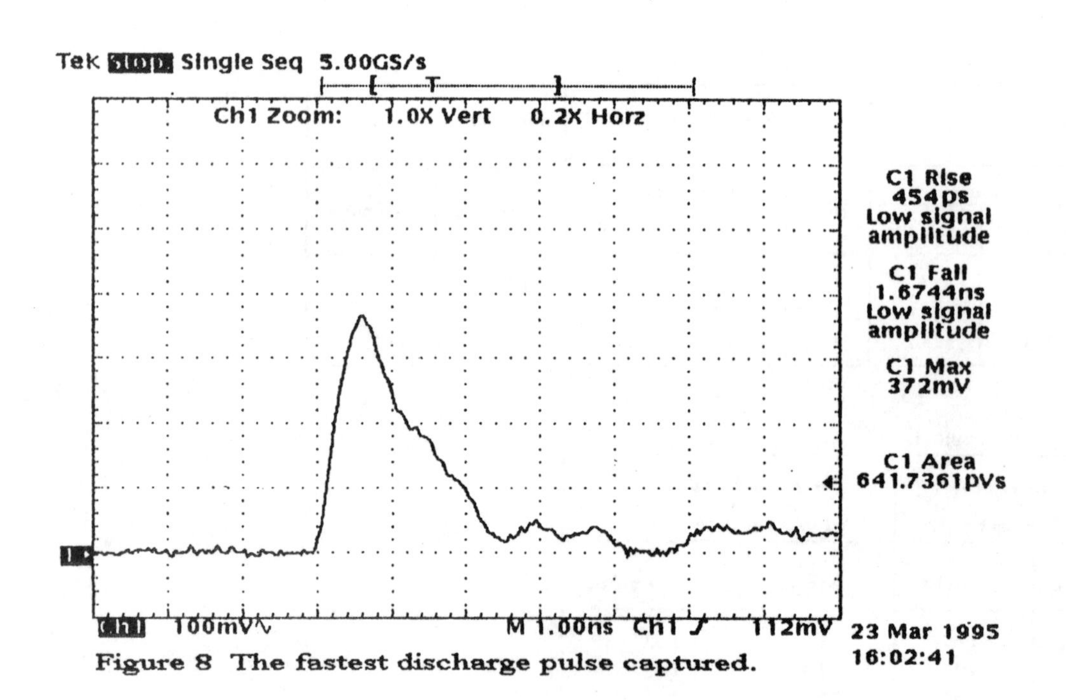

Figure 8 The fastest discharge pulse captured.

Conference Record of the 1996 IEEE International Symposium on Electrical Insulation, Montreal, Quebec, Canada, June 16-19, 1996

Partial Discharge Degradation and Breakdown Studies in Polypropylene Films

Benny Thomas and R. S. Nema
Department of High Voltage Engineering
Indian Institute of Science, Bangalore - 560 012
INDIA.

Abstract: Internal partial discharge characteristics are investigated on impregnated polypropylene films containing artificial cavity of well-defined dimensions, immersed in oil. Electrical breakdown studies are carried out at step-wise rising stress to evaluate constants of inverse power law model. Partial discharge pulse distribution patterns are acquired using PC interfaced multichannel analyser and statistical analysis of the discharge pulse spectrum is done by using 3- parameter Weibull distribution function. The results are compared with that for unimpregnated samples in air.

INTRODUCTION

Polypropylene films are extensively used as dielectric in high voltage capacitors. Internal partial discharges cause long term degradation of polymeric insulation. Eventhough, an in-depth study on partial discharges started as early as 1920's, the basic problem of measurement, evaluation, localisation and coherence between partial discharge quantities and the residual life-time remain unsolved. All investigations show that statistical interpretations of partial discharge data is to be considered, since on the one hand, the partial discharge processes are stochastic in nature and, on the other hand, a large amount of data is involved. Statistical methods can be employed, only if equipments capable of collecting and handling the information faster and more efficiently, are available. One of the advantages of multi-channel analyser based computer-aided measuring system is it's ability to process and transform the information into an understandable output. An attempt is made to study the partial discharge degradation and breakdown characteristics in polypropylene-film samples, having a single artificial cavity of known dimensions.

EXPERIMENTAL PROCEDURE

Electrical grade biaxially oriented polypropylene films of thickness 20 μm are cut as circular pieces of 110 mm diameter. Cylindrical hole of 0.9 mm diameter is drilled at the centre, using a micro-processor controlled, high speed drilling machine. The electrodes used are $\frac{2\pi}{3}$ Rogowski profile uniform field brass electrodes of 58 mm overall diameter and 19 mm radius of curvature, embedded in epoxy (to form 75 mm overall diameter and 20 mm overall thickness) except for the flat portion of the electrode surface (exposed effective diameter of the electrode being 48 mm). Six drilled films are kept in between two single plane films to form a dielectric bounded artificial cavity of 0.9 mm diameter and 120 μm depth. The test sample is placed in between the grounded bottom electrode and the upper high voltage electrode. A thin film of impregnating oil is manually pasted on each layer to form the impregnated sample and, the electrode-sample assembly is completely immersed in oil. Step-wise rising stress is used to obtain the breakdown characteristics of the sample. The voltage is first increased from zero to 1.1 times the inception voltage and, is maintained at this level for a time interval t_s minutes. The voltage is then raised in steps of 500 (rms) volts and maintained for t_s minutes at each step stress, till the sample breakdowns. Time to failure and the applied voltage level at the instant of breakdown are recorded. The time interval (t_s) is chosen to be 1 minute, 3 minutes, 5 minutes and 10 minutes.

Computer aided measurement system is used to monitor the partial discharge pulse distribution patterns. The sample is coupled to a pre-amplifier (a narrow band tuned amplifier with mid-band frequency 500 kHz and band width +/- 5 kHz) through a band-pass filter, a 1000 pF (100 kV) discharge - free coupling capacitor and a tuned detection impedance. The output of the pre-amplifier is amplified using a linear-pulse amplifier (gain : 0.1 – 1.0 and time constant : 1 – 5 μs and, is fed to an IBM PC interfaced multichannel analyser *(EG & G ORTEC, USA)*, where the partial discharge pulses are graded into various channels according to their height, digitized by the analogue to digital converter and, then transmitted to the personal computer (PC). To acquire the pulse distribution pattern, the same stepwise rising stress is applied across the sample, but the final step -voltage level is limited to 2500 V, in view of safety of measuring instruments. Two to five pulse distribution patterns with data acquisition period of 20 seconds each, are recorded in computer floppy diskettes at step voltages of 1.1 Vi, 1000 V, 1500 V, 2000 V and 2500 V, for different step-time intervals of 1, 3, 5 and 10 minutes. Each spectrum contains the data of number of discharges pertaining to 512 different apparent-charge magnitudes. Since the data processing is complicated and, a large capacity of computer memory is needed for on-line analysis, processing of data is done off-line by using another personal computer.

RESULTS AND DISCUSSIONS

The breakdown stress as well as time to failure for unimpregnated samples in air is decreased by about 50 percent in presence of the cavity (Figure 1), whereas the voltage endurance coefficient (n) is reduced from 10.51 to 7.1 (Table 1). For impregnated sample immersed in oil, the breakdown strength and time to failure is decreased by about 68 percent and the n- value is reduced from 12.57 to 9.07. The results show that the internal partial discharges degrade both the unimpregnated and impregnated polypropylene films, for the later, damage is more predominant. The impregnation reduces the amount of long term degradation as indicated by the n value, by an order of 2 for both polypropylene samples with and without cavity, and increases the breakdown strength by 44 percent in the absence of cavity, while by only 10 percent in the presence of cavity. Partial Discharge induced physical and chemical degradation processes on insulating materials are well documented in the literature.

Starr and Endicott [1] proposed the applicability of inverse- power law (IPL) model for linearly increasing stress. Shobha and Nema [2] in their work on internal partial discharge aging on polypropylene films suggested that step-wise rising stress can be used to estimate the constants of IPL model. The step-wise stress considerably reduces the time to failure [3]. The constant (K) of the IPL model is tabulated in Table 1, which is in the same order of magnitude. Since the internal partial discharge is the major aging mechanism in the present experiments and, the K - value indicates a measure of damage to the insulation, it implies that the cumulative damage is independent of the step interval time (t_s). Hence, step-wise rising stress method can be used to estimate the n - value of IPL model. Figures 2 - 4 depict variation of average discharge magnitude (Q_a), number of pulses per second (N) and total discharge magnitude (Q_t) with overvoltage, for step- time intervals of 1, 3, 5 and 10 minutes, for both the unimpregnated samples in air and unimpregnated samples in oil, submitted to internal partial discharges. For unimpregnated samples in air, average discharge magnitude steeply reduces with overvoltage and finally, reaches a stabilized value as shown in Figure 2, whereas number of pulses per second sharply increases and then, reaches a stabilised value (Figure 3). The total discharge magnitude also reaches a stabilised value (Figure 4), independent of the step-time interval. In Figures 2 & 3, it can be seen that the average discharge magnitude is highest and, the pulse repetition rate is lowest at inception voltage level, which implies that discharges of larger magnitudes predominate at lower voltage and subsequently, number of lower magnitude discharges increase. Increase in the number of lower magnitude discharges is an indication of aging of the insulation.

For impregnated samples immersed in oil, the pulse rate, average discharge magnitude and total discharge magnitude are much lower, in comparision with the corresponding values for unimpregnated samples in air and, these discharge quantities substantially remain within a small range of magnitude over the entire range of applied voltage levels. It is to be noted that the magnitude of these discharge quantities is comparable with the stabilised values of respective quantities for unimpregnated samples in air. Impregnation slows down the detrimental aging processes at the begining, but later, aging becomes comparable with or more than that in unimpregnated samples. It may be due to structural or chemical changes in the polymer, promoted by the impregnating liquid.

The Weibull cumulative probability distribution is frequently used to fit the electrical breakdown data. The 3-parameter Weibull distribution is given by

$$lnln\left[\frac{1}{1-F(q)}\right] = \beta\, ln\,(q-\gamma) - \beta\, ln\,\alpha \qquad (1)$$

where α is the scale parameter, which represents the discharge magnitude below which lie 63.2 percent of discharge magnitudes, β is the shape parameter which is a measure of the dispersen of discharge magnitudes, γ is the location parameter which indicates the discharge magnitude below which no discharges occur and q (the discharge magnitude) is the random variable. The pulse distribution data fits 3-parameter Weibull distribution. The variation of statistical parameters α & β with overvoltage are depicted in Figure 5 for unimpregnated sample in air and, in Figure 6 for impregnated sample in oil. The α -values are initially maximum and then, decreases rapidly to a stabilized value, for unimpregnated samples in air. The step-interval time does not affect α except for inception voltage or unaged samples (Figure 5). Figure 6 also shows a similar trend for impregnated sample immersed in oil, where α - values having wide variation with step-time interval at inception voltage level tend to converge and finally, stabilize to a representative value for aged sample, as applied voltage level increases. On the one hand, the scatter of α - values with step-time interval and, on the otherhand, a stabilized value with increase in applied voltage level seems to be a good measure to assess the status of an aged sample, for impregnated samples as well as unimpregnated samples

The β -values are independent of the step-interval time, t_s. Since the β - value is invariable with applied voltage level and step - interval time, the conclusion drawn is that a single breakdown mechanism only is present in the sample. Three parameter Weibull distribution can be used to identify the void discharges which occur in solid insulating materials.

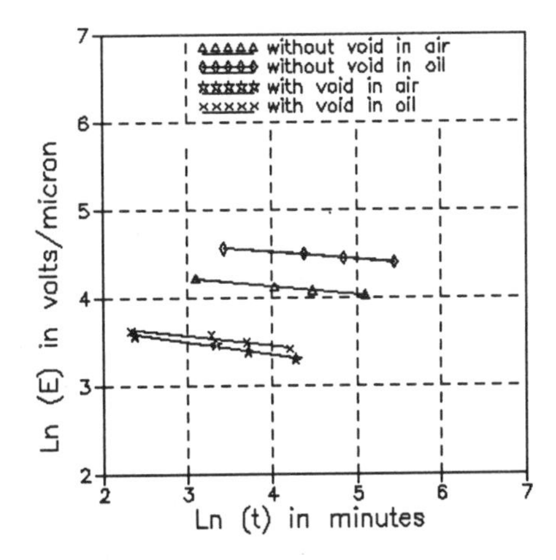

Figure 1. E-t Characteristics

CONCLUSION

Internal partial discharges considerably reduce the life of impregnated and unimpregnated polypropylene films. Step-wise rising stress method can be used to evaluate constants of inverse power law model. The average discharge magnitude (Q_a), total discharge magnitude (Q_t) and pulse repetition rate (N) stabilize to representative value which indicate the status of aged sample. Number of larger magnitude discharges decrease whereas number of smaller magnitude discharges increase with aging. Impregnation slows down the aging processes. Weibull parameters α and β can be used to identify the internal partial discharges and to monitor the status of aging.

REFERENCES

1. Starr,W.T and Endicott,H.S. "Progressive Stress – A New Accelerated Approach to Endurance". AIEE Trans PAS. Vol. 80, 1961, p. 575.

2. Shobha,R and Nema,R.S. "Internal Partial Discharge and Breakdown Characteristics of Thin Polypropylene Films". IEEE CEIDP, 1993, p.408.

3. Laghari,J.R Cygan,P and Khechen,W. "A Short Method of Estimating Life-time of Polypropylene Film using Step- stress Tests". IEEE TransEI. Vol 23(6),1990,p.1180.

4. Benny Thomas and Nema,R.S. "Internal Partial Discharge Characteristics in Thin Polypropylene Films". Conference on High Voltage Engineering, IISc, Bangalore, India, 1995,pp.209-212.

Table 1
E − t Characteristics data

Sample	n (#)	t_s (*min.*)	Time to Failure $t(min.)$	Field Strength $E(V\mu^{-1})$	Constant $K(\#)$
A1	10.51	1	21.977	68.03	4.015 X 10^{20}
		3	56.260	61.56	3.595 X 10^{20}
		5	87.357	59.15	3.666 X 10^{20}
		10	164.022	56.26	4.069 X 10^{20}
A2	7.10	1	11.516	35.52	1.173 X 10^{12}
		3	23.424	32.14	1.173 X 10^{12}
		5	41.338	29.67	1.173 X 10^{12}
		10	78.320	27.11	1.173 X 10^{12}
B1	12.57	1	30.630	95.83	2.475 X 10^{26}
		3	79.870	89.58	2.765 X 10^{26}
		5	126.16	85.42	2.402 X 10^{26}
		10	232.75	81.25	2.362 X 10^{26}
B2	9.07	1	12.071	37.34	2.193 X 10^{15}
		3	18.997	35.52	2.193 X 10^{15}
		5	39.244	32.79	2.193 X 10^{15}
		10	81.068	30.27	2.193 X 10^{15}

n : Endurance Coefficient
t_s : Step-time Interval
A1 : Unimpregnated Sample in Air Without Void
A2 : Unimpregnated Sample in Air With Void
B1 : Impregnated Sample in Oil Without Void
B2 : Impregnated Sample in Oil With Void

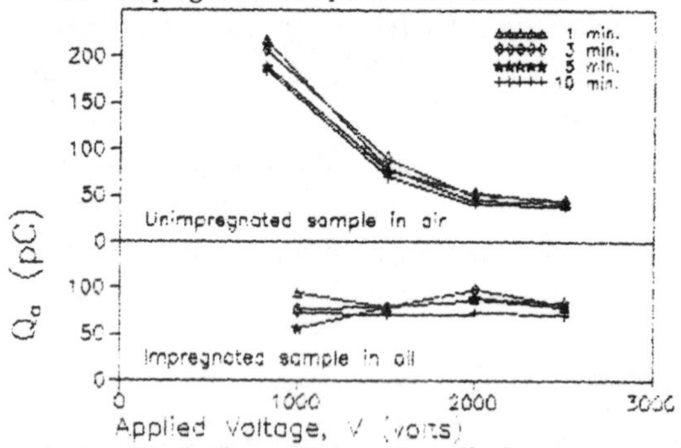

Figure 2. Variation of average discharge magnitude with applied voltage

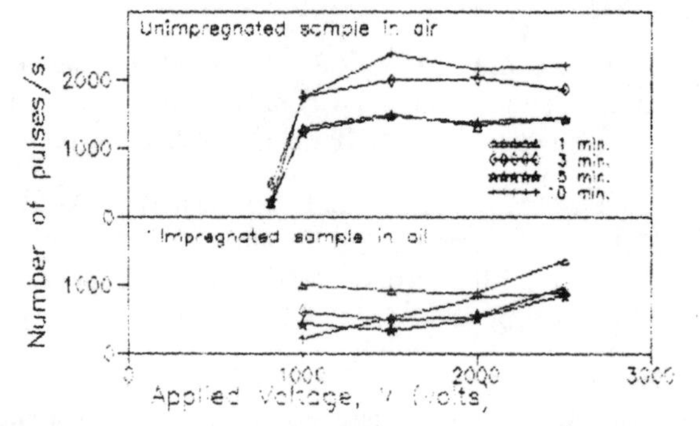

Figure 3. Variation of pulse repetition rate with applied voltage

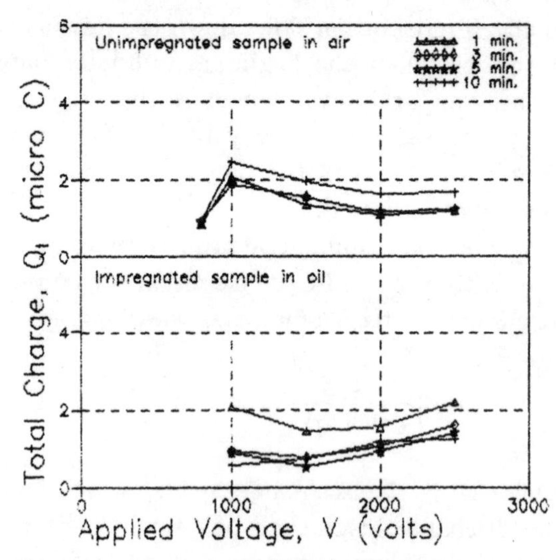

Figure 4. Variation of total discharge magnitude with applied voltage

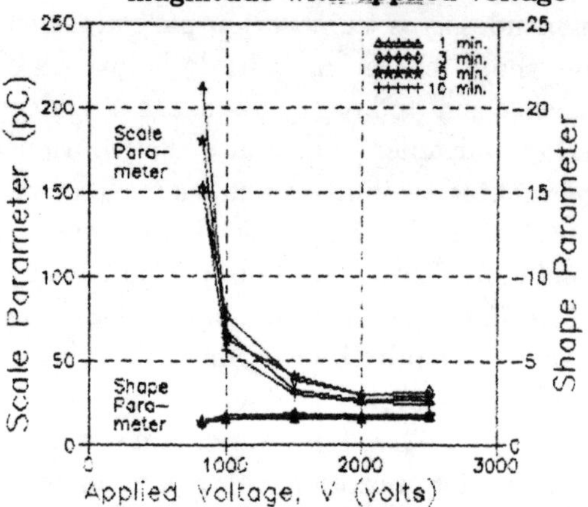

Figure 5. Variation of scale parameter with applied voltage for umimpregnated sample in air

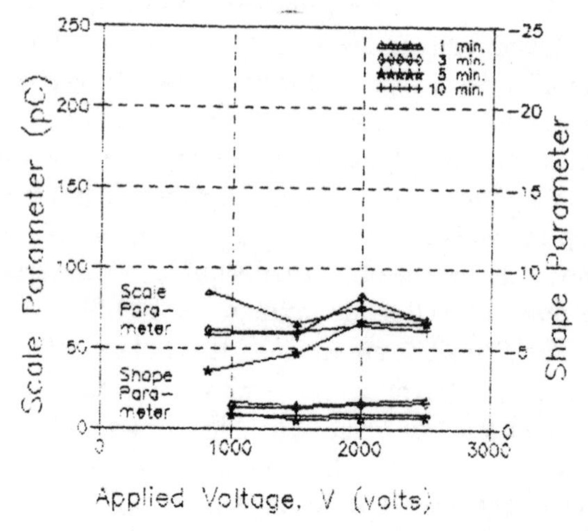

Figure 6. Variation of total discharge magnitude with applied voltage

Conference Record of the 1996 IEEE International Symposium on Electrical Insulation, Montreal, Quebec, Canada, June 16-19, 1996

Effect of Elongation on the Electrical Properties and Morphology of Polypropylene

Dong-Young Yi, Seung Hwangbo, and Min-Koo Han

Department of Electrical Engineering, Seoul Natonal University, Seoul, Korea

Dae-Hee Park

Department of Electronic Materials, Wonkwang University, Iri, Korea

ABSTRACT: Variations of electrical properties of polypropylene film due to the elongation have been investigated. The conductivities of PP films have been decreased with the increase of elongation ratio(λ) and the activation energy(ϕ) of elongated PP films were slightly higher than that of non-elongated PP film. From TSC experimental data, we have found, in the elongated PP films(λ=6, 8), that trap density shows the slight increase trend with the increase of elongation ratio for peak P_1, but trap density for second P_2 decreased largely in the elongated samples. We have also observed the increase trend of AC breakdown strength with the increase of elongation ratio. We could conclude that the decrease of conductivity and increase of dielectric strength with elongation were originated from the increase of trap density and trap depth for TSC peak P_1, and also may be attributed to the dominance of dipolar process at room temperature.

INTRODUCTION

Polypropylene(PP) films are widely used in the power capacitor due to their excellent electrical and mechanical properties. Many studies have been carried out on polypropylene resins and non-oriented films, and the work on the thermal aging behavior of BO-PP films has been reported recently.[1] It has also been reported that a supermolecular structure of polyethylene has a strong influence on the electrical properties.[2]-[4] We have been reported previously that the residual voltage and the surface potential decaying characteristics were strongly affected by the high elongation ratio.[5] From these results, we have found that the trap density and/or trap depth should be changed owing to the elongation. The effects of mechanical drawing or draw ratio(λ) on the dielectric properties and their correlation with morphology in the commercially available non-oriented PP films have been investigated scarcely. It is very important to improve the dielectric properties of PP films more effectively because of an increase of the rated electric stress of power capacitors. The purpose of this work is to investigate the effect of mechanical drawing or elongation on the electrical properties, such as conductivity, thermally stimulated current and dielectric strength, of polypropylene (PP) films.

EXPERIMENTAL

We have prepared the elongated polypropylene films with the use of roll elongation method at 135[°C]. All samples were 50 [μm] thick. Draw ratios(λ) or elongation ratios of elongated PP films were 1(non-elongation),4, 6, 8 and these were determined by the ratio of the length of elongated film to that of non-elongated film. After the elongation process, samples were cleaned with ethanol and ultrasonic cleaner, and dried in a desiccator for 24 hours or more before the preparation of test specimens for each measurement in order to minimize the influence of contamination. We have performed the X-ray diffraction (XRD) analysis to evaluate the structural or morphological changes of polypropylene due to elongation. We made use of vacuum evaporated gold electrode to assure the intimate contact and to eliminate the interfacial problems. We have measured the conduction properties of virgin and elongated samples by logging and monitoring the real time current data with computer. Before the measurement of conduction current, test specimens were preconditioned by short-circuiting the measurement setup for 24 hours or more in an oven to eliminate the memory effect.

Thermally stimulated current(TSC) was measured by a standard technique as follows.[6] An electric field of 0.2[MV/cm] was applied to the sample at room temperature for about 40 minutes. And the sample was cooled down to 77[K] while keeping an electric field of 0.2[MV/cm] : it was kept at this temperature for about 30 minutes under the same applied field. After short-circuiting the sample until the discharge current became negligible, the sample was heated from 77 to 360[K] at a constant rate of 4 [℃/min]. These experiments were carried out under a vacuum of 10^{-5} Torr. Elongation dependence of dielectric breakdown strength(E_b) has been measured by applying voltage pulse of 6 μs width and 0.1 μs rise time.

RESULTS and DISCUSSIONS

We have performed the X-Ray Diffraction(XRD) analysis to evaluate the structural or morphological changes due to elongation. We have found that the crystalline form of elongated and non-elongated PP films used in our experiment are highly isotactic and monoclinic α-form from XRD results as shown in Fig.1. It has been reported that the isotactic polypropylene can be highly crystallized only by the small mechanical drawing because of its steric regularity.[1] We could observe the disappearance of some peaks($2\theta \approx 21.5, 22$) from XRD results, which may be attributed to the orientation of crystalline and amorphous phase, in the elongated PP films. We have investigated the relation between these morphological and the electrical properties, such as conductivity, dielectric strength and TSC(thermally stimulated current), of polypropylene films.

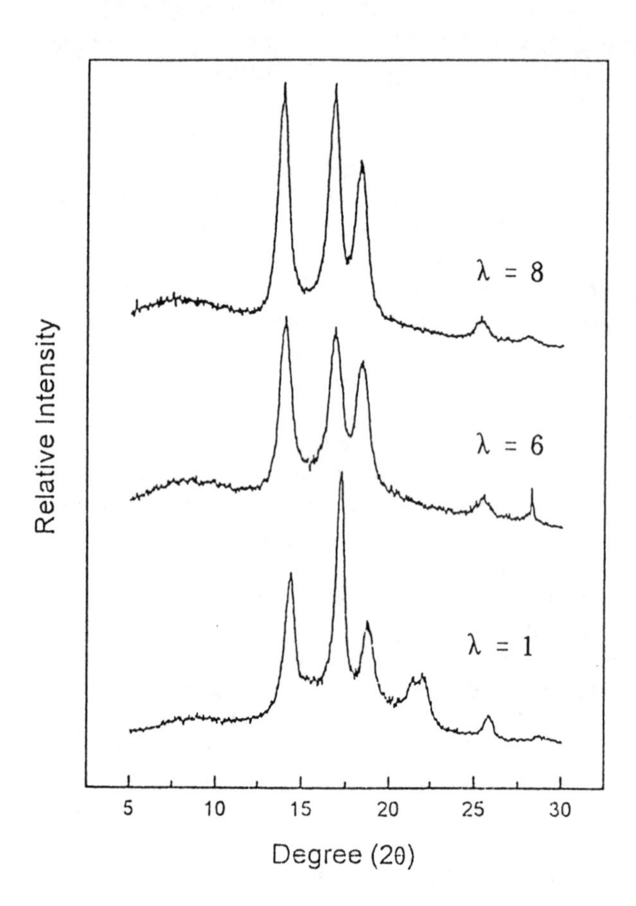

Fig. 1. XRD Results for virgin and elongated PP

Fig 2. shows the relation between the electrical conductivity and the elongation ratios of PP films. We have observed that the conductivities of PP films decreased with the increase of elongation ratio(λ).($\sigma \approx 4 \times 10^{-18}$ [S/cm] for λ=1, $\sigma \approx 6 \times 10^{-19}$ [S/cm] for λ=6, $\sigma \approx 4 \times 10^{-19}$ [S/cm] for λ=8 at room temperature under 0.2 [MV/cm]) We have also calculated, from the temperature dependence of conductivity(Fig 3.), the activation energy(ϕ) of PP films. We have found that the activation energy(ϕ) of elongated PP films were a little higher than that of non-elongated PP film and conduction current of the non-elongated PP films were larger than those of elongated PP film at both the high and low temperature regions.($\phi \approx 1.53$ [eV] for λ=1, $\phi \approx 1.65$ [eV] for λ=6, 8 at low temperature region, $\phi \approx 2.4$ [eV] for λ=1, $\phi \approx 2.75$ [eV] for λ=6, 8 at high temperature region) Transition temperatures of high and low temperature region, at which the dipolar process of conduction is almost over and ionic hopping process begins, have been increased with the increase of elongation ratio. The decrease of conductivity may also be attributed to the dominance of dipolar process in the elongated polypropylene films at room temperature.

It has been well known that conduction of PP is governed by the dipolar process in the amorphous phase or at interfaces between the crystalline and amorphous phases in low temperature region, and in the high temperature region dominated by the ionic hopping through the amorphous phase.[1] Therefore, we could conclude that the orientation effect due to the roll elongation may be closely related to the conduction behavior of elongated PP films such as the decrease of conductivity, increase of activation energy, and the increase of transition temperatures at which the ionic hopping begins.

We have also performed the TSC(thermally stimulated currents) measurements in order to investigate the above changes of conduction behaviors of PP with elongation. Therefore, we have performed the TSC(thermally stimulated currents) measurements to evaluate the elongation dependence of trap density. Fig 4. shows TSC plots of the elongated and non-elongated samples. We could observe the new peak P_1 at 261[K] in the elongated PP films(λ=6, 8), and found that the peak temperature of the peaks P_1 , P_2 was not influenced by the elongation. We have analyzed peak P_1 with the initial rising method for the evaluation of the elongation dependence of effective trap depth.[7] Conduction behaviors may be closely related to the trap depth or trap density of the elongated PP. Each peaks was separated by the thermal cleaning method.[8] The trap depths of the elongated samples, evaluated from peak P_1 at room temperature range, are about 0.3[eV] for both the 6 and 8 times elongated samples. We have also evaluated the relative changes of trap density between the elongated and non-elongated by the integration of each peaks. Trap density shows the slight increase trend with the increase of elongation ratio for peak P_1. However, we have found that the trap density for P_2 decreased largely with elongation and there were little differences between the elongated samples. It has been reported that the trap density increases slightly with a low elongation range and again decreases with further elongation.[9] We have calculated the net TSC charge as a function of the elongation ratio. Fig 5. shows the relation between the net TSC charge and elongation ratios. TSC charge decreased with the increase of elongation ratio and reduced to about one order of magnitude compared with non-elongated PP in the 8 times elongated PP film(λ=8). Therefore, we may conclude that the increase of trap density and trap depth for TSC peak P_1 are mainly responsible for the decrease of conductivity and increase of activation energy with elongation.

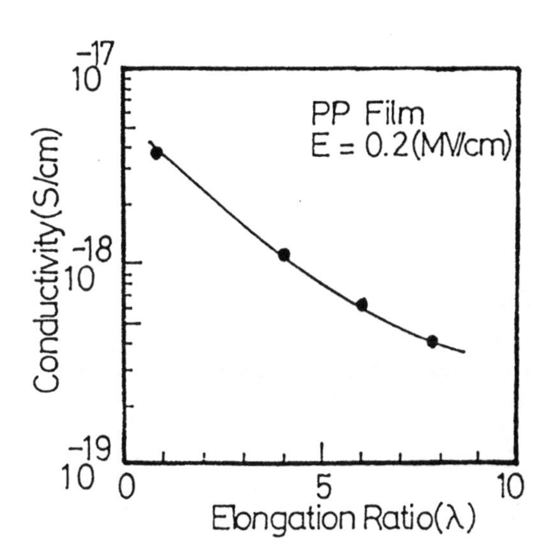

Fig. 2. Elongation dependence of conductivity

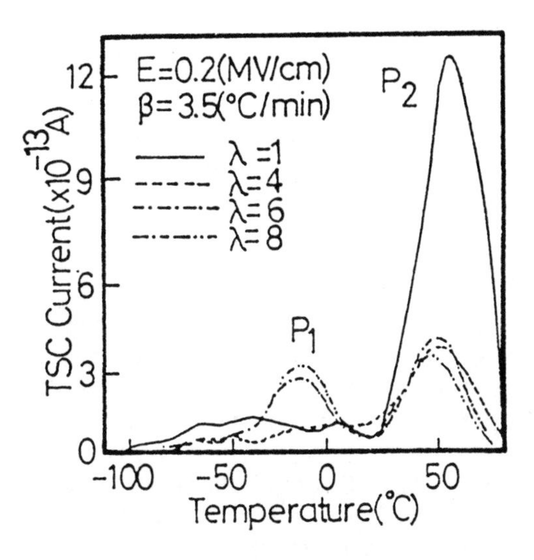

Fig. 4. TSC curve as a function of elongation ratio

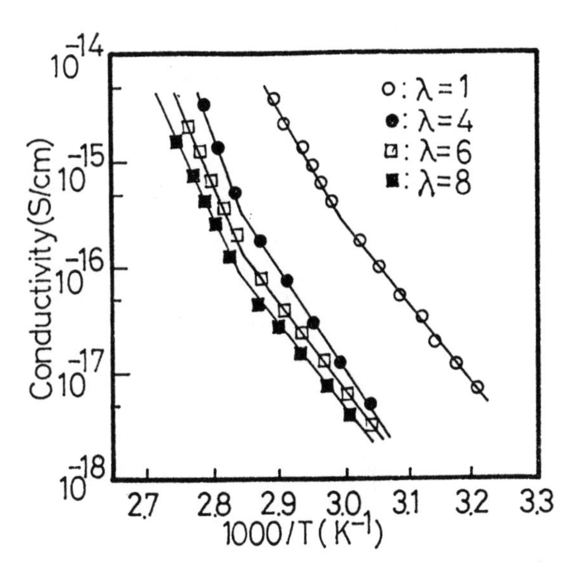

Fig 3. Arrhenius plot from conduction behavior

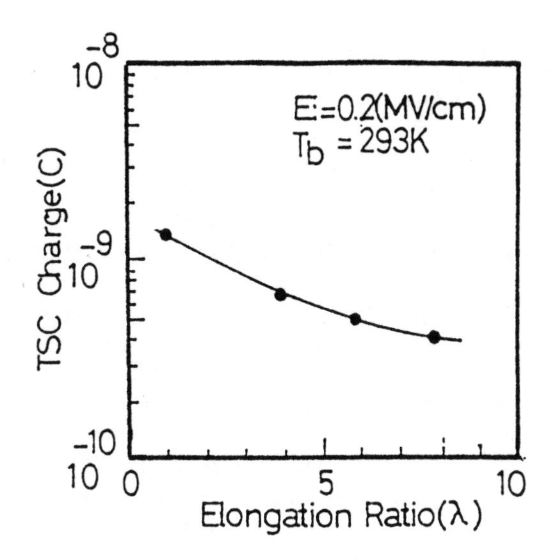

Fig. 5. TSC charge as a function of elongation ratio

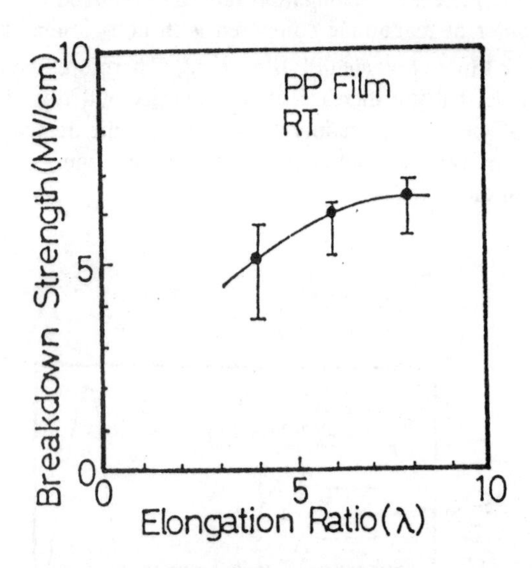

Fig. 6. Elongation dependence of Breakdown strength

We have also observed the increase trend of AC breakdown strength with the increase of elongation ratio.($E_b \approx 3.5$[MV/cm] for $\lambda=1$, $E_b \approx 5$[MV/cm] for $\lambda=4$, $E_b \approx 5.8$[MV/cm] for $\lambda=6$, $E_b \approx 6$[MV/cm] for $\lambda=8$) These increase of dielectric strength with elongation may be due to the same reason for the conduction behaviors such as the domination of dipolar process in conduction of elongated PP at room temperature and the increase of trap depth or density as shown in TSC peak P_1.

CONCLUSION

From the above experimental results and analysis, we could observe that the conductivities of PP films have been decreased with the increase of elongation ratio(λ) and the activation energy(ϕ) of elongated PP films were a little higher than that of non-elongated PP film. From TSC results, we could observe, in the elongated PP films(λ=6, 8), the increase of peak P_1 at 261[K], and that trap depths at room temperature range are about 0.3[eV]. Trap density shows the slight increase trend with the increase of elongation ratio for peak P_1 , but trap density for P_2 decreased largely in the elongated samples. TSC charge have been decreased with the increase of elongation ratio. We have also observed the increase trend of AC breakdown strength with the increase of elongation ratio. Crystalline form of PP films are highly isotactic and monoclinic α-form and we have found the disappearance of some peaks from XRD results, which may be attributed to the orientation of crystalline and amorphous phase, in the elongated PP film.

We could conclude that the decrease of conductivity and increase of dielectric strength with elongation were originated from the increase of trap density and trap depth for TSC peak P_1 , and also may be attributed to the dominance of dipolar process at room temperature. The orientation effect due to the roll elongation may be closely related to the above electrical behavior of elongated PP films

ACKNOWLEDGEMENT

This work was supported by the Korea Electric Power Company(KEPCO).

REFERENCES

1. Umemura, T, K. Abe and K. Akiyama. IEEE Trans EI. Vol. 22., No.6, December 1987, pp. 735-742.
2. Yoshino, K and Y. Inuishi. Oyo Butary. 49, 1980, p212.
3. Kolesov, S. N. IEEE Trans EI. Vol. 15, 1980, p382.
4. Yahagi, K. IEEE Trans EI. Vol. 15, 1980, p241.
5. Park, D.H, J. Kyokane and K. Yoshino. Jpn. J. Appl. Phys. 20, 1987, L65.
6. Nishitani, T, K. Yoshino and Y. Inuishi. Jpn. J. Appl. Phys. 14, 1975, p721.
7. Creswell, R. A and M. M. Perlman. J. Appl. Phys. 41, 1970, p2365.
8. Cowell, T. A. T and J. Woods. Brit. J. Appl. Phys. 18, 1967, p1045.
9. Kobayashi, S and K. Yahagi. Jpn. J. Appl. Phys. 16, 1977, p2053.

Conference Record of the 1996 IEEE International Symposium on Electrical Insulation, Montreal, Quebec, Canada, June 16-19, 1996

Characterization of Reconstituted Mica Paper Capacitors used in High Voltage Electronics Applications

John S. Bowers, PE, CIRM
Engineering Manager
Custom Electronics, Inc.
Browne Street
Oneonta, NY 13820-1096

Abstract - **The purpose of this paper is to describe the characteristics of reconstituted mica paper capacitors that are designed and manufactured for use in high voltage electronics applications. The applications; design and construction; electrical and environmental characteristics; and reliability of this type of capacitor will be described.**

INTRODUCTION

High voltage electronics systems that are designed for commercial, aerospace, and military applications require highly reliable components. These types of electronics circuits and systems include, or can include, the use of reconstituted mica paper capacitors. Reconstituted mica paper capacitors are particularly suited for operation where high ambient temperatures exist [1] and are an excellent choice for these types of systems.

APPLICATIONS

Reconstituted mica paper capacitors are typically used for energy storage, filtering, coupling, etc. in high voltage applications where radiation resistance, corona resistance, high volumetric efficiency, physical durability, and capacitance stability (with respect to temperature, voltage, frequency, or mechanical stresses) are required. These types of applications include, but are not limited to, airborne or surface radar systems, ECM power supplies, high voltage transmitters for missile applications, high voltage TWT power supplies, ignition systems, power transmission systems, laser devices, and gas and oil exploration equipment.

Small, high voltage electronic modules can be designed and manufactured to include these types of capacitors in conjunction with other high voltage components (i.e. resistors, diodes, spark gaps, strip lines, inductors, etc.).

DESIGN AND CONSTRUCTION

The dielectric material used in the design and construction of these types of capacitors is reconstituted mica paper which is impregnated with a liquid polymer resin (i.e. polyester, epoxy, or silicone). The National Electrical Manufacturers Association defines mica paper as flexible, continuous, and uniform layers of mica reconstituted into a paper-like, electrical insulating material composed entirely of small, thin, overlapping flakes or platelets, and which has sufficient strength to be self-supporting and to be capable of being wound into roll form for commercial use [2]. Capacitor grade mica paper does not contain binders, adhesives, foreign matter, or coloring agents, and is substantially free of any substance which will adversely affect its performance.

Capacitor grade, reconstituted mica paper is manufactured from natural muscovite mica ($K_2Al_4Al_2[Si_6O_{20}](OH,F)_4$). An Energy Dispersive X-Ray Spectrum (EDX) for muscovite mica is shown in Fig. 1 [3].

The muscovite mica is subjected to a process in which it heated to approximately 870°C. This heat causes the mica crystal to partially dehydrate and release a portion of the water which is bonded naturally in the crystal. When this occurs, the mica partially exfoliates and smaller particle sizes result. The heated mica is then quenched in a mild alkaline solution. The mica is cooled drained and then subjected to a weak sulfuric acid solution.

The chemical reaction between the caustic and the acid generates a gas between the laminae which causes the mica to expand greatly. The particle size is further reduced by mechanical methods. The mica is screened in the presence of large amounts of water on separating screens to select the desired particle size distribution to produce the paper desired. This "pulp" is then transferred to a specially modified paper machine for sheet forming and drying. The particular design of the head box and drying belts and/or drums are generally proprietary to the individual mica paper manufacturers.

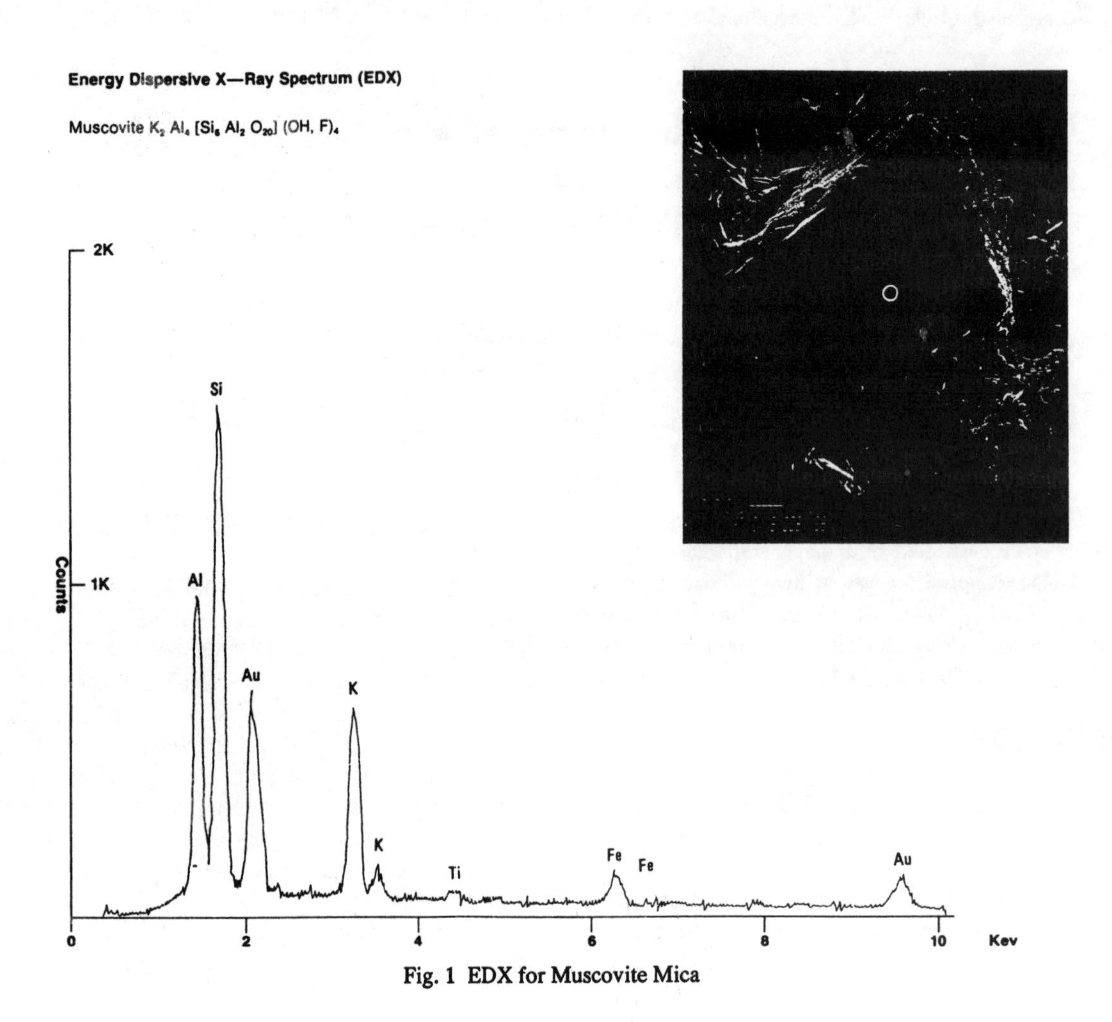

Fig. 1 EDX for Muscovite Mica

The van der Waals' forces between the crystal surfaces of the mica platelets in close proximity hold the layer together. Reconstituted mica paper can range in thicknesses from 12.7μ m (0.0005") to 50.8μm (0.002").

The mica paper is wound with capacitor grade aluminum foil electrodes. Nickel leads are used in "buried foil" construction to make a connection with the anode and the cathode foils. With "extended foil" construction, nickel leads are not required. Usually, two or more layers of mica paper are used in the mica paper capacitor winding. The windings are then lightly pressed into a flat configuration. The windings are then subjected to a vacuum drying cycle and a Vacuum Pressure Impregnation (VPI) process with a liquid resin material (i.e. epoxy, polyester, or silicone) of low viscosity. After the VPI process, the windings are removed from the liquid resin and are then polymerized with a heat curing process. The "cured" windings are now referred to as sections and are cleaned and tested for basic electrical properties (i.e. Capacitance, Dissipation Factor, Dielectric Withstanding Voltage, etc.). The sections are then soldered and assembled into the required configuration and packaged in one of several methods which include tape wrap and end fill, potting in a metal or fiberglass tube, or encapsulated with any one of several types of epoxy. The capacitors are then electrically tested and inspected.

Depending on the type of packaging, capacitance, voltage rating, terminations, etc., various dimensions can be achieved.

ELECTRICAL AND ENVIRONMENTAL CHARACTERISTICS

Reconstituted mica paper capacitors are well known for their outstanding electrical, environmental, and physical characteristics. Most notably, these parts exhibit long life, a very low capacitance drift over the entire temperature range, they can withstand high voltages, they are naturally resistant to the effects of partial discharges, and they exhibit low radiation-induced conductivity caused by the absorption of ionizing radiation such as X-rays, gamma rays, and neutrons. In addition, they exhibit a fractional voltage or charge loss as a function of the absorbed dose.

Electrical Characteristics

Capacitance Range: Generally, the capacitors can be designed in the 100 pF to several μF range (depending on the voltage rating). Capacitance tests are generally conducted in accordance with MIL-STD-202, Method 305.

Capacitance Tolerance: The general capacitance tolerance is ±10%. Other tolerances are available.

Dielectric Constant (K): The dielectric constant of reconstituted mica paper is approximately 6.5 to 8.5 [1,4,5]. As a result of the impregnants used with the reconstituted mica paper, the "effective" K of the reconstituted mica paper capacitors is approximately 5 to 6.

It is interesting to note that the K value for most polymer films that are used as the dielectric material for high voltage capacitors is in the range from 2 to 3.5 [4].

Voltage Range: Voltage ratings are from 1 to 50 kVdc. Typically, each capacitor is designed and manufactured for a specific application.

Dielectric Withstanding Voltage (DWV): This test (also referred to as a high potential, high pot, over potential, or dielectric strength test) is conducted at a voltage higher than the rated voltage for a specific amount of time and is used to prove that the capacitor can operate safely at its rated voltage and that it can withstand a momentary overvoltage condition due to switching, surges, and other similar phenomena. DWV tests for reconstituted capacitors are typically conducted for 15 seconds at 110% to 200% of rated voltage (depending on the design stress of the capacitor) in accordance with MIL-STD-202, Method 301.

Dissipation Factor (DF): The DF is defined as the ratio of the effective series resistance to the capacitive reactance at a frequency [4]. It is also the inverse of the Quality Factor (Q). The DF of reconstituted mica paper capacitors increases with temperature and frequency. Typical DF values range from 0.5% maximum at 25°C to 1% maximum at 125°C. DF measurements are typically conducted in accordance with MIL-STD-202, Method 306.

Insulation Resistance (IR): The insulation resistance of reconstituted mica paper capacitors is 5000 megohms (MΩ) x microfarads (μF) minimum or 100,000 MΩ minimum (whichever is less) at 25°C and 25 MΩ x μF minimum or 1,000 MΩ minimum (whichever is less) at 125°C. Insulation resistance is typically measured at 500 Vdc after two minutes of electrification in accordance with MIL-STD-202, Method 302, Test Condition B.

Thermal Coefficient of Capacitance: The thermal coefficient of capacitance or reconstituted mica paper capacitors is approximately + 5% maximum from 25°C to 125°C and - 3% maximum from 25°C to -65°C.

Inductance: The inductance of reconstituted mica paper capacitors can range from approximately 5 nH for "extended foil" designs to approximately 100 nH for "buried foil" designs.

Energy Density: Typical energy densities range from 0.0061 J/cc (0.1 J/in^3) for a low capacitance, low voltage capacitor to 0.061 J/cc (1.0 J/in^3) for a capacitor designed for short life applications. The individual design will determine the energy density value.

% Capacitance Change, dissipation factor (in %), and insulation resistance (in MΩ x μF) from -55°C to 125°C for typical reconstituted mica paper capacitors are shown in Fig. 2.

Environmental Characteristics

Operating Temperature: Depending on the impregnant, reconstituted mica paper capacitors can withstand the following temperature ranges:

- Epoxy: -65°C to 125°C
- Polyester: -55°C to 200°C
- Silicone: up to 260°C

Thermal Shock: With the exception of unencapsulated reconstituted mica paper capacitors, reconstituted mica paper capacitors are designed and manufactured to pass the thermal shock tests defined in MIL-STD-202, Method 107, Test Condition B.

Immersion: With the exception of unencapsulated reconstituted mica paper capacitors, reconstituted mica paper capacitors are designed and manufactured to pass the immersion tests defined in MIL-STD-202, Method 104, Test Condition B. "Wrap and Fill" units will not meet these requirements unless encapsulated in an assembly or provided with some other form of protection from moisture.

Humidity Resistance: Moisture in the capacitor dielectric can cause a large reduction in leakage resistance and increases in the DF [4]. Reconstituted mica paper capacitors can be subjected to and pass humidity resistance tests that are conducted in accordance with MIL-STD-202, Method 103, Test Condition B. This type of test is not applicable for unencapsulated capacitors.

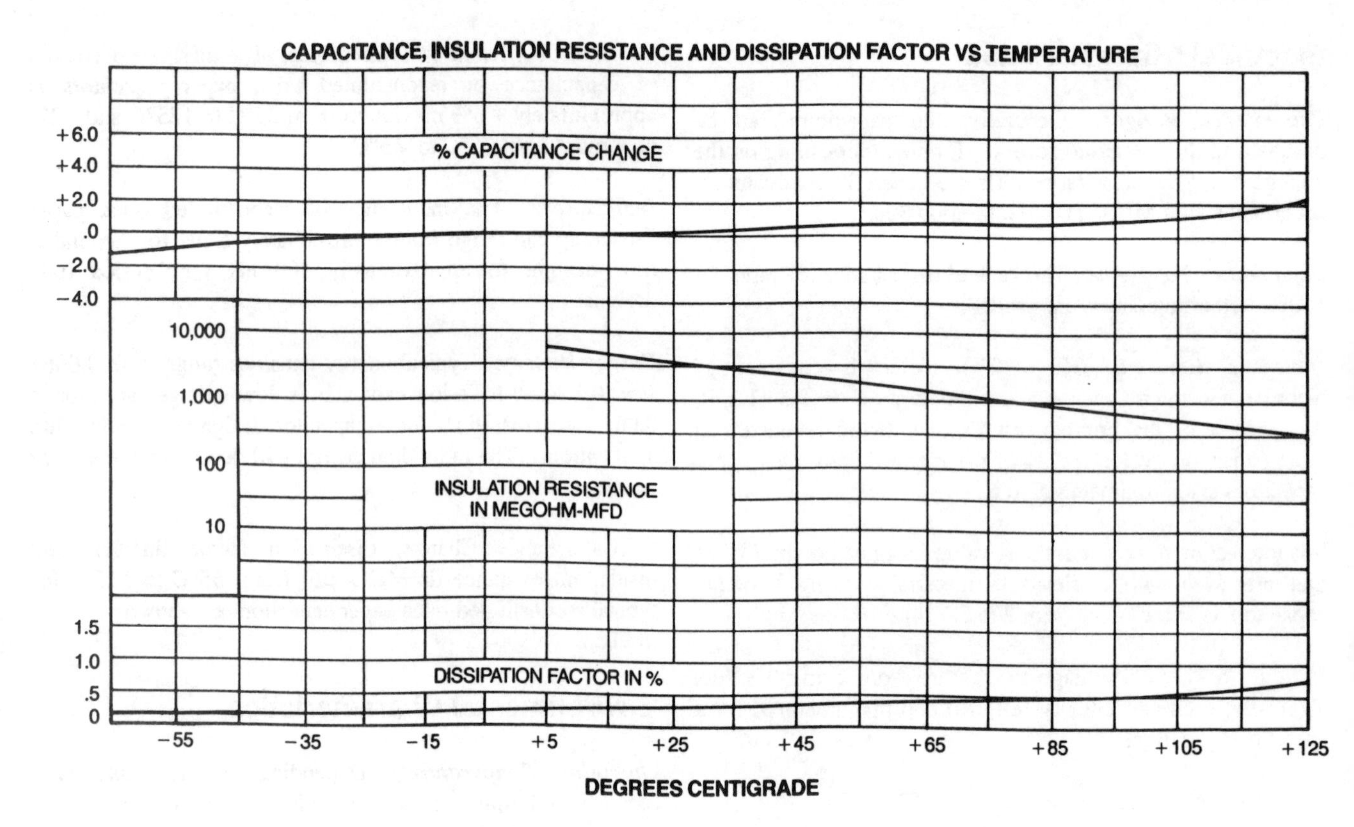

Fig. 2 Capacitance, IR, and DF vs. Temperature

RELIABILITY

High reliability is the number one "strength" of reconstituted mica paper capacitors. A complete understanding of the customer's requirements, a proper design, the selection of highly reliable materials, and a tight control of the manufacturing and testing processes lead to the reputation that these types of capacitors have.

Studies are currently being conducted to determine the voltage and temperature acceleration factors for the dc life of reconstituted mica paper capacitors. A voltage acceleration factor of 7 to 10 is typically used for reconstituted mica paper capacitors.

CONCLUSION

The applications, design and construction, electrical and environmental characteristics, and reliability of this type of capacitor have been described.

High reliability reconstituted mica paper capacitors provide outstanding characteristics when properly designed, manufactured, tested, and applied to high voltage electronics systems.

REFERENCES

[1] Sarjeant, W. J., "Capacitor Fundamentals", Proceedings of the 1989 IEEE Electrical Insulation Conference, Chicago, IL, September 24-28, 1989.

[2] NEMA FI 2 - 1992, "Untreated Mica Paper Used for Electrical Insulation", National Electrical Manufacturers Association, Washington, DC., 1992.

[3] Welton, J.E., SEM Petrology Atlas, The American Association of Petroleum Geologists, Tulsa, Oklahoma, 1984.

[4] Sarjeant, W. J. and Dollinger, R.E., High-Power Electronics, Blue Ridge Summit, PA: TAB Books, 1989.

[5] Kaiser, C.J., The Capacitor Handbook, Olathe, KS: CJ Publishing, 1990.

Conference Record of the 1996 IEEE International Symposium on Electrical Insulation, Montreal, Quebec, Canada, June 16-19, 1996

Physicochemical Characterization of the Thermal Aging of Insulating Paper in Power Transformers

M.C. Lessard, L. Van Nifterik and M. Massé
Institut de recherche d'Hydro-Québec,
1800 boul. Lionel Boulet,
Varennes, Québec, Canada, J3X 1S1

J.F. Penneau
Direction des Études et Recherches d'Électricité de France,
Les Renardières BP 1,
77250 Moret sur Loing, France

Robert Grob
Institut National Polytechnique,
École Nationale Supérieure de Chimie de Toulouse,
118 route de Narbonne,
31077 Toulouse, France

Abstract: **Paper is a low-cost base material with outstanding mechanical and electrical properties, which is why it is still a key element in the insulation of electrical apparatus. Under the effect of a variety of factors including temperature, paper can substantially lose its properties, thus jeopardizing the service life of costly equipment. To remedy this situation, new so-called thermally upgraded papers are being made by certain manufacturers. A study carried out jointly by Hydro-Québec, Électricité de France and the École Nationale Supérieure de Chimie de Toulouse has allowed researchers to qualify the thermal resistance of three different types of thermostable paper. These papers have been selected as being representative of what is available on the market today. The paper samples were subjected to a thermal aging test in the presence of mineral oil to represent normal conditions of operation (150°C). The thermal degradation of the paper insulation is characterized by various physicochemical methods including measurement of the degree of polymerization, determination of 2-furfural in mineral oil by HPLC as well as determination of various sugars (monosaccharides, polysaccharides and anhydrosugars) in the paper using ion chromatography. This last method allows us to verify the formation of cellobiose, which is the real repeat unit of cellulose, as well as that of levoglucosan, which is an anhydrosugar and a precursor of 2-furfural. The evolution of all of these parameters, measured as a function of time, has allowed us to compare the thermal resistance of various insulating papers. The results of this study seem to show that, compared to traditional kraft paper, certain paper types are more susceptible to being thermally upgraded than others. This study also allowed us to demonstrate that the use of an inhibitor in mineral oil (DBPC) does not seem to influence the thermal degradation of these papers.**

INTRODUCTION

Predicting the lifetime of power transformers by monitoring the thermal degradation of their insulating paper is of significant economic importance considering the volume of such equipment with paper-oil insulation used by modern power generation and transmission companies. Many research projects have been undertaken in the past, mainly to study the thermal aging of cellulose and kraft paper [1-6].

Recently, however, thermostabilized papers have become available as a means of increasing the service life of power transformers. In an attempt to assess the thermal behavior of the most commonly used papers to insulate the conductors forming the windings of oil-insulated transformers, Hydro-Québec's research institute, the Direction des Études et Recherches d'Électricité de France and École Nationale Supérieure de Chimie de Toulouse launched a joint project to study the aging of kraft paper and different so-called thermostable papers under the influence of thermal stresses close to those under normal operating conditions. In this context, the main objective of the study reported here was to qualify different thermoresistant papers. A second objective was to identify among the various degradation products any compounds that could be used as new nondestructive tracers for diagnosing the aging of power transformers. Lastly, an interesting comparison was made of the influence of an inhibited oil of North American origin and a European non-inhibited oil on the thermal degradation mechanisms of insulating papers.

TEST PROCEDURE

The thermal resistance of four papers was characterized and compared using aging models which allowed us to analyze various degradation by-products of paper such as furanic by-products and sugars. The paper degradation was characterized by measuring the degree of polymerization, the 2-furfural content in the oil as well as the papers' cellobiose and levoglucosane content after extraction.

Two types of oil were examined, a North-American naphthenic oil with 628 ppm DBPC added (type A) and a European paraffinic oil containing no additives (type B). The oils and papers were dried and conditioned in accordance with the standard protocols used by transformer manufacturers. Many papers were examined and classified into four basic groups:

- Type 1: conventional kraft paper
- Type 2: highly purified kraft-based paper
- Type 3: Insuldur type, chemically modified kraft-based paper
- Type 4: Insuldur type, chemically modified kraft and Manila hemp-based paper

Types 2, 3 and 4 are marketed as thermostable papers.

The following methods of analysis were used: ion chromatography on a CarboPac PA1 or MA1 column for determining sugars and other tracers, inverse-phase high-performance liquid chromatography (HPLC) on a Hamilton PRP1 column for determining the furanic by-products [7] and the additive (DBPC) [8], the degree of viscosimetric polymerization of the paper (ASTM D4243) and Karl-Fischer titration for determining trace quantities of water in the oil (for quality control of the aging imposed on the models).

BLOCK DIAGRAM OF ANALYSES PERFORMED

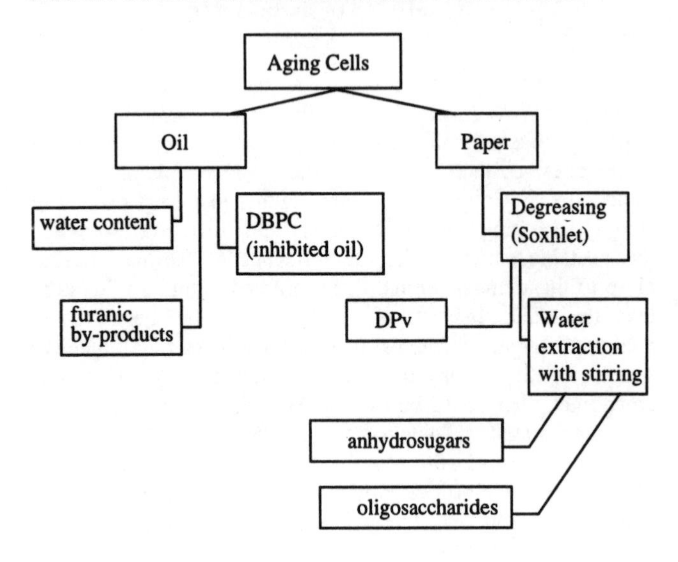

Accelerated thermal aging under normal conditions

The primary objective of this research was to study the degradation of cellulose under normal thermal conditions, in other words thermal conditions allowing aging to be accelerated without interfering with the degradation mechanisms that characterize cellulose aging in a transformer under normal thermal conditions [1].

The models consisted of glass vials with a capacity of 40 mL. Each flask containing 20 mL oil and 1 g paper corresponded to one sampling. The aging cells were conditioned in an inert atmosphere (under nitrogen) and heated to 150°C. The maximum aging time was set at 28 days. The samples were taken from the aging cells at regular intervals, i.e. one every day for the first four days then one every three days to make a total of 12. A standard was prepared for each paper-oil complex, as a non-thermally aged sample for comparaison.

RESULTS AND DISCUSSION

To ensure clarity of the figures throughout the paper, the pictograms for identification of all the paper-oil complex curves will remain the same.

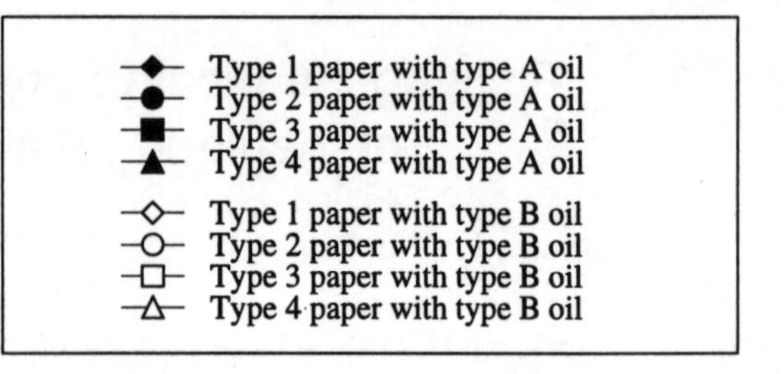

Measurement of degree of paper polymerization

The variation in the degree of polymerization that takes place during thermal aging of the four selected insulating papers at 150°C in the two mineral oils is presented in Figure 1. It can be seen that the cellulose degradation kinetics in terms of the decrease in the degree of polymerization is very fast from the outset of each thermal aging test. This effect is most striking in the case of paper types 1, 2 and 3, somewhat less in the case of type 4. After that, the cellulose degradation kinetics slows down for all four papers. This observation reveals that we first have a break in the most fragile glycosidic bonds, those in the amorphous regions of the cellulose, causing a rapid decrease in the degree of polymerization. The glycosidic bonds in the crystalline regions of the cellulose, for their part, are more resistant to thermal degradation. They will therefore tend to deteriorate during the second period, corresponding to the slow kinetics [9].

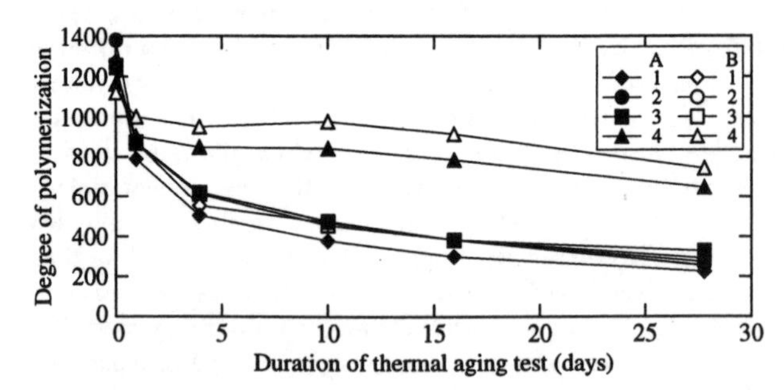

Figure 1. Degree of polymerization of the papers vs. thermal aging at 150°C

The type 4 paper has a good resistance to the imposed thermal stress but the other three are more difficult to differentiate: type 1 seems to be the most sensitive at 150°C but, on the basis of these results, types 2 and 3 could not really be classified as thermostable papers.

It should be noted that, whichever oil is used, type A or type B, it does not modify the performance of paper types 1, 2 or 3 in response to thermal stress, which confirms the expected results, considering that thermal aging takes place in a non oxidizing medium. For type 4, on the other hand, a slight improvement in the thermal resistance could be observed with the paraffinic oil not containing any DBPC. From this result, we can assume that the products formed during DBPC degradation would have some effect on the thermal resistance of type 4 paper.

Analysis of furanic by-products

The major compound resulting from the thermal degradation of paper is 2-furfural, although significant amounts of 5-hydroxymethyl 2-furfural can also be found as well as smaller amounts of 5-methyl 2-furfural. The high concentrations of these compounds reveals considerable thermal degradation of both the cellulose and hemicelluloses. On the other hand, in the case of paper type 4, the concentrations obtained for 2-furfural and 2-furfuryalcohol are very low. It should be noted, however, that the latter is the compound found in the largest quantities in the case of thermal aging of paper type 4, whereas it was not detected at all during the thermal aging of the other three papers. To judge from the traces observed, 2-acetylfuran is a difficult compound to identify and, consequently, to determine.

Since 2-furfural is considered to be a characteristic tracer of the thermal degradation of cellulose [2,3], this section of the work reported will be focused on studying the variations in the 2-furfural content during thermal aging at 150°C. Figure 2 presents a semi-log plot of the variation in the 2-furfural content during thermal aging at 150°C of the four insulating papers in the two mineral oils investigated.

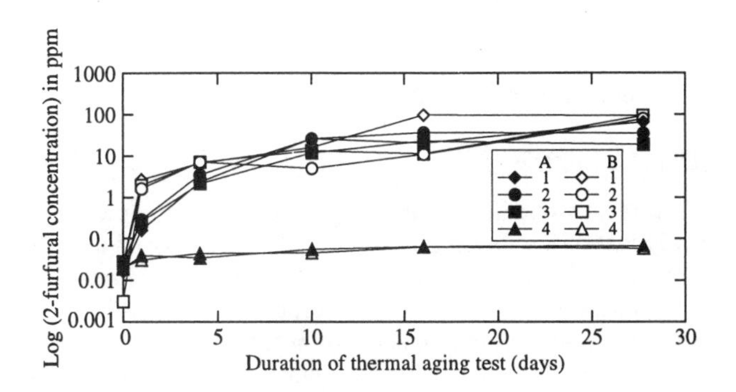

Figure 2. Furfural-2 content (μg/g of paper) vs. thermal aging at 150°C

It is interesting that at this stage of the study the results obtained for the degree of polymerization are very similar to those obtained for the 2-furfural content, in agreement with the expected results, considering that 2-furfural is a tracer of the thermal degradation of cellulose [2,3]. In terms of 2-furfural production, the type 4 paper is a lot less degraded under the influence of the temperature, contrary to the other types. Meanwhile, there is no particular difference between the non thermostable paper (type 1) and paper types 2 and 3 which are sold as thermostable products. Lastly, the type of oil did not seem to make much difference in the thermal degradation of cellulose.

Correlation between the degree of polymerization and the 2-furfural concentration

Figure 3 shows the correlation between the degree of polymerization and the log of the 2-furfural concentration during thermal aging at 150°C for the four insulating papers in the two mineral oils.

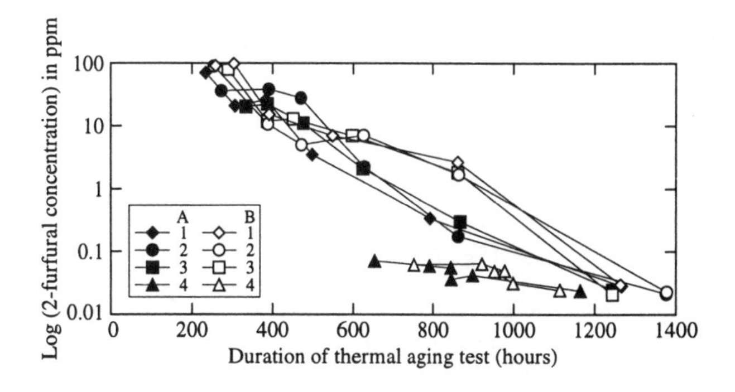

Figure 3. Correlation between the degree of polymerization and the furfural-2 concentration (μg/g of paper) at 150°C

It is clear that the 2-furfural concentration varies considerably for a same paper as it undergoes thermal degradation. Nevertheless, when we plot the log of the 2-furfural concentration versus the degree of polymerization, we obtain a relatively good correlation between these two factors for all four papers, which shows that the concentration is effectively related to the thermal degradation mechanisms of cellulose. Here again, we find no significant difference between the nature of the oil and the paper types 1, 2 and 3 whereas for type 4 a very strong difference in the performance is noted.

Analysis of sugars

Use of ion chromatography made it possible to determine the cellobiose content in aqueous extracts of the paper samples. Cellobiose is an oligosaccharide, the real repeat unit of cellulose [10]. The content evolutions of cellobiose during the thermal aging at 150°C are shown in Figure 4 where it can be seen that the production kinetics of this compound is linear over most of the range studied.

This new analytical approach provided further confirmation of the good resistance to thermal stress demonstrated by type 4 paper. In fact almost no oligosaccharides were detected during aging at 150°C. For the other papers whose cellobiose rate could be detected, type 1 seems the least resistant, followed very closely by type 2. The paper that offered the best resistance in terms of cellobiose production is type 3. Nonetheless, the increase in cellobiose is rapid in all three papers. These results lead to the following classification:

Type 4 >> Type 3 >> Type 2 > Type 1

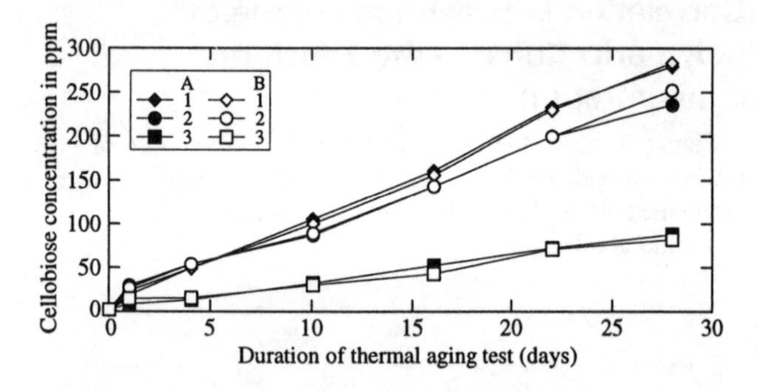

Figure 4. Correlation between the degree of polymerization and the cellobiose concentration (μg/g of paper) at 150°C

Again these analyses do not show any striking differences between papers aged in oil type A and those aged in type B.

Since the real precursors of cellulose and hemicellulose degradation are anhydro-sugars [11], the last phase of this study was devoted to an investigation of the changes that take place in these compounds during the aging process. The presence of 1,6-anhydro-glucose, or levoglucosane, in the aqueous extracts of the paper samples was a strong focal point because it is a known precursor of 2-furfural. Determination of this compound in the four different paper types with type A oil yielded a curve of the variation in quantity during thermal aging (see Figure 5). As in the case of the analysis of cellobiose, analysis of the levoglucosane allowed us to rank the four papers from the point of view of their stability in terms of levoglucosane production as follows:

Type 4 >> Type 3 > Type 2 >> Type 1

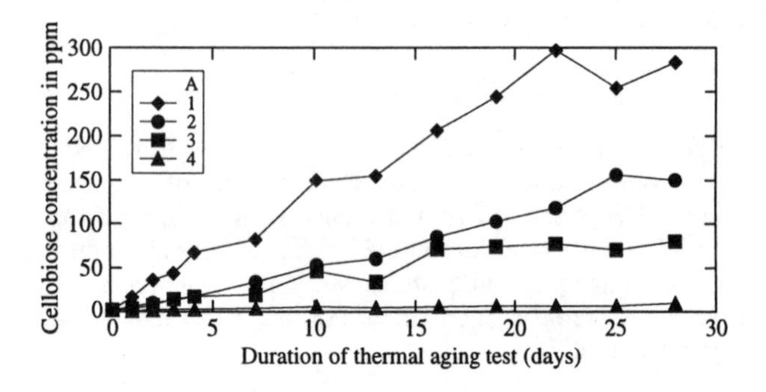

Figure 5. Correlation between the degree of polymerization and the levoglucosane concentration (μg/g of paper) at 150°C

Correlation between the degree of polymerization and the sugar concentration

The degree of polymerization of the four papers and their cellobiose and levoglucosane contents in an aqueous extract (Figures 6 and 7 respectively) reveal an excellent correlation, which is a clear indication that this parameter is directly related to the depolymerization kinetics of insulating papers during thermal aging.

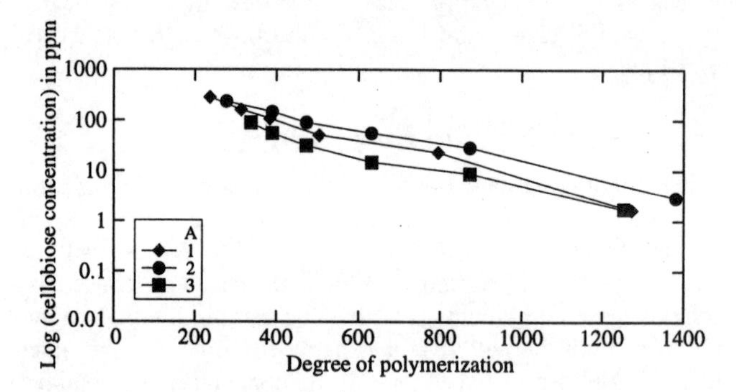

Figure 6. Correlation between the degree of polymerization and the cellobiose concentration (μg/g of paper) at 150°C

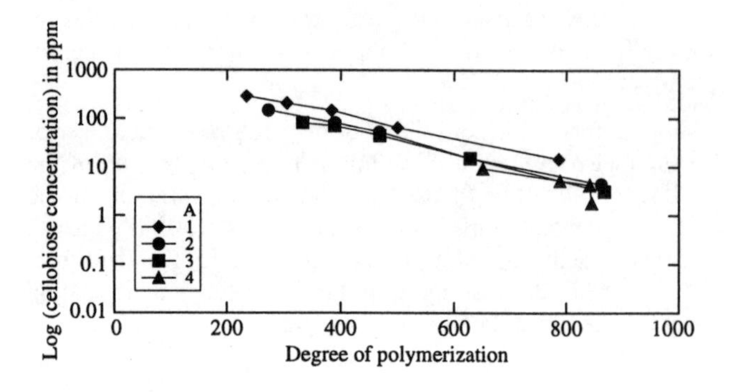

Figure 7. Correlation between the degree of polymerization and the levoglucosane concentration at 150°C

CONCLUSIONS

The results obtained in this experimental work show quite clearly that of the four papers subjected to accelerated thermal aging at 150°C only one, type 4 offers a good resistance. As far as papers types 2 and 3 are concerned, further qualification tests are needed before a final decision can be made as to their effectiveness. Actually, these two papers, despite their being labeled thermostabilized, behaved very similarly to kraft paper which, it should be recalled, is a conventional insulating paper known to be non thermoresistant.

The comparison of two oil types revealed that they had no significant impact on the degradation kinetics of the four papers studied.

On the other hand, this work provided an opportunity to improve our knowledge of the thermal degradation kinetics of cellulose by monitoring, on aqueous extracts of paper, the cellobiose and levoglucosane contents. The production of these sugars seems to be far more constant and reproducible than that of 2-furfural. Consequently, if either of these sugars has adequate solubility in the oil, it could be considered a reliable tracer of thermal aging and used at a later stage as part of a new analysis technique for the diagnosis of power transformers. The solubility of these compounds in oil is currently the focus of a specific study at our laboratories. The work reported here was conducted under what is accepted as normal accelerated aging conditions and further studies are already under way to take account of aging under thermal fault conditions (200°C and 250°C).

ACKNOWLEDGMENTS

The authors offer their sincere thanks to Dr. C. Lamarre, researcher and A. Gendron, senior technician at IREQ, Hydro-Québec, for their scientific advice and, also, to G. Gagné and L. Lam, Engineering and Services, Hydro-Québec, for their financial support.

REFERENCES

1. Vergne, J., Doctoral thesis, INP Toulouse, January 16, 1992.

2. Shroff, D.H., Stannett, A.W., IEE Proceedings, Vol. 132, No. 6, November 1985, p. 312.

3. Burton, P.J., CEGB internal report, 1982.

4. Griffin, P.J., et al. 61th Annual International Conference of Doble Clients, 1994, Sc. 10D.

5. Fabre, M.J., Revue Générale de l'Electricité, Vol. 66, No. 1, January 1957, p.17.

6. Shafizadeh, F., Ellis Horwood, Chemical Science, John Wiley and Sons, Cellulose Chemistry and its Applications, 1985, Chapter 11.

7. Lessard, M.-C., Lamarre, C., Massé, M., Gendron, A., CEIDP, 1995 Annual Report, Virginia Beach, session 4B.

8. Lamarre, C., Gendron, A., IEEE Trans., Dielectrics and Electrical Insulation, Vol. 2, No. 3, June 95, p.413.

9. Broido, A., Javier-Son, A.C., Barrall, E.M., Journal of Applied Polymer Science, Vol. 17, 1973, p. 3627.

10. Monties, B., Les Polymères Végétaux, ed. Gauthier-Villars.

11. Nevell, T.P., Ellis Horwood, Chemical Science, John Wiley and Sons, Cellulose Chemistry and its Applications, 1985, Chapter 10.

Conference Record of the 1996 IEEE International Symposium on Electrical Insulation, Montreal, Quebec, Canada, June 16-19, 1996

Graft Polymerization and Other Methods to Reduce the Hygroscopic Nature of Cellulose Insulation

T. V. Oommen
ABB Transmission Technology Institute
1021 Main Campus Drive
Raleigh, NC 27606

T. L. Andrady
Research Triangle Institute
Camille Dreyfus Lab
RTP, NC 27709

Abstract: **Cellulosic insulation materials are widely used in power transformers and other high voltage devices, but require intensive drying out to remove all moisture. This paper describes three methods used to modify cellulose to reduce its hygroscopicity, of which graft polymerization was the most successful method. The modified paper sheets, however, lost some mechanical strength properties, and showed a tendency to be thicker than the starting material.**

INTRODUCTION

Electrical apparatus such as power transformers use cellulosic materials, mostly kraft paper and pressborad, as the chief insulation material due to low cost and availability. Several tons of paper and pressboard are used in a large power transformer.

In all the above applications the paper has to be thoroughly dried out to remove residual moisture which could be as high as 8%, and bring it to less than 0.5%. This requires elaborate drying methods such as air blast and vacuum. After the dry-out, the paper is impregnated with an insulating fluid such as mineral oil to remove all air pockets and to prevent moisture reabsorption. In dry-type transformers, a resin is forced into the paper to form a moisture barrier.

An alternative to using expensive synthetic materials is by modifying cellulose to make it less hygroscopic. This does not mean that process-free cellulose can be developed; there is always some activity due to the hydroxyl groups present. The modification methods described are attempts to block the hydroxyl groups from free absorption of external moisture. Grafting and acetylation are chemical methods of modification; polymer coating is a surface treatment by hydrophobic polymers.

METHODS OF MODIFICATION

The methods previously used for electrical grade papers are: (1) alkylation, (2) thermal upgrading. In category (1) we may mention cyanoethylation and acetylation by esterification of the hydroxyl groups[1, 2]. Thermal upgrading involves dipping the paper in an aqueous solution of an amine /amide compound such as dicyandiamide and drying out. These compounds, present only in 2-4%, appear to prevent premature aging of cellulose by hydrolytic breakdown at elevated temperatures. The use of upgraded paper has enabled operation of power transformers with better loading capabilities. A survey of thermal upgrading compounds is given in an EPRI report [3]. It must be pointed out that thermal upgrading does not reduce the hygroscopicity of cellulose.

Methods Used in the Study

We used the following methods of cellulose modification: (1) Graft polymerization, (2) Alkylation, (3) Polymer Coating. Previous work on alkylation was already mentioned. Polymer deposition has also been attempted before, but we used new formulations. Graft polymerization has been known for some time to textile chemists, but has never been attempted on electrical grade paper.

Graft Polymerization

Grafting involves attachment of a polymer chain, usually synthetic, to the backbone of a given polymer. It was first applied to cellulose in 1953. An excellent review article by Habeish and Guthrie provides much information [4]. Stannett [5,6] explains the grafting process as follows:

$$\text{Polymer A } (P_A) \rightarrow P_A^{\cdot}$$
$$P_A^{\cdot} + nM_S \rightarrow P_AP_S$$

where $P_A^{\cdot}$ is the free radical generated from polymer P_A. M_S is the monomer of the grafted polymer. In cellulose grafting, P_A is cellulose; P_S is typically a vinyl polymer of acrylonitrile, ethyl acrylate, methyl methacrylate, divinylbenzene (styrene) and diallylphthalate. In the case of cellulose, the free radical is formed by removal of one or more hydrogens attached to the carbon atoms in the ring (see Figure 1 for the molecular structure of cellulose).

Free radical polymerization may be achieved by (1) chain transfer in presence of a catalyst, e.g., ferrous ion (Fe^{2+}), (2) direct oxidation using ceric ion (Ce^{4+}) etc. (3)using chemical initiators such as peroxide, (4) forming cellulose derivatives first, (5) irradiation with UV, gamma rays etc.

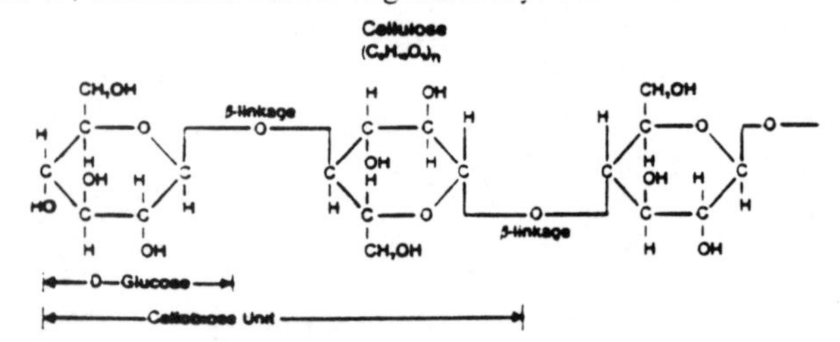

Figure 1. Molecular Structure of Cellulose

Since lignin in paper would retard the grafting process, only low-lignin paper should be used. Electrical grade kraft paper has about 5% lignin, so may be used as such. In our tests, 3 mil kraft paper (Manning 250) was used. Kraft papers have high alpha-cellulose content (80-90%) the desirable type of cellulose. Small and large paper samples were used in the grafting experiments. The small samples, 2" x 2", were soaked in 0.02M ferrous ammonium sulfate solution for 10 minutes, then treated at 70°C for 60 minutes in 5% monomer solution in a 3-necked flask with a reflux condenser, thermometer and a nitrogen inlet tube, see Figure 2.

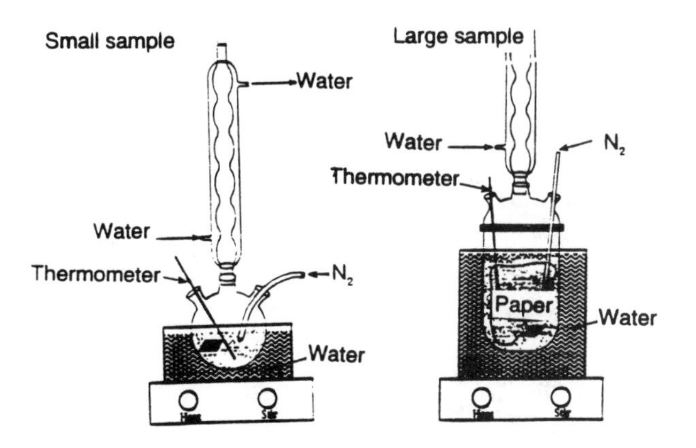

Figure 2. Apparatus for Graft Polymerization

The samples were then removed and washed overnight with a suitable solvent to remove homopolymer, followed by water washing, drying and weighing. The samples were treated for 10 minutes above the glass transition temperature for the grafted polymer (140°C for acrylonitrile grafting). Larger samples were 8" x 10" size, and were similarly prepared.

The different monomers used could be grouped as follows:

5% acrylonitrile in aq. medium (Fe^{++}/H_2O_2)
5% ethyl acetate (" ")
5% acrylonitrile (5% methanol, ")
2.5% styrene + (" ")
2.5% acrylonitrile

The best moisture reduction was obtained for the 5% ethyl acetate and the 5% acrylonitrile (methanol) systems, with water content less than 2% (the others had double the amount of moisture left). The percent of grafting by weight estimates was 74% to 208% for the small samples, and 82-91% for the large samples. The 2% moisture level was found for samples ranging from 74% to 208% grafting. Typically, a high percent moisture was related to a low percentage of grafting.

The grafted sheets were thicker than the base paper, had a lighter color, and appeared to be more brittle.

Polymer Deposition

This was accomplished by dipping an oven-dried sheet of paper in the latex or polymer solution of required concentration in a glass dish for about 30 seconds. The sheet was removed, gently wiped, and then hung in an oven at 45°C for several hours.

The following is a partial list of emulsions and solutions used:

DRARX 110L (carboxylated styrene-butadiene-acrylonitrile)
DARATA 3631 (poly-vinylidene chloride polymer)
DARAN 8600c (vinylidene chloride-vinylidene acrylate copolymer)
POLYVINYLIDENE CHLORIDE solution
POLYSTYRENE solution

Concentrations ranged from 3% to 10% in some cases; 5 to 15% in other cases.

The results showed no significant water reduction. The lowest water content was about 5%. The polymer coating was not adequate to prevent water reabsorption.

Alkylation

Chemically modifying the hydroxyl groups by acetylation etc., was mentioned earlier. Acetylation was done by reacting the paper sample with acetic anhydride; other anhydrides such as propionic and butyric anhydrides were also tested. The reaction temperature was typically 120°C and the reaction time was three hours. Moisture levels at 50% RH in alkylated samples ranged from 4.2 to 6.6%, not a great improvement. Much lower moisture levels from acetylation reported in the literature resulted from a higher level of acetylation which made the paper brittle.

EVALUATION OF MODIFIED PAPERS

Several key properties were studied to evaluate the modified papers which included physical and electrical testing.

Moisture Absorption

Samples were first dried out and then oil impregnated. The moisture pickup with time at 50% RH is shown in Figure 3.

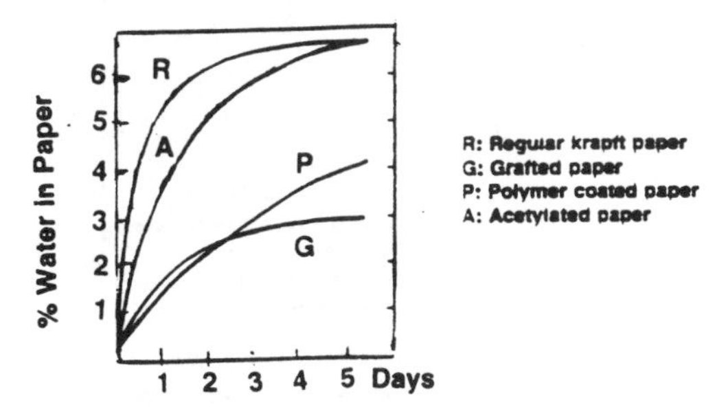

Figure 3. Moisture pickup at 50% RH of Oil Impregnated Paper Samples

It is seen that the grafted paper showed the lowest moisture absorption tendency which peaked below 3%. The polymer coated paper was comparable to the grafted paper but eventually rose to a higher moisture level with no tendency to level off. The acetylated paper was only slightly better than the unmodified paper.

Thermogravimetry

TGA is used to determine the thermal stability of materials. Figure 4 shows TGA of the various papers. The least stable paper was the polymer coated sample. The grafted paper was less stable than the regular and acetylated papers but was much better than the polymer coated paper.

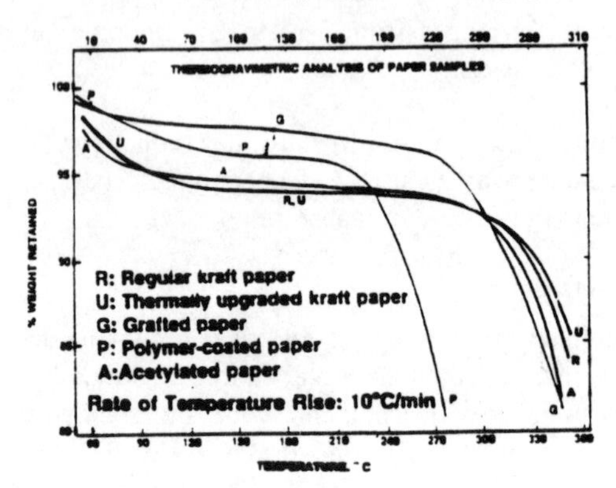

Figure 4. TGA Curves for Paper Samples

Tensile Strength

Tensile strength is an important mechanical strength property for conductor tapes. Tensile strength was measured in machine and cross machine direction on 1" wide samples with an Instron Tester. The breaking load is reported in English and metric units. Figure 5 gives tensile strengths in the machine direction.

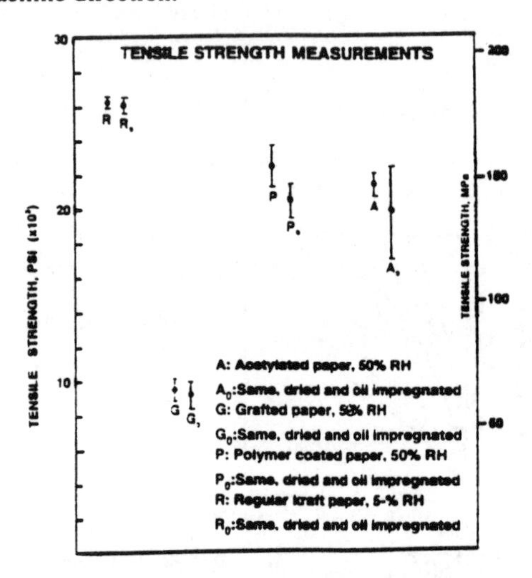

Figure 5. Tensile Strength of Paper samples

The grafted paper had the lowest tensile strength; it was also the most brittle. It was also the thickest paper, with a thickness gain of 80% over the base paper. The polymer coated paper had about 40% thickness gain; the acetylated paper had only 5% gain. The weight gain was in the same range as thickness.

Dissipation Factor

The paper samples were dried and oil impregnated before measuring dissipation factor at different temperatures. The results are shown in Figure 6.

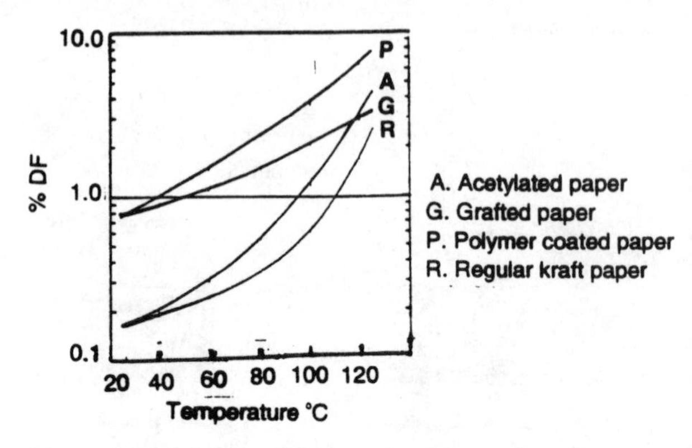

Figure 6. Dissipation Factor (%) of Paper Samples

It will be noticed that compared with the regular paper, the polymer coated and the grafted papers have a much higher initial DF; however, the DF of the grafted paper rises much slower, and the DF is comparable to that of regular paper at 120°C. At 100°C the DF of the grafted paper is at least twice as high as of the regular paper.

Breakdown Voltage

The breakdown voltage of oil impregnated papers are shown in Figure 7. It may appear that the grafted paper has the highest breakdown strength. But it must be remembered that it

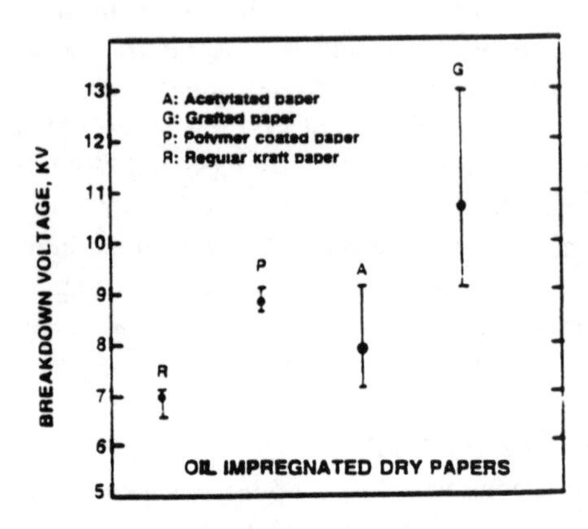

Figure 7. Breakdown voltage of the Modified Papers

has the highest thickness also. If we normalize the thickness, the grafted paper would fare no better than the other papers. The higher variability in the grafted paper results from the uneven grafting of the sheet due to the limitations of the reaction vessel.

Oil Compatibility

This test is usually done to make sure the sample does not contaminate hot transformer oil. Dried samples of identical size were placed in similar volumes of degassed oil and heated at 105°C for 5 days under nitrogen atmosphere. The oil was then tested for contamination. Table 1 lists the properties measured:

Table 1
Oil Properties After Treatment with Hot Oil

	Regular	Grafted	Polymer-coated	Acetylated
Appearance	Clear	Clear	Yellowish	Clear
Dielectric Strength, KV ASTM D 1816	48.8	48.2	37.4	50.0
Diss. Factor, % 60 Hz., 100°C	0.0920	0.0750	0.93%	0.0630
Interfacial Tension, dynes/cm	43.4	43.4	39.7	43.2
Neutr. No. mg KOH/g	0.0064	0.0063	0.0306	0.0064

The coated paper alone shows unsatisfactory DF values, possibly due to disintegration of the coating into the oil.

COMPARISON OF PAPERS

Table 2 summarizes the desirable and undesirable features of the various papers:

Table 2
Desirable and Undesirable Features of Modified Papers

	Desirable	Undesirable
Grafted Paper	1. Lowest moisture absorption 2. Highest diel. str.	1. Highest weight and thickness gain 2. Lowest tensile strength
Polymer-coated paper	1. Moderate moisture abs.	1. Lowest thermal decomp. temp. 2. Highest DF 3. Poor oil compatibility
Acetylated paper	1. Satisfactory thermal and electrical prop.	1. Severe modifi-- cation proc. 2. Moisture abs. high.

CONCLUSION

The study revealed that it is possible to modify cellulose by different methods to reduce its hygroscopicity. Grafting appears to be the most effective method, but results in a thickening of paper and weight gain. The incorporation of a hydrophobic chain appears to lower the fiber to fiber adhesion. Thus any chemical modification comes with a price. A superficial treatment like surface coating appears to have little lasting value.

ACKNOWLEDGMENT

The modified papers were prepared at the Camille Dreyfus Lab of RTI. Guidance from Dr. Vivian Stannett, Prof. Emeritus of NC State University, is gratefully acknowledged.

REFERENCES

1. Kang, B.P., "Some Considerations for Using Alkylated Cellulosic Paper for High Voltage Insulation", *Insulation*. Feb 1977, pp.30-36.
2. Tillman, Anne Marie, "Chemical Modification of Lignocellulosic Materials", M. S. Thesis, Department of Chemistry, Goteborg, Sweden, 1987.
3. EPRI Report EL-4935, "Improved Cellulosic Insulation for Distribution and Power Transformers", March 1987.
4. Habeish, A. and Guthrie, J. T., *The Chemistry and Technology of Cellulosic Copolymers*. Springer-Verlag, New York, 1981.
5. Stannett, V. T.,"Cellulose Grafting: Past, Present and Future", from S DOE Report, *Assessment of Biobased Materials*, Chum, H.L., Ed., December 1989.
6. Schwab, E., Stannett, V., and Hermans, J.J.,"Grafting Onto Cellulosic Fibers", TAPPI, Vol. 44, No.4, April 1961, pp. 251-256.

Conference Record of the 1996 IEEE International Symposium on Electrical Insulation, Montreal, Quebec, Canada, June 16-19, 1996

Fundamental Investigations on the Influence of Temperature and Water Content on the Electrical Behavior of Fluid Impregnated Insulating Papers

K. Dumke H. Borsi E. Gockenbach
Schering Institute of High Voltage Technique and Engineering,
University of Hannover, Callinstr. 25A, D-30167 Hannover, Germany

Abstract: **This paper deals with fundamental investigations on the electric strength of liquid impregnated insulating papers namely aramid paper and cellulosic paper. These investigations point out, that the breakdown voltage at 20 °C of the two papers differs not significantly. However their electric strength differs significantly at higher temperatures. Best results regarding the lifetime can be achieved by using ester liquid impregnated aramid paper. Therefore the combination of aramid and ester liquid proves out to be more than an adequate substitute for mineral oil impregnated cellulosic paper. Finally the correlations between the water content of the impregnating liquid and the water content of the paper show, that using ester liquid diminishes the water inside the insulating paper.**

INTRODUCTION

To distribute electric energy liquid insulated transformers are mainly used and the insulating liquid is for the most part mineral based transformer oil. There is sufficient knowledge of the durability and of the behavior of mineral oil insulations in case of ageing, especially in combination with cellulosic paper. Mineral based transformer oil possesses the disadvantage of combustibility and the disadvantage of endangering the environment in case of a transformer leakage.

Other insulating liquids do not possess these disadvantages, nevertheless these liquids have similar electric properties referred to mineral based transformer oil. Two of these liquids are organic ester liquid and silicon liquid. The thermal limit of both liquids is clearly higher than the limit of mineral oil. To benfit from the advantage of high temperatures it is necessary to use an insulating paper of high thermal resistance, too.

Aramid paper possesses a significant higher thermal resistance than mineral oil and therefore the combination of both, liquids and papers with high temperature durability, should be taken into account for future transformer design.

The influence of the temperature on the electric strength is rarely investigated. Furthermore the correlation between the water content of the insulating paper and the water content of the liquid is unknown.

For this reason investigations were made with the following aims:

- Determination of the influence of the temperature on the electric strength of insulating papers impregnated with different insulating liquids.
- Determination of the correlation between the water content of the insulating liquids and the water content of the impregnated papers.

TEST MATERIALS AND PROCEDURES

Test Materials

The investigations were carried out using two different insulating papers. Important properties of the papers can be taken from table 1.

Table 1
Investigated Insulating Papers

Properties	Paper 1 Cellulosic Paper	Paper 2 Aramid Paper
thickness (one layer)	100 μm	80 μm
max. temperature	105 °C	180 °C
water saturation (at 20 °C)	approx. 16 %	approx. 12 %

Three different insulating liquids were investigated in combination with the two insulating papers:

- mineral based transformer oil (Shell Diala D)
- organic ester liquid (MIDEL 7131)
- silicon liquid (Baysilone M 50 EL).

The water saturation $w_{saturation}$ of the insulating liquids is given in figure 1 [1,2]. Some properties of the investigated liquids are shown in table 2.

Table 2
Investigated Insulating Liquids

Properties	Ester liquid	Silicon liquid	Mineral oil
breakdown voltage according DIN 0370	> 55 kV	> 50 kV	> 60 kV
rel. permittivity ε_r	3.3	2.7	2.2
water saturation (at 20 °C)	2700 ppm	200 ppm	44 ppm

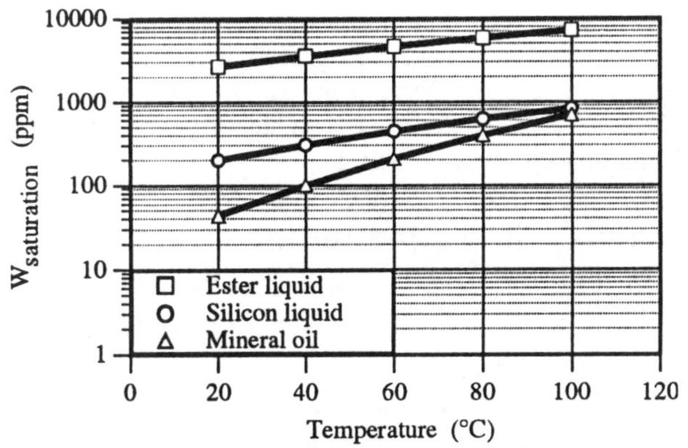

Figure 1. Water saturation in ppm (10^{-6}) of different insulating liquids versus temperature in the range from 20 °C to 100 °C [1,2].

To achieve the same initial conditions the insulating liquids were prepared in separated two-stage processing plants. This preparation guarantees a very low water content for all liquids used in the experiments. To perform some investigations with a water content higher than that after the preparation, samples of the liquids were exposed to an atmosphere of a definite humidity for a certain time. The water content was controlled after the exposition via a coulometric unit.

To achieve after the impregnation of the insulating paper an equilibrium of the water content of the liquids and the water content of the paper, these paper was exposed to the same atmosperic conditions covered by the insulating liquid.

Determination of Water Content

To determine the water content of the insulating liquids, samples were given into a titration vessel and via a coulometric titration (Karl-Fischer-Titration) the water content was measured automatically with a total unaccuracy of 5 %. For the determination of the water content of the impregnated insulating papers, these papers were heated up to 100 °C in an atmosphere of dried nitrogen (N_2), where the nitrogen collected the emitted water. Finally the nitrogen flowed through the titration vessel and the water content was determined using the same equipment mentioned before.

The investigations with the higher water content were performed after the papers were exposed for a certain time to a definite atmosphere and the condition of equilibrium was achieved. During this time the papers were covered with the insulating liquid. Thus different conditions regarding the water content could be realized.

Test Specimen

The investigations on impregnated insulating papers were carried out on test specimen with a thickness of approximately 300 µm (aramid paper) or 320 µm (cellulosic paper), each specimen consists of three layers of insulating cellulosic paper or four layers in case of aramid paper. The test specimen were dried for 120 h under vacuum $p < 1$ mbar and a temperature of 80 °C. After this they were impregnated with the insulating liquids under the same conditions and dumped in a post-impregnation for 48 h under temperature and vacuum.

Test Vessel

The test vessel used for the determination of the breakdown voltage U_D is shown in figure 2. It consists of a glass cylinder, equipped with electrodes mounted on the bottom and the top with an adjustable gap space. The pressure of the electrodes on the test specimen could be controlled via a spring placed in the upper electrode holder.

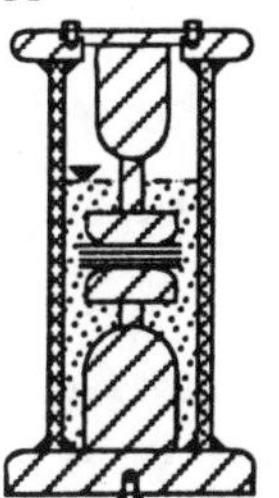

Figure 2. Test vessel with Rogowoski-shaped-electrodes for the determination of the breakdown strength of impregnated insulating papers. A Pt_{100}-controlled-heating system allowed to carry out investigations in a temperature range from 20 °C up to 150 °C.

Electric Test Circuit

The determination of the breakdown strength was executed in a ramp-test, the electric test circuit is given in figure 3.

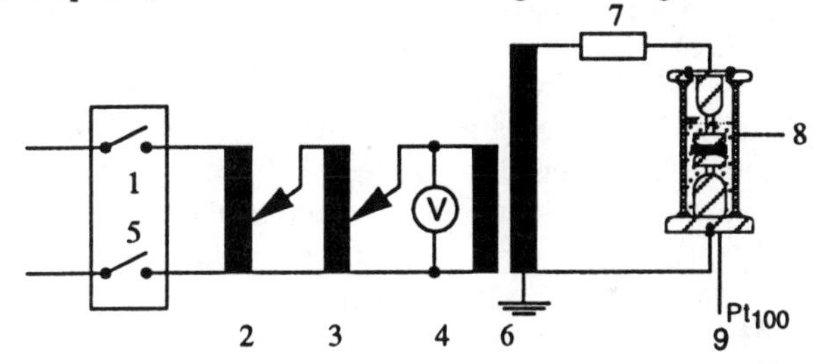

Figure 3. Electric test circuit for the determination of the break-down strength E_D of paper test specimen in a ramp test setup. (1,5) protection circuit, (2) motor actuating auto transformer, (3) auto transformer, (4) voltmeter, (6) high voltage test trans-former (220V/100kV/5kVA), (7) damping resistor, (8) test vessel, (9) thermal control unit with Pt_{100}.

Ageing of Impregnated Paper

The impregnated paper specimen were aged for a period of 1000 h at high temperatures, simultaneously an electric pre-stressing of 20 kV/mm was performed. For the ageing of the specimen the same test vessels, described above, were used. To accelerate the ageing metalic catalysts (each 3 g/l zinc, copper, aluminium, and iron in form of cuttings) were added to the liquids during the ageing.

Test Procedure

Breakdown strength

The determination of the electric strength was performed in a ramp-test as mentioned before, the voltage gradient was $\Delta U = 2$ kV/s. Measurements were carried out at temperatures of 20 °C, 60 °C, and 100 °C using specimen in unaged condition as well as specimen after ageing. Each test was repeated for a collective of 6 samples. At the beginning of the test procedure the water content of the liquids was adjusted. Additionally the water content of the impregnated papers was measured and controlled likewise.

RESULTS AND DISCUSSION

Correlation of the Water Content of Papers and Liquids

The correlation of the water content of the insulating liquids and the impregnated papers is shown in figure 4A exemplary for the investigated cellulosic paper and the three insulating liquids. A comparison of the aramid and cellulosic paper is given in figure 4B, using ester liquid for the impregnation.

As can be seen in figure 4A there is no significant difference between the silicon liquid and the mineral oil. This behavior is due to the water saturation of the liquids, which is almost the same for the investigation temperature of 20 °C.

The curves show that using an insulating liquid with a high water saturation (ester liquid) is combined with a lower water content in the paper compared to other insulating liquids with the same absolute water content but a lower water saturation. This behavior is due to the lower relative humidity of insulating liquids with a higher water saturation at the same absolute water content, however there exists no common curve for the investigated liquids, which combines the three curves to a single correlation of water content of the paper versus relative humidity of the insulating liquids.

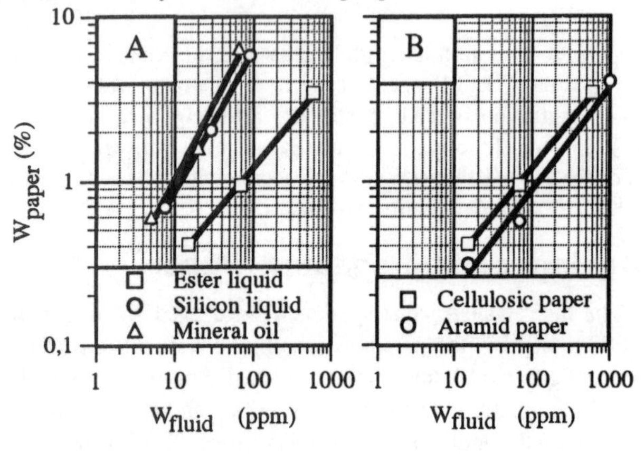

Figure 4. Water content of impregnated paper versus water content of insulating liquids. Temperature $\vartheta = 20$ °C

(A) Cellulosic Paper with different impregnating liquids

(B) Aramid paper and cellulosic paper impregnated with ester liquid

The comparison of the two insulating papers points out, that the water content of the aramid paper is lower than the water content of the cellulosic paper (figure 4B). This effect is directly due to the lower water content of the unimpregnated aramid paper, given in table 1.

Electric Investigations

Influence of temperature, insulating paper and insulating liquids on the electric strength

The breakdown strength of the impregnated insulating papers is given in figure 5A for cellulosic paper and in figure 5B for aramid paper. As can be seen the two insulating papers possesses a different dependence of E_D versus the temperature. For cellulosic paper the breakdown strength decreases for higher temperatures, whereas the curves of the breakdown strength of aramid paper show rather constant values.

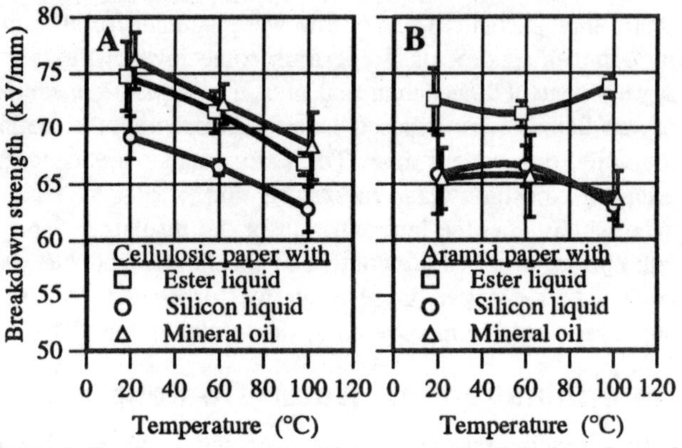

Figure 5. Breakdown strength of impregnated cellulosic and aramid paper versus temperature in the range from 20 °C to 100 °C for different insulating liquids.

Furthermore it can be noticed, that the breakdown strength E_D of cellulosic paper does not differ significantly using ester liquid or mineral oil as impregnating liquid. For aramid paper the highest values of E_D can be achieved by an impregnation with ester liquid.

Influence of water content on the electric strength

As can be seen the breakdown strength decreases for higher water contents of the insulating liquids. This correlation can also be found out for ester liquid and silicon liquid in the same dimension as shown in figure 6.

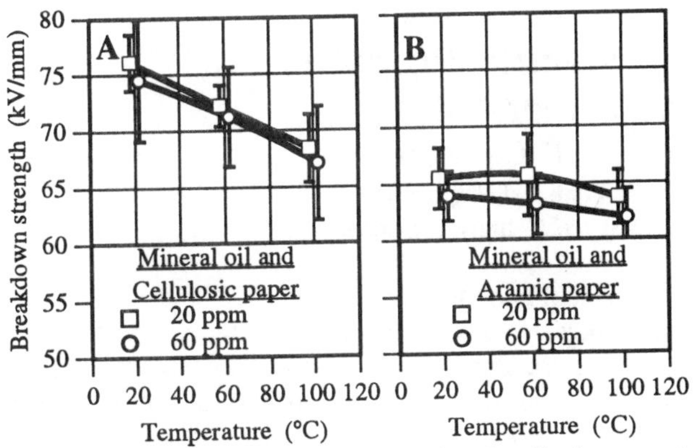

Figure 6. Breakdown strength of mineral oil impregnated cellulosic and aramid paper versus temperature in the range from 20 °C to 100 °C for two different water contents.

Influence of ageing

Figure 7 presents the results of breakdown investigations before and after the ageing, exemplary shown for ester liquid impregnated papers. For the purpose of comparison unaged mineral oil impregnated papers are presented too. As can be seen the ageing reduces the breakdown strength of the ester liquid impregnated aramid paper from approximately 72 kV/mm to 62 kV/mm (at 20 °C), this makes a reduction of 15 %. Nevertheless the breakdown strength reaches nearly the same value compared to unaged mineral oil impregnated aramid paper.

Ageing the cellulosic papers does diminsh the values of E_D for approximately 5 kV compared to the unaged condition.

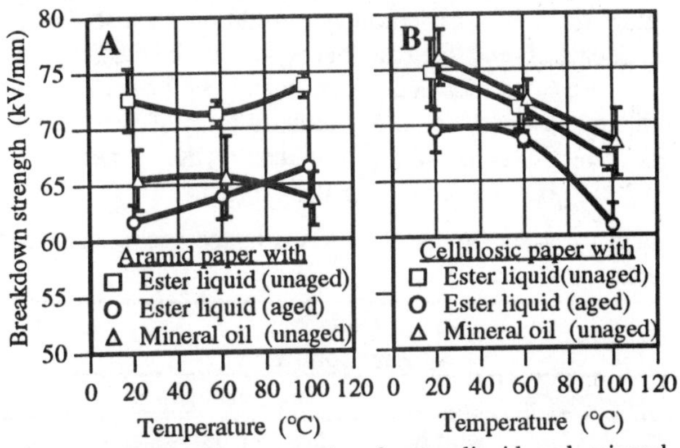

Figure 7. Breakdown strength of ester liquid and mineral oil impregnated cellulosic and aramid paper versus temperature in the range from 20 °C to 100 °C in unaged and aged conditions.

Ageing Conditions:

(A) 1000 h at 150 °C with catalysts in presence of oxygen under an electric stress of 20 kV/mm

(B) 1000 h at 100 °C with catalysts in presence of oxygen under an electric stress of 20 kV/mm

CONCLUSIONS

The main results of the investigations on liquid impregnated insulating papers can be concluded as follows:

- In comparison with mineral oil and silicon liquid the ester liquid shows a much higer water saturation at different temperatures.
- Aramid paper possesses less water than cellulosic paper at the same water content of the insulating liquid.
- The breakdown strength versus temperature of impregnated cellulosic paper differs from that of impregnated aramid paper.
- Ester liquid impregnated aramid paper shows the highest breakdown strength over the whole temperature range from 20 °C to 100 °C.
- The water content of the impregnating liquid does not influence the breakdown strength of impregnated cellulosic paper in a range from 20 ppm to 60 ppm.
- An ageing of the ester liquid impregnated aramid paper causes a reduction of the electric strength of less than 15 %.

ACKNOWLEDGMENT

The authors thank the Deutsche Forschungsgemeinschaft for the financial support and Dr. Beck, Hamburg, for providing with test materials.

REFERENCES

1. Beyer, M. et al., "*Hochspannungstechnik*", Springer Verlag, New York, Zürich, Tokio, 1986.
2. Schröder, U., "Das Teilentladungs- und Gasungsverhalten von Transformatorenöl", Ph. D., University of Hannover.

Conference Record of the 1996 IEEE International Symposium on Electrical Insulation, Montreal, Quebec, Canada, June 16-19, 1996

Influence of Ageing on the Dielectric Properties of Carbonized Transformer Oil

M. Krins H. Borsi E. Gockenbach
Schering Institute of High Voltage Technique and Engineering
University of Hannover, Callinstr. 25A, D-30167 Hannover, Germany

Abstract: **This paper reports about experimental investigations dealing with the influence of ageing processes on the dielectric dissipation factor, the volume resistivity, and the PDIV of carbonized mineral based transformer oil. The oil was carbonized by stressing it with defined numbers of lightning impulses of a known energy and was then subjected for a period of 1000 hours to an accelerated ageing with metallic catalysts at 20 and 90 °C in the presence of oxygen. The analysis of the dielectric properties reveals that an ageing of carbonized oil samples leads to a faster and more intensive degradation of the fluid compared to uncarbonized oil.**

INTRODUCTION

Machine and network transformers are to a high extend supplied with an on-load tap changer (OLTC) which enables in dependence of the load a variation of the transformer ratio without deenergizing the transformer [1]. The arcing in the diverter switch cubicle during the change-over between two windings leads to a degradation and carbonization of the transformer oil which is used as an embedding and insulation medium. Moreover a gradual thermal ageing of the degraded insulating liquid takes place. This ageing can be accelerated by metallic materials present in the oil, such as contact metals or leads which act as catalysts. These effects are generally considered in the dimensioning of the insulation materials and distances.

The described combination of thermal and electrical stresses can result in a decrease of the dielectrical properties of the fluid. So in order to maintain the dielectric strength of the diverter switch the oil is exchanged during regular inspections depending on service time and number of operations [2]. Between the maintenance intervals the OLTC has to perform reliably irrespective of the amount of contamination or the degree of ageing. The foregoing must be considered in the present discussion regarding the prolongation of inspection intervals.

With regard to the outlined problems extensive experimental investigations on the ageing processes of carbonized mineral based transformer oil samples have been performed. In this respect the dielectric dissipation factor tan δ, the volume resistivity ρ_{90}, and the partial discharge inception voltage were referred to for they seemed to be the suitable characteristics of an oil which are sensitive enough concerning the impact of carbon particles and degradation processes due to ageing [3].

TEST PROCEDURES

Carbonization and ageing of the oil

The experimental investigations were performed on the naphthenic mineral based transformer oil Shell Diala D which has a water solubility of 44 ppm at 20 °C and 400 ppm at 80 °C. In order to ensure equal initial conditions for each test series the fluid had to be prepared before starting the measurements, i. e. the transformer oil was filtered, degassed, and the moisture content was reduced in a two-step-oil-preparation-plant [4]. Then the water content of the samples was adjusted to a level of 20 ppm and controlled by using a coulometric Karl-Fischer-Titration device with an accuracy of approximately 1 ppm.

For the carbonization 1.1 l of the prepared oil was filled in a test vessel which contained two vertically arranged homogenous copper electrodes. The carbonization was carried out by using a four-step-impulse-generator which allowed us to produce negative lightning impulses of different voltage levels with a shape of 1.2/50 µs in no-load operation. The oil was stressed at room temperature with a defined number N of lightning impulses which was varied between 0, 1000, 5000, and 10000. The single impulse energy E covered a range of 32.7 and 65.4 Ws.

After the carbonization the oil samples were subjected to different accelerated ageing conditions over a period of 1000 hours. With regard to the operating temperature of OLTCs the influence of ageing at room temperature and at an elevated temperature of 90 °C was scrutinized. Additionally also the catalytic impact of metallic materials present in the OLTC-cubicle was investigated. For these experiments 3 g/l of iron, aluminium, copper, zinc, tin, and chrome in form of very fine cuttings of nearly equal size were put in a filter paper of less than 3 µm pore size which was hung into the oil.

Table 1 gives a survey of the investigated ageing conditions which were all performed in the presence of oxygen.

Table 1
Ageing Conditions

Temperature	Ageing Parameters
20 °C	aged for 1000 h with catalysts and oxygen
90 °C	aged for 1000 h with oxygen
90 °C	aged for 1000 h with catalysts and oxygen

Before starting the measurements which are described in detail in the next chapter the humidity of the carbonized and aged oil samples was adjusted to the required water contents.

Experimental setups and measurements

The measurements of the dielectric dissipation factor tan δ were performed in a temperature range from 20 to 90 °C. According to VDE 0370 and IEC 296 the test cell consisted of a guard-ring-(three-)electrode configuration with an electrode distance of 2 mm [5]. The 50-Hz-test-voltage of 2 kV was supplied over an auto transformer to a high voltage testing transformer. On the high voltage side the test cell was connected to a Schering-bridge with an auxiliary Wagner-branch. The latter was used to eliminate parasitic earth capacities and thus guaranteed a measuring sensitivity of better than $1 \cdot 10^{-6}$.

The volume resistivity ρ_{90} of the carbonized and aged oil samples was determined according to VDE 0303 and IEC 93 in the above mentioned guard-ring-configuration [6]. When the steady-state-temperature of 90 °C was reached a positive DC voltage of 1000 V was applied to the oil and after stressing it for one minute the volume resistivity was determined by a sensitive voltage-current-measurement.

For the determination of the partial discharge inception voltage (PDIV) at room temperature an electrode arrangement (Fig. 1) according to VDE 0380 and IEC 1294 was used [7]. It consisted of a wolfram needle electrode with a tip radius of 3 μm and a steel sphere electrode with a diameter of 12.5 mm. The gap distance was adjusted to 50 mm. The partial discharges (PD) were decoupled by a combination of a narrow-band interference-meter and a decoupling unit which was connected in series to the test cell. The band-width of the PD detection system was 9 kHz, the mid-frequency was chosen to 150 kHz.

Starting from 0 V the voltage on the needle electrode was increased with a constant rate of 1 kV/s till the first steady PD impulses with a charge of 100 pC or more could be detected. Then the arrangement was deenergized and after a period of one minute the oil was stressed anew. From the results of ten successive measurements the mean value and the standard deviation of the PDIV were determined.

RESULTS AND DISCUSSION

Dielectric Dissipation Factor tan δ

In figure 2 the influence of different numbers of impulse stresses on the dielectric dissipation factor of unaged oil samples is illustrated in a temperature range from 20 to 90 °C. The diagram shows for new oil between 20 and 40 °C a slight decrease and for higher temperatures till 90 °C a constant increase of the tan δ which obeys to an exponential dependence on the temperature. In this temperature interval the tan δ is essentially determined by ionic conduction processes [8]. For

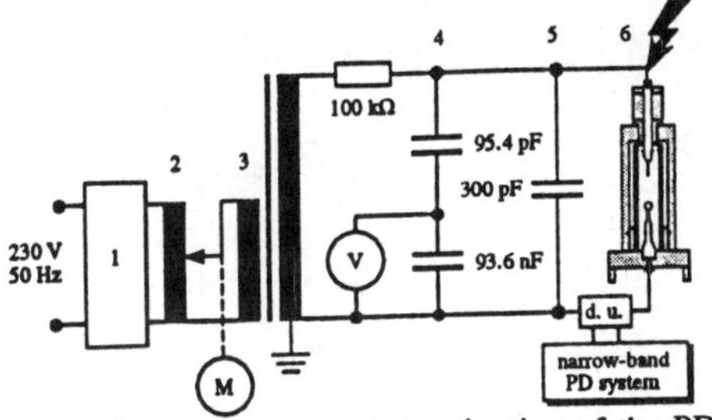

Figure 1. Measuring setup for the determination of the PDIV (1) control unit, (2) motor transformer, (3) high voltage transformer, (4) capacitve voltage divider, (5) coupling capacitance, (6) test cell with PD detection system

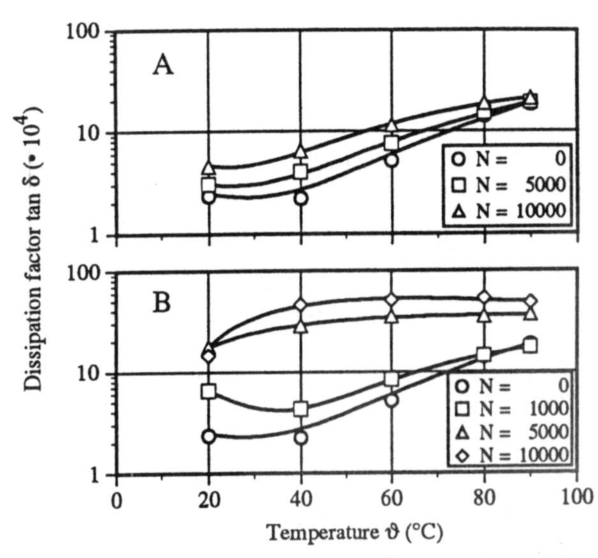

Figure 2. Dielectric dissipation factor tan δ of unaged oil (20 ppm) vs temperature with the number N of impulse stresses as parameter (A) E = 32.7 Ws, (B) E = 65.4 Ws

oil samples which were carbonized with 5000 or 10000 impulses of 32.7 Ws single energy (Fig. 2A) the tan δ curves are shifted to higher values. This can be explained by the increase of the oil conductivity which is caused by the escalating number of impulse discharges from 0 over 5000 to 10000. Moreover between 20 and 60 °C the curve characteristics are nearly the same compared to uncarbonized oil whereas for higher temperatures the curves of the carbonized oil samples are retarded and turn into a nearly horizontal branch. This effect is even stronger when the single energy of the impulses is doubled to 65.4 Ws (Fig. 2B). Here the tan δ values of the investigated oil samples are not only on a 2-12 times higher level compared to uncarbonized oil but it can also be perceived that with an increasing number of impulse stresses the transition into the horizontal characteristic starts at lower temperatures. So besides the number of impulses also the single impulse energy seems to govern the tan δ behaviour. The observed retardation may be explained by the fact that for increasing numbers of impulse stresses the tan δ is predominantly determined by the carbon content and the part of the tan δ rise being due to the temperature is thrust into the background.

The influence of different water contents on the dielectric dissipation factor is presented in Fig. 3. The tan δ of new oil (Fig. 3A) is not significantly effected by different water contents as long as the water is dissolved. At low temperatures an increase of the tan δ can be observed which is especially pronounced for rising water contents. This phenomenon can be related to a so-called "water-maximum" which can be expected between 0 and 10 °C [9]. When the water solubility of the oil is exceeded at low temperatures the water precipitates in form of small water droplets. They take part in the conduction process and are responsible for a local maximum of the tan δ at low temperatures. With increasing water content of the samples the water solubility limit is reached at higher temperatures and

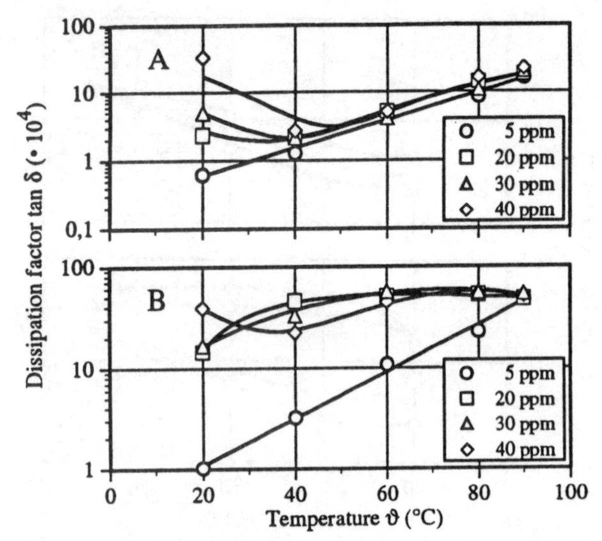

Figure 3. Dielectric dissipation factor tan δ of unaged oil vs temperature with the water content of the oils as parameter
(A) N = 0, (B) N = 10000/E = 65.4 Ws

so the transition from the water maximum into the interval of ionic conduction is shifted to higher temperatures. Thus for high water contents the water maximum affects the tan δ values even up to a temperature of 50 °C.

As can be derived from Fig. 3B the basic statements concerning the influence of water on the tan δ are also valid for carbonized oils. The only exception is the fact that the water maximum at low temperatures is not so much developed for water contents between 20 and 40 ppm. For the carbonized oil samples the influence of the water seems to be drawn to the background compared to the impact of the carbon particles on the tan δ values.

The influence of the investigated ageing processes on the dissipation factor is shown in figure 4 for different degrees of carbonization. The figure reveals that after subjecting the oils to an ageing the tan δ of all samples strictly obeys to an exponential dependence of the temperature. The formation of a water maximum can no longer be made out. This may be explained by the formation of sludge which results from a molecule enlargement due to the ageing. The sludge is expected to adsorb water molecules so that the effective water solubility is increased. This leads to a shifting of the water maximum to lower temperatures with the result that the branch of the maximum does no longer occur in the examined temperature range. Furthermore the bending of the tan δ curves which could be determined for the unaged carbonized oil samples (Fig. 2) has vanished after the ageing.

A comparison of Fig. 4A-D points out that with a rising number of impulse pre-stresses the tan δ of the oil samples is more and more reduced by an ageing at 20 °C with catalysts (20 °C w. c.). This behaviour finds its corresponding analogy in an ageing conditioned mounting of the volume resistivity ρ_{90} (Fig. 5). In contrast to this an ageing at a temperature of 90 °C leads for all scrutinized samples to a significant increase of the dissipation factor compared to the unaged state. Depending on the degree of carbonization the tan δ_{90} values experience a rise by factors between 5 and 10 due to the thermal ageing. Moreover the impact of the carbon content on the tan δ is now more pronounced. After the ageing an increase of the impulse number from 0 over 1000 and 5000 to 10000 leads to a rise of the tan δ_{90} from 99 over 178 and 204 to 448 • 10^{-4} compared to 18, 17, 37, and 48 • 10^{-4} for the unaged samples. The reason for this behaviour can be seen in the fact that the chemical structure of the oils is more degraded for higher numbers of impulse pre-stresses. So the ageing processes can impair more effectivly on an oil with a high carbon content and can thus develop more polar degradation products of high conductivity (Fig. 5). This means that the period of induction, that is the time after which the ageing of an oil starts, is reduced for carbonized oils compared to new fluids. The result is

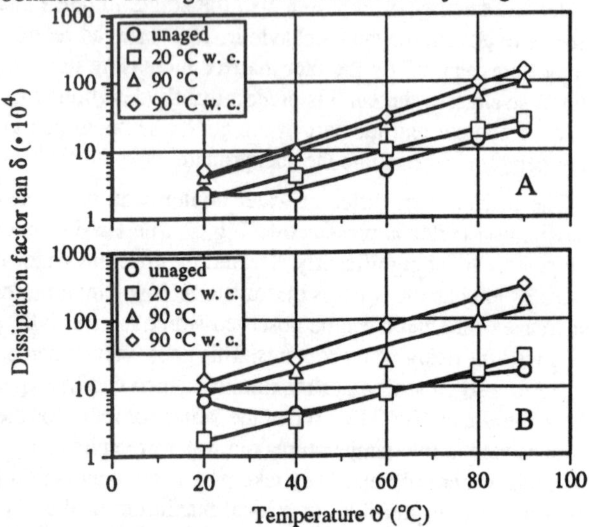

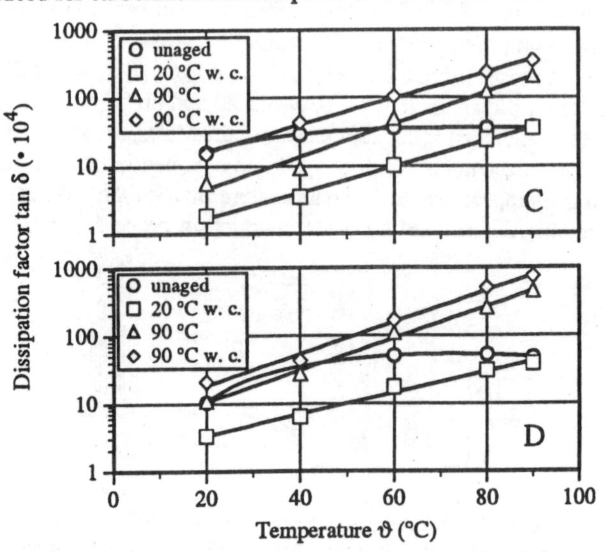

Figure 4. Comparison of dissipation factors tan δ of various oils (20 ppm) vs temperature with the ageing conditions as parameter
(A) N = 0, (B) N = 1000/E = 65.4 Ws, (C) N = 5000/E = 65.4 Ws, (D) N = 10000/E = 65.4 Ws

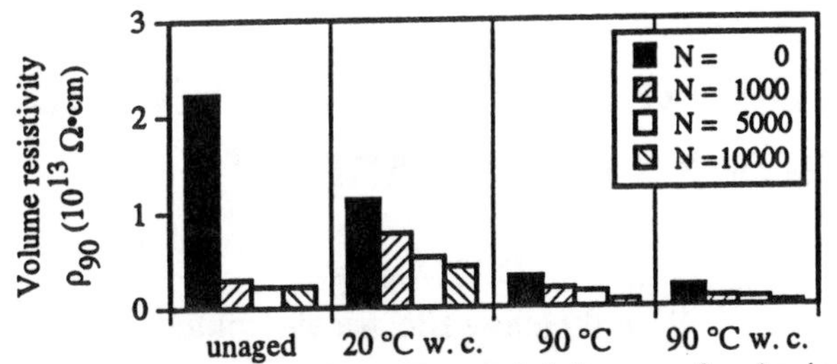

Figure 5. Comparison of the volume resistivity ρ_{90} of carbonized oil samples (20 ppm) for different ageing conditions

a faster ageing of the transformer oil.

An additional ageing with catalysts (w. c.) results in a further parallel shifting of the tan δ curves. As can be derived from Fig. 4 the tan δ_{90} values of oils which have been pre-stressed with 0, 1000, 5000, and 10000 impulses amount to 148, 325, 349, and 749 • 10^{-4}, respectively. Comparing this with the effect which is due to the elevated ageing temperature the presence of metalic catalysts is responsible for a further mounting of the tan δ by factors between 1.5 and 1.8.

Partial discharge inception voltage

In Fig. 6 the impact of the sum energy of various impulse pre-stresses on the PDIV is illustrated for different ageing conditions. The results reveal that the PDIV of unaged oil (Fig. 6A) rises from a level of 26 kV with increasing degree of carbonization till it reaches a maximum of 36 kV at 327 kWs. For higher energies a decrease of the PDIV can be made out to a level of 27 kV at 654 kWs. This behaviour results from two opposing factors. The first is the radiusing of the needle which is due to the conductive carbon particles. They lead to a reduction of the effective electric field strength at the needle tip. Thus the inception of PDs is shifted to higher values. If the carbon content is further increased the carbon particles may on the other hand effectively bridge a part of the gap distance which leads to an earlier inception of PDs and thus counteracts to the first factor.

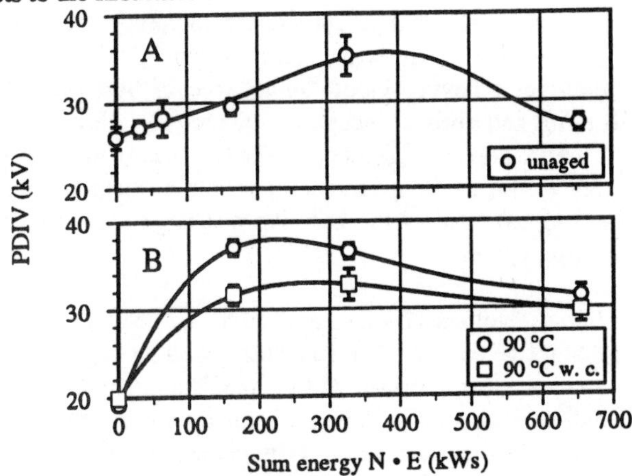

Figure 6. PDIV of carbonized transformer oil (20 ppm) vs sum energy of the impulse pre-stresses with the ageing conditions as parameter
(A) unaged, (B) aged at 90 °C with/without catalysts

An ageing of the carbonized oil at 90 °C (Fig. 6B) results in a shifting of the PDIV maximum to lower impulse energies. This effect can be explained by the above mentioned formation of conductive sludge so that the effective reduction of the gap distance already determines the dielectric strength at a smaller amount of carbon. Thus it is also understandable that the maximum value is not significantly effected by the ageing.

An additional ageing in the presence of catalysts leads to a parallel shifting of the PDIV curve to a lower level whereas the curve characteristic remains the same (Fig. 6B).

CONCLUSIONS

The experimental investigations on carbonized and aged transformer oil samples have revealed the following results:

- Compared to new oil carbonized fluids experience a faster and more intensive ageing process. The higher the carbon content of the oil the more pronounced is the tan δ increase after the ageing and thus the dielectric losses. This has to be considered as an important factor with regard to the permissible prolongation of maintenance intervals.
- Referring to the examined energy range the presence of carbon particles in transformer oil does not significantly reduce the PDIV even after an accelerated ageing process.

ACKNOWLEDGMENT

The authors would like to thank Maschinenfabrik Reinhausen, Regensburg (Germany) for their support. Moreover the authors like to express their gratitude to Mr Stefan Klinger for performing parts of the experiments.

REFERENCES

1. Bleibtreu, A. et al., "Present State and Trends in Developing Tap-Changers of the High Speed Transition Resistor Type with Special Regard to Operating Reliability", Cigré Session, 1976, Report No. 12-07.
2. Baehr, R. et al., "Diagnostic Techniques and Preventive Maintenance Procedures for Large Transformers", Cigré Session, 1982, Report No. 12-13.
3. Lakshminarayanan, T., "Effect of transformer materials on the degradation of transformer oil under semiactive conditions", Int. Conference on Transformers, Bombay, 1986, pp. 10-18.
4. Schröder, U., "Das Teilentladungs- und Gasungsverhalten von Transformatorenöl", Ph. D. thesis, U Hannover, 1994.
5. DIN VDE 0370, Isolieröle, "Neue Isolieröle für Transformatoren, Wandler und Schaltgeräte", Teil 1, December 1978.
6. DIN VDE 0303, Prüfungen von Werkstoffen für die Elektrotechnik, "Messung des elektrischen Widerstandes von nichtmetallenen Werkstoffen", Teil 3, May 1983.
7. DIN VDE 0380, Isolierflüssigkeiten, "Bestimmung der Teilentladungseinsatzspannung (PDIV)", Teil 5, February 1995.
8. Beyer, M. et al., *Hochspannungstechnik*. Springer Verlag New York, Tokio, Berlin, 1986, p. 161.
9. Holle, K.-H., "Über die elektrischen Eigenschaften von Isolierölen, insbesondere über den Einfluß von Wasser auf deren Temperaturverhalten", Ph. D. thesis, TU Braunschweig, 1967.

Conference Record of the 1996 IEEE International Symposium on Electrical Insulation, Montreal, Quebec, Canada, June 16-19, 1996

INFLUENCE OF ELECTRODE TEMPERATURE ON THE CHARACTERISTICS OF A NON UNIFORM FIELD SYSTEM WITH A BARRIER

M.C. Siddagangappa
Department of High Voltage Engineering
Indian Institute of Science
Bangalore 560 012, India

D.K.Mandal
Bharat Heavy Electricals limited
Transmission Business Group
Lodhi Road, New Delhi 110 003, India

Abstract: **Electrode temperature controlled ac corona discharge inception voltages of a non uniform field electrode system and the characteristics change of CIV in the presence of a barrier is investigated. Reduction in the breakdown voltages of the gap for negative standard impulse applications was observed as the point electrode temperature raised upto 200° C and the characteristic occurrence was repeated in the presence of press board barrier.**

INTRODUCTION

When a composite insulation system of gas and solid is subjected to high electrical stresses, charged particles produced in the localised discharges either penetrates or deposits on the solid dielectrics altering the field distribution. Thus, the reliability and the characteristics of a composite insulation system are controlled by combination of electrode field conditions, gaseous medium and of course on the solid dielectric material used together with surface conditions. Invariably, solid dielectric barriers are used in high voltage switchgear systems, whereas the dielectric barrier discharge is finding application as a source of atmospheric pressure non thermal plasma to be used for environmental pollution control [1]

Important investigation by Van Brunt and co-workers [2] reported on the fundamental characteristics behavior of corona discharges in air influenced by solid dielectric barrier. Awad and Castle [3} have studied the electrode temperature controlled corona currents using a co-axial wire tube geometry. Though, temperature influences the swarm data in synthetic air, systematic breakdown voltage measurements in nearly homogenous field condition do not support the temperature dependence as reported by Zaengl et.al [4], wherein both electrodes and synthetic air were maintained at the same temperature in a confined volume.

The present preliminary investigation was conducted to examine the influence of electrode temperature on the characteristics of composite non uniform field system having a solid insulation barrier.

EXPERIMENTAL DETAILS

A point and sphere electrodes were mounted on a common axis on an insulator base and the required gap length could be adjusted. The stainless steel point electrode was of 6 mm dia tapered at one end as a cone forming a tip surface of 1 mm dia. The copper sphere of 62.5 mm was connected to the high potential end of the power source. The point electrode terminated to ground via current measuring circuit, and was insulated with a mica tube from a 750 watt a.c. heater. A grounded metal shield was provided between heater and the point electrode to avoid leakage currents due to the heater supply voltage. A temperature controller with a shielded thermocouple sensor attached was used to maintain the electrode tip at specified temperatures with a variation of ± 1 °C. The high voltage a.c power was derived from a 230 V/115 kV, 10 KVA, 50 Hz transformer having a voltage measuring winding ratio of 1:1000. The input to the transformer was controlled by a voltage regulator.

The corona inception currents (CIC) were obtained by measuring the potential drop across 1 kΩ resistor (± 1 % tolerance). The voltage drop across the resistor due to the current flowing in the circuit was measured using a broad band (2 Hz to 1 MHz) millivoltmeter having input impedance of 300 MΩ. The waveform and corona inception voltage (CIV) levels were also observed on an oscilloscope. The CIV recorded are those, relating to the first appearance of minimum detectable gap current (1 μA) and of the negative discharge sensed by the measuring system.

The impulse voltages used were obtained by 165 kV, 1.5 kJ, 4 stage Marx generator. The standard 1.2/50 μs negative impulse voltages were measured by a precision damped capacitance divider coupled to a peak voltmeter and recorded on a surge oscilloscope (Hafeley). Fifty percent breakdown voltages (V_{50}) determined by up and down method.

Repeated experiments on the composite insulation system

revealed that the area of the circular barrier placed perpendicular to the electrode axis have significant effect on CIV levels. However, the CIV data reaches a stabilization when the barrier diameter was two times that of sphere electrode. Further increase in the area of the barrier have had no effect on the corona inception characteristics of the composite insulation system. Finally the experiments were conducted with pressboard barriers of 150 mm dia and of 1.5 mm thickness.

RESULTS AND DISCUSSION

With a simple experimental setup and conventional measuring method adopted, CIV data for point electrode temperatures of 27, 50, 100, 150 and 200 °C and for gap distance of 20, 30 and 40 mm are shown in figure 1. The corona inception voltages decreased with increase in point electrode temperature. The discharge currents recorded at inception. were of the same order (5 to 7 μA), even when the electrode temperature was at 200 °C , are not shown in the figure. CIV data of a non uniform composite insulation system was obtained for gap lengths of 20, 30 and 40 mm and are plotted in figure 2., against press board barrier position measured with reference to the point electrode. The maximum CIV was observed for the barrier positions around the axial centre of the electrode gap and higher than CIV recorded with no barrier. However, when the barrier moves towards any of the electrodes in the system CIV continue to decrease and falls below the value of simple electrode gap. The CIC showed no significant variations for different barrier positions in the gap.

Further, the effect of electrode temperature on CIV of composite insulation was studied by maintaining the point electrode at 50, 100, 150 and 200 °C. CIV data for gap lengths 20, 30 and 40 mm with constant electrode temperature of 200 °C are plotted in figure 3. The reduced CIV values of heated pointed electrode follow the same trend of the data obtained for studies at room temperature in the presence of a barrier which is shown for comparison in the figure. The investigations were also conducted to assess the change in breakdown strength with electrode temperature, by applying standard negative impulse voltages to the sphere electrode. Negative impulse breakdown data (V_{50}) obtained for various gap distances with electrode temperatures in the range of 20 to 200 °C , are plotted in figure 4. As is evident from the figure, the negative impulse breakdown strength has fallen by about an average of 25% for the gaps studied when the temperature of the point electrode reaches 200 ° C.

Impulse breakdown experiments were extended for the composite system and for different barrier positions. The breakdown of the system if occurred was always through the centre of the barrier and no extraneous breakdown was observed. After each impulse breakdown of the gap, barrier was replaced by new one. The data for 10 mm gap length is plotted in figure 5 for various barrier positions and for constant electrode temperatures. Similar characteristics were observed for 20 to 30 mm gap lengths investigated. By comparing the data of figures 3 and 4 , calculated relative breakdown strength of the gap in the presence of a barrier is always higher and is consistent with the earlier studies [5].

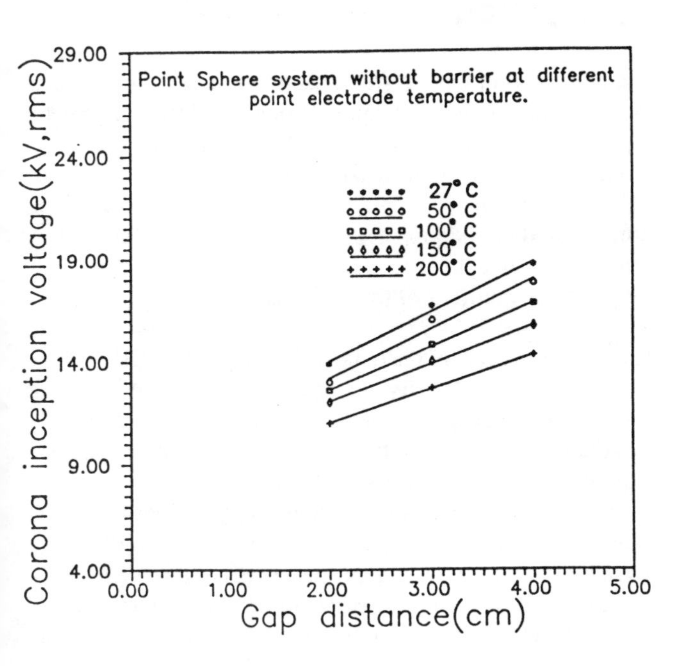

Figure 1. Corona inception voltages as a function of gap length for various point electrode temperatures.

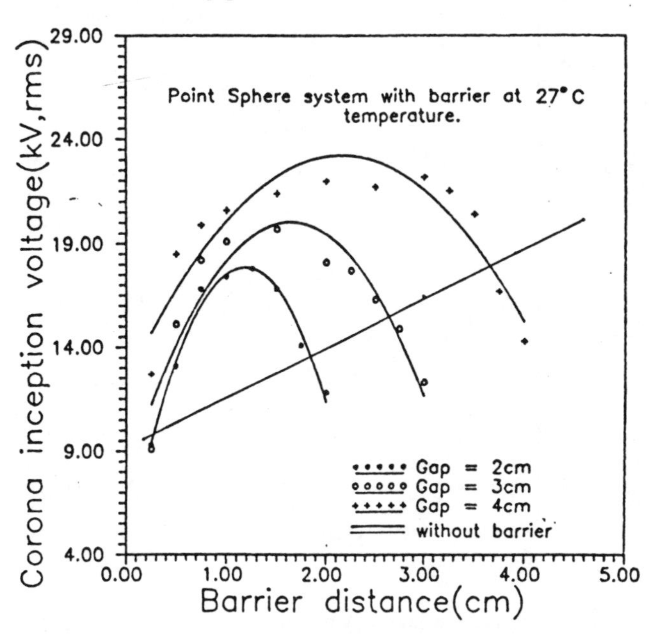

Figure 2. Corona inception voltages as a function of barrier distance with reference to the point electrode.

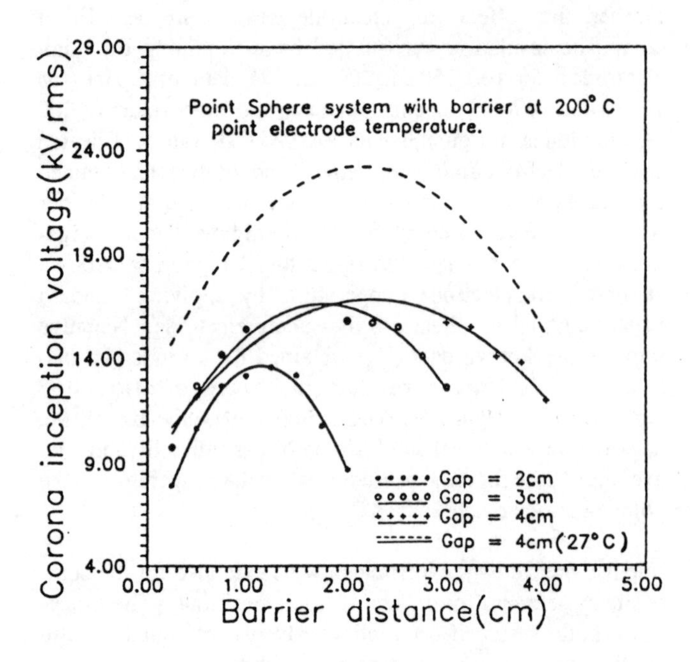

Figure 3. Corona inception voltages as a function of barrier distance at different point electrode temperatures.

Figure 5. Fifty percent breakdown voltages as a function of barrier position for various point electrode temperatures.

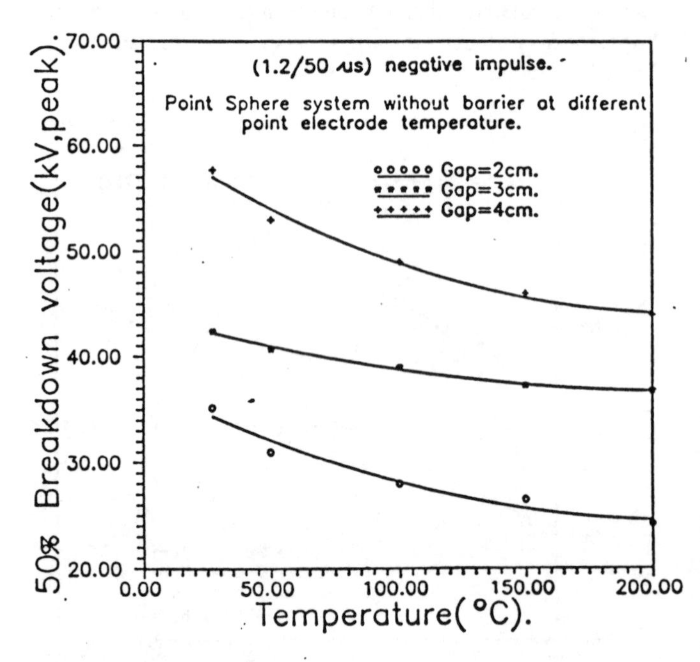

Figure 4. Fifty percent breakdown voltages as a function of point electrode temperatures.

CONCLUSIONS

In a non uniform field system, higher electrode temperature causes a decrease in the magnitude of CIV while there are no appreciable changes in CIC. In the composite insulation system, CIV data is controlled by electrode temperature and as well by the presence of a barrier. Relative impulse breakdown strength of the insulation system is observed to be reduced as a function of temperature, indicating a major role played by the thermal conditions of the electrode.

The classic answer for the temperature dependent CIV data is the change in air density factor δ [6], because of thermal gradient just close to the surface of the point electrode, which constitutes a small percentage of gaseous insulation volume. Early work of Alston [7] indicated pronounced effect of hot spot in uniform fields reducing dc flashover values in air. In light of the recent studies by Zaengal and co workers [4] revealing no temperature dependence of homogeneous field breakdown in synthetic air, reasonable explanation is yet to be evolved for electrode temperature dependence on the dielectric system behaviour.

ACKNOWLEDGMENTS

The authors wish to thank Mr. B.Nageshwar Rao, Mr. Jagannatha and Mr. Rajashekar for the timely help received in this work.

REFERENCES

1. Bijan Pashie, Shirshak K Dhali and Frank I Honea. "Electrical characteristics of a coaxial dielectric barrier discharge". J. Phys. D: Appl. Phys. Vol.2, 1994, pp. 2107 -2110

2. Van Brunt, R. J, Misakian, M, Kulkarni. S. V. and Lakdawala. V.K. "Influence of a Dielectric barrier on the stochastic behaviour of trichel -pulse corona". IEEE Trans. E.I. Vol.26 No.3, June 1991, pp- 405-415.

3. Awad, M.B and Castle, G.S.P. " Breakdown streamers in coronas with heated Discharge Electrode". IEEE Trans E.I. Vol.2, No.3, June 1977, pp. 234-236.

4. Zaengal, W.S, Yimvuthikul and Friedrich, G. " The Temperature Dependence of Homogeneous Field Breakdowns in Synthetic Air". IEEE Trans. EI. Vol.26, No.3, June 1991, pp. 380 -390.

5. Teruya Kouno. " Breakdown of composite Dielectrics: The Barrier effect". IEEE Trans EI. Vol. 15, No.3, June 1980, pp. 259 - 263

6. Cobine, J. D. *Gaseous Conductors*. Dover Publication. N. Y. 1958, p.255

7. Alston, L.L. " High-Temperature effects on Flashover in Air". Proceedings IEE, Vol. 105, Pt A, No.24, December 1958, pp. 549 -553.

Conference Record of the 1996 IEEE International Symposium on Electrical Insulation, Montreal, Quebec, Canada, June 16-19, 1996

STUDIES ON THE DIELECTRIC STRENGTH OF TRANSFORMER OIL UNDER OSCILLATORY IMPULSE VOLTAGES

C. Venkataseshaiah, R. Sarathi and M.V. Aleyas
Department of Electrical Engineering
Indian Institute of Technology
Madras - 600 036, INDIA.

Abstract: **In this paper, the results of investigations carried out on the breakdown characteristics of transformer oil subjected to oscillatory impulse voltage (unidirectional and bidirectional) stresses in the frequency range of 10-100kHz using needle-plane electrode configuration are presented. For the purpose of comparison, experimental results obtained under AC, lightning and switching impulse voltages are also provided. The effect of gap spacing, frequency, field non- uniformity and polarity on the breakdown characteristics are discussed.**

INTRODUCTION

The insulation of power transformers is subjected to unidirectional and bidirectional oscillatory impulse voltage stresses due to switching operations, line faults, lightning etc. The frequency of such oscillatory impulse voltages (OIV) is found to lie in the frequency range of several hundred Hertz to 200 kHz [1]. A study of the dielectric strength of transformer oil insulation under OIV in the above stated frequency range is therefore necessary for optimum design of transformer insulation.

Studies on the breakdown characteristics of transformer oil under AC, DC, lightning and switching impulse voltages have been reported in literature in considerable detail [2,3]. However, limited literature is available on such studies under OIV. Couper et al. [4] and Guanzhicheng et al. [5] studied the breakdown characteristics of air gaps with rod-sphere configuration under OIV in the frequency range of 2 to 60 kHz to examine the effect of polarity, decay rate and frequency. Tsunekarau et al [6] have conducted breakdown studies on transformer insulating materials in the frequency range of 8-240 kHz under uniform field condition. In practical situations highly non-uniform electric field conditions can exist in transformer insulation, due to protrusions in the current carrying conductors or floating metallic particles in the insulating oil. Therefore, there is a need to generate a credible database with regard to breakdown characteristics of transformer oil insulation under OIV with non-uniform field conditions.

In this paper, the results of investigations carried out on the breakdown characteristics of transformer oil subjected to OIV (unidirectional and bidirectional) in the frequency range of 10-100 kHz, under highly non-uniform field conditions are presented. For the purpose of comparison, the results obtained under AC, lightning and switching impulse voltages are also presented.

EXPERIMENTAL PROCEDURE

Fig. 1 shows the circuit arrangement used for the generation of oscillatory impulse voltages. The required frequency and decay rate of OIV were obtained by suitably choosing the values L_S, C_L and R_t [4]. Needle-plane electrode configuration was used for the study as it is representative of the highly non-uniform electric field conditions which exist in the transformer insulation. Two types of needles (Type-1 of 10 μm tip radius and Type-2 of 50 μm tip radius) were used. In the present work, gap spacings of 5 mm, 10mm and 13mm were used. The transformer oil was obtained from a transformer manufacturer and it was used without any further treatment. A capacitor divider and a Nicholet-pro 90 model digital storage oscilloscope were used for measuring the applied voltage.

The following methodology was adopted to determine the breakdown voltage. The oil was allowed to rest for atleast two minutes after the test cell was filled with oil. Initially three impulses were applied at intervals of 30 seconds at a crest voltage which is considerably below the expected breakdown voltage. The sequence was repeated by increasing the applied voltage by 2 to 5 kV each time until first sparkover occurred. The corresponding voltage was taken as one of the readings. This experimental procedure was repeated for thirty readings. The average of the thirty breakdown voltage values is taken as the breakdown voltage for the given configuration. Care has been taken to ensure that, before the application of the impulse, the region between the electrodes was free from bubbles, as well as carbon tracks formed, if any, because of previous breakdowns. The needle electrode was replaced after every three breakdowns. A fresh oil sample was used for obtaining each breakdown voltage value. Whenever the oil sample was replaced, the bottom electrode was cleaned and polished.

The breakdown voltage under AC was determined in the following manner. First the voltage was raised at the rate of 3 kV /second upto seventy percent of the expected breakdown voltage and then at the rate of 1 kV/second until breakdown occured. The average of 10 breakdown voltage values was taken as the breakdown voltage for the given configuration.

RESULTS AND DISCUSSION

Fig. 2 shows the breakdown voltage variation as a function of gap spacing for different voltage waveforms viz. AC, lightning impulse (LI), switching impulse (SI) and unidirectional oscillatory (10 kHz and 100 kHz) impulse voltages (UOIV) for both polarities. Among AC, LI and SI breakdown voltages it is seen that the AC breakdown voltage is the lowest for a given electrode spacing. Also it is seen that, the 10 kHz unidirectional impulse breakdown voltage is nearly the same as that of the switching impulse (100/1000 μs) breakdown voltage. The breakdown voltage under 100 kHz unidirectional oscillatory impulse voltages is always higher than that under 10 kHz UOIV. This observation is found to hold good even for negative polarity. Also, it is seen that the breakdown voltage increases with increase in gap spacing as well as frequency of the OIV (unidirectional and bidirectional). It is observed that the oscillatory impulse breakdown voltage is always higher than the standard lightning impulse breakdown voltage. Similar observations were reported by Couper et al. [4] in their studies on air breakdown with rod-sphere configuration.

The variation of oscillatory impulse ratio with gap spacing is shown in Fig. 3 for three different frequencies viz. 10,50 and 100 kHz. The oscillatory impulse ratio is defined as the ratio of breakdown voltage under OIV to the breakdown voltage under AC for a given configuration. The variation of impulse ratio with different gap spacings under LI is also indicated for the purpose of comparison. The oscillatory impulse ratio of OIV (for both unidirectional and bidirectional) decreases with increase in gap spacing as in the case of LI for both the polarities. The value of the impulse ratio for unidirectional and bidirectional OIV is always higher under negative polarity. Compared with unidirectional OIV, the oscillatory impulse ratio is always higher for bidirectional OIV.

Fig 4 shows the variation in breakdown voltage as a function of gap spacing for two different types of needle electrodes at different frequencies. It is observed that the breakdown voltage with type-1 needle is less compared to tthe breakdown voltage with ype-2 needle. This tendency is the same at different frequencies and gap spacings. Similar phenomena has been observed in the case of bidirectional OIV also.

CONCLUSIONS

The following conclusions can be drawn based on the above investigations.

* The breakdown voltage under OIV is always higher than that under standard lightning impulse voltages. The 10 kHz unidirectional oscillatory impulse breakdown voltage is almost the same as that of switching impulse (100/1000 μs) breakdown voltage. In the frequency range studied it is observed that the breakdown voltage increases with increase in frequency.
* The breakdown voltage under bidirectional OIV is higher than the breakdown voltage under unidirectional OIV for a given oscillatory frequency.
* The negative polarity oscillatory impulse breakdown voltage (unidirectional or bidirectional) is higher than the positive polarity oscillatory impulse breakdown voltage. The effect is more pronounced with higher field non-uniformity.
* The breakdown voltage increases with increase in gap spacing and the effect is more pronounced at higher frequencies of OIV.

REFERENCES

1. Degeneff R.C et al., IEEE Trans. in Power Apparatus and Systems, Vol. PAS-101, No. 6, pp. 1457-1459, June 1982.
2. Sharbaugh A.H et al., IEEE Trans on Electrical Insulation, Vol. EI-13, No. 4, pp. 249-276, Aug. 1978.
3. Forster E.O et al., IEEE Trans. in Dielectrics and Electrical Insulation, Vol. 1, No. 3, pp. 440-446, June 1994.
4. Couper I.D et al., IEEE international conference on properties and applications of dielectric materials, Vol. 1, pp. 156-159, 1988.
5. Guan Zhicheng et al., Paper No 13-01, Sixth International Symposium on High Voltage Engineering, New Orleans, LA, USA, Aug. 28-Sept. 1, 1989.
6. Tsuneharu Teranishi et al., Electrical Engineering in Japan, Vol. 104B, No. 1, pp. 66-73, 1984.

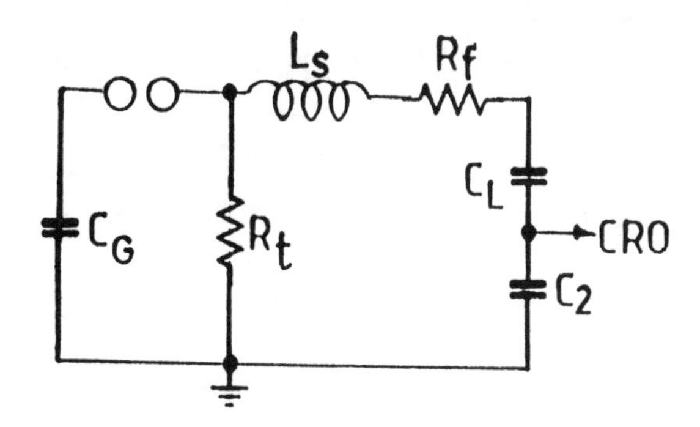

Fig. 1 Circuit arrangement for generation of OIV.

C_G	:	Generator capacitance
L_S	:	Additional series Inductance
R_f	:	Front resistance
R_t	:	Tail resistance
C_L	:	Load/measuring capacitance

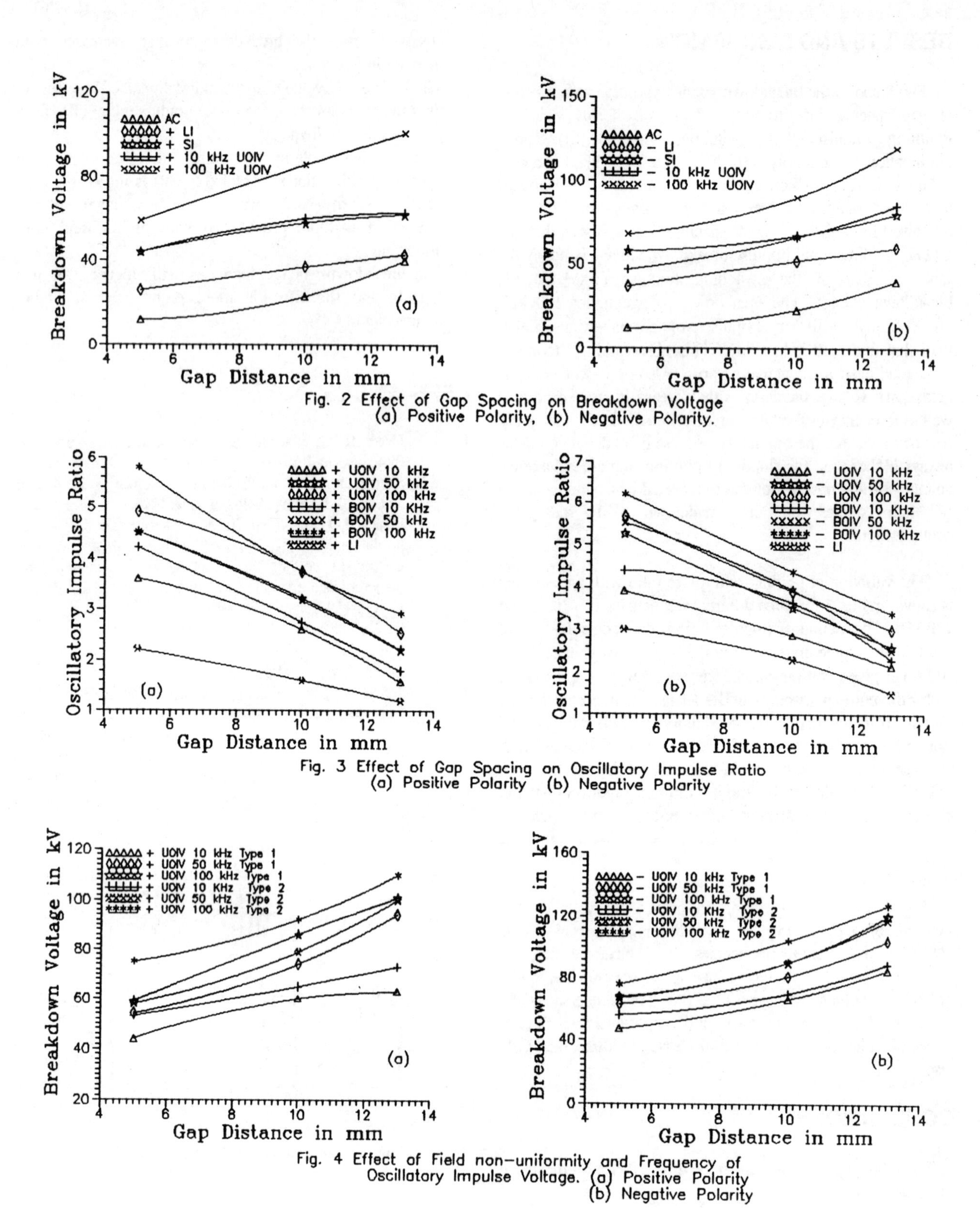

Fig. 2 Effect of Gap Spacing on Breakdown Voltage
(a) Positive Polarity, (b) Negative Polarity.

Fig. 3 Effect of Gap Spacing on Oscillatory Impulse Ratio
(a) Positive Polarity (b) Negative Polarity

Fig. 4 Effect of Field non-uniformity and Frequency of Oscillatory Impulse Voltage. (a) Positive Polarity
(b) Negative Polarity

Conference Record of the 1996 IEEE International Symposium on Electrical Insulation, Montreal, Quebec, Canada, June 16-19, 1996

Arcing Resistance of High Fire Point Dielectric Liquids

Edgar Howells
Cooper Power Systems
Thomas A. Edison Technical Center
11131 Adams Road
Franksville, WI. 53126

Garrett P. McCormick
Cooper Power Systems
Thomas A. Edison Technical Center
11131 Adams Road
Franksville, WI. 53126

Abstract:: **High Fire Point (HFP) dielectric liquids have been in use since the demise of Askarels (PCB/TCB) in the mid - 1970's. Having fire points of at least 300°C, their main application has been in the role of transformer impregnants particularly in units located in, on, or near buildings. This study was aimed at evaluating the suitability of such liquids for application in arcing environments.**

At present, most HFP liquids are produced from one of three different chemical bases. These are : a) Polyol Ester, b) High Molecular Weight Hydrocarbon (HMWH) and c) Dimethyl Silicone . Samples of these liquids were evaluated in a typical oil switch primarily utilized in capacitor switching applications . Although not an HFP liquid, Transformer Oil was included in the test matrix to provide a base line to which the other liquids could be compared.

Each liquid was exposed to 250 cycles of switching 8,000 Volts at 50 Amps. (a duty well within the capability of the switch). The results show that both Polyol-Ester and High Molecular Weight Hydrocarbons compare favorably to Transformer Oil in this application. However, the dielectric breakdown value of Dimethyl Silicone was so rapidly degraded that its use under such conditions requires considerable caution.

INTRODUCTION

The use of Askarels (PCB/TCB) as dielectric fluids was discontinued in the 1970's due to serious environmental concerns. High fire point (HFP) liquids (also known as "less-flammable" liquids) were then developed and the major application of these was as replacement impregnants in transformers formerly filled with Askarel. The success of this procedure then led to the use of these liquids in new transformers, particularly those intended for use in high fire risk areas.

Their primary application has been in padmount transformers, especially those which are located in, on, or close to buildings. Some of these are equipped with immersed loadbreak switches and therefore expose the insulating liquid to arcing activity during switch operation. This investigation was initiated to determine what effect, if any, this arcing had on the various liquids. This would enable us to gauge their suitability for use in this and other types of switchgear.

While some of the high fire point liquids have been used in switching applications in the past, there was little or no data available on the relative dielectric performance of the three main types of high fire point liquids under these conditions. For this reason, it was decided to conduct a test where each liquid was subjected to uniform arcing conditions in identical switching apparatus.

HIGH FIRE POINT LIQUIDS

By definition , a high fire point fluid must have a minimum fire point of 300° C. The dielectric liquids that currently fit into this category can be divided into three groups based on their chemical structures.

High Molecular Weight Hydrocarbons (HMWH) are chemically similar to conventional, mineral oil-based Transformer Oil. However, they have higher molecular weight components and therefore possess much higher fire points. They have good dielectric and lubricating properties and reasonable heat transfer capabilities. HMWH-based fluids have been used in transformers fitted with load break devices such as load tap changers (LTC).

Silicone fluids or more specifically, Polydimethyl Siloxanes (PDMS) are one of the liquids used as an askarel replacement due to their high fire points. These fluids have acceptable electrical properties and generally their heat transfer properties are adequate. They have not been used extensively for applications involving mechanical switches, principally due to the fluid's poor lubricity [1].

Polyol Esters are synthetic fluids. They have good dielectric and heat transfer properties but their relatively high cost has limited their use to special transformer applications. In general they have good lubricating properties and form the main component in many types of synthetic lubricants.

TEST DESCRIPTION

The evaluation was conducted using an oil switch widely used in capacitor switching applications [2]. The tank which normally houses this switch holds a relatively small quantity of oil and liquid samples were to be taken during the tests. Therefore the volume was increased so that the performance of the switch would not be compromised by a shortage of liquid. In addition, the larger tank minimized volume effects on the test results so that liquid samples could be taken periodically. To this end, a larger, rectangular tank designed to hold 41.6 liters (11.0 gallons) was constructed and the head casting, mechanism and contact assembly from the switch (See Fig. 1) were fitted to it.. The same operating mechanism was used throughout the test but fresh contact assemblies were used for each of the liquids. From Fig. 1 it can be seen that the contact assembly consisted of two fixed contacts which were bridged by a bar-like movable contact. This movable contact was mounted on a pivoting arm which was driven by the unit's spring and motor driven operating mechanism.

Three commercially available HFP liquids were used in the test series, a Polyol Ester [3], HMWH [4] and PDMS [5]. Transformer Oil [6] was also included to provide baseline dielectric data for comparison. purposes. Each fluid was evaluated as supplied from the manufacturer in 55 gallon drums.

Gaskets were provided between the head casting and the lid of the rectangular tank. Pressure tests indicated that slow venting would occur around the operating handle at approximately 34.5 kPa above atmospheric pressure (5 psig). The unit was also fitted with a pressure relief valve rated to operate at 69.0 kPa above atmospheric pressure (10 psig) in the event of an over-pressurization.

TEST PROCEDURE

Prior to testing a liquid, the following parameters were determined: a) Flash and Fire Point [7], b) Dissipation Factor [8], c) Dielectric Breakdown [9] and d) Moisture Content [10]. Moisture content was considered to be of importance because it can affect a liquid's dielectric strength. Liquid samples were collected through two valves that were fitted to the rectangular tank, approximately 7.6 cm (3.0 in) from the tank bottom.

Each liquid was then subjected to 250 switching cycles, where a single cycle involved one open and one close operation. The same electrical circuit was used for each test , which provided 50 A (rms) at 8 kV (rms) with a power factor of zero. This was well within the rated capability of the particular switch used. A programmable logic controller (PLC) was used to control the triggering and timing of the switch. It also acted as a safety device by monitoring the switch operations and being capable of shutting down the test if the switch failed to operate.

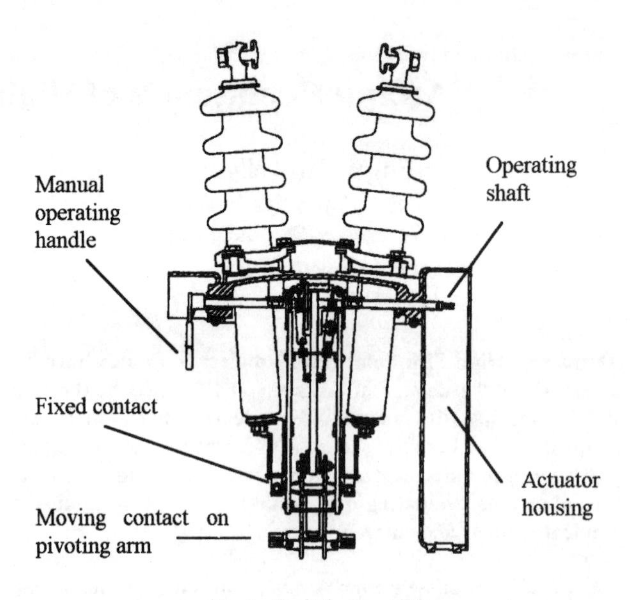

Fig.1 Switch Assembly

During each test the voltage, current and arc voltage were periodically measured and arc energies calculated. These data were used to verify that test conditions remained reasonably consistent for each liquid. Each test was halted after every 50 cycles so that liquid samples could be taken for evaluation. Other samples taken prior to and after the test were subjected to the whole battery of tests described earlier. Intermediate samples only underwent the dielectric breakdown and dissipation factor tests.

At the end of each test, the contacts were examined for signs of undue wear and erosion due to arcing. The contact weights were compared with weights measured prior to the test. The contact assembly was then replaced and the entire apparatus thoroughly cleaned and rinsed prior to filling with another liquid.

RESULTS

Initial Comparisons

As mentioned previously, the primary goal of this study was to compare the performance of these HFP liquids under uniform arcing conditions. Arcing times and voltages were recorded during every fifth cycle and Table I summarizes the results. It can be seen that the arcing times and calculated arc energies for all the HFP liquids are quite similar. It should also be noted that for each liquid there were sizable differences in recorded values on successive cycles, as evidenced by the standard deviations (Std. Dev.). However, it is clear that on average all the HFP liquids experienced similar arcing conditions. Examination of the switch contacts indicated that there was very little arc erosion of the contacts in any of the liquids tested. In addition, the contact weights changed very little throughout the test and they followed essentially the same trend in each liquid.

The arcing data for Transformer Oil indicate that it typically cleared quicker than the HFP liquids. This can be attributed to the fact that Transformer Oil has a lower viscosity and therefore offers less resistance to the movement of the switch. HFP liquids are inherently more viscous.

Table 2 lists the dielectric breakdown results from the liquid samples taken during the test series and it illustrates the degrading effect of arcing. While all the liquids experienced a progressive decrease in dielectric breakdown voltage, there was an alarming drop in the PDMS's performance after the first 50 switching cycles and later samples confirmed that the liquid did not recover. The instrument used to measure the breakdown value has a lower limit of 4 kV and it is reasonable to assume that the values for PDMS continued to drop below that level. The other fluids did not decay rapidly and their breakdown values after 250 switching cycles were 25 kV or greater. The Polyol Ester maintained the highest breakdown value while the values and trend observed for both the transformer oil and HMWH were similar.

Dissipation factor values increased in all cases as the liquids became contaminated with arc by-products, including carbon. Some of the readings from later samples were erratic as arcs would develop within the test cell of the measuring instrument. This phenomenon is noted on the later readings from the HMWH and it also precluded any readings being taken for the later PDMS samples. After 250 switching cycles, the HMWH had the most favorable dissipation factor followed by the Polyol Ester, Transformer Oil and finally the PDMS.

Table 1 Arcing Time and Arc Energy

Liquid	Arcing Time (ms)		Arc Energy (kJ)	
	Average	Std. Dev.	Average	Std. Dev.
Transformer Oil	13.8	4.6	0.901	0.366
HMWH	19.9	5.2	1.317	0.448
Polyol Ester	18.1	3.8	1.113	0.349
PDMS	18.2	4.7	1.009	0.289

Table 2 Dielectric Properties

Dielectric Breakdown Voltage (kV)

Number of operations	Mineral Oil	HMWH	Polyol Ester	PDMS
0	52	49	47	45
50	41	43	36	4*
100	38	35	40	4*
150	33	30	38	4*
200	32	30	43	4*
250	30	25	40	4*

* 4 kV is the minimum measuring limit of the instrument

Dissipation Factor

Number of operations	Transformer Oil	HMWH	Polyol Ester	PDMS
0	<0.0001	<0.0001	0.0003	<0.0001
50	0.0014	<0.0001	0.0005	0.0001
100	0.0190	0.0002	0.0006	arcing
150	0.0373	0.0200	0.0013	arcing
200	0.0500	0.0014	0.0029	arcing
250	0.0540	0.0100	0.0189	arcing

One factor which can adversely affect dielectric strength is a high moisture content and these values were examined to determine if the poor performance of the PDMS fluid was due to moisture ingress. The initial values for all liquids were low, most being less than 25 ppm. The one exception was the Polyol Ester, which had a value of 97 ppm. However, with the exception of PDMS, esters have much higher moisture saturation levels than the other liquids discussed, so this value is still considered to be low. Evaluation of samples of each liquid taken at the end of the test indicated that there was no significant change in moisture content. Therefore, it was concluded that this was not a factor in the PDMS fluid's rapid decay in performance.

The final parameters that were examined were the flash and fire points of the various liquids. These all fell within the ranges specified by the manufacturers and all the HFP liquids had fire points above 300°C. The values recorded after the switching test were almost identical to the initial values and there was no evidence of degradation of this parameter in any of the fluids.

Further Investigation of the PDMS Fluid

The PDMS fluid had degraded dielectrically prior to the first test point at 50 switching cycles. Therefore, it was decided to re-test the fluid in order to better profile the decay in dielectric breakdown voltage. This test involved 25 switching cycles with fluid samples being collected after every fifth cycle. This revealed that the breakdown voltage dropped below the instrumentation limit after 25 cycles.

In an effort to identify the cause for this phenomenon, the degraded PDMS fluid was passed through a series of very fine filters. The filtered fluid was found to have a breakdown voltage that was comparable to the fluid in the virgin state. This indicates that the particulate arc by-products formed in the fluid adversely affect its dielectric performance. Since every fluid was liberally loaded with carbon by the end of the test, it was assumed that the poor performance of the PDMS fluid had to be due to some other arcing by-product.

One possible explanation is that the sizable quantity of particulate by-product formed in the PDMS fluid has a dielectric constant which is much greater than that of the base liquid. To check this, some of this material was dried and compressed to form solid pellets. These were found to have a dielectric constant of 223 as opposed to the value of 2.71 that is quoted for the liquid itself. Computer models of this scenario confirmed that the electrical stress on the liquid is greatly intensified by the presence of such particulates.

For comparison purposes, the dielectric breakdown performance of each of the tested liquids is shown in Fig.2. There it can be seen that the rate of degradation in the PDMS fluid is quite different to that exhibited by the other liquids.

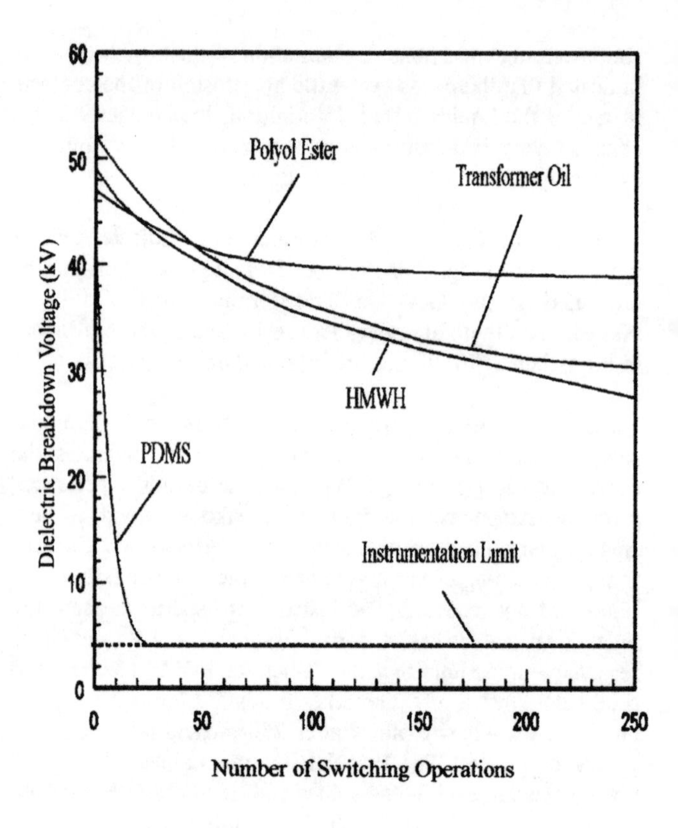

Fig.2 Dielectric Breakdown vs Number of Switching Cycles

CONCLUSIONS

It has been demonstrated that under uniform arcing conditions there are significant differences in the arcing resistance of the three most common types of high fire point dielectric liquids. The results show that both Polyol Ester and High Molecular Weight Hydrocarbons compare favorably to conventional Transformer Oil in this application. However, the dielectric breakdown value of the PDMS fluid was so rapidly degraded that extreme caution is recommended when using it under these conditions.

REFERENCES

1. Dow Corning® 561 Silicone Transformer Liquid User Manual, Dow Corning® Corporation, 1991, pp. 6.1 - 6.3.

2. Kyle® NR Switch, Cooper Power Systems, South Milwaukee, Wisconsin.

3. Envirotemp® 200 Fluid, Cooper Power Systems, Waukesha, Wisconsin.

4. R-Temp® Fluid, Cooper Power Systems, Waukesha, Wisconsin.

5. Dow Corning® 561 Silicone Transformer Liquid, Dow Corning® Corporation, Midland, Michigan.

6. Univolt® 61, Exxon Company, U.S.A., Houston, Texas.

The following references are all contained in the 1994 Annual Book of ASTM Standards, Section 10, Volume 10.03:

7. ASTM D-92, "Standard Test Method for Flash and Fire Points by Cleveland Open Cup", pp. 7 - 11.

8. ASTM D-924, "Standard Test Method for Dissipation Factor (or Power Factor) and Relative Permittivity (Dielectric Constant) of Electrical Insulating Liquids", pp. 96 - 102.

9. ASTM D-877, "Standard Test Method for Dielectric Breakdown Voltage of Insulating Liquids Using Disk Electrodes", pp. 75 - 78.

10. ASTM D-1533, "Standard Test Methods for Water in Insulating Liquids (Karl Fischer Reaction Method)", pp. 174-180.

Conference Record of the 1996 IEEE International Symposium on Electrical Insulation, Montreal, Quebec, Canada, June 16-19, 1996

A STUDY ON THE BREAKDOWN CHARACTERISTICS OF THE FLUOROCARBON

Chang-Su Huh
Dep. of Electrical Engineering
Inha Uni.,Inchon,Yonghyundong
253 Zip Code 402-751,South Korea

Jae-Bok Lee
Korea Electrotechnology Research Institute
Seongjudong 28-1 ,Changwon , Korea

Abstract **: In this paper, we investigated physical properties and electrical characteristics of the fluorocarbon that used as coolants for large power gas-insulated transformer. Volume resistivity of the fluorocarbon is $\rho = 1.87\times10^{15}$ $[\Omega\cdot cm]$ at 1 atm, 27℃. Dielectric constant is 1.86 and decreases as temperature increase. The breakdown voltage at 1 atm is higher than that of transformer oil. The breakdown voltage of fluorocarbon vapor is about 18 kV when pressure in a test chamber increases over 1 kg/cm^2. When fluorocarbon is mixed with SF_6 gas, breakdown voltage of the mixed is higher than that of fluorocarbon. Then fluorocarbon leads to increase over 4 kg/cm^2 in pressure as temperature increase. Therefore, when a gas-insulated transformer is manufactured, the design must be taken into consideration a high-pressure .**

1. Introduction

These day SF_6 gas transformer is suitable to compare with the trend of power substation system which is changed to total gas insulation. But in large power transformer design area , though there is no limit in electrical insulation class with increas of gas preassure but design of power transformer is limmited by cooling machanism when use only SF_6 gas as a insulator and coolant. So hybirid insulating transformer which use SF_6 gas as a insulating material and FC(fluorocarbon) as a seperated cooling material recommended. So we can increase the capacity of transformer many times. FC is injected by nozzle to the coil and core, then evaporated with heat and heat move out by cooling fan in circulation of FC[1,2,3,4]. This mathod is widely studied in many countries but not in our country. So we investigated the AC breakdown voltage characteristics and dielectric constant, volume resistivity of FC and also preassure increase and variation of insulating charateristics were investigated with the temperature variation to be used on design of large power transformer.

2. Experiment

2.1 Experiment of breakdown characteristics of FC

Breakdown test is one of the fundamental and direct method to find a electrical insulating characteristics of material and this method could be classified according to applied voltage kinds. we studied on the AC breakdown characteristics of FC instead of phenomena on anode or cathod which is well known to use as a basic design parameter of hibrid cooling transfermer . Friquency of test voltage is 60 Hz and 50 KV, 500 VA test transformer used Needle-plain electrode and sphere-sphere (12.5mm) electrode is used and the gap was regulated by micrometer in test chamber. We circulated the liquid for uniformity after one by one. Also we suppressed the moving not to take place the variation of breakdown characteristis by flowing of liquid on near of electrode. The breakdown characterristics of FC + SF_6 gas was investigated in preassure test chamber with 12.5 mm sphere - sphere electrode in various gap length which can be regulated by micrometer in outside.1500cc of FC is filled following of vacuum the chamber and SF_6 gas was charged to 1.2 Kg/cm^2 which is considered to restrict maximum pressure increase with temperature rise for not to recive the regulation of high preassure vessel law .

2.2 Measurement of dielectric characteristics of FC

Dielectric constant was meassured with shering bridge ,test cell was cleaned with aceton and distilled water and dried before infilled the test liquid. It's amount is about 40 cc. The dependence of dielectric constant on tempetature was studied on 60 Hz, 1000V with increase the temperature from 25 ℃. Also variation dielectric constant was studied with changing applied voltage on step 250 V

2.2 Measurement of volume resistivity of FC

The volume resistivity was meassured by standard method after one minite with applied voltage 250V.Because volume resistivity was effected by temperature and humidity this parameter was controlled continuously.

3. Experimental result and discussion

3.1 Breakdown chracteristics of FC

The result of Breakdown chracteristics of FC was followed in Fig.1. Breakdown voltage was increase with the electrode gap until 3 mm in needle-plain electrode configure.But we can see in the Fig that the increase rate of breakdown voltage with the gap length is saturated sharply in some elctrode gap range .And then if the gap length is increase continuously, the breakdown voltage is increase again .This mean that electron emission could take place in near the needle elcectrode by high electric field,and it could be thought as a extension of the electrode gap. So in some high electric field range, increase of breakdown voltage is not viewed. But finally the gap breakdowned by high electric field because this effect could not take place continuously in very high voltage [9]. Altough it is prescribed in the rule of the association that the breakdown voltage of new minaral oil is upper 30KV in 2.5 mm gap length at sphere-sphere electrode. But finely purified new oil have higher breakdown voltage about upper 50 KV was published oftenly. we can see the AC breakdown voltage of FC is higher than the minaral oil and breakdown voltage is increase with the gap length in fig 1.

3.2 AC Breakdown chracteristics of FC with SF_6 gas

FC + SF_6 gas mixtured vapor is used in real power transfermer. FC evaporated little in room temperature even though it's boiling point is 97 ℃.Preasure of tank is highly increased with temperature rise when the FC + SF6 gas mixture used as a insulator and coolant.So we studied·

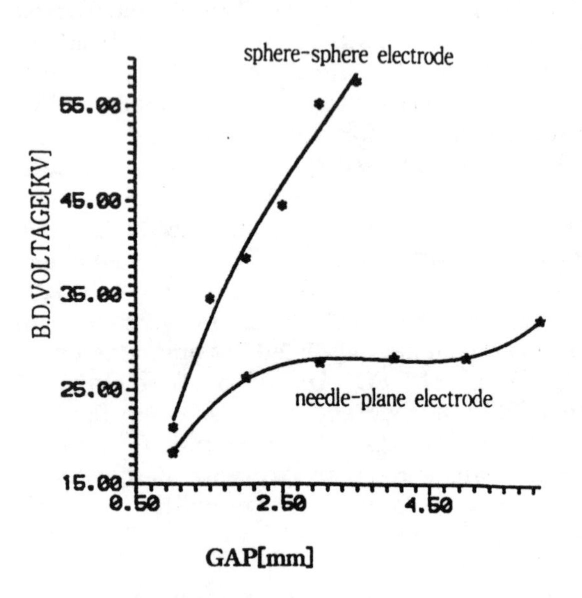

Fig.1 AC breakdown voltage characteristics of FC

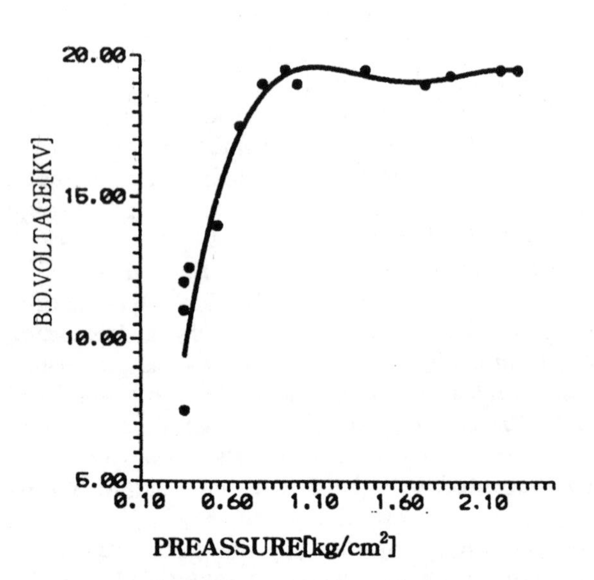

Fig.2 Breakdown voltage of the FC vapor (at sphere electrode ,1.25 mm gap)

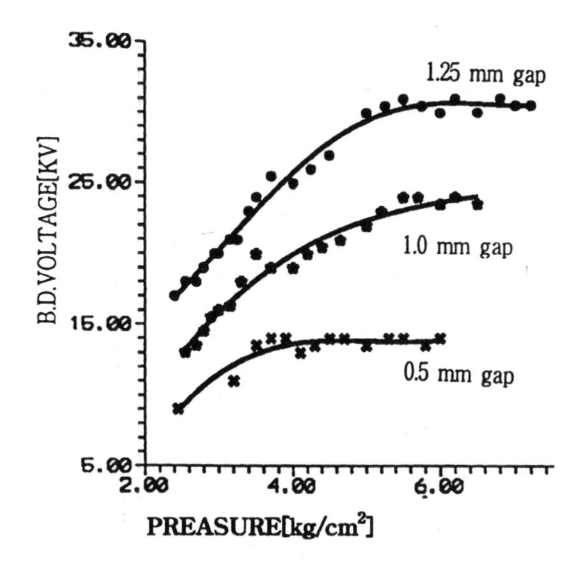

Fig.3 Breakdown voltage of the mixtures of FC vapor with SF6 gas (at sphere electrode)

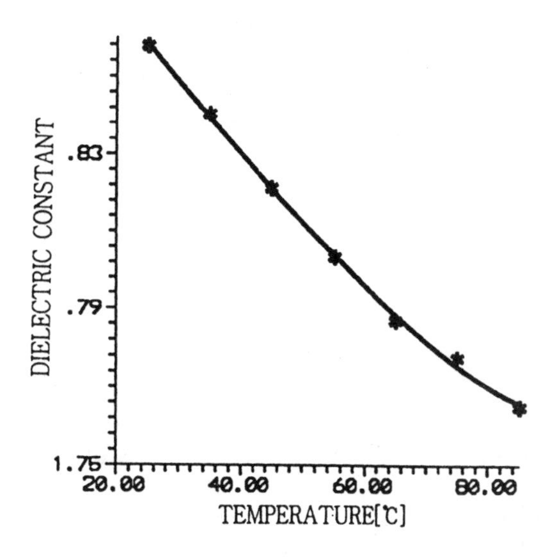

Fig.4 Dependence on temperature of dielectric constant

about the variation of preassure in tank and breakdown voltage with temperature rise. Results are fullowing and it was used as a design factor when we make a model gas transfermer [10].

Breakdown voltage of the FC vapor is increase with the temperature by the preassure rise by evaporation of the FC. This phnomina is saturated at about 0.8 atm . Because there is no increase of density in gas by saturated vapor preassure of FC in closed tank, Also increase of breakdown voltage of the mixtures of FC vapor with SF6 gas (at sphere electrode) is saturated It's result is followed in Fig. 4.

3.3 Measurement of dielectric constant of FC

The capacitance of standard cell of bridge is Co =56.4762 (pF),and after filled the cell with the FC ,capacitance was meassured and caculated by ε_r=Co/C.The characteristics of dielectric constant with the variation of temperature are figured in Fig.5. We know the dielectric constant is in inverse proportion to the temperature because the capacitance connected to the dissipation factor. So to speak,when temperature is increase, orientation of all molecular in applied electric field is disturbed by thermal agitation. So dielctric constant is decrease. Also we found dielectric constant is not much changed with voltage variation .

3.4 Measurement of volume resistivity

Volume resistivity is meassured by ratio of applied voltage by current per unit area after one minute after voltage applied to dismissed the absorption current effect[10].Current is changed with time by many effect. Usally in liquid this effect could be thought as a dipole orientation and charge carrier density variation in the gap and space charge formation in near electrode.But we can find that there is no dipol orientation in high electrical resistivity materiral like FC, by the fact that the rverse current is not viewed when the applied voltage is cut off shortly . Also it is very difficult to thought that there is a some electricric field variarion by charge formation in near electrode.Volume resistivity FC could have been caculated by ρ=3.6 πCRx in our standard test cell. Rx is resist measured in circuit and 3.6πC,637 is electrode constant of test cell in our system.ρ is 1.87exp15 at room temperature.Volume resistivity is decrease with increase of temperature and have high value in increased voltage.

4. Result

In this paper we investigated physical property and electrical characteristics of the fluorocarbon to used in hibrid cooling power transformer,and results are followed.

1.volume resistivity of the FC is 1.87exp15 at room temperature.Volume resistivity is decrease with rise of temperature and have high value in increased voltage.And at same temperature dielectric constant is 1.86.

2.AC breakdown voltage FC is 30 KV in the gap length,1.25mm and 18 KV at vapor of FC in 1 atm.Vessel presssure is increase over 2 kg/cm^2 in operating temperature of transformer.

3.Breakdown voltage of the mixtures of FC vapor with SF6 gas is superior then the liquid FC or FC vaper those data used as a design factor when we design the power transfermer and have a good results.

This work was supported by Electrical Engineering and Science Research Center by fund of KEPCO in 1994.

Reference

1.R.H Hollister " Gas vapor and fire resistant transformers" CH 1510-7/79, pp. 239-242, 1979

2.John R .Mopies, Sanborn F. Philip "A new concept for a compressed gas insulated power transformer" 7th IEEE PES Trans And Distri. Conference and exposition, April 6, pp 176-183, 1979

3.K.Goto, "Development of insulation technology for high-voltage gas insulated transformer "IEEE Trans.on PD, Vol 4,No.2, pp 1096-2004, 1989

4.K.Tokoro,"Development of 77KV 40 MVA gas-vapor cooled transformer" IEEE Trans. on PAS,Vol PAS-101,No11,pp 4341-4349, November 1982

5.Y.Harumoto et al, "Development of 275KV EHV class gas-insulated power transformer" IEEE Trans on PAS,Vol PAS-104, No 9, pp 2501-2508 september 1985

6.K.Goto, T.Yamazaki "Studieson structural integrity for seperate cooling/ sheet-winding gas insulated transformer" IEEE Trans.on PD, Vol 4, No 2, pp 1079-1085 ,April 1989

7. J I S C 2101-1982, pp 37-43

8.T.Takagi et al, "Basic insulation characteristics of perfluorocarbon for large power transformers"IEEE Trans. on PD, Vol 3, No 4, pp 1809-1815,October 1988

9.A.H. Sharbaugh et al, "Progress in the field of electric breakdown in dielectric liquids" IEEE Trans,Electr. insul, Vol EI-13, No 4, pp 249-276, August 1978.

10.A.H.Sharbaugh,P.K.Watson."Breakdown strengths of a perfluoro carbon vapor (FC-75) and mixtures of the vapor with SF_6" IEEE Trans. on PAS,Vol PAS-83, No2, pp 131-136, February 1964

Conference Record of the 1996 IEEE International Symposium on Electrical Insulation, Montreal, Quebec, Canada, June 16-19, 1996

SYNTHESIS AND EVALUATION OF OLEIC ACID ESTERS AS DIELECTRIC LIQUIDS

P.Thomas, S.Sridhar, K.R.Krishnaswamy

Materials Technology Division
Central Research & Testing Laboratory
Central Power Research Institute
Bangalore, India-560 094

Abstract : The worldwide depletion of petroleum crude and the search for technologically and ecologically alternative to PCB's has led to the development of synthetic liquid dielectric which are non-flammable, non-toxic and bio-degradable. Esters like Methyl oleate and Pentaerthritol tetra oleate were synthesised in the laboratory. These esters were evaluated for its Physical, Chemical, Electrical and Ageing properties and the values obtained compare well with those of commercially available liquids like Midel 7131 and RTemp fluid. Further accelerated ageing studies were performed on these esters and the data obtained reveal that the esters prepared in the laboratory has very good chemical stability towards oxidation without inhibitor when compared to Midel 7131 and RTemp fluid which are having the phenolic inhibitors.

INTRODUCTION

Mineral oil is commonly used liquid dielectric for transformer application. Mineral oils are used since 1897, but the flammability aspects were not over looked in the case of certain special applications. It is evident that all hydrocarbon were fulfilling all the properties, except flammability. Eventhough , PCB having excellent chemical stability and high fire resistance, it has been concluded that they were not friendly towards environment. The greater R&D activities has so far been concentrated upon the provision of non-halogenated, non-toxic liquids to replace PCB's to improve the safety of electrical equipments. (1)

Polyol esters fluids have been used as a substitute for PCB's in dielectric applications in the United states, Europe and in the Far east . These oils have excellent electrical properties in addition to fire-resistance characteristics and are non-toxic and bio-degradable.(2). Therefore work has been taken to develop certain esters with the intention of evaluating their application as a replacement for PCB's in special application like indoor transformer and traction transformers.

This report highlights the preparation of certain Oleic acid esters like Methyl oleate and Pentaerthritol tetra oleate and comparing of these with other various insulating liquids.These esters were studied for Physical, chemical and Electrical properties. The Pentaerthriol tetra oleate has very high viscosity and water solubility. These properties are given in the Table.1. It is found that these properties are comparable with other synthetic liquids except resistivity which is lesser compared to the values reported for Midel 7131 an RTemp fluid. The esters are very dependent on the final manufacturing treatment. Accelerated ageing studies were performed on these esters indicates that ester viz. Pentaerthritol tetra oleate and Methyl oleate were having good stability towards oxidation without the addition of oxidation inhibitor. The esters developed have high flash point and is anticipated to replace PCB.

EXPERIMENTAL

Materials : Pentaerthritol (nitration grade), Oleic Acid (Lopour grade), Thionyl chloride, Methanol, Xylene(sulphur free), Sulphuric acid, Triethylamine, Sodium carbonate, Aluminium oxide,Korvi earth, Methanol (spectra grade).

Preparation of Esters:

1. Methyl Oleate: The oleic acid and methanol were taken in 1:10 ratio and refluxed using acid catalyst. The reaction is carried out at 60-70 deg.C for 10 hrs. Then the excess methanol distilled off and the resulting product is washed with water and sodium carbonate. Then the product purified by passing thro' the mixed bed of Korvi earth, Alumina and Mol.sieve.

2. Pentaerthritol tetra Oleate : The first step involved in the preparation of ester is that the acid chloride of oleic acid prepared by refluxing it with thionyl chloride. Then the acid chloride is distilled off and made to react with the monopentaerthritol at the rate of 10 ml per minute using triethylamine as base . The crude products washed well with water and 2% sodium carbonate solution. The product is purified using mixed bed at elevated temperature.

RESULTS AND DISCUSSION

The esters prepared viz. Pentaerthritol tetra oleate and Methyl oleate were hot filtered thro' G4 sintered crucible to avoid contamination and moisture present. The physical properties like Viscosity, Pour point, Flash point, Density and chemical properties like Neutralisation value , Corrosive Sulphur , presence of Oxidation inhibitor and Electrical properties like Electric strength, Tan delta, Resistivity and Dielectric constant were evaluated as per IS:335-1993 "Specification for New Insulating oils" Bureau of Indian Standards. These parameters were compared with other commercially available liquids like Midel 7131 and RTemp fluid and given in the Table 1. Ageing studies performed on these esters shows that these Ester are comparable with those of Midel 7131 and RTemp fluid.

Heat Transfer Properties:

The physical properties like Density , Viscosity, Flash point and Pour point are the properties which determines the heat transfer properties of equipments and its performance under low temperature condition. It is noted that the viscosity of these liquids at operating temperatures are comparable to the conventional oils. The viscosity-temperature relationship is given in the figure below. The further depression of pour point can be best achieved by addition of dodecyl benzene, a single ring aromatic hydrocarbon with a C12 side chain. The flash point values obtained indicates that they can be used for higher temperature application.(4)

Chemical Stability:

Stability is the most important chemical property of the ester, although their water absorption properties are also important because these affect stability. These esters were subjected to oxidation stability in accordance with IS:335-1993, Appendix B.

The neutralisation number is the measure of the concentration of acidic components of the fluid. The concentration of acid obtained are only marginal for Pentaerthritol ester and methyl ester when compared to Midel 7131 and RTemp fluid which is 0.01 and 0.014 respectively. No sediment or precipitable sludge observed for any of these liquids. The values are given in the table 2(1). This shows that these two esters viz Methyl Oleate and Pentaerthritol tetra oleate are chemically stable without the addition of oxidation inhibitor. The ester linkage is exceptionally stable one, bond energies determinations predict that the ester linkage is more stable than the C-C bond.(4) The chemical stability in case of Pentaerthritol may be due to the β-carbon atom, thus preventing the degradation via ring formation. It was observed that the chemical stability can be enhanced by the addition of oxidation inhibitors.(5)

Electrical and Ageing Properties :

Electrical properties like Electric strength , Tan delta and Resistivity were compared with Midel 7131 and RTemp fluid before it is subjected to ageing test. All the electrical properties except Resistivity are comparable with Midel 7131 and RTemp fluid. These values are given in Table 1. Ageing studies were performed on these ester as per ASTM D-1934. The Values obtained were compared with Midel 7131 and RTemp fluid .

Table.1 Comparison of the Dielectric Properties of Esters with that of other Liquid Dielectrics.

Sl. No	Dielectric Properties	PETO	Methyl Oleate	RTemp Fluid	Midel 7131
1.	Density, @29.5C	0.91	0.875	0.87	0.95
2.	Kinematic Viscosity cst @ 27 C	128	9	253	53
3.	Flash Point, C	255	188	253	233
4.	Pour Point, C	-12	-6	-18	-20
5.	Acidity, mg KOH/gm	NIL	NIL	NIL	NIL
6.	Electric Strength	55	68	65	75
7.	Power Factor, DDF @ 90 C	0.0020	0.0021	0.00025	0.0023
8.	Dielectric Constant @ 27 C	3.1	3.2	2.2	3.2
9.	Spec. Resistance x10E-12, ohm-cm	25	25	40	30
10.	Oxidation Inhibitor	NIL	NIL	Present	Present
11	Corrosive Sulphur	Non-Corrosive	Non-Corrosive	Non-Corrosive	Non-Corrosive

Table.2 Results of accelerated ageing studies by Glass tube method and by open beaker method .

Sl. No	Dielectric Properties	PETO	Methyl Oleate	RTemp Fluid	Midel 7131
2(1)	Oxidation stability (Glass tube)				
a.	Neutralization Value after oxidation, mg KOH/gm	0.001	0.001	0.014	0.010
b.	Total Sludge, % by weight	Nil	Nil	Nil	Nil
2(2)	Accelerated ageing (open beaker)				
a.	Power Factor, DDF @ 90 C	0.01	0.01	0.05	0.064
b.	Dielectric Strength	35	35	52	69
c.	Spec. Resistance,@90C x10E-12, ohm-cm	2.5	2.6	0.37	7
d.	Total Acidity mg KOH/gm	0.022	0.022	0.045	Nil
e.	Total sludge, % by weight	Nil	Nil	Nil	Nil

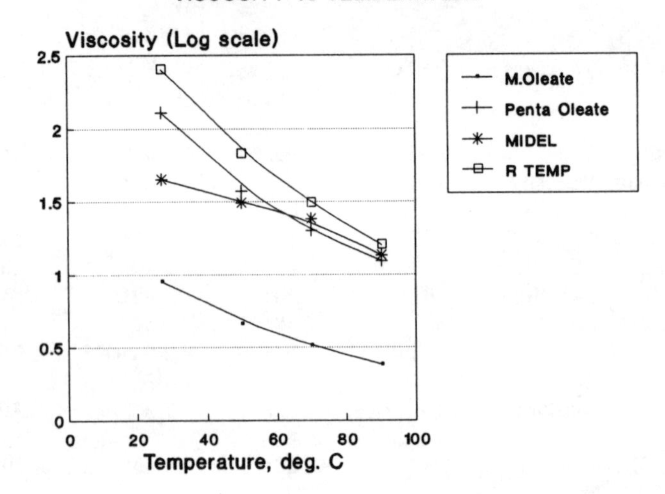

and given in Table 2(2). The dielectric dissipation factor and volume resistivity after ageing test found that the values are better than the RTemp fluid but poorer than the Midel 7131 which is also a pentaerthritol ester of vegetable origin. The marginal decrease in the Electric strength of the aged ester was observed . The electric strength of an any aged fluid is the function of the concentration of moisture, particulate matter and product of oxidation of the esters. These factors alone or in combination will reduce the resistance to Breakdown strength of oil . As anticipated , the particulate in the aged fluid is only a trace quantities.

CONCLUSION

The physical , chemical and electrical properties are comparable with the other commercially available liquids like Midel 7131 and RTemp fluid.
Ageing parameters indicates that the esters are chemically stable without the oxidation inhibitor where as Midel 7131 and RTemp fluid are containing phenolic inhibitor.

ACKNOWLEDGMENT

The authors are thankful to the CPRI management for according permission to publish this paper

REFERENCES :

1. K.W. Plessner and E.H.Reynolds, "Novel single and mixed liquids for high voltage dielectrics ", CIGRE , 15- 07,p 1-12, 1976.

2. David sundin,"The service history of Ester based fluids for railway transformers ", IEEE. International conference on Electrical Insulation, p31-34, Toronto, Canada, June 3-6, 1990

3. P.L. Gervason, "Survey of flammability methods applied to New liquid dielectric-related by-products, International conference on Dielectric Materials , Measurement and application, September 1984

4. Atkins,D.C., Baker,H.R., Murphy,C.M and Zisman,W.A., Industrial Eng.Chem,39,491(1947)

5. P.R.Krishnamoorthy, S.VijayaKumari, K.R.Krishnaswamy and P.Thomas, "Effect of BTA and DBPC on the accelerated oxidation of New and Reclaimed Tr.oil- A comparative study", IEEE 3rd International conference on properties and Dielectric Materials, p 732-735, July 8-12, 1991, Japan.

6. F.M.Clark, Insulating Materials for Design and Practice Miscellaneous Insulating Liquids, Pare 211-248.

7. A.C.M.Wilson, Insulating liquids, their use, manufacture and properties, Peter peregrinus Ltd, UK & NY, 1980 IEE,London

Conference Record of the 1996 IEEE International Symposium on Electrical Insulation, Montreal, Quebec, Canada, June 16-19, 1996

Influence of Co-field and Cross field Flow of Mineral Oil with and without Additives on Conduction Current and Breakdown Voltage in Highly Nonuniform Fields.

I.Y. Megahed
Electrical Engineering Dept.
University of Alexandria
Egypt

A.A. Zaky
Arab Academy for Science
and Technology, Alexandria
Egypt.

M.A. Abdallah
Electrical Engineering Dept.
University of Alexandria
Egypt

Abstract: The paper presents the results of the effect of enforced co-field and cross-field oil flow on the conduction current and breakdown voltage of degassed oil, oil saturated with O_2 and with SF6 and oil containing 1-methylnaphthalene (MN) and dimethylaniline (DMA) as additives. Direct voltage and a point-to-plane electrode geometry were used and results were obtained for both polarities of the point electrode . A general conclusion from all experiments is that oil flow, whether co-field or cross-field, raises the breakdown voltage and lowers the conduction current. The results also show that with the exception of DMA, all additives both gaseous (O_2 and SF6) or solid (MN) raised the breakdown voltage and reduced conduction current, compared with degassed oil, for both polarities of the point electrode. These effects are attributed to the electron-trapping properties of the additives.

INTRODUCTION

The effect of enforced motion on conduction current and breakdown in insulating liquids is of practical importance in all applications in which fluids are used as insulants and coolants which circulate either by natural or by forced convection. However, work in this area has been rather limited [1-3], and the results reported are an extension of those previously reported by the authors on the effect of flow on the breakdown voltage and conduction current in mineral oil [4].

Enforced motion, as distinct from self or electrohydrodynamic motion (EHD), is produced by external mechanical means and can be either parallel to the direction of the field (co-field) or perpendicular to it (cross-field). The effect of both modes of motion on conduction current and breakdown voltage in transformer oil has been investigated using direct voltages and point-to-plane electrode geometry. Tests were carried out on degassed oil , O_2-saturated oil , SF6-saturated oil and oil containing 1-methylnaphthalene (MN) and dimethylaniline (DMA) as additives.

O_2 and SF6 have large electron-trapping cross sections and high ionization potentials (O_2 : 12.1 ev , SF6 : 10-15 ev), MN has a large electron-trapping cross section but a low ionization potential (7.96 ev) DMA does not trap electrons and has a low ionization potential (7.14 ev). These additives were chosen because their effects on the propagation velocities of both positive and negative streamers have been reported by several authors [5,6] and it is interesting to determine whether there is any correlation between breakdown voltage and streamer velocity.

EXPERIMENTAL

All tests were carried out on transformer oil (Shell Diala B) complying with BS 148:1959. The liquid samples were filtered through a 0.3 μm Millipore filter and degassed at a pressure of 0.1 Pa for 10h. For experiments carried out on liquid samples saturated with oxygen or SF6 dried gas was allowed to bubble slowly through the degassed oil which then remained in contact with the gas at atmospheric pressure for 16 h. The concentration of the polyaromatic additives used were 0.1 Molar for 1-methylnaphthalene (MN) and 1.0 Molar for dimethylaniline (DMA).

In the vertical electrode arrangement the point electrode was a nickel-plated sewing needle of tip radius 25 microns held along the central axis of the cylindrical oil flow duct by collect-type spring jaws. The profiled Ni-Cr plane electrode had a diameter of 2 cm. Direct voltage and a diverter circuit were used in all tests and the live electrode was always the point. Breakdown and conduction measurements were carried out for gaps of 200, 400, 700 and 900 μm . For measurements under flow conditions the pressure applied varied between 1.7 and 17 kPa. For breakdown voltage measurements each gap was stress conditioned for two hours at 60% of the expected breakdown voltage. The breakdown voltage was taken as the average of ten breakdowns. New electrodes and a fresh liquid sample were used for each gap setting.

For conduction current measurements a new sample was admitted into the cell and a voltage corresponding to 60% of the breakdown voltage measured previously for the same polarity and gap length was applied for 30 minutes (conditioning period) before starting the current measurements. The current was measured by means of a Keithley 617 programmable electrometer.

RESULTS

a) *Breakdown*

Figure 1 shows a set of typical breakdown voltage/gap length characteristics for the different additives used under co-field flow for both polarities of the point electrode. For both point polarities SF6 was the most effective with the breakdown voltage about 6 kV above that of degassed oil. Oxygen and 1-methylnaphthalene raise the breakdown voltage but to lesser extent. Dimethylaniline has practically no effect. In all cases the breakdown voltages with the point negative are significantly higher than those with the point positive.

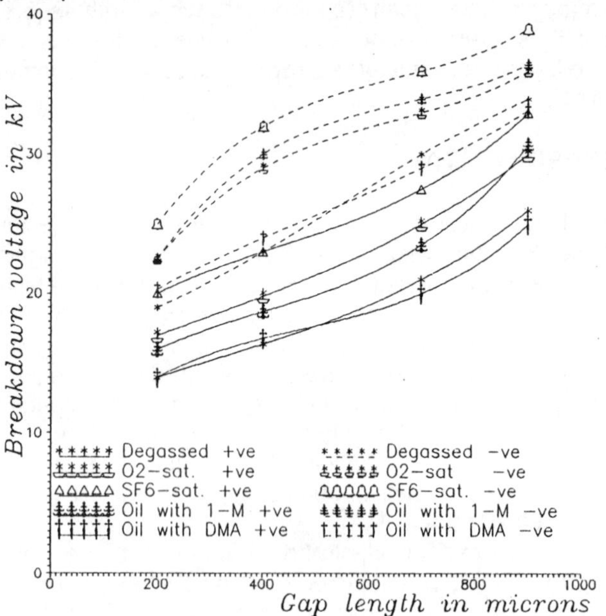

Fig.1 Breakdown Voltage/gap length characteristics for oil with different additives and co-field velocity 52cm/s. full lines positive point,broken lines negative point

Figure 2 shows the same characteristics as Fig.1 except that in this case the flow is cross-field. The relative effect of the various additives remains the same as for co-field flow.

Figures 3 and 4 show the effect which the direction of flow has on the breakdown voltage for both point polarities. It is evident that flow direction has a significant effect only when the point is positive. In this case the increase in breakdown voltage is much greater with cross-field flow than with co-field flow. With the point negative both flow directions produce a similar increase in the breakdown voltage. Similar results were obtained with degassed oil and with oil containing the other additives.

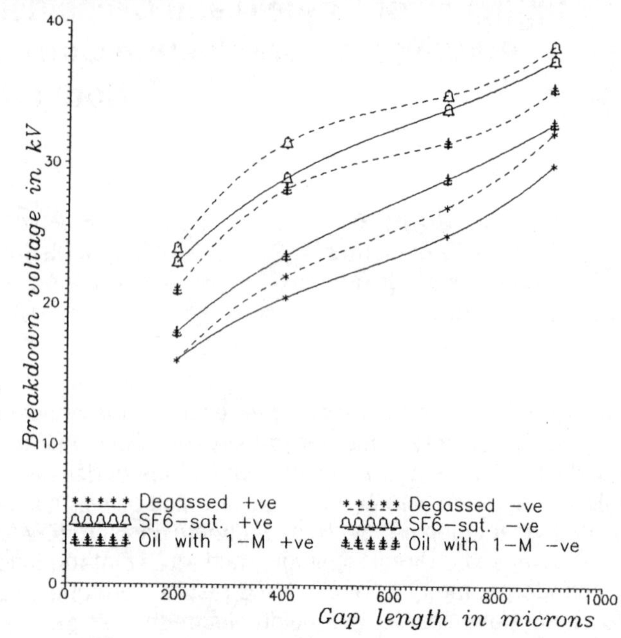

Fig.2 Breakdown Voltage/gap length characteristics for oil with different additives and cross-field flow velocity 52cm/s.full lines positive point,broken lines negative point

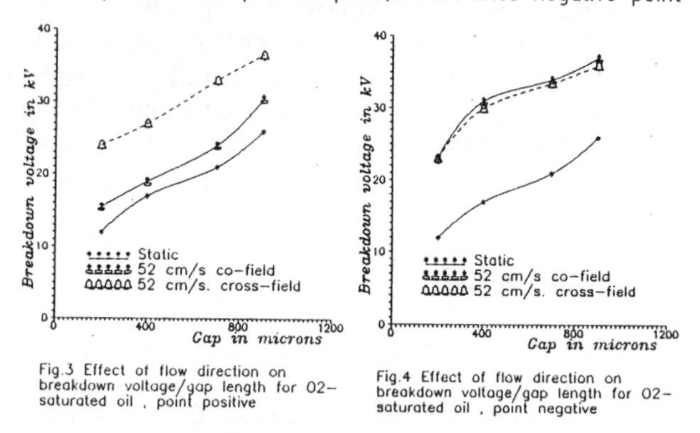

Fig.3 Effect of flow direction on breakdown voltage/gap length for O2-saturated oil , point positive

Fig.4 Effect of flow direction on breakdown voltage/gap length for O2-saturated oil , point negative

Figures 5 and 6 and 7 and 8 show the effect of flow direction on the breakdown voltage/gap length characteristics for SF6-saturated oil for different flow rates and for both point polarities. Once again a difference in the characteristics obtained with the two flow modes is apparent only when the point is positive. With this point polarity and co-field flow the breakdown voltage increases very little with increasing flow rate. However, with cross-field flow the breakdown voltage increases significantly with increasing flow rate.

It should be mentioned that characteristics similar to the ones shown here for SF6-saturated oil , were obtained for O_2-saturated oil and oil with (MN) as additive. All three additives produced differences in the breakdown voltage/gap length characteristics under co-field and cross-field flow for the point positive as described above (Figs. 5 and 7). With DMA as additive, however, the characteristics were similar to those obtained with degassed oil with the breakdown voltage increasing with increasing flow rate for both co-field and cross-field flow.

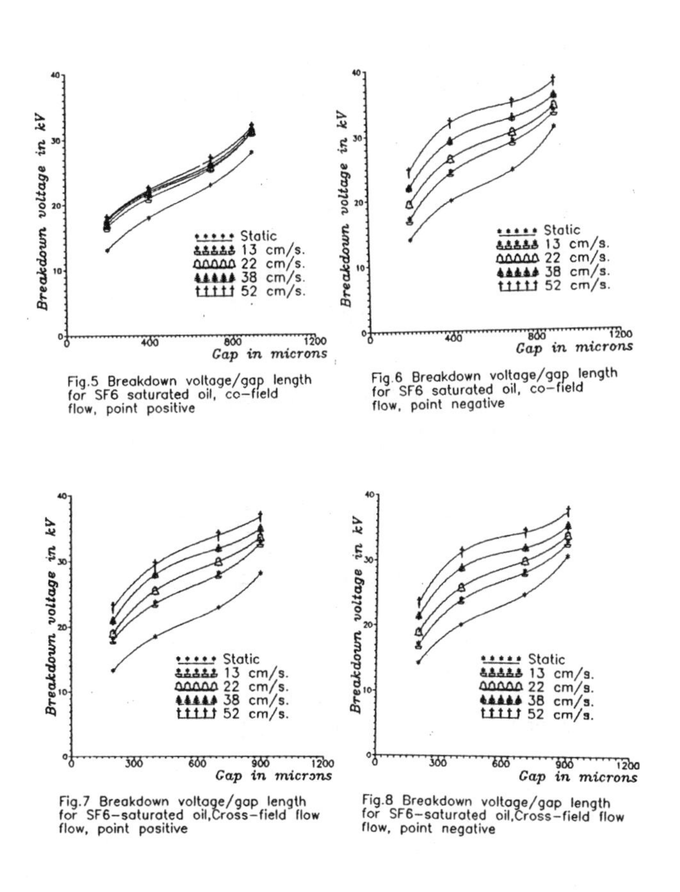

Fig.5 Breakdown voltage/gap length for SF6 saturated oil, co-field flow, point positive

Fig.6 Breakdown voltage/gap length for SF6 saturated oil, co-field flow, point negative

Fig.7 Breakdown voltage/gap length for SF6-saturated oil, Cross-field flow, point positive

Fig.8 Breakdown voltage/gap length for SF6-saturated oil, Cross-field flow, point negative

b) *Conduction*

Figure 9 shows typical current / voltage characteristics for degassed oil under stationary conditions, under co-field flow and under cross-field flow for both point polarities. In general, currents with the point positive are higher than those with the point negative for stationary oil as well as for both types of oil flow. Oil flow produces a marked reduction in current but the reduction with cross-field flow is slightly greater than with co-field flow .

Figure 10 and 11 show the effect of additives on the conduction current for stationary oil and both polarities.

Figures 12 and 13 show the same characteristics with cross-field flow and are very similar to those obtained with co-field flow.

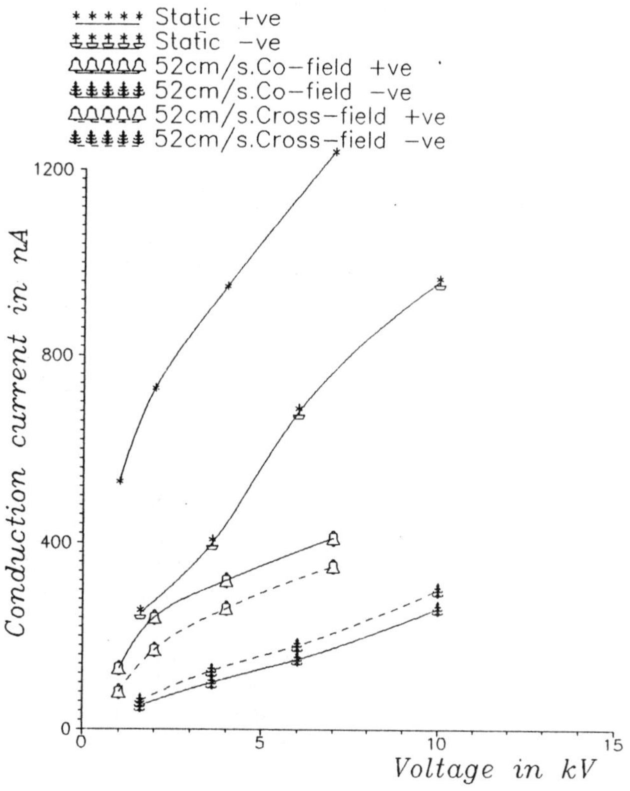

Fig. 9 Effect of flow direction on current/voltage, for both point polarities Degassed oil, gap length 400 microns

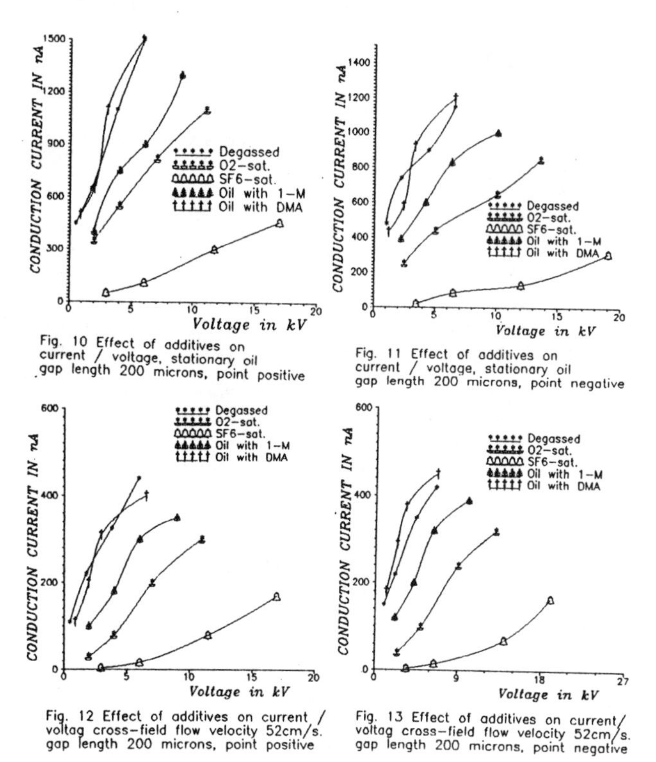

Fig. 10 Effect of additives on current / voltage, stationary oil gap length 200 microns, point positive

Fig. 11 Effect of additives on current / voltage, stationary oil gap length 200 microns, point negative

Fig. 12 Effect of additives on current / voltag cross-field flow velocity 52cm/s. gap length 200 microns, point positive

Fig. 13 Effect of additives on current/ voltag cross-field flow velocity 52cm/s. gap length 200 microns, point negative

The conduction current results indicate that, the same as for breakdown, SF6 is the most effective additive and DMA has no effect.

The present results indicate that the most effective additives are those which have electron-trapping properties viz. SF6, O_2 and MN. The non-electron trapping DMA had no effect on either breakdown or conduction properties of the oil. The fact that additives which have a large electron-trapping cross section accelerate negative streamers and additives which have a low ionization potential (e.g. DMA) accelerate positive streamers [6] indicates that there is no correlation between breakdown voltage and streamer propagation velocity.

The increase in breakdown voltage and decrease in conduction current brought about by oil flow can in general be attributed to two factors: (i) the formation of space charges at needle tip and the resulting formation of an intense local field there becomes more difficult under oil flow conditions and,(ii) the flow will have a cooling effect at the point tip retarding the formation of vapour bubbles and the establishment at the point electrode of the low density warm regions which precede breakdown [7].

REFERENCES

1. Boone,W. And Vermeer,J. "The Influence of Liquid Circulation on its Electric Strength". 4th International Conference on Dielectric Liquids, Dublin, 1972, pp.214-217.

2. Theodossiou,G., Nelson,K.K., Lee,M.J. and Odell,G.M. "The Influence of Electrohydrodynamic Motion on the Breakdown of Liquid Dielectrics". J.Phys. D., Vol. 21 1988, pp. 45-50.

3. Zaky,A.A., Megahed,I.Y. and El-Awa,M. "Effect of Liquid Flow Velocity on the Breakdown of Mineral Oil under Highly Nonuniform Fields". 10th ICDL, Grenoble, 1990, pp. 564-568.

4. Zaky,A.A., Megahed,I.Y. and Abdallah,M.A. "Influence of Co-field Motion on the Conduction Current in Oil under Nonuniform dc Fields". IEEE Trans. EI, Vol.1, No.4, August 1994, pp. 734-740.

5. Devins,J.C., Rzad,S.J. and Schwabe R.J. "The Role of Electronic Processes in the Electrical Breakdown of Insulating Liquids". Can. J. Chem., Vol. 55, No. 11, 1977, pp.1899-1905.

6. Nakao,Y., Itoh,H., Hoshino,S., Sakai,Y. and Tagashira,H."Effects of Additives on Prebreakdown Phenomena in n-Hexane" . IEEE Trans. EI, Vol. 1, No.3, June 1994, pp. 383 -390.

7. Allan,R.N. and Hizal,E.M. "Prebreakdown Phenomena in Transformer Oil Subjected to Nonuniform Fields". Proc. IEE, Vol. 121 1974, pp. 227-321.

Conference Record of the 1996 IEEE International Symposium on Electrical Insulation, Montreal, Quebec, Canada, June 16-19, 1996

The Degradation Properties in the Corona Charged Epoxy Composites by TSC Measurement

J. S. Kim, S. K. Lee, W. S. Choi, G. H. Park and J. U. Lee
Dept. of Electrical Eng., Kwangwoon University
447-1, Wolgye-dong, Nowon-ku, Seoul, Korea

***Abstract*: Corona degradation phenomena of epoxy composites to be used for transformers were studied. A formation of corona electrets was first observed by applying high voltages to five different kinds of specimens for a given mixing rate. And then Thermally Stimulated Currents(TSC) were measured in a temperature range of -160~200(℃). A behaviour of carrier and its origin in epoxy composites were examined. Various effects of corona degradation on epoxy composites were also discussed in the present article.**

INTRODUCTION

As the scale of a power delivery system is increased, the insulating design techniques such as materialization of high electric field and security of reliability in insulating composition are required accordingly.

Previous works have been performed to develop new composites whose insulating properties are better than existing materials and to examine the impact of the degradation phenomena on electrical properties in epoxy composites which are generally employed for mold materials in transformers.

A main purpose of this research is to study the TSC values in corona charged epoxy composites and interrelationship between corona stress and material degradation. The TSC values associated with the determination of currents which were released from poled specimens were measured as a function of temperatures. Poling was accomplished by placing the specimen under a high DC stress and followed by cooling with liquid nitrogen to freeze the molecular motion of the polymer, resulting in immobilization of the charges.

When the specimens are warmed up slowly, the trapped charges are released. By measuring the amount of released charges and their characteristic release temperatures, inferences on the space charges which is formed in the polymer can be made.

It has been noticed that peak temperatures for a given heating rate depend on a variety of factors such as an age of the specimen, thermal history, and charge density.

To investigate the effect of corona degradation on the properties of epoxy composites, the TSC spectra were measured as a function of temperatures and the various aspects of corona degradation effects were analyzed.

EXPERIMENTATION

Specimen Preparation

The specimens used in this study were designed to have a constant mixing rate in the epoxy resin of Bisphenol-A type and the MeTHPA(Methyl Tetra Hydro Phthalic Anhydride). Bisphenol-A type is in liquid phase at room temperature, and the MeTHPA is a hardner of acid hydro-anhydride system.

The DY-040 and the SiO_2 with amount of 5(wt · phr) are then added to improve both the impact strength and the machinery strength. The silane coupling agents, which belong to an amino silane system(N-(N-(β-Aminoetyl)-Aminopropyl-trimetoxy-Silane)) have been used for a surface treatment of fillers.

Design of Mixing Ratio

Since the electrical and physical properties of epoxy composites depend greatly on the mixing ratio among resin, hardner and filler, as well as the hardening conditions[1]~[2], these aspects need to be considered in manufacturing specimens. The design parameters and their conditions are summarized in Table 1.

Table 1. Mixing Ratio (wt · phr)

Samples	Epoxy	Hardener	DY -040	Filler	Curing Condition
H 70FN	100	70	10	0	▶1st Curing; 100(℃) ×4(hr) ▶2nd Curing; 140(℃) ×10(hr) ▶Deg. Curing; 140(℃) ×10(hr)
H100FN	100	100	10	0	
H130FN	100	130	10	0	
H100F60	100	100	10	60	
SH100F60	100	100	10	60	

Experimental Method

After the corona electrets[3]~[4] were formed by appling the high electric field, 300(kV/cm), to epoxy composites, the TSC spectra were measured in the temperature range of -160~200(℃) at a heating rate of 5(℃/min). A Schematic drawing of the experimental apparatus for the TSC measurement is given in Fig. 1.

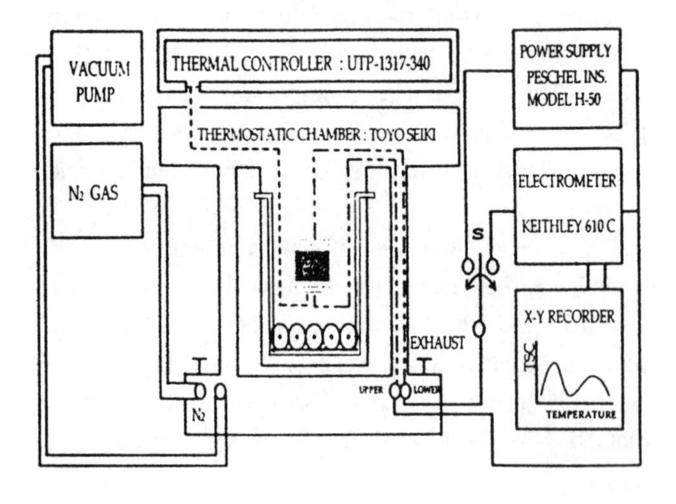

Fig. 1. TSC Apparatus

RESULTS AND DISCUSSION

TSC Spectra due to Electrical Strength

Fig. 2 shows the TSC spectra formed at the electric fields of 5, 10, and 30(kV/cm). They were observed from electrets for the 1st cured(120(℃)×4(hr)) H100FN specimens. Similarly, Fig. 3 display the TSC spectra of the second cured(120(℃)×4(hr)) H100FN specimens at 5, 10, and 30(kV/cm).

From the experimental study, the five peaks of δ, γ, β, α_1 and α_2 are obtained at temperatures of -100(℃), -40(℃), 20(℃), 100(℃) and 130(℃), respectively. The first three peaks which are observed at low temperatures below glass transition temperature(T_g) are relaxation ones due to the action of side chain or terminal radical[5]~[6]. Also, both the α_1 peak near 100(℃) and the α_2 peak above T_g are the ones caused by ionic space charge. A reason causing these peaks are as follows:

(1) Since the δ peak has been suggested by either a methyl radical($-CH_3-$) or a non-acted epoxy radical, the overall size of the peak is not so big. In the molecular structure, it is expected that the methyl radical is independently operated in proportion to the level of hardener added and it also affects to the curing reaction.

(2) The γ peak is produced by the mixture of hydroxyether radical, methoxy radical(CH-OH), aliphatic ether radical(-CH-O-), and aromatic ether radical(-COO-CH-) which is related to hardner. Also, as the amount of hardner is increased, the size of peak is getting bigger likewise the δ peak.

(3) The β peak appears to be resulted from the aromatic ether radical (-COO-CH-) which is thermally oxided through a procedure of maximum reaction.

(4) The α_1 peak is believed to be the effects from the virtue of molecule which is related to glass transition temperature.

(5) The α_2 peak which is observed at the highest temperature, 130(℃), is generated by the ionic space charge which is accumulated by impurities or inside defects.

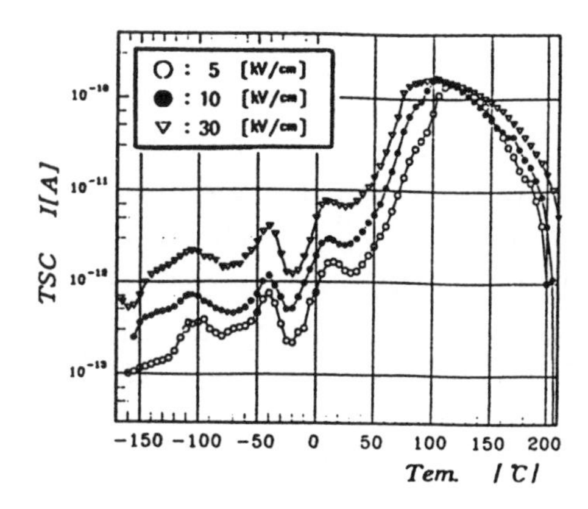

Fig. 2. The dependence on forming electrical strength in the 1st cured H100FN samples

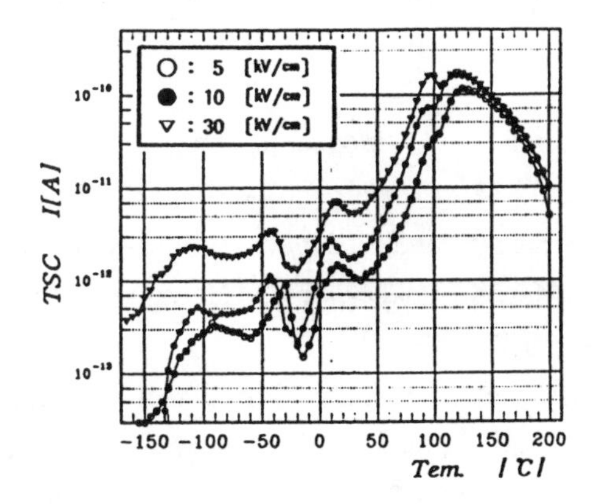

Fig. 3. The dependence on forming electrical strength in the 2nd cured H100FN samples

TSC Spectra due to Curing Time

The TSC spectra classfied according to the curing time and the ones observed from electrets formed by charging corona are shown in Fig. 4, 5, 6, 7, and 8. It is noted that in the case of the longest curing time, 240(hrs), the size of the α peak which is existing above 100(℃) is smaller than those of the 1st cured specimen and the 2nd cured specimen. Particularly, both the α_1 peak and the α_2 peak move to higher temperatures. The gap between these two peaks is rather significant and this result implies that space charge is accelerated by the corona charge[7]~[8]. On the other hand, as the curing time is longer, the size of TSC spectra is getting smaller. This feature also indicates a degradation due to the curing time and corona. In case of H100F60 and SH100F60 specimens, the α_1 peak and the α_2 peak at high temperatures are not separable because of the effect due to the filler and silane coupling agents.

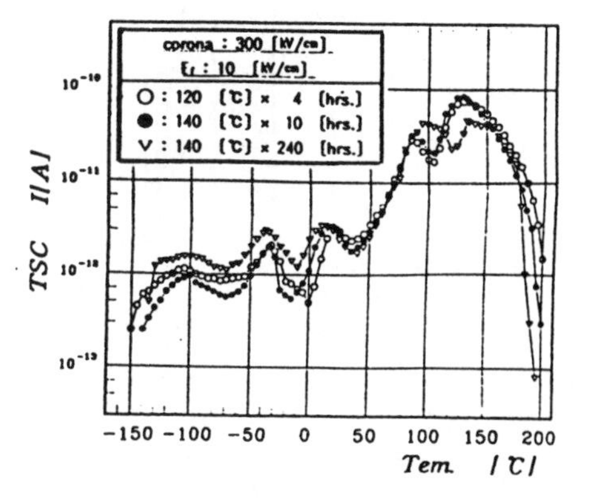

Fig. 4. TSC spectra obtained from the different curing time in corona charged H70FN samples

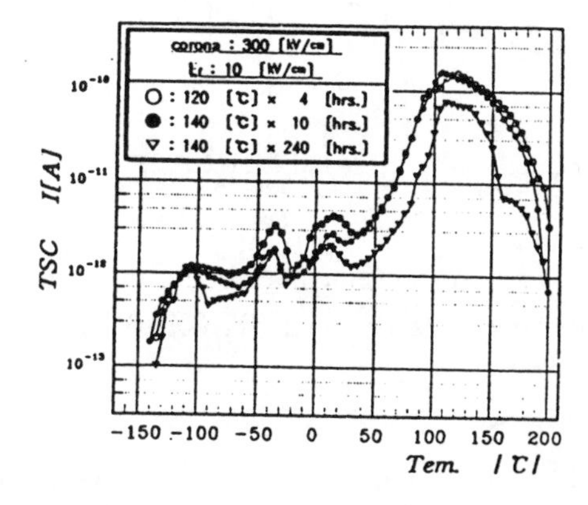

Fig. 5. TSC spectra obtained from the different curing time in corona charged H100FN samples

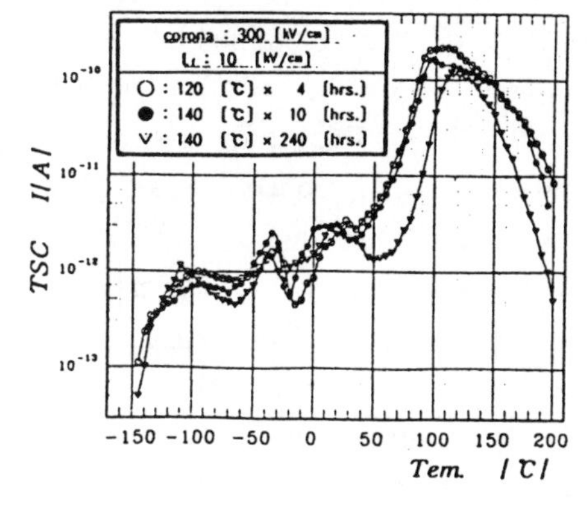

Fig. 6. TSC spectra obtained from the different curing time in corona charged H130FN samples

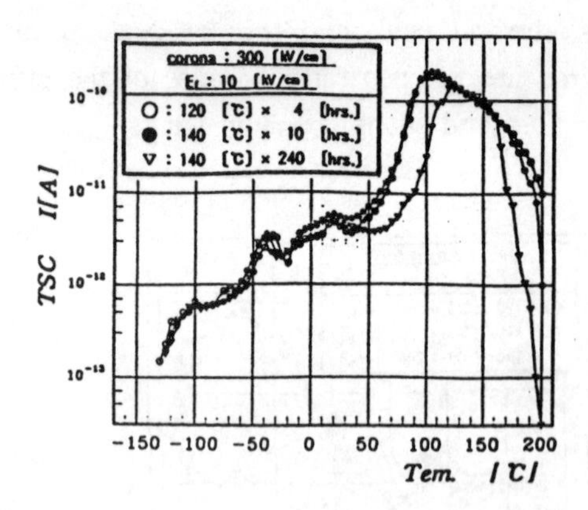

Fig. 7. TSC spectra obtained from the different curing time in corona charged H100F60 samples

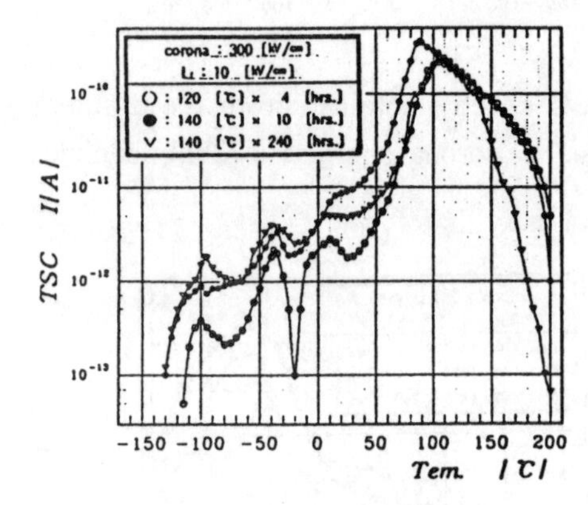

Fig. 8. TSC spectra obtained from the different curing time in corona charged SH100F60 samples

CONCLUSION

In order to understand the degradation characteristics of the epoxy composites due to corona discharge, the TSC spectra were measured in the temperature range of -160~200(℃). From a series of experiments, the five peaks including δ, γ, β, α_1 and α_2 were obtained at temperature of -100(℃), -40(℃), 20(℃), 100(℃), and 130(℃), respectively. Results of this study indicate that (1) the δ peak is resulted from the methyl radical; the γ peak is produced due to the mixture by hydroxyether radical, methoxy radical, aliphatic ether radical and aromatic ether radical; the β peak is obtained from aromatic ether radical; the α_1 peak is possibly due to molecule which is related to the glass transition temperature; and the α_2 peak seems to turn up by the ionic space charge, and (2) in case of the longest curing time, the size of the α peak at 100(℃) or above is smaller than those of the first cured and the second cured specimen; particularly, the α_1 peak and the α_2 peak tend to locate at higher temperature and the gap between two peaks is significant and these results suggest that the space charge is accelerated by the corona charge; and the curing time and the corona voltage are also acted as a basis of degradation.

REFERENCES

1. T. Takahama, O. Hayashi, F. Sato, "Electric Strength of Epoxide Resins and its Relation to the Structure", J. Appl. Poly. Sci., Vol.26, pp.2211-2220, No.5, 1981.

2. D. H. Kaelble, J. Moacanin and A. Gupta, "Physical and Mechanical Properties of Epoxy Resins, Chemistry and Technology, Clayton A. May Edi." Marcel Dekker, Inc., New York and Basel, pp.603-652, 1973.

3. J. V. Duffy and G. F. Lee, "The Effect of Steric Hinderance on Physical Properties in an Amine-Cured Epoxy", J. Appl. Poly. Sci., Vol.35, pp.1367-1375, 1988.

4. M. Ochi, M. H. Okasaki, M. Shimbo, "Mechanical Relaxation Mechanism of Epoxide Resins Cured Aliphatic Diamines", J. Poly. Sci., Phys. Ed., Vol.20,pp.689-699, 1982.

5. O. Delatycki, J. C. Shaw and J. G. Williama, "Viscoelastic Properties of Epoxy-Diamine Networks", J. Poly. Sci., A2, Vol.7, pp.753-762, 1969.

6. L. Simoni, "A General Approach to the Endurance of Electrical Insulation under Temperature and Voltages", Colloid & Polymer Sci., 260, pp.297-302, 1982.

7. S. S. Sastry, G. Satyanandam, "Effects of Fillers on Electrical Properties of Epoxy Composites", J. Appl. Poly. Sci., Vol.26, pp.1607 -1615, 1988.

8. J. J. O'Dwyer, "The Theory of Electrical Conduction and Breakdown in Solid Dielectrics", Clarendon Press. Oxford, 1973.

Conference Record of the 1996 IEEE International Symposium on Electrical Insulation, Montreal, Quebec, Canada, June 16-19, 1996

ELECTROSTATIC ENHANCED HEAT TRANSFER IN A DIELECTRIC LIQUID

Yang Jia-Xiang Chi Xiao-Chun Liu Ji Ding Li-Jian
Department of Electrical Materials
Harbin University of Science and Technology
Harbin, 150040, P.R.China

Yang He
Harbin Institute of Technology, P.R.China

Abstract: **In this paper, the phenomena of electrostatic enhanced heat transfer in a dielectric fluid have been investigated experimentally using a horizontal model pipe.**

The saturation pressure in the model pipe, the heat flux both in heated and cooled ends are all measured under high dc voltage. Experimental results show that the heat transfer properties of dielectric working liquids can be improved by the electrostatic field.

INTRODUCTION

With the development of high voltage technique, the effects of high electric field on insulating materials become more and more significant, and it has been widely applied to electrostatic enhanced heat transfer in highly insulating liquid.

It has been reported that the enhancement of boiling heat transfer with an electrostatic field is about 8.5 times of that without the electric field in the dielectric liquid Freon-11 mixed with ethylalcohol [1], and the enhancement of condensation heat transfer has been attained up to 6 times by using C_6F_{14} (perflorohexan) with the helical wire electrode [2]. Therefore, much attention is paid to this effective method in recently years [3][4][5].

In this paper, some experiments on enhanced heat transfer by the high electric field in the dielectric liquids have been done using a model pipe, and the experimental results are reported. According to these results, it is found that the wick structure can be substituted for the electrode structure in a conventional heat pipe.

EXPERIMENTAL SYSTEM

Figure 1 is the schematic diagram of a model pipe. The pipe is composed of a tube of oxygen-free copper which is 330 mm long, outer diameter is 32 mm, and the thickness of tube wall is 2.5 mm. Two grooved flanges have been welded on either end of the tube with sealing wash for vacuum and pressure sealing respectively. The electrode structure consists of a single polished copper wire (high voltage electrode) and the copper tube (grounded electrode). The wire is stretched by the two plexiglass end ports and runs along the axis of the tube about 3 mm above the bottom of the inner tube wall. The condensed fluid flows axially from the cooled section (as a condenser) to the heated section (as an evaporator) through the electrode structure. There is no thread both in the inner walls of the cooled section and the heated section.

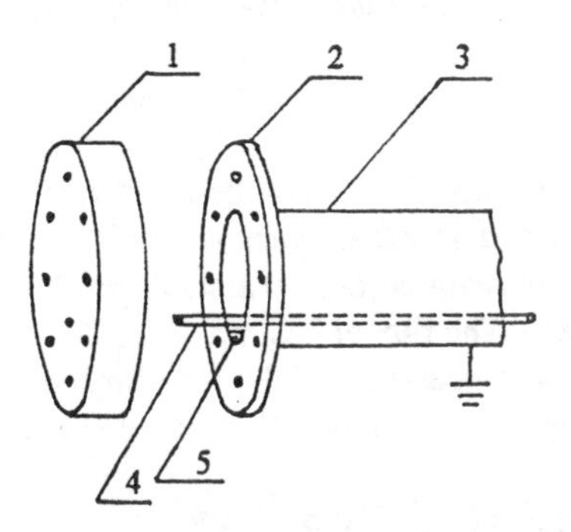

1.plexiglass end port 2.flange 3.copper tube
4.polished copper wire 5.working fluid
Fig.1 Schematic diagram of model pipe

The model pipe is divided axially three parts: heated section, cooled section and adiabatic section. The heated section (130 mm) is heated by a hot water jacket (HWJ). The cooled section (130 mm) is separated from the evaporator by a 20 mm adiabatic section, and it is cooled by a coolant water jacket (CWJ). Two temperature bath circuit systems supply the CWJ and the HWJ with the cooling water and the heating water separately.

The experimental system is shown in Figure 2. It consists of four parts: a model pipe, a circuit cooling system, a circuit heating system and a temperature measurement system. In either of circuit cooling and heating system, there is a throttle valve and a stop valve, therefore, the cooling water flux and the heating water flux can be adjusted in the light of the experimental requirements.

Considering the convenience of the experiments, Freon-11, which boils at 24.3 ℃, is selected as the experimental working fluid.

PROCEDURE

When the experiment starts, the two temperature baths are firstly brought up to the desired temperature, one is for the cooling circuit system, the other is for the heating circuit system. And then the circuit systems turn on to establish axial liquid communication. It usually takes about 2 hours to reach the

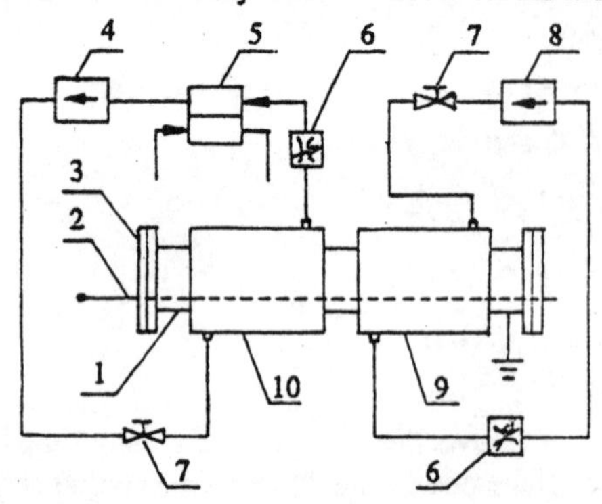

1.copper tube 2.polished copper wire 3.plexiglass end port
4.cooling water bath 5.heat exchanger
6.throttle valve 7.stop valve 8.heating water bath
9.hot water jacket 10.coolant water jacket
Fig.2 Schematic diagram of experimental system

thermal equilibrium. Then the high dc voltage is applied to the electrode structure in 2.0 kV step, and only about 30 minutes waiting time is required between stable reading at different applied voltage. The flux of heating water and cooling water, the inlet/outlet temperature of the two water jackets and the saturation pressure of working fluid Freon-11 are measured.

EXPERIMENTAL RESULTS

During the experimental procedure, it is observed clearly that the working fluid is tilted as the high voltage is applied to the electrode structure, as shown in Figure 3. It is also found that

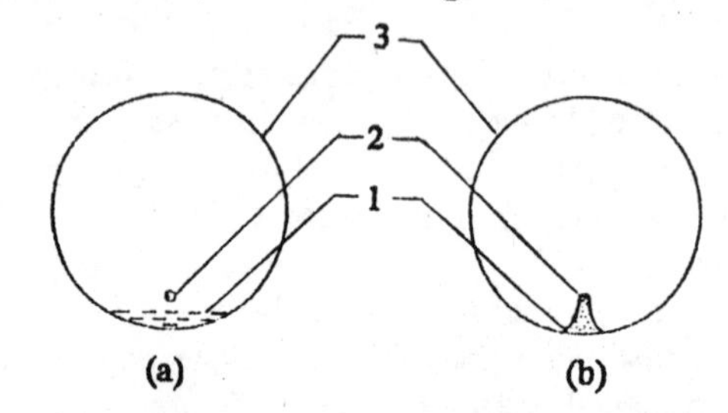

1.working fluid 2.polished copper wire 3.copper tube
(a) no voltage is applied (b) voltage is applied enough
Fig. 3 Cross-section view of the model pipe

the working fluid boils more intensely, and more bubbles escape from the working liquid.

Based on the measured data of the cooling water flux, the heating water flux, the temperature difference between inlet and outlet of the water jacket, and the saturation pressure of the working fluid Freon-11, the heat flux both in the cooled and the heated section can be obtained according to the definition:

$$q=\frac{C\cdot Q\cdot \Delta T}{S}$$

where C is the heat capacity of water, D is the density of the water, Q is the flux of the cooling water or the heating water, ΔT is the mean temperature difference, and S is the area of heat-exchange surface.

With the following test conditions: the average water flux in CWJ is 4.5 ml/sec, the inlet temperature of CWJ is 11.0 ℃, the water flux in HWJ is 9.5 ml/sec, the inlet temperature of HWJ is 59.0 ℃, the room temperature is 19.6 ℃, experimental results are obtained, as shown in Figure 4, Figure 5 and Figure 6.

Figure 4 shows the relationship between the saturation pressure of working fluid Freon-11 and the applied voltage on

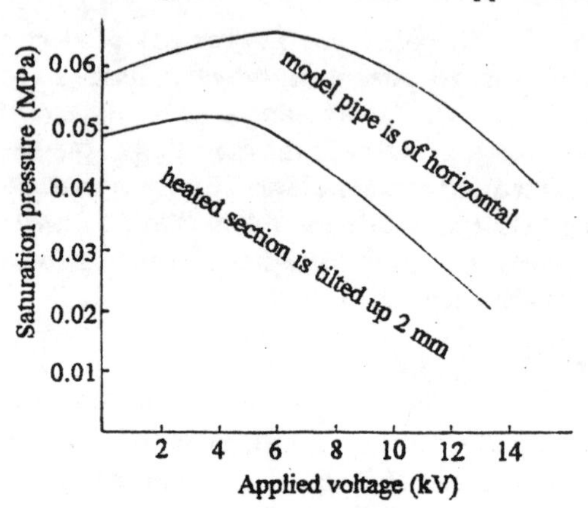

Fig. 4 The relationship between saturation pressure and applied voltage

the electrode structure. It is obvious that the applied voltage has effect on the saturation pressure of Freon-11. As the applied voltage increases, the saturation pressure increases firstly and then decreases.

There has similar changing tendency of the saturation pressure whether the model pipe is horizontal or not (the heated section is tilted up 2 mm). But there is still a slight difference that the extreme point is at around 6 kV when the pipe is horizontal, while the extreme point is at about 4 kV when the heated section of the pipe is tilted up 2 mm.

Figure 5 is the dependence curve of the heat flux in the cooled section on the applied voltage. From the curve it can be found that the changing tendency of the heat flux in the cooled section with the applied voltage is similar to that of the saturation pressure.

Figure 6 shows the dependence curve of heat flux in the heated section on the applied voltage. It can be seen that the changing tendency of heat flux in the heated section with the applied voltage reverses to that of heat flux in the cooled section. As the applied voltage increases, the heat flux in the heated section is decreased at first, and then increased. But, the extreme point is still at the same position, that is to say, the extreme point of heat flux is at around 6 kV and 4 kV when the model pipe is horizontal or not, respectively.

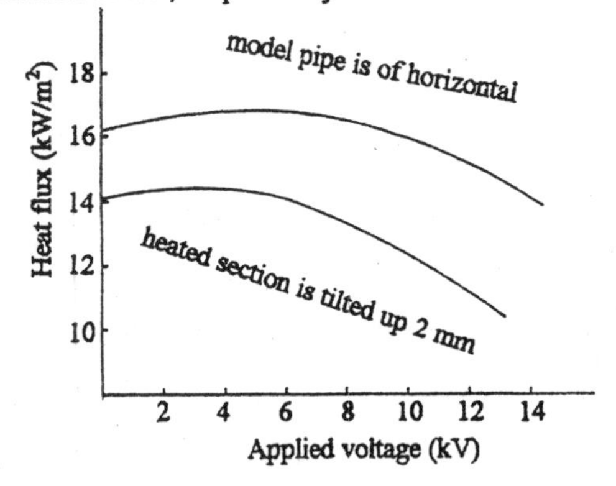

Fig.5 Relationship between the heat flux in the cooled section and the applied voltage

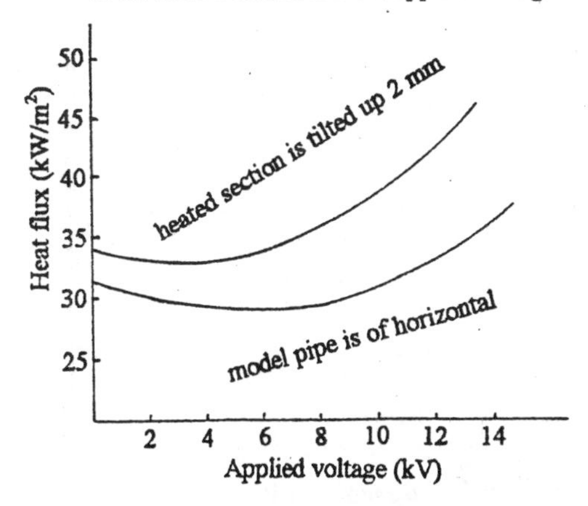

Fig.6 Dependence curve of heat flux in the heated section on the applied voltage

According to the figures listed above and the other experimental results obtained, it can be learned that the changing tendency of the saturation pressure of working fluid is always similar to that of the heat flux in the cooled section, but always reverses to that of heat flux in the heated section, regardless varied test conditions.

DISCUSSION

It is clear from above results that the poor heat-transfer properties of the dielectric liquid are improved, and higher applied voltage, much more increase. This is due to heat transfer enhanced by high electric field.

(1) Augmentation of boiling and condensation heat transfer

It is known that the electrohydrodynamic force in highly insulating dielectric liquids can be described as the following equation [6][7].

$$\overline{f^e} = \rho_f \overline{E} - \frac{1}{2}E^2\nabla\varepsilon + \nabla[\frac{1}{2}\rho E^2(\frac{\partial\varepsilon}{\partial\rho})_T]$$

where, the first term is the force per unit volume on a medium containing free charges ρ_f.

the second term is the polarization force exerted on a dielectric liquid due to a nonuniform electric field.

the last term is the electrostrictive force, which can be neglected for incompressible fluid.

In heated section of the model pipe, the nonuniform distribution of temperature field will lead to the nonuniform distribution of conductivity σ and dielectric permittivity ε, that is, $\nabla\sigma \neq 0$ and $\nabla\varepsilon \neq 0$. If $\nabla\sigma$ is nonzero in the liquid bulk, free charges can be acted upon by the body force $\rho_f \overline{E}$. And if $\nabla\varepsilon$ is nonzero in the liquid, the polarization force $-\frac{1}{2}E^2\nabla\varepsilon$ will tend to cause liquid motion. In two-phase heat transfer, both $\rho_f \overline{E}$ and $-\frac{1}{2}E^2\nabla\varepsilon$ are in existence [5]. Therefore, under the applied voltage enough, the electrohydrodynamic force $\overline{f^e}$ will increase the liquid flow, so the bubbles can easily escape from the working liquid, and the boiling point of working fluid is decreased. Consequent to the lower boiling point, the latent heat of vaporization of working liquid Freon-11 goes down, and the heat flux in the heated section is reduced, as shown in Figure 6. On the other hand, because the bubbles escape from the working fluid easily, there will be more vapor in the model pipe, as a result, the saturation pressure of working liquid goes up, as shown in Figure 4.

As increasing the applied voltage, the electric field intensity between the electrode structure increases, and the electrohydrodynamic force $\overline{f^e}$ become more stronger, so the bubbles escape more easily, and the boiling point is lower. As a result, the heat flux remains decreasing, meanwhile, the saturation pressure increases more. When the applied voltage reaches a certain value (around 6 kV when the model pipe is horizontal), the electrohydrodynamic force $\overline{f^e}$ holds stable because the working fluid polarized completely, so the boiling point is the lowest, the heat flux is the smallest, and the saturation pressure is the topmost.

In condensed section of the model pipe, the corona discharge around high voltage wire electrode leads to the ionization of the vapor, and forms the ionic wind. When the ionic wind impacts on the inner heat-transfer surface of the model pipe, it will affect film condensation by destabilizing the liquid film to promote pseudodropwise condensation. The resultant heat flux in the cooled section is increased along with the increase of applied voltage, as shown in Figure 5.

As the applied voltage increases more, the experimental results become a little contradictory, that is, when the heat flux in heated section increases with applied voltage, the saturation pressure of working fluid decreases, but the heat flux in cooled section decreases too.

When the applied voltage is brought up further, the effect of electrostatic cooling accelerates the heat exchange between the vapor and the heating water, due to the intense ionic wind makes the flow of steam chaos, and hence the heat flux in heated section is increased, as shown in figure 6. In the condensed section, however, the effect of electrostatic cooling will make the vapor condense directly in the tube, so the saturation pressure is reduced, as shown in Figure 4. Even the heat-exchange between the vapor and the heating water is augmented due to the ionic wind, the reduction of saturation pressure lowers the heat-exchange between the vapor and the cooling water, consequently, the heat flux in cooled section is dropped, as shown in Figure 5.

(2) Enhanced liquid flowing from the condenser to the evaporator

As shown in Figure 4, Figure 5 and Figure 6, the heat transfer is promoted when the pipe is tilted up (the evaporator elevation above the condenser).

While the pipe is tilted up, the evaporation capacity of the heated end will decrease sharply and cause ' dry heat ' of the model pipe, eventually, the heat transfer properties of the model pipe will be completely destroyed. But not so does when utilizing the dc high voltage. Due to the ' rise ' effect of polarization force, as shown in Figure 7. The polarization

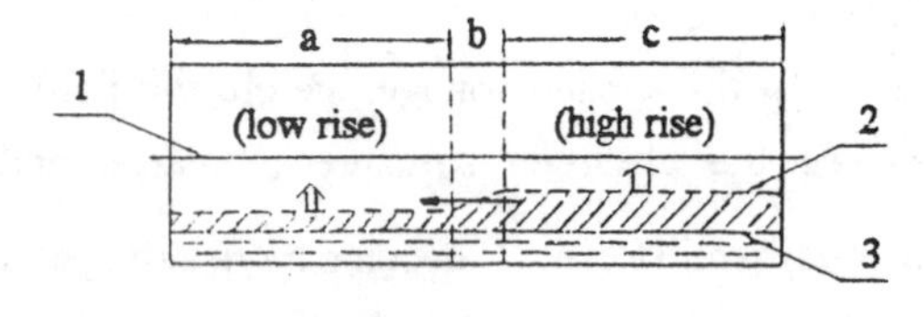

1.high voltage electrode 2.liquid surface with enough voltage
3.liquid surface without voltage
a.heated section b.adiabatic section c.cooled section
Fig.7 The rise effect of polarization force

force in cooled end is higher than that in the heated end, and the resultant lifting height of cooled end is above that of the heated end. Hence, it will transport the dielectric working liquid from the cooled end to the heated end. It indicates that the method of electrostatic enhanced heat transfer is preferable under the condition of inclined position.

CONCLUSIONS

Based on the experimental results and the discussion mentioned above, the following conclusions can be obtained:

(1) The thermal transfer properties of dielectric liquid are promoted by the electrostatic field.

(2) Electrostatic enhanced heat transfer of dielectric liquid has an effect apparently on the inclined pipe (the evaporator elevation above the condenser).

(3) An electrode structure can replace the capillary wick in conventional heat pipe, which can gather condensed liquid and transport it from condenser to evaporator.

Further experiments are necessary to study the working performance of the model pipe, and to find desired working fluids used in heat and mass transfer engineering.

REFERENCES

1. Ogata J, et al. ' Augmentation of Boiling Heat Transfer by Utilizing the EHD Effect'. JSME International Journal, Series B. Vol. 56. July 1990, pp. 2044-2051.
2. Yamashita K, et al. ' Heat Transfer Characteristics of an EHD Condenser ' . JSME International Journal, Series B. Vol. 57. May 1991, pp. 1727-1733.
3. Atten P, et al. ' Electroconvection and its Effect on Heat Transfer ' . IEEE Trans EI. Vol.23. No.4, 1988, pp. 659-667.
4. Ohadi, M. M, et al. ' Heat Transfer Enhancement of Laminar and Turbulent Pipe Flow via Corona Discharge' Int. J. Heat Mass Transfer. Vol. 34, 1991, pp. 1175-1187.
5. Poulter R and Allen, P. H. G.' Electrohydrodynamically Augmented Heat and Mass Transfer in the Shell/Tube Heat Exchanger ' . Proc. Eighth Int. Heat Transfer Conference. Vol.6, 1986, pp. 2963-2968.
6. Castellanos A. ' Coulomb-driven Convection in Electro-hydrodynamics '. IEEE Trans EI. Vol.26. No.6, December 1991, pp.1201-1215.
7. Chang, J.S. ' Stratified Gas-Liquid Two-Phase Electro-hydrodynamics in Horizontal Pipe Flow ' . IEEE Trans IA. Vol.25. No.2, March/April 1989, pp. 241-247.

Conference Record of the 1996 IEEE International Symposium on Electrical Insulation, Montreal, Quebec, Canada, June 16-19, 1996

Influences of Electrical Field on Boiling-Condensation Heat Transfer System

Yang Jia-Xiang Ding Li-Jian Chi Xiao-Chun Liu Ji

Department of Electrical Materials Engineering

Harbin Institute of Electrical Technology

Harbin, 150040, P. R. CHINA

Yang He

Department of Thermal Energy Engineering

Harbin Institute of Technology

Harbin, 150001, P. R. CHINA

***Abstract*: In this paper, the influences of electrical field on boiling-condensation heat transfer system have been investigated using a cylinder heat transfer model. Freon-11 is selected as working fluid. The condensation heat transfer coefficient, the boiling heat flux and the saturation pressure are measured in this investigation. According to the experimental results, it is found that the electrical field can influence heat transfer system. The boiling heat transfer is enhanced by the applied voltage, and the saturate vapor of working fluid is condensed on the high voltage electrode directly when the applied voltage is higher than 6 kv. The experimental results have been discussed, and it is considered that the high electrical field strength change the thermal properties of working fluid.**

INTRODUCTION

Effects of electrical field on heat transfer have been studied extensively [1-7]. A. Yabe et al. [2,3] demonstrated the influences of corona wind on the augmentation of convection and vaporization. Electric coupled pool boiling has been investigated by several authors [1,4,5]. Influences of electrical field on condensation heat transfer have been studied too [1,6,7]. Some useful laws on electric coupled heat transfer have been achieved through those studies mentioned above. However, these laws usually limited by the experimental conditions. Especially, it was only considering a single phase (boiling or condensation) when the laws on the electric coupled boiling and condensation heat transfer were concluded. Therefore, some other effects of electrical field may be neglected.

Recently, an investigation of electric coupled heat transfer has been carried out using a boiling-condensation heat transfer model. This investigation studied the effects of electrical field on both boiling and condensation heat transfer at the same time. The relevant parameters, such as condensation heat transfer coefficient, heat flux in the experimental model, saturation pressure, etc., are measured in the investigation. In this paper, some experimental results based on the investigation are presented and discussed.

DESCRIPTION OF EXPERIMENTAL SYSTEM

Figure 1 is the schematic diagram of the experimental system. Used in the investigation. This experimental system is made up of five major parts: a boiling-condensation heat transfer model, a heating section, a cooling section, a dc high voltage source and a refilling section.

A glass cylinder (outside diameter: 200 mm, inside diameter: 180 mm, height: 200 mm) with two end-covers is the major

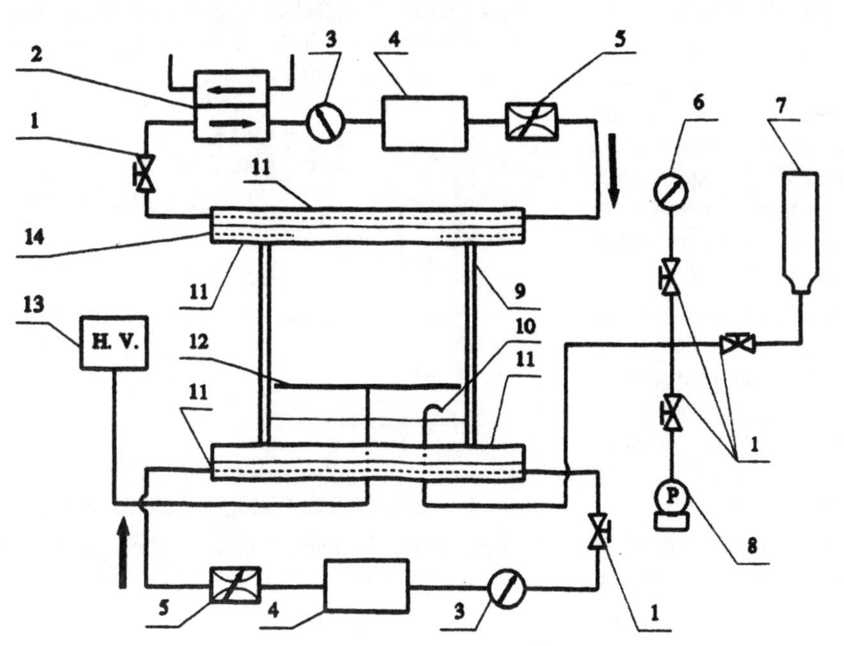

Figure 1 Schematic diagram of experimental system

part in the boil-condensation heat transfer model. The upper cover is used as condensation plate, and the bottom cover is functioned as heating plate. There are two small channels (its diameter is 2 mm) milled in the upper cover, which is used for venting non-condensable gases remaining in the model and dissolved in the working fluid. The two covers are attached to identical covers with the milled heating and cooling channels. A copper mesh is set into the experimental model as a high voltage electrode. It is placed parallel to the upper and bottom covers, and the distance to the upper or bottom cover is adjustable. In this experiment, the distance between the copper mesh and the bottom is 20 mm, and it is in the vapor of working fluid. The copper mesh electrode is connected with high voltage source.

The upper cover is cooled with cool water from the cooling section. The cooling section consists of a temperature bath (with a pump), a flow-meter, a throttle valve, an stop valve and a heat exchanger. The temperature of cooling water can be controlled by the temperature bath and the heat exchanger. The water flow rate can be adjusted through the throttle valve and stop valve. The bottom cover is heated with hot water from the heating section. The heating section is similar to the cooling section, just no heat exchanger. This section can provide a constant hot water flow for the experimental model.

The refilling section is composed of a pressure manometer, a vacuum pump, a working fluid conservator and three stop valves. This section is used for refilling the working fluid to the experimental model. In this experiment, Freon-11 is selected as working fluid.

PROCEDURE AND MEASUREMENTS

Before each test the experimental model is connected to the vacuum pump. The working fluid is refilled into the model from the conservator until a vacuum of 300 Pa is reached.

When the experiment starts, the states of heating and cooling sections are first set to the desired condition (including water temperature, water flow rate). Experiments are always carried out with a higher saturation pressure than atmospheric pressure, which ensures that no air can penetrate the experimental model. At the beginning of test, non-condensable gases remaining in the cylinder or dissolved in the working fluid are vented out through the two small channels in the upper cover.

After the thermal equilibrium is reached, measurements will be recorded. It usually takes about 3 hours to reach the thermal equilibrium at the first time. Then the dc voltage is applied to the copper mesh electrode. Only about 40 minutes waiting time is required between stable measurements at different voltages. Measurements are made for five different voltages: 0, 6, 9, 12 and 15 kv. The copper mesh electrode is always negative.

The temperature differences at inlet and outlet of the water flow are measured in both cooling section and heating section. In the experiment, the water flow rates are recorded in heating section and cooling section respectively. The wall temperature of condensation surface is obtained by extrapolating the temperature measured at three different layers in the condensation plate. The saturation pressure and temperature are gauged too.

The condensation heat transfer coefficient (h_c) is calculated from the measurements of cooling water flow rate, temperature difference of cooling water, wall temperature and saturation temperature. It can be expressed as

$$h_c = \frac{q}{t_s - t_w};$$

$$q = \frac{\rho \cdot C_p \cdot Q \cdot \Delta t}{F},$$

where, q is the heat flux; t_s is the saturation temperature; t_w is the wall temperature of condensation surface; ρ is the density of water; C_p is the heat capacity of water; Q is the cooling water flow rate; Δt is the temperature difference of cooling water; F is the heat transfer area.

In order to evaluate the effects of electrical field, the saturation pressures, the heat fluxes in heating section and the condensation heat transfer coefficients are compared under different voltages respectively. The rate of condensation heat transfer coefficient (h_c/h_0) is also calculated (where h_c is the coefficient with different voltages, h_0 is the coefficient with no applied voltage).

EXPERIMENTAL RESULTS

The experiment has been carried out according to the procedure mentioned above. The experimental results are described as following ("T" in figure refers the temperature at inlet of hot water):

The influence of applied voltage on heat flux in heat section is shown in figure 2. It is found that the heat flux is increased with the increase of applied voltage.

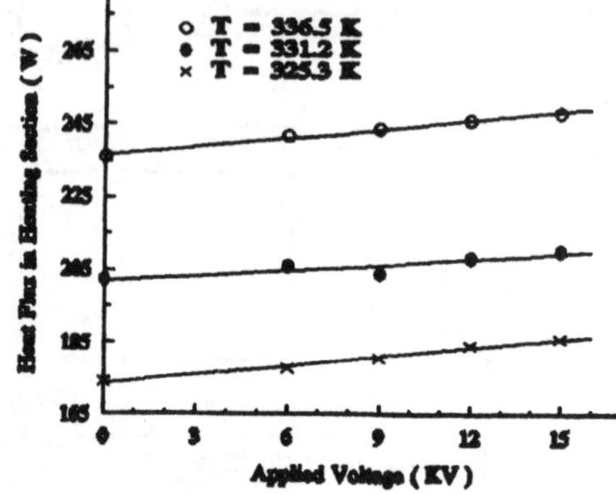

Figure 2 Influence of applied voltage on heat flux in heating section

Figure 3 gives the relation between the saturation pressure and the applied voltage. From this figure, the saturation pressure increases slightly with 6 kv applied voltage, and then it decreases along with the increase of applied voltage.

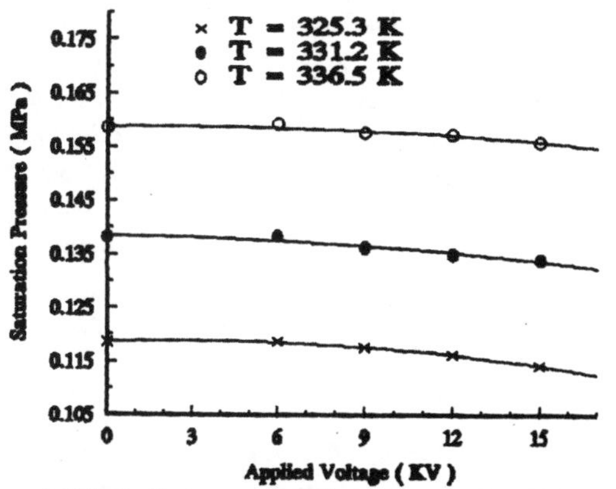

Figure 3 Relation between saturation pressure and applied voltage

Figure 4 illustrates the relation between the condensation heat transfer coefficient and the applied voltages. The condensation coefficient is the lowest with the 6 kv applied voltage, and then it increases obviously under a higher applied voltage.

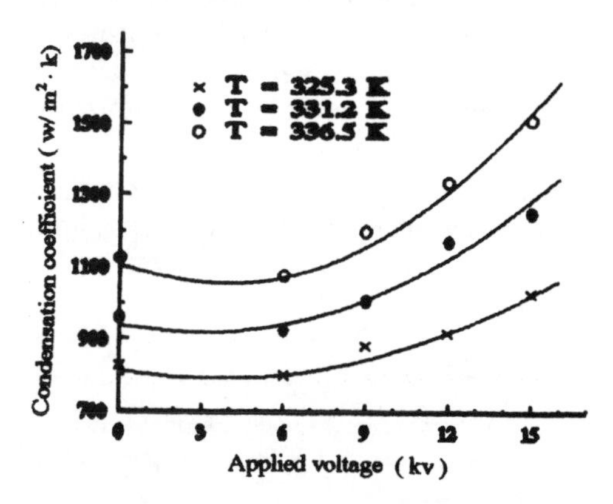

Figure 4 Relation between condensation coefficient and applied voltage

Figure 5 is the ratio of condensation coefficient h_c/h_0 under different applied voltages. The changing tendency of h_c/h_0 is similar to that of condensation coefficient (refer to figure 4).

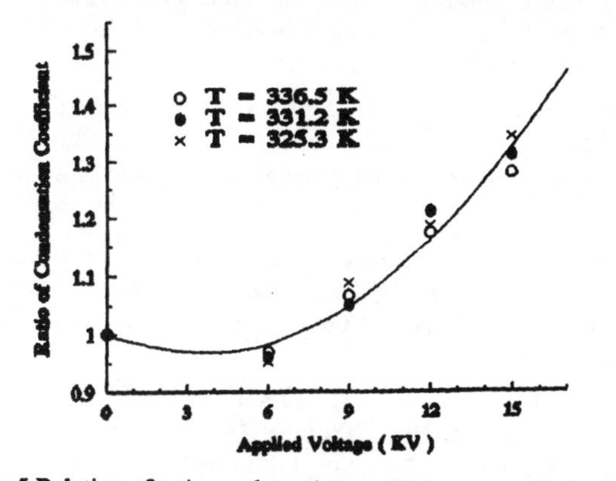

Figure 5 Relation of ratio condensation coefficient to applied voltage

In the experiment, the following phenomena are observed. If the other conditions are kept as constant, along with the increase of applied voltage, the departure sizes of bubbles from the heating surface become smaller, and the escaping speed of bubbles from the heating surface is increased. Above phenomena are more obvious when the applied voltage is increased more. In addition, it is found that there is some liquid on the copper mesh electrode when the voltage is higher than 6 kv, and the higher voltage, the more liquid on the electrode. According to the observation, the liquid on the electrode is neither the condensed liquid dropped from the condensation surface, nor the liquid "jetted" from the boiling liquid.

DISCUSSION

Because the distance between the copper mesh electrode and the heating surface is only 20 mm in this investigation, the electrical field strength between the electrode and the heating plate is much greater than that in the other area. Therefore, the major effects of electrical field should be on boiling heat transfer, and the condensation heat transfer could be influenced by the boiling heat transfer because of the combined boiling-condensation heat transfer system. In other words, the change of condensation heat transfer should be subordinated to the change of boiling heat transfer.

Enhancement of boiling heat transfer

Figure 2 reveals that the heat flux of heating section is influenced by the applied voltage, and the heat flux is increased almost proportional to the applied voltage. This result has been obtained by several authors [1,4,5], and it is easy to understand. When an electrical field is applied to the working fluid, there are two types of electrostatic forces in the working fluid. One is dielectrophretic force, which reflects the motion of the dielectric liquid in a non-uniform electrical field. The other is the coulomb attraction owing to the electrical field in the test system.

Consequent to the dielectrophoresis, the higher dielectric permittivity phase is accelerated towards the area of higher field strength. In case of the boiling liquid, the conductivity dielectric permittivity of liquid phase is higher than that of vapor phase. Therefore, the electrical field strength in vapor (bubbles) is stronger relatively. As a result, the bubbles on the heating surface tend to be displaced by the accelerated liquid, and then the release rate of bubbles from the heating surface is increased. Consequently, the heat transfer is enhanced.

On the other hand, there are some charges at the interface between the liquid and the bubbles due to the friction, so there is a coulomb attraction force under the electrical field. As a result, the bubbles could be drawn out by this force. Consequent to the electrical field force and buoyancy, the departure size of bubbles from the heating surface will be reduced, and the bubble release rates will be increased. Therefore, the heat transfer is increased too. Figure 2 and the phenomena observed in the experiment support above analyses.

High voltage condensing

From figure 3 it can be found that the saturation pressure is influenced by the applied voltage. In this experimental system, three cases maybe influence the saturation pressure. The first is the boiling heat transfer, the next is the condensation heat transfer, and the last is the high voltage copper mesh electrode.

As described above, electrical field can enhance the boiling heat transfer. Corresponding to the enhancement of boiling heat transfer, the vaporization of working fluid will be more excited. Therefore, the saturation pressure should be improved with the enhancement of boiling heat transfer because of the sealed system. However, from figure 3 the saturation pressure is increased only with the 6 kv applied voltage, and the saturation pressure is decreased along with the increase of applied voltage after the applied voltage is higher than 6 kv. So, the enhancement of boiling heat transfer is not the only reason.

On the other side, Condensation heat transfer and saturation pressure can be influenced each other. From figure 4 and figure 5, it is found that the condensation heat transfer has a relation with the applied voltage, and the condensation heat transfer coefficient is increased obviously after the applied voltage is higher than 9 kv. Usually, the enhancement of condensation heat transfer can be observed on the condensation surface. If the condensation heat transfer was enhanced, it would have much more condensed liquid on the condensation surface. In this experiment, the electrical field on the condensation surface is so weak that we do not observe any changes on the condensation surface with the varied applied voltages. Therefore, the variation of condensation heat transfer coefficient is not the reason for saturation pressure's change, but the result of saturation pressure's change. Considering this point, the other reason for the change of saturation pressure lies in the high voltage electrode.

As mentioned before, there is some liquid on the copper mesh electrode when the applied voltage is higher than 6 kv, and the higher applied voltage, the more liquid on it. In other words, the saturate vapor is condensed on the high voltage electrode directly after the applied voltage is higher than 6 kv. This phenomenon has never been reported before. Because the saturate vapor condenses on the high voltage electrode directly, some saturate vapor changes to liquid. As a result, the saturation pressure would be decreased, and then the condensation heat transfer coefficient would be increased because the temperature difference goes down between the condensation surface and the saturate vapor. Therefore, the experimental results are in agreement with each other.

However, how does the saturate vapor condense on the high voltage electrode directly? What about the heat released from the condensation process? From the experiment, it is found that the amount of liquid on the electrode has a relation to the applied voltage, and this phenomenon is observed only after the applied voltage is higher than 6 kv. Hence, it is assumed that the high voltage change the thermal properties of working fluid (Freon-11) vapor around the electrode, and improve the Freon-11's latent heat of vaporization. Consequently, some "saturate vapor" is condensed on the electrode directly, and some "saturate vapor" remains the vapor phase after taking up the heat released from the condensation. This procedure is named as high voltage condensing. Therefore, the saturation pressure is dropped although the boiling heat transfer is enhanced. In addition, this effect is more obviously under the higher voltage. Using above hypothesis can explain the experimental results reasonably. However, it needs further study to confirm.

CONCLUSIONS

The influences of electrical field on boiling-condensation heat transfer system have been investigated experimentally. According to the experimental results and above analyses, the following conclusions can be made:

1. The electrical field can influence the boiling heat transfer. The boiling heat transfer is enhanced by a high dc voltage.

2. The Freon-11 vapor can be condensed on the high voltage electrode. It is considered that the thermal properties of Freon-11 vapor be changed when the electrical field strength is high enough, and the Freon-11's latent heat of vaporization would be increased.

This investigation is helpful for the application of high voltage technique to the other fields. Further experiments with other working fluids are necessary to confirm the hypothesis achieved in this paper.

REFERENCES

1. T. B. Jones, "Electrohydrodynamiclly Enhanced Heat Transfer in Liquid-A Review", Advanced in Heat Transfer, Vol. 14, 1978, pp. 107-148.
2. A. Yabe, Y. Mori and K. Hijikata, "EHD Study of the Corona Wind Between Copper and Plate Electrodes", AIAA, J., Vol. 16, 1978, pp. 340-345.
3. A. Yabe, Y. Mori and K. Hijikata, "Heat Transfer Augmentation around a Downward-Facing Flat Plate by Non-uniform Electric Fields", Proc. 6th Int. Heat Transfer Conf., Toronto, 1978, Vol. 3, pp. 171-176.
4. P. Cooper, "EHD Enhancement of Nucleate Boiling", Trans. Of ASME, Journal of Heat Transfer, Vol. 112, 1990, pp. 458-464.
5. J. Berghmans, "Electrostatic Fields and the Maximum Heat Flux", Int. J. Heat Transfer, Vol. 19, 1976, pp. 791-797.
6. A. B. Dkovsky and M. K. Bologa, "Vapour Film Condensation Heat Transfer and Hydrodynamics Under the Influences of an Electric Field", Int. J. Heat Transfer, Vol. 24, No. 5, 1981, pp. 811-819.
7. A. Yabe, K. Kikuchi, T. Taketani, et al., "Augmentation of Condensation Heat Transfer by Applying Non-uniform Electric Fields", Proc. 7th Int. Heat Transfer Conf., Vol. 5, 1982, pp. 189-194.

Conference Record of the 1996 IEEE International Symposium on Electrical Insulation, Montreal, Quebec, Canada, June 16-19, 1996

Movement of Charged Conducting Particles Under A.C. Voltages in Insulating Liquids

S.M. EL-Makkawy
Electrical Engineering Department
Faculty of Engineering
Suez Canal University
Port-Said,Egypt

Abstract: **The paper presents a computer oriented analysis and procedure for computation of the dynamic motion of free conducting particles in insulating liquiids under alternating voltage at different frequencies. The analysis contains the charge simulation technique to compute the electric lifting field for any conducting particle shape with parallel plate electrode arrangement. Numerical solution of the non-linear dynamic equations of the particle motion has been carried out and graphically visualized. The opposing viscous force is the most significant component and is dissipated in medium as fluid drag.**

This study includes the influence of the most important design parameters on the particle motion, such as applied voltage magnitude and frequency, particle density and dimensions, liquid density and restituation coefficient.

1. INTRODUCTION

One of the major failure mechanisms in insulating liquids is the existence of free particles in the liquid[1]. The presence of impurity particles in oil-insulated power equipment is unavoidable and such contamination affects the insulating properties of the oil and hence the insulating integrity of the equipment itself [2,3]. Particles that are most likely to be found in transformers are composed of iron, copper, aluminum and cellulose; and are introduced into the tank by contact of the oil with coils, the core and solid insulating structures, both during manufacturing and in service conditions [3]. Carbon particles are produced by electrical discharges in oil and their presence is unavoidable in the oil circuit breakers and of transformers [4]. Particle concentration in insulating liquids is reduced by filteration, however, filtering the liquids can only reduce the number and can't completely remove the parrticles[2,5]. The electric strength of oil is a controlling factor for the power frequency strength requirements for oil-paper insulation system. The dielectric properties of the oil-paper insulation system are good when the paper is dried and impregnated, and when the oil is dry and essentially free of contaminating particles. However, the dielectric strength can be significantly reduced if the water content of the paper is high and/or if the oil is contaminated by water or particles [3].

Some investigations have involved the calculation and measurment the motion of the particle in fluid dielectrics under the influence of direct and alternating voltage [1,6,7]. However, no theoretical-mathematical model exists to predetermine the breakdown voltage of contaminated under these conditions. Such a theoretical treatment is very complicated, since a great number of time-dependent variables affect the breakdown process. One side of this difficulty is the dynamic behaviour simulation of the moving particles as well as the calculation of the distorted electric field strength under the influence of particle motion.

A resting particle at the bottom of an electrode will be lifted, if a sufficient high voltage is applied. After levitation, the position of the particle in the gap is continually changing with time. The maximum height reached by a particle is thereby suspected to be the important parameter lowering the breakdown voltage by particle contaminated in fluid dielectric.

In this paper, our interest is devoted to the simulation of conducting particle motion in insulating liquid under A.C. voltages of different frequencies. Hereby, the study includes the influence of the important design parameters on this motion. The numerical solution of the non-linear dynamic equations of the particle motion resulted in a detailed analysis of particle maximum height, influence of applied voltage magnitude and frequency, influence of particle density and dimensions, influence of restituation coefficient for different type of insulating liquids.

2. Particle Movement Under A.C. Voltage

When the field at the electrode on which the particle is resting reaches a certain value E_L, the particle will be levitated. The initial motion of a particle of mass *m* acquiring a charge *q* after it has been lifted is then described by the equation,

$$m \cdot \ddot{x} = q \cdot E(x,t) - B(\dot{x}) \cdot \dot{x} \pm mg \qquad (1)$$

Where *E* is the instantaneous electric field strength value, *x* is the distance crossed by the particle at time *t*, $\ddot{x}$ is the particle acceleration, $\dot{x}$ is the particle velocity and $B(\dot{x})$ is the coefficient of friction.

Since for an unchanged geometrical electrode system the electric field strength is directly proportional to the applied voltage, the instantaneous electric field strength *E* can be expressed for an alternating sinusoidal applied voltage as :

$$E = E_m \operatorname{Sin}(\omega t+\phi) \qquad (2)$$

Where E_m is the electric field at the distance *x* corresponding to the maximum applied voltage, ω is the angular velocity and ϕ is the phase angle at the lifting instance (t=0) and can be computed as :

$$\phi = \operatorname{Sin}^{-1}(E_L/E_m) \qquad (3)$$

2.1 Drag Forces Calculation

In this work the particle under study is spherical particle. The drag force is for small relative velocities proportional to the velocity and can be expressed as :[1]

$$F(\dot{x}) = 6\pi\eta R \, . \, \dot{x}(t). Kd \quad (4)$$

Where η is the dynamic viscosity (N.m), R is the radius of ball or particle under study (m).

$$kd = 0.1673 + [0.7198 \ln(R_e)] \quad (5)$$

Is the correction factor for non-linearity in drag, if $R_e \geq 2$. If $R_e < 2$ then $C_f = 1$. R_e is the Reynolds number

$$R_e = 2R \, \dot{x}(t) \frac{\rho_s(\rho_s - \rho_l)}{\eta(\rho_s + .5\rho_l)} \quad (6)$$

Where ρ_s the density of the ball under study (Kg/m^3) and ρ_l the density of the insulating liquid (Kg/m^3).

At higher values of R_e ,(4) should be replaced by:

$$FD = Cd(\tfrac{1}{2} \dot{x}^2 A \frac{\rho_s(\rho_s - \rho_l)}{(\rho_s + .5\rho_l)}) \quad (7)$$

Where Cd is the drag coefficint and A the projected area. The coefficient Cd is empirically given by

$$Kd = Cd\,(R_e/24) \quad (8)$$

2.2 Particle Charging and Electric Field Calculation by Charge Simulation Method

The charge simulation method makes use of mathematical linearization and expresses the laplace's equation as a superposition of particular solution due to discrete ficticious charges (for example point-line and ring charges). These charges replace the physically distributed surface charges on the electrode. They must be placed outside the space in which the field is to be computed . The magnitude of these charges have to be calculated so that their integrated effect satisfy the boundary conditions exactly at a selected number of contour points on the boundary equal to the number of the simulating charges. This can be represented by the following linear system of equations, Which must be the solved for the unknown column of the amounts of the charges (Q).

$$[P] \, . \, [Q] = [\phi] \quad (9)$$

Where [P] is the matrix of the potential coefficients of the charges related to the contour points and $[\phi]$ is the column of boundary conditions, which must be satisfied. For more details about the C.S.M. and its use in particle contaminated field calculations it's reffered to [8].

i) The particle is in contact with the first electrode:

In this case the particle is considered as a part of the electrode configuration is simulated by *n* charges and the particle is simulated by *m* charges, then the potential coefficient matrix [P] to be calculated and inverted will be an (m+n)*(m+n) matrix.

ii) The particle is moving in the gap:

The electric charge acquired by the particle and calculated in the excceding underpoint will be assumed to be constant during the gap crossing. The change in the potential portion $(\phi_o)_n$, caused by the particle simulating charges due to particle motion, at the *n* contour points of the electrodes should be taken into consideration after every displacement interval. Thus, the *n* simulating charges corresponding to the electrodes must be recalculated after each displacement interval to ensure correct electrode potential values $(\phi)_n$ at the *n* contour points representing the electrodes. Thus the following system of equations should be solved at every displacement interval to recalculate the simulating charge system $[Q]_n$

$$[P]_{nn} \cdot [Q]_n = [\phi_n] - [\phi_o]_n \quad (10)$$

iii) The particle is in contact with the other electrode:

Like the case in under point (i), an (m+n)*(m+n) potential coefficient matrix is to be errected and inverted to calculate the *n* simulating charges corresponding to the electrodes as well as the *m* simulating charges representing the new charge acquired by the particle at this second electrode.

2.3 Numerrical solution of the differential equation of particle motion

The numerical solution of the differential equation of particle motion as formulated by (1) is undertaken in this work as follows [9]:

The equation will be transfered in a system of linear differential equations

$$\dot{x} = V$$

$$\ddot{x} = (1/m).[Q.E(x,t)-B(V).V)] \pm g \quad (11)$$

For small time intervals Δt , it's approximately valid that:

$$\Delta x = \Delta t \, . \, V \quad \text{and}$$

$$\Delta V = \Delta t \, [(1/m).(Q.E(x,t) - B(V).V) \pm g] \quad (12)$$

Begining from the start pont (t_o , x_o , V_o), one obtains forr the distance (x_i) and the velocity (V_i) after the *i* th time interval ($t_i = t_o + i.\Delta t$)

$$x_i = x_{i-1} + \Delta t \, . \, V_{i-1} \quad \text{and}$$

$$V_i = V_{i-1} + \Delta t \, [(1/m).(Q.E(x_i , t_i) - B(V_{i-1}).V_{i-1}) \pm g] \quad (13)$$

According to these formulee , the particle motion can be calculated . If at time *t* the particle touched the electrode, so the reflection process will be discribed through the restituation coefficient R_C . As start values for the succeeding motion on gets:

$$t_o = t_n \;, \quad x_o = x_n \quad \text{and} \quad V_o = -R_c \, V_n$$

3. RESULTS AND DISCUSSION

As discussed before, the particle motion process under the influence of electric field is of great particle interest. From the field calculations it's already known that, the nearer the particle can reach toward the inner electrode, the more is the field disturbance caused by the particle. Thus, the maximum height, which the particle can reach during the simulation time, is of main interest. Consequently, all factors affecting the maximum height reached by a particle should be investigated.

Fig.1 illustrates a typical behaviour of conducting particle motion under A.C. voltage in silicone oil. The figure displays the distance travelled by a spherical aluminum particle in a parallel plane gap as a function of time . The figure displays also the applied sinusoidal voltage wave, to anable a comparsion between the particle motion and the instantaneous voltage values.

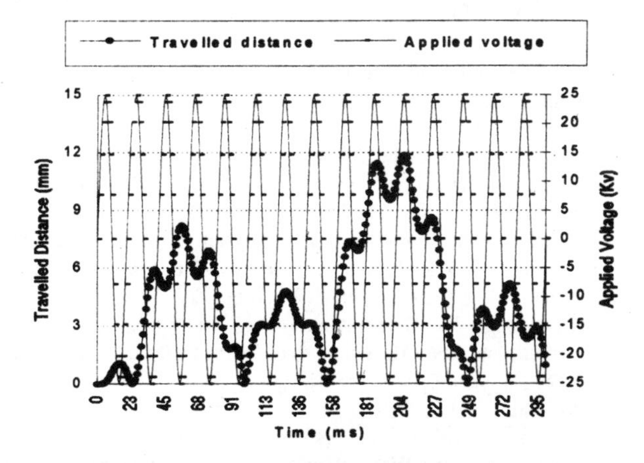

Fig.1. Distance travelled by spherical aluminum particle in parallel plate gap as well as the applied AC voltage wave (50 Hz) as a function of time. V=25Kv , R=.5mm.

Since the driving force isn't constant under A.C. voltage, the motion is more complex than under D.C. voltage. The particle at lift off move a relatively short distance during a half cycle, so that it bounces on the bottom electrode and don't cross the entire gap until higher voltages are applied or enough energy is accumulated. The particle may remain in midgap during one or more cycles and it can accumulate kinetic energy over several bounces. Nearly the same behaviour of particle motion is given in Fig.2. The latter case describes the particle motion in the same parallel plate gap as in Fig.1, but for a different two types of insulating liquids, at the same applied voltage. It's clear from the figure that, in the case of transformer oil the particle remain in midgap during the periode and it can't be cross the gap, and in the case of Freon TF the particle could accumulate enough energy to cross the gap and impact the upper electrode. This mean that, the higher density of the insulating liquid the heigher is its travelled distance. The summary of typical liquid properties is shown in table I.

To clearify the factors affecting the maximum height of a conducting particle in insulating liquids, the influence of the different parameterrs should be studied.

Table I

Summery of Typical Liquid Properties [5]

Property	Freon TF	Silicone Oil	Transformer Oil
$\varepsilon_2/\varepsilon_0$	2.39	2.29	2.29
ρ_m(Kg/m^3)	1565	816-949	900
η(Kg/m.s)	(7.32-7.64)E-4	8.16E-4-.019	1.58-2.1E-2

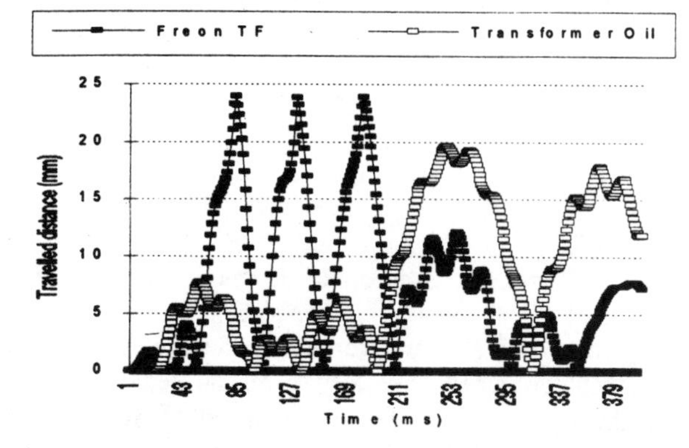

Fig.2. Distance travelled by spherical aluminum particle in parallel plate gap as a function of time. at different insulating liquids. V=25Kv , R=.5mm , R_c=.5 , d=25mm , R_e=50.

Fig.3 illustrates, for example, the effect of the particle radius on the maximum particle height in a parallel plate gap for two different insulating liquids, namely transformer oil and silicone oil. It's clear from the figure, that the maximum particle height decreases directly proportional to the particle radius. It's clear also that the maximum height is higher in case of silicone oil than in case of transformer oil.

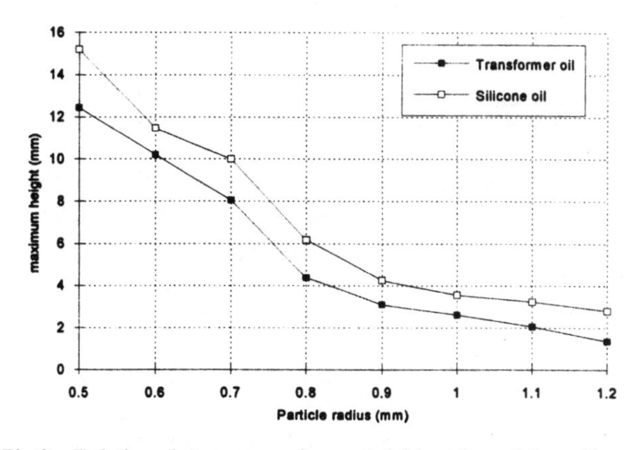

Fig.3. Relation between maximum height and particle radius at different insulating liquid. V=20Kv , R_c=.6 , Spherical aluminum particle of R_e=50.

Fig.4 displays the maximum height of a steel spherical particle in parallel plate gap as a function of restitution coefficient for three different insulating liquids, namely transformer oil, silicone oil and freon TF. It's seen from the figure, that the maximum height of the particle increase with restitution coefficient increase.

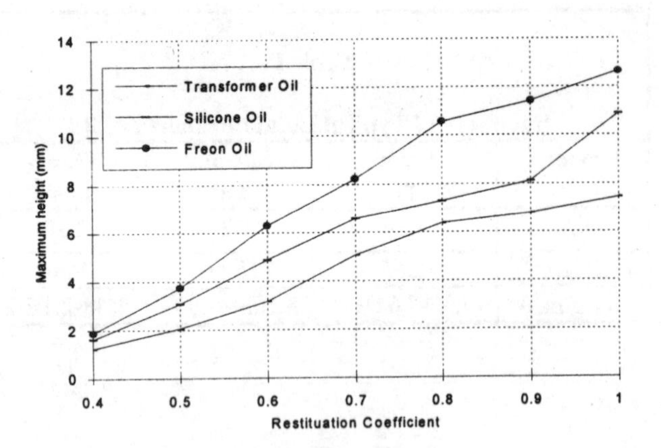

Fig.4. Relation between maximum height and restitution coefficient at different insulating liquids. V=30Kv, Steel particle of R=.5mm, R_e=50.

Fig.5 illustrates the maximum height of the particle as a function of the frequency for the same arrangement as by Fig.4, but for a 50Kv A.C. voltage and for two different insulating liquids, namely, transformer oil and freon TF. It's seen from the figure, that, the maximum height of the particle decays exponentially with frequency increase.

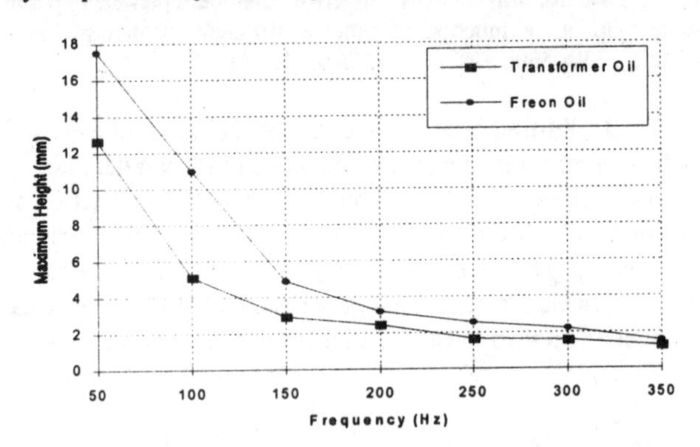

Fig.5. Relation between maximum height and frequency at different insulating liquids. Steel particle of R=.5mm , R_c=.5 , R_e=50 and V=50Kv.

Fig.6 displays also the maximum height of the particle as a function of frequency for two spherical particles of different materials, namely, steel and copper particles in transformer oil. The lighter particle(steel) travells of course higher distances than the copper one.

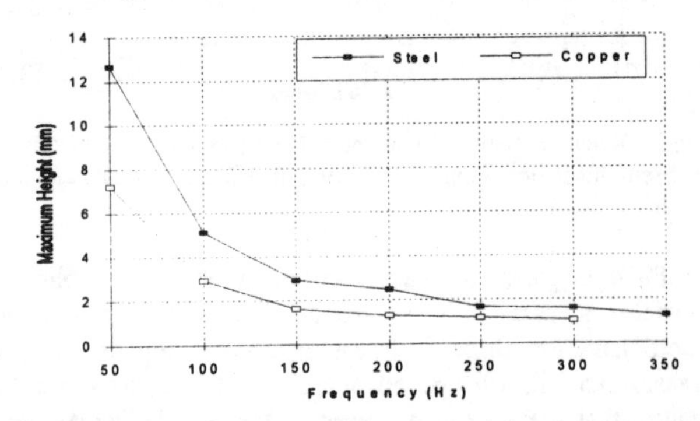

Fig.6. Relation between maximum height and frequency at different particle materials. R=.5mm , V=50Kv , R_c=.5 , R_e=50 in Transformer oil.

4. CONCLUSION

The paper describes an algorithm for the numerical calculation of free conducting particle motion in insulating liquids under A.C. voltage of different frequencies. A computer program is written and the simulation results of the particle motion are introduced and discussed.

The important parameter suspected to be the main reason of lowering the breakdown voltage by particle contaminated insulating liquids is the maximum height reached by the particle. This maximum height is found to be directly proportional to the applied voltage, inversly proportional to the supply frequency and also inversly proportional to the weight of the particle. The increase of the insulating liquid density and restitution coefficient causes increasing of the maximum height of conducting particle.

5. REFERENCES

1. S.Birlasekaran."The Movement of A Conducting Particle in Transformer Oil in AC Fields".IEEE Trans EI.Vol.28, No.1, 1993, pp.9-17.
2. Sheshakamel Jayaram."Influence of Impurities on Electro Convection in Insulating Liquids".IEEE Trans EI-27.No.2, 1992, pp.255-270.
3. Kamal Miners."Particles and Moisture effect on Dielectric Strength of Transformer Oil Using VDE Electrodes".IEEE Trrans EI.Vol.PAS-101, No.3, 1982, pp.751-756.
4. M.Darveniza."A Demonstration of The Effect of Carbon Particles on The AC Strength of Transformer Oil".Proc. of the IEEE Int. Conf. on Properties and Applications of Diel. Mate. , 1985, pp.154-156.
5. A.P.Washabaugh,M.Zahn and J.R.Melcher."Electro Hydro Dynamic Traveling-Wave Pumping of Homogeneous Semi-Insulating liquids".IEEE Trans EI.Vol.24, No5, 1989, pp.807-834.
6. G.B.Denegri,G.Liberti,G.Molinari,A.Viviant."Field-Enhanced Motion of Impurity Particles in Fluid Dielectrics Under Linear Conditions".IEEE Trans EI.Vol.12, No.2, 1977,pp.114-124.
7. J.F.Roach,M.Rosada and H.F.Levy."Liquid and Particle Motions in Transformer Oil Under 60Hz Stress".Proc. of the IEEE Int. Symp. on Elect. Ins., NewYork, NY, 1980, pp.234-238.
8. M.Abdel-Salam."On The Computation of Lifting Voltages for Free Conducting Particles".IEEE, PES.Winter Meeting, NewYork, NY, February 1979.
9. R.M.Radwan,Fathi M.H.Youssef,,S.S.El-Dessouky and S.M.El-Makkawy."An Approach to Dynamic Behaviour Simulation for Particle Contaminants in Compressed SF6 Gas Insulated Systems".5th ISH, Braunschweig 1987, p.12-11.

Conference Record of the 1996 IEEE International Symposium on Electrical Insulation, Montreal, Quebec, Canada, June 16-19, 1996

The Influence Of SF6 And SF6/N2 Gas Pressure To The Breakdown Performance Of Polyester Film

Zhang peihong Gong Guoli Dong Guangyu
Department of Electrical Material
Harbin University of Science & Technology
Harbin, 150040, P.R.China
Dong Zhenhua
Shenyang Research Institute of Transformer
Shenyang, P.R.China

Abstract: **SF6 has been widely used as the gas insulating medium in gas insulated transformer, and polyester film used as the turn insulation and other insulating materials. In this paper, the insulation strength of turn insulation of SF6/N2-film is tested when the SF6 is replaced by SF6/N2 mixed gas, and also compared with that of SF6-film. The results show that the power frequency breakdown voltage and breakdown stress of SF6/N2-film is lower than that of SF6-film in the same pressure and the same film thickness, the mean value of the former is about 91% of the latter. In order to reach the same level of turn insulation strength in the operation range, the gas pressure must be increased by 0.05 Mpa.**

INTRODUCTION

At present, SF6 is used as the insulating medium in the gas insulated transformer (GIT). But it is expensive and its liquefying temperature is high. Reference materials show, under the operating pressure, when the temperature is elow -18'C, it has the possibility of liquefying. Compared with SF6, SF6/N2 mixed gas has the advantages of better performance in low temperature, less influence by conducting particle, lower sensitivity to rough electrode, lower cost, etc. When the pressure proportion of SF6/N2 is 60/40, its insulation strength is 92% of SF6. If SF6/N2 is used as the gas insulating medium in GIT to replace SF6, it is possible to improve some performance and reduce cost. In this paper, the power frequency insulation strength of SF6/N2-film, used as turn insulation of GIT, is tested in the different gas pressure (in usual use range) and different film thickness. At the same time, the insulation strength of SF6-film is also tested in order to compare with each other.

EXPERIMENT

The turn insulation of GIT is a gas impregnated film insulation.

Test Model

In this test, the flat copper wires wrapped by polyester film and clamped by a clamping device are used as the turn insulation test model, shown in Fig.1. The clamping device is made by epoxy resin plate, its function is to sustain the samples and uniform the force between the turns. The influence to the distribution of the electrical field can be

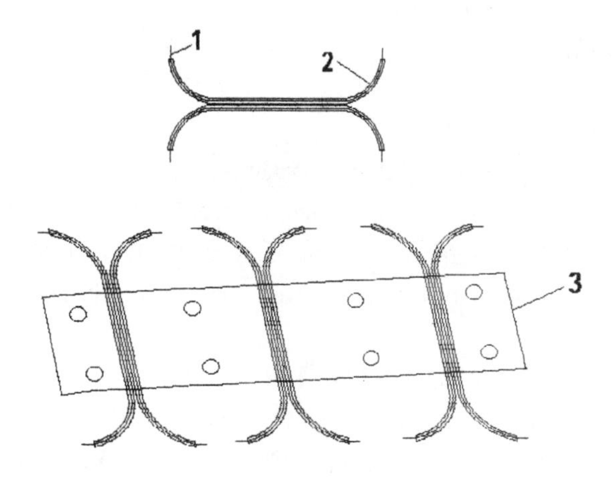

Figure 1. The test model
1.--flat copper wire 2.--polyester film 3.--clamping device made by epoxy resin plate

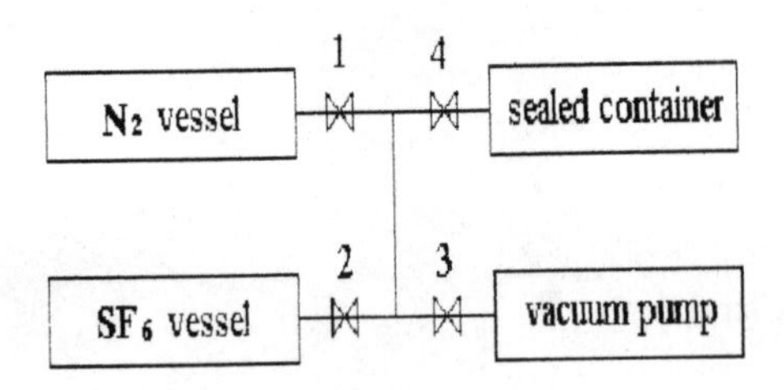

Figure 2. The schematic diagram of gas aerate-deflate equipment
1,2,3,4--valve

neglect testified by way of experiment. In order to study the relation between the insulation strength of gas-film and the film thickness, the film thickness wrapped on copper wire is 0.04x1, 0.04x2, 0.04x3, 0.04x4, 0.04x5, 0.04x6 respectively.

Gas Aerate-Deflate Equipment

A gas aerate-deflate equipment is used to fill the different pressure gas in a sealed container, the schematic diagram is shown in Fig.2. The gas filling process is as below: first, open the valve 3, 4, after pumped vacuum, close 3, open 2, when reach the partial pressure as demand, close 2, open 1, until gas filling finished, close 1. The test is done at six different gas pressure, that is 0.02, 0.05, 0.10, 0.12, 0.15, 0.20 Mpa. The pressure proportion of SF6/N2 is 70/30.

Test Procedure

First, put the model in a container, connect the wires, seal the container. Then, fill the gas according to the above process, lay aside twelve hours to make the two kinds of gas full mixed, then, test with power frequency voltage.

RESULTS AND DISCUSSION

The breakdown voltage and breakdown field stress of SF6/N2-film insulation in different gas pressure and different film thickness are shown in Fig.3, Fig.4, Fig.5, Fig.6. From the curves, we can see that when the film thickness is constant, the breakdown voltage and breakdown stress increase with the increasing of gas

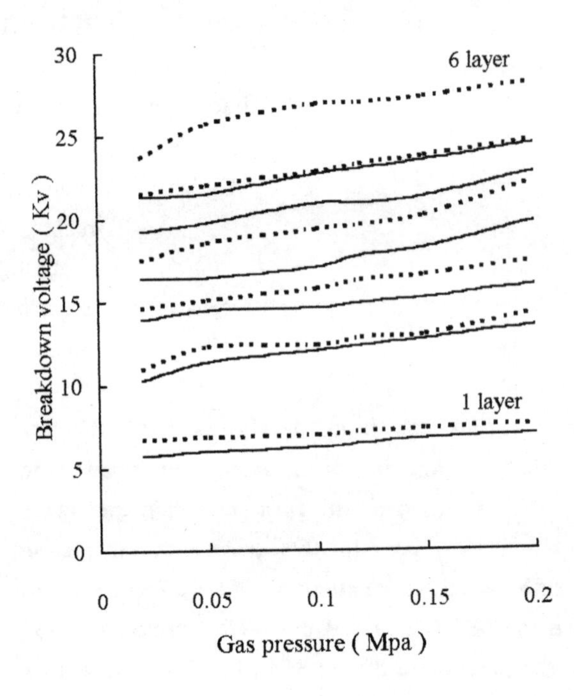

Figure 3. The breakdown voltage of different thickness gas-film insulation to the gas pressure

---- SF6 —— SF6 /N2

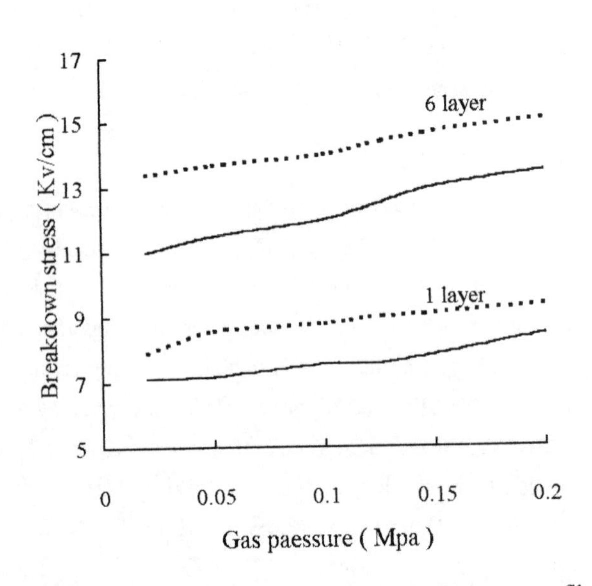

Figure 4. The breakdown stress of different thickness gas-film insulation to the gas pressure

---- SF6 —— SF6 /N2

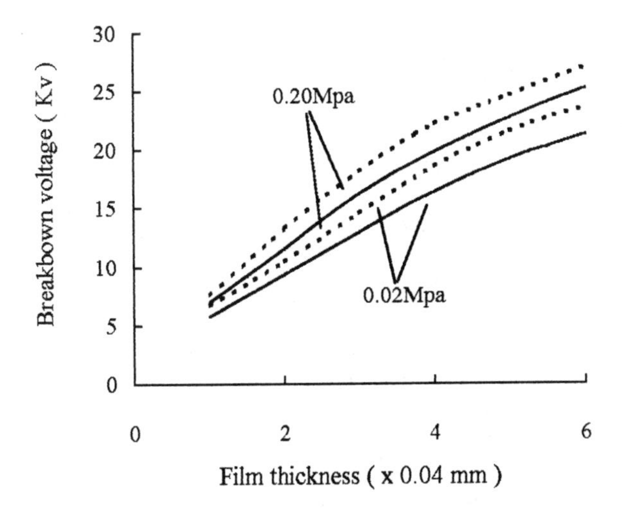

Figure 5. The breakdown voltage of different gas pressure to the film thickness

---- SF_6 —— SF_6/N_2

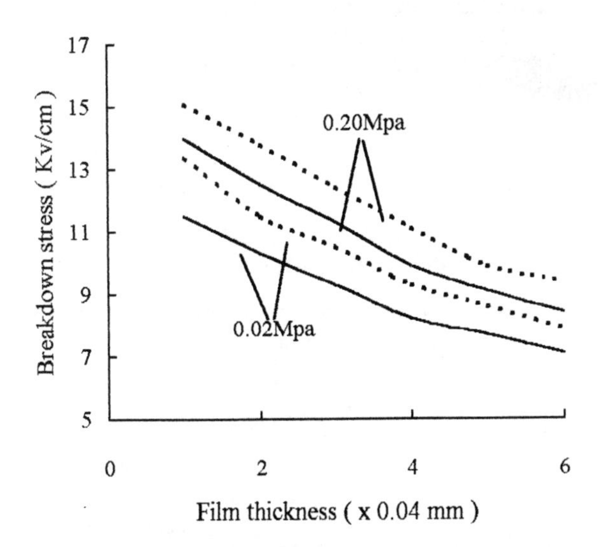

Figure 6. The breakdown stress of different gas pressure to film thickness

---- SF_6 —— SF_6/N_2

pressure; when the gas pressure is constant, the breakdown voltage increases with the increasing of the film thickness, the breakdown field stress decreases with the increasing of the film thickness. These tendencies are consistent with SF6-film.

The results also show that at the same gas pressure and the same film thickness, the insulation strength of SF6/N2-film is lower than that of SF6-film, the mean value of the former is about 91% of the latter.

The increase of insulation strength with the gas pressure is because when the gas pressure increases, the insulation strength of gas itself increases also. At the same gas pressure, the insulation strength of SF6/N2 is lower than that of SF6, so, under the same gas pressure and at the same film thickness, the breakdown voltage and the breakdown stress of the SF6/N2-film are lower than those of the SF6-film.

Now, in the GIT, the operating gas pressure is 0.10-0.15 Mpa, the turn insulating thickness is 2-3 layer polyester film. If SF6/N2 is used as the gas medium to replace SF6, shown in Fig.3, the gas pressure must be increased by 0.05 Mpa in order to reach the same turn insulation strength demand.

CONCLUSION

In this paper, only the power frequency insulation strength of turn insulation when SF6/N2 is used as the gas insulating medium of GIT to replace SF6 is concerned. On the basis of the experimental results described above, the following conclusions are drawn:

1. The change tendencies of power frequency breakdown voltage and field stress of SF6/N2-film to the gas pressure and film thickness are consistent with those of SF6-film.
2. In the same gas pressure and same film thickness, the mean value of the insulation strength of SF6/N2-film is about 91% of that of SF6-film.
3. If SF6/N2 is used as the gas insulating medium in GIT, the gas pressure must be increased by 0.05 Mpa in order to reach the same insulation strength.

REFERENCE

1. Ren Zhongyi, Gong Guoli, Gas Insulation Harbin Institute of Electrical Engineering, 1993.
2. Shenyang Research Institute of Transformer, Gas Insulated Transformer, 1987.

Conference Record of the 1996 IEEE International Symposium on Electrical Insulation, Montreal, Quebec, Canada, June 16-19, 1996

"Evaluation of MV-XLPE Cables"

J.T.Benjaminsen and H.Faremo
Norwegian Electric Power Research Institute (EFI)
Norway

J.A.Olsen
Federation of Norwegian Utilities
Norway

N.P.Mikkelsen
Elselskabet BHHH
Denmark

L.Olsson
Gøteborg Energi AB
Sweden

***Abstract*: An extensive investigation of the long term wet ageing performance of MV-XLPE cables was started by Nordic (Norway, Sweden, Denmark and Finland) utilities and cable manufacturers in 1990. Representatives from the manufacturers and utilities were members of the project Steering Committee. The investigations were performed by the Norwegian Electric Power Research Institute (EFI).**

The work was divided into two parts, a three years research part and a two years accelerated long term wet ageing test.

In the research part, a total of 5 km cable core of old and new constructions were tested. 3.8 km of this cable core were aged under wet conditions in EFI's laboratories over a period of three years. The aim was to identify the effect of different ageing parameters in order to establish test procedures and evaluation criteria for an accelerated long term test of MV-XLPE cables.

Based on the results from the research work, a two years accelerated long term test was proposed, and in the summer of 1993 the wet ageing of cables from 15 European cable manufacturers was started. In the autumn of 1995 these cables aged two years were tested to breakdown with an AC step test. 14 of the 15 cables met the requirements set by the Steering Committee.

INTRODUCTION

In the Nordic countries polymeric cables for 12 and 24 kV were introduced more than 20 years ago. Since the introduction several improvements with respect to materials, production processes, constructions, etc. have been introduced. One of the most important changes is probably the changeover from a painted-graphited to an extruded insulation screen. The painted/graphited cable construction showed an increased failure rate after 10-15 years of service. The reason for this was in most cases enhanced water tree growth from the painted/graphited insulation screen.

However, problems have also been experienced with one of the early cables with strippable extruded insulation screen containing a special additive. This cable type had suffered some breakdowns which have been found to have the same cause of failure as the old cables with painted/graphited insulation screens. Also the time to failure for this early extruded strippable insulation screen with the special additive was the same as for the cables with painted/graphited insulation screens, i.e. 10-15 years. Service experience with other cables having strippable extruded insulation screens have been good since they were first introduced (1975 - 1981).

When this Nordic investigation started, several countries had already started work in order to develop a long term ageing test in a wet environment. International groups (for instance CIGRE WG 21-11) had also been working in this field for several years, but with limited success [5].

RESEARCH WORK

Test Objects

The main part of the research work was done on XLPE cables produced in 1990, and included both strippable and fully bonded insulation screens. Cables from two different manufacturers were used in some of the tests, in order to study if the test results were affected by this.

An interesting feature was the inclusion of three old, not previously energized XLPE cables, which could be compared with the new constructions. These old cables, which were all produced in the mid seventies, were of a construction which had suffered several failures in service.

Table 1 and 2 show the different cable types included in the tests.

Table 1: New cables: 12 kV - 150 mm^2 Al with swelling powder in the conductor.

Cable Code	Manu-facturer	Insulation Screen	Comments
X1	A	Strippable	
X2	A	Strippable	Other material batches
X3	A	Strippable	Other cross section (50 mm^2)
Y1	B.	Strippable	
Y2	B	Strippable	Other production
F	A	Fully bonded	

Table 2: Old cables: 12 kV - different cross sections.

Cable Code	Manu-facturer	Insulation Screen	Comments
G1	C	Tapes and paint	35 mm^2 Al
G2	D	Early strippable	95 mm^2 Al
G3	E	Tapes and graphite	185 mm^2 Al

The IEC denomination of the cables is 6/10 (12) kV; nominal insulation thickness is 3.4 mm.

Test Program

A large scale test program (see Table 3) was set up, and the aim was to identify the effect of different testing parameters in order to establish test procedures and evaluation criteria for an accelerated long term wet ageing test of MV-XLPE cables. The test parameters were: Preconditioning, ageing voltage, ageing temperature and ageing time.

These parameters were chosen from EFI and the Steering Committees experience together with [6] and [7].

Table 3 summarizes the test program, where altogether 19 different comparisons were performed.

Table 3: Test program

Parameters		Preconditioning		
Voltage	Temp. [°C]	No	Wet, 55°C for 1 month	Dry, 90°C for 200 h
$3\ U_0$	Low (25±5)	X1 Y1	X1, X2, X3 Y1, F G1, G2, G3	X1 Y2
	High (50±5)		X1 F	X1
$4\ U_0$	Low (25±5)		X1	X1
	High (50±5)			X1 Y2

Three different preconditioning procedures were applied prior to the ageing. These were *wet preconditioning, dry preconditioning and no preconditioning*. The ageing was performed at $3U_0$ and $4U_0$ and at two different temperatures (room temperature and 50°C). All external coverings (sheaths and earth screens) were removed from all cables before preconditioning started.

In order to perform statistical analysis after each ageing interval (0, 1, 2 and 3 years of ageing) an AC breakdown step test was performed on ten cable samples from all 19 cable groups, each cable sample having an active test length of 5m. Water treeing analysis was performed on ~10 cm^3 of the cable insulation of each cable including all breakdown sites. These procedures are in agreement with the proposal of CIGRE WG 21-11, "Characterization test" [4].

Results

The test program proved that XLPE cables manufactured in the mid seventies failed to pass a two years wet ageing test performed at $3U_0$ - room temperature - wet preconditioning before ageing. The reason for the breakdowns during ageing has proven to be long vented water trees from the insulation screen. This is ageing mechanisms similar to those observed in service aged cables (for instance showed in [3]). Also other characteristic features for these cables were maintained at these ageing conditions (see Table 4). As a background for a long term test these observations were important when the parameters of a long term test that copies features from service as closely as possible should be decided upon.

Table 4: Old cables; comparisons of some typical features of service and laboratory experiences.

Type of Water Trees	Service Aged	Laboratory Aged
Bow-tie trees	Low density. Longest trees initiated from inhomogenities.	Higher density. Longest trees initiated from inhomogenities.
Vented trees from the cond. screen	Low density. Variable lengths.	Variable density as a function of cable length. Variable lengths.
Vented trees from the insul.screen	G1 and G3 types: High density. G2 type: Low density. Long lengths.	Variable densities. Long lengths

The cables with today's construction (1990) aged at $3U_0$ - room temperature - wet preconditioned before ageing did all pass three years of ageing. In Figure 1 is shown the breakdown results for"old" and "new" cables tested according to the same conditions. The differences in breakdown levels are significant between "old" and "new" cables.

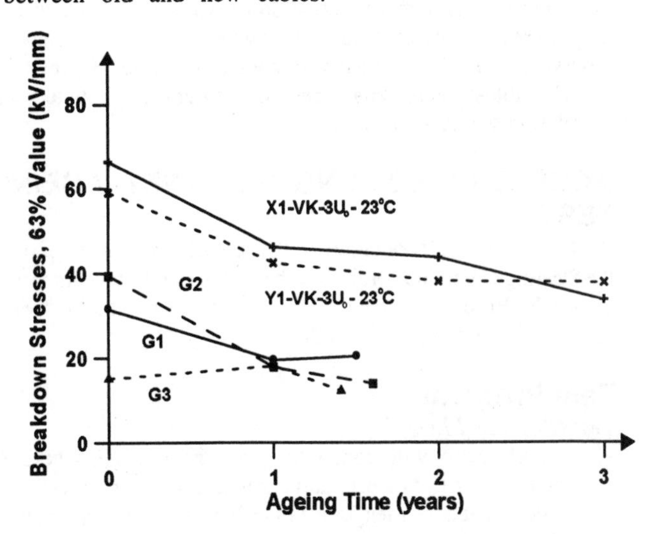

Figure 1:
Comparison of residual breakdown strengths ($E_{max\ 63\%}$) between "old" and "new" cables. The "old" cables broke down after approximately 1.5 years of ageing. VK = wet preconditioned.

Ageing at $4U_0$ combined with an ageing temperature of 50°C resulted in breakdowns of one of the "new" cables, Y2, after 2.3 years of ageing. The cause of the failure was enhanced growth of vented water trees from the conductor screen. The reason for this water tree growth seems to be severe corrosion of the aluminium conductor due to the actual ageing parameters. This phenomenon was also observed on the other cables aged at 50°C (X1 and F), however, not to the same extent

Conductor corrosion was not observed in any of the cables aged at 23°C. Ten years of experience at EFI performing investigations of service failures of XLPE cables from the Nordic countries, has not indicated that water tree growth from the conductor screen is a problem for these cable types. This brings up the question whether a test where both ageing voltage and temperature are increased ($4U_0$ and 50°C) results in representative ageing mechanisms compared to service performance. As a worst case such a test could turns a cable down that in service might have had an excellent service performance.

The influence of the different test parameters are discussed in [1] and [2] and can be summarized as follows:

Preconditioning has an influence upon the breakdown field strength for the reference values and after one year of ageing. However, after two and three years of ageing no differences appeared on any of the cables in our tests.

Increasing the ***ageing voltage*** from $3U_0$ to $4U_0$ did not affect the residual breakdown strength significantly.

Increasing both ***ageing voltage and temperature*** at the same time result in a significant reduction of the residual breakdown strength.

The correlations between breakdown values and water tree lengths are inconsistent. "Old" cables where many water trees are observed, seem to give a relatively good correlation. However, in "new" cables where only a few water trees are initiated, the correlation seems to be poor.

ACCELERATED LONG TERM WET AGEING TEST

Based on the results obtained in the research part of the project the Steering Committee proposed the following two years test of 12 kV XLPE cables intended for use in wet surroundings valid for the voltage range of 12 to 24 kV.

Test Program

Test Object and Length

- 12 kV cable with conductor size 150 mm^2 Al, where all outer coverings (sheath and earth screen) are removed.
- Active cable test length 60 m (12x5 m active test lengths). Total test length approximately 100 m.

Preconditioning

- All external coverings are removed before the preconditioning starts.
- In order to remove the cross linking byproducts (acetophenone, cumyl alcohol, etc) and to wet both insulation and screen materials the test objects shall be immersed in a water tank for 45 days at 55±5°C. After both one and three weeks of preconditioning the water shall be renewed.

Note: *Based on the service experience in the Nordic countries EFI did not expect enhanced vented water tree growth from the conductor screen in any of the cables tested in the research part. When testing new cable types this feature is, however, not known. The Steering Committee therefore decided to keep the wet preconditioning as a part of the test. By doing this no special requirement for cable construction (i.e. stranded conductor without swelling powder) is necessary. All cables will be wet throughout the whole insulation system already at the start of the test.*

Electrical Test before Ageing

- Before ageing starts the whole test length shall be exposed to an AC-test at $10U_0$ (60 kV_{rms}) for one minute.

Ageing Parameters

- Voltage: $3U_0$ (18 kV) i.e. maximum electric field stress at the conductor screen of approximately 5 kV/mm.
- Frequency: 50 Hz.
- Temperature : 30±10°C on the ageing water. The ageing is performed isothermally.
- Quality of the ageing water:
  Tap water added 0.03g NaCl per litre of water. The pH-value of the water shall be between 6.5-7.5, controlled every month.
- The ageing shall be performed with water on the conductor and on the outside of the cables.
- Test duration: 17500 hours (2 years).

Evaluation after Ageing

The aged cable length shall be cut into 12 test samples of 5 m active lengths (required cable for terminations in addition). Each test sample shall be tested to breakdown with an AC step test after the following procedure:

- Start at $5U_0$ for 5 minutes.
- Voltage increasing in steps of U_0 every 5 minutes to breakdown occurs.

Note: *Only cables totally submerged during the whole ageing period shall be used.*

Approval Criterion

A: All 12 test objects shall withstand the step of $10U_0$ for 5 minutes

or

B: All 12 test objects shall withstand the step of $9U_0$ for 5 minute
8 of the 12 test objects shall withstand the step of $11U_0$ for 5 minutes.
4 of the 12 test objects shall withstand the step of $13U_0$ for 5 minutes.

For the time being it is not set any requirements to the longest allowable water tree after ageing. One of the reasons are that the labs are not always able to stop the breakdown before the reason (the water tree) is completely burnt away. Therefore no certain value can be given for the most critical water tree of the cable. The distance to the next long water tree in "new" cables is often too long to be detected by general visual inspections.

Results

In the summer of 1993 a test according to this Nordic proposal for test procedure was started on cables from 15 different European cable manufacturers. With one exception, all the tests were performed on 12 kV cables (6/10 (12)). Both cables with peroxide curing, silane curing and WTR insulation materials were included. 10 of the cables had fully bonded insulation screens, while 5 had easy strip insulation screens. 12 of the cables had water blocking in the conductor or solid conductor.

During the two years of ageing no breakdown occurred in any of the cables. After the two years ageing of the cables were completed, 12 test objects from each cable were tested to breakdown with an AC step test. One out of the 15 cables did not pass the approval criterion set by the Steering Committee.

The twelve withstand voltages for each cable were sorted in increasing order; test object no 1 representing the lowest observed voltage for each cable and test object no 12 representing the highest.

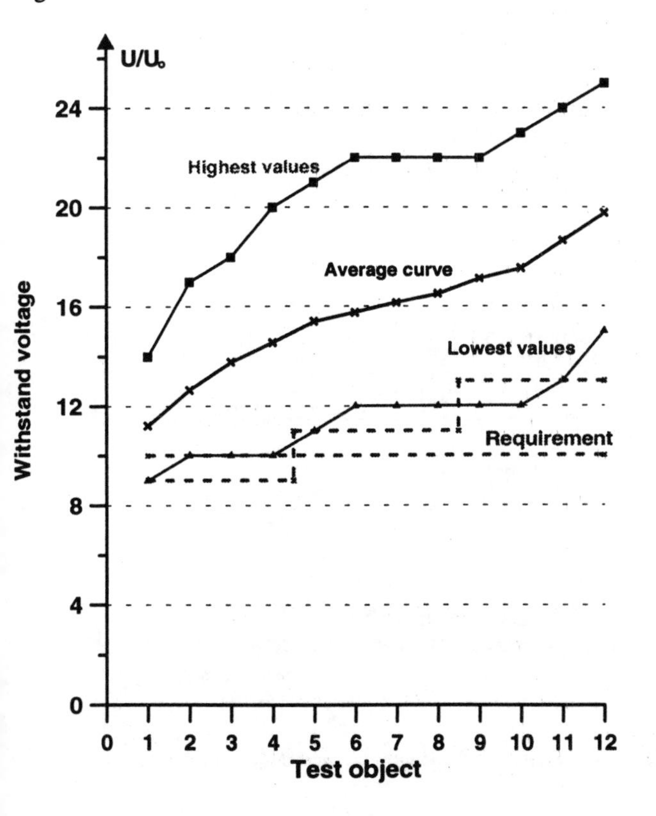

Figure 2:
Curves for the highest and lowest withstand voltages observed. In addition the curve for average withstand voltages for all the cables and the requirements to pass the test are shown.
*Note: In this figure the **withstand** values are used. The **breakdown** values will be **one** U_0 higher.*

In Figure 2 is shown a curve for the average withstand voltages for the 14 cables which passed the test. The curve is drawn by taking the average value of the lowest observed withstand voltage for each of the 14 cables (test object no 1), then the next lowest up to the average value of the highest withstand voltage for each cable (test object no 12). Also curves through the highest and lowest withstand voltages observed are drawn to show the scatter in the results. The requirements to pass the test is also shown in the same figure.

CONCLUSIONS

Ageing at $3U_0$ - room temperature - wet preconditioning - two years, clearly differentiates between "new" and "old" XLPE cables.

Features from service are reproduced at these test conditions.

Based on these results a two years accelerated long term test was proposed. 15 cables from different European XLPE cable manufacturers have been exposed to this test. One of the cables did not pass the test

ACKNOWLEDGEMENT

The Steering Committee consists of utilities and manufacturers of XLPE cables in the Nordic countries. This project highly appreciates both the technical and financial contributions from all members.

REFERENCES

[1] Benjaminsen, J.T, H.Faremo, P.Soelberg: "Long Term Testing of MV-XLPE Cables". CIRED 95, Brussels, 8-11 May 1995, Paper no 3.1.

[2] Faremo, H, J.T.Benjaminsen, J.A.Olsen, N.P.Mikkelsen, L.Olsson: "The Effect of Preconditioning on the Performance of MV-XLPE Cables in Long Term Water Treeing Tests". JICABLE 95, Versailles, France, 26.-28. June 1995.

[3] Faremo, H, B.E.Knutsen, J.A.Olsen: "12 and 24 kV XLPE Cables in Norway. Faults Related to Water Treeing". CIRED 1993, Birmingham 17.-20. May 1993.

[4] Steennis, E.F, A.M.F.J. van der Laar: "Characterisation Test and Classification Procedure for Water Tree Aged Medium Voltage Cables". Electra No. 125. July 1989, pp. 89-101.

[5] Steennis, E.F, H.Faremo: "State of the Art of Water Tree Testing on Cables. The Development of an Accelerated Ageing Test" on behalf of CIGRE WG 21-11 and TF 15-06-05. CIGRE 92, Paris 31. Aug. - 5. Sept. 1992.

[6] Faremo, H, R.Naybour, M.S.Papadopulos, E.F.Steennis: "A Water Treeing Assessment Test Specification for European Harmonization". IEEE-ICC, Birmingham 20-23. July 1992.

[7] Merschel, F: "Langzeitprüfungen an VPE-isolierten Mittelspannungskabeln - Ergebnisse, Erfarungen, Normung".Elektrizitätswirtschaft, Jg. 93 (1994) Heft 19, pp 1135-1143.

Conference Record of the 1996 IEEE International Symposium on Electrical Insulation, Montreal, Quebec, Canada, June 16-19, 1996

Diagnosis and Restoration of Water Tree Aged XLPE Cable Materials

Hallvard Faremo
Norwegian Electric Power Research Institute (EFI)

Erling Ildstad
Norwegian University of Science and Technology (NTNU)

N-7034 Trondheim, Norway

***Abstract*: One main purpose of this work was to study the effect of silicone restoration liquid on water treed XLPE cable insulation materials. Evaluation parameters used were AC and impulse breakdown strengths as well as water tree examinations. In addition a non-destructive diagnostic method was used to evaluate the degree of water treeing and restoration. The laboratory experiments were performed on water tree aged Rogowski type test objects with semiconducting screens and an insulation thickness of 0.7 mm.**

The results show that the short term effect of restoration is to strongly increase the AC and impulse breakdown strength. The long term performance, however, has to be further examined.

Return voltage measurements may become a useful diagnostic tool for evaluating the degree of water treeing and restoration of XLPE cable insulation.

INTRODUCTION

Retoration of water treed XLPE cables was first introduced in USA by using the procedure of drying and acetophenone impregnation [1]. Later a silicone based technique without drying was introduced [2]. This latter CableCure® method is considered to give a more durable restoration because the restoration liquid is said to oglomerize in contact with water.

In a previous paper we have shown that the effect of CableCure-silicone restoration is to increase the AC-breakdown strength of water tree aged test objects to a value higher than that of degassed and not-aged test objects [3]. This high improvement seemed to be for permanent nature if the test objects were kept in direct contact with the restoration liquid during the continued water tree ageing. However, if the surplus restoration liquid was removed water trees continued to grow and the improvement in AC breakdown strength was reduced to a value slightly higher than that of non-restored test objects. One objective of this paper was to find out whether restoration has a similar effect on residual impulse breakdown strength.

Today replacement or restoration criteria for old cables are very rough. Based upon failure statistics only, it is very likely that cables are replaced or restored earlier than necessary. By correct diagnosis of the condition great savings can be made by replacing or restoring an old cable at the right moment. Thus there is a fundamental need for diagnostic test methods to detect cables with a high probability of failure. A second objective of this paper was to demonstrate the use of dielectric response measurements as a diagnostic tool to evaluate the degree of water treeing and restoration.

EXPERIMENTAL

Test samples

All tests were performed on Rogowski type test objects with an insulation thickness of 0.7 mm, pressure moulded between 0.5 mm thick semiconducting layers [4] (see Figure 1). The materials used were NCPE 4201S for the insulation and NCPE 0592 for the semiconductors.

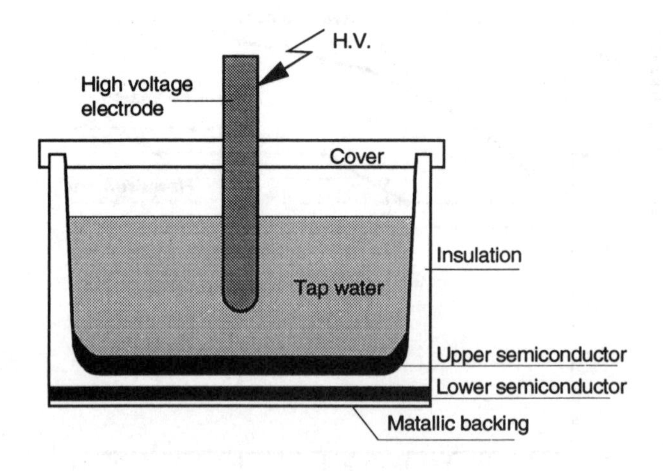

Figure 1: Sketch of the Rogowski type test object.

Ageing conditions and test procedures

56 test objects were filled with water energized at a 50 Hz AC electric stress of 10 kV/mm and subjected to temperature cycling between 90°C (16 hours) and 10°C (8 hours). The effect of this temperature cycling is to enhance water tree initation and growth.

Before ageing and at the time intervals indicated in Figure 2 we performed diagnostic testing, water tree examination and 1.2/50 μs impulse breakdown strength measurements.

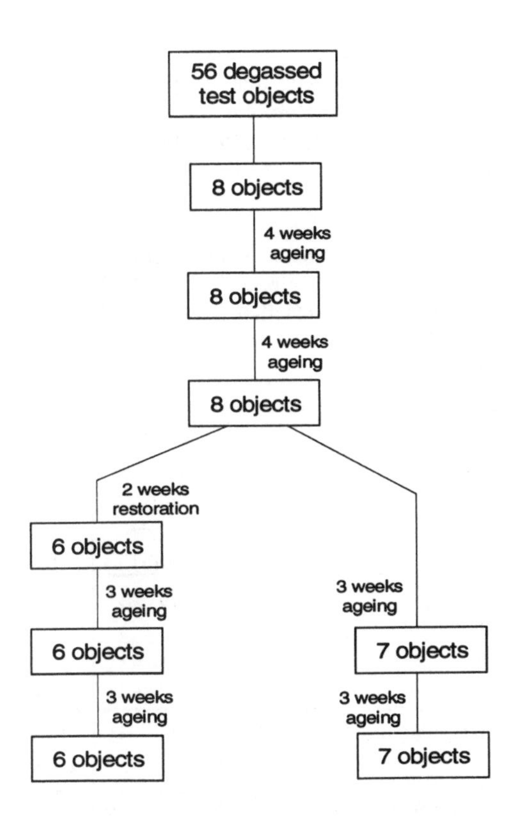

Figure 2: Test procedure. Water tree ageing of 56 Rogowski type test objects at 10 kV/mm. Temperature cycling between 90°C (16 hours) and 10°C (8 hours). Restoration by CableCure-silicone liquid.

Restoration

After 8 weeks of ageing restoration was performed using a CableCure silicone based restoration liquid supplied by UTILX. No drying was performed before restoration, only the surface of the test objects were wiped dry using paper towels. Then approximately 20 ml of the restoration liquid was filled into each test object which then was placed in an enclosed metallic box for 2 weeks at room temperature (23°C). Typically each test object gained 1.5% in weight during this impregnation.

After restoration the surplus liquid impregnant was removed and the surface of the upper semiconductor was wiped dry with a paper towel before the cup was refilled with tap water and reenergized at 10 kV/mm AC.

Diagnostic testing

Time domaine dielectric response measurements are characterized by measure of either depolarization current or voltage return caused by application of a DC voltage across the insulation material. In case of these small test objects, with a capacitance of about 50 pF, we found it most practical to measure the voltage return. All measurements were performed using the computerized experimental arrangment schematically shown in Figure 3.

A high voltage amplifier (Trek 20/20) was used to charge the test objects. Measurements of depolarization current and return voltage were performed by an electrometer (Keithly 617).

The test objects were charged for 15 minutes at different DC levels up to 10 kV before grounding for 3 seconds, followed by response measurments during 15 minutes.

All mesurements were performed at room temperature.

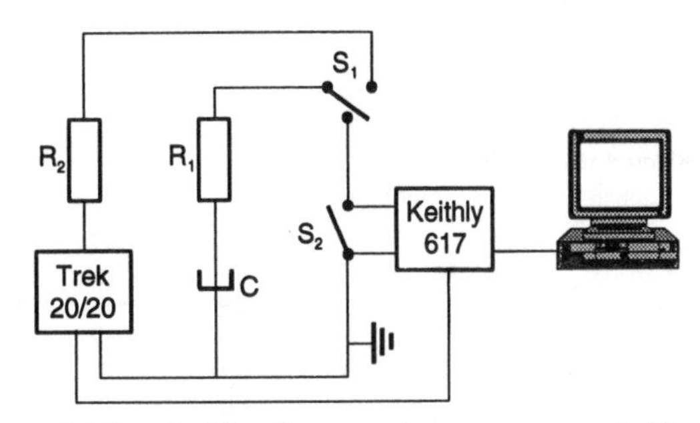

Figure 3: The principle of return voltage measurements. R_1 and R_2 are charging and discharging resistors respectively, C represents the test object and S_1 and S_2 are PC-controlled switches.

Impulse breakdown testing

During breakdown testing the samples were soaked in silicone oil to avoid external flashover. At each voltage level five lightning impulses (1.2/50 µs) were applied at intervals of 30 s. The voltage was increased at steps of 10 kV until electrical breakdown occurred. The lowest test voltage was 50 kV and all tests were performed at room temperature.

Water tree examinations

Microscopy water tree examinations were performed on 0.5 mm thick microtomed slices stained by the methylene blue dye procedure using a magnification of 100 times. Five slices from each of the aged test objects were examined with respect to the longest water tree in each slice.

RESULTS AND DISCUSSION

Breakdown strength

Results presented in Figure 4 and 5 demonstrate that the effect of restoration was to increase the impulse breakdown strength to a value about 20% higher than that of degassed non-aged objects. When removing the restoration liquid and continuing the water tree ageing a significant reduction of breakdown strength was observed. At the termination of the experiments, six weeks after restoration the impulse breakdown strength of the CableCure restored test objects was not significantly higher than that of the non-restored objects.

For comparison previous results from examination of residual AC breakdown strength are also presented in Figures 4 and 5 [3]. It is seen that in general the impulse breakdown strength is about two times higher than the AC strength.

After restoration the residual breakdown strength drops relatively fast down to almost the same levels as if no treatment had taken place. This rapid reduction of breakdown strength indicates that oglomerization is not a dominant process. However, it can be argued that two weeks of impregnation is insufficient because the oglomerization process is very slow, and may take several years to become effective [5].

As expected continued ageing with a reservoir of restoration liquid gave a more permanent high AC breakdown strength [3].

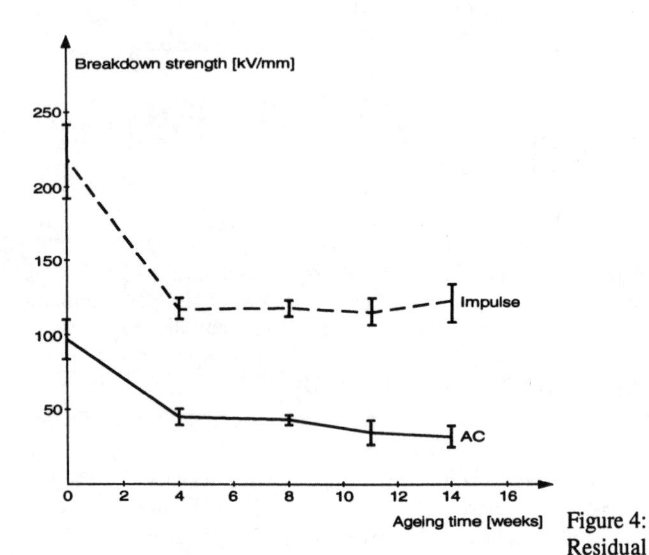

Figure 4: Residual AC and impulse breakdown strengths ($E_{63\%}$)without any restoration as a function of ageing time for water tree aged test objects. The 95% confidence levels are indiacted.

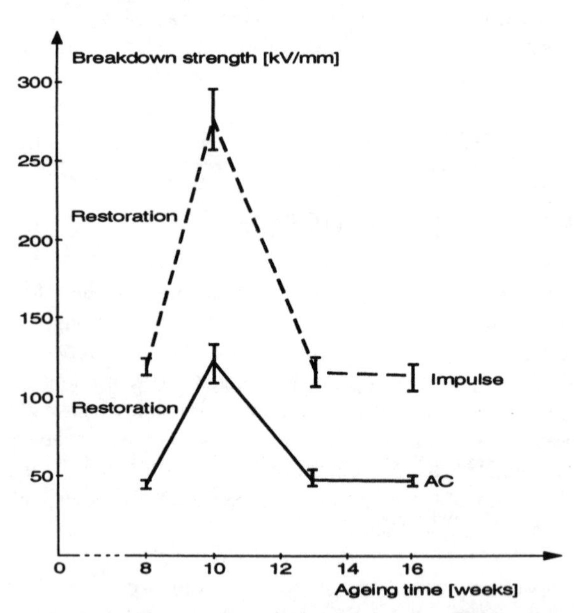

Figure 5: Residual AC and impulse breakdown strength ($E_{63\%}$) as a function of ageing time. CableCure silicone restoration during weeks 8-10. The 95% confidence levels are indicated.

Water tree lengths

Results from microscopy water tree analysis are presented in Table 1. Due to the temperature cycling, facilitating condensation of water during cooling of the insulation from 90 to 20°C, the largest water trees had reached a relatively stable length after about four weeks of ageing.

Table 1: Results from the water tree investigations performed on Rogowski type test objects. Ageing at 10 kV/mm with cyclic temperature variations between 90°C for 16 hours and 10°C for 8 hours. Two weeks of restoration after 8 weeks of ageing.

Ageing time [weeks]	Treatment	Bow-ties $\bar{\ell}_{max} \pm s$ [μm]	Vented upper semicon $\bar{\ell}_{max} \pm s$ [μm]	Vented lower semicon $\bar{\ell}_{max} \pm s$ [μm]
4	None	175±28	178±69	264±84
8	"	194±47	229±52	217±74
11	"	190±46	251±53	247±41
14	"	187±57	218±64	243±76
8	CableCure	182±52	271±95	280±133
11	"	199±43	249±83	243±100
14	"	218±58	269±56	264±109

This observation is in agreement with the residual breakdown values presented in Figure 4. After an initial rapid reduction the first 4 weeks the breakdown strength is reduced at a slow rate. This result demonstrate the strong correlation between long lengths of water trees and low residual breakdown strengths.

At the termination of the tests it was not possible to observe any difference in water tree lengths between restored and non-restored test objects.

Voltage return measurements

Results from measurements of return voltages are presented in Figure 6 and 7. It is clearly demonstrated that both water tree ageing and restoration causes increased magnitude of return voltage. In agreement with previous observations on service aged cable samples [6], the effect of increasing the applied poling voltage was to cause a more than linear increase of return voltage values.

Results presented in Figure 7 show that after restoration, water tree ageing resulted in a strong increase of return voltage. This increased slow depolarisation may be caused by increased content of polar groups and higher ability for water absorption.

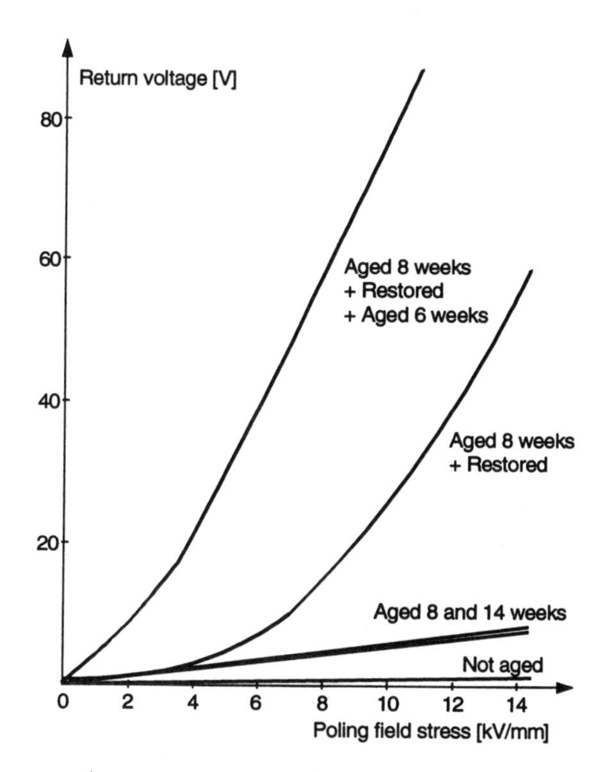

Figure 6: Measured return voltage values (after 90 s) as a function of applied DC poling voltage.

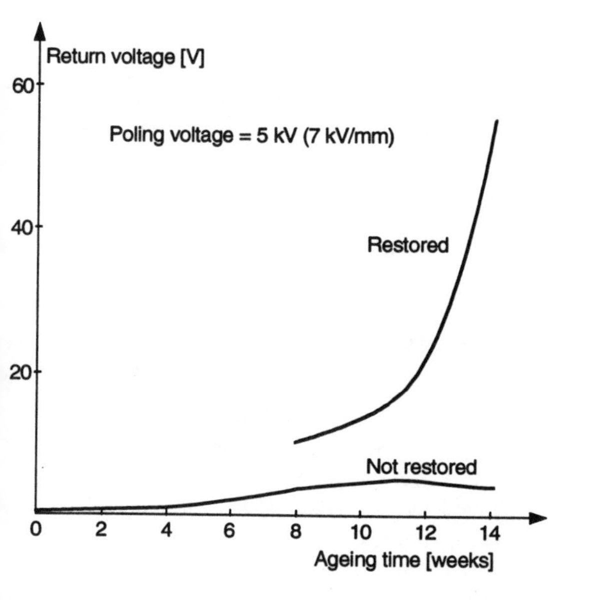

Figure 7: Return voltage versus ageing time. Readings were taken after 15 min poling at 5 kV, 3 s short circuit and 90 seconds return voltage measurements.

CONCLUSIONS

Water tree analysis shows that reduction of residual impulse breakdown strength is closely related to the length of the longest water trees.

It is demonstrated that CableCure silicone restoration strongly increase the residual 1.2/50 μs lightning impulse breakdown strength of water tree aged XLPE cable insulation.

The results confirm previous observations that it is important to keep the restoration liquid in contact with the insulation for a long time in order to obtain a permanent effect.

Dielectric response measurements may become a valuable diagnostic tool to evaluate the degree of water treeing and restoration of XLPE cable insulation.

ACKNOWLEDGEMENT

The authors would like to express their thanks to R.R.Myklebust who carried out most of the experimental work. We would also gratefully acknowledge the supply of CableCure restoration liquid from UTILX.

REFERENCES

[1] G.S.Eager, C.Katz, B.Fryszczyn, F.E.Fischer, E.Thalmann: "Extending Service Life of Installed 15-35 kV Extruded Dielectric Cables". IEEE Trans. PAS, Vol. 103, Aug. 1984, pp. 1997-2005.

[2] P.R.Nannery, J.W.Tarpey, J.S.Lacenere, D.F.Meyer, G.Bertini: "Extending the Serivice Life of 15 kV Polyethylene URD Cable Using Silicone Liquid". IEEE Trans. on Power Deliv., Vol. 4, No. 4, Oct. 1989, pp. 191-195.

[3] H.Faremo, E.Ildstad: "Rehabilitation of Water Tree Aged XLPE Insulation". IEEE Int. Symp. on El. Insul., Pittsburg USA, June 5-8 1994, Conf. Proceedings pp. 188-192.

[4] H.Faremo, E.Ildstad: "The EFI Test Method for Accelerated Growth of Water Trees". IEEE Int. Symp. on El. Ins., June 1990, Toronto, Conf. Record: pp. 191-194.

[5] G.J.Bertini, D.L.Kleyer: "The Extension of CableCure Dielectric Enhancement Technology to Feeder and 35 kV Cables". ICC, St.Petersburg, Florida, Nov. 1992.

[6] S.Hvidsten, E.Ildstad, B.Holmgren, P.Werelius: "Correlation between AC Breakdown Strength and Low Frequency Dielectric Loss of Water Tree Aged XLPE Cables". Stockholm Power Tech., June 18-22, 1995, pp. 42-47.

Conference Record of the 1996 IEEE International Symposium on Electrical Insulation, Montreal, Quebec, Canada, June 16-19, 1996

Dielectric Behaviour of AC Aged Polyethylene in Humid Environment

P.C.N.Scarpa, A.T.Bulinski*, S.Bamji* and D.K.Das-Gupta

School of Electronic Engineering and Computer Systems
University of Wales, Bangor
Dean Street, Bangor, Gwynedd, LL57 1UT, U.K.

*National Research Council
Ottawa, Canada, K1A 0R6

Abstract: **The present paper reports the results of a study of electrical ageing of low density polyethylene (LDPE) aged in humid environment (0.1M NaCl) at an AC stress of 6kV/mm, 1kHz, at room temperature (RT) and at 65°C, and cross-linked polyethylene (XLPE) AC aged in humid environment (water)at an AC stress of 6kV/mm, 50Hz, at RT, for an extended period of time. For this study the dielectric spectroscopy data in the frequency range of 10^{-5}Hz to 10^{6}Hz and their comparative analysis, have been used to provide electrical analog models of the ageing.**

INTRODUCTION

Polyethylene, one of the most used electrical insulating materials, despite its excellent physical and chemical properties, is known to exhibit structural degradation when submitted to a continued AC stress in humid environment, which may lead to a breakdown of the polymer [1]. It has been shown that the polarization and dielectric behaviour of polymers subjected to a stress in an adverse environment, may be expected to change as the time progresses, due to an enhanced conductivity originating from ionic products and injected and subsequently localized space charges [2,3].

In the present work the dielectric relaxation behaviour as a function of time, of polyethylene aged at constant alternating electric field in a humid environment, has been used as a measure of the ageing.

EXPERIMENTAL

The samples used in this work were (i) flat sheet samples of low density polyethylene (LDPE) moulded in a hot press to a thickness of approximately 0.6mm, and (ii) pieces of cross-linked (XLPE) cable (15kV rating)

The preparation of the flat LDPE samples has been described elsewhere [4]. The LDPE samples were aged for up to 454 hours at room temperature (RT) and at 65°C, with an AC field of 6kV/mm at 1kHz. The ageing was carried out with the samples in a specially constructed cell, in which one side of the sample with a 30mm vacuum deposited aluminium electrode was connected to earth and the other side, previously sandblasted, was kept in contact with a 0.1M NaCl aqueous solution. After the ageing the samples were removed from the cell, cleaned carefully with distiled water, and then both surfaces of each sample were provides with vacuum deposited aluminium electrodes of 12mm in diameter. The samples were then located in a stainless steel chamber where the dielectric measurements were performed. The dielectric data in the frequency of 10^{-5}Hz to $5*10^{-2}$Hz were calculated from the transient discharging current, via Hamon approximation, whereas bridge measurements provide the data in the frequency range of 10Haz to 10^{6}Hz.

The ageing of the XLPE cables was carried out with the samples in a specially designed stainless steel tank, filled with tap water, at RT, with an AC field of 6kV/mm at 50Hz, for up to 6000 hours. The dielectric measurements were performed with the cables in the water tank. The outer semicon layers of the cables were in contact with water and formed one electrode, the other electrode being the central metallic cores. Further details of the ageing and measurements set-up, techniques and procedure have been described elsewhere [3,5], and will not be repeated here.

RESULTS AND DISCUSSION

Figures 1 and 2 show the dielectric behaviour of the real and imaginary parts [$\chi'(\omega)$ and $\chi''(\omega,)$ respectively] of the complex susceptibility $\chi(\omega)$ of the unaged and AC aged for 6000 hours in humid environment, XLPE cable samples. The data presented in figures 1 and 2 were obtained from the measured and calculated frequency response of the real [$\varepsilon'(\omega)$] and imaginary [$\varepsilon''(\omega)$] parts of the complex permittivity $\varepsilon(\omega)$, in the same manner described in details elsewhere [3]. The data in these figures have been fitted to show a presence of three relaxation processes, in parallel, in which each process may be represented by a susceptibility function. Taking into account the asymptotic behaviour below and above the peak frequency ω_p, the imaginary part of the Cole-Cole, Davidson-Cole, Havriliak-Negami, and Jonscher, Dissado-Hill susceptibility functions [6], may be represented, respectively, by the following expressions:

$$\chi''(\omega) \propto \left[(\omega/\omega_p)^{\alpha-1} + (\omega/\omega_p)^{1-\alpha}\right]^{-1} \quad (1)$$

$$\chi''(\omega) \propto \left[(\omega/\omega_p)^{-1} + (\omega/\omega_p)^{\beta}\right]^{-1} \quad (2)$$

$$\chi''(\omega) \propto \left[(\omega/\omega_p)^{\alpha-1} + (\omega/\omega_p)^{\beta(1-\alpha)}\right]^{-1} \quad (3)$$

$$\chi''(\omega) \propto \left[(\omega/\omega_p)^{-m} + (\omega/\omega_p)^{1-n}\right]^{-1} \quad (4)$$

where α, β, m and n fall in the range (0,1). As can be seen from figures 1 and 2, better fittings were obtained with the use of the Havriliak-Negami (3) and Jonscher, Dissado-Hill (4) susceptibility functions, in comparison with the fittings obtained with the susceptibility functions with only one parameter, i.e., the Cole-Cole (1) and Davidson-Cole (2) functions.

It is suggested that the high frequency loss, which does not appear to change significantly with the ageing, may be attributed to clustered water. Also, the medium frequency peak, which becomes broader with continuing ageing, may be possibly due to an enhancement of oxidation products including carboxylic groups, together with polar moieties of additives and cross-linking by-products diffusing out of the cables. It is worthwhile to notice that the low frequency loss peak changes significantly with ageing time. This peak may, initially, be attributed to the glass transition temperature (T_g) of the polymer [7], and it may be modified by localized space charges arising from injection from the field and subsequent trapping. It is suggested that with further ageing a Quasi-DC (QDC) process [8] is brought about, due to movement of real charges as the XLPE cable becomes more conductive. a QDC process may be represented by [6],

$$\chi'(\omega) \propto \chi''(\omega) \propto \omega^{n-1} \quad (5)$$

Further evidence of the change of the nature of the dielectric response from dipolar to a space charge origin, may be seen in figures 3 and 4. These figures show the dielectric behaviour of the real [χ'(ω)] and imaginary [χ"(ω)] parts of the complex susceptibility [χ(ω)] of the unaged and AC aged for 100 hours and 454 hours in humid environment at RT and at 65°C, respectively. Due to the presence of QDC processes, the data in figures 3 and 4 were fitted only with the fractional power law (4) e (5) (i.e., the Jonscher "universal" law).

As may be seen from figure 3, the QDC process observed in the low frequency range, gives a progressively flatter response as the ageing goes on. Also, the amplitude of the high frequency loss peak does not appear to change significantly with ageing time, a broadening of that peak being observed. It may be suggested that the formation of oxidation products and the injected and subsequently trapped space charges give origin to a competitive process, in which the space charge component becomes more dominant as the ageing progresses.

Similar situation may be observed with ageing at 65°C, as shown in figure 4. It is suggested that the thermal annealing would assist a release of the space charges possibly introduced during the manufacturing process. and it promotes a redistribution of the topographical impurities. in the polymer. As a result single QDC process spanning from the low frequency range up to ~10^2Hz in the medium frequency range, instead of the two QDC processes observed in the unaged sample. The effect of continued AC stress at 65°C is to enhance this process, as evidenced by the increase of the values of χ'(ω) and χ"(ω) with ageing time.

ACKNOWLEDGEMENT

One of the authors (PCNS) is grateful to LAC-COPEL/UFPR (Curitiba-PR - Brazil) and ÇNPq (Brazil) for a maintenance grant.

REFERENCES

1. Dissado, L.A. and Fothergill, J.C. *Electrical Degradation and Breakdown in Polymers*. Peter Peregrinus Ltd, London, 1992.
2. Das-Gupta, D.K.; Svatík, A.; Bulinski, A.T.; Densley, R.J.; Bamji, S. and Carlsson, D.J. "On the Nature of AC Field Aging of Cross-linked Polyethylene Using Liquid Electrodes". J. Phys. D: Appl. Phys. Vol. 23, 1990, pp. 1599-1607.
3. Scarpa, P.C.N.; Svatík, A. and Das-Gupta, D.K. "Dielectric Spectroscopy of Polyethylene in the Frequency Range of 10^{-5}Hz to 10^6Hz". Accepted for publication in Polym. Eng. Sci., 1995.
4. Bulinski, A.; Bamji, S.; Densley, J.; Gustafsson, A. and Gedde, U.W. "Water Treeing in Binary Linear Polyethylene Blends - The Mechanical Aspect". IEEE Trans. Diel. Elect. Insul. Vol.1 No.6, 1994, pp. 949-962.
5. Scarpa, P.C.N.; Bulinski, A.T.; Bamji, S. and Das-Gupta, D.K. "Dielectric Spectroscopy of AC Aged Polyethylene in the Frequency Range of 10^{-5}Hz to 10^6Hz". Annual Report of the 1995 Conference on Electrical Insulation and Dielectric Phenomena. IEEE Publication 95CH35842, pp. 81-84.
6. Jonscher, A.K. *Dielectric Relaxation in Solids*. Chelsea Dielectric Press, London, 1983.
7. Williams, G. "Molecular Aspects of Multiple Dielectric Relaxation Processes in Solid Polymers". Adv. Polym. Sci. Vol. 33, 1979, pp. 60-92.

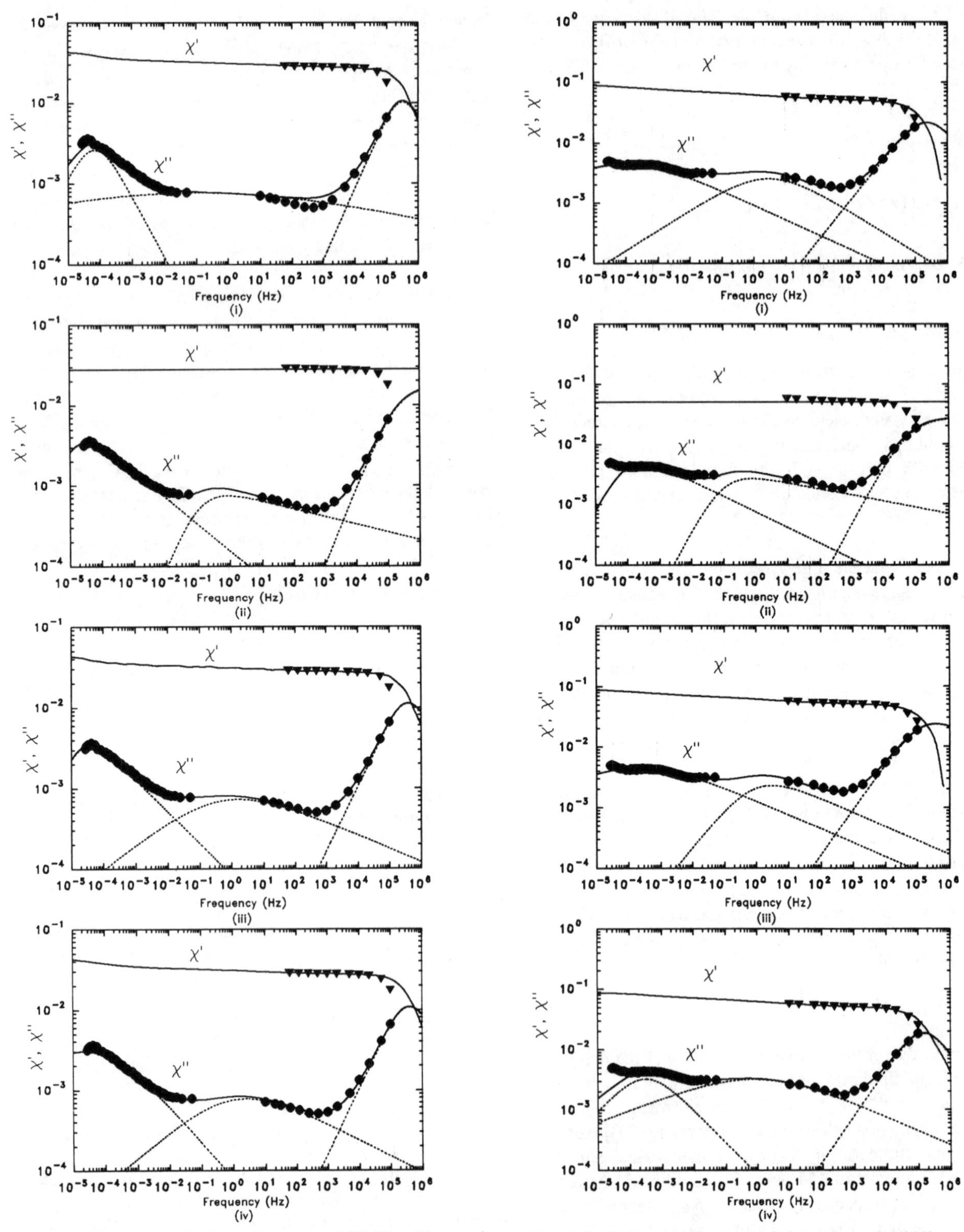

Figure 1: Dielectric behaviour of the unaged XLPE cable sample, after 48 hours of immersion in tap water, fitted with (i) Cole-Cole, (ii) Davidson-Cole, (iii) Havriliak-Negami and (iv) Jonscher , Dissado-Hill functions

Figure 2: Dielectric behaviour of the AC aged (6kV/mm, 50Hz, RT) XLPE cable sample, after 6000 hours of ageing in water, fitted with (i) Cole-Cole, (ii) Davidson-Cole, (iii) Havriliak-Negami and (iv) Jonscher , Dissado-Hill functions

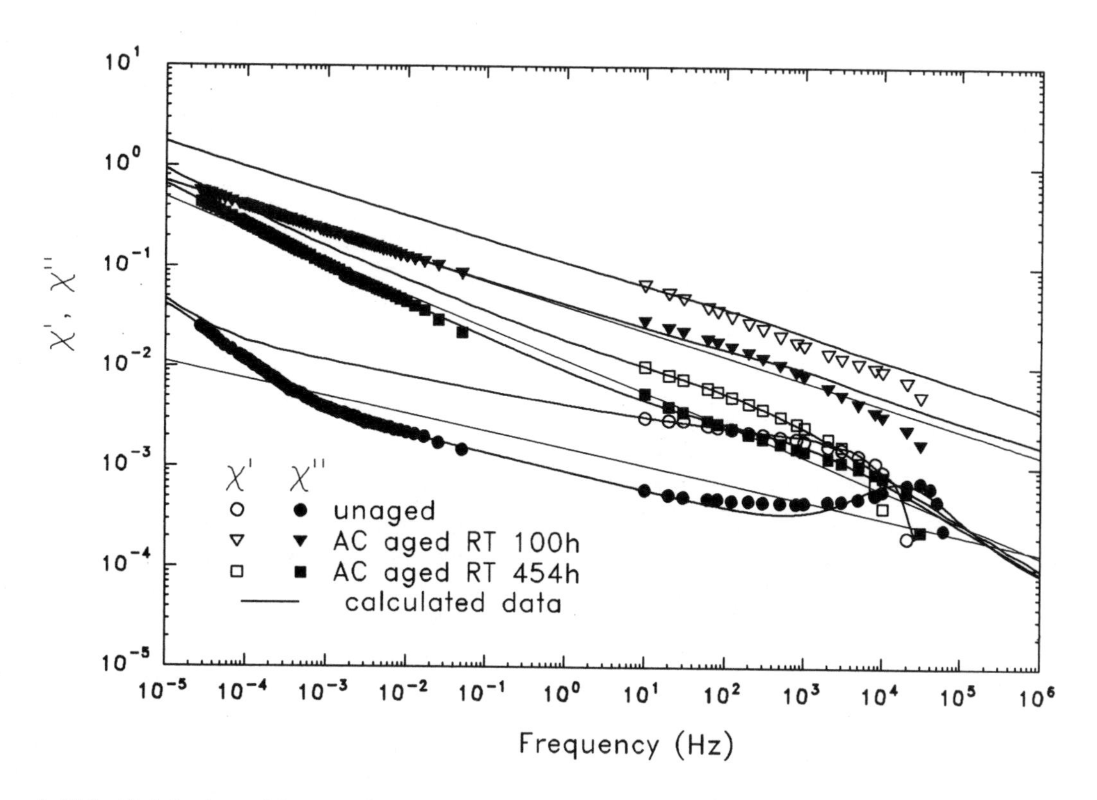

Figure 3: Dielectric behaviour of the unaged and the AC aged (6kV/mm, 1kHz, RT, 0.1M NaCl) for 100 hours and 454 hours LDPE samples.

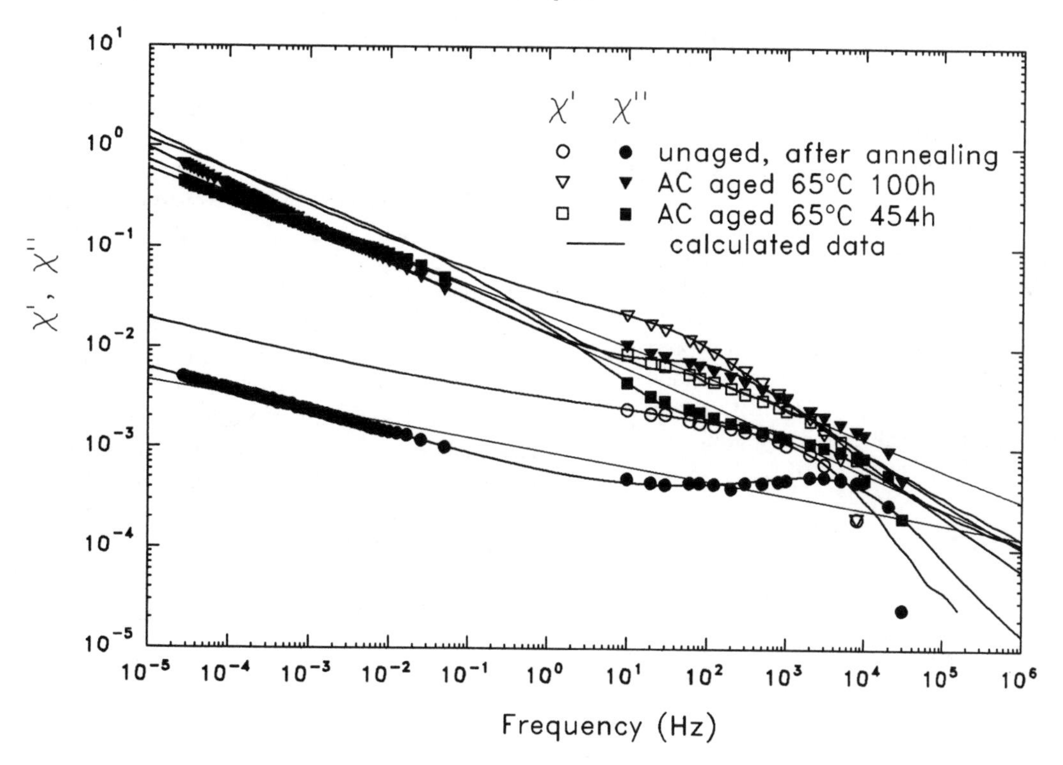

Figure 4: Dielectric behaviour of the unaged, after thermal annealing (vacuum, 50°C, 96 hours), and the AC aged (6kV/mm, 1kHz, 65°C, 0.1M NaCl) for 100 hours and 454 hours LDPE samples.

Conference Record of the 1996 IEEE International Symposium on Electrical Insulation, Montreal, Quebec, Canada, June 16-19, 1996

THE INFLUENCE OF WATER IN XLPE CABLE CONDUCTOR ON XLPE INSULATION BREAKDOWN VOLTAGE AND PARTIAL DISCHARGE

S.V.Nikolajevic, B.B.Stojanovic
The Electric Power Distribution Company
Beograd, SR Yugoslavia

ABSTRACT This paper presents the results of a continuing investigation into degradation of the crosslinked polyethylene (XLPE) cable insulation. The paper deals with the changing of water absorption of various types of XLPE cable insulations: steam and nitrogen-dry cured crosslinked polyethylene (XL) and steam and nitrogen-dry cured water tree retardant crosslinked polyethylene (WTR-XL). The results of the study into effect of water absorption on breakdown stress (AC BDS) and partial discharge for different XLPE cable insulations are also given. During the aging tests, the cable conductor was poured with the tap water and the cable ends were properly closed.

INTRODUCTION

The water (or water vapor) pressure in the cable conductor is influenced by the quantity of water in the cable conductor, the temperature and the electric field. The water absorption in the cable insulation depends on water quantity in the cable conductor, temperature, electric field in the cable insulation and water pressure in XLPE cable conductor. The results obtained in this investigation were discussed in the framework of Maxwell's and Fick's lows [2], [3], [4], [5], [9].

The aim of this investigation is to present the influence of water on breakdown voltage and partial discharge of XLPE insulation and the changing of water (or water vapor) pressure in XLPE cable conductor. Multiple stress aging data emphasize a synergistic effect of water, pressure of water (or water vapor), electric field and temperature. The aging test was performed on XLPE cable models. The samples taken from these cable models were subjected to electric stress and heating.

EXPERIMENTAL

Investigation was carried out for cable models having the following construction: conductor of aluminium stranded wires, 50 mm^2 in cross section, 8.8 mm in diameter; conductor screen from semiconducting layer, type HFDS 0592, 0.5 mm thick; insulation with and without water tree retardant crosslinked polyethylene, types 4208 XL and HFDS 4201S XL respectively, 2.4 mm thick; insulation screen from semiconducting layer, type HFDS 0592, 0.6 mm thick, metallic screen from copper tapes (2 tapes) dimension (0.15 x 20) mm and polyethylene sheath 1.9 mm thick. Great care was taken for ensuring homogenous compounding and technologies in actual production, for every of four types of cable models. The steam and nitrogen-dry cured cable insulation was cooled with water.

Cable samples were chosen in such a way to cover a wide range of XLPE cables, having therefore, the same construction, materials and technology as cables used in the actual transmission and distribution networks.

From each reel of cable models of these manufacturing runs, samples were cut and sequentially numbered. Cable samples were prepared for the tests so that the insulation screen was carefully removed from both ends, leaving (5 - 10)m between stress cones as electrodes during the aging. Each cable sample was poured with the tap water and cable ends were properly closed. The aging conditions for all samples were: temperature 90°C, ambient temperature (17°C-25°C), frequency 50 Hz and the electrical field on conductor screen was varied 0 kV/mm, 4.2 kV/mm, 8.4 kV/mm and 15.1 kV/mm.

RESULTS AND DISCUSSION

The initial unaged cable characteristics were determined by measuring: water content and AC BDS. Periodically five samples of each group were removed from the aging test. The insulation specimens for measuring of water content were taken from the places on cable samples close to where the break down occurred. The pressure of water (or water vapor) in the cable was measured daily. It can be observed on Fig. 1 that with the increased electric field the water (water-vapor) pressure in the cable decrease. This could be attributed to the greater diffusion of water vapor and condensation to liquid water upon increased electric field, which eventually leads to the faster propagation of water in the insulation system.

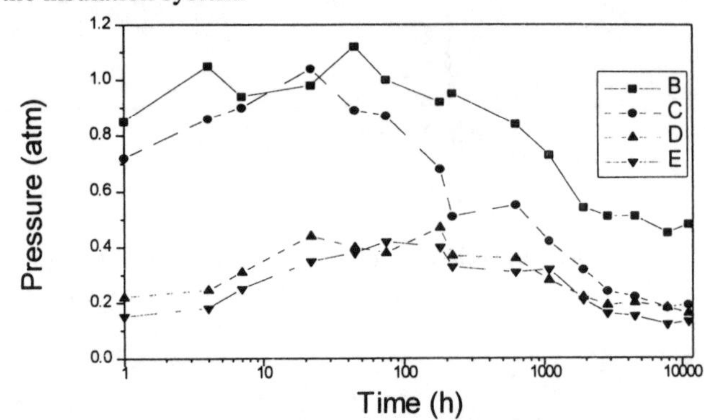

Figure 1. Pressure in XLPE cable as a function of aging time at $90C^0$, B-0 kV/mm, C-4.2 kV/mm, D- 8.4 kV/mm, E- 15.75 kV/mm.

At the beginning of the aging process, the pressure increase rapidly from the atmospheric (as the referent), and the highest value of 1.1 atm occurs after 50 hours of aging in conditions 90^{0}C with no electric field. Maximum values for the cases with the electric field of 8.4 kV/mm and 15.75 kV/mm were found to be 0.4 and 0.3 atm respectively, after approximately 2000 hours of aging.

The crosslinked polyethylene is characterized by activated absorption at which the molecules of water permeate between the crosslinked polyethylene molecules. Due to the relative thermal or electrical (or together thermal and electrical) movement of the crosslinked polyethylene molecules and their building blocks the water molecules easily permeate the crosslinked polyethylene insulation as the water molecules are very small (2.7 x 10 E-10 m), 100 - 10000 times smaller then XLPE molecules [1], [4], [7]. The permeate of water molecules into the crosslinked polyethylene cable insulation along the radial direction (r) (Fig. 2) is determined by Fickes laws as:

$$P = -D\left(\frac{\partial c}{\partial r}\right) \qquad (1)$$

$$\frac{\partial c}{\partial t} = D\frac{\partial^2 c}{\partial r^2} \qquad (2)$$

where P is the moisture gradient, D is the absorption coefficient, C is the moisture concentration, t is the aging time.

It is reasonable to assume that the moisture gradient (P) and the absorption coefficient (D) greatly depend on the chemical constitution of crosslinked polyethylene cable insulation (XLPE) and temperature and can be expressed by the following equations:

$$P = Po \exp(-Q_P/RT) \qquad (3)$$

$$D = Do \exp(-Q_T/RT) \qquad (4)$$

where Q_P and Q_T express the absorption activation energy, Po and Do are the values of P and D at the initial temperature, R is the universal gas constant and T is the absolute temperature.

Rates of permeate for steady and non steady state water flow through the unit area of cable insulation are illustrated by Eq. (1) and (2), respectively.

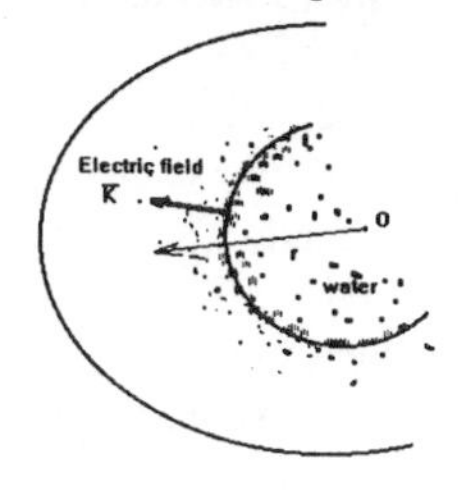

Fig. 2. Cross-section part of cable insulation with the cable conductor filled with the water

Steam cured XL cable insulations have many microvoids, 50-100 μm in diameter, formed by water penetration, whereas dry cured XL cable insulations have voids about 10 times smaller [11]. WTR XL cable insulations have microvoids about the same size as those in dry cured XL. However, the presence of ethylene copolymer as a filler, substantially reduces the number of voids, contributing to good WTR-XL water retardant characteristics . Therefore, the WTR XL is selected as the water retardant material [11].

When the cable conductor was poured with the tap water and the cable ends were properly closed, cable samples were subjected to electrical stress and heating. That means that the inside surface of the cable insulation was left in continuous contact with water during the testing. When molecules of water are exposed to electrical field, they take orientation in the direction of electrical field towards the surface of XLPE insulation. According to the electrical theory of absorption (Maxwell's law) these molecules are treated as electrical dipoles, [3], [5] and [9]. Under the influence of temperature gradient and electrical gradient water molecules move in the direction of the heat and electrical flow. The process of absorption will not be stationary until the flow has attained a steady-state value. After some time , the processes of water absorption for certain type of XLPE insulation for the determined aging conditions are finished. Considering that steam cured XL insulation contains voids of greater size, the conditions for rapid penetration of water into the insulation system exist. Increased temperature and electrical field contribute to water absorption and faster degradation of cable insulation. Water absorption rates of change for various types of XLPE insulations and for various aging conditions are shown in Fig-s 3-6.

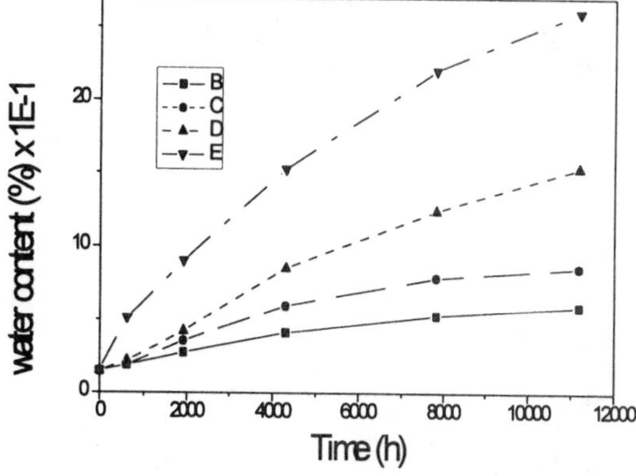

Figure 3. Water content in steam cured XL as a function of aging time at 90°C B-0 kV/mm, C- 4.2 kV/mm, D-8.4 kV/mm, E-15.1 kV/mm

As it was expected the worst results were obtained for steam cured XL insulation with the highest value of water content for electric field E=15.1 kV/mm. Significantly lower values (2-5 times) were obtained for dry cured XL, due to different manufacturing process (see Fig. 3). Water absorption for steam cured WTR-XL is more than three times lower than that for steam cured XL (Fig. 3) for the same aging conditions. That difference is not that significant for dry cured XL and WTR-XL but it still exists (Fig. 4 and Fig. 6). Tests carried out on dry cured WTR-XL insulation showed the best results comparing to other types of XLPE insulations (Fig. 6).

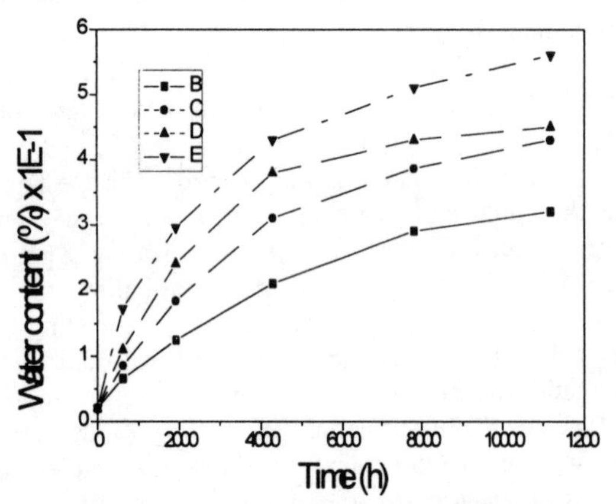

Figure 4. Water content in dry cured XL as a function of aging time at 90°C B-0 kV/mm, C- 4.2 kV/mm, D-8.4 kV/mm, E-15.1 kV/mm

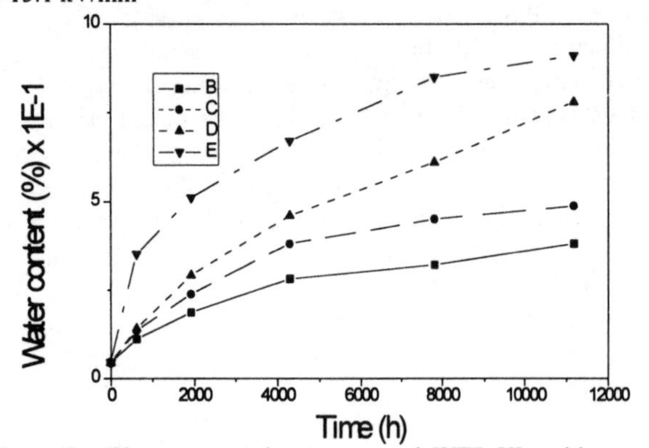

Figure 5. Water content in steam cured WTR-XL cables as a function of aging time at 90^{0}C, B-0 kV/mm, C- 4.2 kV/mm, D-8.4 kV/mm, E- 15.1 kV/mm.

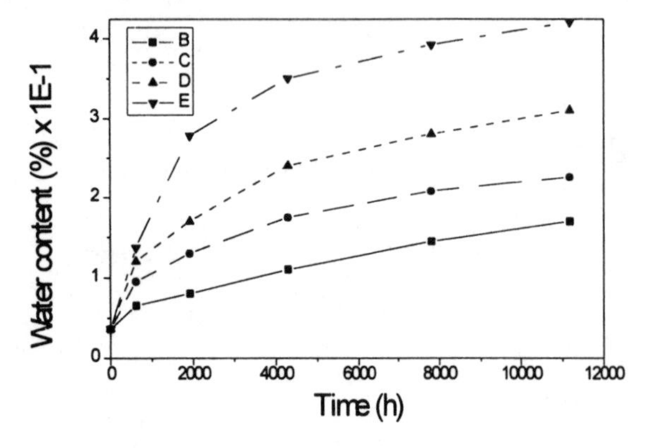

Figure 6. Water content in dry cured WTR-XL as a function of aging time at 90°C B-0 kV/mm, C- 4.2 kV/mm, D-8.4 kV/mm, E-15.1 kV/mm

The changing of AC BDS as a function of aging time for different XLPE insulations is presented in Figures 7-10. The initial AC BDS values for steam and dry cured XL with WTR-XL are lower than the corresponding initial values for steam and dry cured XL without WTR-XL, due to various types of XLPE insulations and different manufacturing technologies. For dry and steam cured XL insulations AC BDS decreases rapidly within the aging time (see Fig. 7 and Fig. 8), whereas steam and dry cured WTR-XL insulations have more stable characteristics (Fig. 9 and Fig. 10). Dry cured WTR-XL has a nearly constant AC BDS characteristic after 2000h of aging.

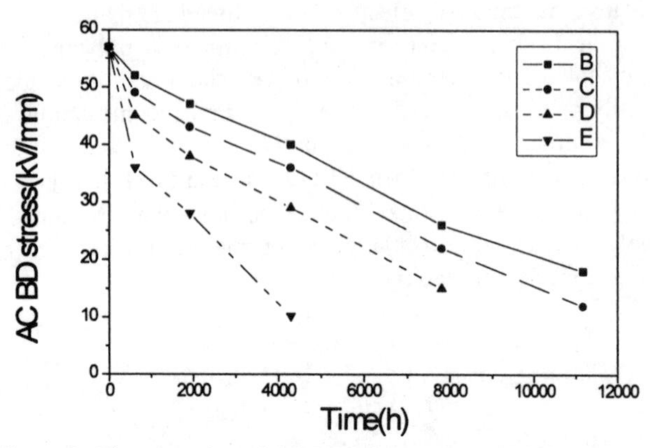

Figure 7. The changing of AC BDS for steam cured XL cables as a function of aging time at 90°C, electrical stress B-0 kV/mm, C-4.2 kV/mm, D-8.4 kV/mm and E-15.1 kV/mm.

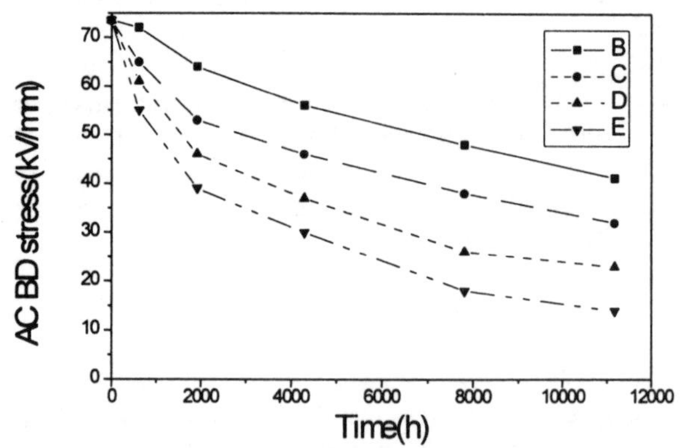

Figure 8. The changing of AC BDS for dry cured XL cables as a function of aging time at 90°C, electrical stress B-0 kV/mm, C-4.2 kV/mm, D-8.4 kV/mm and E-15.1 kV/mm.

Dry cured WTR-XL has shown the best properties regarding the maximum electric field when the partial discharge begins, whereas the worst characteristics were recorded for steam cured XL insulation (see Fig. 11).

The partial discharge characteristics as a function of aging time given in Fig. 12 show that the worst characteristic was obtained for steam cured XL insulation, whereas partial discharge measured in dry cured WTR-XL can be neglected. The partial discharge measured at 5kV for dry cured XL and steam cured WTR-XL exists but it is still significantly lower than the one measured in steam cured XL insulation, after 3500h of aging. Fig. 11 and Fig. 12 refer to aging conditions comprising cable conductor with water and cable ends properly closed , temperature 90^0 and electrical field on conductor screen 4.2 kV/mm.

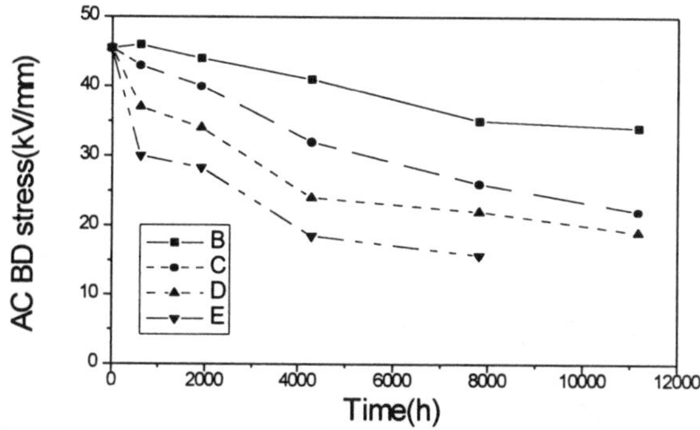

Figure 9. The changing of AC BDS for steam cured WTR-XL cables as a function of aging time at 90°C, electrical stress B-0 kV/mm, C-4.2 kV/mm, D-8.4 kV/mm and E-15.1 kV/mm.

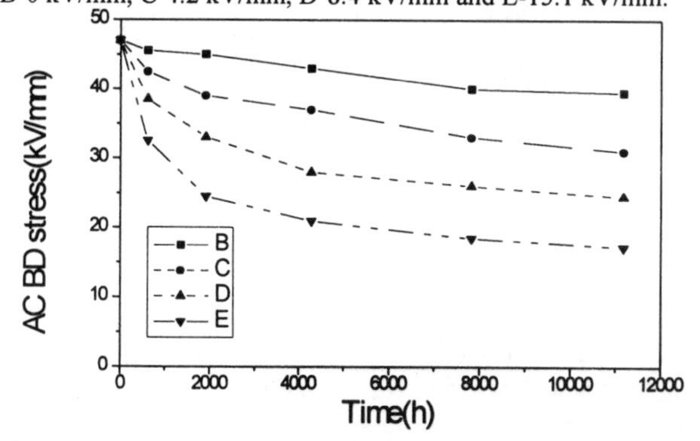

Figure 10. The changing of AC BDS for dry cured WTR-XL cables as a function of aging time at 90°C, electrical stress B-0 kV/mm, C-4.2 kV/mm, D-8.4 kV/mm and E-15.1 kV/mm.

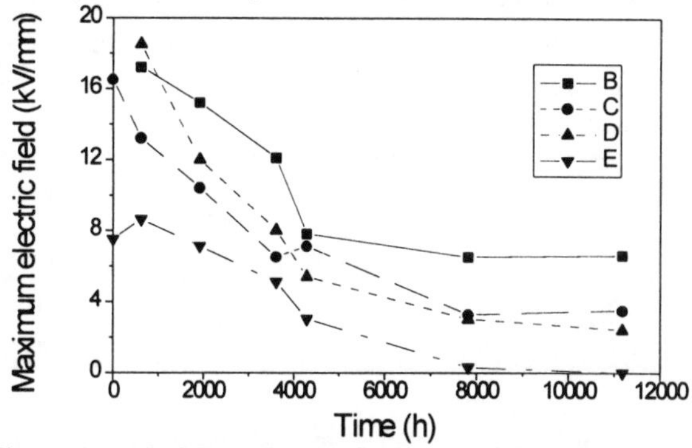

Figure 11. Maximum electric field in XLPE cable insulations when the partial discharge begins. B-dry cured WTR-XL cables, C- steam cured WTR-XL cables, D- dry cured XL cables, E- steam cured XL cables

CONCLUSIONS

The results obtained in this investigation are summarized as follows:

-There is a significant pressure in the cable when the tap water is injected in the XLPE cable conductor with the cable ends properly closed. The water ingress into the cable can be due to the production, transportation, stocking, layering, etc.

-The water absorption in XLPE insulation is affected by aging time, technology of crosslinking, type of insulation and electrical stress of insulation.

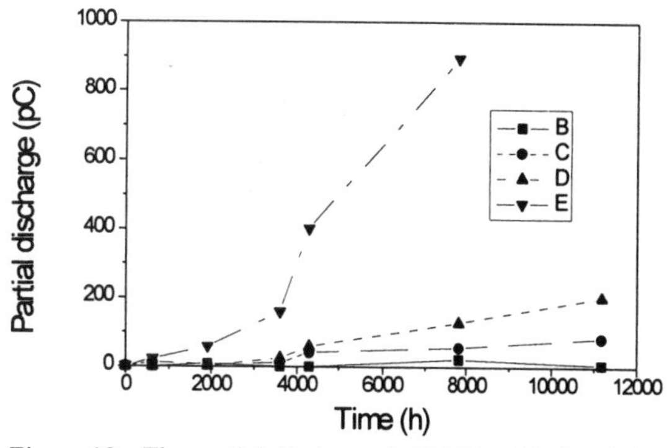

Figure 12. The partial discharge in XLPE cable insulation: B-dry cured WTR-XL cables, C- steam cured WTR-XL cables, D- dry cured XL cables, E- steam cured XL cables

-Steam and dry cured XL insulations have AC BDS characteristics depending greatly upon the electrical stress on conductor screen, whereas WTR-XL has more stable AC BDS characteristics

- The voltage stress prior to insulation breakdown rapidly decrease within the aging time especially in steam cured XLPE insulations.

- Partial discharge for dry cured WTR-XL aged in electrical field and with water in cable conductor is very low

REFERENCES

1. Walton, M. D et al. "Accelerated aging of extruded dielectric power cables ",IEEE Transactions on Power Delivery, April 1992.
2. Tanaka J. and R. Luter "Analysis of cables with visible halos", IEEE 1982. Electrical insulation, Philadelphia, 1982.
3. Nikolajevic, S.V. "Investigation of water effects on degradation of crosslinked polyethylene (XLPE) insulation", IEEE Trans.on Power Delivery,October 1993.
4. Malic, D. "Termodinamika i Termotehnika", Beograd, 1972.
5. Maslov, V. "Moisture and resistance of electrical insulations", Moscow, 1975.
6. Favrie, E. and H. Auclair "Effect of water on electrical properties of extruded synthetical insulations application on cables", IEEE Trans. on Power Apparatus and Systems, Vol. PAS-99, No. 3, 1980.
7. Wilkens, W.D. "A 2^5 factorial experiment investigating the effects of operating conditions on cable insulation life", IEEE Trans. on Elect. Insul.,Vol. EI-12, No.1, 1977.
8. Eager, G.S et al. "Extruding service life installed 15-35 kV extruded dielectric cables", IEEE Trans. on Power Apar. and Syst., Vol. PAS-103, No.8, 1984.
9. Surutka, J. "Elektromagnetika", Beograd, 1975.
10. Lawson, J. H. and J. W. Wahlstrom: "Investigation of insulation deterioration in 15 kV and 22 kV Polyethylene Cables removed from service IEEE Transactions on Power Aparatus and Systems March/April, 1972.
11. Otani, K. et al. "Development of materials for high performance XLPE cables", Wire Journal In., March 1985.

Conference Record of the 1996 IEEE International Symposium on Electrical Insulation, Montreal, Quebec, Canada, June 16-19, 1996

The Influence of the Material Conditions on the Breakdown Behaviour of XLPE-Samples at Voltages of Different Shapes

E. Gockenbach and G. Schiller
Schering Institute of High Voltage Technique and Engineering
University of Hannover, Callinstraße 25A, D-30167 Hannover, Germany

Abstract: **Breakdown investigations at XLPE model cables were worked out at voltages of different shapes. The breakdown voltages are quite different. The highest breakdown voltage was measured at DC voltage with negative polarity and the lowest at AC voltage with 250 Hz. The DC strength is mostly influenced by the water content. While dry model cables have a very high breakdown strength up to 400 kV/mm, the value of wet model cables is approximately 80 % of the former. If the model cables contain water trees, the DC breakdown strength is reduced further. The most sensitive voltage form in order to discover water trees is clearly the 0.1-Hz-AC-Voltage.**

INTRODUCTION

Although XLPE cables normally operate at power frequency (50/60 Hz) it is essential to know their breakdown behaviour at other voltage forms. By using voltages of different shapes, the breakdown voltage can evaluate the insulation stress at transient overvoltages. Concerning the electrical field distribution the best voltage form is given by the power frequency. However, large test equipment is necessary because of the high load of the cable. Alternatives have been discussed for some years, the Very Low Frequency Voltage (sinusodal or cosine-rectangular) and Oscillating Voltages (impulse voltages with frequencies from 1 kHz up to 12 kHz). In order to use such voltage shapes it is necessary to know the behaviour of the material depending on the composition and the condition of the material. For on-site testing it is additionally important to know the sensitivity of the breakdown voltage to the voltage shape for typical cable faults. Therefore breakdown investigations were carried out at continuous and impulse voltages. Model cables were used to diminish the size of test equipment, to perform a large quantity of measurements and to reduce the testing costs. The investigations were worked out at model cables at the state of delivery, after thermal conditioning and containing water and water trees.

TEST MATERIAL AND PROCEDURE

Model Cables

Figure 1 shows cross sections of the two investigated model cables. Model cable 1 consists of a tinned copper conductor, an inner semiconductive screen, an insulation of XLPE and an outer semiconductive screen. Model cable 2 has only a third of

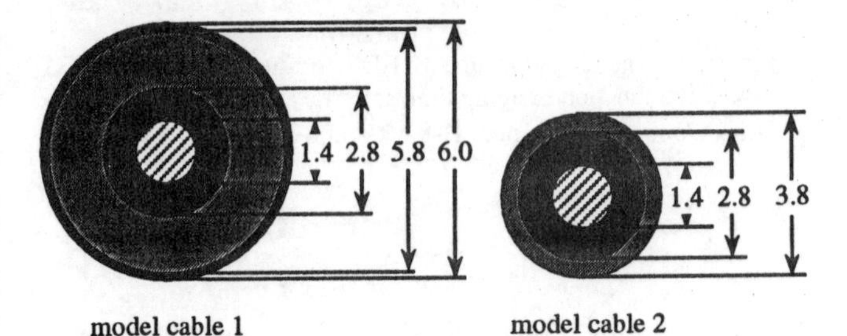

Figure 1. Cross sections of model cables

the insulation thickness of model cable 1 and the outer semiconductive layer is missing. The investigated insulation compound was a copolymer. In the case of model cable 1 a very smooth compound with low surface protusions was selected for the inner conducter layer and a standard compound for the outer layer. The conductor layer of model cable 2 is made of the standard material.

The field enhancement factor as well as the capacitance per unit are in same order than the values of real medium voltage cables. Table I shows these values:

Table I
Data of Model Cables

	Type 1	Type 2
capacitance per unit length	175.6 pF/m	418.8 pF/m
field enhancement factor	116%	147%

The model cables were steam-cured /1/. Therefore the results are mostly useful for steam cured cables, but in many cases they can also used for dry cured cables.

Non-conditioned model cables of type 1 have a water content of approximately 500 ppm and model cables of type 2 about 200 ppm. The breakdown measurements took place two months or more after cross-linking, when the diffusion processes of cross-linking by-products or water are very slow compared to the first days after cross-linking.

A part of the investigations were carried out on vacuum-dried model cables. After the conditioning (one week at 90 °C and 1 mbar for type 1 and three days for type 2) the model cables contain no cross-linking by-products or water, which could influence the breakdown voltage.

Typical cable faults are caused by water and water trees. In order to find the best voltage form for revealing this fault, one part of the samples was only conditioned and another part additionally electrically prestressed. The inner conductor of these samples was replaced by a thinner one, so that water with a temperature of 70 °C could penetrate into the insulation. The outer conductor layer was contacted with normal water, which was grounded. The water had a temperature of about 40 °C. In the case of electrical prestressing, the model cables were connected with 15 kV at 250 Hz over 250 hours. The aging at 250 Hz was chosen, because optical measurements had shown clearly longer and tighter water trees compared to 50 Hz.

Test circuits

The breakdown investigations were carried out at continous voltages,

- negative DC voltage and
- sinusodal AC voltage (0.1 Hz, 50 Hz and 250 Hz)

and at impulse voltages,

- negative lightning impulse with 1.2/50 μs (LI) and
- bipolar impulse voltage with a frequency of 10 kHz (OV).

The DC test circuit consists of a voltage controlled 50-Hz-unit connected with a Greinacher-cascade. The maximum output voltage is 800 kV at a current of 50 mA.

The 0.1-Hz-AC-voltage was produced by a test circuit described in /2/. A voltage doubler unit for each phase (capacitor and diode) was added in order to get a higher output voltage (peak voltage approximately 300 kV).

The investigations at 50 Hz and 250 Hz were performed with the test circuit described in /3/.

The test circuit for LI is a four-step-impulse-generator with a maximum output voltage of 300 kV. In case of OV only three steps were used and an air coil was connected to the model cable. The maximum peak value is about 250 kV. The breakdown voltage values at bipolar impulse voltage refer to the positive maximum (= first peak) of the impulse.

Test procedure

The breakdown voltage of the model cables was measured by using the step test at 20 °C. The initial voltage was approximately 50% of the expected breakdown voltage. The step time was always 5 minutes or 10 impulses respectively per step at the impulse voltages. The nominal value of the Weibull distribution (63%) and 95%-confidence-intervals were calculated out of a collective minimum of 6 test samples.

A water termination system was used for all AC measurements and the DC measurements at model cable 2. At DC voltage, the termination consists of a metal deflector and a tube filled with mineral oil. Because of the low currents at DC voltage it was sufficient to ground the outer conductor screen at some points. The outer conductor screen controls the electrical field at the impulse voltages. Therefore both ends of the conductor screen and both inner conductor ends of the model cable were connected with high voltage. The active part of the cable was put in water connected to ground. The active cable length was 2 meters in all tests.

RESULTS AND DISCUSSION

Model cables of type 1

Figure 2 shows the breakdown voltage of model cables of type 1 at different voltage forms and at different states of the material. After vacuum drying the breakdown voltage at DC reaches the highest value of all combinations. The nominal value is approxmately 360 kV. The lowest breakdown voltages were measured at 0.1 Hz and 50 Hz at model cables containing water trees. These values are only a quarter of the maximum breakdown voltage at DC.

At DC and LI the breakdown voltages rise after the vacuum conditioning. At DC by-products and water are dissociated and cause heteropolar charges and an increase in field strength in front of the electrodes (Fig. 3). This effect is much stronger than the electron injection at the cathode. After vacuum drying only electrons are injected by the cathode. They reduce the field strength and cause an increase of the breakdown voltage. The vacuum drying leads furthermore to a higher material density reducing the mean free path of the electrons leading to the increase of breakdown voltage at LI after vacuum drying.

The breakdown voltage at 50 and 250 Hz is reduced after vacuum drying. The reason are the missing by-products being known as voltage stabilizers at AC voltages.

The breakdown voltage at 0.1 Hz and OV is not influenced by the vacuum drying. It is assumed that the destabilizing effect of heterocharges and the stabilizing effect of the by-products are in balance.

The influence of water on the breakdown voltage is very small. Regarding all the stack columns of model cables conditioned in water, there is no clear difference between the breakdown voltage of model cables in the state of delivery and after a vacuum drying. This result is evident because of the shrinking of microvoids during every conditioning at higher temperatures. It does not matter for the water content, if the model cables are vacuum-dried or not before the water conditioning. At vacuum-dried model cables, a higher content of water causes only at DC and LI voltage a significant drop of breakdown strength. This can be explained with the mechanism mentioned above.

If the model cables contain water trees the breakdown voltage is reduced at every voltage shape and it is in every case lower compared to the state containing only water. The most significant influence can be established at the 0.1 Hz voltage. The breakdown voltage at this frequency is only one third of the state without water trees. The DC strength is lowered to about 50% after the water tree conditioning. At 50 Hz the breakdown voltage reduction is 15 to 40% and at 250 Hz up to 10%. With increasing frequency the breakdown voltage reduction decreases. Responsible for this effect is the frequency-dependent electrical treeing. While at 0.1 Hz tree-like-tree structures with a fast growth rate can be observed, at 50 Hz and especially at 250 Hz the slow bush-like-treeing predominates.

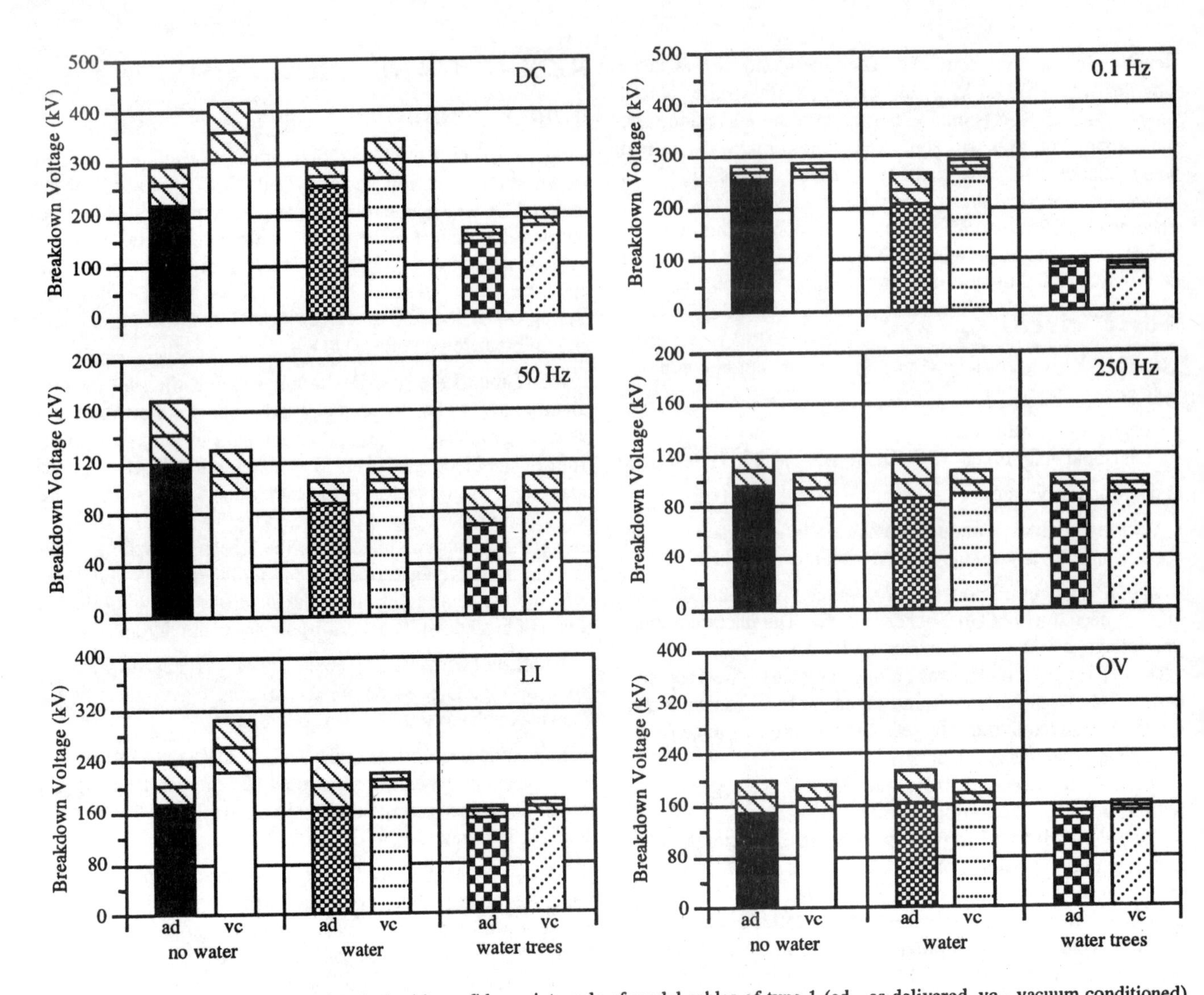

Figure 2. Breakdown Voltage (peak value) with confidence intervals of model cables of type 1 (ad - as delivered, vc - vacuum conditioned)

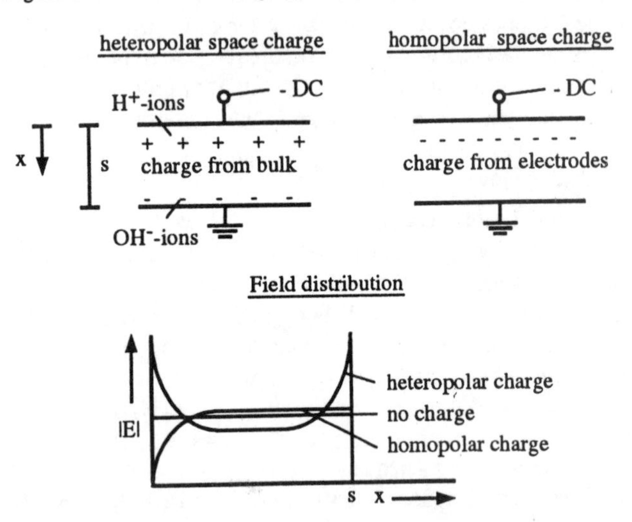

Figure 3. Influence of the polarity of the charge on the electrical field at DC voltage

The impulse voltage is much more reduced at LI (25 to 40%) than at OV (about 10%). This can be also explained by the different electrical treeing forms. Optical investigations have shown that electrical trees at LI grow very straigtht without branching while at OV some branches come out of the main tree delaying the growth.

Although after vacuum drying the density of water trees is clearly larger with an average increase of the water tree length of approximately 50%, the breakdown values do not differ from the non-conditioned and electrically prestressed values. This result shows that breakdown measurements are a good dectection method for singular irregularities like water trees.

Table II shows the breakdown voltage after water tree aging and its relation to the breakdown voltage of non-aged model cables. It can be established that the 0.1-Hz-voltage has the lowest breakdown value and the highest sensitivity for detecting water trees.

Table II Breakdown Voltages (peak values) of vacuum-dried model cables with water trees		
Test voltage	Breakdown Voltage (kV)	Rel. Breakdown Voltage (%)
DC	191	53
0.1 Hz	86	34
50 Hz	94	85
250 Hz	96	101
LI	167	64
OV	157	91

Model cables of type 2

To find out if the reduction of the DC- and 0.1-Hz-electrical strength after water tree aging is caused by the water or if it is the field disturbance due to the water trees, breakdown investigations were worked out at model cables of type 2 in dependence on the water content. The breakdown voltages at DC, 0.1 and 50 Hz are shown in Figure 4. In the state of delivery (A) all breakdown values decrease at a temperature of 20 °C which is caused by a supersaturation with water (Figure 5) due to the microvoids. After vacuum drying (B) the DC breakdown value decreases with an increasing water content of up to 56%. At 0.1 Hz the breakdown voltage is reduced by 31% at most and at 50 Hz there is no clear influence of the water content on the breakdown voltage. This means that the DC strength is mainly influenced by the water content and not directly by the field disturbance due to the water trees. The breakdown voltage at 0.1 Hz is more sensitive to the field disturbance.

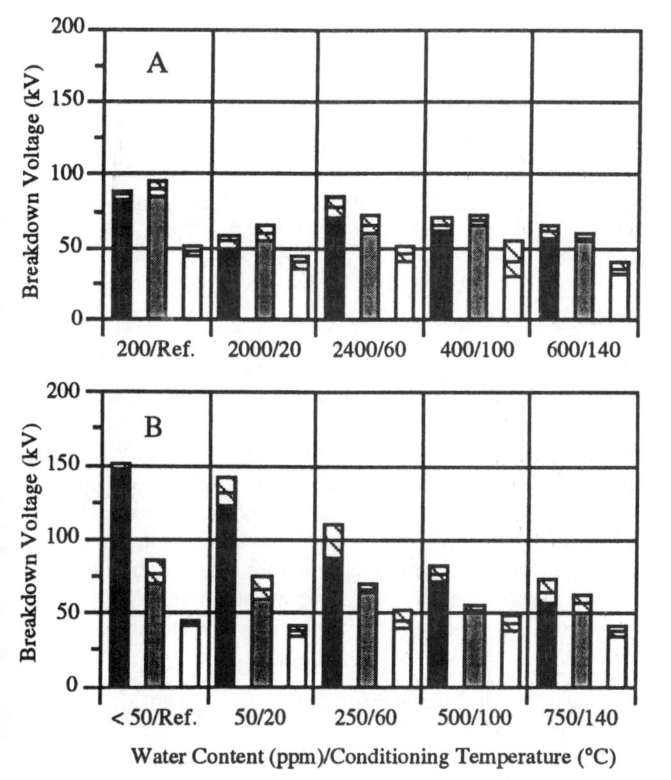

Figure 4. Breakdown voltage with confidence intervals of model cables of type 2 vs water content in the insulation respectivly conditioning temperature of the water at different frequencies
(A) in the state of delivery
(B) after vacuum conditioning
with: ■ DC, ■ 0.1 Hz, □ 50 Hz

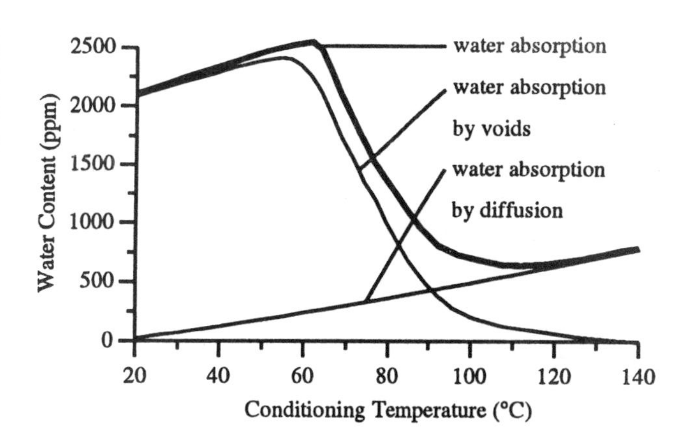

Figure 5. Water absorption vs conditioning temperature of steam cured XLPE insulation

CONCLUSIONS

The breakdown investigations at steam cured model cables have shown the following results:

- After vacuum drying the breakdown voltage increases at DC- and LI voltage, it decreases at 50 and 250 Hz and it shows no influence at 0.1 Hz and OV.
- Model cables conditioned in water have a slightly reduced breakdown voltage at DC, LI, 50 and 250 Hz.
- Unconditioned model cables can take up much water at low temperatures because of the microvoids in the insulation.
- Model cables containing water trees have a low breakdown voltage at 0.1 Hz and DC voltage. The low DC voltage after water tree aging is mainly caused by the high local content of water. The breakdown voltage at 0.1 Hz is very sensitive due to the field disturbance.

ACKNOWLEDGMENT

The authors thank Alcatel Kabelmetal Hannover for providing the test materials.

REFERENCES

1. Schädlich, H., H. G. Land and C. D. Ritschel, "Investigations on Dielectric Strength of Model Cables with PE and XLPE Insulation", 7th Int. Sym. on High Volt. Eng., Paper No. 23.12
2. Grönefeld, P. "Elektrische Prüfung von Kunststoffen mit Spannungen unterschiedlicher Form", doctoral thesis, TU Hannover 1986
3. Gockenbach E. and G. Schiller, "Effect of Frequency on Electrical Strength of XLPE Insulating Materials", 7th Intern. Sym. on High Volt. Eng., Paper No. 23.05

Conference Record of the 1996 IEEE International Symposium on Electrical Insulation, Montreal, Quebec, Canada, June 16-19, 1996

Prequalification tests of 345-kV XLPE cable system at IREQ

J.-L. Parpal, R. Awad, M. Choquette
Hydro-Québec

J. Becker, L. Hiivala, S. Chatterjee
Alcatel Cable

T. Kojima
Fujikura

R.D. Rosevear, O. Morelli
Pirelli

Abstract: **Extruded polymeric cables and accessories are an alternative to self-contained fluid filled (SCFF) cables for Extra-High-Voltage (EHV) systems. Crosslinked polyethylene (XLPE) insulated cables have many advantages over the traditional fluid-filled pressurized cable system with the elimination of the hydraulic system and the associated equipment and complications during the installation and operation/maintenance of such systems. Other advantages of the extruded cable system arise from the new accessory technologies and the use of splices and terminations that are prefabricated and pretested in the factory and require less time to install on site. Concerns over the long term reliability of high voltage cable systems, in particular the accessories and the lack of service experience above 300 kV led to the decision of Hydro-Québec to carry out a prequalification (long-term) test program to assess the reliability of the cable materials and to verify the cable and accessory installation methods to be employed. The cables were installed in duct banks and manholes simulating actual installation conditions used in Hydro-Québec underground cable network. This paper describes the program and results of the prequalification tests of 345-kV XLPE cables and accessories performed at Hydro-Québec's Research Institute (IREQ) in partnership with three international cable manufacturers, Alcatel, Fujikura and Pirelli.**

INTRODUCTION

Since 1989, 120-kV XLPE insulated cables and accessories have been used for three major circuits in Montréal [1]. In view of the increasing electrical demand in downtown Montréal, new underground 315-kV circuits are planned to be installed in the years to come. However, at the present time there are no XLPE cable systems in service anywhere in the world using prefabricated (premolded) joints above 275 kV. The design of prefabricated joints at 315 kV (or 345 kV) required to operate at high electrical stress levels still presents the greatest technical challenge toward achieving consistent performance and acceptable field assembly times. Since the reliability of the cable and accessories is required to be at least on a par with that of fluid-filled cables, and since there is no service experience worldwide at that voltage level for XLPE cables with premolded joints installed in manholes, there was definitely a need for a prequalification program for these materials.

The main objectives of the prequalification testing program were:

- to assess the long-term reliability of the cable systems and in particular the premolded joints
- to ensure that the cable, premolded joint design and field installation techniques were compatible with standard Hydro-Québec duct and manhole dimensions
- evaluate the different cable clamping systems

Although the voltage level of the Hydro-Québec system is 315 kV, the test program was designed for 345 kV (U_0 = 200 kV), which is the standard operating voltage for many electrical utilities in North America. Thus the long-term test, which is the first part of the test program, was performed at 345 kV (phase-to-ground) so that these cables and accessories would be qualified for the entire North-American market. This implies that all type test voltages (based on Hydro-Québec technical specification SN-49.1 for 120 kV, 230 kV and 315 kV) were those required for 345 kV cables and accessories. The type tests of the cable, joints and terminations were performed by the cable manufacturers in their respective laboratories prior to the installation of the materials for the long-term testing.

This paper describes the results of the prequalification test program of 345-kV XLPE cables and accessories that was completed, between November 1995 and January 1996, at Hydro-Québec's Research Institute (IREQ) in partnership with three cable manufacturers, Alcatel, Fujikura and Pirelli.

THE TEST PROGRAM

The tests selected for this program were based on Working Group 21.03 recommendations, of the Conférence Internationale des Grands Réseaux Électriques (CIGRE) in which the aging conditions are designed to verify the life expectancy of the cable system and reveal any signs of degradation [2]. The more stringent conditions in the Hydro-Québec program are aimed at obtaining accelerated aging, which will provide useful data on the long-term dielectric performance of the various components and, also on the thermomechanical behavior of the entire cable system.

Three test loops of cable installed in ducts as well as a joint in a manhole (rigid clamping point or free-expansion) with 2 outdoor and 2 SF_6 potheads was set up at the Hydro-Québec's research facility in Varennes in the same fashion as of the utility's standard installations (Figure 1). This cable testing facility is unique in Canada and was specially designed to test extra high voltage cables rated from 120 kV up to 1200 kV AC or DC. Cables ranging in length between 100-200 meters with joints can be tested in conditions similar to those in service, with thermal cycling, over periods of months. Experienced research personnel have successfully performed the delicate testing of the 500-kV DC paper-oil insulated (1989-90) [3] and 800-kV AC PPLP-insulated oil-filled (1992-93) [4] cable systems.

Current transformers were used to heat the cable and thermocouples, placed along the surface of the cable, monitored the temperature. The conductor temperature was extrapolated every minute from the measured surface temperatures along the cable loop using a software developped at IREQ for the 500 kV and 800 kV testing.

Type Tests

The type tests on the cable, joint and terminations were performed by the cable manufacturers before the beginning of the long-term testing. The type tests of the 345-kV cable systems were performed according to Hydro-Québec specification SN-49.1 in general, to International Electrotechnical Commission (IEC) Standard 885-2 for partial discharges and IEC 230 for impulse testing. A summary of the type test results is presented for all cable manufacturers in Table I.

Electrical Tests After Installation

Upon the completion of the cables and accessories installation at IREQ, an AC voltage of 400 kV (phase-to-ground) between the conductor and the metallic shield was applied to the cable assembly for 15 min and the dielectric losses were measured.

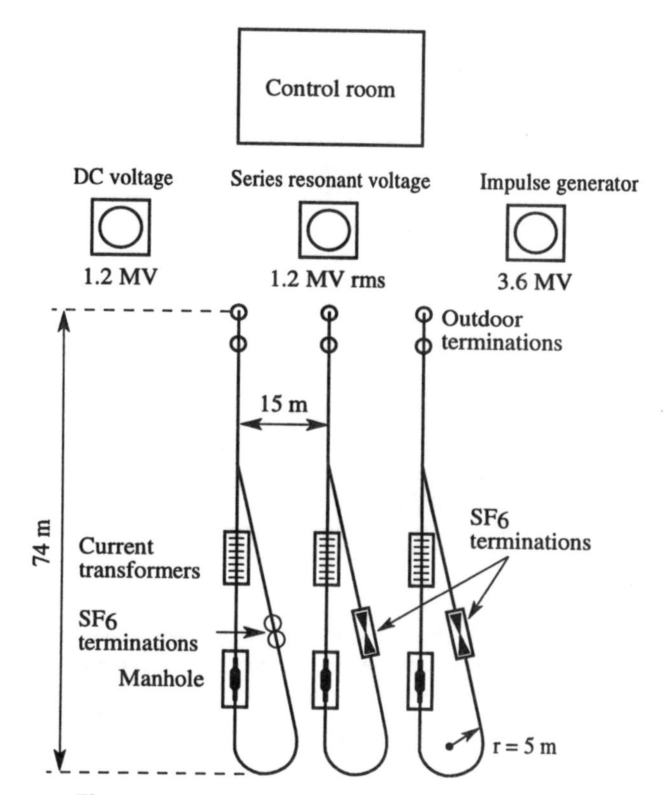

Figure 1: Installation scheme of the cable test loops

Table I: Type Test Results (U_o = 200 kV)

Article SN-49.1	ALCATEL	FUJIKURA	PIRELLI
Bending test	Passed	Passed	Passed
Partial-discharge test on complete loop	Passed	Passed	Passed
Measurement of power factor vs. temperature and voltage	***T = 18°C*** $3{,}95 \times 10^{-4}$ at U_0 $7{,}93 \times 10^{-4}$ at $2U_0$ ***T = 95°C*** $3{,}68 \times 10^{-4}$ at U_0 $4{,}81 \times 10^{-4}$ at $2U_0$	***T = 21°C*** $4{,}5 \times 10^{-4}$ at U_0 $4{,}7 \times 10^{-4}$ at $2U_0$ ***T = 95+5°C*** $3{,}9 \times 10^{-4}$ at U_0 $4{,}3 \times 10^{-4}$ at $2U_0$	***T = 19°C*** $2{,}8 \times 10^{-4}$ at U_0 $2{,}9 \times 10^{-4}$ at $2U_0$ ***T = 99°C*** $8{,}3 \times 10^{-4}$ at U_0 $7{,}5 \times 10^{-4}$ at $2U_0$
Load cycling test (30 cycles at 2 U_0)	Passed	Passed	Passed
Partial-discharge test (repeat)	Passed	Passed	Passed
Switching impulse test	Passed	Passed	Passed
Lightning impulse test	Passed	Passed	Passed
Power frequency test	Passed	Passed	Passed
Partial-discharge test (repeat)	Passed	Passed	Passed

Long-Term Test

The long-term test in the prequalification program was considered to be the best means of establishing a precise indication of the reliability of the proposed cable system. The accelerated aging resulting from more stringent test conditions not only provides information on the dielectric performance of the different cable components but on the thermomechanical behavior of the cable system as well. Upon successful completion of the 400-kV test, each cable loop was subjected to voltage for the long-term test under the temperature and durations conditions listed in Table II.

Limited Type Tests

Following the long-term test, the cable and accessories were subjected to lightning and switching impulses and AC voltage (Table II). These tests were intended to give precise indications of any degradation in the insulating materials that could have resulted from the long-term test.

THE MATERIAL TESTED

The respective characteristics of the Alcatel, Fujikura and Pirelli cables and accessories are summarized in Table III. The cables were pulled into 200 mm diameter Epoxy Reinforced fiber glass (FRE) ducts. The joints were installed in standard Hydro-Québec manholes having the following dimensions: 11 m long, 3 m wide and 3 m high. Two types of prefabricated joints (sectionalised) were tested: an Ethylene-Propylene-Diene-terpolymer (EPDM) rubber monoblock slip-on type (Figure 2a) and a combination of premolded EPDM rubber, a compression device and an epoxy unit (Figure 2b).

Table II: Long-term and limited type test conditions

Long-term test	
Voltage	345 kV (phase-to-ground)
Duration (continous)	6000 h (over a period of 8.3 months)
Daily thermal cycling	8 h heating (95°C max.) / 16 h cooling
Limited type tests	
Dielectric losses	400 kV at 95°C
Lightning impulse tests	1300 kV, 10 positive and 10 negative polarity at 95°C
Switching impulse tests	900 kV, 10 positive and 10 negative polarity at 95°C
AC withstand test	450 kV during 1 h

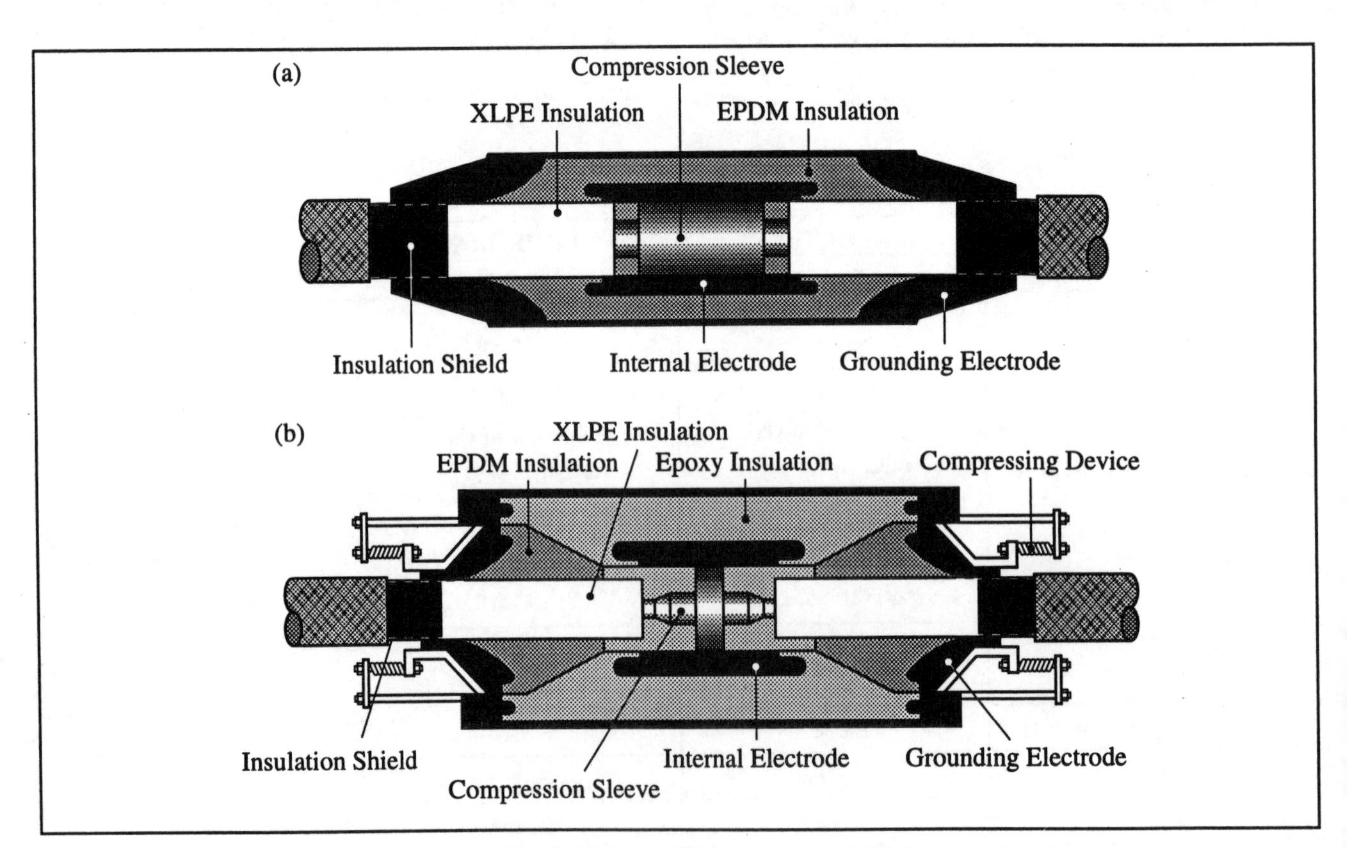

Figure 2: Type of premolded joints: (a) Slip-on-type and (b) Compression Type

Table III: Summary of cable and accessories characteristics

Component	ALCATEL	FUJIKURA	PIRELLI
Conductor*	Cu 1600 mm^2	Cu 1000 mm^2	Cu 1600 mm^2
XLPE insulation	26.6 mm	27 mm	25 mm
Metallic sheath	Lead alloy	Corrugated Aluminum	Lead alloy
Outer jacket	Polyethylene	Polyethylene	Polyethylene
Overall diameter	130 mm	140 mm	130 mm
Prefabricated joint	EPDM slip-on type	EPDM/Epoxy compression type	EPDM slip-on type
Outdoor terminations	EPDM stress cone in SF_6	Condenser-cone (paper/aluminum) in silicone oil	EPDM stress cone in silicone oil
SF_6 terminations	EPDM stress cone in SF_6/SF_6	Condenser-cone (paper/aluminum) in silicone oil/SF_6	EPDM stress cone in silicone oil/SF_6
* The actual conductor size will be based on the transmission capacity required for future projects.			

The cable and joint installation in the manhole allowed for the expansion and thrust of the cable as a result of the conductor temperature rise due to the load current. Two types of installation were tested:

- rigid clamping mode (joint and cable rigidly clamped)
- free expansion loop (cable laid on supports, forming unconstrained expansion loop on each side of the joint)

The manufacturers provided the SF_6 enclosures together with the SF_6 terminations and two types of configurations were tested:

- side-by-side (vertical)
- back-to-back (horizontal)

CONCLUSION

The long-term and limited type tests of the Alcatel, Fujikura and Pirelli 345-kV cable systems (XLPE cable, prefabricated EPDM joint, two outdoor and two SF_6 terminations) were successfully completed. The success criteria used in the prequalification test program were to complete the 6000 h (or the equivalent) time test at 345 kV (phase-to-ground) and pass the limited type tests without any electrical failure of the cable or accessories.

Fujikura and Pirelli completed 6 000 h at an applied voltage of 345 kV. Due to a later start of the tests, Alcatel completed 3300 h at 345 kV and 800 h at 400 kV. The combine duration and voltage level of Alcatel's long-term test were considered to be equivalent to 6000 h at 345 kV when the power law is used to extrapolate the lifetime.

ACKNOWLEDGMENT

We gratefully acknowledge the support and collaboration of our colleagues from Alcatel, Fujikura, Pirelli and Hydro-Québec who helped us realize this project.

REFERENCES

1. Awad R., Choquette M., "Hydro-Québec Experience with 120 kV XLPE Insulated Cables", CEA Meeting 1993.
2. CIGRE Working Group 21.03, "Recommendations for Electrical Tests Prequalification and Development on Extruded Cables and Accessories at Voltages > 150 (170) kV and < 400 (420) kV", Electra 1993, pp. 15-19.
3. Bell N., Bui-Van Q., Couderc D., Ludasi G., Meyere P., Picard C., "450 kV DC Underwater Crossing of the St-Lawrence River of a 1500 km Overhead Line with Five Terminals", CIGRE 1992, Paper 21-301.
4. Couderc D., Bui-Van Q., Vallée A., "Development and Testing of a 800 kV PPLP-Insulated Oil-Filled Cable and its Accessories", CIGRE 1996, Report No. 131 (21/22).

Conference Record of the 1996 IEEE International Symposium on Electrical Insulation, Montreal, Quebec, Canada, June 16-19, 1996

Electrical Breakdown Strength of 5 kV XLPE Cable with Imperfections Under Combined ac-dc Voltage

S. Grzybowski, J.Fan
High Voltage Laboratory
Electrical and Computer Engineering Department
Mississippi State University
Mississippi State, MS 39762

Abstract: **This paper presents data on electrical breakdown strength of 5 kV XLPE cable with imperfection under ac voltage, dc voltage and superimposed ac upon dc voltage (pulsate voltage). In the later case, the measurement was taken over a wide range of pulsate ratio, which is defined as p=Vac/Vdc. Comparison with the former measurement results for sound cables shows that the pulsate voltage can be used to detect the imperfections inside XLPE cable. Based on past research, the results obtained under combined ac plus dc voltage follow an expression which relates the breakdown voltage to the sum of the arctangent and arccotangent function of the parameter "p".**

INTRODUCTION

Since the introduction of the PE and XLPE insulated cable, it has been a common practice for the industry or utility to employ dc voltage to detect the gross imperfections before energized or deterioration after the in-service aging. However, experience shows that in some cases, a dc test is believed to further deteriorate the service-aged cable. In a recent review, B.S.Berntein [1] described the projects initiated by EPRI which show that a dc test will reduce the remaining life of the aged cable. Other studies [2] show that dc test in accordance with AEIC specifications without flashover does not appear to affect the remaining life of the service-aged cable while it does if flashover occurs. A lot of research effort has already been taken to find an alternate test method to the dc test in the last few years, such as the study of the XLPE cable breakdown characteristic under PRW (Pulsed Resonant Wave)[3] and VLF (Very Low Frequency ac)[2] excitation.

The fact is that dc test has its undeniable advantages in testing cables: Lower power is required during the test, which means a smaller size of test equipment. A dc leakage current can be detected during the test, which is another indication of cable insulation deterioration. So it will be nice if we can keep these advantages when looking for an alternative method to dc test. On the other hand, dc test has its disadvantages when used to test XLPE cables. Study shows that, unlike paper-oil insulation, XLPE is able to trap space charges when stressed, especially when dc stress is applied. These trapped space charges will distort the field distribution inside the cable insulation. In addition, the absence of the partial discharge may develop different test results compared with the ac test.

This study presents ac, dc and pulsate breakdown characteristics of XLPE cable with artificial imperfections. Two types of cable samples are employed; these include cable samples with artificial holes and samples with sharp needles.

LABORATORY MEASUREMENT

A schematic diagram of the test set up used in this study is sketched in Fig. 1. The dc source used here is a portable SF_6 insulated unit and can be assembled in modules. Two modules are used in the present test with each rated at 175 kV, giving a total output of 350 kV. An ac transformer rated at 40 kV is used as the ac source. A pair of sphere gaps is employed as a protection system of the ac transformer to avoid overvoltages caused by the breakdown of the cable sample. The 5 kV XLPE cable sample with insulation thickness of 2.286 mm (90 mils) is connected in series with the dc and ac source, so a pulsate voltage waveshape as indicated in Fig. 2 is applied to the cable sample.

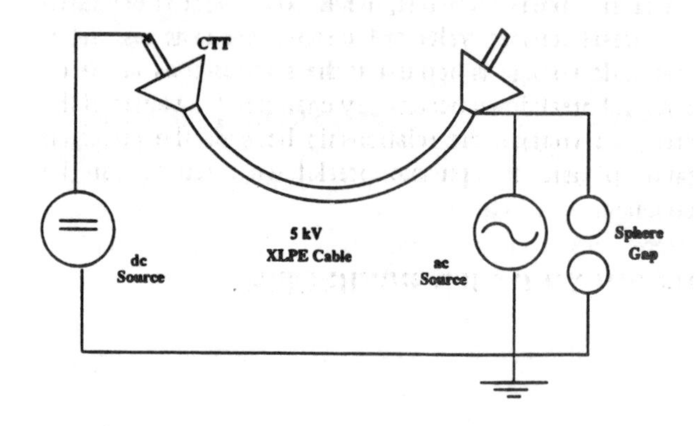

Fig. 1 Schematic diagram of equipment set up

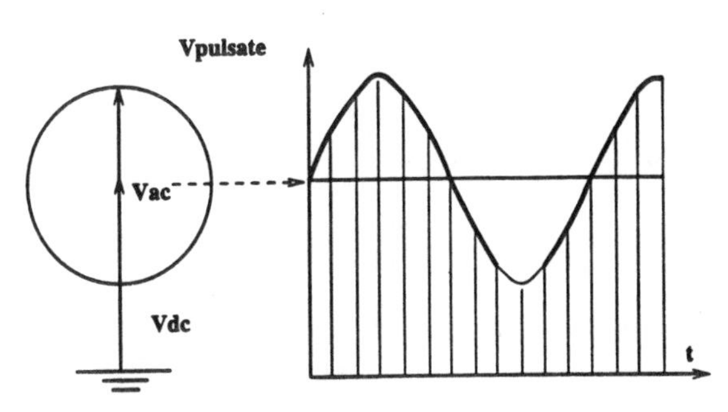

Fig. 2 Waveform of the pulsate voltage applied to the cable samples

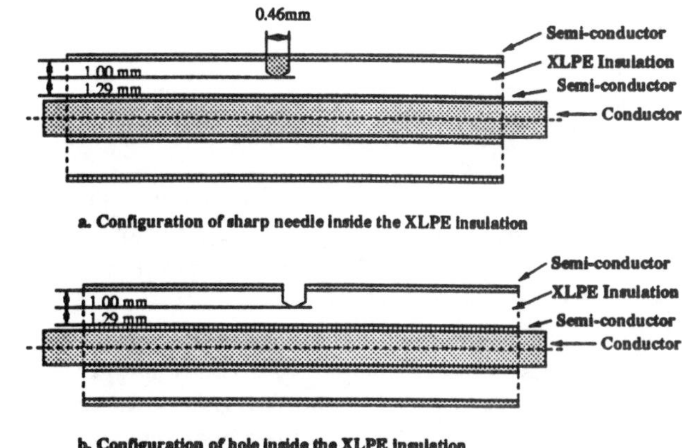

Fig. 3 Imperfection configuration inside the cable sample

Newly produced 5 kV XLPE cable provided by the manufacturer is cut into samples. Each sample is about 3.5 metres long. Both ends of the cable sample is stripped and the semiconductor layer at the end is removed with extreme care to avoid any damage. A convenient oil-insulated cable test termination is used. It can avoid abnormal field stress concentration at the end of the stripped cable samples. No flashover or failure at the end of the semiconductor layer was observed during the test with the use of these terminations.

Two types of imperfections as shown in Fig. 3 are studied:

a. Sharp needles; a sharp metal needle is embedded radially inside the cable insulation, which is a simulation of the metallic contamination in the cable insulation.

b. Small holes; a small hole as indicated in Fig. 3 was made radially by using a drill. This is a simulation of the voids or holes inside a cable.

Prior to the pulsate test, ac and dc breakdown strength of the cable samples are measured separately. The voltage applied to the cable samples will start from zero and is increased at a rate of 5 kV per minute for ac breakdown test. For dc test a rate of 10 kV per minute is used. During the pulsate breakdown test, the ac component is increased to a preset constant value at the same rate as ac test. After that, a dc voltage is applied at the same rate as in the dc test until breakdown occurs. By changing the values of the preset ac voltage, the relationship between the pulsation ratio 'p' and the pulsate breakdown strength can be obtained.

RESULTS OF INVESTIGATION

The results of the ac and the dc breakdown test are shown in Table 1 for both types of imperfections. From the measured results, one can see that the dc breakdown strength is much higher than the ac breakdown strength. A comparison with the breakdown strength of sound cable shows that the percentage decrease of ac strength is much higher than that of the dc strength.

Table 1
Average ac and dc breakdown strength of 5kV XLPE cable

Type	ac (kV)	dc (kV)
Sound cable	68.0	260.0
Hole inside	26.2	192.0
Needle inside	23.7	172.0

For pulsate voltage, three samples are used for each preset ac value. The relationship of the ac component with the dc component in a pulsate breakdown is shown in Fig. 4 and Fig. 5 for both types of imperfections. A sharp reduction of the dc component in the vicinity of the ac breakdown strength is observed. Different curve shapes when the ac component is less than 10 kV are also observed for these two imperfections.

According to the previous study of the paper-oil insulation system [5] and the new XLPE cable [6], the pulsate breakdown voltage should be related to the sum of the arctangent (ac component) and arccotangent (dc component) function of the pulsation ratio 'p'. The previous developed equations are shown as follows:

$$\mathbf{Vb = Vdc + Vac} \quad (1)$$

$$\mathbf{Vdc = \frac{2}{\pi} \cdot A \cdot arccotan(\frac{p}{\alpha})} \quad (2)$$

$$\mathbf{Vac = \frac{2}{\pi} \cdot B \cdot arctan(\frac{p}{\beta})} \quad (3)$$

Where: A = dc breakdown voltage
B = ac breakdown voltage
α = the value of p when dc component is 0.5 A
β = the value of p when ac component is 0.5 B

Based on the previous analysis technique, the characteristic pulsation parameters α and β are calculated with the present test results of XLPE cable samples. For the cable samples with artificial holes, α=0.2154 and β=0.0787 are calculated, while for the cable samples with embedded needles, α=0.212 and β=0.104. The results are presented on Fig. 6 and Fig. 7. Good fitness is observed for the hole imperfection in Fig. 7, while a poor fitness is shown in the needle imperfection, case of small p, Fig. 6.

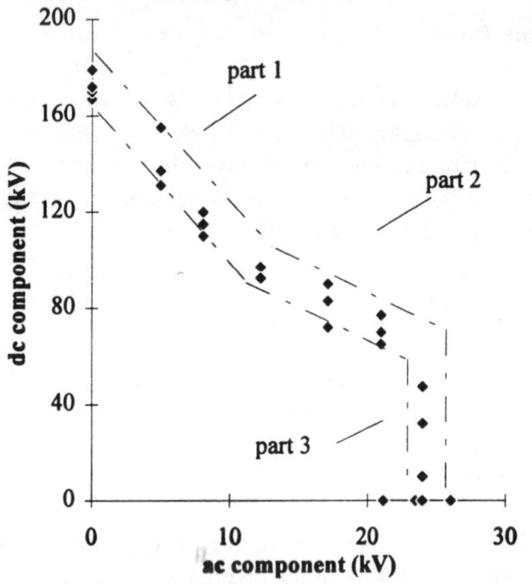

Fig. 4 Pulsate breakdown of 5 kV XLPE cable with needle inside : dc component vs. ac component

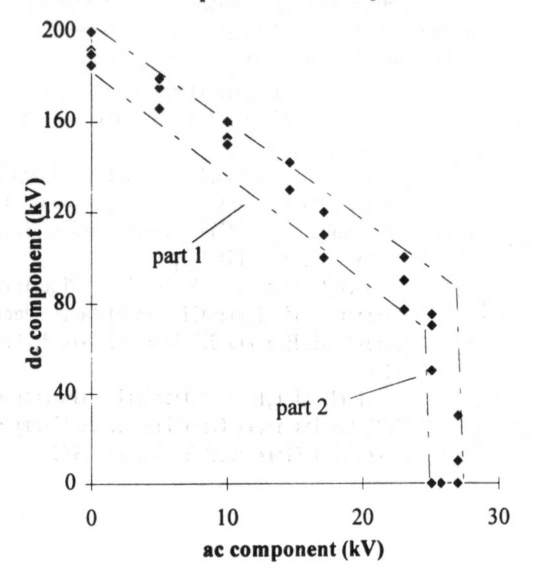

Fig. 5 Pulsate breakdown of 5 kV XLPE cable with hole inside : dc component vs. ac component

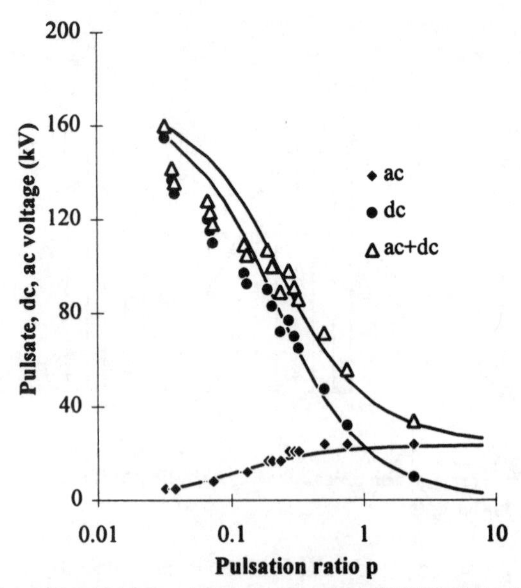

Fig. 6 Electrical breakdown strength of 5 kV XLPE cable with needle inside vs. pulsation ratio p

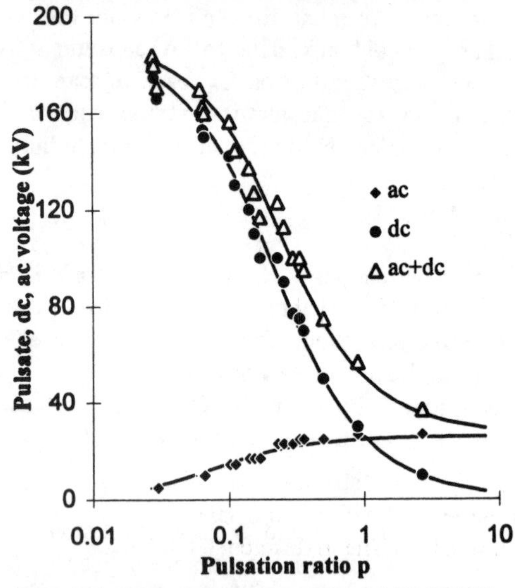

Fig. 7 Electrical breakdown strength of 5 kV XLPE cable with hole inside vs. pulsation ratio p

DISCUSSION OF RESULTS

From the curve given in Fig. 4, which is for the needle embedded samples, one can see that there are three parts, which are indicated as part 1, part 2 and part 3, while for the cable samples with artificial holes, only two parts are observed. The different curve shapes in Fig. 4 and Fig. 5 cause the different fit of the measured results to the theoretical results. There is an abnormal drop of the dc component with the introduction of the ac component in Fig. 4. This causes the measured dc component, and therefore pulsate breakdown voltage, to be less fit to the theoretical results, obtained from equations(1,2 and 3),

Fig. 6. Former study shows that for the oil-paper insulation and new XLPE cable, the test results fit well to the theoretical data.

It is known that XLPE dielectric has the ability to trap space charges. In the case with imperfections inside the cable, there is a space charge accumulation at the tip of the needle electrode or hole under dc stress. These space charges might cause severe field distortion inside the cable insulation. From Table 1, a high dc breakdown strength is observed for both imperfect cable samples. This means that the field distortion helps raise the dc breakdown strength. On the other hand, ac breakdown strength is relatively much lower, indicating that for ac breakdown, space charge accumulation plays a less important role.

In Fig. 4, the abnormal drop of the dc component with the introduction of ac voltage can be explained like this: At part one, the ac component causes the needle electrode to inject electrons to the cable insulation, so space charge at the tip of the needle caused by the dc stress is partially neutralized. This means that the field distortion is corrected, so a severe drop of the dc component is observed in part one. This confirms the previous study [4] showing a dc component current under ac stress in aged cable (needle plays partly the role of water trees), which is believed to be caused by the electron accumulation at the tip of the water trees. With further increase of the ac component, shown as part two in Fig. 4, field correction caused by the electron injection becomes less important in the cable breakdown, so the curve has less slope. Finally, in part three, ac stress starts to dominate the breakdown process, so the value of dc component is more unpredictable. For the cable samples with holes, no electron injection is expected as in the needle electrode case, so only two parts are observed in Fig. 5. Further study is necessary to clarify the detail activity of the space charge inside the cable in different situations.

In the case of aged cable, water trees can play the role of needle electrode in this study, so if a pulsate voltage is employed to test the XLPE cable instead of a dc voltage, then less space charges are expected to be trapped inside the aged cable after the dc test, because it may be neutralized by the introduction of ac voltage.

From the curves given in Fig. 6 and Fig. 7, and the previous study, another interesting observation is that in any case, measured data of ac component are always fit well to the theoretical curves.

CONCLUSIONS

Several conclusions are obtained based on the present study and analysis:

a. Breakdown characteristics of 5 kV XLPE cable samples with two types of imperfections are studied under ac, dc, and pulsate excitation. A much higher dc breakdown voltage is observed than ac breakdown voltage.
b. Imperfections inside the cable can be detected by using pulsate voltage.
c. A space charge is believed to play an important role in the dc and pulsate breakdown of cable samples with imperfections while it has less effect on the ac breakdown strength.
d. A good fit is found between the measured data and the theoretical curves for cable samples with artificial holes, while less fit is observed in the case of needle embedded cable samples.
e. Further study is necessary to deal with the pulsate breakdown characteristic of aged cable .

ACKNOWLEDGMENT

Special thanks are due to Southwire Company for providing 5kV XLPE cable and especially to Mr. Darrell Jeter for his help and support.

REFERENCES

1. Bernstein,B. "Cable Testing: Can we do better?". IEEE Electrical Insulation Magazine, Vol. 10, No.4, July/August.1994, pp. 33-38.
2. Eager,G.S. Fryszczyn,B. Katz,C. Elbadaly,H.A. and Jean,A.R. "Effect of DC Testing Water Tree Deteriorated Cable and A Preliminary Evaluation of VLF as Alternate". IEEE Trans PWRD, Vol. 7, No. 3, July 1992, pp. 1582-1591.
3. Ward,B.H. Steiner,J.P. "An Alternative to DC Testing of Installed Polymer Power Cables". 7th ISH. Paper No. 75.06, Dresden, Germany, August, 1991.
4. Oonishi,H. Urano,F. Mochizuki,T. Soma,K. Kotani,K. Kamio,K. "Development of New Diagnostic Method for Hot-Line XLPE Cables with Water Trees". IEEE Trans PWRD, Vol.-2, No. 1, January 1987, pp. 1-7.
5. Nowaczyk,H. Grzybowski,S. Kuffel,E. "Electrical Breakdown Strength of Paper-Oil Insulation Under Pulsating Voltages". IEEE Trans EI, Vol. 22, No. 3, June 1987, pp.249-253.
6. Grzybowski,S. McMellon,R. "Electrical Breakdown Strength of XLPE Cables under Combined ac-dc Voltage". 9th ISH, Paper No. 1086,Graz, Austria, August, 1995.

Conference Record of the 1996 IEEE International Symposium on Electrical Insulation, Montreal, Quebec, Canada, June 16-19, 1996

EXPERIMENTAL INVESTIGATIONS OF OVERVOLTAGES IN 6kV STATION SERVICE CABLE NETWORKS OF THERMAL POWER PLANTS

Petar I. Vukelja, Radomir M. Naumov, Goran V. Drobnjak and Jovan D. Mrvić
Nikola Tesla Institute
Belgrade, Yugoslavia, Koste Glavinića 8A

Abstract: **The paper presents the results of experimental investigations of overvoltages on 6kV isolated neutral station service cable networks of thermal power plants. On the basis of these investigations, certain measures are proposed for limiting overvoltages and for the reliability of station service of thermal power plants.**

1.INTRODUCTION

Nikola Tesla Institute investigated overvoltages in five 6kV station service cable networks of thermal power plants [1,2,3]. The networks have similar characteristics. Six kV busbars with cable feeders are supplied from 220/6/6 or 110/6/6 or 15.75/6.3/6.3 or 21/6.3/6.3 kV power transformers. The loads at the other cable feeders end are either a high voltage motor (HV motor) or a 6/0.4kV power transformer. The power transformer is also connected to the busbar by cable. The number of cable feeders is usually more than 10 and they are from 10 to 400 m long. The cables are shielded and have three leads with cross-linked insulation. The shield was grounded at both ends. The overvoltages were recorded with capacitive voltage measurement systems made at the Nikola Tesla Institute. Wideband capacitive voltage measurement systems recorded a flat response from below power frequencies to 10MHz. Investigations of overvoltages were performed for:

- appearance and interruption of metal earth faults,
- intermittent earth faults,
- switching operation of HV motors switchgear,
- switching operation of transformers switchgear,
- transfer of the network supply from one transformer to another.

2. EXPERIMENTAL INVESTIGATIONS OF OVERVOLTAGES

2.1. Overvoltages for earth faults

Overvoltages were investigated for closing and opening of metal earth faults and for intermittent earth faults [1]. Earth faults are just signalised in station service networks of thermal power plants and networks continue to operate normally under conditions of earth faults lasting up to 8h.

The metal earth fault was accomplished on one of the phase conductors of a chosen cable feeder of the network by connecting that conductor to the earthing system. The connection was made through a current transformer (12kV, 60A/5A), which also served for earth-fault current recording. The appearance and interruption of the metal earth fault was accomplished by closing and opening the circuit breaker to which the earthed cable feeder was connected. The number of switching cycles (closing and opening operations of the circuit breaker) for the metal earth fault ranged from 5 to 20.

The intermittent earth fault was accomplished by moving an earthed metal rod tied to a wooden bar towards and away from one the phase conductors. When the distance would decrease sufficiently, an electric arc would appear which could be broken or re-established by moving the metal rod on the bar away from or towards the phase conductor. For intermittent earth faults, tens and even hundreds of instances of repeated occurrences and interruptions of earth faults were achieved.

A capacitance potential divider, connected to the 6kV busbars, was used to register the phase-to-earth transient voltages at the instant of appearance and interruption of metal earth faults, as well as for intermittent earth faults. The resulting overvoltages were not excessively high; the highest measured at the instant of appearance of a metal earth fault was 2.45 p.u. and, for an intermittent fault 2.8 p.u. Their frequencies and steepness were also relatively low. When earth faults appeared, the frequency of their oscillation was at most several kilohertz, while at the instant of interruption the frequency remained under 50Hz.

The switching out of a metal earth fault and the interruption of an intermittent one caused 2nd subharmonic ferroresonances. The voltages to earth were of the order of

1.7-2.1 p.u. and continued indefinitely in the network. For as long as they last, the phase-to-earth voltages, apart from the 50Hz fundamental harmonic, have a significant second 25Hz subharmonic and frequently one additional subharmonic whose frequency varies from less than one hertz to several hertz.

The earth-fault current differed in amplitude and form from network to network, but it never exceeded the value of 30A RMS. In addition to the fundamental harmonic (50Hz), earth-fault currents contained higher-order harmonics, but their amplitudes did not exceed 10% of the fundamental. The existing Yugoslav technical recommendations state that, if the earth-fault current exceeds the value of 30A RMS, the network should be divided or the power system should be earthed at a neutral point.

2.2. Overvoltages during switching operations of HV motors switchgear

The switching overvoltages were measured on the insulation of 33 HV motors ranging from 160kW to 6.5MW [2]. Transient voltages were recorded for switchgear closing and opening during normal HV motor operation and for their opening during aborted start (when the motor starting current is interrupted). Switchgear opening during aborted start is possible and does occur in practice, usually when protection circuits react due to the absence of corresponding technical conditions for normal operation. The switchgear used for HV motor switching operations included: oil minimum circuit breakers, air magnetic switching devices (circuit breakers and contactors) and vacuum contactors. About 150 switching cycles (closing and opening operations) of 19 HV motors (160kW to 6.5MW) were performed using oil minimum circuit breakers, 75 switching cycles of 8 HV motors (200kW to 2MW) using air magnetic switching devices and 40 switching cycles of 4 HV motors (350kW to 500kW) using vacuum contactors.

The closing operations with oil minimum circuit breakers and their opening during normal HV motor operation did not cause high overvoltages (max. 2.25 p.u.). Oil minimum circuit breaker opening during HV motor aborted start generally caused overvoltages (max. 4.45 p.u.). Such high overvoltages occur due to current chopping before it naturally passes through zero and due to reignitions; results show that higher overvoltages occur for opening of older oil minimum circuit breaker models.

The closing of air magnetic switching devices, their opening during normal HV motor operation and during aborted start did not cause excessively high overvoltages (max. 2.55 p.u.)

The closing of vacuum contactors generally caused high overvoltages (max. 3.5 p.u.). Multiple prestrikes appeared in all three phases. The corresponding overvoltages resembled a column of several dozen cut voltage waves with steepness of up to several dozen kV/µs; rise times of chopped waves are under 1µs. Overvoltages with rise times in the nanosecond region are unevenly distributed in the motor stator windings. They significantly stress the line-end coils and turns. It is considered that overvoltages with rise times above 1µs are generally evenly distributed across the insulation between turns of the coils. The presence of unearthed star capacitor banks at the switchgear-cable terminations reduces the number of prestrikes as well as the overvoltage rise times. The rise times which were under 1µs without capacitor banks (3 x 60kvar) increased to at least 5µs when the banks were introduced. The opening of vacuum contactors in normal HV motor operation and during aborted start generally caused no overvoltages, or only insignificant overvoltages (max. 1.85 p.u.).

2.3. Overvoltages during switching operations of transformers switchgear

The switching overvoltages were measured on the 6kV insulation of three transformers 6/0.4kV, one transformer 220/6/6kV and two transformers 110/6/6kV [3]. Transient voltages were recorded for switching in and switching out unloaded transformers with oil minimum circuit breakers. Thirty five switching cycle operations of three 6/0,4kV transformers were performed; of these, 10 were made during an earth fault. The closing of 6/0.4kV transformers did not cause excessively high overvoltages (max. 2.1 p.u.); with earth faults, overvoltages were higher (max. 3.25 p.u.). During opening operations, multiple reignition in the circuit breakers appeared and current chopping before it naturally passes through zero occurred; there were high overvoltages (max. 2.3 p.u. in normal operation and max. 3.85 p.u. with earth fault).

The closing of 220/6/6kV and 110/6/6kV unloaded transformers, which supply the 6kV station service networks of thermal power plants, caused 2nd and 3rd harmonic ferroresonance on the 6kV side; the voltages to earth exceeded 3.7 p.u. and continued indefinitely in the network.

In most cases, a very simple but very efficient measure was applied for the removal of ferroresonance. This was done by connecting a resistor between the ends of the open delta of the secondary windings of the inductive-voltage-transformer set. Sometimes the resistors had to be more than 100Ω, but usually 20 - 40Ω sufficed.

2.4 Overvoltages for transfer of the network supply from one transformer to another

The transfer of station service cable networks supply in thermal power plants from one transformer to another is a normal, albeit rare occurrence. This can be performed either with automatic switching devices or manually. A hundred and fifteen automatic transfers of station service cable network supply from one transformer to another were performed without interrupting operation [3]; the load transfer occurred very quickly in intervals ranging from 15 to 50ms, practically without overvoltages (max. 1.2 p.u.). The manual transfer of load from one supply to another (performed 8 times) caused only moderate overvoltages (1.9p.u), but resulted in large transient currents in the station service network and a decrease of voltage at 6kV busbars (over 20%).

3. CONCLUSIONS

Investigations have shown that high overvoltages can occur in station service cable networks in thermal power plants. They appear frequently due to the relatively large number of switching operations. The shapes and duration of these overvoltages differ widely from one case to another. They can last from several μs to tens and even hundreds of ms, sometimes remaining indefinitely in the network (ferroresonance). Wave-front rise times of overvoltages range from hundreds of ns to several ms, i.e. steepness of overvoltages reach tens of kV/μs. For apparatus with maximum voltages of 7.2kV - the type used in station service cable networks of thermal power plants (excluding motors), the rated short duration power frequency withstand voltage is 20kV, while the rated lightning impulse withstand voltage is 60kV. The apparatus switching impulse withstand voltage is not known, but is assumed to be above 90% of the amplitude of the short duration power frequency withstand voltage $0.9\sqrt{2}*20 \approx 25$kV. The 6kV motor stator winding insulation withstand voltages to earth, for switching overvoltages, have not been sufficiently investigated. According to earlier studies [4, 5], the stator insulation withstand voltage to earth reaches values of up to 5 p.u, while the turn-insulation withstand voltage is upwards of 1.5 p.u, depending on the rise time of the voltage wave front. Recent investigations [6,7,8,9] indicate that these withstand voltages are much higher.

On the basis of the results of overvoltage studies and in order to increase the reliability of operation of station service cable networks in thermal power plants, the following is recommended:

- The earth fault should be eliminated as quickly as possible. High quality and reliable relay protection circuits should be used for determining the earth fault location.
- The number of switching operations during earth faults should be reduced to a minimum.
- 20Ω resistors should be placed at the ends of the open delta secondary windings of the inductive voltage transformer set on 6kV busbars and on the 6kV side of transformers feeding the network.
- The supply from one transformer to another should not be transferred without the use of automatic devices.
- The insulation to ground of the stator winding of HV motors which are frequently switched using old types of oil minimum circuit breakers and which frequently switch out during aborted start, should be protected from overvoltages with surge arresters.
- Motor manufacturers should be consulted on the necessity of protecting the turn-insulation of windings in motors switched by vacuum switching devices. Unearthed capacitor banks used for reactive energy compensation or R-C circuits connected in series (usually R = 100Ω, C = 0.1 to 0.2μF) can serve as protection devices.

New and refurbished networks should be equipped with high quality switchgear without pre- and re-ignitions of the electric arc and without current chopping before it passes through zero.

REFERENCES

1. Vukelja, P.I., Naumov, R.M., Vučnić, M.M. and Budišin, P.B.: "Experimental investigations of overvoltages in neutral isolated networks". IEE Proceedings-C, Vol. 140, N°.5, September 1993, pp.343-350.

2. Vukelja, P.I., Naumov, R.M., Vučinić, M.M. and Budišin, P.B.: "Experimental investigations of high-voltage motor switching surges". IEE Proc.-Gener. Transm. Distrib., Vol.142, N°.3, May 1995, pp. 233-239.

3. Vukelja, P.I.:"Overvoltages in station service networks of thermal power plants, measures and means for their limitation"., Report of 21th Conference JUKO CIGRE, Vrnjačka Banja, October 1993, (in Serbian), paper 33.01.

4. IEEE Working Group progress report :"Impulse voltage strength of AC rotating machines". IEEE Trans., August 1981, Pas-100, (8), pp. 4041-4053.

5. Working Group 13.02 of Study Committee 13:"Interruption of small inductive currents: Chapter 3, Part B." Electra, 1984, (95) pp. 31-45.

6. Gupta, B.K., Nilsson, N.E. and Sharma, D.K.: "Protection of motor against high voltage switching surges." IEEE Trans., March 1992, EC-7(1), pp. 139-147.

7. Gupta, B.K., Lloyd, B.A., Stone, G.C., Sharma, D.K. and Fitzerald, J.P.: "Turn insulation capability of large AC motors, Part 2-Impulse strengtht." IEEE Trans., December 1987, EC-2(4), pp. 666-673.

8. Gupta, B.K., Lloyd, B.A., Stone, G.C., Sharma, D.K., Nilsson, N.C., and Fitzerald, J.P.: "Turn insulation capability of large AC motors, Part 3-Insulation co-ordination." IEEE Trans., December 1987, EC-2(4), pp. 674-678.

9. Narang, A., Gupta, B.K., Dick, E.P. and Sharma, D.K.: "Measurements and analysis of surge distribution in motor stator windings." IEEE Trans., March 1989, EC-4(1), pp. 126-134.

Conference Record of the 1996 IEEE International Symposium on Electrical Insulation, Montreal, Quebec, Canada, June 16-19, 1996

INVESTIGATIONS ON THE AGING OF CROSS-LINKED ETHYLENE VINYL SILANE COPOLYMER

Haridoss Sarma, C.Mahabir and A.Shaikevitch

AT Plastics Inc.
Brampton - Ontario
Canada L6W 3G4

Abstract: **Cross-linked ethylene vinyl silane copolymer (EVS) used as an electrical insulation in low voltage secondary cables was examined for its oxidative aging behaviour. Thermal aging of the copolymer was found to consist of three stages. The first stage consisted of an increase in the silane cross-linking induced thermally, the second stage was defined by the consumption of antioxidants with a subsequent increase in oxidation leading to brittle failure as the third stage. The type and concentration of the antioxidants determined the extent of the second stage, influencing the target application of the copolymer. For a copolymer with selected antioxidant concentration, the brittle failure occurring in the third stage could be related to an increase in cross-link density of the copolymer. Aging of the copolymer under UV conditions, resulted first in oxidation, characterised by carbonyl species and polymer unsaturation, leading to brittle failure. There was little or no increase in cross-linking. Under either aging conditions the electrical breakdown strength of the coplymer was retained to above 50% until the samples became brittle. Practical implications of the results from these investigations are discussed.**

INTRODUCTION

Cross-linked polyethylene (XLPE) is widely used as an electrical insulation. The most common technique to cross-link PE is through free radicals initiated by peroxide or electron irradiation. A second method that is rapidly gaining acceptance in the wire and cable industry is through the reaction of water with an organic silane grafted to or copolymerized with the ethylene main chains. The cross-linking of these ethylene vinyl silane copolymers (EVS) is essentially controlled by diffusion of water in the polymer [1]. There is experimental evidence to show its increased resistance to moisture induced degradation in the presence of an electric field attrinuted to its reactivity with water [2]. Mechanical measurements on EVS copolymer [3] have also indicated that more cross-links are formed within an already existing gel in an extended cross-linking time. Thermo-oxidative processes during the time of cross-linking have been shown to contribute to this behaviour. The kinetics of degradation of silane grafted PE has been compared to peroxide cross-linked PE and shown to exhibit increased thermal stability attributable to its polyfunctional network structure [4].

The results of some investigations on thermally and ultra-violet (UV) aged EVS are presented in this paper. Fourier transformed infra-red (FTIR) spectroscopy has been extensively used to identify and interpret the chemical changes occurring in the copolymer as a result of the aging process. Polymer oxidation, changes in the characteristic silane absorption, silane cross-linking and polymer unsaturation have been measured and correlated. Solvent extraction and tensile measurements have also been used to confirm the aging induced cross-linking process, which distinguishes silane cross-linked PE from a peroxide cross-linked. The EVS copolymer was investigated both in its cross-linked and uncross-linked state. The presence and absence of antioxidants in the copolymer helped to focus on the role played by antioxidants in the aging process. The study of oxidation induction by non-isothermal measurements and analysis by thermogravimetry were found to be of great benefit as ancillary methods. The a.c. breakdown strength of selected aged and virgin copolymer films were also measured as an electrical diagnostic. Additional cross-linking of the copolymer induced thermally either without or with a co-operative effect of oxidation during thermal aging and a purely oxidative degradation during UV aging were evident from these investigations. The copolymer retained its a.c.breakdown strength until a mechanical brittle failure occurred. For the purposes of secondary cables insulated with EVS for which the predominant aging mechanism is via oxidation, dielectric constant or loss measurements will be more useful than the electrical breakdown strength measurements.

EXPERIMENTAL

The EVS copolymer used in this investigation was of a melt flow index of 0.65g/10min with 2.0% by weight of copolymerized vinyl trimethoxy silane. Both films and insulated wire samples of this copolymer were studied. In the case of films, either a pressed or an extruded film was employed. An organic tin catalyst masterbatch with and without antioxidants (AO) was used to cross-link the coplymer. The samples were cross-linked by exposing them to ambient humidity and temperature or to water at 90°C. It has to be noted that in either case, the melting point of the crystalline fractions could never have been reached. This is in contrast to peroxide cross-linking wherein the polymer is taken well above

TABLE 1

Experimental Samples with their Cross-linking and Aging Conditions

Samples	Conditions for cross-linking	Conditions for thermal aging
EVS copolymer	uncross-linked	100μm films, 130°C and 180°C for several hours
EVS copolymer + catalyst	water at 90°C for 8 hours	100μm films, 130°C and 180°C for several hours
EVS copolymer + catalyst + 0.2% AO	water at 90°C for 168 hours	100μm films, 150°C for up to 10 days
EVS copolymer + catalyst + 0.45% AO	ambient temperature and humidity	375μm insulation on 20 AWG wire, 150°C for up to 5 weeks

its crystalline melting point. The cross-linked network in the case of EVS copolymer thus takes place below its crystalline melting point. This has strong implications in the aging behaviour of the EVS copolymer.

The sample identification with their corresponding conditions under which they were cross-linked and aged are given in Table 1. Thermal aging of the copolymer in its cross-linked and uncross-linked state without any antioxidant was performed in an oven at 130°C and 180°C. In the case of samples containing 0.2% by weight of phenolic esters the aging temperature was at 150°C. Wire samples insulated with EVS containing 0.45% by weight of antioxidants were aged at 150°C in an oven with a controlled air flow of 100 - 200 air changes per hour. Oxygen consumption at these ventilation rates is expected to be very great resulting in a very severe thermal aging condition.

The experiments on UV aging were performed in a UV weatherometer (xenon arc) with a UV source of 6000 watt and a radiation level of 0.35 watt/m^2 at 340 nm. The black panel temperature was 63°C which decreased to 40°C during a periodic water spray (102 min light only and 18 min light and water) on the samples. Only the samples with 0.25 % of AO were used for this testing.

The aged samples were characterised using an FTIR spectrometer, the spectra being recorded at 2 wavenumber resolution. Some measurements using a micro beam were made on the wire samples to detect the radial changes in antioxidant concentration, oxidative species and polymer chain unsaturation. The changes in the characteristic IR absorptions of the methoxy and silane groups (1090, 1260 and 800 cm^{-1}) [2], of the olefinic double bonds (R.CH=CH$_2$ at 908 cm^{-1}, R.CH=CH.R' at 965 cm^{-1} and RR'C=CH$_2$ at 888 cm^{-1}), of the carbonyl groups at 1720 cm^{-1} were monitored. Solvent extraction as per procedure ASTM D2765 was performed on all aged samples. Tensile measurements were also carried out in the case of wire samples.

Thermal analysis by methods such as differential scanning calorimetry (DSC), and thermogravimetry (TGA) were performed on wire samples alone to study the stabilisation and degradation of the EVS copolymers. In the case of the DSC method, a Perkin Elmer Model DSC7 instrument was used to determine the oxidation induction themperature of the sample (5 mg wt.) encapsulated in standard aluminum pans covered by a wire mesh, purging it with oxygen at the rate of 50 ml/min while increasing its temperature beyond its crystalline melting point at a constant rate of 10°C/min. The oxidation induction temperature reported as T_{ox}, was determined by the point of intersection of the extrapolated base line and the tangent at a point on the enthalpy curve deviating by 10% from the base line. This method had been very well documented by earlier workers who had studied the influence of experimental conditions such as mass of the sample, oxygen flow rate, antioxidant type and concentration [5]. It was also shown that this method can be used as an integrated measure for the concentration of remaining antioxidant activity in thermally aged XLPE [5,6]. Thermogravimetric analysis was carried out in a Perkin Elmer Thermal Analyzer TGA7. The heating ratewas 20°C/min and the flow rate of either nitrogen or air was kept constant at 20 cc/min throughout the study. The first derivative of the TG curves was calculated using the equipment software. It should be noted that DSC measures the onset of oxidation in terms of the change in enthalpy whereas TGA measures the degradation by the change in mass. These two techniques were complimentary to FTIR and solvent extraction methods in identifying and confirming the co-operative effect of oxidation and thermal cross-linking during the copolymer aging process. Electrical breakdown strength measurements were made on films by clamping them between two brass electrodes in a cell containing silicone oil which was replaced as and when necessary. A ramp voltage (200 v/min) was applied to the sample until failure occurred. The breakdown voltages of different specimens of the same kind were then analyzed using Weibull statistics and Weibull α values (62.5% probability values) reported as a percentage retention after aging.

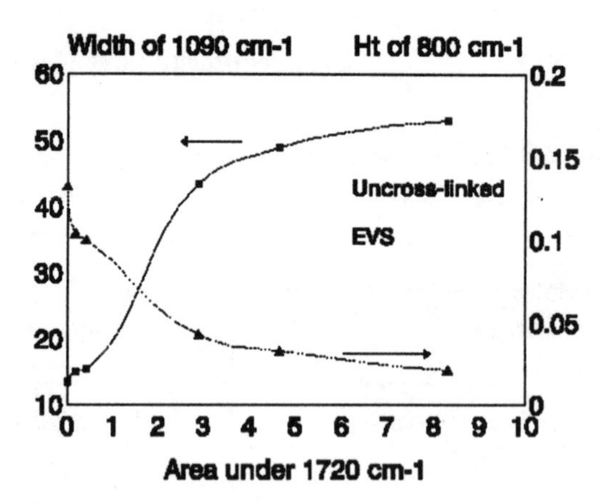

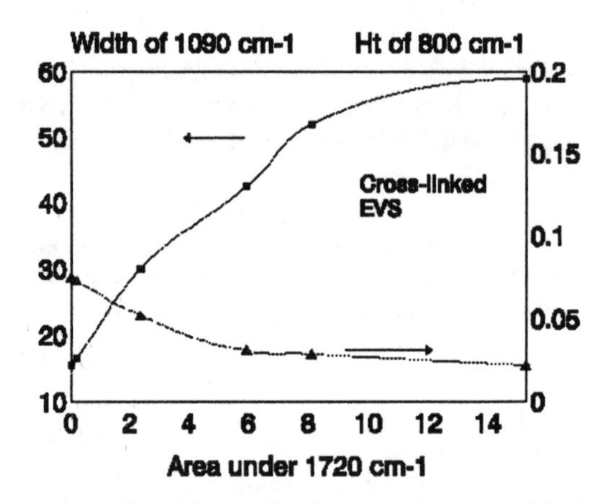

Figure 1. Spectroscopic changes in thermally aged EVS (a) uncross-linked (above) (b) cross-linked (below)

RESULTS AND DISCUSSION

EVS Copolymer without any antioxidants

Both cross-linked and uncross-linked copolymer films when aged at 130°C and 180°C exhibited excessive oxidative degradation as demonstrated by the increase in area under 1720 cm^{-1} absorption. Accompanying this, was a decrease in the methoxy groups as identified by the CH_3 vibration associated with Si in the methoxy group (800 cm^{-1}). There was no sharp increase in 1025 cm^{-1} (Si-O-Si) absorption. However, the width of the main Si-O absorption (from the methoxy silane) at 1090 cm^{-1} was found to increase with an overall reduction in its intensity. The relationship between these infra-red absorptions in the case of films aged at 130°C is given in Figure 1a for uncross-linked polymer and in Figure 1b for cross-linked polymer. The aging at 180°C was too excessive in altering the characteristic absorptions of silane, possibly overlapped by those of short chains from chain scission.

The increase in the width of 1090 cm^{-1} absorption can be explained as arising due to the contributions from the Si-O-Si vibrations resulting from different cross-link moieties. In the case of the uncross-linked copolymer a thermal cross-linking mechanism can be postulated. The very patent [7] on EVS copolymer indeed claims its cross-linkablity by a thermal process alone. It is also possible that the trace moisture in the sample could hydrolyse the methoxy groups which can then condense to form a cross-linked network. Because the aging is carried out at temperature above the crystalline melting point of the polymer, it can also be expected that molecules already part of the network form even more cross-links. The increased mobility of the silane groups at aging temperatures can add to newer cross-links. Other intermediate reactions between the polymer hydroperoxide and methoxy silane groups which contribute to the network are also possible. The same arguments will hold true for the observations with the cross-linked polymer. Thus a substantial amount of gel is formed when an uncross-linked polymer is thermally aged (Figure 2). Even for a fully cross-linked polymer, moderate increases in gel could result from one of the above routes. Chain scission through further oxidative degradation decreases the gel. Because neither of these samples contained antioxidant for stabilisation, this result indicates that cross-linking is very much controlled by oxidation. The co-operative effect between cross-linking and thermo-oxidation implies a proper stabilisation is required to optimise the copolymer for an intended application.

An increase in polymer trans unsaturation (RCH=CHR' at 965 cm^{-1}) was the other main chemical change identified after oxidative aging. The effect of sample thickness in limiting oxygen uptake leading to polymer degradation was also confirmed.

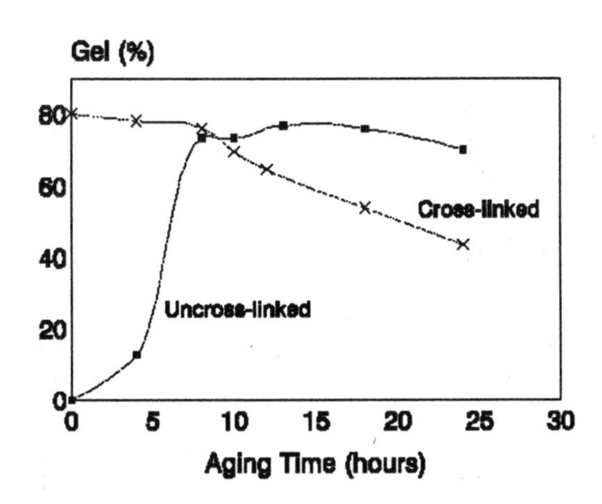

Figure 2. Gel content variation in thermally aged EVS

EVS Copolymer with 0.2% antioxidants

Conventional phenolic ester antioxidants were used for these experiments. Thin films of the copolymer were extruded with a tin catalyst/AO masterbatch and cross-linked in water at 90°C. The samples were then cut into small pieces and oven aged at 150°C for up to 10 days. Films from the same batch were also aged in a UV weatherometer.

Thermal cross-linking stood out to be an important feature of these experiments as well. The measured increase in gel content while thermo-oxidative aging the polymer and a decrease while photo-oxidative aging are shown in Figure 3. An increase in gel during the initial phases of thermal oxidation quickly dropping off to a low value in the final phase of extensive aging was observed.

The results from FTIR investigations on these aged samples support the above measurements on gel content. The absorption at 1025 wavenumbers arising from the Si-O vibrations of the cross-link increased more after thermal aging than after UV weathering (Figure 4). Although solvent extraction failed to detect any apparent increase in gel content after UV weathering, spectroscopically, a small but noticeable, increase in this shoulder was apparent indicating the co-operative effect of silane cross-linking and oxidative degradation. An increase in the $Si\text{-}OCH_3$ absorptions (1261 and 800 cm^{-1}) is however surprising. Figure 4 also reveals that ester carbonyls (1218 cm^{-1}) and terminyl vinyl $RCH{=}CH_2$ groups (909 cm^{-1}) were produced under UV exposure conditions

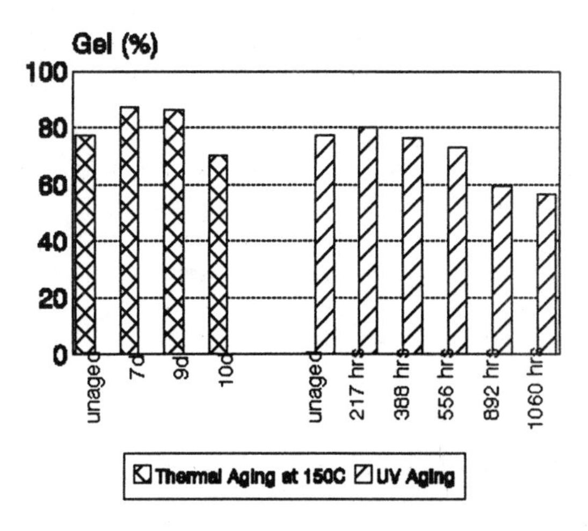

Figure 3. Gel content variation in aged EVS (0.2% AO)only.

Further analysis of spectra for chain unsaturation showed that there was an increase in trans unsaturation under both thermo-oxidative and UV aging. A comparison of oxidative degradation between the two aging conditions is given in Figure 5. Only a small but measurable increase in the carbonyl (ketone C=O) stretching vibration at 1720 cm^{-1} was observed in the case of thermal aging. On the contrary, UV aging, under the conditions chosen, resulted in a substantial increase in this absorption after 388 hours of irradiation and beyond this time a more pronounced degradation leading to an ester carbonyl with a saturated absorption at 1740 cm^{-1}. This assignment is validated by the presence of 1218 cm^{-1} absorption seen in Figure 4. The presence of 1640 cm^{-1} absorption after UV weathering could be related to C=C vibration resulting from the chain unsaturation $RCH{=}CH_2$ (seen

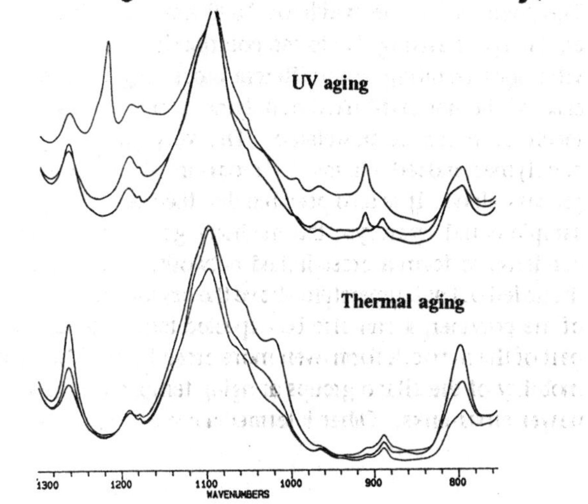

Figure 4. Spectroscopic changes in aged EVS (0.2% AO)

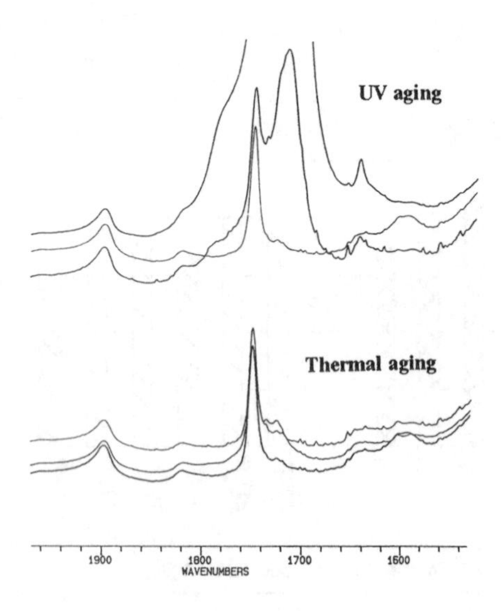

Figure 5. Spectroscopic changes in aged EVS (0.2% AO)

in Figure 4) the C-H linkages about the double bond absorbing at 909 cm^{-1}.

This experimental data showed that the antioxidants chosen for this experiment were not efficient against UV weathering. Trace amounts of oxidation of the copolymer occurring during initial stages of UV aging could contribute to an increase of cross-linking. Oxidation beyond this stage leads to chain scission resulting in a decrease in gel content. There is no added contribution to an increase of network density as the mobility of silane groups are not increased under UV aging. The temperature at which the samples are generally exposed to UV under common practice, is lower than the cross-linking temperature, which again is lower than the crystalline melting point. The deterioration of mechanical properties (tensile strength and elongation) of the copolymer under thermo-oxidative aging could be more due to a localised increase of cross-link density and under UV aging it will be due to oxidative degradation marked by polymer chain scission. The choice of the antioxidant for a specific application is therefore very critical in determining the useful life of the copolymer insulation.

This feature of the EVS copolymer may even be a distinct disadvantage when comparing it to peroxide cross-linked PE. Temperatures greater than the PE crystalline melting point are usually employed in the latter which will facilitate the formation of a network throughout the polymer matrix before service installations. The aging temperature for most of the intended applications is well below the peroxide cross-linking temperature. True oxidative degradation leading to chain scission and hence a decrease in gel is a frequent occurrence

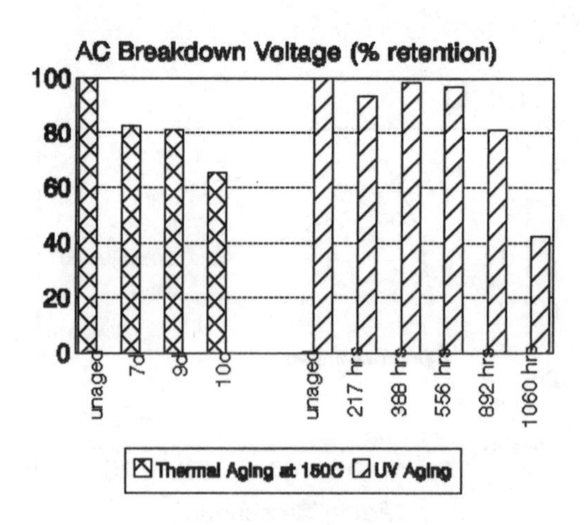

Figure 6. Electrical breakdown strength of aged EVS (0.2% AO)

here. This was verified using a standard XLPE insulation containing antioxidants.

The electrical breakdown strength of the aged EVS samples were measured using a ramp a.c.voltage and the values are reported in Figure 6 as % retention of the Weibull α values. It can be seen that in either case the breakdown voltages decreased only at the onset of mechanical degradation exhibited by the brittle failure of the samples. There was a measurable decrease in the breakdown strength of the insulation with an increase in gel content (network density) of the copolymer during the initial stages of thermal aging. UV weathering, which increased the oxidation and decreased the gel content of the polymer (Figures 3 and 5) did not result in a decrease in electrical strength of the insulation. Independent measurements on the copolymer cross-linked to various degrees did not show any influence of the amount of gel on breakdown strength. These results indicate that the morphology of the gel with reference to the solubles in the polymer is a critical property. Electrical breakdown strength, a key indicator of the performance of the insulation and the most widely used by end-users is thus not a very accurate predictor of the aging of EVS as a secondary cable insulation. Because the insulation chemically oxidises first, leading to a mechanical failure, either tensile property or its resistance to mechanical indentation (hardness) is a better diagnostic method to characterize aging and to predict the remaining life of the insulation. In addition dielectric constant and dissipation measurements would be useful.

EVS Copolymer with 0.45% antioxidants

Having examined the aging behaviour of unstabilised and

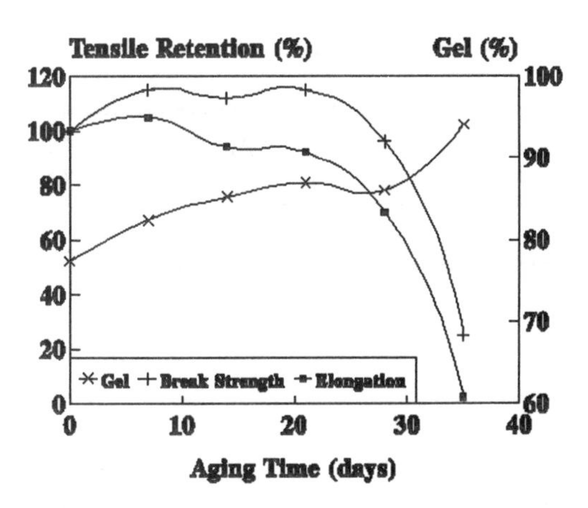

Figure 7. Evaluation of thermally aged EVS (0.45% AO)

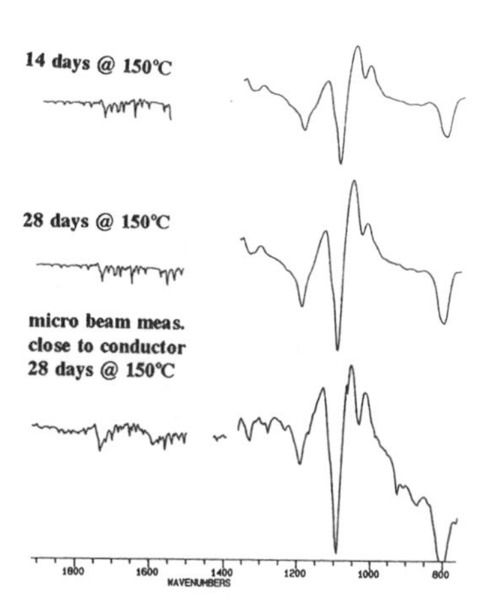

Figure 8. Spectroscopic changes in thermally aged EVS (0.45% AO)

sufficiently stabilised copolymer, the experiments were extended to one containing 0.45% antioxidant, primarily to focus on the role played by the latter in controlling thermal cross-linking.

The gel content and the tensile break strength and elongations of extruded wire insulation were measured as a function of thermo-oxidative aging at 150°C and results are shown in Figure 7. The gel content increased initially slowly and with a rapid rise beyond 21 days of aging. The initial increase in cross-link density resulted in a moderate increase in tensile break strength and a decrease in tensile elongation; any further increase in gel content resulted in a significant decrease in tensile properties.

The spectral differences between the unaged and aged insulation are shown in Figure 8. Increase in 1025 cm^{-1} absorption (Si-O-Si) with corresponding decreases in 1090 cm-1 (Si-O..CH_3), 1260 and 800 cm^{-1} (CH_3 associated with Si) absorptions is evident from the figure. Within the limited sensitivity of FTIR, oxidation of the sample was observed only when the sample became amber in colour. There was only a very small loss of antioxidant concentration (1740 cm^{-1}) due to consumption. There were no changes observed in any silanol intermediate groups.

Because of the spectral overlap of the infra-red absorptions of antioxidants, an alternate thermal analysis method was used to determine the residual antioxidant activity at different stages of aging. The oxidation temperature T_{ox} as measured from the dynamic DSC showed a steady decrease as a function of aging time (Table 2). The consumption of antioxidants during aging of the copolymer was reflected in lower measured values for oxidation temperature, the value for the sample at the end of the aging scheme (brittle failure) corresponding to that of the virgin copolymer with no antioxidant. This is in sharp contrast to homopolymer PE for which the T_{ox} value for an oxidised sample was always less than that for an unstabilised virgin. This could be due to an inherent thermal resistance offered by an increase in Si-O-Si cross-link moieties [4].

The increased resistance of EVS copolymer to thermal degradation was also evident from TGA studies. The measured temperature onset for thermal decomposition of the copolymer in nitrogen and air along with the corresponding maximum temperature for decomposition and the thermal residue left at 600°C are listed in Table 3. Firstly, a measurable solid residue left after complete decomposition of the copolymer is derived from the silicious (SiO_2) material. This residue was increased after polymer aging with a possible increase in Si-O bonds. Secondly, the temperature onset for degradation was a good measure of antioxidant activity in the polymer. It was lower for polymer without antioxidant and shifted to higher temperatures with the presence of antioxidants. As the antioxidants were consumed during aging, it shifted back to lower temperature. However, when the oxidation was complete leading to brittle failure, there was an apparent increase in the onset, possibly indicating an increase in Si-O-Si cross-links providing thermal resistance. Thirdly, the maximum temperature for decomposition was greater for cross-linked EVS copolymer reflecting the inherent thermal resistance offered by the cross-link moieties [4]. It stayed relatively constant during aging with a sharp increase after 21 days of aging, once again corresponding to an increase in gel content explained above. Thus the dynamic DSC and TGA methods lend support to the molecular events identified by infra-red spectroscopy and act as ancillary methods to confirm the oxidative degradation process of EVS copolymer insulation. Surface studies such as XPS (x-ray photo-electron spectroscopy) may throw further light on these events at a microscopic scale.

TABLE 2

Oxidation Temperatures (°C) for Aged and Unaged EVS Copolymer

PE	EVS un-XL	EVS XL	Aged 7 d	Aged 14 d	Aged 21 d	Aged 28 d	Aged 35 d
201.7	204.2	241.0	230.7	222.9	219.1	213.0	217.0

*un-XL : uncross-linked; XL : cross-linked; Aging at 150°C

TABLE 3

Parameters from TGA Measurements on EVS Copolymer Aged at 150°C

Sample	T (°C) onset		T max (°C)		residue at 600°C (%)	
	nitrogen	air	nitrogen	air	nitrogen	air
PE	256.7	260.8	391.3	386.9	0	0
EVS (uncross-linked)	252.8	251.9	462.5	461.7	1.1	1.3
EVS (cross-linked) + 0.45% AO	288.8	271.0	468.1	473.8	1.5	1.8
Aged 7 days	266.8	260.9	470.2	473.8	1.2	1.2
Aged 14 days	269.1	261.0	474.3	473.4	1.4	1.5
Aged 21 days	258.4	260.3	475.6	479.6	1.3	2.1
Aged 28 days	271.3	253.4	472.7	485.7	1.1	2.4
Aged 35 days	322.8	301.5	474.0	490.6	3.7	9.4

CONCLUSION

The present investigations on EVS copolymers show that they are inherently more resistance to thermal and oxidative degradation than homopolymer PE. Although cross-linked EVS copolymers could be oxidatively more stable than peroxide cross-linked PE, this advantage is overridden by the co-operative effect of oxidation and thermal cross-linking. As proposed by Hjertberg etal. [3], it is highly likely that the polymer hydro peroxide resulting from the oxidation can interact with the methoxy silane via hydrogen bonded silanols to produce a silane cross-linked network, although the present spectroscopic investigations did not show any evidence for this intermediate reactions to occur. What it shows, however, is a definite increase in cross-link network as a function of thermal aging. For an unstabilised copolymer the relationship between oxidation, reduction in methoxy groups and cross-linking is very clear. In the case of a stabilised copolymer, the antioxidants control this inter-dependence extending the useful temperature range of its application. For a very well stabilised polymer, increased cross-link density within the gel is still a possibility with oxidation occurring at a microscopic level. The energy differences between C-C bond (628 kJ/mol) in the case of peroxide cross-linked PE and Si-O-Si bond (779 kJ/mol) in EVS copolymer [8] indicates a better electrical and thermal stability of the silane cross-linked network. However, the lower C-Si bond energy (430 kJ/mol) could be conducive for scission of the silane side group making it more mobile and hence giving rise to additional cross-links. The loss of mechanical properties of a well stabilised EVS copolymer after thermo-oxidative aging may very well be due to this increase in cross-link density. The differences in temperatures of cross-linking (below crystalline melting point) and aging (above crystalline melting point) accentuates this process of deterioration of tensile properties. The picture for UV weathering is, however, different. Polymer oxidation leading to chain scission without a measurable increase in three dimensional network, is the predominant mechanism here, resulting in a loss of properties.

The electrical breakdown strength of EVS copolymer is found to be fairly constant until the insulation becomes brittle during aging. There is no direct correlation between the amount of cross-link density (or the concentration per volume of Si-O-Si) and breakdown strength. However, it may depend on the regularity with which the network is formed. It may also depend on the localisation of Si-O-Si bonds with respect to the crystalline morphology. More work is needed here to identify the morphological location of silane groups in the polymer. This could be of value in understanding the oxidative process and more specifically the moisture induced degradation of EVS copolymer. For the purposes of secondary cables insulated with EVS for which the predominant aging mechanism is via oxidation either by thermal or UV means, dielectric methods will be more useful as diagnostic techniques than the electrical breakdown strength measurements.

In conclusion, it can be said that thermal aging of EVS copolymer consists of three stages. The first stage consists of an increase in the silane cross-linking induced thermally, the second stage is defined by the consumption of antioxidants with a subsequent increase in oxidation leading to brittle failure as the third stage. The type and concentration of the antioxidants determine the extent of the second stage, influencing the target application of the copolymer. The third stage essentially resulting in brittle failure is accentuated by excessive oxidative degradation. For a copolymer with selected antioxidant concentration, this brittle failure could be related to an increase in cross-link density of the copolymer. On the contrary, aging of the copolymer under UV conditions results first in oxidation, characterised by carbonyl species and polymer unsaturation, leading to brittle failure. There is only a small or no increase in cross-linking. Under either aging conditions the electrical breakdown strength of the coplymer is retained significantly until the samples become brittle. The type and concentration of antioxidants are critical while deriving the full use of EVS copolymer for different applications.

REFERENCES

1. Sarma, H. and Shaikevitch, A. "Cross-linking of Ethylene Vinyl Silane Copolymers", Int.Wire Assoc. Tech. Conf., Mexico City, Oct 1994, pp.140-148.
2. Shaikevitch, A. and Sarma,H. "Chemical Nature of Water Trees in Cross-linked Ethylene Vinyl Silane Copolymer", IEEE Int.Symp. on Elect.Insu. June 1994, pp.360-363.
3. Hjertberg, T., Palmlof, M. and Sultan, B.A. "Chemical Reactions in Cross-linking of Copolymers of Ethylene and Vinyltrimethoxy Silane", J.Appl. Polym. Science Vol.42, 1991, pp.1185-1192.
4. Sen, A.K., Mukherjee, B., Bhattacharya, A.S., De, P.P. and Bhowmick, A. "Degradation of Silane and peroxide Cross-linked Polyethylene and Ethylene Propylene Rubber", Polym.Deg.Stability, Vol.36, 1992, pp.281-289.
5. Karlsson, K., Assargren, C. and Gedde, U.W. "Thermal Analysis for the Assessment of Antioxidant Content in Polyethylene", Polym. Testing, Vol.9, 1990, pp.421-431.
6. Karlsson, K., Smith, G.D. and Gedde, U.W. "Molecular Structure, Morphology and Antioxidant Consumption in Medium Density Polyethylene Pipes in Hot-Water Applications", Polym. Eng. Science, Vol.32, 1992, pp.694-697.
7. Zutty, Nathan.L and Charleston, W. "Heat Curing of Ethylene Vinyl Silane Copolymers" US patent 3,225,018, 1965
8. Cartasegna, S. "Silane Grafted Moisture Curable Ethylene Propylene Elastomers for the Cable Industry", Amer. Chem Soc, Rubber Div. Meet. Vol.50, April 8, 1986, pp.722-739.

Conference Record of the 1996 IEEE International Symposium on Electrical Insulation, Montreal, Quebec, Canada, June 16-19, 1996

A comparison of analytical and conventional techniques for thermal endurance characterization of cable insulation

A. Motori and G.C. Montanari
University of Bologna
Viale Risorgimento 2
40136 Bologna, ITALY

L. Centurioni and G. Coletti
University of Genova
Via All'Opera Pia 11a
Genova, ITALY

Abstract: **The thermal endurance characterization of ethylene-propylene rubber (EPR) insulated cables for medium voltage applications is performed by both conventional long-term life tests and resorting to an analytical technique, which is based on measurements of the oxidative stability of the insulation. It is shown that the Temperature Index and the Halving Interval provided by the two methods are very close. Comparison are finally made with the results previously obtained applying the same procedures to high voltage EPR cables.**

INTRODUCTION

The need to provide thermal endurance characterization of insulating materials in times shorter than those required by IEC 216 Standard [1] (about 1 year) is forcing to investigate new techniques. These techniques should provide estimates of the indices generally used for thermal endurance characterization, that is, TI, temperature index, and HIC, the halving interval, in times as short as possible, but compatible with the aim to achieve index estimates which are close to those provided by conventional tests, made according to IEC 216 Standard (let us recall that TI is the temperature corresponding to the life of 20000 hours, while HIC is the temperature interval, referred to TI, in which life halves or double).

IEC 1026 Standard [2] suggests criteria for the determination of TI and HIC, resorting to the so-called analytical techniques for the determination of the slope of the endurance line and to a conventional life test for its location. In this paper, the results obtained applying to EPR cables for medium voltage applications these techniques are presented and discussed. Only a careful choice of the measured property used to derive the activation energy of the degradation process, as well as of the diagnostic property used for the determination of the location of the endurance line, can provide index estimates which are close to the values derived by conventional tests. It is shown that oxidation time measurements provide estimates of the activation energy which fit well those derived by conventional tests. The results obtained are also compared with those elsewhere derived for EPR used in high voltage insulation [3].

PRINCIPLES OF THE ANALYTICAL METHOD

IEC 1026 Standard suggests criteria for the determination of the thermal endurance line and the relevant indices (TI and HIC) for endurance characterization [2]. The slope of the line can be derived by analytical techniques, which enable short-time estimation of the activation energy of the degradation process prevailing in the temperature range of the tests. The line location is provided by an experimental point, which is derived by a conventional life test at a chosen temperature, and establishing a suitable end-point for the selected diagnostic property (the life time should exceed 300 hours).

It must be observed that the values of TI and HIC can remarkably depend on the chosen diagnostic property and life time; therefore, this selection should be related to the expected service conditions of the insulation.

For cross-linked polyethylene (XLPE) and EPR high voltage cables, dielectric strength, tensile strength and tensile modulus were considered as diagnostic properties, and their noticeable decrease with aging time as end-point were selected [3, 4].

Moreover, as oxidation is the prevailing degradation reaction for these materials in the temperature range of life tests and design, the analytical technique adopted for the estimation of the life line slope was the measurement of oxidative stability. Unlike other techniques suggested by IEC 1026 Standard (e.g., thermogravimetric analysis), oxidative stability measurements can be performed at temperatures very close to those of the life tests. Performing these measurements at three or more temperatures until a characteristic time (e.g. the oxidation maximum time) is reached, the slope of the thermal endurance line can be estimated by the activation energy (E) of the oxidation process, according to the Arrhenius model, that is:

$$t_o = A \exp(E/kT) \qquad (1)$$

where t_o is the oxidation time at the temperature T.

The equation of the thermal endurance line is, therefore [1, 2]:

$$\log t_L = a + b/T \qquad (2)$$

where t_L is the life time and $b = E/2.3k$ is the slope. The slope is determined according to (1), while the line location is provided by the life time, t_{LC}, derived by the conventional life test at the temperature T_C, that is:

$$a = \log t_{LC} - b/T_C. \qquad (3)$$

Alternatively, the Eyring model, which is based on the activation free energy of the degradation process and provides comparable results, can be applied [3, 4].

MATERIAL AND PROCEDURES

The tested cables, used for medium voltage applications (EPR_{MV}), had copper conductor 1.4 mm in diameter and EPR insulation 1.0 mm thick, made by an ethylene-propylene diene terpolymer. Dicumylperoxide and 1,2-dihydro-2,2,4-trimethyl-quinoline were used as cross-linking agent and antioxidant, respectively. EPR was also filled with calcined kaolin.

Conventional thermal life tests were performed in an oven at 135, 150 and 170°C, according to IEC 216 Standard. After removing the conductor, the mechanical properties (tensile strength and elongation at break) of the insulation were measured as a function of aging time. Measurements were made at room temperature and 50% r.h.. Five specimens were tested for each aging time.

Oxidative stability measurements were performed by differential scanning calorimetry and oxidation maximum time (OMT) was chosen as the property for the application of the analytical technique. Measurements were made on specimens cut from the unaged cables at 175, 190, 200 and 210°C with the procedure described elsewhere [5]. Five tests were performed for each temperature.

APPLICATION OF THE ANALYTICAL METHOD

Figures 1-3 shows the elongation at break (EL) as a function of aging time for the three test temperatures (the values are referred to the initial one, ELo, measured on the unaged specimen).

According to IEC 216, considering the elongation at break as the diagnostic property and its decrease to 50% of the initial value as the end-point, the thermal endurance graph reported in Figure 4 is obtained. This graph provides 104.3°C and 7.5°C for TI and HIC, respectively.

The oxidation maximum times are reported in Figure 5 as a function of the reciprocal absolute temperature. The activation energy of the oxidation process of the insulation is provided by the slope of the regression line of the points.

The thermal endurance graph, obtained resorting to the analytical procedure, i.e. plotting the line which has the slope determined by the OMT tests and crosses a life point derived by a conventional life test, is reported in Figure 6. In the Figure, the three endurance lines provided by the conventional thermal life tests at 135, 150 and 170°C, property elongation at break, end-point 50%, are shown.

The values of TI and HIC derived by the conventional life tests (Figure 4) and the analytical procedure (Figure 6) are reported in Tables 1 and 2, respectively. In the same Tables, the indices previously obtained on an EPR insulation for high voltage cables (EPR_{HV}) by both methods and two different diagnostic properties are also reported for comparison.

With reference to EPR_{MV}, it can be observed that the values of TI and HIC derived by both the analytical technique at the three test temperatures and the conventional procedure are in good agreement. In Table 2, the results obtained considering the conventional tests at 135 and 170°C are omitted for the sake of brevity; however, they provideTI=107.0, HIC=6.7 and TI=110.5, HIC=6.8, respectively. Only the results derived considering the conventional test at 135°C provide differences in the estimate of TI slightly larger than HIC, but still acceptable thinking that much more significant variations are obtained considering conventional characterization in which diagnostic properties and/or end-point is varied [3, 4].

Similar conclusions can be drawn considering the thermal endurance characterization previously carried out on HV EPR grade (see again Tables 1 and 2). Also in that case the values of TI and HIC obtained from conventional and analytical techniques were very close (even closer than for the MV EPR grade).

These results support the use of the proposed analytical method for short-time thermal endurance characterization of insulating materials for which oxidation is the leading degradation reaction in the test and extrapolation temperature range. Such a method can be seen as an alternative to the conventional procedure, in particular for the comparison of similar insulating materials, candidate for the same applications. Of course, one must be well aware that changes of antioxidant and/or other chemical constituents can affect the accuracy of the method, providing results which become as far from those derived by the conventional procedure as the diagnostic properties used for the latter are influenced by the above-mentioned changes. In fact, it must be underlined that the very good agreement discussed above derives also from the appropriate selection of diagnostic property and end-point. For example, results obtained by a similar comparison made for different types of LV EPR grade provided contrasting indications, depending on the kind of additives.

CONCLUSIONS

The proposed analytical technique for short-time thermal endurance characterization can be used to estimate the Temperature Index and the Halving Interval of insulating materials for which oxidation is the leading degradation reaction in the test and extrapolation temperature range.

Such a method can be considered as an alternative to the conventional long-term life tests, mainly when similar insulating materials, candidate for the same applications, must be compared. In fact, although a reference conventio-nal thermal aging would still be appropriate for one reference material, at least to check the most suitable diagnostic property, the time saved to obtain the thermal characterization of other similar materials would be considerable, causing both economic savings and a shortening of the "time to the market" of improved products.

REFERENCES

[1] IEC 216 Standard, Fourth issue (1990-1994).

[2] IEC 1026 Standard, First issue (1991).

[3] G. Mazzanti, G.C. Montanari, A. Motori, *J. Phys. D: Appl. Phys.* **27**, 2601-2611, 1994.

[4] G.C. Montanari, A. Motori, *J. Phys. D: Appl. Phys.* **24**, 1172-1181, 1994.

[5] G.C. Montanari, A. Motori, A. Bulinski, S. Bamji, J. Densley. Proc. IEEE Int. Symp. on Electr. Insul., Baltimore (USA) 1992, 44-48.

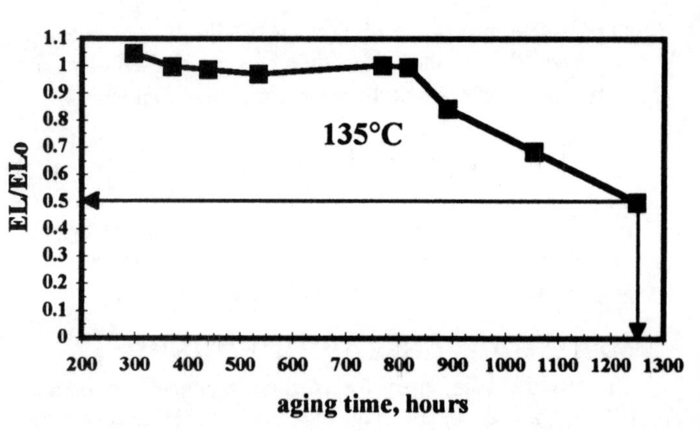

Fig. 1: Elongation at break as a function of aging time (aging temperature: 135°C).

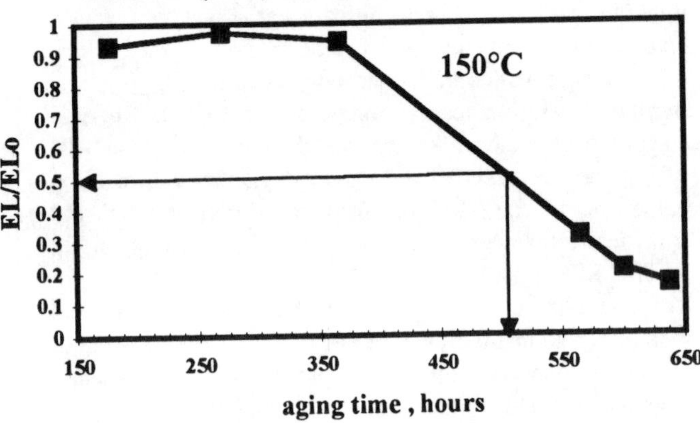

Fig. 2: Elongation at break as a function of aging time (aging temperature: 150°C).

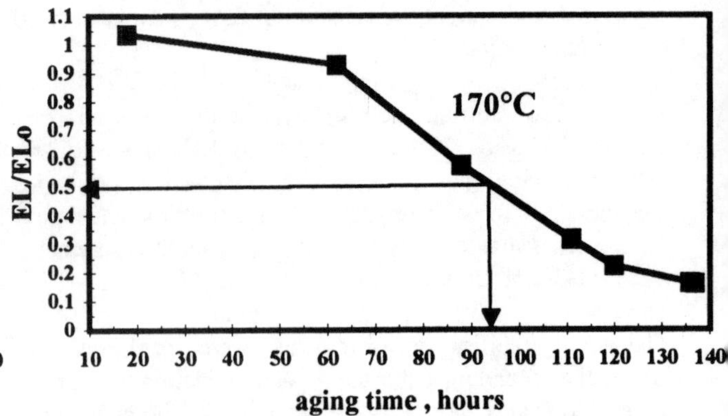

Fig. 3: Elongation at break as a function of aging time (aging temperature: 170°C).

Table 1: Values of TI and HIC derived by the conventional procedure.

Material	Property	End point	Temperature, °C	Life time, h	TI (°C)	HIC (°C)
EPR_{HV}	ES	0.5	160	550.0	111.1	8.4
			150	953.0		
			130	3754.5		
	TS	0.5	160	548.0	110.9	8.5
			150	982.4		
			130	3789.0		
EPR_{MV}	EL	0.5	170	94.4	104.3	7.5
			150	506.0		
			135	1250.0		

Table 2: Values of TI and HIC derived by the analytical technique (L_{150} is the life time derived by the conventional test at 150°C).

Material	Property	End point	L_{150} (h)	TI (°C)	HIC (°C)
EPR_{HV}	ES	0.5	953.0	115.5	7.4
	TS	0.5	982.4	115.8	7.4
EPR_{MV}	EL	0.5	506.0	111.0	6.8

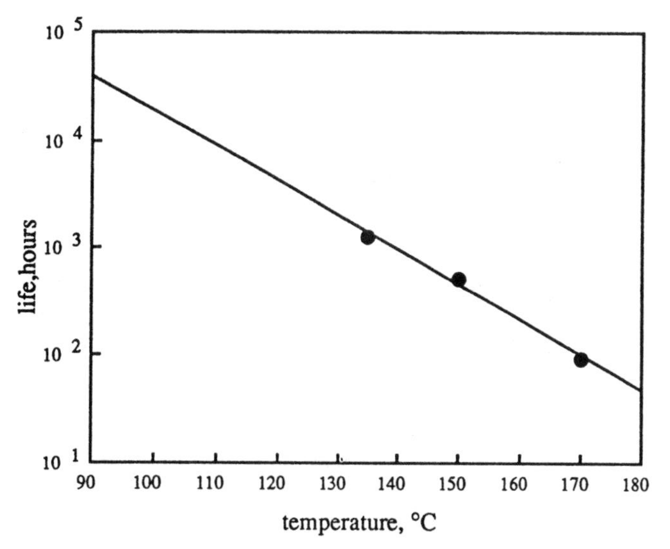

Fig. 4: Thermal endurance graph according to IEC 216 (diagnostic property elongation at break, end-point 0.5).

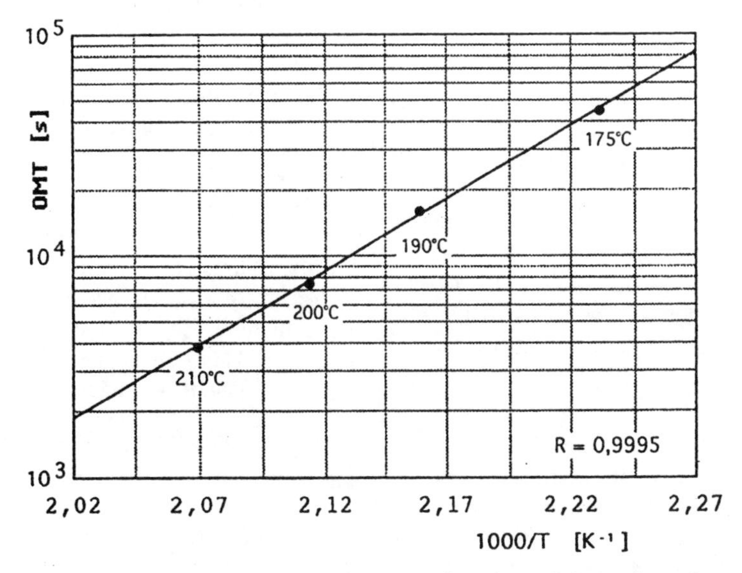

Fig. 5: Oxidation maximum time as a function of the reciprocal absolute temperature.

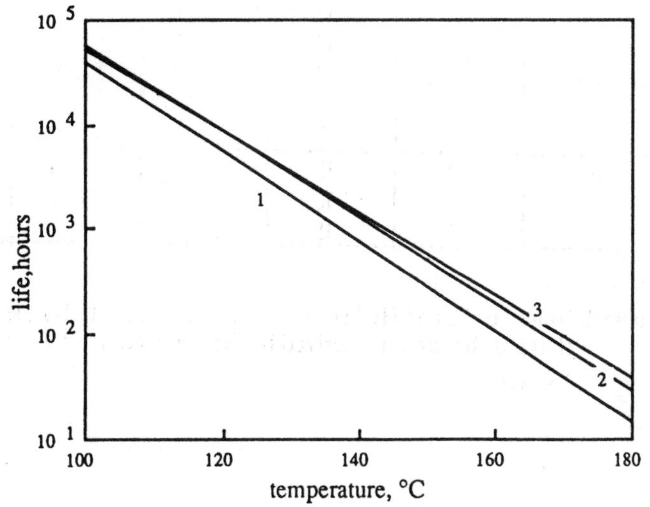

Fig. 6: Thermal endurance graph according to IEC 1026 (diagnostic property elongation at break, end-point 0.5; conventional life test at (1) 135°C, (2) 150°C and (3) 170°C).

Conference Record of the 1996 IEEE International Symposium on Electrical Insulation, Montreal, Quebec, Canada, June 16-19, 1996

The Accelerated Ageing of a LDP under Simultaneously Action of Heat and Electrical Field

Authors: George Mares
Gabriel Constantinescu
Ovidiu Dumitrescu

EUROTEST S.A.
Research, Equipment Testing, Industrial Engineering & Scientific Services
Splaiul Unirii 313, sect. 3, P.O. Box 4-77,
Bucharest 73204, Romania

Abstract: **The LDP (low density polyethylene) subject at two stress factors - heat and electrostatic field, it needs, for interpretation of the ageing data, of one equation with a explicitly dependence of simultaneously action of the stress parameters. For this, the equation used determine a value of the activation energy in a good concordance with the value of energy showed in literature.**

INTRODUCTION

Your accelerated ageing tests under simultaneously action of heat and electrostatic field of a LDP lead to more difficulties for determination the lifetime. In this paper is showed a statistical model for the interpretation of experimentally data and obtained the lifetime for using of this material in specific conditions. The activation energy calculated with this model it has an average of on tree curves of isopotential. In conformity with this model the activation energy is adequate at scission of polymeric chains.

EXPERIMENTAL

The sheets of LDP with 100 x 100 mm dimensions of the surface and 0.05 mm thickness were aged in forced air circulation oven. Each sample was put between two electrodes (in conformity with ASTM) in order to apply the electrostatic field. On each sample the potential was applied from the individual source. The individual sources were controlled of the slave computer connected with the master computer. The rate in order to grow the potential was of 300 V/s. When a sample is break, automatically the individual source of power supply is interrupting and time until break is recording. The maximum errors of measurements and control of high voltage are under 1.5%.

RESULTS AND DISCUSSION

The LDP sheets were accelerated ageing, simultaneous under action of heat at the temperatures of 80, 90 and 100°C (under softening temperature of 104°C) and at high voltage of 3, 4 and 5 kV (equivalent one intensity of electrostatic field of 60, 80 and 100 kV/mm). For each temperature were used for ageing voltages of 3, 4 and 5 kV.

In order to determine the lifetime in this experimentally conditions most adequate is the Weibull repartition function.

The Weibull repartition function has following form:

$$F_t(t,\theta,k) = 1 - \exp(-t^k \cdot \theta^{-k}) \quad (1)$$

where t is time, θ is scale factor and k is form factor.

The average time $\bar{t}$ is:

$$\bar{t} = \theta \cdot \Gamma(1+1/k) \quad (2)$$

where $\Gamma(t)$ is the Gamma function.

If the variation coefficient depend not of voltage and temperature, then the form coefficient k is a constant of materials.

On the other hand, the held mean time for a specification held voltage is:

$$\bar{t} = C(T) \cdot U^{-\gamma} \quad (3)$$

where $C(T)$ is a factor depending of absolute temperature T, U is the voltage and γ is a material constant.

Generally, exist a voltage for which under this voltage the important degradation phenomena are of thermo-oxidation. This voltage we named lower voltage U_l. The equations (3) will be used in following form:

$$\bar{t} = C(T) \cdot M^{-\gamma}(U - U_l) \quad (4)$$

where

$$M(U-U_l) = \begin{cases} U & for U \geq U_l \\ 1 & for U < U_l \end{cases}$$

is a numerical function.

From equations (2) and (4) the scale factor will be:

$$\theta = \frac{C(T)}{\Gamma(1+1/k)} \cdot M^{-\gamma}(U - U_l) \quad (5)$$

and for k = material constant then

$$\theta = \tilde{C}(T) \cdot M^{-\gamma}(U - U_l) \quad (6)$$

where $\tilde{C}(T) = C(T) / \Gamma(1+1/k)$

On the other hand for the thermal accelerated ageing lifetime is:

$$\ln \bar{t} = A + E/RT \quad (7)$$

where E is the activation energy, R is the gas constant while A is a constant.

For the Weibull partition function we obtain following expression:

$$F(t,\theta,k,T,U) = 1 - \exp\left[-C \cdot e^{-\frac{kB}{T}} \cdot M^{\gamma \cdot k}(U - U_l) \cdot t^k\right] \quad (8)$$

where $B=E/R$ while C is a material constant.

For the case when the simultaneous ageing under action heat and electrostatic field will head at synergetic effects then equation (8) will be modified.

Rewriting equation (8) in following form:

$$\ln\left|\ln(1-F(t,\theta,k,T,U))\right| = \ln C - \frac{kB}{T} + \ln t + \\ +\gamma \cdot k \cdot \ln M(U-U_l) \quad (9)$$

it is easy using in order to calculate lifetime under this form:

$$\ln t = k^{-1}\ln\left|\ln(1-F(t,\theta,k,T,U))\right| - k^{-1}\ln C + \\ +BT^{-1} - \gamma \cdot \ln M(U-U_l) \quad (10)$$

or

$$\ln t = \tilde{C} + BT^{-1} - \gamma \cdot \ln M(U-U_l) \quad (11)$$

when

$$\tilde{C} = k^{-1}\ln\left|\ln(1-F(t,\theta,k,T,U)\right| - k^{-1}\ln C$$

For one curve of constant voltage the equation (11) may be rewriting in following way:

$$\ln t = \tilde{C}_U + BT^{-1} \quad (12)$$

when

$$\tilde{C}_U = \tilde{C} - \gamma \cdot \ln M(U-U_l)$$

Using the equation (12) for the linear regression in order to obtain the value of B constant, the results are showed in table I.

Using the equation (11) in final form

$$\ln t = 6.953 + \frac{27903}{T} - 9.89 \cdot \ln U \quad (14)$$

will give lifetime, but for low voltage the lifetime will have unrealistic values. For this experiment the apparent energy is around of 56 Kcal/mol adequate at scission polymeric chains. In the case when the ageing is at low voltage and usual temperatures the activation energy will decreasing because for long time the degradation phenomena will be due thermo-oxidation reactions.

CONCLUSIONS

In order to determine lifetime behind accelerated ageing under simultaneously action of heat and electrostatic field is necessary to rewrite equation (8) in synergetic conditions, else obtained lifetime for using condition will be unreal. For low voltages and long time used the apparent activation energy will be adequate more to thermo-oxidation phenomena. For this experiment is very important that put in evidence a activation energy grater than thermo-oxidation energy. This is natural because thermal ageing time was short in order to induce a thermo-oxidation degradation. In other hand, the equation (8) may be used if U_l voltage will be correct calculated and the activation energy will be a average depending of emergence probability of two degradation phenomena: scission of polymeric chains and thermo-oxidation, or in other hand investigated the synergetic effect.

Table I

Voltage [V]	B	$\tilde{C}_U$	The error of B with a significant level of 5%	Correlation coefficient [%]	Standard deviation of $\tilde{C}_U$
3000	28677.55	-71.893	1.63	99.99	<< 1
4000	29673.24	-77.848	3144.15	98.20	15.749
5000	25359.47	-67.770	3815.25	99.40	7.35

Table II

Temperature [°C]	γ	$\tilde{C}_T$	The error of γ with a significant level of 5%	Standard deviation of $\tilde{C}_T$	Correlation coefficient [%]
80	10.1073	90.2799	0.0503	0.29	99.99
90	10.4055	90.2013	2.4279	14.069	98.70
100	9.1778	78.3527	1.3527	7.61	99.50

For the isotherm curve equation (11) may be rewriting in following way:

$$\ln t = \tilde{C}_T - \gamma \cdot \ln M(U-U_l) \quad (13)$$

where

$$\tilde{C}_T = \tilde{C} + BT^{-1}$$

In conformity with equation (13) the linear regression will have following values of $\tilde{C}_T$ and γ, shaved in table II.

For this linear regression $M(U - U_l) = U$.

The estimations of parameters given at averages will be $\hat{B} = 27903$, $\hat{\gamma} = 9.89$ and $\hat{\tilde{C}} = 6.953$.

For the cases when accelerated ageing under simultaneously action and the intensity of electrostatic field has small values, when the experiments of ageing will be on long time, we assume that with a high probability the form parameter k will be a function of temperature. This function of temperature will included explicitly the glass temperature.

Conference Record of the 1996 IEEE International Symposium on Electrical Insulation, Montreal, Quebec, Canada, June 16-19, 1996

Modeling of Thermal and Radiative Aging of Polymeric Cable Materials

Montgomery T. Shaw and Yong-Ming Liu
Polymer Science Program and Chemical Engineering Department
University of Connecticut
Storrs, CT 06269-3136

Abstract: **It is of critical importance to safety in the nuclear power industry to estimate the lifetime of polymeric materials in an environment featuring elevated temperatures and radiation. To this end we are examining the aging of polymeric materials, in cables and other components in nuclear reactor containments by comparing aging processes for a variety of materials under natural conditions with those under accelerated laboratory conditions typical of those used in qualification. The physical property data derived from the naturally and artificially aged specimens are compared using three methods: (1) traditional method, (2) activation energy from natural data alone, and (3) internal standards method. All three methods work with the properties of the material in its aged, but usable state, and not with a failure time of the material.**

INTRODUCTION

While the prediction of long-term aging performance has been practiced for years [1], it is a practice that attracted considerable controversy. The use of accelerated thermal aging based on the Arrhenius theory has many limitations and the significance of dose-rate effects on accelerated radiation aging has been examined extensively [2, 3]. To make up for uncertainty in the extrapolation of performance to a time far in the future, accelerated testing conditions are made purposefully conservative. Experience has suggested they may be unrealistically conservative, leading to expensive replacements and use of materials which many not be optimal.

This approach is to compare the aging processes for a variety of materials under natural conditions with those under accelerated laboratory (or qualification) conditions. Specifically, we wish to test the null hypothesis that the normally specified accelerated aging produces the same degradation in material properties as the corresponding natural aging.

ANALYSIS

We use three methods to compare the properties of naturally and artificially aged materials:

1) Traditional method
2) Activation energy from natural data
3) Internal standard method

All three methods work with the properties of the material in its aged, but usable state, and not with a failure time of the material.

Traditional Method

With this method we start with the results of the artificial experiments and attempt to predict the results of natural aging. We then compare the predicted property with the actual property. If the two agree, then the hypothesis of equivalence cannot be rejected. Otherwise we use parametric or non-parametric methods to reject the null hypothesis. Furthermore we can then test hypotheses concerning the conservative aspects of artificial aging.

The predicted properties are arrived at in the following manner: First we need the time, temperature and dose for the particular natural aging conditions chosen for the comparison. This might be, for example, 36600 h at 133 °F, along with 1.76 krad. Then we must pick a published activation energy: it might be 1.1 eV if we are considering an EPR insulation on a cable. The conditions for natural aging are then converted to an arbitrary standard temperature of 140 °F, becoming 23000 h at 140 °F and 1.76 krad. This point is placed numerically on a plot of the properties vs. time for the artificial aging brought to the same activation energy. Of course the 23000 h for the shifted natural point will never exactly coincide with any of the shifted times for the accelerated data; moreover, the radiation doses will never exactly coincide. Thus the accelerated data must be interpolated, first with respect to shifted time at each total dose and then with respect to dose. For trials we chose the Akima minimum-curvature interpolation method [4]. The result of this interpolation will be the predicted property.

On finishing the interpolation we plot the actual property vs. the predicted property. If the agreement were perfect, all the points would lie on a 45-degree line in such a plot. If the points fall below the line, then the natural properties are falling faster than what one would predict from the artificial aging data. With the example shown in Figure 1, we cannot reject the hypothesis that the artificial and natural aging are equivalent. One problem with this interpretation is that it does not give a picture of how the data at early times, such as is the case in Figure 1, will be predictive of what comes later.

Activation Energy from the Natural Data Set

The second method does not rely on the artificial aging results at all, but works entirely with the natural data set. The logic is that the natural set can provide an activation energy and this activation energy can be compared with that expected. For example a reasonable activation energy for EPR insulation is 1.1 eV; this value would be used for predicting natural aging. If analysis of the natural aging data gives the same number, within experimental error, then the two are consistent. If the activation energy is different, then we can reject the hypothesis that the two are equivalent. Unfortunately, we don't have complete natural aging curves; the data are discrete, and correspond to different and changing conditions.

Because the shapes of the property decay curves are not known, they might be approximated by a Taylor series in log time [5]. We generally use up to third order. We also assume that time shifting process is of the form:

$$t_{shifted} = t \; / \; a_\sigma \tag{1}$$

where t is the time the specimen has actually been exposed, $t_{shifted}$ is the shifted time and a_σ is the time shift factor which depends on dose rate and temperature. The expression we use for a_σ follows from earlier work by Burnay [3]. This expression is

$$a_\sigma = \left(1 + \frac{\dot{D}}{a_1}\right)^{-a_2} \exp\left[\frac{a_3}{R}\left(\frac{1}{T} - \frac{1}{T_0}\right)\right] \tag{2}$$

where $\dot{D}$ is the dose rate, T is the absolute temperature, T_0 is the reference temperature, R is the gas constant in the appropriate units, and a_i are parameters. The activation energy is identified with a_3, while a_2 provides information about the influence of dose rate on the degradation process.

Once having assumed a functional form for the property decay function *f(t)* and the shift factor a_σ, the task remains to find the parameters for both. The concept of the method is simple: the three parameters in the shift law described in Eqn. 2 are varied until the shifted data fall along a smooth curve with the smallest possible variation.

An example result is shown in Figure 2. Should a_3 be found significantly different than the published activation energy, the hypothesis that artificial and natural aging are the same can be rejected. Likewise, if a_2 is significantly different than 1.0 and/or a_1 is large compared to the normal natural dose rates, then the equal-dose-equal-damage premise might be questioned. Values of a_3 much less than the published values would be of particular concern because this can mean that the predicted results for natural aging would be overly optimistic. However, the possibility exists that our materials are somewhat different than those reported in the literature, so the actual comparison would be done with our accelerated data. This comparison is under investigation.

Internal Standards Method

To reduce the effect of inaccurate environmental data, we have proposed the "internal standards" method. The idea is that a test material itself provides an indicator of the aging severity. We can take one (or all) of the materials as the metric, i.e., the internal standard. Thus, the properties can be correlated in terms of this internal reference to monitor the level of exposure. The properties of a preferred internal standard should change monotonically during aging with minimum random variation. At the simplest level, one examines the properties of two materials, A and B, drawn from the same bundle, and thereby the same natural aging conditions. If material A shows more damage than B, but if they have been predicted to be the same based on accelerated aging tests, then we might reject the hypothesis that the two methods of aging are the equivalent.

To set up the relation between the standard and a material to be tested, and to describe the role of the activation energy in the internal standards method, we can examine synthetic property decay data for any material by assuming an equation such as

$$f_i\,(t;\,T) = exp\left(-\,M_i\,t\,e^{-E_i/RT}\right) \tag{3}$$

where M_i is a material-dependent parameter, E_i is the activation energy, and T is the absolute aging temperature. While this form for *f*, the relative property value, is vastly oversimplified, it serves to illustrate the method. We then generate property decay data for material *i* and a standard *s* under both artificial and natural aging conditions, as presented schematically in Figure 3. If we plot $log(f_i)$, the property of material *i*, vs. $log(f_s)$, the property of the standard *s* for a fixed temperature *T*, the "slope" of this relation, *n*, (we take a linear regression) can be derived as:

$$n = \frac{M_i}{M_s} \exp\left[\frac{E_s - E_i}{RT}\right] \tag{4}$$

Clearly, the "slope" *n* depends on both the activation energy difference and the temperature, as well as the fundamental properties of the material as expressed by the ratio $M_i \; / \; M_s$. *n* is the slope for a linear relation only if the data are from the same temperature; if there are a variety of temperatures in the data set, the points will scatter somewhat (unless $E_s = E_i$).

To compare natural and artificial aging for the "unknown" relative to the standard, we apply Eqn. 4 for natural *N* and artificial *A* aging conditions; the ratio of the two slopes is then:

$$\frac{n_N}{n_A} = \exp\left[\frac{E_i - E_s}{R}\left(\frac{1}{T_N} - \frac{1}{T_A}\right)\right] \tag{5}$$

In Eqn. 5 we have lumped the artificial aging temperatures together into T_A and natural aging temperatures together into T_N. In this equation the ratio M_i / M_s has canceled. Also, we anticipate that Eqn. 5 will be the first term of expansions of other property decay functions $f(t)$.

The application of Eqn. 5 as follows: The artificial and natural property decay data for two materials are plotted using log-log scales and the two slopes n_A and n_N are found. The ratio of the slopes between natural and the artificial aging should be given by the right-hand side of Eqn. 5. Applying this to real data, as shown in Figure 4 and Figure 5, we have an expected ratio n_N / n_A of 2.3 ± 0.7 with E_i = 0.86 eV (material R) and E_s = 1.2 eV (material K) and a found ratio of 2.2 ± 1.5. Thus we cannot reject the hypothesis that the artificial and natural aging are equivalent. Should the found ratio have been higher, then we could conclude that the prediction of natural aging from artificial aging data is flawed for at least one of the materials. The method cannot tell which one with a single comparison, but by running several two-way comparisons, this should become evident.

Clearly, the internal standards method, while of great potential utility, is still a primitive technique. In developing an expression for the relative behavior of two materials, we need to assume a property-decay function $f(t)$. One improvement would be to use functions derived from artificial data; these would in general be different for the two materials. Also the question of radiation needs to be addressed; of the two materials, one might have far more radiation sensitivity than the other. These considerations suggest that the method should be used mainly to compare two similar materials, e.g., two EPR insulations from two manufacturers, that are anticipated to behave the same in the natural environment.

CONCLUSION

All three of the methods have as their foundation the assumption that the property decay function is shifted in some simple manner with changes in the environment. Some differences exist in the details. The equal-dose-equal-damage premise means that $f = f(D / a_T)$, i.e., time is not an argument and temperature works through the dose D via a temperature shift a_T, and not through time. The second assumption is $f = f(t / a_\sigma (D; T))$; that is, dose rate and temperature cause only multiplicative time shifts and do not interact with time in some more complicated (non-separable) fashion.

Comparing the internal standards method with the conventional method is more straightforward. The conventional method looks at the aging conditions in more detail, while the internal standards method lumps these together. On the other hand, the conventional method suffers by requiring these conditions to be known accurately, something that proves to be very difficult to do.

The accuracy of the environmental data in a natural aging experiment is of great concern. We can reduce the error with better monitoring devices. Recognizing that the material reacts to the instantaneous environment, e.g., $dt_{shift} = dt / a_\sigma (t)$, we must have a complete environmental history for each specimen from the moment it was received from the supplier to the moment it was tested. With enough accurate environmental information, we should be able to reach reliable conclusions.

REFERENCES

1. S. Carfagno and R.S. Gibson. *A Review of Equipment Aging Theory and Technology*. Electronic Power Research Institute, Palo Alto, CA: September 1980. Report NP-1558.
2. K.T. Gillen and R.L. Clough, "A Kinetic Model for Predicting Oxidative Degradation Rates in Combined Radiation-Thermal Environments", *J. Polym. Sci., Polym. Chem. Ed.*, Vol. 23, p2683 (1985).
3. S.G. Burnay "A Practical Model for Lifetime Prediction of Elastomeric Seals in Nuclear Environments", ACS Polymer Preprints, Vol. 31, No. 2, p.339 (1990).
4. H. Akima, "A new Method of Interpolation and Smooth Curve Fitting Based on Local Procedures", *J. ACM*. Vol. 17, p.589 (1970).
5. C.S. Shen, G.V. Gordon and M.T. Shaw, "Modeling of Multiple Physical Properties of Aged Polymer Materials", SPE ANTEC Conf. Proc. Vol. 35: p.668. (1989).

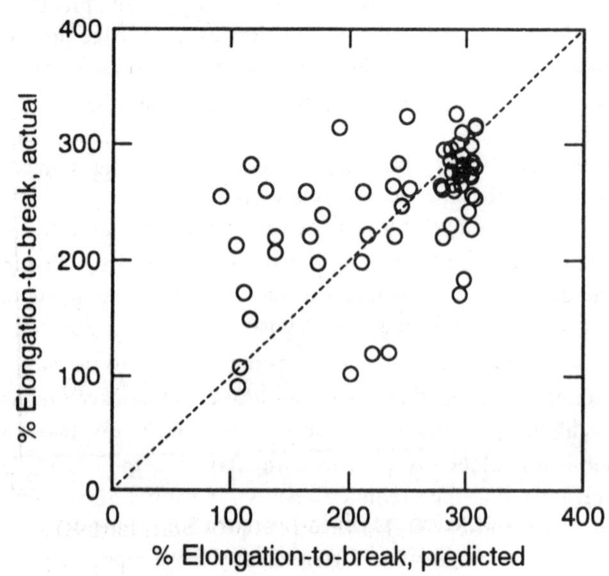

Figure 1. Example results using the traditional method, a Neoprene Rockbestos jacket with an assumed activation energy of 0.86 eV [1].

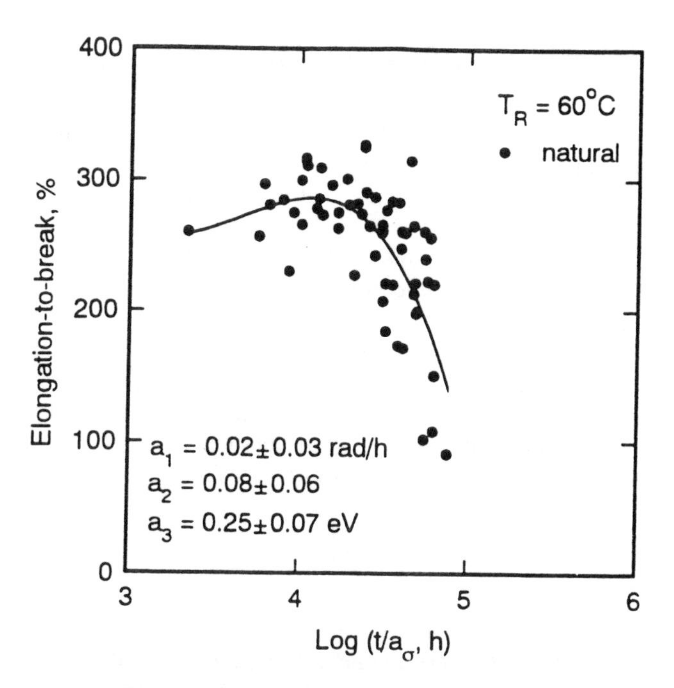

Figure 2. Modeling of the Elongation-to-Break of material R. Reference conditions are T=60°C and $\dot{D}$=0.

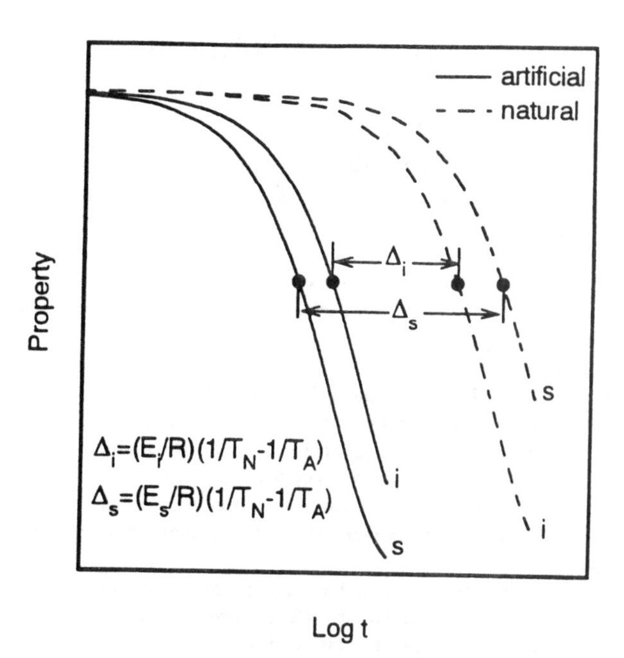

Figure 3. Property decay curves for a standard and test material under natural and artificial aging conditions.

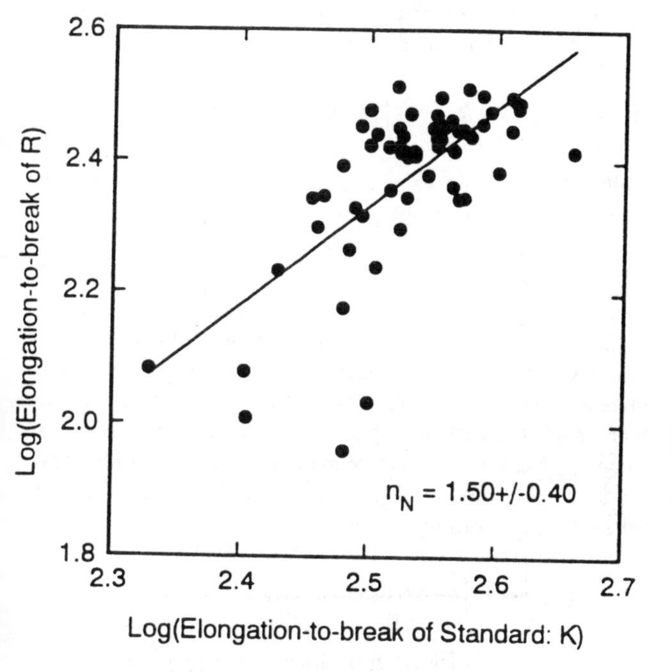

Figure 4. Elongation-to-break of Material R vs. the same property of the standard - Material K under the natural aging condition.

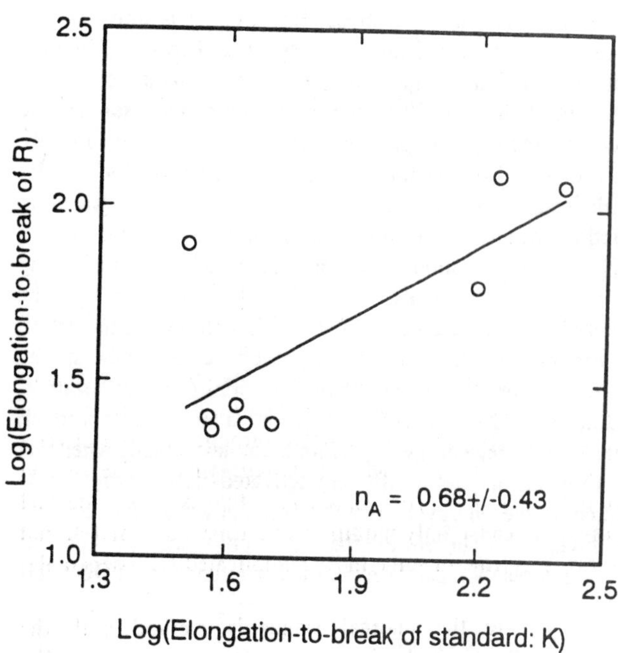

Figure 5. Elongation-to-break of Material R vs. the same property of the standard - Material K under the artificial aging condition.

Conference Record of the 1996 IEEE International Symposium on Electrical Insulation, Montreal, Quebec, Canada, June 16-19, 1996

A New Thermo-Electrical Life Model Based on Space-Charge Trapping

L. Dissado
Department of Engineering, University of Leicester, Leicester, U.K.

G. Mazzanti
Dipartimento di Ingegneria Elettrica, University of Bologna, Bologna, Italy

G. C. Montanari
Dipartimento di Ingegneria Elettrica, University of Bologna, Bologna, Italy

Abstract-A new thermo-electrical model which relates the time-to-formation of microcavities or crazes in insulating materials to the applied dc electrical and thermal stresses is proposed.

The model is based on the Eyring law, but factors are introduced in the expression of the state free energy, which are functions of the electrostatic and electromechanical energy associated with space-charge trapped within the material.

The presence of both electrical and thermal thresholds characterizes the model which can, therefore, describe properly typical behaviours of polymeric insulating materials reported in literature.

In the paper, the model is applied to the results of multi-stress life tests performed, at different levels of temperature and electric stress, on flat specimens of Polyethylene-Terephthalate (PET). It is shown that the model can satisfactorily describe the behaviour of the experimental data, providing insights into the chemical-physical characteristics of the insulation.

INTRODUCTION

Electrical treeing is the phenomenon that generally causes breakdown of insulating materials after a period of application of electrical and thermal stress. But, treeing is only the last stage of the whole aging process. The period of treeing induction is, indeed, generally, the predominant component determining lifetime. The thermodynamic approach is one way of tackling the problem of modelling such process. According to Eyring's theory, applied by Dakin [1] to the life of dielectrics, temperature accelerates chemical degradation reactions occurring in polymeric insulation, at a rate which depends on reaction rate constants of the prevailing degradation reaction (e.g. oxidation, chain scission), till the material is no more able to satisfy design requirements. In this mainframe, degradation is regarded essentially as an irreversible process. Considering an initial (reactant) state 1, characterized by free energy G_1, and a degraded (product) state 2, with free energy G_2, degradation takes place whenever the reactants acquire a sufficient activated-state energy, G_a, to overcome the energy barrier (i.e. the activation free energy) $DG\#=G_a-G_1$. Therefore, the rate constant, R, for the prevailing reaction can be expressed as a function of DG# [2]:

$$R = \frac{kT}{h}\exp(-DG\#/kT) \quad (1)$$

where k and h are Boltzmann's and Planck's constant respectively, T the absolute (thermodynamic) temperature. Some thermal and thermo-electrical life models have been derived from (1), on the basis of an inverse-proportional relationship between time-to-failure, i.e. life, and reaction rate constant [2, 3].

On the other hand, it is not unlikely that chemical reactions behave as reversible processes, as discussed in [4]. In this case, both forward reaction rate constant R_f, and backward reaction rate constant R_b, must be considered, which can be succinctly expressed, on the basis of (1), as follows

$$R_i = \frac{kT}{h}\exp\left[-\left(G_a - G_j\right)/kT\right] \quad (2)$$

with i=f when j=1, and i=b when j=2. In this context, failure occurs when a suitable minimum amount of reactant states is converted into product states. Under these hypotheses, a thermal reversible life model can be set up, expressed by [4]:

$$L = \left(R_f + R_b\right)^{-1}\ln\left[A/\left(A-A^*\right)\right] \quad (3)$$

In (3), A is the equilibrium value of the rate of conversion of polymer sites from initial to degraded state, i.e.:

$$A = R_f/\left(R_f + R_b\right) = 1/\left[1+\exp\left(DG/kT\right)\right] \quad (4)$$

while A* is the limit value of A (end-point) over which failure takes place, and DG is given by:

$$DG = G_2 - G_1 \quad (5)$$

Equation (4) shows that A depends on material properties (via DG) and on temperature, and that A increases as T increases. For a given material, if a temperature exists such that $A \le A^*$, life will never end at that temperature, even at $t=\infty$ Then, the highest temperature matching such a condition is the thermal threshold, T_{t0}, that satisfies the following relationship, according to (4):

$$A^* = A(T_{t0}) = 1/\left[1+\exp\left(DG/kT_{t0}\right)\right] \quad (6)$$

and hence:

$$T_{t0} = DG/\left\{k\ln\left[(1-A^*)/A^*\right]\right\} \quad (7)$$

Therefore, the reversible thermal life model features a thermal threshold.

SPACE-CHARGE ROLE IN ELECTRO-THERMAL MODELING

In the light of the thermodynamic aging theory, it was assumed [3, 4] that the effect of electric field, E, consists of a reduction of the activation energy of the degrading reaction. According to the reversible approach, this could be due to a field-induced rise in the ground state free energy of the reactants, which becomes $G_1(E)$, being $G_1(E)>G_1$. Assuming that the activated and product state energies are less raised (or unraised) by the field, this involves a lowering of the energy barrier of the forward reaction (which is accelerated with respect to the action of temperature only) and thus a rise of R_f, which becomes $R_f(E)$. Also, DG changes to DG(E), according to (5), being DG(E)<DG.

Under these hypotheses, a thermo-electrical model can be derived from model (3) replacing G_1 with $G_1(E)$, R_f with $R_f(E)$ and DG with DG(E). Then, it is obtained:

$$L = \left[R_f(E)+R_b\right]^{-1}\ln\{A(E)/[A(E)-A^*]\} \tag{8}$$

The field effect can be addressed to trapping of space-charge in the presence of an electric field [4]. Space-charge trapping causes local space-charge concentrations, and hence a storage of electrostatic and electro-mechanical energy; thus, local forces are induced, able to promote cracks and plastic deformations in the polymer structure. The action of these forces could be particularly effective in the case of ac voltage, i.e. when the forces have a fatigue nature; however, also in dc conditions they can be able to accelerate material degradation.

In thermodynamic terms, space-charge trapping involves a dependence of the free energies of reactant, activated and product state (and thus of model (8)) on the electromechanical and electrostatic energies stored in a space-charge centre, W_{em} and W_{es} respectively. These are related, in turn, to the applied field. In order to establish such relationship, the following assumptions are made:

- dc voltage is applied;
- space-charge is injected from electrodes, protrusions, or void discharges, but the injection currents are insufficient to cause damage;
- the trapped space charge comes into equilibrium in a short time (compared with aging time);
- local space-charge centres of equal charge q are formed;
- close to the space-charge centre, the external field is negligible compared to the space-charge field;
- the charge q is a function of the applied field E, via a power law expression, i.e.:

$$q = C\,E^b \tag{9}$$

Finally, the problem is further simplified assuming that a space charge centre occupies a spherical region of radius r_0, with uniform density ρ_0.

Consequently, W_{em} and W_{es} can be expressed as a function of the applied field. The resulting expressions of W_{em} and W_{es} are [4]:

$$W_{es} = C_1 q^2 = C_1' E^{2b} \tag{10}$$

$$W_{em} = C_2 q^4 = C_2' E^{4b} \tag{11}$$

with

$$C_1 = 3/(20\pi\varepsilon r_0) \tag{12}$$

$$C_2 = 3\alpha^2/(4480\,\pi^3 Y \varepsilon^4 r_0^5) \tag{13}$$

where α is the electrostriction coefficient, Y is Young's modulus; moreover, $C_1' = C_1C^2$ and $C_2' = C_2C^4$.

It can be assumed further that $G_1(E)$ is proportional to both W_{em} and W_{es}, namely:

$$G_1(E) = G_1 + AW_{es} + BW_{em} \tag{14}$$

and that the electromechanical energy is released (i.e. the relevant free energy is not raised) in the activated and product state, whereas the electrostatic energy is not (this seems realistic for chain scission reaction, which promotes the formation of free volume inside the material). Then, the electrostatic energy contribution to reaction rate constants is zero, since it raises the energy of all states equally; only the electro-mechanic energy is active in reducing DG and raising R_f. Consequently, DG(E), A(E) and $R_f(E)$ become, from (15):

$$DG(E) = G_2 - G_1 - BW_{em} = DG - BC_2'E^{4b} \tag{15}$$

$$A(E) = 1/\{1+\exp[DG(E)/kT]\} = 1/\{1+\exp[(DK - C_2''E^{4b})/T]\} \tag{16}$$

$$R_f(E) = \frac{kT}{h}\exp\left[-\frac{G_a - G_1 - BW_{em}}{kT}\right] = \frac{kT}{h}\exp\left[-\frac{G_a - G_1 - BC_2'E^{4b}}{kT}\right] \tag{17}$$

where DK is defined as:

$$DK = (G_2 - G_1)/k \tag{18}$$

and $C_2'' = BC_2'/k$. Note that R_b remains unchanged.

Now, the thermo-electrical model based on space-charge trapping can be fully expressed in terms of electric field and temperature. Defining the parameter DDK as:

$$DDK = [G_a - (G_1 + G_2)/2]/k \tag{19}$$

model (8) can be rewritten on the basis of (15) and (17), as:

$$L = \frac{h}{kT}\frac{\exp(DDK/T)\exp\left(-C_2'' E^{4b}/2T\right)\left\{-\ln\left[(A(E)-A^*)/A(E)\right]\right\}}{\exp\left\{-\left[DK - C_2'' E^{4b}\right]/2T\right\}+\exp\left\{\left[DK - C_2'' E^{4b}\right]/2T\right\}} \tag{20}$$

where A(E), given by (16), has not been fully expressed for the sake of simplicity. Note that model (20) is characterized by 5 parameters, i.e. DDK, DK, b, A* (or T_{t0}) and C_2''.

Note that from previous considerations, life in (20) is actually not time-to-breakdown, but time to formation of cavities or crazes, i.e. approximately the time to partial discharge (PD) inception. Moreover, the model does not account for contaminants, protrusions or voids initially present in the material, which is essentially regarded as homogeneous.

Model (20) also features a thermal threshold (like model (7)) that is field dependent, $T_t(E)$. Its expression can be derived from (7), giving:

$$T_t(E) = DG(E)/\left\{k\ln\left[(1-A^*)/A^*\right]\right\} = T_{t0} - C_2'' E^{4b}/\left\{\ln\left[(1-A^*)/A^*\right]\right\} \tag{21}$$

Thus, $T_t(E)$ decreases as E increases and may eventually become zero.

Each electrical field value corresponding to the field dependent thermal threshold is in turn an electrical threshold. Then, from (21) an electrical threshold, E_t, can be derived:

$$E_t(T) = \left\{\frac{DK - T\ln\left((1-A^*)/A^*\right)}{C_2''}\right\}^{1/4b} = \left\{\frac{(T_{t0}-T)\ln\left((1-A^*)/A^*\right)}{C_2''}\right\}^{1/4b} \tag{22}$$

Analogously to the thermal threshold, the electrical threshold decreases as temperature increases.

FITTING THE SPACE-CHARGE MODEL TO EXPERIMENTAL RESULTS

Model (20) has been applied to the results of electro-thermal life tests realized under dc voltage on flat specimens of PolyEthylene Therephthalate (PET). The tests were carried out at various voltage levels, at three different temperatures, namely 180, 160 and 110°C. The specimen thickness was 36 μ m; test procedures are fully described in [5].

Time to breakdown instead of time to PD inception was measured in the tests; nevertheless, the data can be regarded as suitable for the use of model (20), on condition that the specimens are homogeneous, and that both the time to formation of space-charge centres and the time of treeing growth are negligible when compared to the treeing induction time. This is expected to hold particularly when the values of applied electrical field are low or moderate.

Table I
Values of the parameters of the space-charge reversible model, calculated for dc PET data [5].

Parameter	dimension	value
DDK	K	$1.79 \cdot 10^4$
DK	K	465.5
C_2''	$K\,(kV/mm)^{-4b}$	1.59
b	adimensional	0.40
A*	adimensional	0.40

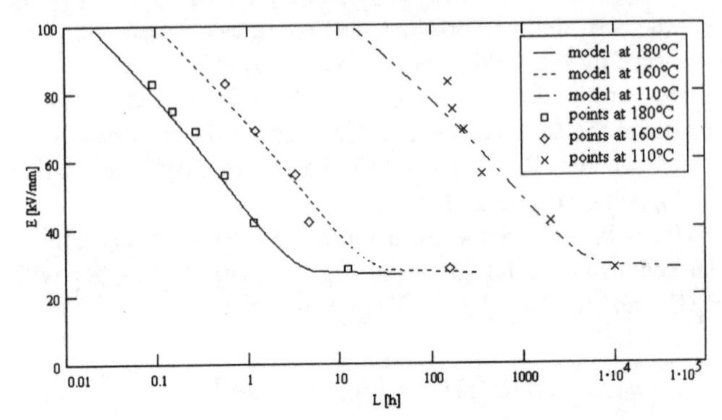

Figure 1. Electrical life lines at different temperatures derived from model (21) (on the basis of the parameter values reported in Table I). Experimental life points are also displayed.

The values of model parameters, listed in Table I, have been derived from PET data by a computer routine that exploits the symplex algorithm. Correctly, they are not function of temperature, since model (20) is fully explicited in terms of both stresses, as previously remarked. The activation energy value associated to DDK is 1.54 eV, and corresponds to quanta of an *intra*-molecular vibration, which is meaningful for a bond breaking process. The activation energy value associated to DK is 0.04 eV, i.e. in the order of an *inter*-molecular ('phonon') vibration. It could correspond to the displacement of free polymer chain ends after chain scission.

Figure 1 shows the electrical life lines at different temperatures derived from model (20) (on the basis of the parameter values reported in Table I), together with the experimental life points. It can be noted that the agreement between experimental data and model predictions is good. In particular, the electrical threshold has been accurately inferred at the different temperatures (actually, it is quite constant with temperature, and its value is close to 28 kV/mm). Hence, both the requirements needed for the application of the space-charge model to time-to-breakdown data, and the hypotheses on which the space-charge theory is based, are likely fulfilled for this data set.

In figure 2, the thermal life lines derived according to model (20) are plotted in Arrhenius graph, together with PET data. In this case, the agreement is satisfactory, but not so good as in the case of the electrical-life plot. This happens since the thermal-life plot emphasizes the differences between model estimates and test results in the two situations when they are significant, namely: at high fields (see curve at 83 kV/mm), and close to the threshold. In the first case, at high fields, errors consist in an underestimation of time-to-failure. This could be addressed to limits of applicability of the space charge model at high fields, where the contribution of electrical failure, due to localized phenomena (e.g. imperfections, protrusions), can prevail in the thermodynamic approach and, moreover, the treeing growth period may be no more negligible with respect to the treeing induction period. In the second case, even a very small shifting of the electrical threshold estimated by the model from that suggested by experimental results in the electrical-life plot, can produce a large deviation between life estimated by the model and experimental evidences in thermal-life plot, close to the threshold field.

The electrical threshold is plotted as a function of -1000/T, according to (22), in figure 3. The so-called threshold curve is obtained; its points have as coordinates the electrical and thermal stress values that yield infinite life.

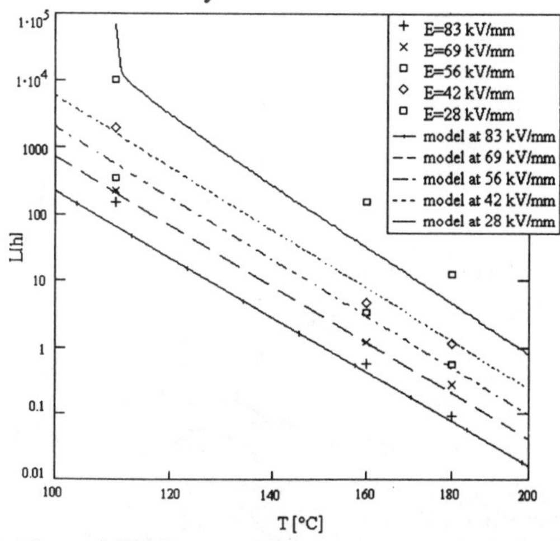

Figure 2. Thermal life lines at different temperatures derived from model (20) on the basis of the parameter values reported in Table I. Experimental life points are also displayed.

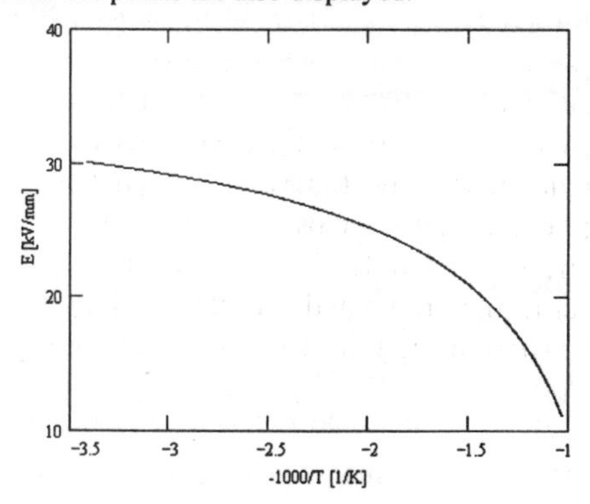

Figure 3. Electrical threshold vs -1000/T, according to (22).

The curves reported in figures 1-3 are similar to the corresponding curves relevant to other insulations which showed a tendency to threshold, even if test procedures were different (see e.g., the shape of the curves obtained from ac multi-stress life tests on XLPE cable models [6]). Moreover, in [7] the model herein described was shown to be able to fit the results of ac electrical life tests performed on XLPE cable models, even if in this case the model can be seen as a phenomenological one. These evidences suggest the possibility that the thermo-electrical space-charge model be successfully extended to systems, like cables, and to ac conditions, provided that the changes in the space-charge aging mechanisms are correctly accounted for.

CONCLUSIONS

The thermo-electrical life model based on space-charge trapping satisfactorily fits dc life test data relevant to PET flat specimens. Model parameter calculation leads to activation energy values that indicate chain scission as a possible main aging reaction, in agreement with the hypotheses that led to the model. The model also reproduces the shape of electrical and thermal life lines coming from test data that are not fully compatible with model applicability hypotheses, such as ac tests on cable models. This is encourageing in the perspective of a further development of the space-charge model.

REFERENCES

1. T.W. Dakin, "Electrical Insulation Deterioration Treated as a Chemical Rate Phenomenon", AIEE Trans., pp.113-122, 1948.
2. G.C. Montanari, G. Mazzanti, "From Thermodynamic to Phenomenological Multistress Models for Insulating Materials without or with evidence of Threshold", Journal of Physics D: Applied Physics, Vol.27, pp.1691-1702, 1994.
3. J.P. Crine, J.L. Parpal, G. Lessard, "A Model of Solid Dielectric Aging", Proc. IEEE Symp. on El. Ins. pp.347-351, Toronto, June 1990.
4. L. Dissado, G. Mazzanti, G.C Montanari, "The Incorporation of Space Charge Degradation into the Life Model for Electrical Insulating Materials", IEEE Trans. on Diel. and El. Ins., Vol. 2, n. 6, pp.1147-1158, December 1995.
5. S.M. Gubanski and G.C. Montanari, "Modelistic Investigation of Multi-stress AC-DC Endurance of PET Films", ETEP, Vol. 2, n.1, pp. 5-14, 1992
6. G. Mazzanti, G.C. Montanari, "A Comparison between XLPE and EPR as Insulating Materials for HV Cables", IEEE/PES Wint.Meeting, Baltimore (Maryland) 1996.
7. L. Dissado, G. Mazzanti, G.C Montanari, "The role of trapped space charge in the electrical ageing of insulating materials", atti del congresso 2nd Int. Conf. on Space Charge in Solid Dielectr. (CSC'2), pp.189-194, Antibes, 2-7 aprile 1995.

ACKNOWLEDGEMENTS

The authors gratefully acknowledge Dr. Stanislav Gubanski for having provided data.

Conference Record of the 1996 IEEE International Symposium on Electrical Insulation, Montreal, Quebec, Canada, June 16-19, 1996

Electrical Aging of Extruded Dielectric Cables: A Physical Model

J.-P. Crine, C. Dang, J.-L. Parpal,
Institut de recherche d'Hydro-Québec(IREQ)

Abstract **A model is proposed to describe the electrical aging of polymeric cables. It is based on simple thermodynamics concepts in the Eyring theory and supposes that the first step in electrical aging is essentially a molecular process. Our model of electrical aging under ac fields supposes that the molecular-chain deformation which will generate submicrocavities in the amorphous region of the insulation, is essentially a fatigue process. Above a critical field F_c, there is an exponential relation between time and field, whereas below F_c, the breakdown strength of the insulation varies very little with time; in other words, there is very limited aging. The model confirms that there is a relation between cable endurance and the insulation morphology, and that the size of submicrocavities is ultimately limited by the amorphous-phase thickness.**

INTRODUCTION

The electrical aging phenomena of extruded cables are still poorly known. Although there are many models and theories [1- 4], few are reliable mainly because they are unable to describe all the interactions between the various parameters (field, time, insulation, morphology, etc.). Whatever the theory, it appears that electrical-aging results obtained at high fields are described by an exponential relation between time and field [1]. At low fields, the breakdown field values vary very little with time [1], suggesting some sort of threshold value. This general behavior was predicted years ago by Dakin [2] and the cable-aging models giving the best fits to experimental data are those of Simoni and Montanari [3, 4] and of Crine *et. al.* [1, 5]. The process is known to be thermally activated and Montanari's and Crine's models are somehow based on the rate theory associated with the name of Eyring [6].

The proposed model described in this paper relies on simple thermodynamics arguments [5], the free-volume breakdown theory of Artbauer [7]; the submicrocavity formation observed by Zhurkov during mechanical aging of polymers [8] and on the molecular model of aging of Crine and Vijh [9]. An exponential relation exists between field and time at high electric fields, electrical aging is a thermally-activated mechanism, high frequencies accelerate aging, below a given critical field there is limited aging and, thus, the breakdown field varies very little with time. This model is the only one based on a simple equation containing no adjustable constant, it depends on two physical parameters which can be easily applied to any material.

THEORETICAL BACKGROUND

It is assumed that, the van der Waals bonds can be readely broken above a given critical ac field F_c, and "holes" are thus generated. After bond breaking, molecular chains can move against each other and holes may aggregate together to form larger defects, called submicrocavities, Figure 1. The maximum length of the submicrocavities (*i.e.* accumulation of connected "holes") corresponds to λ_{max} which is shown to correspond to the thickness of the amorphous phase in the insulation [15]. This concept is essentially the same as the submicrocavity formation first reported by Zhurkov [8] and observed by others during the mechanical aging of various polymers [10].

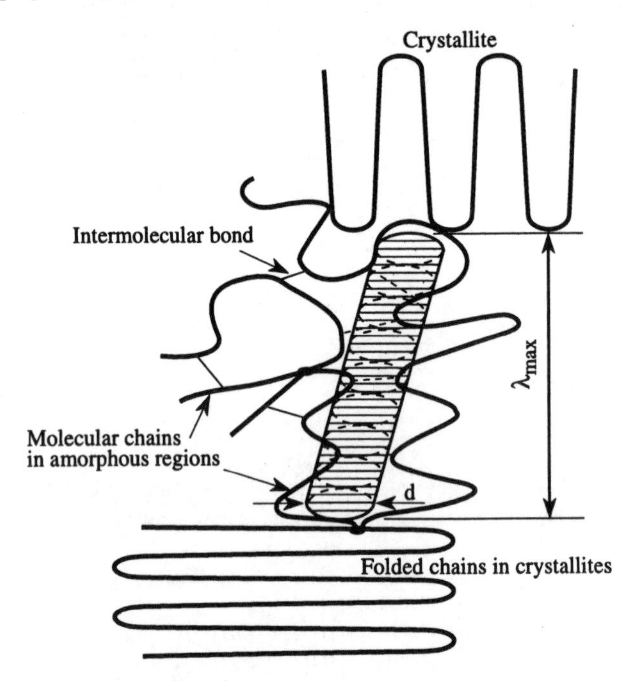

Figure 1. Schematic representation of bond breaking induced above a critical F_c by an ac stress.

It is supposed that the interchain bond breaking process is controlled by an energy barrier whose height between the original and final state is the activation energy $\Delta G_o^{\ddagger}$, as depicted Figure 2. When a field F is applied, molecular chains are deformed over a distance λ and $\Delta G_o^{\ddagger}$ is reduced by an amount equal to the work of deformation applied on the barrier, $W=e\lambda F$. For the sake of simplicity, in the rest of the paper the term ΔG_o is used instead of the conventional $\Delta G_o^{\ddagger}$ usually reserved for the activation energy. Note that a more suitable definition for W would be $W = \Delta V\sigma$, where σ is the electromechanical stress generated by the ac field and ΔV is the activation volume of the process. This is similar to the electromechanical energy W_{em} recently introduced by Dissado *et. al.* [11] and by Blaise [12]. Note that the electrostatic energy (Maxwell stress) is negligible at fields lower than ~2×10^8 V/m [11].

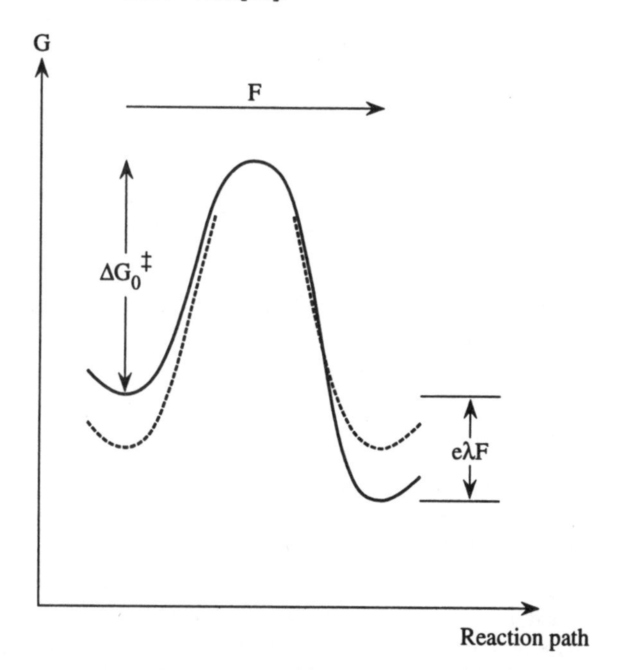

Figure 2. Schematic representation of the free energy barrier controlling aging.

It is important to insist on the fact that this energy barrier is not typical of charge trapping (and detrapping) but rather of bonds breaking (and hence, of bond strength). Of course, the final breakdown process is electronic, and with its high-energy release, it will induce some intramolecular breaking [13] but this paper is concerned only with aging, *i.e.* pre-breakdown processes.

The time to reach the state of bonds disruption, which is the final state of aging, is given by [15]

$$t \simeq \left(\frac{h}{2kT}\right)\exp\left(\frac{\Delta G_o}{kT}\right)\operatorname{csch}\left(\frac{e\lambda F}{kT}\right) \qquad (1)$$

where k and h are the Boltzmann and Planck constants, respectively, and T is the temperature (in K). At high fields, Eq. 1 predicts an exponential relation between field and time. Thus ΔG_o and λ can be readily deduced from the intercept and slope of a F vs. log t plot, provided that at high fields ($F > F_c$) λ has a constant value, λ_{max}. At low fields, the field-dependence of the time-to-breakdown is more complex than a simple exponential relation. The transition between the exponential and non-exponential regimes occurs at a so-called critical field F_c. It should be noted that the critical field is not a threshold field since there is still some aging at lower field values.

As already stated, the critical field in this model should not be confused with a threshold field. In fact, there is no demonstration of the possible existence of such a field, below which there is no aging, whereas actual results provide contrary experimental evidence. Recently published results of Katz and Walker [14], show clearly that the breakdown strength of EPR cables aged for up to 13 years in service decreases exponentially with time. As can be seen in [14] or in Figures. 16 and 21 of [1], there is no sign at all of any threshold, even after 13 years in service.

SUBMICROCAVITY FORMATION

It was assumed that the voltage endurance of polymers somehow depends on the strength of their van der Waals bonds, hence on the energy of cohesion E_{coh} of the material. It was indeed observed by Crine and Vijh [9] that the dielectric strength of various polymers depended on their cohesive energy density, which is the ratio of $E_{coh} / \overline{V}$, where $\overline{V}$ is the molar volume. It can be speculated that bond breaking occurs when the work of bond deformation induced by the field is higher than the energy of cohesion of the polymer. Since the work of deformation $W = e\lambda F$, this implies that bond rupture occurs for

$$e\lambda F \geq E_{coh} \qquad (2)$$

which corresponds to $F \geq F_c$ and $\lambda = \lambda_{max}$. Thus, the value of F_c is

$$F_c \simeq E_{coh} / e\lambda_{max} \qquad (3)$$

By definition, $E_{coh} = \Delta H_{vap} - kT$, where ΔH_{vap} is the heat of vaporization of the polymer. Substituting in Eq. 3 gives

$$F_c \simeq (\Delta H_{vap} - kT) / e\lambda_{max} \qquad (4)$$

When the calculated F_c values are compared with those deduced from experiment, the calculated F_c values are mostly 2 to 4 times higher than the experimental values depending on the test conditions [15]. Considering all the experimental uncertainties and the crudeness of our assumptions, this can be taken as a fair prediction of the upper limit value of F_c.

At this point, it is important to stress the differences between our submicrocavities (whose size is typically in the 10^{-8} m range) and the much larger microcavities (typical diameters of 10^{-6}-10^{-5} m) generated during the manufacturing process. Obviously, electrons injected into the latter microcavities may gain much larger kinetic energies than when injected into the submicrocavities but their number is extremely small and any degradation effect associated with them would be necessarily small at moderate fields (< 20 kV/mm) [16].

We assume that at fields higher than the critical field, F_C, submicrocavities are formed. This assumption is based on evidence of small-angle X-ray scattering (SAXS) measurements performed on polymers mechanically aged [8, 10]. Zhurkov et al. [8] gave the name of submicrocavities to the free volume rearrangement occurring between molecular chains, since the size is typically in the 8-40 nm range. With time of aging, the submicrocavities tend to coalesce to form microcavities, which eventually become fracture initiation sites. The other experimental techniques allowing the detection of "defects" in the 8-40 nm range are the small-angle neutron scattering (SANS) and the positron annihilation [17]. Interestingly, positron annihilation measurements performed years ago on various polymers [17] subjected to electrical fields showed that the intensity of positron lifetime in PE varies abruptly in the 10-15 kV/mm range, *i.e.* at ~ F_C.

ΔG_o and λ_{max} DEPENDENCE

In addition to the concept of critical field for submicrocavity formation, the other two main parameters of our model are ΔG_o and λ_{max}. It has been observed that the values of ΔG_o and λ_{max} deduced from existing cable aging data replotted on F vs. log time graphs are linearly related [1]. This linear relation between ΔG_o and λ_{max} can be predicted by simple thermodynamic arguments. From the activated entropy ΔS and enthalpy ΔH of a given process, the activation energy ΔG_o can be writen as [18]

$$\Delta G_o = \Delta V\,(\alpha / \beta)\left[(\kappa\alpha)^{-1} - T\right] \tag{5}$$

where ΔV is the activation volume, κ is a constant, α is the thermal expansion coefficient and β is the static compressibility. It is assumed that the size of ΔV is equal to the sum of the volume of "holes" created in the polymer under high stress. Hirai and Eyring [19] had speculated that "holes" have to form prior to the displacement of viscous molecular chains subjected to mechanical strains. The holes creation proposed by Hirai and Eyring leads to a deformation of the energy barrier highly similar to the deformation depicted in Figure 2 and equal to $e\lambda F$ in our model. For the sake of simplicity, let us suppose that each hole is a sphere of diameter d, which is the average intermolecular chain distance (*i.e.* the available space between two chains). We also suppose that, after further interchain bond breaking, neighbouring elementary holes located between chains can be connected together to form submicrocavities (see Fig. 1). As a first approximation, we may suppose that this additional free volume ΔV has a cylindrical shape of length λ_{max} and diameter d. It can be shown [15] that

$$\Delta G_o \simeq \lambda_{max} \frac{\pi\, d^2}{4} \frac{\alpha}{\beta}\left(\frac{1}{\kappa\alpha} - T\right) \tag{6}$$

and that ΔG_o and λ_{max} are linearly related.

We assumed that submicrocavities more readily form in the amorphous phase where there is more free volume available for chain rearrangement and because the energy of cohesion (controlling F_C, see Eq. 4) is slightly lower than for the more rigid crystalline phase. Thus, λ_{max} values are expected to be of the same order of magnitude as the amorphous phase thickness suggesting a crude correlation between λ_{max} and the polymer crystallinity. Note also that the size of submicrocavities experimentally detected by Zhurkov *et. al.* [8] for various mechanically aged polymers agrees well with the size of their average amorphous large thickness. Thus, the more amorphous EPR polymer, has a value of λ_{max} (~ 25 nm) larger than the more crystalline high-density PE with λ_{max} ~ 5 nm [15].

CONCLUSION

Extruded cable lifetime can be reasonably well predicted by Eq. 1 and the main features of our simple model can be summarized as follows:

- The energy required to break the intermolecular (van der Waals) bonds must be greater than the energy of cohesion of the polymer, which may occur at fields higher than the critical field F_C. This corresponds to the lower field limit of the exponential relation between the time-to-breakdown and the field.
- The two major parameters describing the aging process are the activation energy of interchain bond breaking ΔG_o and the length of the submicrocavities thus formed, λ_{max}. Their value can be evaluated from the exponential regime in F vs. log t plots.
- G_o and λ_{max} are linearly related by the force required to pull molecular chains apart.
- At $F > F_C$, submicrocavities (free-volume rearrangement composed of tiny "holes" of diameter equal to the interchain distance) form with a maximum total length equal to the amorphous-phase thickness of the polymer.
- It seems that submicrocavity formation is not an instantaneous process and fields applied for periods of time shorter than the formation time may possibly not initiate submicrocavities.
- After long times of aging or application of high fields, submicrocavities may coalesce to eventually form microcavities. Electrons injected in these submicrocavities can acquire enough kinetic energy to

become hot electrons and generate further damage. In other words, the onset of electrical aging is essentially a molecular process and electrons play a major role only near the final breakdown (*i.e.* fast) process.

Our model not only is able to describe the physical process but can also set the limits of the experimental conditions for accelerated aging. Finally, it should be noted that this model has been applied to other insulating polymers (epoxy, polypropylene, polyimide, etc.) used in the insulation of satellite equipment [20], suggesting that the model may apply to all sort of polymers and is not restricted to cable insulation.

ACKNOWLEDGMENTS

Part of this work was done with EPRI funding (RP 2713-1) and we thank B.S. Bernstein, A.K. Vijh, G.C. Montanari, L.A. Dissado and E. Brancato for their comments. The text revision by L.K. Regnier was greatly appreciated.

REFERENCES

1. Dang, C., Parpal, J.-L., Crine, J.-P. "Electrical Aging of Extruded Dielectric Cables - Review of Existing Theories and Data," accepted for publication, IEEE Transactions on EI, 1996.

2. Dakin, T.W. "Electrical Insulation Deterioration Treated as a Chemical Rate Phenomenon," AIEE Transactions, vol. 67, pp. 113-122, 1948.

3. Simoni, L., Mazzanti G. and Montanari, G.C. "Life Models for Insulating Materials in Combined-Stress Conditions," Proceedings of 1994 ICPADM Conference, pp. 827-832, 1994.

4. Montanari G.C. and Mazzanti, G. "From Thermodynamic to Phenomenological Multi-Stress Models for Insulating Materials without or with Evidence of Threshold," Journal of Physics D., vol. 27, pp. 1691-1702, 1994.

5. Crine, J.-P. "A Model of Solid Dielectrics Aging," Proceedings of IEEE International Symposium on EI, pp. 25-27, 1990.

6. Glasstone, S., Laidler K.J. and Eyring, H. "The Theory of Rate Processes," McGraw Hill, New York, 1941.

7. Artbauer, J. "Zur Temperaturablangigkeit der electrischen Durchschlagfestigkeit amorpher Polymerenmit Relaxation under der Glastemperatur ," Journal of Polymer Science C, vol. 16, pp. 477-485, 1967.

8. Zhurkov, S.N., Zakrevskyi, V.A., Korsukov V.E. and Kuksenko, V.S. "Mechanism of Submicrocrack Generation in Stressed Polymers," Journal of Polymer Science A2, vol. 10, pp. 1509-1520, 1972.

9. Crine J.-P. and Vijh, A.K. "A Molecular Approach to the Physico-Chemical Factors in the Electric Breakdown of Polymers," Applied Physics Communications, vol. 5, pp. 139-163, 1985.

10. Wendorff, J.-M. "Submicroscopic Defects in Strained Polyoxymethylene," Progress in Colloid and Polymer Science, vol. 66, pp. 135-141, 1979.

11. Dissado, L., Mazzanti G. and Montanari, G.C. "The Role of Trapped Space Charges in Electrical Ageing of Insulating Materials," Proc. of 2nd Int. Conference on Space Charge in Solid Dielectrics, pp.189-94 , 1995.

12. Blaise, G. "Charge Trapping/Detrapping Induced Lattice Polarization/Relaxation Processes," Proc. of 2nd Int. Conference on Space Charge in Solid Dielectrics,pp.164-69 , 1995.

13. Zeller, H.R. "Breakdown and Prebreakdown Phenomena in Solid Dielectrics," Proceedings of 2nd ICSD Conference, pp. 98-109, 1986.

14. Katz, C. and Walker, M. "An Assessment of Field Aged 15 and 35 kV EPR Insulated Cables," IEEE Trans. Power Delivery, vol. 10, pp.25-33, 1995.

15. Parpal J.-L., Dang C. Crine J.-P., "Electrical Aging of Extruded Dielectric Cables Part II-A Physical Model", submitted to IEEE Trans. Dielectrics and Elect. Insul., 1996.

16. Novak J.P., Bartnikas R., "Ionization and Excitation Behavior in a Microcavity", IEEE Trans. Diel. Electr. Insul., 1995, **2**, pp.724-728.

17. Brandt, W. and Wilkenfeld, J. "Electric Field Dependence of Positronium Formation in Condensed Matter," Physical Review B, vol. 12, pp. 2579-2587, 1975.

18. Crine, J.-P. "A Thermodynamic Model for the Compensation Law and its Physical Significance for Polymers," Macromolecular Science B, vol. 23, pp. 201-219, 1984.

19. Hirai N. and Eyring, H. "Bulk Viscosity of Liquids," Journal of Applied Physics, vol. 29, pp.810-816, 1958.

20. Arens, I. "The Stress-Lifetime Relation of Electrical Insulation,"Proceedings of European Space Power Conference, Graz (Austria), August 1993.

Conference Record of the 1996 IEEE International Symposium on Electrical Insulation, Montreal, Quebec, Canada, June 16-19, 1996

On the Use of Doped Polyethylene as an Insulating Material for HVDC Cables

M.Salah Khalil
Dept. of Electrical &Electronics Engineering
College of Engineering
Sultan Qaboos University
P.O.Box 33, Al-Khod 123, Muscat , Sultanate of Oman.

***Abstract*: The merits of hvdc cables with polymeric insulation are well recognized. However, the development of such cables is still hampered due to the problems resulting from the complicated dependence of the electrical conductivity of the polymer on the temperature and the dc electric field and the effects of space charge accumulation in this material. Different methods have been suggested to solve these problems yet none of these methods seem to give a conclusive solution. The present report provides, firstly a critical review of the previous works reported in the literature concerning the development of hvdc cables with polymeric insulation. Different aspects of those works are examined and discussed. Secondly, an account is given on an investigation using low density polyethylene (LDPE) doped with an inorganic additive as a candidate insulating material for hvdc cables. Preliminary results from measurements of dc breakdown strength and insulation resistivity of both the undoped and the doped materials are presented. It is shown that the incorporation of an inorganic additive into LDPE has improved the performance of the doped material under polarity reversal dc conditions at room temperature. Moreover, the dependency of the insulation resistivity on temperature for the doped material appears to be beneficially modified.**

INTRODUCTION

Polyethylene has been sucessfully used for high voltage ac(hvac)and extra hvac cables. To-day crosslinked polyethylene (XLPE) is used as an insulating material for 400 kV ac cables. Attempts to use this highly insulated material for hvdc cables have failed, so far, and mainly cables with oil-impregnated paper are still used for this purpose. The main reason behind this state of affairs is the distinct difference between the ac and dc insulation phenomena. In ac cables, the radial stress distribution is capacitive and depends on permittivity of the dielectric which is slightly affected by temperature variations in the working range. In this case the maximum electric field always appears at the internal conductor screen surface. In dc cables the electric field distribution depends on the resistivity of the insulation which is sharply dependent on temperature, electric field strength, time of electricfication and load conditions. In this case, and according to load conditions, the maximum stress may be multiplied several times and may occur at the conductor screen surface or at at the boundary of the dielectric near the insulation shield, a phenomenon known as stress inversion which sets a limit to the load rating in hvdc cables. On the other hand, the non-uniform dc resistivity of polymeric materials and its dependence on the temperature variation and electric field distribution results in the accumulation of space charge in the material under dc conditions [1]. Such space charge causes a non-uniform distribution of the electric field in the insulating material so that the dc breakdown strength is reduced specially with polarity reversals and/or temperature variations. Two coefficients were suggested to be used in the evaluation of the performance of hvdc cables with polymeric insulation and to take care of the mentioned above effects [2]. Those coefficients are the temperature rise coefficient K_{TDC} and the polarity reversal coefficient K_{PDC} respectively where :

$$K_{TDC} = \frac{\text{dc breakdown strength at H.T.}}{\text{dc breakdown strength at L.T.}} \%$$

and

$$K_{PDC} = \frac{\text{dc breakdown strength under polarity reversals}}{\text{dc breakdown strength at unipolar voltage}} \%$$

It is expected that K_{TDC} will be sensitive to polarity reversals. Thus K_{TDC}(D.P) and K_{TDC}(R.P) will be used to take care of the effects of unipolar and polarity reversal voltage applications respectively. Similarly, K_{PDC} is expected to be sensitive to temperature variations, thus two values K_{PDC}(L.T.) and K_{PDC}(H.T) will be used to evaluate the effects of low temperature and high temperature performance respectively. It is evident that the larger the coefficients K_{TDC} and K_{PDC} the better is the performance of the cable. A value of 100% for each coefficient will indicate the independence of the cable performance on both temperature variations and polarity reversals.

DEVELOPMENT TRENDS OF HVDC CABLES WITH POLYMERIC INSULATION

Various methods have been attempted to develop hvdc cables with polymeric insulation. Those methods can be broadly categorized according to the development approach as follows :

- Development of cables with limited stress inversion.
- Development of cables with suppressed space charge .
- Development of cables using LDPE, XLPE or HDPE without any modifications.

The two suggested coefficients: K_{TDC} and K_{PDC} will be used to analyse the published test results of those works as shown in Table I.

Development of cables with limited stress inversion

The aim of this work was to establish the principles of a stress inversion free or limited stress inversion dc cable [2 and references therein,4]. For this purpose special polymeric compounds were developed by incorporating mineral fillers into different polymers including XLPE. Those compounds exhibited an increased thermal conductivity and reduced the dependence of the the electrical resistivity on temperature thus rendering the possibility for designing a stress inversion free cable. It is interesting to point out here that no mention has been made to space charge effects in this work. Measurements made on cable models made of these materials indicated that although the high temperature performance of these cables was improved, their dc breakdown strength under polarity reversals are low compared to the unipolar dc breakdown strength (low K_{PDC})as shown in Table I.

Development of cables with suppressed space charge

Different methods have been suggested to suppress space charges in PE [2,3 and references therein]. Only the method of the modification of the bulk characteristics will be discussed here as it is the most widely used method. Two main techniques have been suggested: (i) the blending method [5,6], (ii) the additives method [8-11].

The Blending Method : In these works , ionomer has been blended with XLPE and HDPE respectively [5,6]. Results indicate that blending of XLPE with ionomer improves the impulse breakdown strength with or without superposing dc voltage. As for HDPE, dc breakdown strength with unipolar and polarity reversals at high temperature were improved. However, in an independent report by EPRI, it was recommended not to use ionomers for extruded dc cable insulation [7] .

The Use of Additives: Using an unspecified additive, Ando et al. [8] have shown that the incorporation of that additive has modified the space charge distribution in the filled material, increased the dc breakdown strength and improved its temperature dependency rendering a constant dc breakdown value within a temperature range from 20°C to 90°C However,. the dc breakdown strength of the filled material at polarity reversals and high temperature was still low (low K_{PDC} at H.T).

In another attempt, Khalil et al.[9,10] have investigated the effects of different inorganic additives on the space charge formation in LDPE using small test samples. The results of their works indicate that: i) the addition of TiO2, BaTiO3 and iodine to LDPE seems to suppress the space charge and modify its distribution patterns in the doped material ii) the dc breakdown strength of the TiO2 doped LDPE seems to become insensitive to polarity revarsals. iii)the addition of BaTiO3 and iodine to LDPE considerably reduces the dc breakdown strength of of the doped material[3] .

Ogata et al. [11] used an unspecified conductive inorganic additive to XLPE. As shown in Table I, the results of this work indicate that the performance of the cable with the doped material at high temperature and under direct polarity has been remarkably improved (higher K_{TDC}) yet the effects of polarity reversals at low temperature are still considerable (low K_{PDC} at L.T).

The Use of LDPE, XLPE and HDPE for HVDC Cable Insulation without any Modifications:

In this work attempts have been made to evaluate the performance of ac cables with LDPE, XLPE, and HDPE insulations under hvdc conditions. No modifications have been made either to the materials or to the cable design [10.11]. As shown in Table I, the results of this work indicate that the dc breakdown strength for the three materials is considerably reduced by temperature rise (low K_{TDC}). This work also shows that XLPE and HDPE exhibited thermal instability when loaded. The results of this work indicate clearly that the insulating material should be modified in order to develop a polymeric hvdc cable with accepted performance.

EXPERIMENTAL DETAILS

The main objectives of the present experimental work were to investigate the effects of doping LDPE with an inorganic additive on i) the dc breakdown strength of the doped material under both direct and polarity reversal conditions ii) the dependency of the insulation resistivity on temperature.

Preliminary tests were made using two materials: undoprd LDPE and LDPE doped with an inorganic additive. 3 types of tests were conducted: unipolar dc breakdown tests, dc breakdown tests with polarity reversals and insulation resistance evaluation test. Breakdown test samples were cylinders with hemispherically tipped electrodes made by moulding. The distance between the electrode tips was 0.25 mm. Breakdown tests were made on 4 groups of samples with an average of 16 samples for each group. All breakdown tests were made at room temperature.. Insulation resistance evaluation tests were made using discs of about 2 mm thick and 170 mm diameter. The samples were provided with conventional gold electrodes and guard rings made by vacuum evaporation. The experimental apparatus used for these measurements is described elsewhere [12]. Insulation resistivity was evaluated at 4 temperatures: 50°C,60 °C,70 °C and 80 °C. The voltage used for the present resistivity tests was 50 kV dc.

RESULTS AND DISCUSSION

Breakdown Tests

Figure 1. shows the Weibull plots of dc breakdown tests for both the undoped and doped doped materials. For the undoped material, the plots show that the dc breakdown strength under direct polarity can be described by a Weibull

Table I
Effects of Temperature rise, Polarity Reversals, Insulating Material and Additive Content on the dc Breakdown Strength of HVDC cables with Polymeric Insulation

Ref.	Sample Details		dc Breakdown Strength of the Material at R.T	Effect of Temperature Rise		Effect of Polarity Reversals	
	Material	Thickness		K_{TDC} (D.P)	K_{TDC} (R.P)	K_{PDC} (L.T)	K_{PDC} (H.T)
Epri Project [4]	XLPE+A XLPE	5.4 mm 5.2 mm	94 kV/mm 69 kV/mm	85% 79%	150% 125%	44% 31%	79% 39%
Suzuki et al. [6]	Blended HDPE	3.5 mm	-	-	82%	-	-
Ando et al [8]	XLPE+A XLPE	6 mm 6mm	100kV/mm 87kV/mm	100% 68%	- -	- -	54% 44%
Ogata et al. [11]	XLPE+A XLPE	2.5 mm 9 mm 2.5 mm 9 mm	140 Kv/mm 122 Kv/mm 164kV/mm 105kV/mm	85% 82% 46% 61%	114% 106% 102% -	72% 64% 49% -	97% 83% 108% 104
Pays et al. [12-13]	LDPE XLPE HDPE	6.5 mm " "	140kV/mm 150kV/mm 130kV/mm	64% 65% 38%	- - -	- - -	- - -

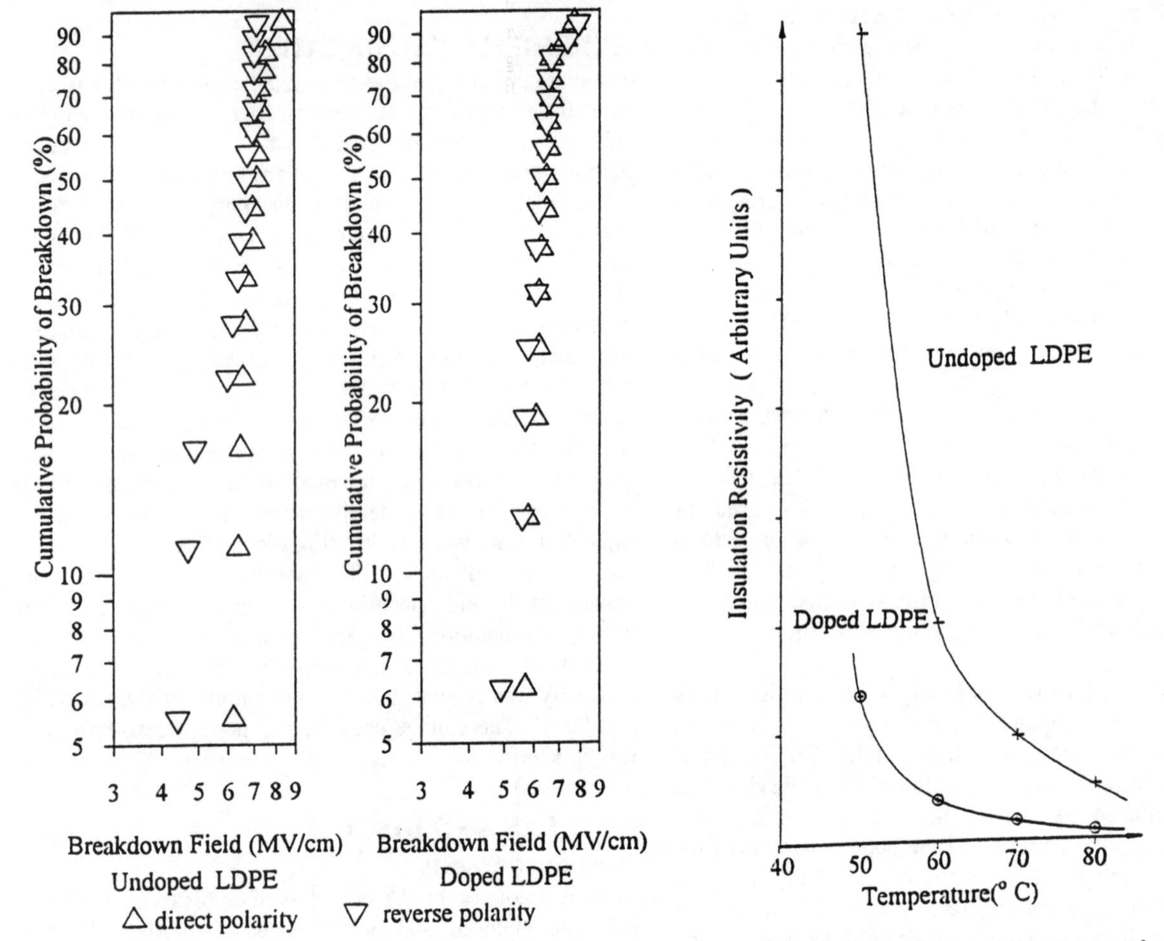

Figure 1. Weibull plots of dc breakdown data.

Figure 2. Variation of resistivity versus temperature for undoped and doped LDPE at an average dc electric field of 0.25 MV/cm.

distribution (β=12.2 and $E_{63\%}$ =7.41Mv/cm). Under polarity reversals, the distribution appears to be divided into two distriburtions: the main distribution (80% of the samples) remains unchanged within statistical uncertainty and a secondary distribution (~20% of the samples) which have reduced their failure breakdown strength. These low probability failures lie outside of the 90% confidence limits of the main distribution and therefore can be considered independent of it. For the doped material, the distributions for both direct and polarity reversals are slightly different from each other ; in case of direct polarity β=12.8,and $E_{63\%}$ = 6.75 Mv/cm while under polarity reversals β=8.9 and $E_{63\%}$ = 6.65 Mv/cm. These results indicate that, for the doped material, the characteristic stress does not change much by polarity reversals while β is reduced indicating larger dispersion in the latter case. Comparison between the characteristic stresses of the undoped and doped materials at direct polarities shows that the incorporation of the additive reduces the characteristic stress of the main distribution by about 9 % if compared with the corresponding value for the doped material. However, the incorporation of the additive has caused the breakdown strength to be almost insensitive to polarity reversals. In this case K_{PDC} = 98.5%

Evaluation of Insulation Resistivity

Figure 2. shows the variation of the insulation resistivity versus temperature for the undoped and doped LDPE. The insulation resistivity is evidently reduced by the incorporation of the additive. It can also be seen that the dependence of the insulation resistivity of the doped material on temperature in the range from 60°C to 80 °C has been reduced if compared with the corresponding value for the undoped material. Such a reduction is considered beneficial for the performance of hvdc cables with polymeric insulation[2 and references therein]

CONCLUSION

The present preliminary study indicates that although the incorporation of an inorganic additive into polyethylene has slightly reduced the dc breakdown strength of the doped material, yet it improved the polarity reversal effect. The additive also seems to have reduced the insulation resistivity of the compounded material. However, the dependency of insulation resistivity on temperature has been beneficially reduced. Further work should be continued using different additives with different concentrations. The use of additives to polyethylene can be the solution to the problem of hvdc cables with polymeric insulation. However, the effects of additives on the cable performance should be carefully investigated .

ACKNOWLEDGEMENT

The author is grateful to NKT Power Cables, Denmark for supporting the present work and permission to publish some of the above results. Thanks also go to the Danish Electrical Research Institute (DELRI), ELSAM, ELKRAFT and the Technical University of Denmark for their support. Special thanks go to the faculty members of the Electric Power Engineering Department, Technical University of Denmark: Dr. P.O.Henk. for useful discussions and Dr. M.Henriksen and his team for their contributions to the experimental facilities used in this work.

REFERENCES

1. McAllister,I.W.,G.C.Crichton and A.Pedersen "Charge Accumulation in DC Cables:A Macroscopic Approach". Conference Record of the 1994 IEEE International Symposium on Electrical Insulation, pp.212-216.
2. Khalil,M.Salah. "High Voltage Direct Current Cables with Polymeric Insulation: the State of the Art". Conference Record of the 1992 Nordic Insulation Symposium.ISSN 1102-4925, report No. 6.1
3. Khalil,M.S. "On the Methods of Suppression of Space Charge Effects in Polyethylene". Conference Record of 1990 Nordic Insulation Symposium, report No.4.2.
4. Extruded Dielectric DC Cable Development. EPRI Project 7828-1 final report, 1980.
5. Fukagawa, H, Y. Nitta and Y.Sekii. "Development of a New Insulating Material for DC XLPE Cables". Conference Record of 1987 Jicable, pp. 283-287.
6. Suzuki,T,T.Niwa, S.Yoshide,T.Takahaski and M.Hatada. "New Insulating Materials for HVDC Cables". Proceedings of the 1989 3rd International Conference on Conduction and Breakdown in Solid Dielectrics, pp. 442-447.
7. Space Charge Measurument in DC Cable Materials. EPRI report EL-7301, Research Project 7897. 1992.
8. Ando,N, F.Numajiri, K.Tomori,K.Muraki and T.Kumaga. "Development of 250 kV DC Crosslinked Polyethylene Cable".1976 Hitachi Review. Vol.25.
9. Khalil, M.S and A.A. Zaky. "Influence of Iodine on the Distribution of Space Charge in LDPE". Intern.J. Polymeric Mater. Vol. 11, January 1985, pp. 1-8.
10. Khalil, M.S., A.A. Zaky, and B.S. Hansen. "The Influence of TiO_2 and $BaTiO_3$ Additives on the Space Charge Distribution in LDPE". Annual Report of the 1985 Conference on Electrical Insulation and Dielectric Phenomena, pp. 143-148.
11. Ogata, S,Y. Maekawa, K.Terashima, R.Okiai,S.Yoshida,H. Yamanouchi and S.Yokoya. "Study on the Dielectric Characteristics of DC XLPE Cables". IEEE Trans.on Power Delivery.Vol.5. No. 3, July 1990, pp. 1239-1247.
12. Pays, M and M.Louis. "Extruded Synthetic Insulations for Development of HVDC Energy Cables". Conference Record of 1987 Jicable, pp. 111-121.
13. Pays, M. "New Developments in the Field of High Voltage and Extra High Voltage Cables:The User Point of View". IEEE.PES 1989, Transmission and Distribution Conference. Paper No.89 TD356-7 PWRD.
14. Henk, P.O. M.Henriksen, B.S.Hansen, and H.S. Johanesson. "DC Cable Insulation Conductivity as a Function of Temperatur and Time". Conference Record of 1986 Nordic Insulation Symposium, report No.16.

Conference Record of the 1996 IEEE International Symposium on Electrical Insulation, Montreal, Quebec, Canada, June 16-19, 1996

Effect of Barium Titanate ($BaTiO_3$) Additive on the Short-Term DC Breakdown strength of Polyethylene

M.Salah Khalil
Dept. of Electrical & Electronics Engineering
College of Engineering. Sultan Qaboos University
P.O. Box 33, Al-Khod 123
Muscat, Sultanate of Oman

P.O. Henk
Department of Electric Power Engineering
Technical University of Denmark
Bldg. 325, DK-2800 Lyngby, Denmark

***Abstract*: The use of additives to insulating materials is one of the methods to improve certain properties of these materials. Additives can also be used to provide more insight into some precesses like conduction, space charge formation and breakdown under certain conditions of field application.**

In the present paper, the effect of the addition of fine particles 1wt % $BaTiO_3$ to plain low density polyethylene (LDPE) on the short-term dc breakdown strength of LDPE at room temperature was investigated. The characteristics of the used polyethylene are as follows: density 0.925 g/cm^3, melt index 0.25 g/10 min. The $BaTiO_3$ used was laboratory grade with particle size less than 7 µm. Special cylindrical test samples of both undoped and doped materials were used in this investigation. Stainless steel hemispherically tipped electrodes were embedded in the material by moulding. The mean value of the gap length between the electrodes was 0.25 mm. The design of the test sample allows for determining the intrinsic breakdown strength of the material. The Weibull plots were used to analyse the breakdown test results. Analysis of the results indicate that the addition of $BaTiO_3$ to LDPE has reduced the short term dc breakdown strength of the doped material by about 16% if compared with the corresponding value for the plain LDPE.

An attempt is made to correlate between the present results, and earlier published results about the effect of $BaTiO_3$ on dc conductivity and space charge formation in LDPE.

INTRODUCTION

In polymeric materials additives or fillers are usually used to improve a certain property of the polymer, be it chemical, mechanical, thermal or electrical. Additives have been used in cable manufacture to reduce the possibility of tree generation and polymer oxidation. Additives have also been used for making emmission shields [1]. Lately, the use of additives to modify certain electrical properties of the polymer has attained som importance. Additives, including $BaTiO_3$ have been used to improve the dielectric properties of polymers (increase the dielectric constant) as well as in attempts to develop new polymeric insulating materials for HVDC cables [2-6]. In the latter case the aim of using the additives was to suppress the build-up of space charge in LDPE under dc conditions and modify the dc resistivity dependence on the temperature and field in this material [4-6]. Results of these investigations have shown that the incorporation of this highly polar additive in LDPE has a remarkable beneficial effect by reducing the density of the remanent space charge and modifying its distribution in the polymer [8]. In order to guarantee that the incorporation of the additive in the host material will not impair its desirable properties, the effects of such incorporation on these properties should be investigated. The objective of the present investigation was to investigate the effect of the addition of $BaTiO_3$ to LDPE on the short term dc breakdown strength of this material at room temperature.

EXPERIMENTAL

The polyethylene used for the present experiments was of ATO Chemie 1002 CJV. Its characteristics are as follows: density 0.925 gm/cm^3 at 20°C and melt index 0.25 gm/10 min. The barium titanate was laboratory grade made by Merck. Its particle size is less than 7 µm. Test samples were cylinders of 36 mm diameter and about 50 mm long. Two hemispherically tipped, 20 mm diameter stainless steel electrodes were embedded in the investigated material by moulding. The special mould design allowed the adjustment of the gap between the tips of the electrodes. The breakdown test sample is shown in Fig. 1.

The measurement of breakdown strength of insulating materials is always beset with difficulties. Accuracy of results depends on sample preparation, experimental conditions and sample and electrode configurations. The electrical strength of a material may be affected by a multitude of factors such as

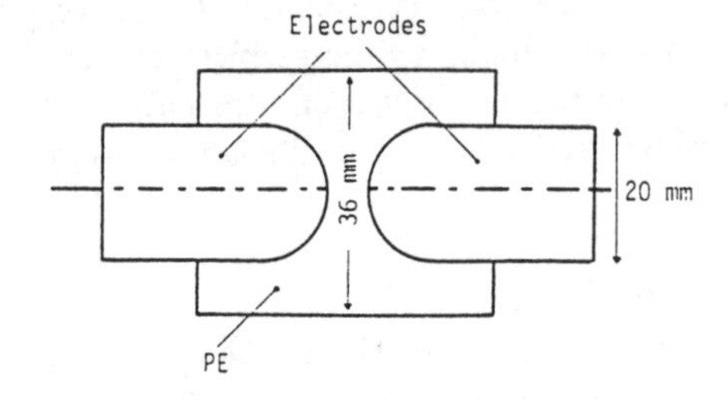

Fig. 1. Configuration of breakdown test sample.

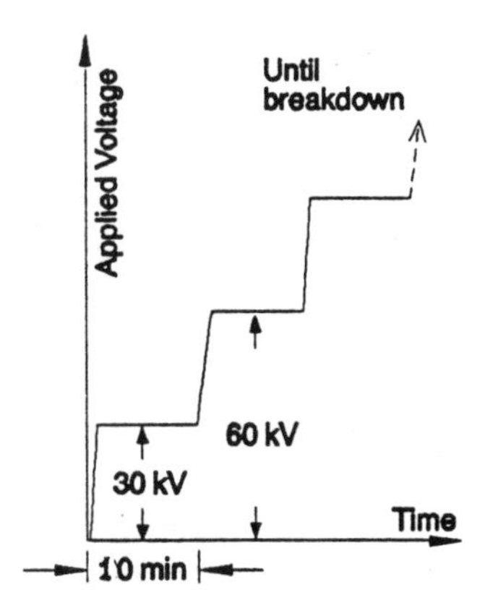

Fig. 2. Profile of Voltage Application

the frequency, the waveform and rate of rise of application of the voltage, the thickness and homogeneity of the sample, presence of gaseous inclusions, moisture or other contamination, previous conditioning of the specimen, the ambient temperature, thermal conductivity of test electrodes, intensity of surface discharges prior to breakdown, the area and volume between electrodes under maximum voltage stress. In the present investigation, the use of the described above electrode/material system, ensures perfect adherence of electrodes to the polyethylene, thus eliminating the undesirable effects of partial discharges. Certain precautioms were made such as preparing the samples under ultraclean conditions, providing clean and smooth electrode/polymer interface thus excluding the effects of any external factors on the breakdown results. The use of such electrode arrangements in breakdown tests has been previously reported by different authors [5,7]. In the present work a total number of 33 samples was used: 17 samples of undoped LDPE and 16 samples of 1 wt% $BaTiO_3$ doped LDPE. The gap length in the test samples was measured by X-ray. The mean value of the gap length was found to be 0.25 mm with a standard deviation of 0.03 mm in all the samples. In order to avoid flashovers, the samples were immersed in insulating oil using a special test cell where breakdown tests were made. Breakdown tests were performed using an increasing stepwise voltage where the voltage was raised in steps, each step consisting of 30 kV and the sample was left at the level for 10 minutes before the voltage was raised agein to the next step until breakdown occurred. The profile of the voltage application is depicted in Fig. 2. All breakdown tests were made using dc voltage at room temperature.

RESULTS AND DISCUSSION

Figure 3 shows the Weibull plot of the undoped LDPE and the 1 wt% $BaTiO_3$ doped LDPE. It is clear that the undoped LDPE plot obeys the Weibull function very well (β = 12.2, and the 63% value of E_c = 7.4 MV/cm).

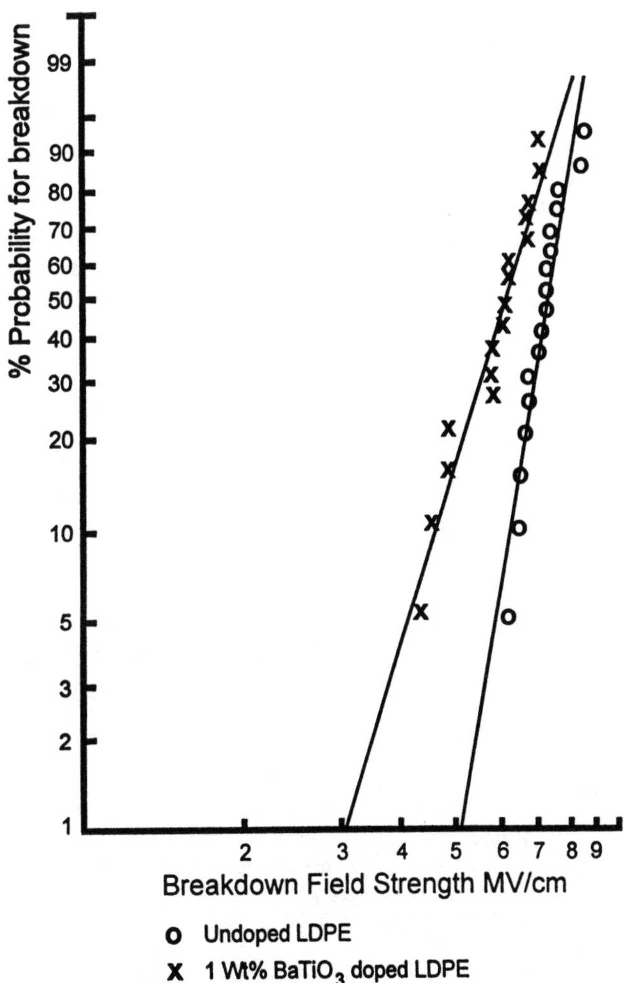

Fig. 3. Weibull plot of results.

For the doped material, β = 6.3 and the 63% value is 6.25 MV/cm. In the doped material the breakdown strength is reduced by about 16% if compared to the corresponding value for the undoped LDPE. The reduction of β is very significant in the doped material indicating the high dispersion in the results. Such a reduction in the breakdown strength and the large dispersion in the breakdown results of the doped material can be attributed to:

i) The effect of the barium titanate particles which can be dispersed in the material with clusters of particles acting as field enhancing defect centers.
ii) The increase of the conductivity of the doped material. Earlier dc conductivity measurements have shown that the addition of 1 wt% $BaTiO_3$ particles to LDPE has increased the dc conductivity of the doped material by a factor of 3 if compared to the corresponding value for the undoped LDPE [8].

iii) The change of the remanent space charge distribution and electric field in the doped material due to the additive. Previous measurements of space charge distributions in $BaTiO_3$ doped LDPE using different techniques have clearly indicated that the addition of such an additive has a considerable effect on both the density of the remanent space charge and its distribution patterns [4,9]. With heterocharge distribution at any of the electrodes, the electric field concentration will be enhanced there increasing the probability of the dc breakdown strength of the material.

ACKNOWLEDGEMENTS

The authors thank NKT A.S. for supplying the materials in the form of cylindrical samples. Thanks also go to dr. M. Henriksen, Dept. of Electric Power Systems, for assisting in setting up the breakdown equipment used. One of the authors (M.S.K.) is grateful to the Dept. of Electric Power Systems, DTU, for supporting this work and providing experimental facilities. The authors are grateful to Dr. L.A. Dissado, Kings College, University of London, for useful discussions of breakdown results.

REFERENCES

[1] S.E. Gleizer, G.I. Meschanov, Yu. V. Obraztsov, I.B. Pershkow and M. Yv Shuvalov, "High Voltage Cable with XLPE Insulation in the USSR-Operational Experience, Investigations of Ageing Processes and Long-Term Tests", CIGRE Rep. 21-109, 1990.

[2] S. Dasgupta, "Polypropylene filled with Barium Titanate: Dielectric and Mechanical Properties", Journal of Applied Polymer Science, Vol. 22, pp. 2384-2386, 1978.

[3] H. Kita and K. Okamoto, "Dielectric Properties of Polymers Containing Dispersed TCNQ Salts", Journal of Applied Polymer Science, Vol. 31, 1383-1392, 1986.

[4] M.S. Khalil, A.A. Zaky and B.S. Hansen, "The Influence of TiO_2 and $BaTiO_3$ Additives on the Space Charge Distribution in LDPE", Conf. on Electr. Insul. & Dielectric Phenomena Ann. Report, pp. 143-148, 1985.

[5] M. Salah Khalil, P.O. Henk, and M. Henriksen, "The Influence of Titatium Dioxide Additive on theShort-Term dc Breakdown Strength of Polyethylene", Conference Record of the 1990 IEEE International Symposium on Electrical Insulation, pp 268-271, Canada, June 1990.

[6] M.S. Khalil, "On the Methods of Suppression of Space Charge Effect Polyethylene", Records of the Nordic Insulation Symposium, NORD-IS 90, Denmark, June 1990.

[7] P.H- Fischer and K.W. Nissen, "The Short-Time Electric Breakdown Behavior of Polyethylene", IEEE Trans.Electr. Insul., Vol. EI-11, No. 2, June 1976.

[8] M.Salah Khalil, P.O. Henk, and M. Henriksen, "Determination of Charge Carrier Mobility in Doped Low Density Polyethylene Using D,C, Transients", Proceedings 3rd Int. Conf. on Conduction and Breakdown in Solid Dielectrics, pp. 192196, 1989.

[9] M. Salah Khalil, A. Cherifi, A. Toureille and J.P. Reboulk, "On Effects of Polarizing Temperature and Additive Content on Space Charge Formation and Electric Field Distribution in Plain LDPE and LDPE doped with $BaTiO_3$: Evidence from Measurements using the Thermal Step Method", Proceedings of the 5th International Conference on Conduction and Breakdown in Solid Dielectrics", ICSD'95, pp. 288-292, UK, July 1995.

Conference Record of the 1996 IEEE International Symposium on Electrical Insulation, Montreal, Quebec, Canada, June 16-19, 1996

Development of Insulation Material for DC Cables -Space Charge Properties of Metallocene Catalyzed Polyethylene-

Shinan Wang, Michitomo Fujita, Gen Tanimoto, Fumio Aida and Yasutaka Fujiwara
Showa Electric Wire & Cable Co., Ltd.
Kawasaki-shi, Kanagawa-ken, 210 Japan

***Abstract*: Applicability of polyethylene polymerized using a metallocene catalyst to an insulation material for DC power cables has been investigated. The accumulation properties of space charge in a range from room temperature to 90°C was examined using the pulse electroacoustic method. The result indicated that hetero-charge was observed in the vicinity of the surface of sample at increased temperature, and that hetero-space charge produced by activation of an antioxidant added to the sample was increased or decreased with the increase of the temperature of sample. A hetero-charge absorber was added for controlling the hetero-space charge. As a result, hetero-charge was not observed on the sample in a high electric field extending from room temperature to high temperature. It was confirmed that the volume resistivity of sample was increased at high temperature.**

IINTRODUCTION

Research and development on DC power cables have progressed so far. The insulation materials of most of DC power cables are conventional polyethylene modified or additive contained. The polyethylene polymerized with Ziegler-Natta catalyst as the base polymer is also proposed. The insulation material is characterized by being superior in space charge properties, DC and impulse breakdown properties[1-3]. A new type polyethylene polymerized using a metallocene catalyst has recently been developed. It is superior to conventional polyethylene in narrow molecular weight distribution, uniform distribution of the comonomer, less amount of catalyst residue and the structure controllable at the molecular level. It is expected to be applied to the insulation material of wires and cables because of various superior characteristics[4]. It is, however, unknown for the present whether or not the newly developed polyethylene will be applied to DC power cables because the characteristics including space charge properties, DC and impulse breakdown properties have not been examined yet.

The authers have investigated the accumulation properties of space charge of the newly developed polyethylene in a range from room temperature to 90°C with the pulse electroacoustic method[5]. Ionic carriers produced by the activation of the antioxidant contained in the newly developed polyethylene are emerged in the form of hetero-space charge in the vicinity of the surface of sample at high temperature. Hetero-space charge disappears and the total amount of space charge accumulated is reduced at high temperature in the newly developed polyethylene modified by adding the hetero-space charge absorber for preventing the emergence of hetero-space charge. This insulation material is expected to be promising as a new type insulation for DC power cables. The experimental results will be described hereafter.

EXPERIMENTAL

Samples

Two types of the newly developed base polyethylene applicable to an insulation material for DC power cables are used in the experiment. The samples used for the experiment are listed in Table I. Samples 1 and 4 are original samples containing an antioxidant and have the densities of 0.915 and 0.935g/cm^3, respectively. Samples 2 and 5 are prepared by adding 1% of the hetero-charge absorber to Samples 1 and 4, respectively, and Samples 3 and 6 are prepared by adding 3% of it to Samples 1 and 4, respectively, for controlling hetero-space charge at high temperature.

Table I Samples

Samples	Thickness (mm)	Density (g/cm^3)	Specifications
Sample1	0.5, 1	0.915	Original, Including anti-oxidation agent
Sample2	0.5, 1	-	Add hetero-charge absorber 1% to Sample1
Sample3	0.5, 1	-	Add hetero-charge absorber 3% to Sample1
Sample4	0.5, 1	0.935	Original, Including anti-oxidation agent
Sample5	0.5, 1	-	Add hetero-charge absorber 1% to Sample4
Sample6	0.5, 1	-	Add hetero-charge absorber 3% to Sample4

Space Charge Measurement

The space charge distribution is measured by the pulse electroacoustic method(PEA) with a space charge measuring apparatus of temperature variable illustrated in Fig. 1. The piezo-electric element for transducing the pressure wave signal to the electrical signal is composed of $LiNbO_3$.

The space charge distribution is measured by putting the sample sheet between aluminum electrodes A and B 45 mm each in diameter. The lower half of the cell including the sample sheet for measuring the space charge is immersed in silicone insulating oil to heat up the sample sheet by increasing the temperature of the silicone oil. The upper half of the cell including the piezo-electric element and BNC for output of the electric signal is exposed to the atmosphere. Since the propagation velocity of the sound wave in the sample is reduced with increasing the temperature, the measurement is started after leaving the sample sheet to stand for further one hour after the temperature indicator shows the preset temperature till exactly reaching the preset temperature of the sample. Thus, the propagation velocity of sound wave becomes stable.

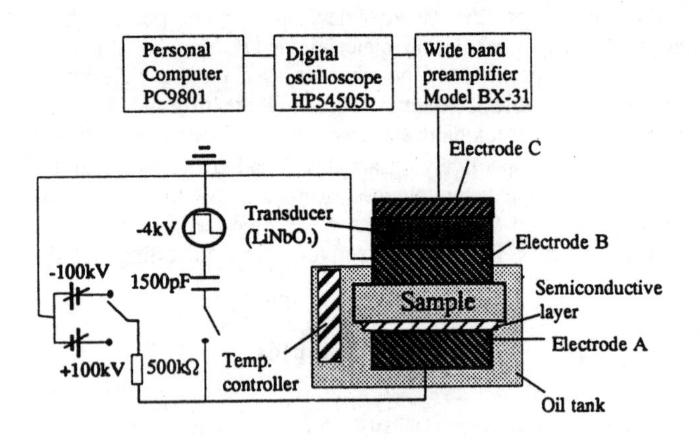

Figure 1. Measuring system for space charge distribution

The sample sheet for measuring the space charge distribution is 0.5 mm thick. The voltage applied to the sample is DC-20 kV and pulse width and voltage for measuring the space charge distribution are 15 ns and -4 kV, respectively, in this experiment. The measurement is carried out under application of voltage at room temperature, 60 and 90°C. The experiment is conducted in such a way that the accumulation of space charge in the sample is saturated approximately three hours after raising up to the specified preset temperatures by continously heating up the sample from room temperature and the saturated space charge distributions at each temperature are measured.

Also the relationship of the sample sheet 1 mm thick between the volume resistivity and the temperature is determined with an ultra-high resistivity measuring apparatus (Advantest R8340).

RESULTS AND DISCUSSION

Space Charge Distributions of Original Samples

The space charge distributions of Samples 1 and 4 were measured in a range from room temperature to high temperature under application of electric field, E, of -40 kV/mm. Figs. 2 and 3 illustrating the space charge distributions three hours after charging at specified temperatures reveal that hetero- and homo-charges are observed in the vicinity of the negative and positive electrodes, respectively, and positive charges are distributed in the center of the sample in both Samples 1 and 4. A lot of hetero-charges in the vicinity of the negative electrode are increased with increasing the temperature of sample. The density of hetero-charge in Sample 1 is lower than that in Sample 4 at 60°C, while that in Sample 1 is approximately double as much as in Sample 4 at 90°C. It is, therefore, inferred that hetero-charges in Sample 4 are decreased at 90°C because many more homo-charges are injected at that temperature.

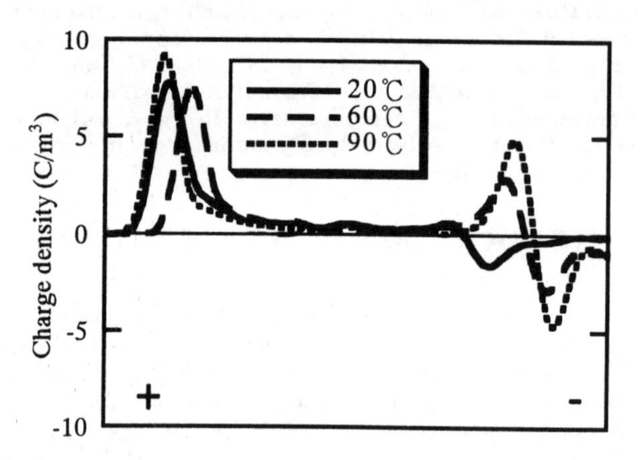

Figure 2. Space charge distribution of Sample 1

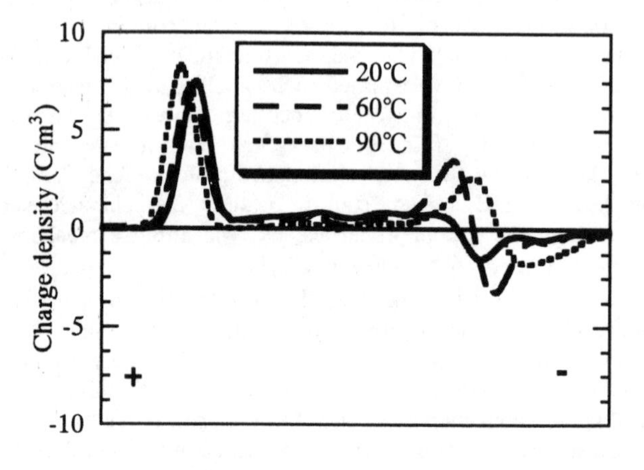

Figure 3. Space charge distribution of Sample 4

Space Charge Distributions of Modified Samples

It is considered that hetero-space charge in the vicinity of the surfaces of Samples 1 and 4 has a large effect on the DC breakdown and volume resistivity of the insulation material. Inorganic porous fine particles are added as the hetero-space charge absorber to Samples 1 and 4 to inhibit the generation of hetero-space charge in this experiment. The accumulation properties of space charge of the samples containing the absorber determined by the same charging application and measurement as in Figs. 2 and 3 are illustrated in Figs. 4 and 5. Homo-charge is observed in place of hetero-charge in the vicinity of the surface of sample on the negative electrode side in Figs. 4(a) and 5(a). Although the width of the waveform of space charge distribution is increased due to the decreasing of the propagation velocity of sound wave in the sample by raising the temperature of sample from room temperature to 90°C, the space charge distribution in the sample is unchanged. Figs. 4(b) and 5(b) reveal that low hetero-charge is observed at 90°C in the vicinity of the negative electrode in the samples containing slightly more hetero-charge absorber than those samples.

Fig. 6 illustrates the volume resistivities of original Samples 1 and 4, and Samples 2, 3, 5 and 6 containing the hetero-space charge absorber in a range from room temperature to 90°C. The figure reveals that the volume resistivities of Samples 1 and 4 are lower than those of the other samples and that the more the additions of the hetero-charge absorber, the higher the volume resistivities at high temperature are.

Effect of Hetero-charge Absorber

Homo-charges injected from the positive and negative electrodes are concentrated into the center of the sample sheet of conventional low density polyethylene without containing the antioxidant by heating up the temperature of it, where the charges are neutralized, thereby decreasing the total amount of charges. The effect of the antioxidant on the accumulation properties of space charge in conventional low density polyethylene has already been reported. The report presents that hetero-charge is accumulated in the vicinity of the surface of sample containing 1% of the antioxidant[6]. Although the neutralization of positive charges with negative ones in the centers of the newly developed Samples 1 and 4 shown in Figs. 2 and 3, respectively, is not observed even though the temperature is increased, a positive hetero-charge layer is formed on the negative electrode side by the effect of the antioxidant. This is because the positive hetero-space charge is accumulated in the vicinity of the negative electrode by the ionic carriers of the antioxidant under the application of electric field at high temperature. No negative hetero-charge is observed in the vicinity of the positive electrode. This is because charges injected from the positive electrode neutralize the negative hetero-space charges.

The existence of hetero-space charge is an important factor detrimentally affecting the DC dielectric breakdown strength. Although hetero-space charges disappear by reducing the additions of the antioxidant as already known, it is undesirable

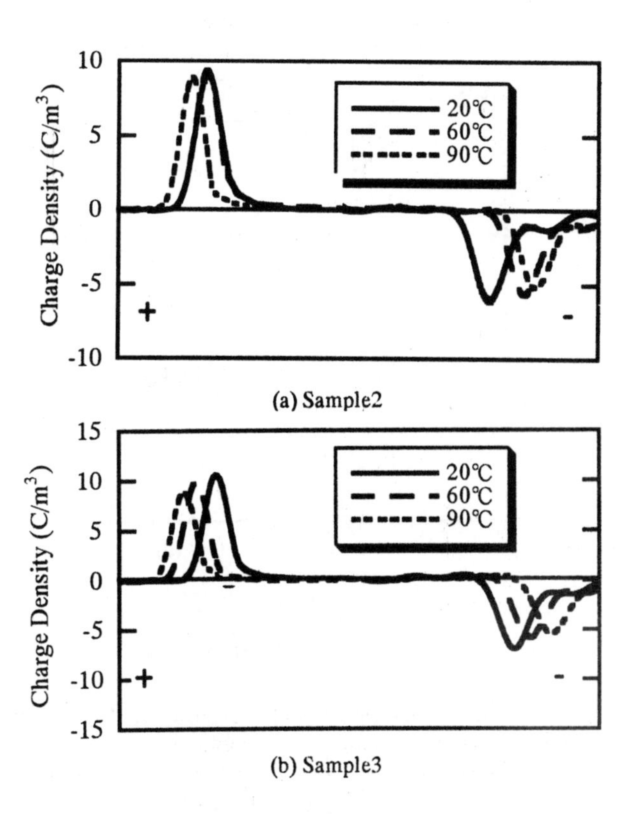

Figure 4. Space charge distribution of Samples 2 and 3

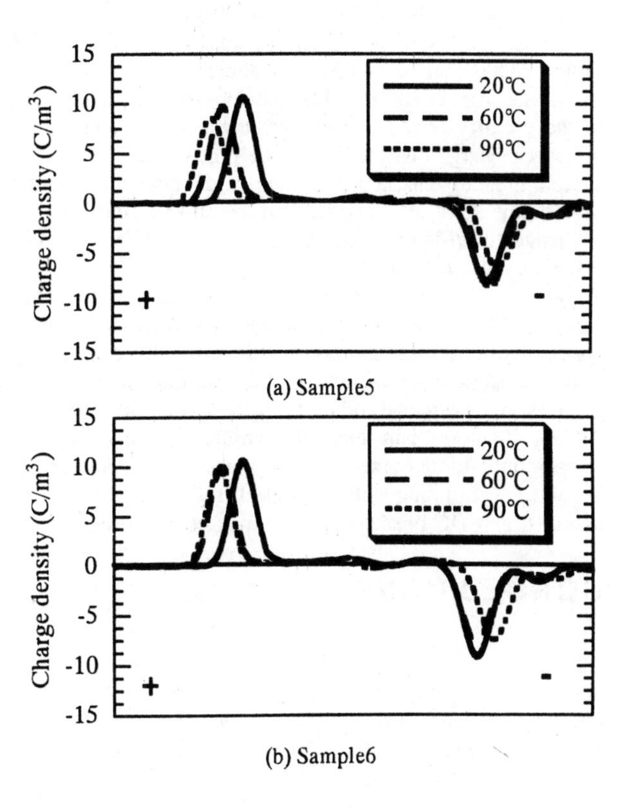

Figure 5. Space charge distribution of Samples 5 and 6

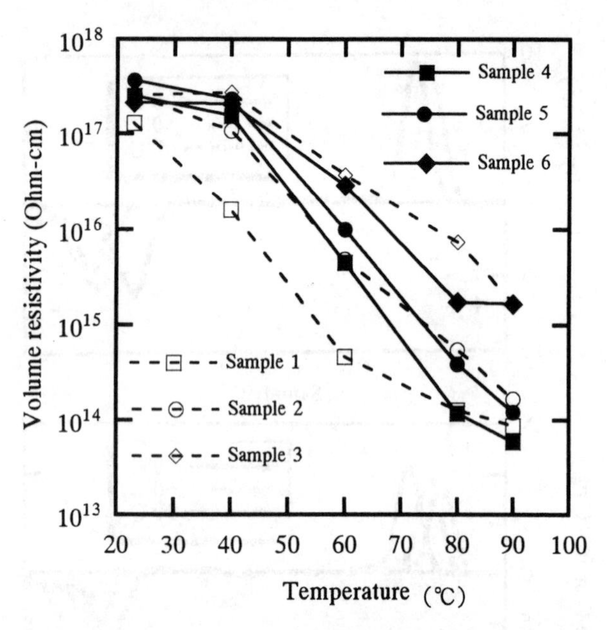

Figure 6. Volume resistivity of new type polyethylene samples

for thermal aging characteristics of DC power cables to reduce the addition of the antioxidant. Accordingly, the addition of the hetero-space charge absorber to the sample is tried in this experiment. The space charge absorber plays a role to remove the hetero-space charge from the insulation material by adsorbing acidic and basic ions generated in it. Comparing the space charge distributions of the samples containing the space charge absorber (see Figs. 5 and 6) with those of the original samples (see Figs. 2 and 3), the effectiveness of the hetero-charge absorber is confirmed because the hetero- space charges in the vicinity of the negative electrode are adsorbed. Since the space charge distribution of the high electric field-applied sample is hardly changed extending from room temperature to high temperature, it is inferred that the effect of the absorber is unchanged even at high temperature. Meanwhile, the volume resistivity of the sample containing the hetero-space charge absorber at high temperature is higher than that of the original sample. This is because the sample hardly becomes ionically conductive since the carriers derived from the antioxidant at high temperature are adsorbed. The optimum addition of the hetero-space charge absorber is 1 to 3% in the experiment. Adding it more than that, the volume resistivity at high temperature further increases. Since the absorber itself becomes a hopping site for electrical conductivity, it is assumed that DC breakdown value may be decreased.

CONCLUSIONS

The accumulation properties of space charge of the newly developed polyethylene and that containing various amounts of the hetero-space charge absorber have been investigated extending from room temperature to 90°C with the pulse electroacoustic method. The results obtained by the investigation are as follows:

(1) Although the amount of space charge accumulated in the newly developed polyethylene polymerized using the metallocene catalyst is smaller than that in conventional polyethylene, hetero-space charge is generated in the vicinity of the surface of the sample on the negative electrode side by the effect of the antioxidant added. It was found that the hetero-charge is increased or decreased with raising the temperature of the sample.

(2) The hetero-charge is not generated in the vicinity of the surface of the sample on the negative electrode side extending from room temperature to 90°C in the newly developed polyethylene containing the hetero-charge absorber. The volume resistivity of it at high temperature is higher than that of the original polyethylene.

Those results indicate that the newly developed polyethylene containing the hetero-charge absorber will be one of the promising insulation material for DC power cables. The applicability of it will be further investigated in detail in the future by examining the properties including DC breakdown and impulse breakdown.

REFERENCES

1. Y.Tanaka, Y.Li, T.Takada and M. Ikeda. "Space charge distribution in low-density polyethylene with charge-injection suppression layers". J.Phys. D: Phys. 28,1995, pp.1232-1238.
2. M.Ikeda, Y.Umeshima, Y.Tanaka and T.Takada."Development of New Cross-linked Polyethylene for DC Power Cable Insulator". Proceedings of 1995 International Symposium on Electrical Insulating Materias,Tokyo, Sept.17-20,1995, pp.403-406
3. K.Ogawa, T.Suzuki, T.Niwa, S.Yosida, T.Takahashi and M.Hatada. "Investigation of Insulating Materials for DC Cables (Part 3)".Proceedings of 21st Symposium on Electrical Insulating Materias,Tokyo, Sept.26-28,1988, pp.297-282
4. Editors. "Polymers of New Olefin-Possibility of New Polymers with Metallocene Catalyst". Polyfile, Oct.,1994, pp.60-63 (in Japanese)
5. T.Takada, T.Maeno and H.Kushibe. "An Electric Stress-pulse Technique for the Measurement of Charges in a Plastic Plate Irradiated by an Electron Beam". IEEE Trans., Vol.EI-22, No.4, Aug.,1987, pp.497-501
6. T.Takada, T.Uozumi, N.Yasuda,T.Fukuiand N.Noda. "Space Charge Distribution in Polyethylene under a High Temperature-the Effect of Antioxidants". Proceedings of 1994 Annual Conference of Electric and Energy Department of IEE Japan, pp.653-654 (in Japanese)

Conference Record of the 1996 IEEE International Symposium on Electrical Insulation, Montreal, Quebec, Canada, June 16-19, 1996

Space Charge Fields in DC Cables

I.W. McAllister, G.C. Crichton and A. Pedersen†
Department of Electric Power Engineering, Building 325
Department of Physics, Building 309
Technical University of Denmark
DK-2800 Lyngby, Denmark

Abstract: **The space charge that accumulates in DC cables can, mathematically, be resolved into two components. One is related to the temperature and the other to the magnitude of the electric field strength. Analytical expressions for the electric fields arising from each of these space charge components are derived. Thereafter, the significance of these field components under both normal operating conditions and immediately following polarity reversal is discussed.**

INTRODUCTION

Previously, the authors have shown that the accumulation of space charge in DC cable insulation is an inherent phenomenon [1]. The presence of such space charge in the bulk of the dielectric constrains the field to be Poissonian. The corresponding field strength $\vec{E}$ can thus be expressed as

$$\vec{E} = \vec{E}_L + \delta\vec{E} \qquad (1)$$

where $\vec{E}_L$ represents the basic Laplace field [2] associated with the applied voltage U, and $\delta\vec{E}$ represents the basic Poisson field [2]. This latter component is established by the space charges in the dielectric together with the associated Poissonian induced charges [3] on the electrodes.

Traditionally, electric fields associated with DC cables are evaluated directly using the functional dependence of the insulation conductivity on temperature and field strength. This circuit theory approach is used in many papers and textbooks on cable technology. In the present study, analytical expressions for the electric field are derived from a knowledge of the inherent space charge distribution: i.e. a field theoretical approach is adopted. This approach allows the significance of $\delta\vec{E}$ under steady-state operation and for conditions immediately following polarity reversal to be discussed in depth.

SPACE CHARGE IN THE INSULATION

Consider a coaxial cable insulated with a macroscopically homogeneous dielectric of constant permittivity. The conductivity γ of this dielectric is a function of both the temperature T and the magnitude of the electric field strength $|E|$. In this study, the γ dependence is represented by an empirical relationship which is valid over the practical range of T and $|E|$ [4]; viz.

$$\gamma = \gamma_a(|E|/|E_a|)^{\nu}\exp[\alpha(T - T_a)] \qquad (2)$$

where γ_a is the conductivity for a reference field strength E_a and temperature T_a. The parameter ν is a material constant which for oil-impregnated paper is approximately zero, but which for polyethylene can be ascribed values in the range 2.1 to 2.4 [4]: α is a constant and for the dielectrics used in cables the value 0.1 K^{-1} is appropriate.

For a loaded DC cable, which is in

† deceased

thermal equilibrium, it has been shown in [1] that the space charge density ρ may mathematically be considered as the sum of two components:

$$\rho = \rho_T + \rho_{|E|} \tag{3}$$

where ρ_T is the component associated with temperature while $\rho_{|E|}$ is that related to the magnitude of the electric field strength. In [1] expressions were derived for these components: viz.

$$\rho_T = \frac{\varepsilon k \beta U}{a^2} \frac{(r/a)^{k-2}}{(b/a)^k - 1} \tag{4}$$

$$\rho_{|E|} = \frac{\varepsilon k(k - \beta) U}{a^2} \frac{(r/a)^{k-2}}{(b/a)^k - 1} \tag{5}$$

where a and b are the inner and outer radii of the dielectric, and r is a cylindrical coordinate. U represents the applied DC voltage while k and β are dimensionless parameters given by

$$k = \frac{\nu + \beta}{\nu + 1} \tag{6}$$

$$\beta = \alpha(T_a - T_b)/\ln(b/a) \tag{7}$$

T_a and T_b are the temperatures at distances a and b from the cable axis. From (3), (4) and (5) it is evident that

$$\rho = \frac{\varepsilon k^2 U}{a^2} \frac{(r/a)^{k-2}}{(b/a)^k - 1} \tag{8}$$

and thus ρ_T and $\rho_{|E|}$ may be simply expressed in terms of ρ: viz.

$$\rho_T = (\beta/k)\rho \tag{9}$$

$$\rho_{|E|} = (1 - \beta/k)\rho \tag{10}$$

FIELD SOURCES

Insulation Space Charge

In evaluating the field distribution as a function of the radial distance r, it is necessary to refer to the total charge Q per unit length enclosed within a cylindrical shell of thickness $(r - a)$:

$$Q = \int_a^r \rho 2\pi r' \mathrm{d}r' \tag{11}$$

where r' is a dummy variable. Upon substitution for ρ, (8), and integrating we obtain

$$Q = 2\pi\varepsilon k U \frac{(r/a)^k - 1}{(b/a)^k - 1} \tag{12}$$

Hence on the basis of (4) and (5) we arrive at the simple relationships

$$Q_T = (\beta/k)Q \tag{13}$$

$$Q_{|E|} = (1 - \beta/k)Q \tag{14}$$

Charge Induced by the Space Charge

For the volume space charge distribution, the Poissonian induced charge q may be expressed as [2]

$$q = -\int \lambda(r)\rho \mathrm{d}\Omega \tag{15}$$

where $\mathrm{d}\Omega$ is a volume element. The parameter λ is the proportionality factor between the space charge accumulated in the insulation and the charge induced on the conductor in question. The λ-function is a solution of the general Laplace equation [2]

$$\vec{\nabla} \cdot (\varepsilon \vec{\nabla} \lambda) = 0 \tag{16}$$

in which ε denotes permittivity. The relevant boundary conditions are $\lambda = 1$ for $r = a$ and $\lambda = 0$ for $r = b$. For the present simple case of the coaxial geometry with a homogeneous dielectric, the appropriate solution of Laplace's equation for the λ-function is

$$\lambda = 1 - \frac{\ln(r/a)}{\ln(b/a)} \tag{17}$$

Owing to the axial symmetry of the space charge distribution, it is however more appropriate to consider q as the charge induced per unit length. Hence for a cylindrical shell of radius r and thickness dr we have

$$q = -\int_a^b \lambda\rho 2\pi r \mathrm{d}r \tag{18}$$

Upon combining (17) with (8) and integrating, we can express q as

$$q = 2\pi\varepsilon k U\left[1 - \frac{(b/a)^k - 1}{k \ln(b/a)}\right] \tag{19}$$

The components of q follow directly: viz.

$$q_T = (\beta/k)q \tag{20}$$

$$q_{|E|} = (1 - \beta/k)q \tag{21}$$

As the induced charge is of *opposite* polarity to the source charge, see (18), this implies that the field components associated with Q and q are in opposition.

ELECTRIC FIELDS

In view of the symmetry of the space charge distribution, the basic Poisson field strength δE at a distance r is given by

$$\delta E = \frac{Q + q}{2\pi\varepsilon r} \tag{22}$$

Owing to symmetry, all vector quantities will be directed either radially away from or towards the axis. We can therefore replace all vector equations with scalar equations. The direction of the field is away from the axis for E positive. Upon introducing (12) and (19) into (22), we obtain

$$\delta E = \frac{U}{r\ln(b/a)}\left[\frac{k \ln(b/a)}{(b/a)^k - 1}(r/a)^k - 1\right] \tag{23}$$

and the associated components of δE are

$$\delta E_T = (\beta/k)\delta E \tag{24}$$

$$\delta E_{|E|} = (1 - \beta/k)\delta E \tag{25}$$

For the coaxial geometry, the basic Laplace field is simply given by

$$E_{\mathrm{L}} = \frac{U}{r\ln(b/a)} \tag{26}$$

Thus on the basis of (1), the net steady state field E can be expressed as

$$E = \frac{U}{a}\frac{k(r/a)^{k-1}}{(b/a)^k - 1} \tag{27}$$

It should be noted that this field expression is identical with that derived in [4] using the traditional approach.

The reversal of the applied voltage polarity can occur on a time-scale which is very short in comparison to the relaxation time constant associated with the accumulation of space charge in the cable insulation. Thus, immediately following such a reversal of polarity, the magnitude and distribution of the space charges are virtually unaffected. Consequently, although the basic Laplace field has changed polarity, the basic Poisson field is effectively unchanged. Hence following polarity reversal, the *initial* field E_{R} may be expressed as

$$E_{\mathrm{R}} = -\frac{U}{a}\left[\frac{2}{(r/a)\ln(b/a)} - \frac{k(r/a)^{k-1}}{(b/a)^k - 1}\right] \tag{28}$$

APPLICATION OF THEORY

To illustrate the order of magnitude of the effects of the accumulated space charge in a coaxial DC cable, we consider the 250 kV, 500 MW XLPE cable proposed by Fukagawa *et al.* [5]. As ν can attain values between 2.1 and 2.4, we assume for expediency a value of 2.25. The parameter β can be evaluated if the power dissipated in the conductor per unit length P_{c} is known [1]. The relationship is

$$\beta = \frac{\alpha P_{\mathrm{c}}}{2\pi\kappa} \tag{29}$$

We assume that the temperature of the conductor has the specified maximum value of 80 ^{o}C, and that the thermal conductivity κ of XLPE is 0.3 W/Km. Thereafter, on the basis of the data given in [5], we obtain $\beta = 4.53$.

Using (6), (8), (9) and (10), the radial variations of ρ, ρ_T and $\rho_{|E|}$ were derived. Results are shown in Figure 1, in which x is the radial distance from the surface of the inner conductor; i.e. $x = r - a$. On letting d denote the insulation thickness, we have $d = b - a$. From Figure 1, it is evident that the space charge distributions are only slightly non-uniform in that these distributions deviate by <5% from their average values. Moreover $|\rho_T|$ attains values of about twice $|\rho_{|E|}|$. However, as these components are of opposite polarity, the net space charge ρ is about $(\rho_T/2)$ such that ρ lies in the range of $12 < \rho/(\mathrm{pC/mm^3}) \lesssim 13$.

The electric fields associated with the space charge were evaluated using (6) (23) to (27) and (29) and the results are illustrated in Figure 2. The behaviour of δE, δE_T and $\delta E_{|E|}$ can be understood with reference to the corresponding space charge distributions, where we recall that the induced charge is of opposite polarity to the space charge. As a consequence of this latter feature, the values of δE, δE_T and $\delta E_{|E|}$ are zero for

$$(r/a) = \left[\frac{(b/a)^k - 1}{k \ln(b/a)}\right]^{1/k} \tag{30}$$

see (23). With respect to Figure 2, we have the relationship

$$(x/d) = \frac{(r/a) - 1}{(b/a) - 1} \tag{31}$$

Upon using the present values of k and (b/a), we find that $(x/d)_0 = 0.47$, see Figure 2. As a result of the space charge field δE changing polarity at $(x/d)_0$, this field opposes E_L for $(x/d) < (x/d)_0$,

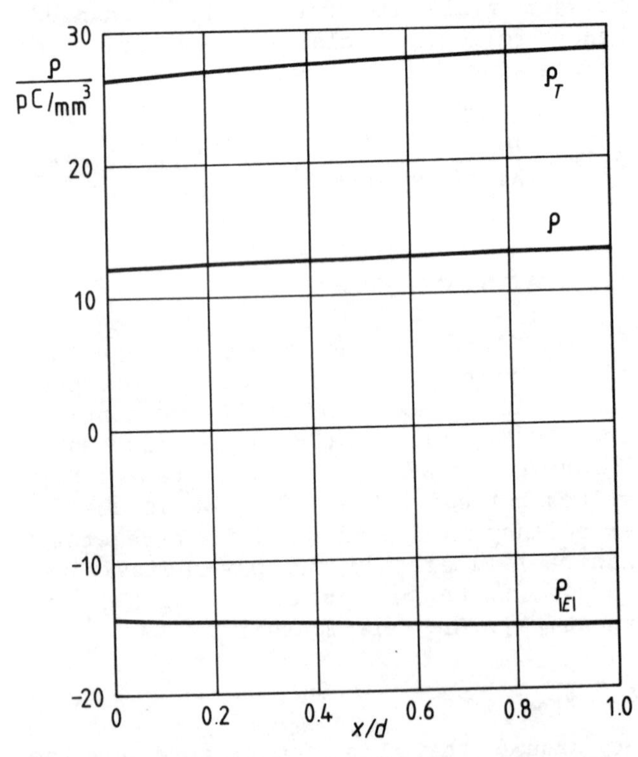

Figure 1. Radial variation of space charge ρ and its components in a loaded DC cable.

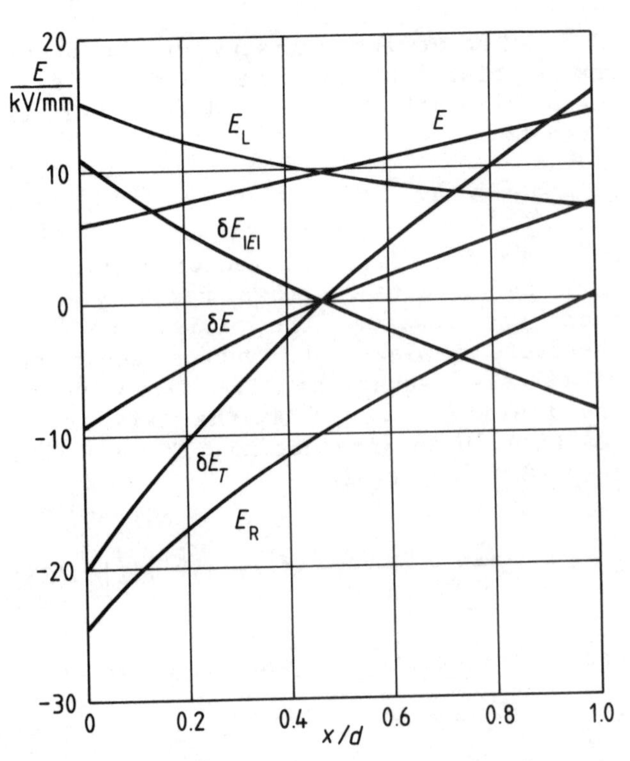

Figure 2. Radial variation of electric field distributions E in a loaded DC cable.

whereas δE augments E_L for $(x/d) > (x/d)_0$ such that we obtain stress inversion: i.e. the greatest stress now occurs at the cable sheath.

For the present ν and β values, the exponent of the radial distance, $(k-1)$, is 1.08. As a result, the steady-state field E varies essentially linearly with the radial distance. Following polarity reversal, the magnitude of the field at the inner conductor is seen to be increased by 61% above that of E_L, see Figure 2.

CONCLUSION

From a knowledge of the inherent space charge distribution in a DC cable, it is possible to identify the influence upon the electric field distribution of the parameters which control the insulation conductivity. For a loaded cable, it is shown that the major influence on the field distribution is associated with temperature, although this is seen to be counteracted partially by that of the electric field. This behaviour suggests that the conductivity parameters could be selected such as to minimise the resultant space-charge-generated electric field. Such an approach would reduce overstressing of the insulation which otherwise would certainly be encountered on polarity reversal.

REFERENCES

1. McAllister, I.W., G.C. Crichton and A. Pedersen. "Charge Accumulation in DC Cables: A Macroscopic Approach". Conference Record of the 1994 IEEE International Symposium on Electrical Insulation. IEEE Publication 94CH3445-4, 1994, pp.212-216.

2. Pedersen, A., G.C. Crichton and I.W. McAllister. "The Functional Relation between Partial Discharges and Induced Charge". IEEE Trans. Dielect. & Elect. Insul., vol.2, 1995, pp.535-543.

3. Pedersen, A., G.C. Crichton and I.W. McAllister. "Partial Discharge Detection: Theoretical and Practical Aspects". IEE Proc. - Science, Measurement and Technology, vol.142, 1995, pp.29-36.

4. Privezentsev, V., I. Grodnev, S. Kholodny and I. Ryazanov. *Fundamentals of Cable Engineering*. Mir Publishers, 1973.

5. Fukagawa, H., H. Miyauchi, Y. Yamada, S. Yoshida and N. Ando. "Insulation Properties of 250 kV DC XLPE Cables". IEEE Trans. Power Appar. & Syst., vol.PAS-100, 1981, pp.3175-3184.

Conference Record of the 1996 IEEE International Symposium on Electrical Insulation, Montreal, Quebec, Canada, June 16-19, 1996

Thermally Stimulated Currents and Space Charge Studies on Field-Aged Extruded Cable Material

Normand Amyot, Serge Pélissou, Alain Toureille*
Hydro-Québec (IREQ)
1800 Montée Ste-Julie, Varennes
Qc., Canada, J3X 1S1
*Université de Montpellier II
Lab. d'Électrotechnique de Montpellier
C.P.079, 34095 Montpellier, France

Abstract: **In the perspective of gaining more knowledge on extruded cable field aging diagnosis, complementary techniques were investigated: thermally stimulated currents (TSC) and space charge measurements, the latter being performed by the thermal step (TS) method. Measurements were taken on 28 kV extruded cable samples of cross-linked polyethylene (XLPE). Samples were peeled-off from three cables; one unaged and two field-aged. Both techniques show differences between field-aged and unaged cable material. Results obtained by TS show that aged material can store more space charges that lead to greater intensity of the electrical field in some sites in the polymer matrix and eventually initiate electrical trees leading to breakdown. Comparison with TSC results show that the origin of space charge formation cannot be attributed uniquely to traps formed by carbonyl groups from polymer oxidation.**

INTRODUCTION

The failure rate of polymer-insulated power cable systems at distribution voltages is still significantly higher than can be accepted by electrical utilities so that serious questions are still raised about the reliability and long-term performance of extruded distribution cables. One of the key issues is the need for reliable and convenient techniques to evaluate the condition (aging and degradation) of cables and accessories. The use of TSC as a diagnosis tool for underground cable material has recently shown that systematic differences can be observed for the intensity of a relaxation peak (β) located near -30°C for field-aged and unaged cables [1,2]. Positive correlation was obtained in preliminary work [2] between the density of water trees in cables and the intensity of this peak. The origin of this low-temperature peak is assumed to be the glass-rubber transition and the relaxation mechanism dipolar from carbonyl groups formed by material oxidation. In this paper, samples from the same cable material have been studied by TSC and TS [3] in order to compare both techniques as diagnosis tools for the detection of field aging. Space charge density and electrical field distributions are compared to TSC spectra in order to gain more insight for their interpretation.

EXPERIMENTAL

The TSC and TS measurements were performed on 200 µm thick ribbons peeled off [4] from extruded XLPE cables. Two field-aged cables and one unaged 28-kV cable were investigated. Table 1 gives the principal manufacturing characteristics of these cables. The field-aged cables had failed in service due to numerous water trees.

Table1: Cable manufacturing characteristics

Cables	PIRV (unaged)	RI5 (field-aged)	RI7 (field-aged)
Age [years]	0	9	14
Rated voltage [kV]	28	28	27
Mean electrical field in service [kV/mm]	0	2	2
Manufacturing year	1989	1978	1975
Conductor	Al, 750 MCM	Al, 500 MCM	Al, 750 MCM
Insulation	XLPE-4201 EC	XLPE-4201	XLPE-4201
Concentric neutral	tin copper	tin copper	tin copper
Curing	steam	steam	steam

Gold electrodes 1 cm in diameter were evaporated on both sides of the ribbons and square samples of 1.5 cm x 1.5 cm were taken. The TSC spectra were obtained by heating the samples with the electrodes shorted at a rate of 10°C/min from T_o = -100°C to T_f = 130°C. The samples were previously annealed at T_a = 120°C for t_a = 15 min, polarization temperature T_p = 110°C for 10 min., the cooling rate was 10° C/min and the freezing temperature (T_o) was -100°C. A SOLOMAT instrument, model 91000, was used for the TSC measurements. All measurements were performed in helium to improve heat transfer to the sample. For the TS measurements, samples are placed horizontally between two isothermal sources from cylindrical brass pieces of same diameter as the evaporated gold electrodes. The hot source is at T = 20°C and the cold sink is at T_o = -10°C. The thermal step ΔT = 30°C propagates through the sample and the charge exchange from one electrode to the other is recorded with an electrometer. Numerical treatment and validation of the method have been discussed elsewhere [5]. Measurements have been carried out before and after conditioning the samples by applying an electrical field E_p = 10 kV/mm field during 4 hours at 70°C.

RESULTS AND DISCUSSION

Figure 1 shows TSC spectra of the field-aged and unaged cables whose manufacturing characteristics are presented in Table 1. In all cases, one well-defined relaxation mode is present at about -28°C. This peak has been previously attributed to the β relaxation corresponding to the glass-rubber transition in XLPE [1,6]. Another peak can be observed at higher temperature (~115°C). Poor reproducibility was noted for this high-temperature region of the spectra when highly reproducible values were obtained for the β peak region. The β peak was assumed to be of dipolar origin since the total integrated charge computed by integration is proportional to the polarization field [1]. These dipoles consist in carbonyl groups from oxidized material as confirmed by FTIR spectroscopy [6]. Figure 2 shows a comparison of the β peaks for all three cables; it can be seen that for both field-aged cables, the β peak maxima show a greater intensity compared to the unaged cable. The total released charges computed from β peak integration are shown in Table 2. Total charge is proportional to peak intensity.

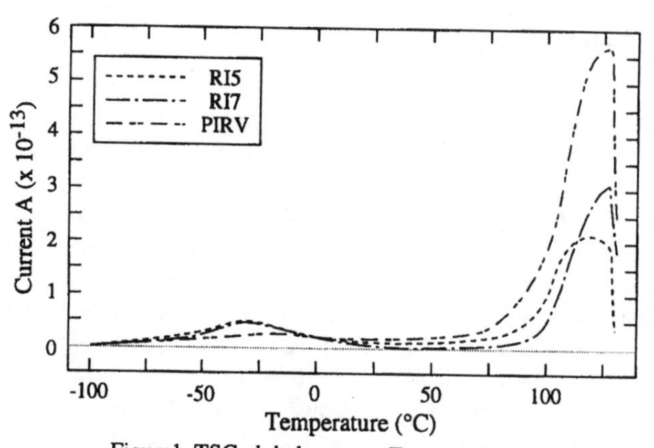

Figure1. TSC global spectra; E_p = 1.6 kV/mm.

Space charge distribution of samples before conditioning are presented in Figure 3. One can see that field-aged cable samples contain more space charges than the unaged one. This is confirmed by quantitative data obtained by peak integration shown in Table 3. The presence of residual charges in field-aged cable samples can be explained by accumulation in service since space charge decay is very slow at room temperature even for short-circuited samples [7-8].

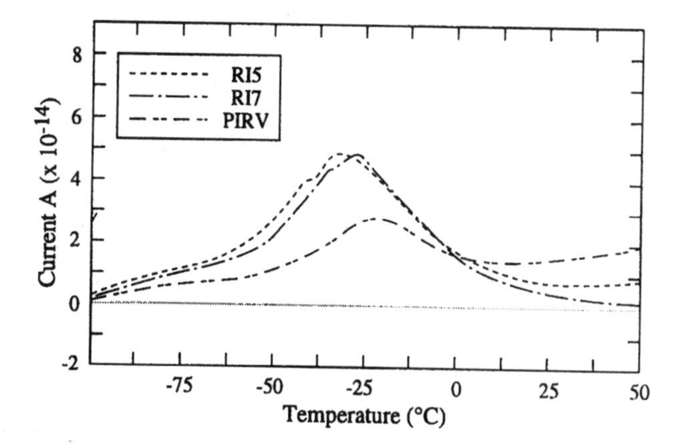

Figure 2. Low temperature (β) peaks from TSC spectra; E_p = 1.6 kV/mm.

Table 2: Low-temperature (β) peak integration

Cable	Total released charge from β peak ($\times 10^{-11}$ C)
PIRV (unaged)	0.61
RI5 (field-aged)	1.27
RI7 (field-aged)	1.27

The space charge distribution in conditioned samples is shown in figure 4. All samples show homocharge accumulation near the cathode whereas homocharge accumulation near the anode is observable only for samples RI5 and PIRV. Space charge density is higher for field-aged samples than for unaged material.

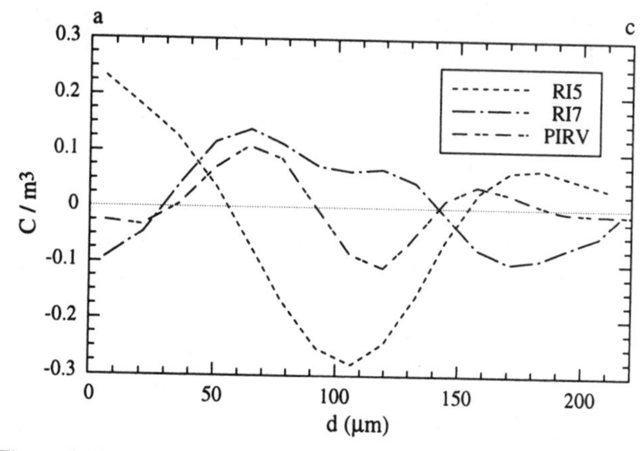

Figure 3. Space charge distribution by TS; before conditioning (E_p = 0).

As shown in Table 3, space charge content has greatly increased after sample conditioning for all samples; this holds for both for negative and positive charges. It is noteworthy that positive and negative space charge content has increased much more for the field-aged material. This can be explained by greater trap density in field-aged cable material. Xinsheng et al. [9] have found that trap density may be taken as a parameter which indicates the aging degree of polyolefin compound subjected to corona discharge. However, because of the complexity of XLPE morphology and structure, the exact origin of these traps is very difficult to determine precisely. Traps can originate from changes in the polymer's morphology (water-trees, amorphous-crystalline boundaries), free radicals, impurities (diffusion from semiconducting shields) and even from polar groups (carbonyl from oxidation). The fact that the relative difference between charge content of new and aged material is not the same in the TSC experiment (released charges from dipole reorientation in Table 2) and the TS (space charges in Table 3), shows that dipoles from carbonyl groups only can not account for all the traps leading to space charge formation. The traps must then have other origin than carbonyl groups. Knowing that our field-aged cables contain numerous water-trees, it is conceivable that treed regions contain other than polar traps. Hegerberg et al. [10] suggest that free radicals from broken chain molecules in water-treed regions may act as traps. Another possibility is that microcavities and microchannels in water trees become sites for interfacial polarization where charge can accumulate.

Table 3: Space charge content

Cable:	PIRV	RI5	RI7
Charge before conditioning ($x10^{-10}$ C)			
Positive (+)	3.64	7.56	7.07
Negative (-)	3.35	12.7	4.79
Residual charge (Positive-Negative)	+0.29	-5.14	+2.26
Charge after conditioning ($x10^{-10}$C)			
Positive (+)	5.58	33.5	17.7
Negative (-)	5.81	40.6	14.9
Residual charge (Positive-Negative)	-0.23	-7.1	+2.8

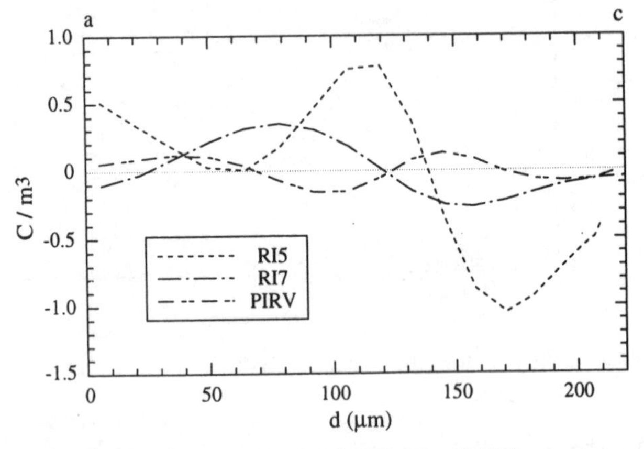

Figure 4. Space charge distribution by TS; E_p = 10 kV/mm; t = 4 hours, T = 70°C.

Figure 5 shows the electrical field distribution after conditioning. It can be seen that the electrical field varies a lot in each sample. It is noteworthy that the maximum electrical field is higher for field-aged samples than for the unaged; fields as high as 1.6 kV/mm are observed for RI5. This is related to the greater amount of space charge in the field-aged cable material. As mentioned above, trap density can be an indication of cable material aging. When conditioned with E_p = 10kV/mm at 70°C for 4 hours, samples with the greatest trap density also accumulate more space charges whether negative or positive. This greater space charge accumulation turns into greater electrical field maxima in the field-aged samples. The occurrence of such high fields near defects in the polymer matrix can in turn give rise to electrical trees and finally lead to breakdown.

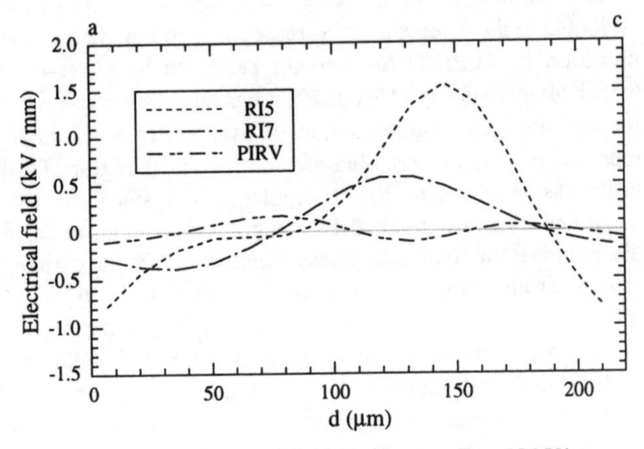

Figure 5. Computed electrical field distribution. E_p = 10 kV/mm, t = 4 hours, T = 70°C.

CONCLUSION

Both TSC and TS measurements show differences between field-aged and unaged cable material. TSC low-temperature peak (β) intensity is greater for field-aged cables. Space charge density distribution obtained by TS show that aged material can store more space charges than the unaged material. We conclude that field-aged cable material contains more trap sites than virgin material. Greater space charge accumulation leads to greater intensity of the electrical field in some sites in the polymer matrix. In these stressed regions electrical tree formation can be promoted by the presence of a defect (impurity, water-tree tip, cavity...). Comparison with TSC results show that the origin of space charge formation cannot be attributed uniquely to traps formed by carbonyl groups from polymer oxidation. Free radicals and microcavities from water-trees could also act as traps. When in service, these traps can be filled with charge carriers and give rise to high enough fields to promote electrical tree formation and lead to breakdown.

REFERENCES

1. Amyot N., Pélissou S. «Thermally Stimulated Currents (TSC) on New and Field-Aged Cable Material ». Annual Report of the Conf. on Electr. Insul. and Diel Phenom. (CEIDP), pp. 489-96, 1994.

2. Amyot N., Pélissou S.,"Diagnosis of Field Aged Extruded Cable Material using the Thermally Stimulated Currents (TSC) Method", 5th Int. Conf. on Cond. and Break. in Sol. Dielectr. (ICSD), pp. 580-4, 1995.

3. Toureille A., Santana J., Joumha A., Vella N., « Measure of Space Charges by the Thermal Step Method », Proc. of the 4th Int. Conf. on Properties and Application of Dielectric Materials (ICPADM), pp. 721-4, 1994.

4. Pélissou S., Crine J.P., Castonguay J., Haridoss S., Bose T.K., Merabet M., Tobazéon R., "Nature et Distribution de l'Eau dans les Câbles Réticulés à la Vapeur", 3rd Int. Conf. Polym. Insul. Pow. Cables (Jicâble), pp. 270-5, 1987.

5. Cherifi A., Abou Dakka M.,Toureille A.., « The Validation of the Thermal Step Method », IEEE Trans. on Electr. Insul., Vol. 27, No.6, pp. 1152-8, 1992.

6. Gubanski S.M., Montanari G.C., Motori A. «TSDC Investigations of Thermally and Electro-Thermally Aged Cable Models», Conf. Rec. of the IEEE Symp. on Electr. Insul., pp. 54-7, 1994.

7. Xinsheng Wang, Demin Tu, « Space Charge in XLPE Power Cable under dc Electrical Stress and Heat Treatment », IEEE Trans. on Dielectrics and Electrical Insulation, Vol. 2, No.3 (June), pp. 467-74, 1995.

8. Sabir A., Toureille A., «Space charge Measurements in High Voltage Cables », Proc. of the 4th Int. Conf. on Cond. and Breakdown in Solid Dielectr., pp. 92-6, 1992.

9. Xinsheng Wang, Yongging Huang, Demin Tu, « Trap Density as an Indication of Electrical Ageing Degree in Polymers », Proc. of the 4th Int. Conf. on Prop. and Appl. of Dielectr. Mat. (ICPADM), pp. 740-3, 1994.

10. Hegerberg R., Lundgaard L., Ilstad E., «Dielectric Relaxation and DC Conductivity of Transformer Pressboard and XLPE Cable Insulation», Proc. of the 4th Int. Conf. on Cond. and Breakdown in Sol. Dielectr., pp. 47-51, 1992.

Conference Record of the 1996 IEEE International Symposium on Electrical Insulation, Montreal, Quebec, Canada, June 16-19, 1996

Effect of water on the space charge formation in XLPE

Hiroyuki Miyata, Ayako Yokoyama, Tohru Takahashi, Syuji Yamamaoto
Fujikura Ltd.
Tokyo, Japan

ABSTRACT: In this paper, we describe the effect of water on the space charge in crosslinked polyethylene (XLPE).
In order to study the effects of water and by-products of crosslinking, we prepared two types of samples. The water in the first one (Type A) is controlled by immersing in water after removing the by-products, and the water in the other type (Type B) of samples is controlled by the water from the decomposition of cumyl-alcohol by heating.

We measured the spacecharge formation by pulsed electro-acoustic(PEA) method.

A large difference was observed between Type A and Type B. In Type A samples (containing only water) the space charge distribution changes from homogeneous to heterogeneous as the water content increases, whereas in Type B (containing water and by-products) all samples exhibit heterogeneous space charge distribution.

However, merely the effect of water for both types was almost same, including peculiar space charge behavior near the water solibility limit.

1. INTRODUCTION

Space charge formation has long been considered to be a factor affecting the insulation performance of power cables and its effect is particularly pronounced in DC cables. A number of studies have been reported[1)-3)] so far on the space charge formation phenomenon due to DC voltage in XLPE and in PE which is the base of XLPE.

In order to grasp the space charge formation phenomenon in DC cables, it is considered most effective to obtain directly the space charge distribution in samples that are several millimeters thick. So far, the authors have measured the space charge distribution in XLPE and PE using the pulsed electro-acoustic method (PEA method) which allows easy measurement of the space charge distribution in thick samples, and have investigated the effects of the measurement conditions, the type of material, and the additive materials, etc., on space charge formation[4)]. In this study, attention was focussed on the water content as a factor affecting the DC characteristic, and the effects of water absorbed from the outside and of the water generated due to the decomposition of cumyl-alcohol on the space charge characteristics were investigated.

2. METHOD OF EXPERIMENTS

2.1 Measuring Equipment

The measurement set up for the PEA method is shown in Fig. 1. Detailed description of the principles of measurement can be found elsewhere in the literature5), 6).

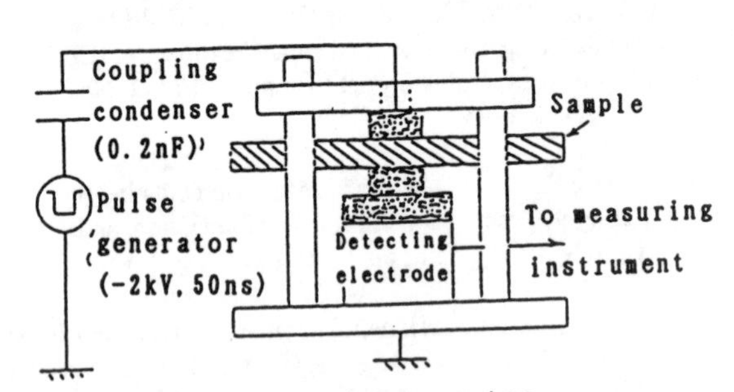

Fig. 1. Experimental apparatus

2.2 Test Samples

The experiments were conducted using the samples shown in Table 1 and Table 2 in order to investigate the relationship between the water content in the samples and the space charge characteristics.
The insulating material in all the samples was XLPE (base PE: LDPE, d=0.92). In the samples of Table 1(type A), the by-products of crosslinking agent(DCP) in XLPE are completely removed by drying for ten days at 50℃ and then the samples are immersed in a constant temperature water tank to cause water absorption. The water content

in the XLPE is controlled by the immersion water temperature , and hence these samples correspond to water absorption XLPE that does not contain any by-products of DCP decomposition. Next, in the samples of Table 2(type B), the water content is adjusted by the water from decomposition of cumyl-alcohol (CA) which is a by-product of DCP decomposition, by heat treating. These samples correspond to XLPE in which the by-products of DCP decomposition and water content are present simultaneously.

Each sample(for typeA and typeB)has the water content as shown in Table1 and 2. The XLPE sample shape with 2mm thickness is shown in Fig. 2.

In order to prevent the variation of water content during PEA measurement, XLPE sample (coated silicone oil on the surface) was directly inserted between metal(Al) electrodes under the pressure of $5kg/cm^2$.

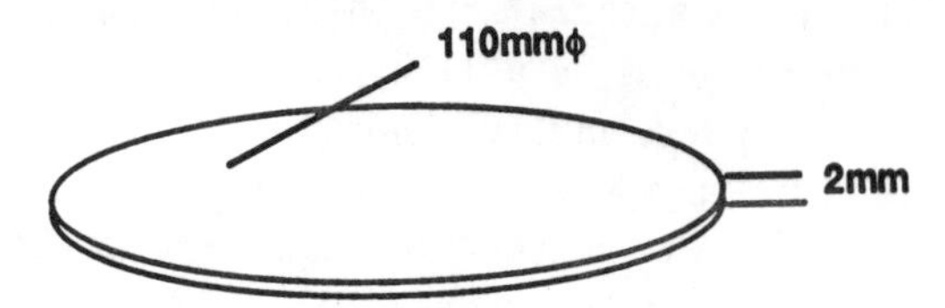

Fig.2 Sample Shape

Table 1 Water content of samples subjected to water absorption. (type A)

No	CONDITION	water content
1	Drying 50℃10days	25ppm
2	5days in water (70℃)	100ppm
3	5days in water (80℃)	220ppm
4	5days in water (90℃)	350ppm

Table 2 Water content of decomposition samples(type B)

No	CONDITION	water content
5	Heat-press only	20ppm
6	Heating 30min. (160℃)	60ppm
7	Heating 30min. (160℃)	75ppm
8	Heating 30min. (160℃)	95ppm
9	Heating 1hr. (160℃)	200ppm
10	Heating 1hr. (160℃)	240ppm
11	Heating 2hrs. (160℃)	330ppm
12	Heating 2hrs. (160℃)	580ppm

2.3 Space Charge Formation Conditions

The conditions of space charge formation are shown in Table 3 below.

Table 3 Space Charge Formation Condition

Electric field	30kV/mm
Temperature	RT
Time	3 hours

3. RESULTS AND DISCUSSION

The results of measuring the space charge distribution are shown in Fig. 3 for the type A samples and in Fig. 4 for the type B samples.

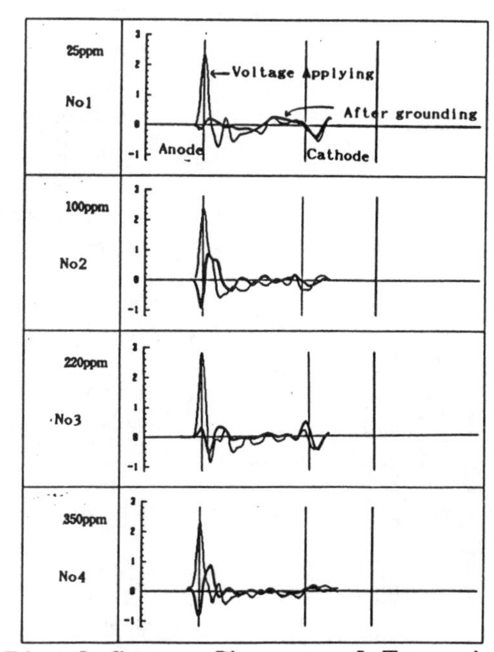

Fig.3 Space Charge of Type A

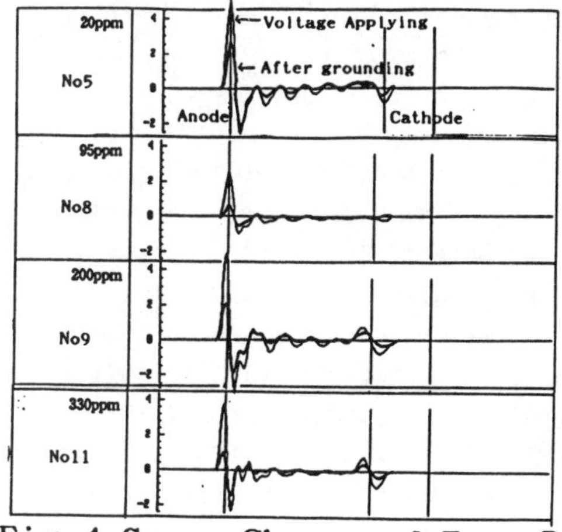

Fig.4 Space Charge of Type B

As for the effect of crosslinking by-products(acetophenone, cumylalcohol) on space charge formation in XLPE, it is said that the distribution changes from heterogeneousto homogeneous depending on whether by-products are present or not[7].

In the results of measurement during this study (shown in Fig. 3), while the space charge distribution is heterogeneous when voltage is being applied, the space charge distribution after grounding was homogeneous. In the case of homogeneous space charge distribution, it is difficult to separate the charge induced by the electrode and the charge accumulated in the sample in the signal when voltage is being applied, and hence the discussion of the space charge distribution in the samples in this study will be made based on the results of measurement after grounding, and the accumulated space charge after grounding in the vicinity of the electrode for each sample is shown in Table. 4.

Table 4 Amount of accumlated space charge in the vicnity of the electrodes

No	water content	Amount of space charge		type
		Anode	Cathode	
1	25ppm	0.036	-0.054	A
2	100ppm	0.163	0.041	
3	220ppm	-0.051	0.050	
4	350ppm	0.172	0.048	
5	20ppm	-0.298	0.030	B
6	60ppm	-0.112	0.043	
7	75ppm	-0.097	0.026	
8	95ppm	-0.090	-0.003	
9	200ppm	-0.265	0.086	
10	240ppm	-0.186	0.043	
11	330ppm	-0.112	0.054	
12	580ppm	-0.029	0.002	

Amount:nC/mm^2

Although no large difference is observed in the amount of accumulated charge depending on the method of adjusting the water content, while the space charge distribution shows that the homogeneous (+) charge accumulation near the anode is significant in the type A samples, accumulation of heterogeneous (-) charge is observed near the anode in the type B samples.

The relationship between the water content and the amount of accumulated charge near the electrode for these samples are shown in Fig. 5.

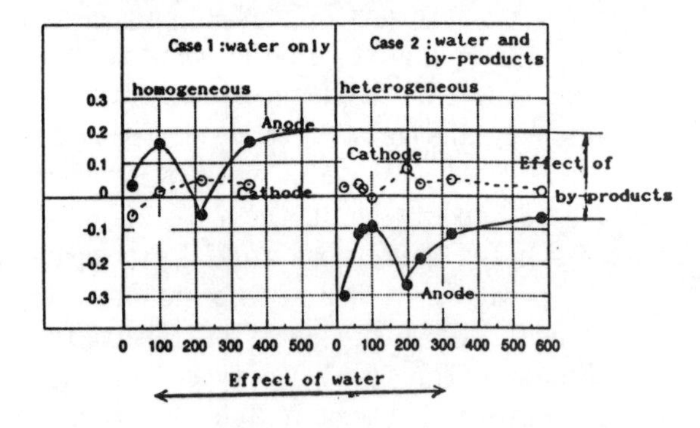

Fig.5 Change of space harge

From Fig. 5, it is clear that the space charge near the cathode is not affected much by the water content irrespective of the method of water absorption. Near the anode, a characteristic point is present near 100~200ppm in all the samples. When the type A samples and the type B samples are compared, difference is observed in the accumulated charge near the anode, and the heterogeneous (-) charge near the anode has apparently increased due to the presence of by-products. However, in our previous studies, the large heterogeneous space charge such as in XLPE is not observed even if reagent grade acetophenone and cumyl-alcohol are absorbed in LDPE[4], and hence this heterogeneous charge near the anode is considered due to the combined effects of the by-products and some other factors (such as crosslinking).

The effect of water content can be considered to be the same in both type A and type B when this heterogeneous charge is excluded.

The space charge characteristics at different amounts of water content are as follows.

- Water content of 100ppm or less

The hole injection from the anode increases. The homogeneous charge increases in type A samples, and the heterogeneous charge decreases in type B samples.

- Water content of 100 to 200ppm

Either the hole injection from the anode decreases or heterogeneous charge accumulates near the anode.
The space charge distribution changes slightly to heterogeneous in type A samples, and heterogeneous spacecharge increases in type B samples.

Water content of 100 to 200ppm means the solibility limit for XLPE (namly, water droplets appear in XLPE insulation above 100~200ppm), and the peculiar space charge phenomina seems to be the behavior associated with water solibility limit.

- Water content of more than 200ppm

The hole injection from the anode increases toward the saturation.
The homogeneous charge increases in type A samples, and the heterogeneous charge decreases in type B samples.

4. CONCLUSIONS

The relationship between the water content and the space charge characteristics was investigated using two types of samples (type A:containing only water,
type B:containing water and by-products) and the following results were obtained.

(1) Effect of by-products

The space charge distribution is homogeneous when there are no by-products (type A) and is heterogeneous when there are some by-products (type B). From this, it is considered that the movement of internalcarriers becomes easy in the presence of by-products.

(2) Effect of water

Hole injection gets enhanced due to increase in the water content. The reason for this is considered to be the presence of water ions (H30+).

However, near the water content level of 100~200ppm(which is the solibility limit for XLPE), peculiar space charge behavior were observed.

REFERENCES

1) Technical Reports of Institution of Electrical Engineers (Part 2) No. 194, p.69 (1986)

2) Technical Reports of Institution of Electrical Engineers (Part 2) No. 304, p.16 (1989)

3) Technical Reports of Institution of Electrical Engineers (Part 2) No. 237, p.7 (1987)

4) Takahashi, et al., JICABLE'95 p168

5) T. Takada, et al., IEEE Trans Electr. Insul., Vol. EI-18, No. 6 (1983)

6) Fukunaga, et al., Electrical Engineering Reports, 110, 647 (1990)

7) Lee, et al., 1991 Annual National Convention of the Institution of Electrical Engineers No. 236 (1991)

8) Moore's Physical Chemistry, Tokyo Chemical Association

Conference Record of the 1996 IEEE International Symposium on Electrical Insulation, Montreal, Quebec, Canada, June 16-19, 1996

Measurement of the trapping and detrapping properties of polymers in relation with their microstructure.

B. VALLAYER, P. HOURQUEBIE, D. MARSACQ
CEA, Le Ripault, 37260 MONTS, France
H. JANAH
Alcatel Cable, 536 quai de la Loire, 62225 CALAIS, France

Abstract : In the field of Space Charge Physics, the role of electrical traps on space charge behaviour and therefore on the breakdown properties has been now well-established. However, the traps in polymers are very difficult to define compared to the case of ceramics for which a lot of studies have been performed.
A new specific method for measuring the trapping and detrapping properties of dielectric materials has been developed. This method allows to characterize the electrostatic state of an insulating sample after irradiation by a high energy electron beam. We will discuss the basis of the method and its general possibilities to measure breakdown relevant parameters as the secondary electron yield for instance.
Moreover, the method has been used on several polymers as HDPE and LDPE. The difference of trapping properties between those materials can be explained by microstructure evolutions (crystallinity ratio) due to a difference of the branching rate.
This difference of trapping and detrapping properties of these two polymers could be connected to the breakdown behaviour of the two materials which is known to be very different.

INTRODUCTION.

In the field of electrical insulation, polyethylene plays a major role due to its great performances. However, polyethylene based devices still have to face reliability problems because a lot of fundamental mechanisms as electrical breakdown or ageing have not yet been well understood. Moreover, it is well known [1] that space charge accumulation and relaxation in polyethylene largely affect its dielectric properties. For instance, it is now well established that charge trapping change some physical constants of a dielectrical material such as the dielectrical constant [2] or the heat capacity [3]. Therefore, it is fundamental to know the electrical trapping properties of dielectrics, and to identify the traps where the electrons are located.
In the case of ceramics, where the nature of the traps is far more easy to understand, the trapping properties have been studied in order to point out some correlation with their breakdown behavior [4].
The necessity to understand the trapping properties of dielectrics has generated the development of several new experimental methods in order to measure the spatial distribution of charges in a given dielectrical sample [5]. The space charge is generally created by setting the sample between two electrodes and applying a high power electrical voltage across it. Then, the experiment consists in measuring the residual quantity of charge trapped in the sample. Therefore, if all of those methods are able to characterize the final state of a charging effect, it is always very difficult to study the overall trapping process and then determine the trapping properties of the materials.
In the same time, Le Gressus [8-9] has proposed an original method, called the Mirror Method, in order to measure the trapping properties of insulators. The principle is to use the electron beam of a Secondary Electron Microscope in order to inject and measure a given quantity of charge in the sample. Therefore, it is possible to characterize directly the trapping properties of the insulating material, with minimized troubles due to the environment, such as electrodes influence, atmosphere etc...
Despite of lot of works [10-11], the first correlations between some structural properties of polyethylene (molecular weight, crystallinity ratio, microstructure) and the dielectrical properties have been cleared only recently by Marsacq and al. [12]. This comes from the fact that these structural properties cannot vary independently and are often studied on polymers coming from different sources.
In this paper, we will show some changes in the trapping/detrapping properties of different materials, using the Mirror Method.

EXPERIMENTAL.

The Mirror Method.

The Mirror Method consists in injecting a given quantity Qi of negative electrical charge beneath the surface of a dielectrical sample and to measure the voltage distribution in vacuum, created by the residual trapped charge Qt.
The primary electrons Qi are injected in the sample with a sufficient energy (30kV) to allow the electron penetration (typical penetration depth is 15µm), without any damage in the material. Therefore a space charge builds up in the bulk of the sample and when the residual trapped charge is high enough, it deflects the low energy scanning beams of a SEM (from 200V to 3000V) and an image of the chamber is produced on the screen. These images can be analysed in

order to determine the cartography of the electrical potential in the vacuum. This cartography is represented by the curves 1/R as a function of the scanning potential Vs (figure 1) where R is the radius of the equipotential surfaces of potentiel Vs. From those curves, it is possible to determine two main parameters :

- Qt : the trapped quantity of charge after irradiation is given by the slope of the curve at the origin. Later, we will present the results refering to a trapping yield Π=Qt/Qi. Therefore, Π is representative of the trapping properties of the sample. In a first approach, Π and Δ (secondary electron emission yield) are linked by the following expression : $\Pi+\Delta=1$.

- Vd : the potential above which the mirror effect disappeared. Actually, as Vd depends on Qt, we will present the results refering to a detrapping ratio DR=Vd/Qt. This ratio is representative of the detrapping properties of the sample.

The experimental conditions are the same for all the presented mirror tests : Qi=30pC, and the primary beam is focused at the surface of the sample.

Materials.

Several different polymers have been studied in order to measure their trapping yield and detrapping ratio. The properties of these materials are reported on table 1. The first two polymers (HDPE and LDPE) came from the same source while the others (LDPE1-5) came from another. All the samples have been compression moulded. The crystallinity (Cryst%), the molecular weight ($\overline{M_w}$) have been measured in our laboratory. As the static dielectric constant is useful in order to calculate the trapping yield, it has also been measured at room temperature.

RESULTS AND DISCUSSION.

The figure 1 displays the curve 1/R=f(Vs) for the first two samples. The results shows that the HDPE has a lower trapping yield but a higher detrapping ratio than the LDPE. At the present time, it is interesting to analyse this behaviour in terms of crystallinity or molecular weight effect since these parameters are easily accessible.

Nevertheless, one fundamental point is that the detrapping ratio is representative of the detrapping properties of the sample. The detrapping of an electrical space charge have been recently recognized to be the driving force of electrical breakdown. Moreover, those interpretations should be consistent with the fact that the HDPE is known to have a higher breakdown strength than the LDPE [1].

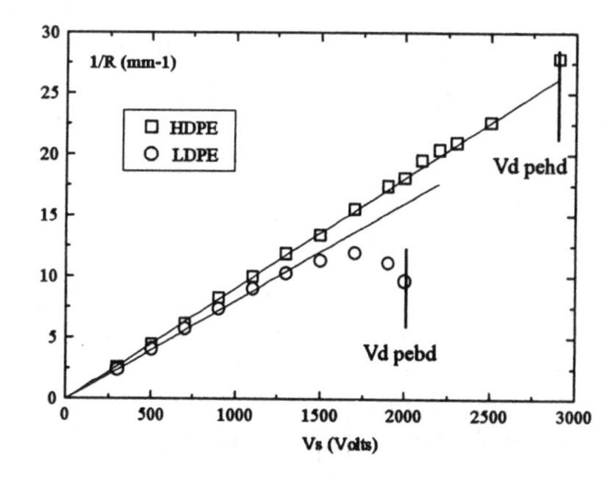

Figure 1 : Curve 1/R=f(Vs) for HDPE and LDPE. For the HDPE, Qt=22pC (Π=74%) and Vd=2900V (DR=134VpC^{-1}). For the LDPE, Qt=25,5pC (Π=85%) and Vd=2000V (DR=82VpC^{-1}).

Sample	Cryst%	$\overline{M_w}$
HDPE	70	391000
LDPE	40	148000
LDPE1	35	154000
LDPE2	25	111000
LDPE3	25	148000
LDPE4	12	164000
LDPE5	25	137000

Table 1 : Crystallinity and Molecular weight of the different samples.

Study of trapping properties.

We have plotted on the figure 2 the trapping yield of the 5 LDPE as a function of a) the crystalline ratio ; b) the molecular weight. Moreover, we have reported in table 2 the trapping yield measured on an atactic (0% crystallinity) and an isotactic (40% crystallinity) polystyrene of same molecular weight.

Actually, these results show that there is no influence of the molecular weight on the trapping properties of the 5 samples as the curve displayed on figure 2-b does not show any significant variation, especially in the case where crystallinity is identical for three different samples (25%). This behaviour is probably related to the range of variation of the $\overline{M_w}$ which is not wide enough in order to have a significant effect. Actually, as it is classical for physical properties of polymers,

the molecular weight effect has been shown to be important in the case of low $\overline{M_w}$ [12].
On the other hand, the effect of the crystallinity ratio on the trapping properties confirms the difference that has been pointed out by the HDPE/LDPE comparison : the higher is the crystallinity ratio, the lower is the trapping ratio.
This is definitly consistent with the results previously obtained [12] and those that are displayed in table 2, showing that, in the case of polystyrene, the same phenomenon is observed.
As a conclusion, these results indicate that the amorphous phase increases the trapping properties of polymers. Moreover, for a given crystalline ratio (25%), the figure 2-a) shows that the trapping properties are able to vary significantly, showing that the crystallinity ratio is not the determining parameter which exclusively controls the trapping properties. Other parameters should be taken into account (morphology, chemical and structural defects, etc...).

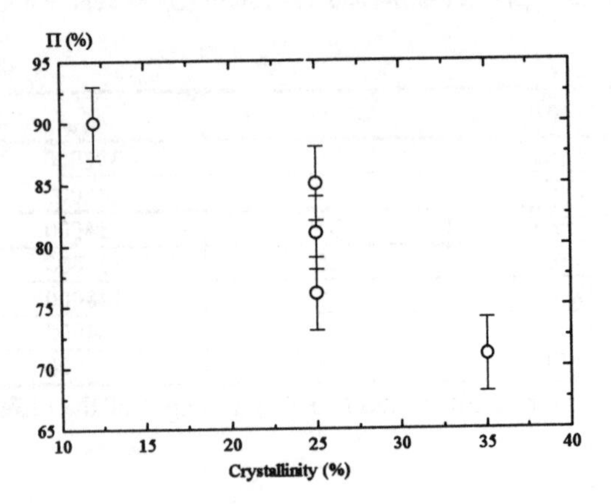

Figure 2-a) : Trapping yield as a function of the crystallinity.

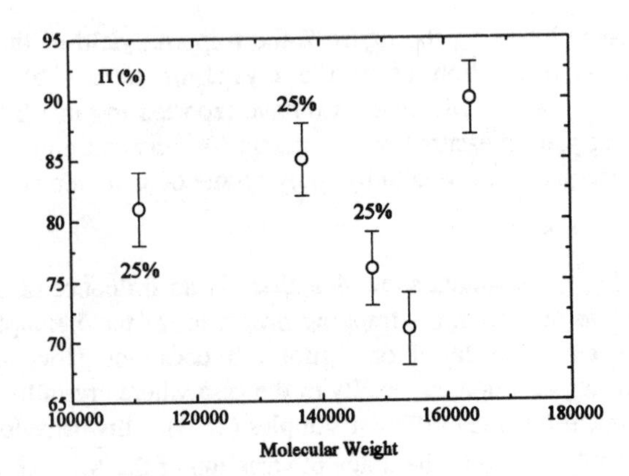

Figure 2-b) : Trapping yield as a function of the Molecular Weight.

	Crystalline Ratio (%)	Trapping Yield (%)
Atactic	0	64,5±3
Isotactic	40	49,5±3

Table 2 : Effect of tacticity on the trapping properties of polystyren.

Study of the detrapping properties.

The curve 3 displays the detrapping ratio of the 5 LDPE as a function of a) : the molecular weight; b) the crystallinity ratio. Unlike the results concerning the trapping properties, these curves show that there is no variation of the detrapping ratio as a function of the crystallinity, while the influence of the molecular weight on the detrapping properties is clear.

These results indicate that the trapping and the detrapping of a space charge (i.e. a single electrical charge) in a polymer does not depend on the same physical phenomenon.
Actually, Blaise has already investigated this fact and has developped a new model for the study of dielectrics, taking into account the polarization effect [13]. When an electrical charge is trapped into a dielectrical material, the polarization that occurs is changing the surrounding of this charge and the binding energy between the charge and the medium raises. In the case of ceramics, a lot of experimental evidences have been obtained showing for example that the critical detrapping temperature of a pure single-crystal of sapphire is higher than the critical trapping temperature [14].

Presently, a higher detrapping ratio indicates that the polarization is easier when the length of the polymer chain is small. This interpretation is in agreement with Deaning [11] who has already pointed out that the cohesive energy between chains increases with the $\overline{M_w}$. An increase of the cohesive energy should lead to a less polarizable material. In terms of traps energy, to a smaller chain length should correspond a higher trap energy in the sample after the polarization has settled down.

Moreover, it seems normal that the detrapping properties do not depend on the crystallinity ratio of the sample, as we have shown previously that the trapping of the electrical charges occurs in the amorphous phase. Therefore the detrapping of the electrical charges is depending on the intrinsic properties of the amorphous phase and is representative of the strength of the polarization surrounding the trapped charges.

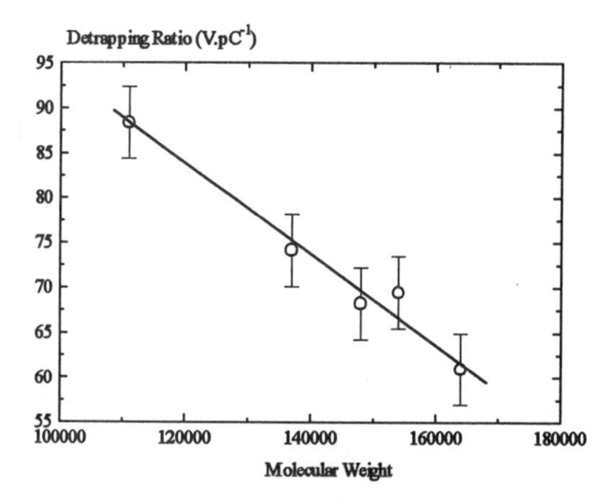

Figure 3-a) : Detrapping Ratio as a function of the Molecular Weight.

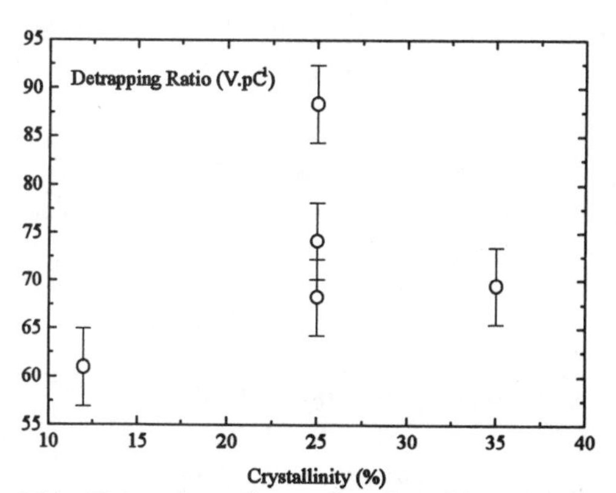

Figure 3-b) : Detrapping ratio as a function of the crystallinity.

CONCLUSION.

In this work, structural parameters (crystallinity, molecular weight) of polyethylene have been correlated to the trapping and detrapping behaviour by using mirror-measurements. First, by increasing the crystallinity ratio the charging capability is decreasing, showing that the trapping occurs in amorphous phase. These results have been shown on several types of polymers. Secondly, the detrapping properties are changing with the molecular weight as the detrapping ratio increases when $\overline{M_w}$ decreases. This result is due to the polarization that occurs in the samples after the trapping of the electrical charges. Moreover, significant changes have been pointed out for several polyethylene samples which are not so different. As a consequence, this work shows the absolute necessity to associate space charge behaviour of materials and their traps nature to a high level of structural characterization.

REFERENCES.

[1] R.Bartnikas
"Influence of space charge on the short and long term performance of solid dielectrics"
Supplément à la revue "le vide, les couches minces", N°275, pp 9 - 33, (1995)

[2] A. Bel Hadj Mohamed, K. Raouadi, C. Cheffi
"Variation of the a-SiO2 single-crystal complex permittivity as function of the trapped charge density"
Proceedings, CSC'2, Avril 1995, Antibes, France.

[3] D. Moya-Siesse, A. Sami, G. Moya
"Caracterisation par des méthodes calorimétriques de l'effet d'une charge d'espace dans les diélectriques solides"
Proceedings, CSC'2, Avril 1995, Antibes, France.

[4] B. Vallayer
"Klystron windows characterisation."
Proceedings CEIDP; IEEE annual report 1994, pp 3561

[5] A.S. De Reggi
"Experimental Methods for Dielectric Polarization Measurements."
Conférence Interdisciplinaire sur les Diélectriques; Supplément n°260; Le Vide, Les Couches Minces; Jan-Fev 1992.

[6] A. Toureille
"Space Charge characterization by coupling Thermal Step Method and Thermostimulated Currents."
Proceedings, CSC'2, Avril 1995, Antibes, France.

[7] H. St-Onge, R. Bartnikas, M. Braunovic, C.H. De Tourreil, M. Duval
"Research to determine the acceptable emergency operating temperatures for extruded dielectrics cables."
EPRI, Report EL-938, Project 933-1, Palo Alto, 1978.

[8] C. Le Gressus, F. Valin, H. Henriot, M. Gautier, J.P. Duraud, T.S. Sudarshan, R.G. Bommakanti, G. Blaise, J. Appl. Phys., 69, 6325, (May 1991).

[9] B. Vallayer, Thesis (95-14), Mai 1995, Ecole Centrale de Lyon, Ecully, France

[10] E.Kuffel, S.Grysybowski, P.Zubiellik
IEEE Transactions on Electrical Insulation, pp 200 - 203, (1980).

[11] R.D.Deaning
"Polymer Structure Properties and Applications"
CAHNERS BOOKS, (1972).

[12] D. Marsacq, P Hourquebie, L. Olmedo, H. Janah.
"Effect of structural parameters of polyethylene on space charge properties"
Proceedings, CSC'2, Avril 1995, Antibes, France.

[13] G. Blaise
"Microscopic and Macroscopic description of the polarization of dielectric materials. The concept of the polaron applied to the conduction and trapping of charges."
Conférence Interdisciplinaire sur les Diélectriques; Supplément n°260; Le Vide, Les Couches Minces; Jan-Fev 1992.

[14] J. Bigarre, C. Rambaut, S. Fayeulle.
"Effet de la température sur le piégeage et le dépiégeage de charge"
Proceedings, CSC'2, Avril 1995, Antibes, France.

Conference Record of the 1996 IEEE International Symposium on Electrical Insulation, Montreal, Quebec, Canada, June 16-19, 1996

The Effect of Space Charges on Conduction Current in Polymer by Modified PEA Method

Seung Hwang-bo, Do-Hong Yun, Dong-Young Yi and Min-Koo Han

Department of Electrical Engineering, Seoul National University, Seoul, Korea

Abstract : **Direct measurement of space charge and conduction current was carried out on low-density polyethylene degraded by ultra-violet using a pulsed electro-acoustic (PEA)method. Dominant hetero-charges were formed near both electrodes by high voltage application and was found to be deeply trapped. In this paper, the effect of temperature and electric field reversal on the detrapping and trapping of space charges was investigated and the role of space charge in electrical conduction was discussed quantitatively. The main mechanism for detrapping and trapping of space charges was Poole-Frenkel model.**

INTRODUCTION

There have been a large number of experimental and theoretical works on electrical conduction and breakdown of polymers, specially polyethylene(PE)[1]. However the physical mechanisms have been not fully understood yet because of the effect of space charge that distorts the interface field on metal-polymer contact and the internal field in polymer. Only a few studies have been done on the change of the conduction mechanisms and the detrapping of space charges, which had been formed previously, based on the quantitative work. The reason is the difficulty in the direct measurement of space charge distribution and conduction current simultaneously.

The probing of space charge distribution in polymer has been usually done by the pressure wave propagation(PWP) method or the pulsed electro-acoustic (PEA) method. The PEA method that was proposed by Takada (1986) utilizes non-uniform pressure wave of uncompensated internal charges in a specimen induced by Lorentz force under short electrical pulse.

In our work, the PEA which was modified to measure a conduction current simultaneously by using a guard in low electrode technique was employed for probing the space charge distribution in polymer. We successfully improved the spatial resolution by shortening the duration time of electric pulse less than 10 ns and increasing the amplifier gain. The distortion in output voltage signals due to impedance-mismathing on a specimen was corrected by deconvolution process.

In this paper, the effect of temperature and electric field reversal on the detrapping and trapping of space charges was investigated in detail and the role of space charge in electrical conduction mechanisms is discussed quantitatively.

EXPERIMENTAL

The specimen was 500μm thick film of low-density PE degraded by Ultra-Violet(UV) for several days at room temperature to increase the density of trap sites which is well known to carbonyl groups due to oxidation.

The experimental set up is shown in figure 1. A DC voltage was applied across a specimen with Al metal electrodes located on both sides of the specimen and the diameter of lower electrode was 3 cm. The gap between guard and lower electrode was 2 mm. The height and width of the applied electric pulse were 1.5kV, 10ns respectively and a feedthrough terminal of 50Ω was used for suppressing the reflection of input electric pulse. A 28μm PVDF film and a 1mm thick PVDF rubber were used as a detecting sensor and a absorber for converting pressure wave to voltage signal respectively. The output voltage signal, which frequency bandwidth was less than 50MHz due to conversion by piezo-material (PVDF), was amplified with BX-31(bandwidth : 0 ~ 75MHz, Gain:10) and HP8447F(bandwidth : 0.1 ~ 1300MHz, Gain : 47dB) and recorded with a digital storage oscilloscope (HP54100A). For improving S/N ratio of signals, the averaging of data and the filtering and deconvolution by S/W were performed.

The current-time characteristic was measured with sampling rate of 0.6 sec with a electrometer(Keithley 617). The charge and discharge current were measured after elapse of enough time, about more than 30 minutes, to protect the effect of previous test. The measurement for space charge and conduction current was carried out alternatively in oven and the temperature of test cell was measured by a thermocouple inserted in it.

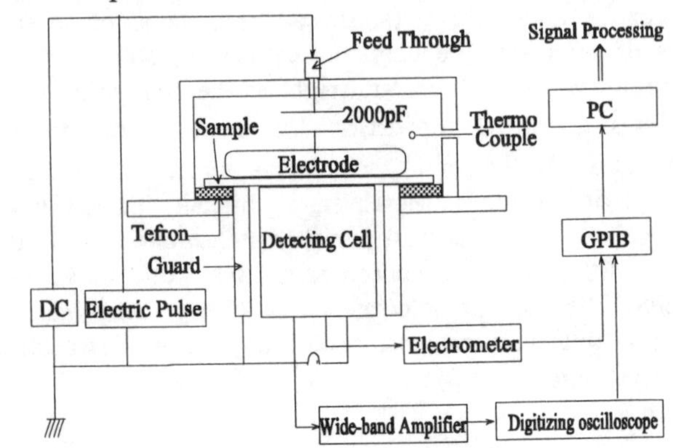

Figure 1. Schematic diagram of experimental set-up for measuring the space charge and conduction current

Since the sensitivity of the PVDF film is changed due to temperature, i.e. the peak height and acoustic velocity of output signals was reduced, the signals were calibrated by comparing with those measured at 20℃(RT).

RESULTS and DISCUSSION

Current and Space Charge under Applied voltage

Figure 2 shows the space charge distribution and current-time characteristics of degraded LDPE under applied electric field of 36 kV/mm at RT, respectively. Figure 2(a) represents the space charge distribution measured per a 20 minute for 200 minutes under voltage application. The difference between negative and positive peaks in profile was induced by the different transmitting path of acoustic waves generated both electrodes. From this figure and the result after discharging(not shown here), we found that the amounts of hetero-space charges formed near both electrodes were almost alike and were increased linearly with time. While it is well known that the homo-charges are formed in pure LDPE and the amount of charges is relatively less than those in this degraded LDPE by UV. One reason for this discrepancy is the increasing of the density of trap sites at bulk that are donor-like or acceptor-like defects generated by UV.

Each points in figure 2(b) represents the mean values of conduction currents measured per a 5 minute under voltage application, respectively. The conduction currents were increased with time as likely as space charge. The major reason for this phenomena is the field enhancement near electrodes due to the increment of hetero-charges rather than that of mobility with time and we can verify this in figure 3, i.e., the fields near both electrodes are enhanced but in bulk reduced with time. Therefore we can conclude that the drift velocity is nearly independent of the applied field due to very low-mobility in degtaded LDPE.

Figure 3 shows the electric field obtained by Poisson's equation (1) in specimen. Although it includes some errors induced by different acoustic wave path, it is not dominant for the trend of profile because it is relative.

$$\frac{\partial E(x)}{\partial x} = \frac{\rho(x)}{\varepsilon} \tag{1}$$

where $\vec{E}$, ρ and ε represent electric field, charge density and the permittivity of specimen, respectively.

The total current density(J) in polymer is made up of the three components of drift, diffusion and displacement and described as follows.

$$j = qn\mu E + qD_q \frac{\partial n}{\partial x} + \varepsilon \frac{\partial E}{\partial t} \tag{2}$$

where D_q, n and μ are the diffusion coefficient , density and mobility for charge carriers. Here, assuming steady-state conditions then $dE/dt=0$ and neglecting the diffusion current, which is justified in the case of low-mobility solids[2], this becomes approximately:

$$j = qn\mu E \tag{3}$$

But, if the homo-charges near the electrodes is formed, the diffusion term must be considerd because it act mainly as potential barrier and has the effect on diffusion, i.e. the filed will be reduced there and the ratio of diffusion to drift current will be enhanced.

The charge density, that is n in Eq. (3), is largely made up of injected charges(n_{inj}) from electrodes and generated ones(n_{gen}) from defects or trap sites in bulk of specimen, i.e. $n = n_{inj} + n_{gen}$. Therfore, the major origin of hetero-charge formation near both electrodes can be explained as two mechanisms. One is the detrapping(in case carrier is electron) or trapping(in case carrier is hole) of charges generated from donor-like or acceptor-like defects and the other is the trapping of injected carriers by the opposite electrode.

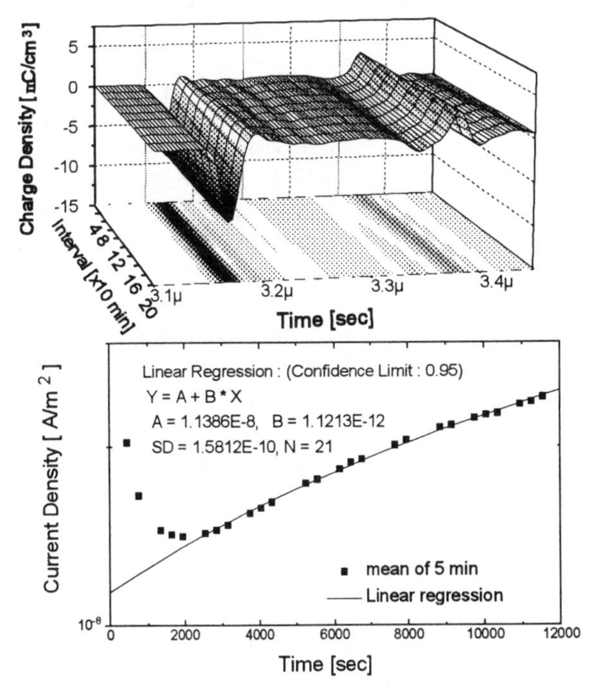

Figure 2. The current-time characteristics(a;upper) and space charge distribution(b;lower) of degraded LDPE under 36 kV/mm.

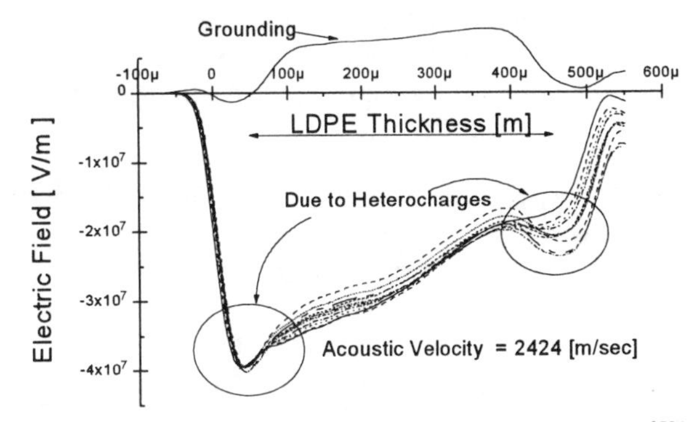

Figure 3. The Field distribution of degraded LDPE under 36 kV/mm

If the barrier heights at both electrodes are almost identical and one carrier is dominant, the formation of hetero-charges by injected carriers from the opposite electrode is difficult in the same manner as likely as at the injecting electrode. However, once hetero-charges are formed, it is possible because hetero-charges act as trap sites or potential barrier. In former case, the relation of $j_{inj} < j_{drift}$ should be satisfied with the formation of hetero-charges.

Current and Space Charge under Reverse Voltage

In the previous section, the hetero-charges had been formed at 36 kV/mm. Here, in order to investigate the effects of space charges on the steady-state current and detrapping of space charges in specimen, we applied positive(same polarity) and negative(reverse one) voltages up to 18 kV/mm again and measured the space charge distribution and current at RT simultaneously. The negative currents were converted to positive ones for comparison.

Figure 4 shows the space charge distribution under applied voltage for 15 minutes and after discharging. From the results after discharging, we found that the space charge distributions were nearly unchanged in both cases. The reason for this is that the trapping or detrapping of space charges did not happened. Accordingly, we could know that the potential barrier at metal-polymer interface is high and the injection current was not dominant in this range.

Figure 5 represents the Log-log plots(log j vs. E) of steady-state current density-field in the same manner as described above. Each point is a mean value for 1 minute after voltage is applied for 15 minutes. The current density measured under negative voltage(homo-charge) is higher than that at positive one(hetero-charge). Therefore we may conclude that the steady-state current density depends on the field in bulk rather than that on the interface of polymer-metal.

The slopes of the curves fitting the negative and positive case are about 1 and 1.8~3 respectively and then this phenomena can be explained that the hetero-charges near both electrodes act as potential barrier at low field. While at high field they enhance the carrier injection rate from electrodes and act as trap sites for injected carriers if mobility is very low and for drifted carriers across the specimen if the mobility is high. The sudden drop of current at high field is considered due to the trapping of injected carriers, however there is no evidence for this.

Current and Space Charge with Variation of Temperature and Time

In order to verify the effect of temperature and time on the steady-state conductivity-field and the detrapping of space charges from trapped sites, we have also performed the experiments at 40 and 60℃. Figure 6 and 7 represent the space charge-field(a) and the typical Poole-Frenkel plots(log σ vs. $E^{1/2}$) of conductivity-field(b) in degraded LDPE.

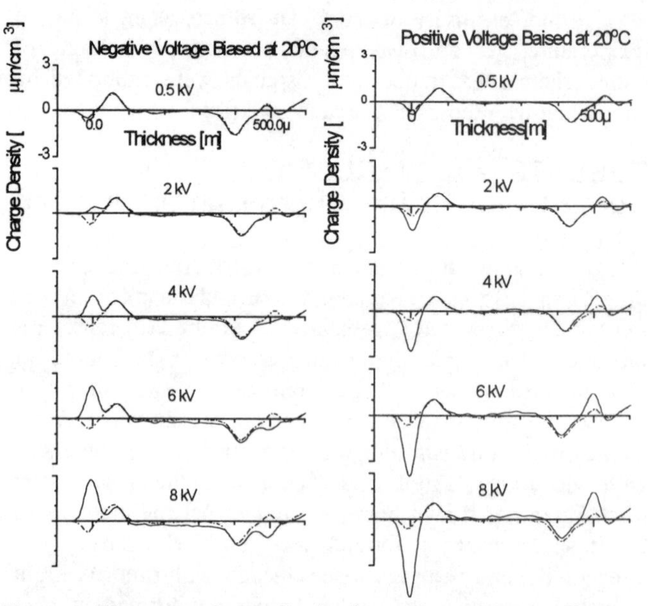

Figure 4. The space charge distribution under voltage application and after discharging in degraded LDPE which hetero-charge had been formed already.(a;left) Reverse voltage (b:right) the sampe polarity of 36kV/mm

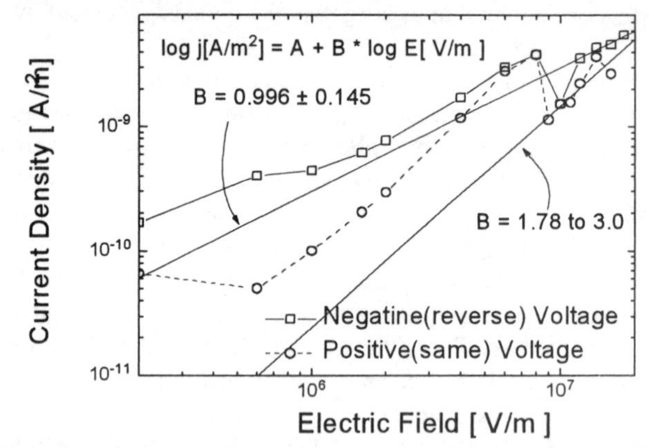

Figure 5. Log-log plot of steady state current density-field in degraded LDPE which was measured with space charge distribution shown in figure 4.

The Poole-Frenkel conduction is a bulk-limited process caused by thermal excitation of the carriers over a field-lowered trap barrier surrounding trapped or donor sites. The conductivity in this case is given by[3],

$$\sigma = \sigma_0 \exp[(\beta_{PF} E^{1/2} - W_t) / kT] \qquad (4)$$

where $\sigma = \sigma_0 \exp[-W_t / kT]$ is the low field conductivity, W_t is the trap barrier, and $\beta_{PF} = (e^3 / \pi\varepsilon_r\varepsilon_0)^{1/2}$ is the Poole-Frenkel coefficient.

From (a) of figure 6, which represents the change of space charge distribution with increasing reverse electric field at 40℃, the reduction of space charges was found at high field. This is due to the detrapping of trapped space charges and agrees well with the result of conductivity data in (b) of figure 6, i.e. the steady-state conductivity was being increased above 3.2×10^3 [$V^{1/2}/m^{1/2}$]. The slope of linear fitting curve in this range gives a relative permittivity ε_r = 30.4. Compared with the accepted value of 2.3 for LDPE and a three dimensional treatment of Poole-Frenkel effect, which yields $2kT$ in denominator of the argument of the exponential term in Equation (4), it is too high. Disagreement is due to the inclusion of low field-effect, the low density and locality of trapped charges and the measuring time, i.e. the slope will be more sharper with elapse of time because the reduction rate at low field is very higher than that at high field.

Figure 7 shows the experimental results carried out at 60℃. A large reverse(hetero-charge) space charges was formed with increasing reverse electric field. This is due to the trapping of generated carriers near the opposite electrode. Also, the conductivity was enhanced with time at high field because of the increment of the interface field on metal-polymer. Accordingly, it requires enough time to fit the slope of the Poole-Frenkel plot exactly. These phenomena coincide with the result of figure 2(b) except that it is happened at low electric field.

The slope of the fitting curve in figure 7(b) yields a relative permittivity ε_r = 4.54. Considering the 3-dimensional treatment of detrapping, that is $kT \Rightarrow 2kT$, this value is good agreement with 2.3. Therefore, we can conclude that the Poole-Frenkel is a proper model for the trapping and detrapping of space charges and as conduction mechanism in degraded LDPE by UV.

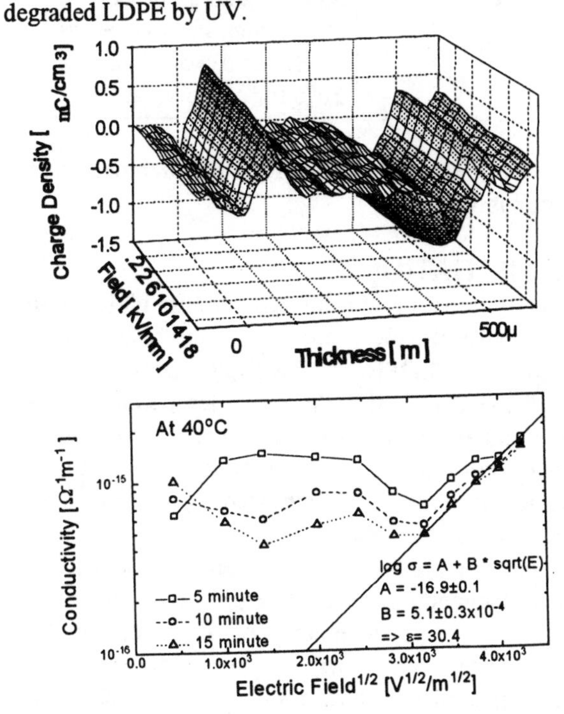

Figure 6. The space charge distribution after discharging(a;upper) and typical Poole-Frenkel plot(b;lower) of steady-state conductivity -field in degraded LDPE at 40℃.

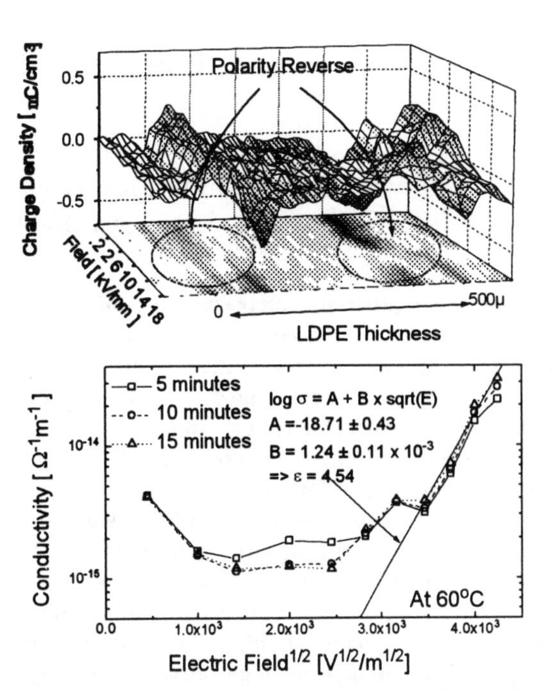

Figure 7. The space charge distribution after discharging(a;upper) and typical Poole-Frenkel plot(b;lower) of steady-state conductivity -field in degraded LDPE at 60℃.

CONCLUSION

The effect of temperature and electric field reversal on the detrapping and trapping of space charges in degraded LDPE by UV was discussed quantitatively through the direct measurement of the space charge distribution and conduction current by PEA method. The conduction can be explained as follows; (1) The trapping and detrapping of space charges can be explained from the Poole-Frenkel model and the effect of Poole-lowed trap depth is enhanced with increasing temperature. (2) The Schottky injection is not dominant in the trapping and detrapping of space charges. (3) Current increases with time because of the enhancement of interface field on metal-polymer due to hetero-charges and then it requires enough time to fit the Poole-Frenkel plot exactly.

ACKNOWLEDGMENT

This work was supported by the Korea Electric Power Corporation.

REFERENCES

1. Suzuoki. Y, Muto. H, Mizutani.T and Ieda. M "Effect of space charge on electrical conduction in high-density polyethylene ". J.Phys.D Appl. Phys. 18.,pp.2293-2302.1985.
2. Nespurek S. And Sworakowski J. "A Differential Method of Analysis of Steady-State Space-Charge-Limited Current-Voltage Characteristics ". Phys. Stat. Sol. (A) 41,pp.619-627, 1977.
3. Nath R., Kaura T. Amd Perlman M.M. "Steady-state Conduction in Linear Low-density Polyethylene with Poole-lowerd Trap Depth". IEEE Trans. Electr. Insul., Vol. 25, pp.419-425,1990.

Conference Record of the 1996 IEEE International Symposium on Electrical Insulation, Montreal, Quebec, Canada, June 16-19, 1996

Electrical Tree Initiation in Polyethylene Absorbing Penning Gas

N. Shimizu, N. Tohyama and H. Sato
Department of Electrical Engineering, Nagoya University
Nagoya, 464-01, Japan

Abstract: **Ac tree initiation voltage was examined in untreated LDPE, vacuum degassed LDPE and LDPE absorbing He gas (He gas was absorbed after vacuum degassing). We have already reported that vacuum degassed LDPE shows much higher tree initiation voltage than untreated one because of absence of oxygen. Therefore we expected that LDPE absorbing He shows the same property with vacuum degassed LDPE. However tree initiation voltage of LDPE absorbing He is as low as that of untreated LDPE. LDPE absorbing Ar gas shows the same tendency. He or Ar gas does not change so much impulse tree initiation voltage. LDPE absorbing He was not well dyed with methylene blue after ac voltage application, which indicates that active oxidation does not occur. Low ac tree initiation voltage in LDPE absorbing He or Ar may be caused by Penning ionization in free volume.**

INTRODUCTION

We have studied extruded solid polymer insulation for superconducting power transmission cable [1]. Recently, we examined tree initiation properties in polyethylene absorbing helium (He) gas taking account of He leakage accident [2]. As the result, we found that He gas decreases ac tree initiation voltage. Argon (Ar) gas also shows even larger effect. In this paper we will report the detailed results and discuss the effect of Penning ionization on tree initiation.

EXPERIMENTAL

Tree testing specimen, a low density polyethylene (LDPE, 0.918 g/cm^3) block to which a needle electrode (tip radius 3 μm) was inserted, was used. The specimens were subjected to several different treatments described in Table 1 before voltage application. In ac experiments, constant ac voltage was applied to the needle electrode in silicone oil (at room temperature) or in liquid nitrogen (at 77 K), and time to tree initiation was measured. Tree initiation was detected by partial discharge (>0.5 pC) and/or observation with optical microscope. In impulse experiments, only one shot of 1.2/50μs impulse voltage was applied to one specimen to avoid space charge accumulation. 15 specimens were used for each condition. Voltages corresponding to 20, 50, 80 % cumulative tree initiation probability were obtained from Weibull probability paper.

Table 1 Specimens and Treatments

Specimen	Treatment
Untreated Specimen	untreated
Degassed Specimen	Degassed by vacuum pump for over 90 h at room temperature.
He Specimen (Ar Specimen)	After degassing process above, kept in He (or Ar) gas (purity >99.9999%) for over 72h at 75 C
N_2 Specimen	After degassing process above, kept in N_2 gas (purity >99.9995%) for over 72h at 75 C
He/air Specimen (Ar/air Specimen)	After degassing process above, kept in mixture gas of He (or Ar) 50% and air 50% for over 40h at 50 C

AC TREE INITIATION

Time to tree initiation at room temperature is shown in Fig. 1. We have already reported that the resistivity to tree initiation of LDPE is drastically improved by removal of oxygen [3,4]. At room temperature, tree initiation voltage of degassed specimen from which oxygen was removed by vacuum pumping is actually much higher than that of untreated specimen. N_2 specimen also shows the same characteristics with degassed specimen. However, He and Ar specimens, which do not include oxygen either, show much lower tree initiation voltage than that of degassed or N_2 specimen. Ar specimen shows even lower value than that of untreated specimen. These results suggest that He or Ar gas in LDPE decreases the resistivity to ac tree initiation.

In He/air or Ar/air specimen, which absorb both oxygen and He or Ar, show lower tree initiation voltage

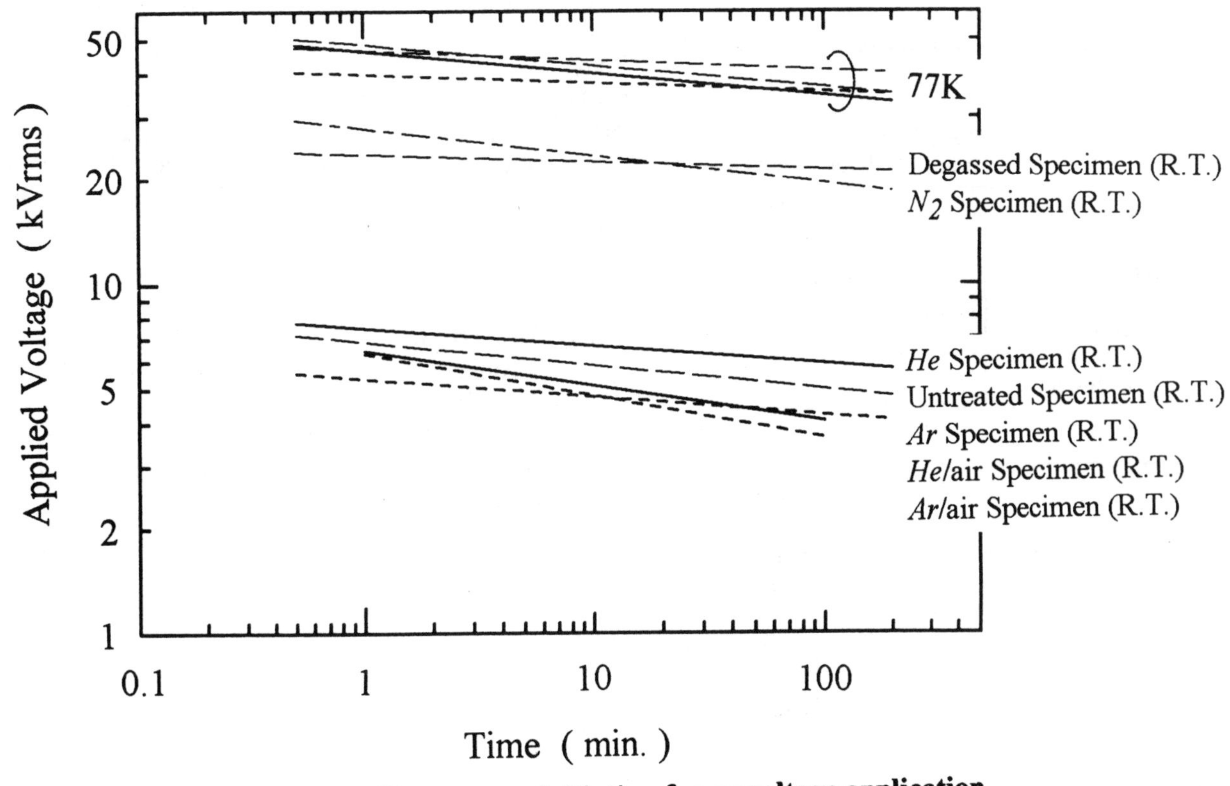

Fig. 1 Time to tree initiation for ac voltage application.

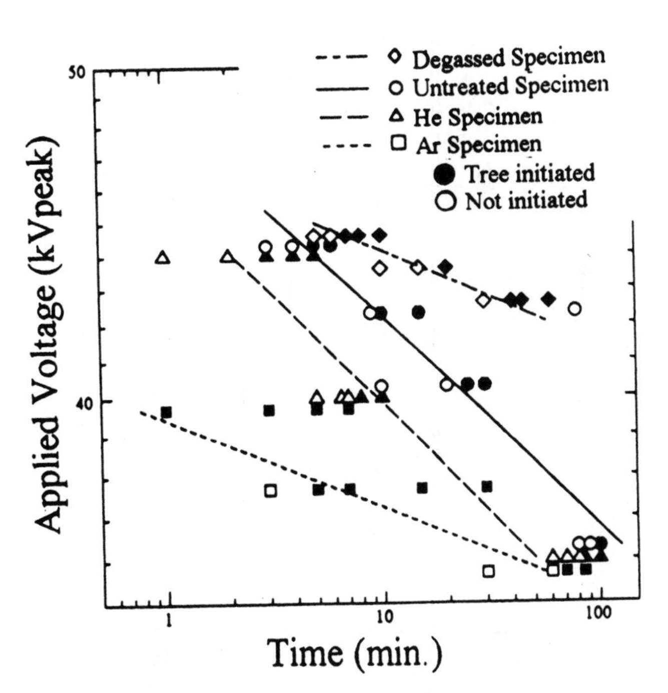

Fig. 2 Time to tree initiation at 77 K.

than that of He or Ar specimen. Especially, the difference between He and He/air specimens is clear.

At 77 K, the difference in tree initiation voltage among specimens are small compared to that at room temperature. However the detailed data in Fig. 2 show that tree initiation voltage decreases in the order of degassed specimen > untreated specimen> He specimen> Ar specimen. Here, untreated specimen at 77 K shows much higher tree initiation voltage than at room temperature. Following things can be considered [5]; at low temperature 1) oxygen gas is frozen and autoxidation process does not occur, 2) field relaxation by homo space charge may be enhanced because of low diffusion and 3) recombination of broken polymer chain may takes place resulting in increasing fundamental electrical strength.

OXIDATION PRODUCTS

Degassed and He specimens are not well dyed with methylene blue after ac voltage application, while untreated specimen well dyed. Table 2 shows dyeing

Table 2 Dyeing Ratio

(3 KVrms was applied for 3 hours at room temperature)

specimen	number of dyed specimens / number of specimens tested
Untreated Specimen	9 / 9
Degassed Specimen	0 / 9
He Specimen	0 / 9
He/air Specimen	3 / 6
Ar/air Specimen	4 / 4

ratio (number of dyed specimens divided by number of specimens tested) in specimens just before tree initiation. Being dyed with methylene blue indicates existing of oxidation products [6.7], so that the above results reveal that active oxidation does not occur in degassed or He specimen, but it does in untreated specimen. In mixture gas specimens, Ar/air specimen shows the same feature with untreated, but He/air specimen shows less active oxidation.

IMPULSE TREE

Fig. 3 compares impulse tree initiation voltage among specimens. He and Ar specimens show a little lower value in negative polarity at room temperature. However the difference is small compared to the results by ac voltage in Fig. 1.

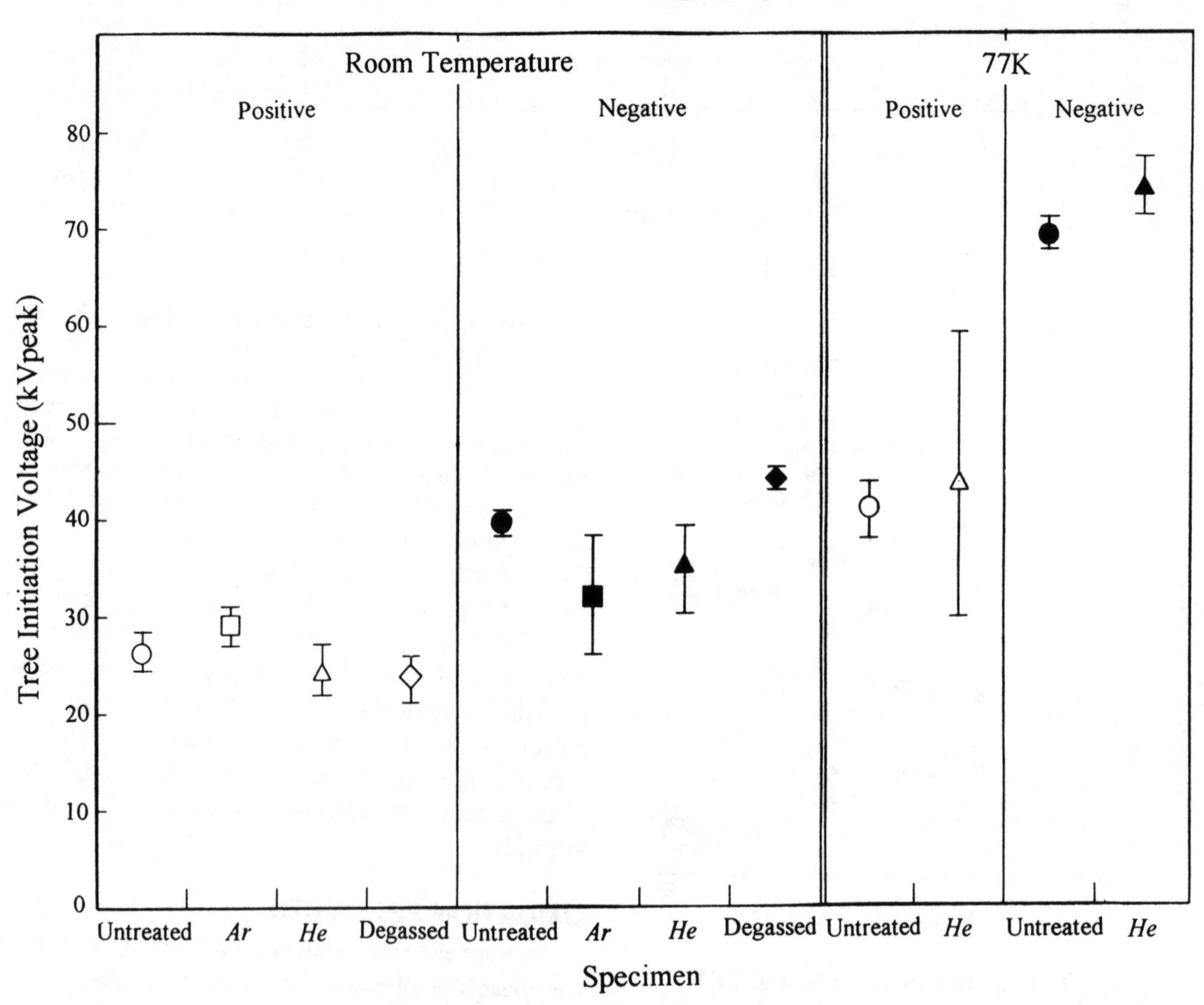

Fig. 3 Impulse tree initiation voltage.

DISCUSSION

We have already made clear that the degassed LDPE shows much higher ac tree initiation voltage than untreated one dose because of absence of oxygen [3,4]. Therefore we expected that He or Ar specimen shows the same property with degassed one. However the experimental results were not so.

He specimen is not well dyed with methylene blue after ac voltage application. This result is the same with that in degassed specimen, and indicates that autoxidation process does not occur in He specimen. We must consider peculiar process to He or Ar gas to explain ac tree initiation voltage as low as, or even lower than, that of untreated specimen.

Penning ionization process is a candidate to explain the present results. Fig. 4 shows the tree initiation mechanism in LDPE absorbing He or Ar gas. He (or Ar) molecules are exited to metastable level by electron bombardment or UV in free volume. These metastable molecules collide with the wall of free volume and cut the polymer chain, and then return to ground state. Ar specimen shows lower ac tree initiation voltage than that of He specimen. This is perhaps because that metastable energy level of Ar (≅ 11.5 eV) is closer to bonding energy of polyethylene molecules (3 ~ 4 eV) than metastable energy level of He (≅ 20 eV).

In this model, electro-chemical process in free volume and on it's wall is important in tree initiation. Here we do not consider electric discharge is the main process, because tree initiation voltage of Ar specimen is lower than that of He specimen although breakdown strength of Ar gas is higher than that of He gas.

At 77 K, bond scission of polymer chain is virtually difficult to occur because of bond recombination and excitation of He or Ar molecule is perhaps less frequent than at room temperature. These may explain the small difference in ac tree initiation voltage among specimens.

In He/air or Ar/air specimen, which absorbs both Penning gas and oxygen, both processes of Penning ionization and autoxidation take place and make ac tree initiation voltage lower.

As to impulse tree, duration period of impulse voltage is < 100 μs, whereas Penning ionization process requires several milliseconds to be completed [8]. Therefore, it is reasonable that impulse tree initiation voltage of LDPE absorbing He or Ar gas is close to that of untreated LDPE.

CONCLUSIONS

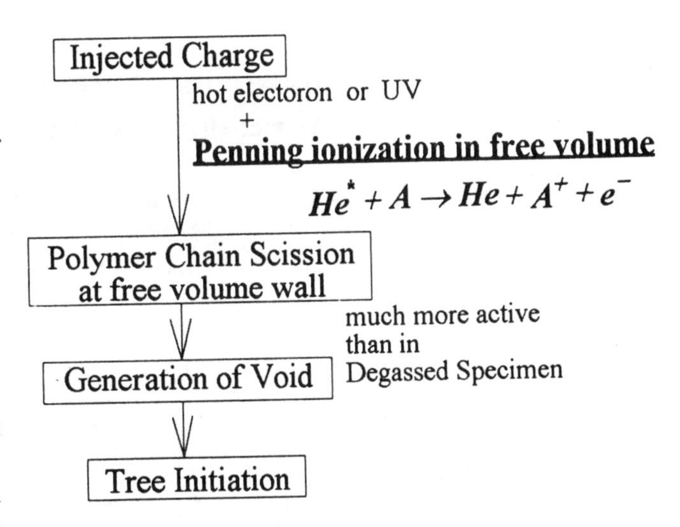

Fig. 4 Ac tree initiation mechanism in LDPE absorbing Penning gas.

1) Ac tree initiation voltage at room temperature of LDPE absorbing He or Ar gas is much lower than that of degassed LDPE although free from autoxidation process.
2) Ac tree initiation voltage of LDPE absorbing He or Ar gas at 77K is a little lower than that of untreated LDPE.
3) LDPE absorbing He gas shows little trace of oxidation after voltage application as well as degassed LDPE, while active oxidation is observed in untreated LDPE.
4) Impulse tree initiation voltage of LDPE absorbing He or Ar gas is not so different from that of untreated or degassed LDPE.
5) The low ac tree initiation voltage in LDPE absorbing He or Ar gas may be explained by Penning ionization process.

REFERENCES

1. Kosaki, M , M.Nagao, Y.Mizuno ,N.Shimizu and K.Horii "Development of Extruded Polymer Insulated Superconducting Cable". Cryogenics, Vol.32, pp.885-894, 1992
2. Tohyama, N , Y.Tateno, H.Sato and N.Shimizu "Tree Initiation in Polyethylene Absorbing Helium Gas" Proc. 1995 Int. Symp. Electrical Insulating Materials pp.455-458, 1995
3. Shimizu, N and K.Horii "The Effect of Absorbed Oxygen on Electrical Treeing in Polymers" IEEE Trans. EI Vol.20, pp.561-566, 1985
4. Shimizu, N "Role of Oxygen in Electrical Treeing" Proc. 1993 Int. Workshop on Elec. Insulation pp.11-16, 1993
5. Uchida, K and N.Shimizu "The Effect of Temperature and Voltage on Polymer Chain Scission in High-field Region" IEEE Trans. EI Vol.26, pp.271-277, 1991
6. Trotman, E.R. *"Dyeing and Chemical Technology of Textile Fiber"* Charles Giffin and Co. England p.52, 1970
7. Nunes, S.L "Dyeing Terminology" IEEE Trans. EI Vol.18, pp.556, 1983
8. von Engel, A *"Ionization Gases"*, Oxford, 1965

Conference Record of the 1996 IEEE International Symposium on Electrical Insulation, Montreal, Quebec, Canada, June 16-19, 1996

Time Resolved Measurements of Electroluminescence in XLPE under Impulse Voltage Conditions

M. Kaufhold, S.S. Bamji and A.T. Bulinski
National Research Council Canada
Ottawa, Ontario K1A 0R6, Canada

Abstract: **Time resolved electroluminescence (EL) measurements were performed on high voltage grade XLPE cable insulation using negative impulse voltages of different shape and duration. To create divergent fields in the insulation, needle electrodes made of semiconductive polymer were used. It is shown that the EL emission probability is independent of the impulse rise time but increases with the impulse fall time and amplitude. Independently of the impulse shape and duration, EL pulses are always grouped into two distinct time intervals between which there is no EL activity. The first EL period begins at the rising portion of the impulse and the second at the falling portion when the impulse voltage drops below 50% of its peak value. EL emission during the first interval is believed to be caused by electron injection and the impact excitation processes, whereas the EL in the second interval is seen to be due to the local electric field reversal and the electron-hole recombination.**

INTRODUCTION

Prior to partial discharge (PD) inception and electrical treeing, insulating polymers subjected to high electric fields usually emit electroluminescence (EL). Crosslinked polyethylene (XLPE) is widely used as the insulation material in underground high voltage distribution as well as in transmission class cables which operate at voltages up to 500 kV. Most of the studies on EL in polyethylene and XLPE have been performed with ac [1, 2] or half-rectified ac [2] voltage. However, an underground high voltage cable is not only subjected to ac voltage but also to electrical impulse stress due to switching and lightning surges. Also, prior to being commissioned into service, as well as during service, the cables are tested with impulse voltages in order to determine their quality and predict their reliability in service [3, 4]; yet, very few EL studies have been performed with impulse voltage [5, 6].

It is imperative to develop impulse tests which can ensure that a particular cable insulation will perform satisfactorily during its service lifetime and which does not cause any adverse effects. This paper describes the behaviour of EL emission in cable grade XLPE subjected to negative high voltage impulses of various amplitudes and shapes. Time resolved measurements of EL under impulse voltage could determine the duration and the effects of various time dependent mechanisms such as charge injection, trapping and decay within the polymer, and could be used to understand how these mechanisms relate to insulation failure.

EXPERIMENTAL

Sample Preparation

XLPE slabs (18×9×4 mm) each having a semiconductor protrusion were used in this study. Semiconductor sheets, 200 μm thick, were first partially crosslinked and obliquely cut to produce protrusions having a tip radius of ~2 μm. These were then sandwiched between 2 mm thick pre-moulded polyethylene slabs. The samples were crosslinked at 170°C for 30 min and each sample had a semicon tip held 9 mm from the bottom surface. To minimise the concentration of the volatile crosslinking by-products, the samples were pre-treated in a rough vacuum at 90°C for 1 week and then kept in vacuum at room temperature till they were electrically tested in a light-tight chamber shown in Fig. 1 and described elsewhere [2].

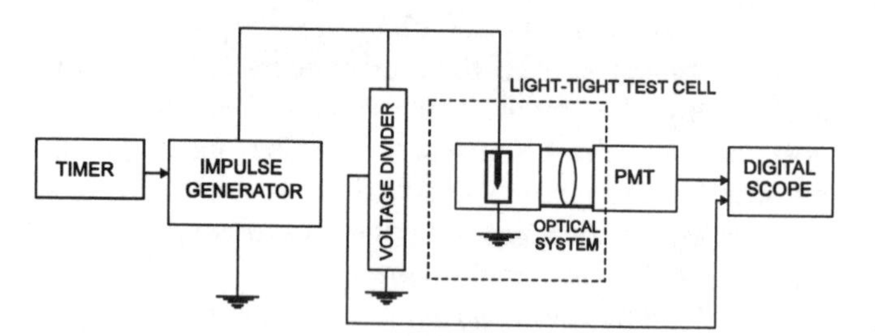

Figure 1. Block diagram of the experimental set-up.

Measurement Technique

Negative impulses were generated with a single stage spark gap generator. An RC network was used to shape impulses with rise times (time between 10% and 90% of the peak value) of 0.06 μs to 160 μs and fall times (time from the peak to 50% of the peak value) of 1μs to 20 ms. For each tests, at least 100 impulses with a repetition frequency of 1 Hz were applied to the sample.

A photomultiplier tube (PMT), operating in pulse counting mode was used to detect the EL pulses. To reduce electromagnetic noise in the output circuit of the PMT, great care was taken to properly shield all the signal lines. Since the PMT output pulses have a duration of only a few ns they are hard to distinguish from the electromagnetic noise generated by the spark gap; hence, to improve the signal to noise ratio and perform time resolved measurements the PMT output was terminated with a high impedance (1 MΩ). This caused the

PMT output pulse to have a long tail time which could be easily distinguished from the electromagnetic noise as shown in Fig. 2.

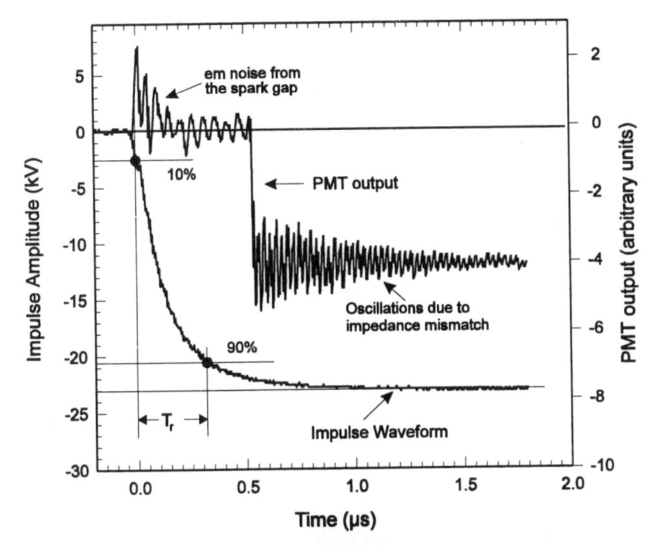

Figure 2. Example of a typical signal from the PMT recorded during impulse application.

The EL pulses detected by the PMT were counted by a 4 channel, 1 GS/s digital scope, having a 500 MHz bandwidth, which was triggered by the spark gap. Time resolved measurements were performed to determine the time between the instant the spark gap was triggered and the emission of the EL pulse. The measured time was compensated by the transit time (33 ns) of the electrons in the PMT and the difference between the travel times (10 ns) of the voltage and the PMT pulses along the length of the cables.

After the impulse tests, the samples were examined under an optical microscope (resolution 2 μm) as well as PD measurements (detection limit 0.1 pC) were performed with ac voltage in order to detect if treeing had occurred. If an electrical tree was detected at the semicon tip the results for such specimen were rejected.

RESULTS

Typical EL emission probability versus the amplitude of the negative impulses having rise times of 0.06 μs and 10 μs but the same fall time of 150 μs are shown in Fig. 3. The EL emission probability was calculated as the percentage of the number of impulses which gave rise to at least one EL pulse to the total number of impulses applied to the sample. The curves show that for both, the fast and slow rise times, EL inception occurs at about 18 kV. The EL emission probability increases with the amplitude of the impulse voltage but it is not effected by the impulse rise time.

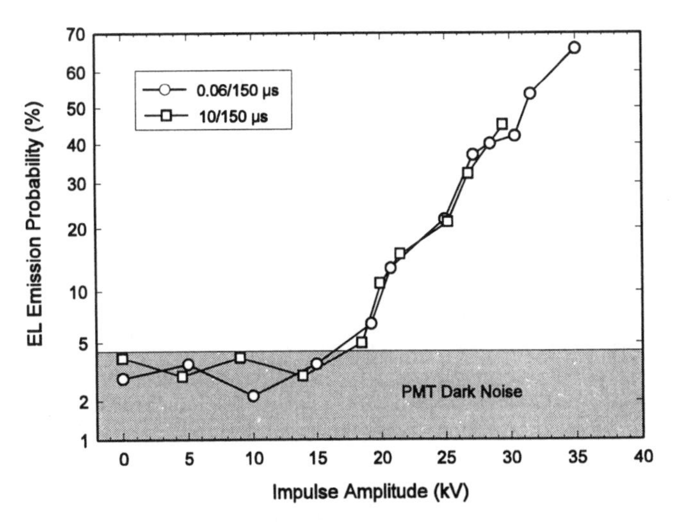

Figure 3. EL emission probability versus impulse amplitude.

Fig. 4 shows the EL emission probability versus the rise time for negative 30 kV impulses. Even for rise times in the range of 0.06 μs to 120 μs, (fall time held constant at 20 ms) no increase in the EL emission probability was observed.

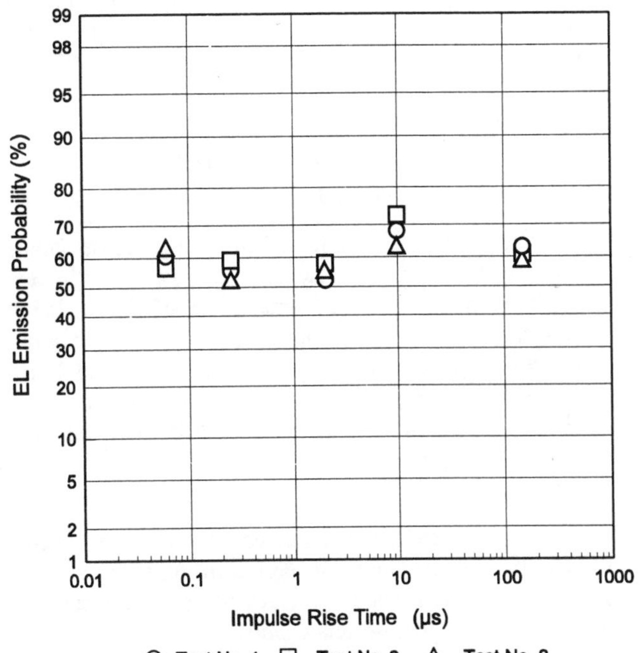

Figure 4. Effect of impulse rise time on the EL emission probability.

On the other hand, Fig. 5 illustrates that the EL emission probability, during the application of negative 25 kV impulses, with a rise time of 0.06 μs, increased when the fall time was increased from 1 μs to 16 ms.

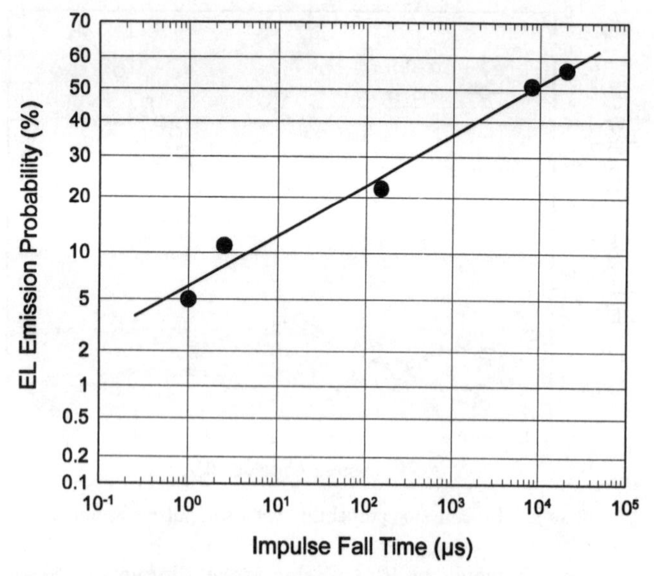

Figure 5. Effect of the impulse fall time on the EL emission probability.

Time resolved measurements were performed to determine the emission of EL pulses during negative impulses having the same amplitude of 35 kV but three different shapes which were as follows:

rise time of 0.06 μs and fall time of 2.5 μs,
rise time of 0.06 μs and fall time of 150 μs,
rise time of 10 μs and fall time of 200 μs.

Figs. 6a-c show the temporal behaviour of these impulses with the time plotted for clarity on a log scale. The dots represent the cumulative density (integration of the probability density function) of the EL pulses that occur during voltage application. For example, Fig. 6c shows that for the 0.06/2.5 μs impulse there is a 10% probability that EL pulses will occur within 0.1 μs after the spark gap is triggered; while, there is a 75% probability that they will occur within 10 μs after the trigger.

Figs. 6a-c show that there are two distinct time periods of EL activity separated by a time period during which no EL activity was observed. The EL pulses first occurred from time t_1 to t_2, shown in Figs. 6a-c, and denoted in this paper as the first EL period. EL pulses did not occur after time t_2 until the applied voltage had decreased to less than 50% of the peak value at time t_3. Then, EL pulses again occurred from time t_3 to t_4, denoted as the second EL period.

EL pulses could be observed within several tens of ns after the spark gap was triggered and the cumulative probability of EL emission in the first EL period was 50% to 80% as compared to 20% to 50% for the second EL period.

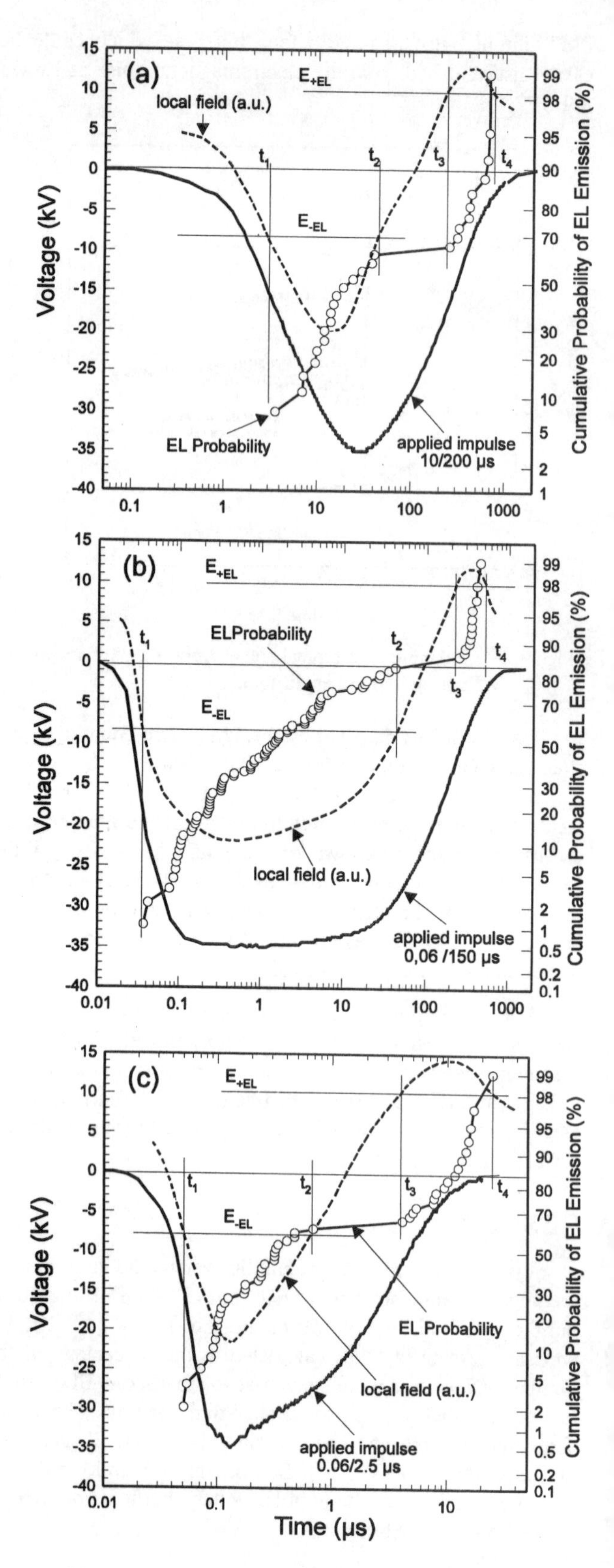

Figure 6. Cumulative probability of the EL emission for impulses of different shape and duration.

As shown in Fig. 3, an EL pulse was not observed for every impulse applied to the sample and sometimes more than one EL pulses were detected during the first or the second EL period. However, an EL pulse occurred during the second EL period only if another EL pulse had previously occurred during the first EL period.

DISCUSSION

Previously, it was shown [1, 2] that for EL emission in polyethylene charge injection is required at points of electrical stress enhancement in the polymer. For negative impulse, above a certain threshold voltage, electrons will be injected into the polymer. These injected electrons can gain energy from the applied field and collide with the molecules of the polymer, additives, antioxidants etc. to cause impact excitation. EL is emitted when the excited molecules return to their ground state or when the injected charges recombine with ionic species or charges of opposite polarity trapped in the polymer.

The charges injected into the polymer will also be trapped in the shallow and deep trapping centers of the polymer and form a space charge which will decrease the field at the injecting semicon electrode and prevent any further charge injection. Since the space charge influences the electric field intensity at the semicon tip, as well as EL emission, the temporal behaviour of EL emission was used to qualitatively describe the field at the semicon tip as schematically shown in Fig. 7.

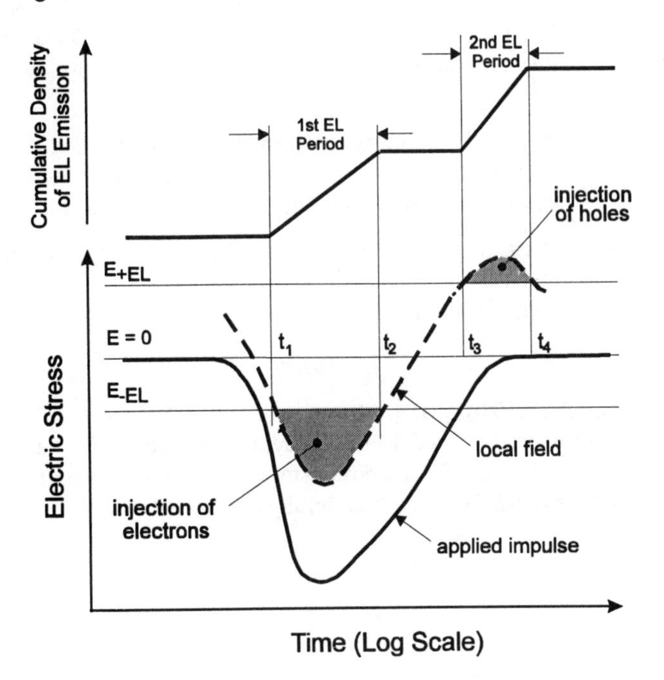

Figure 7. A model for the temporal behaviour of charge injection, local electric field, and El emission with respect to applied field.

Due to the space charge the local electric field at the semicon tip, as depicted by the dashed curve in Fig. 7, is different than the applied field, shown by the solid line. EL pulses would start at time t_1 when the local electric field in the polymer exceeds the EL emission level, E_{-EL}, and stop at time t_2 when the local field falls below this level. Thus, EL emission would occur during the first EL period, from time t_1 to t_2, as long as the local field exceeds the EL emission level, E_{-EL}. After time t_2, as the applied field decreases, the local electric field also decreases, the electrons in the shallow traps detrap but the electrons in the deep traps will remain behind and these trapped electrons will cause a reversal of the local electric field at the semicon tip. EL pulses cannot occur after time t_2 until the field reversal causes the injection of charges of opposite polarity or holes into the polymer. At time t_3, when the local electric field exceeds the EL emission level, E_{+EL}, light would again be emitted due to the recombination of the injected holes with the trapped electrons. As shown in Fig. 7, the EL emission stress during hole injection, E_{+EL}, was set higher than the EL stress during electron injection, E_{-EL}, because it is easier to inject electrons than holes into polymer. As long as the local electric field exceeds the EL emission level, E_{+EL}, light pulses would be emitted during the second EL period from time t_3 to t_4.

The higher the amplitude of the impulse voltage above the EL inception level the greater is the amount of electrons and holes injected into the polymer and the greater is the EL emission probability as shown in Figure 3. Similar model for EL emission was previously presented for half-rectified ac v-oltage [2].

The local electric field, denoted in Fig. 7, is also shown by the dashed lines in Figs. 6a-c. In these figures, EL pulses start at time t_1 when the local electric field exceeds the EL inception level and occur during the first EL period until time t_2 when the local field drops below that level. Comparison of Figs. 6b and 6c shows that increasing the fall time of the impulse voltage significantly increases the duration of the first and second EL periods. On the other hand, comparison of Fig. 6a and 6b shows that increasing the rise time does not increase the duration of the first or the second EL period. This could explain why the EL emission probability increases with the fall time but not with the rise time of the impulse, as shown in Figs. 4 and 5.

After field reversal the electrons trapped in the deep traps can attract the injected holes to cause recombination and EL emission again occurs during the second EL period. During the increasing portion of the subsequent impulse, EL could again be emitted due to the recombination of the injected electrons with the trapped holes. However, since the number of EL pulses does not increase with the rise time of the applied voltage it suggests that EL emission during the increasing portion of the impulse is not likely due to the recombination of the injected electrons with the trapped holes.

It indicates that after injection most of the holes recombine to give rise to EL and only a few holes get trapped in the deep traps. This was verified by varying the repetition rate of the impulses from 1 mHz to 1 kHz during which EL emission probability did not change.

Commercial, cable grade XLPE used in this study has many additives, which could give rise to fluorescence and phosphorescence caused by impact excitation. Fluorescence emission due to electronic transitions between molecular levels of the same multiplicity (singlet to singlet) is usually short lived ($\geq 10^{-8}$ s) while, phosphorescence emission due to electronic transitions between levels of different multiplicity (triplet to singlet) is relatively long lived. Since no EL emission occurs after the applied impulse returns to ground potential it suggests that phosphorescence is not responsible for EL emission.

Due to space charge in the polymer, the local electric field at the semicon tip held at negative 35 kV is ≅300 kV/mm which is at least 80% less [5, 7] than the electric field calculated without the space charge effect. Assuming the maximum mean free path of 20 nm (100 interatomic spacing) [8], the maximum energy gained by the injected charges would be ≅ 6 eV. This energy is sufficient to cause impact excitation and EL emission in the visible and uv ranges but it is too low to cause impact ionisation (7.6 to 9 eV) [8].

CONCLUSIONS

- Time resolved measurements of EL during unipolar negative impulses have shown that EL emission occurs during two distinct periods because of the local electric field reversal in polymer. These emission periods are s-eparated by a time period during which no EL emission is observed.
- EL emission occurs only during the application of an electric field and when the local electric field exceeds the EL emission levels.
- The EL emission probability increases with the impulse amplitude. It is independent of the impulse rise time between 0.06 µs and 160 µs but increases with the impulse fall time.
- EL emission can occur with several tens of nanoseconds from the instant of impulse application.

REFERENCES

[1]. Lebey T. and Laurent C., "Charge injection and electroluminescence as a prelude to dielectric breakdown", J. Appl. Phys. vol. 68, pp. 275-282, 1990.

[2] Bamji S.S., Bulinski, A.T. and Densley R.J., "Degradation of polymeric insulation due to photoemission caused by high electric fields", IEEE Trans. on Electr. Insul., vol. 24, pp. 91-98, 1989.

[3] Kalkner W. and Krage I., "PD detection and localisation using unipolar impulse voltages for on-site testing of polymer insulated cables", Proc. 8th Inter. Symp. on HV Engineering, Aug 23-27, 1993, Yokohama, Japan, pp. 205-208.

[4] Lemke, E. and Schmiegel P., "Complex Discharge Analyzing (CDA) - an alternative procedure for diagnosis tests on HV power apparatus of extremely high capacity", Proc. 9th Inter. Symp. on HV Engineering, Aug 28 - Sept 1, 1995, Graz, Austria, Paper No. 5617.

[5] Shimizu N., "Treeing phenomena of polymeric materials at low temperature", Ph.D. Thesis, Nagoya University, Nogoya, Japan, 1979.

[6] Mary D., Laurent C., Guastavino J., Krause E., and Mayoux C., "Electroluminescence and high-field conduction in poly(ethylene-2, 6-naphthalate)", Annual Report, Conf. on Electr. Insul and Dieletric Phenomena., Oct 23-26, 1994, Dallas, TX, pp. 298-303.

[7] Pietsch, R., Untersuchungen fuer die Bedeutung von Elektrolumineszenz fuer die dielektrische Alterung von Polyethylen. Ph.D. Thesis, RWTH Aachen, 1992

[8] Dissado L.A. and Fothergill J.C., "Electrical Degradation and Breakdown in Polymers", Peter Pelegrinus Ltd., London, 1992.

Acknowledgements

Financial support from the German Academic Exchange Service (DAAD) for the postdoctoral fellowship of one of the authors at NRC Canada is gratefully acknowledged.

Conference Record of the 1996 IEEE International Symposium on Electrical Insulation, Montreal, Quebec, Canada, June 16-19, 1996

The Role of XLPE Type on Electroluminescence and Subseqent Electrical Treeing

F. Kabir, J-M. Braun and J. Densley
Ontario Hydro Technologies
800 Kipling Ave.
Toronto, Ontario, Canada
M8Z 5S4

R.N. Hampton
BICC Cables Ltd
Erith Technology Centre
Church Manorway
Erith, Kent, DA8 1HS
England

S. Verne
BICC plc
Quantum House
Maylands Ave.
Hemel Hempstead, Herts,
HP2 4SJ
England

M. Walton
BICC Cables Corp
Marshall Technology Center
PO Box 430, Scottsville
TX, 75688
USA

Abstract

Extruded polymers used as insulations for high-voltage cables are susceptible, at high stress, to electrical-tree degradation which initiates at stress enhancements. Light is emitted during the tree-initiation phase, prior to the formation of channels and partial discharges. This paper presents data on the light emission and electrical treeing characteristics for two XLPEs and shows that the light inception is directly related to the tree inception, as measured by ramp- or constant-voltage tests.

Introduction

Low density polyethylene (LDPE) and cross-linked polyethylene (XLPE) are now beginning to be used as insulation for high-voltage transmission-class cables up to 500 kV. These materials have excellent electrical properties for cable applications, e.g., low permittitvity, low loss and a high breakdown strength. However, they are susceptible to electrical ageing when subjected to high electrical stresses [1]. One form of electrical ageing is electrical treeing phenomena originating from local stress enhancements. These stress enhancements are considered to be a very important factor in electrical ageing. Possible sources of stress enhancements in extruded cables are protrusions on the semiconducting shield/insulation interfaces or metallic or high permittivity contaminants located either within the insulation or located at interfaces. Thus, the quality of the interfaces is a very important parameter in determining the electrical performance of extruded cables.

Electrical treeing has three phases, initiation, propagation and final breakdown. The mechanisms responsible for the deterioration during tree initiation are not clearly understood. The following have been proposed - Joule heating, impact ionization, damage due to hot electrons, damage due to ultraviolet (UV) radiation, and mechanical fatigue due to Maxwell stress.

Laboratory studies of tree initiation have usually been performed with metallic needle/plane or needle/needle electrodes to model a stress enhancement. The electrostatic field distribution can be calculated for such a geometry so that observed phenomena can be directly related to the applied field. In practice, however, space charge is present near the needle tip so that the actual field in this region is the combination of the applied and space charge fields.

At high applied AC electrical stresses, light is emitted; this light can be caused by any of the above-mentioned mechanisms or by a chemical reaction. Previous studies have shown that [2-6]:

- The inception voltage of the light (electroluminescence, EL) is closely related to the threshold voltage below which there is no electrical ageing.
- EL extends in wavelength from the visible into the uv and this uv may contribute to polymer degradation.
- Gases dissolved in the insulation influence both the EL and electrical tree initiation, e.g. the presence of oxygen reduces the amount of light emitted at all wavelengths, hence increases the light inception, and decreases the tree inception voltage or time-to-tree by about 50% compared to nitrogen.
- The phase relationship of the EL depends on the local field, i.e., the combination of the space charge and applied fields.

Most of the previous studies have used metallic needle electrodes, while in practice, the stress enhancements are more likely to occur at the tips of protrusions of semiconducting compound at the shields. The results presented in this paper will include tests to measure the light inception voltage and electrical tree characteristics for specimens with semiconducting compound electrodes.

Experimental

Each specimen consisted of a needle-shaped semiconducting electrode, made from a commercial-grade compound, embedded in cable-grade XLPEs in such a way that the distance between the tip of the needle and the adjacent ground plane was 10 mm. The tip radii of the specimens varied between 1 and 15 μm. All specimens were treated in a vacuum oven for 24 h at 60°C to remove the cross-linking by-products.

Details of the light-tight test cell and light-collection system have been presented elsewhere [6]. An unique feature of the test cell was that an elliptical mirror was used to increase the efficiency of the light collection. The tip of the needle is located at the focal point of the mirror. To suppress partial discharges, the specimens were either embedded in a transparent epoxy resin or tested with the test cell pressurised to two atmospheres of nitrogen.

The light collected and reflected by the mirror was focused onto the photocathode of a photomultiplier. The output pulses from the photomultiplier were measured by a specially built pulse height analyzer (PHA), which displayed the pulse amplitude distribution (15 channels) and also the phase position of the pulses with respect to the AC waveform. The PHA also recorded the light intensity, defined as the product of the number of pulses per channel times the channel number.

The test procedure adopted was to increase the voltage in 1 kV increments and to measure the EL for a period of 30 s at each voltage until the light intensity increased significantly above the background noise level. The voltage was then removed and the specimen heated for about 40 minutes to reach 70°C and the voltage reapplied in similar steps to re-measure the light inception voltage (LIV). The light intensity at zero voltage, i.e., the background level, was always greater at 70°C than at room temperature, presumably due to increased chemiluminescence at the higher temperature. The voltage was again removed and the specimen allowed to cool to room temperature. The LIV was again measured and a tree initiation test was carried out by raising the voltage in 1 kV steps above the LIV until an electrical tree was initiated. This we define as the tree inception voltage (TIV). Tree initiation was clearly detected by large increases in the number and amplitude of light pulses and the onset of PDs. If the tree did not initiate at the maximum rated voltage, the voltage was kept constant and the time to initiate an electrical tree (TTT) was recorded. The latter tests were performed outside the light-tight test cell and the trees were detected visually using a CCD camera and a video tape recorder.

Results and Discussion

The variation of the light intensity with applied voltage for two XLPE specimens is shown in Figure 1. The light intensity increased rapidly above light inception, with the light emission occurring in the increasing positive and negative quadrants of the applied voltage. This is in agreement with the results of other studies[9]. The peaks of the intensity occurred at phase angles between 40 and 70° in the positive half cycle and between 220 and 250° in the negative half cycle. The phase behaviour is due to the modification of the electric field at the electrode tip due to the space charge set up by the mobile charges. The light intensity was approximately similar in each half cycle at inception but, at higher voltages, the intensity was greater in the negative half cycle, sometimes approaching twice that occurring in the positive half cycle.

Figure 2 shows the effect of the test atmosphere on the light inception voltage. The tests were initially performed at room temperature (Room Temperature 1 in Figure 2), then at 70°C, followed by another test at room temperature (Room Temperature 2 in Figure 2). The results for specimens tested in nitrogen show that the light inception voltage of the second test at room temperature, i.e., after the test at 70°C, was about 25% lower than the initial value at room temperature. The reduction in LIV is due to the removal of oxygen out of the specimen by the increased rate of diffusion during the immersion in nitrogen at 70°C. For the specimens embedded in epoxy resin the diffusion of gases into or out of the specimen, even at 70°C, was negligible with the result that the LIV was independent of the test sequence as can be seen from Figure 2.

Figure 1 represents light intensity data for single specimens of two XLPE types and shows different light backgrounds and light inception voltages. The XLPEs used in this study had the same base polymer, but different stabilizer packages. In common with most high field phenomena the light emission and treeing data required significant numbers of specimens to be tested and analysed using sophisticated statistical techniques. Figures 3, 4 and 5 show light and tree inception voltages and the time to tree at 30 kV, respectively. These figures display the mean and 95% confidence limits for these data. The analyses shown in Figures 3 to 5 clearly demonstrate that there are significant differences between the two types of XLPE's.

Comparison of Figures 3 and 4 shows that light inception occurs at voltages approximately half that of the tree inception. This supports the hypothesis that light emission is a pre-cursor to electrical treeing. Figures 3 and 4 also support the general trend observed by the authors that enhanced tree inception voltages correlate with enhanced light inception: specimens with low light inception voltages give poor electrical treeing performance.

The preceding discussion has focused on the correlation between the LIV's and the TIV's; perhaps the most striking feature is the increase in the time to tree at 30 kV (Figure 5) accompanying the improvements in LIV. Here the improvements in endurance are very large and this should correlate well with improved ageing performance of cable insulations as measured by the ageing parameter n in the life equation:

$$E^n t = \text{constant.}$$

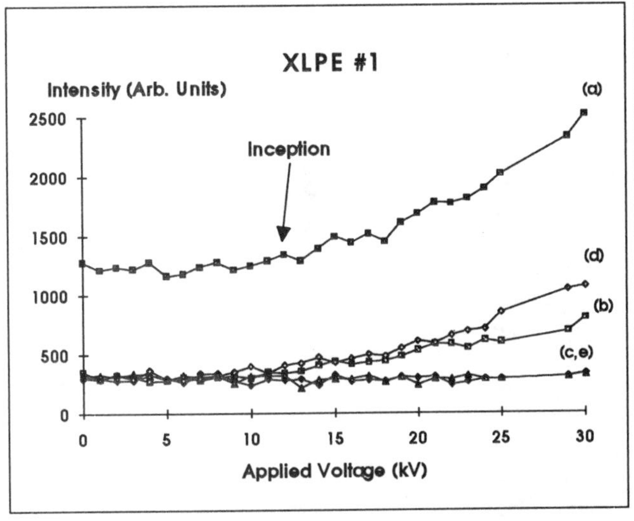

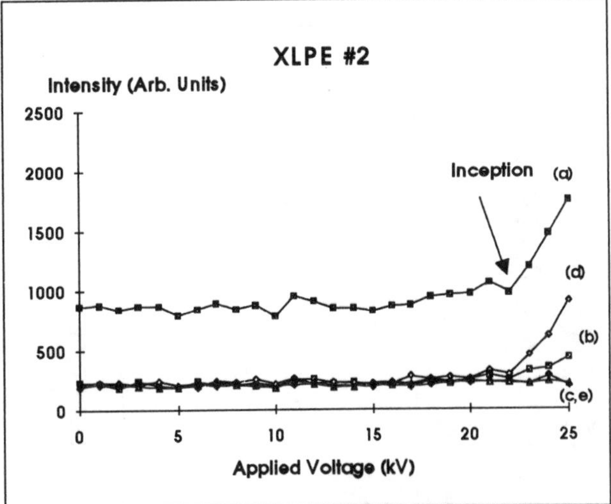

(a) - Full Cycle: (b) - 0 to 90: (c) - 90 to 180: (d) - 180 to 270: (e) - 270 to 360

Figure 1. Emitted Light Vs. Applied Voltage for Two XLPE Specimens

Conclusions

The experimental results show that:

- electroluminescence occurs at protrusions of semiconducting compounds before the onset of partial discharges and electrical treeing.
- the light inception voltage decreases when the oxygen dissolved in the polymer is removed or replaced with nitrogen.
- there is good correlation between the light inception inception voltage, tree inception voltage and time to tree inception, i.e., specimens and materials with a low light inception voltage will exhibit poor electrical treeing performance.
- the resistance of XLPE to electrical treeing can be improved by the careful selection of an appropriate stabilizer package.

Acknowledgements

The authors would like to thank the Directors of BICC Cables for supporting the work described above and for giving permission to publish the paper.

References

1. Densley, R.J. and Bartnikas, R. and Bernstein, B.S., 'Multiple-Stress Aging of Solid-Dielectric Extruded Dry-Cured Insulation Systems for Power Transmission Cables', IEEE Transactions on Power Delivery, Vol. 9(1), pp. 559-571, Jan. 1994.

2. Bamji, S.S., Bulinski, A.T., and Densley, J., 'Evidence of Near-Ultraviolet Emission during Electrical-Tree Initiation in Polyethylene', J. Appl. Phys., Vol. 61(2), pp. 694-699, Jan. 1987.

3. Shimizu, N., Katsukawa, H., Miyauchi, M., Kosaki, M. and Horii, K.,., 'The Space Charge Behaviour and Luminescence Phenomena in Polymers at 77 K', IEEE Trans. Elec. Insul., Vol. EI-14, pp. 256-263, 1979.

4. Laurent, C., Mayoux, C. and Noel, S., 'Dielectric Breakdown of Polyethylene in Divergent Field: Role of Dissolved Gases and Electroluminescence', J. Appl. Phys., Vol. 54, pp. 1532-1539, March 1983.

5. Champion, J.V., Dodd, S.J. and Stevens, G.C., 'Quantitative Measurement of Light Emission during the Early Stages of Electrical Breakdown in Epoxy and Unsaturated Polyester Resins', J. Phys. D: Appl. Phys., Vol. 26, pp. 819-828, 1993.

6. Kabir, F., Braun, J-M., Densley, J. and Hampton, R.N., 'Light Emission from Highly-Stressed Polymeric Cable Insulation', Conf. Rec. of 1994 IEEE Intl. Symp. Elect. Insul., Pittsburgh, pp. 37-40, June 1994.

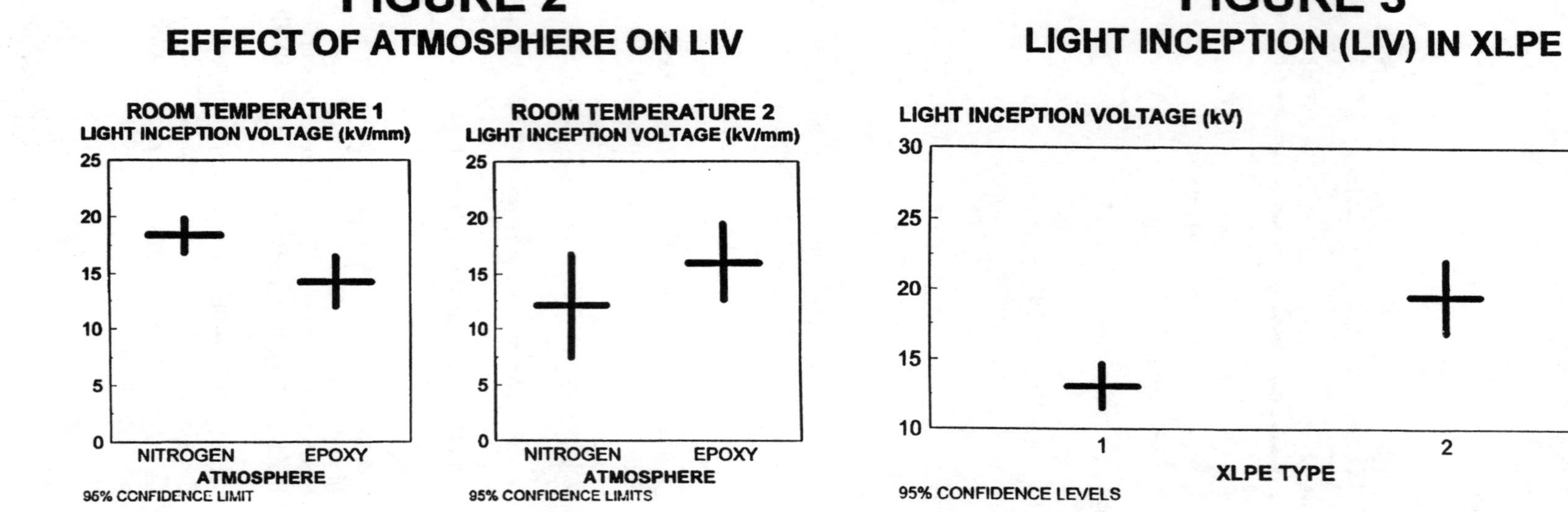
FIGURE 2
EFFECT OF ATMOSPHERE ON LIV
ROOM TEMPERATURE 1
LIGHT INCEPTION VOLTAGE (kV/mm)
25
20
15
10
5
0
NITROGEN
EPOXY
ATMOSPHERE
95% CONFIDENCE LIMIT
ROOM TEMPERATURE 2
LIGHT INCEPTION VOLTAGE (kV/mm)
25
20
15
10
5
0
NITROGEN
EPOXY
ATMOSPHERE
95% CONFIDENCE LIMITS
FIGURE 4
TREE INCEPTION VOLTAGE (TIV) IN XLPE
TREE INCEPTION VOLTAGE (kV)
30
25
20
15
10
1
2
XLPE TYPE
95% CONFIDENCE LEVELS

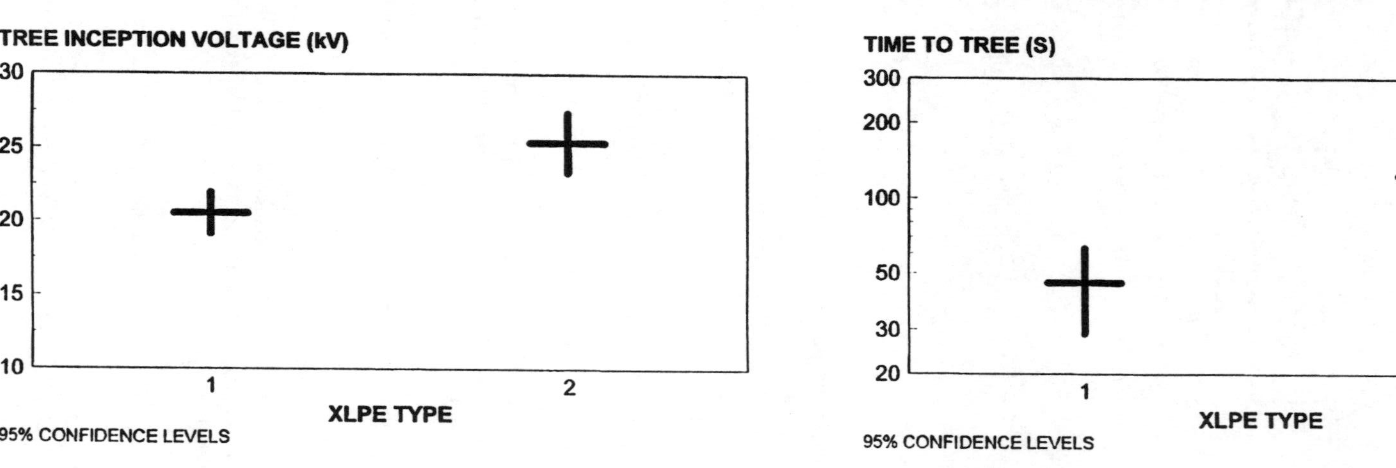
FIGURE 3
LIGHT INCEPTION (LIV) IN XLPE
LIGHT INCEPTION VOLTAGE (kV)
30
25
20
15
10
1
2
XLPE TYPE
95% CONFIDENCE LEVELS
FIGURE 5
TIME TO TREE (TTT) AT 30 kV IN XLPE
TIME TO TREE (S)
300
200
100
50
30
20
1
2
XLPE TYPE
95% CONFIDENCE LEVELS

Conference Record of the 1996 IEEE International Symposium on Electrical Insulation, Montreal, Quebec, Canada, June 16-19, 1996

Effect of the Interfacial Pressure and Electrode Gap on the Breakdown Strength of Various Dielectric Interfaces

Chinh Dang
Institut de Recherche d'Hydro-Québec
Varennes, Québec, Canada J3X 1S1

***Abstract*: Electrical breakdown at the interface of two dielectric surfaces is a major design consideration in underground cable accessories. Our previous studies have shown that the dielectric strength of greased and non-greased interfaces can be greatly affected by their surface conditions as well as their substrate material. These observations were made in both laboratory molded samples and real cable joints.**

The present paper deals with the effect of the interfacial pressure and electrode gap on the breakdown strength of various molded dielectric interfaces. Results showed that greased and non-greased interfaces have each a distinct behavior in both parameters examined. Finally, a plot of the breakdown voltage as a function of the product of pressure and gap was introduced and seems to give a more appropriate representation of both effects.

INTRODUCTION

In cable accessories, a major design consideration is given to the voltage withstand of dielectric interfaces, this is especially critical to premolded cable joints in extra high-voltage applications [1]. Once the insulation materials are selected, both the interfacial pressure and the electrode gap can be adjusted to reach the design goal. However, only a few published studies have examined the influence of these parameters on the dielectric strength of an interface. A previous study by Fournier *et al.* [2] was carried out on an EPDM/EPDM interface with a pair of sharp metallic electrodes which would have a very different charge-injection characteristic than that of smooth semiconducting electrodes found in premolded joints.

The present study uses molded slab samples with embedded semiconducting electrodes to truly emulate the situation existed in premolded cable accessories. Various insulation materials as well as different types of interfaces were included in the investigation.

TEST SAMPLES

Fig. 1 shows a molded slab with a pair of embedded semiconducting electrodes and the test configuration. Each test sample consists of two molded slabs, one with electrodes and the other with no electrode as substrate. The interface of the two slabs can be either a layer of thin grease film or a

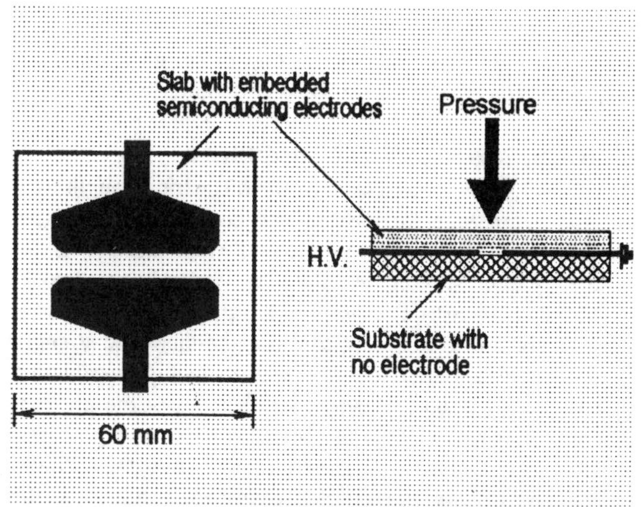

Figure 1 Slab sample and test configuration

layer of 1.5 mm high-K (high dielectric constant) material, or nothing at all. The insulation material of slabs is made of either epoxy, EPR (Ethylene Propylene Rubber compound), or XLPE (Cross-linked Polyethylene with tree-retardant). For a greased interface, a thin layer of grease was first spread on both opposing surfaces of the two molded slabs, then they were inserted into a mechanical press with an overlapped length of 20 mm. A pressure of 75 kPa (20 lbs) was then applied to the overlapping slabs and the final assembly was made by sliding one slab over the other to eliminate the excess grease. This procedure was necessary in order to create a greased interface similar to that observed in cable joints, and was applied to both EPR/EPR and EPR/XLPE greased interfaces. However, in cable accessories with epoxy components such as elbow connectors, no sliding was performed in their assemblage, therefore we carried out the experiment in the case of epoxy substrate without sliding, and a thicker layer of grease was left at the interface. In addition, both sanded (with a 180-grit aluminum oxide emery paper) and non-sanded XLPE slab surfaces were also used.

A fixed weight was applied on the top of the test sample to generate a desired interfacial pressure. The breakdown test was achieved by raising the ac voltage with a ramp speed of 0.5 kV/s at ambient temperature (~ 23°C), and the dielectric strength was calculated from the measurement of the breakdown voltage and the electrode gap. Each type of test sample was identified by three sets of informations. The first letter designates the type of slab with electrodes, which is either R (EPR slab with embedded electrodes) or X (XLPE slab with embedded electrodes). The second set of letters gives the make-up of the interface which can be either greased (G2) or non-greased (NG), either sanded (SD) or non-sanded (NS), or with a high-K layer (K). The third set is for the type of substrate, made of either epoxy (E), EPR (R), or XLPE (X).

TEST RESULTS

Each data point represents the average of six measurements, and its error bar shows the standard deviation.

Effect of the interfacial pressure

Figs. 2 and 3 show the results of the dielectric strength as a function of pressure for greased and non-greased interfaces respectively. All these measurements were taken with an electrode gap of ~5 mm. A saturation is clearly observed in Fig. 2 for the dielectric strength of all greased interfaces at a pressure above 100 kPa. By contrast, the dielectric strength of all non-greased interfaces in Fig. 3, including those with a high-K layer, exhibits a linear dependence with the pressure over the investigated range of 10 to 200 kPa. In both cases, the R/R interface has the highest dielectric strength, followed by R/E and R/X interfaces. Moreover, the sanding on the XLPE surface reduces the dielectric strength of the R/X interfaces; however, the greasing and the high-K layer seem to significantly attenuate this reduction. These observations are consistent with our reported findings in [3]. It should be noted that the relatively higher dielectric strength of the greased R/E interface at low pressures is probable due to the particular greasing method of this interface as described above. With increasing pressure, the thickness of the grease layer at the interface would be reduced to a minimum as in other interfaces.

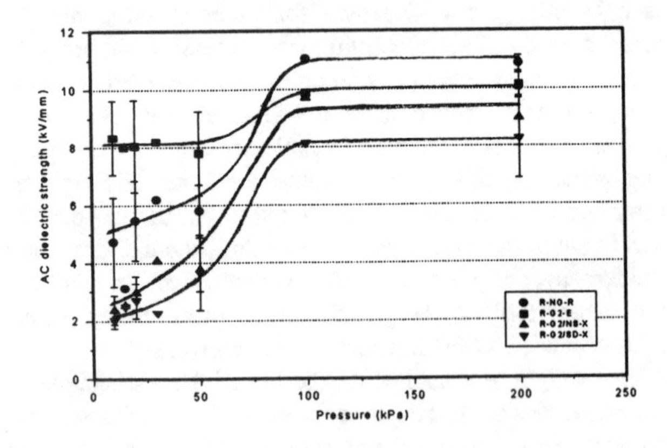

Figure 2 Plot of the dielectric strength as a function of pressure for greased interfaces. Hand-drawn curves show the tendency of each data set.

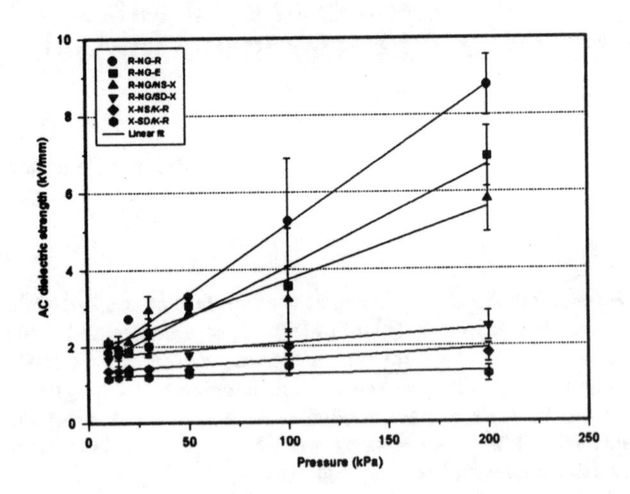

Figure 3 Plot of the dielectric strength as a function of pressure for non-greased interfaces. Straight lines are fitted to each data set.

By comparing the results in Figs. 2 and 3, the dielectric strength of the non-greased interfaces can surpass that of the greased interfaces at higher pressures if the linear relation would hold true. For the R/R interface, the cross-over point will occur at the pressure of 260 kPa.

The high-K layered interfaces have the lowest dielectric strength which remains nearly constant with pressure. This might be due to the controlling resistivity of the high-K layer [4].

Effect of the electrode gap

The dependence of the dielectric strength on the electrode gap for both the greased and non-greased interfaces at a pressure of 100 kPa is shown in Fig. 4 on a log-log graph. It is common to find a power-law relationship between the dielectric strength and the electrode gap [5], and Blaise [6] has recently developped a space-charge model to account for such a relationship. For convenience, our data are divided into six groups. The first group comprises of all the non-sanded greased interfaces (i.e. R-G2-R, R-G2-E and R-G2/NS-X) and is fitted to a thick solid line in Fig. 4. The second group is the sanded greased interface (i.e. R-G2/SD-X) represented by a thin solid line. The third group is made of all non-sanded non-greased interfaces (i.e. R-NG-R, R-NG-E and R-NG/NS-X) fitted to a thick dashed line. The fourth group is the sanded non-greased interface (i.e. R-NG/SD-X) fitted to a thin dashed line. Finally the fifth and sixth group, represented by the thick and thin dotted lines, are the high-K layered interfaces with a non-sanded and sanded XLPE surface, respectively. The coefficients of the power-law fitting are given in Table 1 where we also include the coefficients of the exponential fitting. Except the group 6, the exponential fit seems to yield a better correlation.

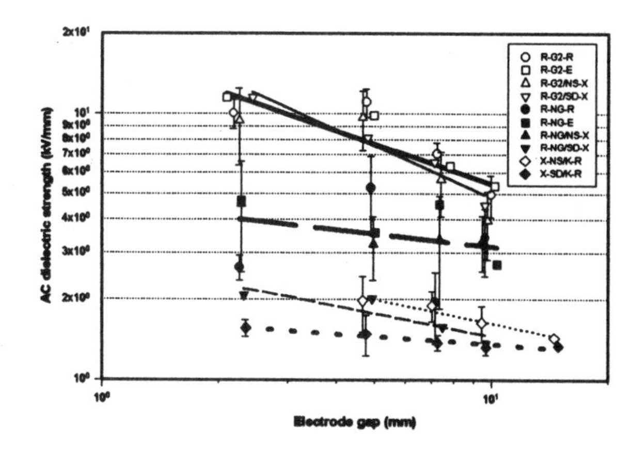

Figure 4 Plot of the AC dielectric strength as a function of the electrode gap for both greased and non-greased interfaces at 100 kPa. Straight lines on this log-log graph represent the power-law fit.

Table 1 - Coefficients of the fitting in Fig. 4

Sample	Coefficients	Power fit*: $y = A\,x^B$	Exponential fit*: $y = A\exp(Bx)$
group 1	A ± std err.	17.3 ± 3.1	14.4 ± 1.5
	B ± std err.	-0.50 ± 0.10	-0.11 ± 0.01
	correl. r^2	0.70	0.83
group 2	A ± std err.	21.4 ± 3.7	15.5 ± 0.8
	B ± std err.	-0.65 ± 0.10	-0.13 ± 0.01
	correl. r^2	0.95	0.99
group 3	A ± std err.	4.56 ± 1.12	4.27 ± 0.83
	B ± std err.	-0.16 ± 0.15	-0.034 ± 0.029
	correl. r^2	0.10	0.12
group 4	A ± std err.	2.81 ± 0.51	2.51 ± 0.21
	B ± std err.	-0.29 ± 0.10	-0.060 ± 0.013
	correl. r^2	0.79	0.92
group 5	A ± std err.	3.20 ± 0.35	2.33 ± 0.10
	B ± std err.	-0.29 ± 0.05	-0.034 ± 0.005
	correl. r^2	0.94	0.96
group 6	A ± std err.	1.70 ± 0.06	1.57 ± 0.06
	B ± std err.	-0.10 ± 0.02	-0.013 ± 0.004
	correl. r^2	0.90	0.76

Note: *x is the electrode gap and y is the dielectric strength.

The exponent B of the power law clearly indicates that the dielectric strength of greased interfaces (i.e. group 1 and 2) drops significantly faster than that of non-greased interfaces (i.e. group 3, 4, 5 and 6) as a function of the electrode gap. This means that at long gaps, a non-greased interface can have a dielectric performance better than a greased interface. However, the large dispersion observed in the data sets of non-greased interfaces suggests that air voids at the interface can play an important role, whereas the presence of a grease or high-K layer would greatly reduce the dispersion [3].

DISCUSSION

The test results obtained in this study are similar to those found in [2] with some noticeable differences probably caused by the electrode material and its shape. Indeed, our measured dielectric strengths were larger than those measured with sharp metallic electrodes for the same gap and type of interface. Moreover, although the relation between the dielectric strength and the pressure for greased interfaces showed saturation in both studies, ours occurred at a pressure of 100 kPa much higher than the value of 20 kPa found in [2].

The "fair" agreement between the power law and the data of the dielectric strength as a function of the electrode gap may perhaps be due to the nonuniform distribution of the electric field in the electrode interval.

In Figs. 2, 3 and 4, the dielectric strength is plotted on the ordinate axis, it increases with the pressure and decreases with the electrode gap. In fact, the measured breakdown voltage has increased with both parameters and has shown a similar trend with both parameters. We can combine both data sets in a single plot by using the breakdown voltage as ordinate and the product of pressure and gap as abscissa. This type of plot is common in gas breakdown studies, and it seems to work as well as for our study. An illustration is given in Fig. 5 with the data sets of a non-greased R-NG/NS-X and a greased R-G2/NS-X interface. Data obtained with different values of pressure and electrode gap are all included, they all follow the same curve for each type of interface. Similar trend is observed for other interfaces.

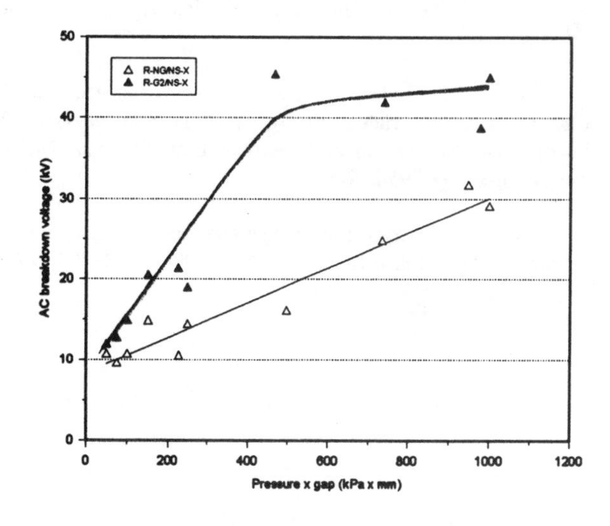

Fig. 5 Plot of the breakdown voltage as a function of the product of the pressure and gap. Data sets include breakdown values at different pressures and different electrode gaps.

CONCLUSIONS

Test results have clearly indicated that the breakdown voltage of greased interfaces shows saturation with the interfacial pressure and the electrode gap. By contrast, non-greased interfaces, including those with a high-K layer, increases linearly with both parameters over the investigated range.

ACKNOWLEDGMENT

The author would like to thank the financial support of both the Canadian Electrical Association (CEA) and Hydro-Quebec in this project. He also wish to thank the technical contribution of D. Lalancette and N. Claveau (technical staff of IREQ), as well as the helpful advices of both CEA project monitors, R. Filter (Ontario Hydro Technologies) and G. Buchanan (Manito Hydro).

REFERENCES

1. Hayashi, M., T. Akiyama, T. Inoue, H. Inoue, K. Fudamoto, T. Nakano and S. Gotoh. "Development of Prefabricated Joints for 154 kV XLPE Cable". Third International Conference on Polymer Insulated Power Cables, Jicable 91, France, pp. 99-105.

2. Fournier, D. and L. Lamarre. "Effect of Pressure and Length on Interfacial Breakdown Between Two Dielectric Surfaces". Conference Record of the 1992 IEEE International Symposium on Electrical Insulation, Baltimore, MD USA, pp. 270-272.

3. Dang, C. "Effect of Surface Conditions on the Breakdown Strength of Various Dielectric Interfaces". 1995 International Symposium on Electrical Insulation and Materials, Tokyo, Japan, pp. 189-192.

4. Blake, A.E., G.J. Clarke and W.T. Starr. "Improvements in Stress Control Materials". The 7th IEEE/PES Conference and Exposition on Transmission and Distribution, April 1979, Atlanta, Georgia.

5. Takahiko, M., K. Haga and T. Maeda. "Creepage breakdown Characteristics of Printed Wiring Board in Silicone Gel". 1995 International Symposium on Electrical Insulation and Materials, Tokyo, Japan, pp. 295-298.

6. Blaise, G. "Space-charge Physics and the Breakdown Process". Journal of Applied Physics, vol. 77, April 1995, pp. 2916-2927.

Conference Record of the 1996 IEEE International Symposium on Electrical Insulation, Montreal, Quebec, Canada, June 16-19, 1996

EFFECT OF THE SURFACE ROUGHNESS ON INTERFACIAL BREAKDOWN BETWEEN TWO DIELECTRIC SURFACES

Daniel Fournier
Institut de recherche d'Hydro-Québec (IREQ)
1800 boul. Lionel Boulet, Varennes, Québec, Canada J3X 1S1
Tel.: (514) 652-8069, Fax: (514) 652-8962

***Abstract*: Cable splices and accessories are the weak link in an underground power distribution system. Investigation of problems related to cable splices and accessories becomes quite intricate once the simpler causes of failures are dismissed to allow more complex phenomena to be examined. The interfacial breakdown between two internal dielectric surfaces represents one of the major causes of failure for power cable joints. In order to better understand this phenomenon, breakdown experiments were performed at interfaces found in cable splices. An experimental jig was designed to induce breakdown between dielectric surfaces longitudinally along their interface. Effects of surface roughness at EPDM/XLPE and EPDM/EPDM interfaces as well as the presence of silicone grease are taken into account.**

INTRODUCTION

The interfacial breakdown between two dielectric surfaces is complex and constitutes one of the principal modes of failure for power cable joints [1]. In previous work [2-4], breakdown experiments were performed on laboratory samples made from two unaged EPDM slices pressed one against the other. More recently, breakdown experiments were performed on interfaces found in both new and field-aged straight joints and T-connectors [5,6]. It was found that the dielectric strength of an interface greatly depends on both the interfacial pressure and the presence of a dielectric grease. The results have also shown that the dielectric strength decreases with aging.

The purpose of this paper is to report some preliminary results on the effect of surface conditions on the dielectric strength of interfaces found in both new straight joints and T-connectors.

TEST SAMPLES AND EXPERIMENTAL SET-UP

The test samples consisted of 24-mm-thick round slices of a straight joint (25 kV class). The slices were cut perpendicular to the joint axis at specific positions (see Fig. 1) to get straight clean samples with the EPDM/EPDM and EPDM/XLPE interfaces. It is worth noting that this procedure produces slices with large and small external diameters, that is, two of each per cable joint. Both new straight joints and T-connectors were used.

The slices were then put into a special breakdown test cell which allows thin electrodes to be inserted at the desired interface. The electrodes were flat stainless steel needles 1.2 cm long and 300 mm thick, with a round tip of 60 mm radius. They were set 4 mm apart at the selected interface. For each test sample, six AC breakdown measurements were taken at each interface. A special sensor with a tiny load cell at its centre (for the force sensing) was designed and constructed to determine the interfacial pressure. A detailed description of the samples and of the experimental set-up is given in [5,6].

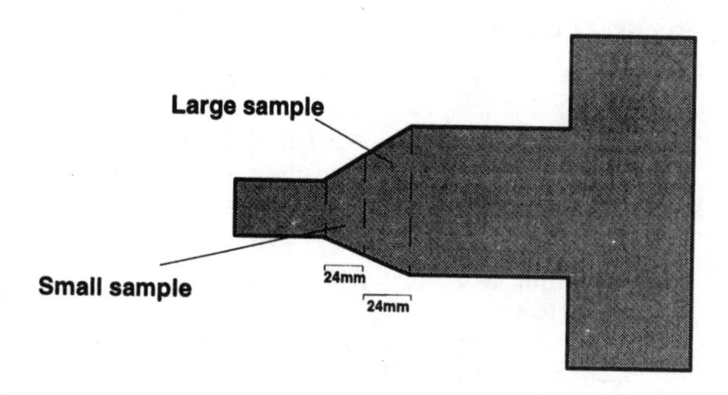

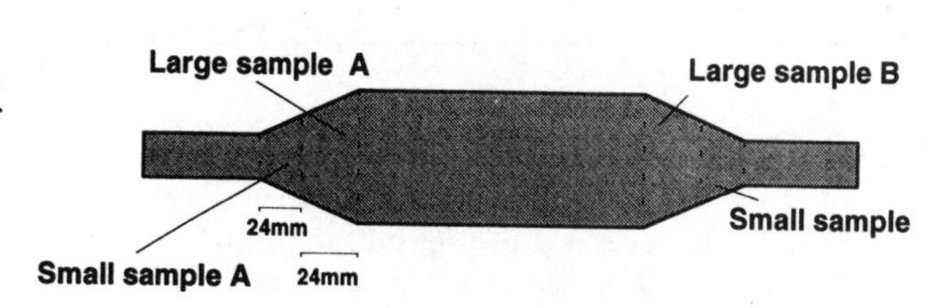

Fig. 1. Location of the test slices on a straight joint and a T-connector

Table 1 summarises the interface and surface conditions of our samples. The cable surface was prepared in different ways to create six conditions: greased/no sanding, greased/normal sanding (150 grit), greased/fine sanding (a 150-grit followed by a 600 grit), ungreased/no sanding, ungreased/normal sanding and ungreased/fine sanding. The sanding was done with an aluminum oxide emery paper. To install the splice on the cable, silicone grease was normally spread over the cable surface, and the splice or T-connector was then slipped on. To create ungreased surfaces, the test slice was disassembled, and both EPDM and XLPE surfaces of the test slice were wiped clean with 111-trichloroethane, and reassembled without adding grease. Two adapters from different manufacturers were used for the T-connectors. Adapters "B" are slightly smaller than adapters "A" and consequently fit more tightly on the cable.

Table 1: Test sample description and identification

Interface &surface conditions		Joint type (750 MCM compact cable)		
		"T" 600 A, adap. A	"T" 600 A, adap. B	Straight 600 A
Greased	no sanding	# 1	# 2	# 13
	fine sanding	# 3	# 4	# 14
	normal sanding	# 5	# 6	# 15
Ungreased	no sanding	# 7	# 8	# 16
	fine sanding	# 9	# 10	# 17
	normal sanding	# 11	# 12	# 18

EXPERIMENTAL RESULTS

Figures 2 to 7 show plots of the dielectric strength against interfacial pressure for all test slices cut from both T-connectors and straight cable joints. Table 2 explains the symbols used in these plots to indicate the type of interface and the sanding conditions .

Table 2: Meaning of the symbols in Figures 2 to 7

Interface	Sanding condition		
	none	normal (150 grit)	fine (150+600 grit)
EPDM/XLPE	●	▲	■
EPDM/EPDM	O	N/A	N/A

Effect of grease

Figures 2 to 7 show that the dielectric strength of EPDM/XLPE interfaces improves by a factor of 3-4 when the XLPE surface is sanded and greased, whiel that of unsanded EPDM/XLPE interfaces improves sligtly when grease is present.
It was also observed that the dielectric performance of ungreased EPDM/EPDM interfaces is as good as that of greased interfaces.

Effect of interfacial pressure

As confirmed by previous work [3-6], the dielectric strength of interfaces tends to increase with the interfacial pressure. This is especially noticeable with ungreased test samples (see Figs. 4 and 5).

Effect of sanding

Some cable joints and T-connectors were dissected to examine the spreading of the grease at EPDM/XLPE interfaces. It was been observed that interfaces with normal sanding had a thin layer of grease all over the surface, although thicker spots of grease were present in some areas. The situation is different in the case of no sanding or fine sanding. In both cases, there were regions with no grease at all, although in some areas, thick spots of grease were present. It should be mentioned that EPDM/EPDM interfaces also showed regions where there was no grease.

It is suspected that the non-uniform spreading of grease at interfaces might cause scatter in the interfacial breakdown test results.

The result have showed that the dielectric strength of greased EPDM/XLPE interfaces in T-connectors is better with a normal sanding than with a fine sanding. Indeed, the greater roughness of the normal-sanded XLPE surface keeps the grease at the interface and allows more uniform spreading. On the other hand, for ungreased interfaces, the dielectric strength of test samples with unsanded XLPE surfaces is much higher by a factor of 2-3 than that obtained when the surfaces are sanded.

Effect of adapter size for T-connectors

Adapter "B", which offers a tighter fit at both EPDM/EPDM and XLPE/EPDM interfaces, generated higher interfacial pressures (see Figs. 2-5). In fact, interfacial pressures are roughly twice as high as those obtained with adapter "A". The higher pressure with adapter "B" seems to slightly improve the interfacial dielectric strength of ungreased test samples for T-connectors.

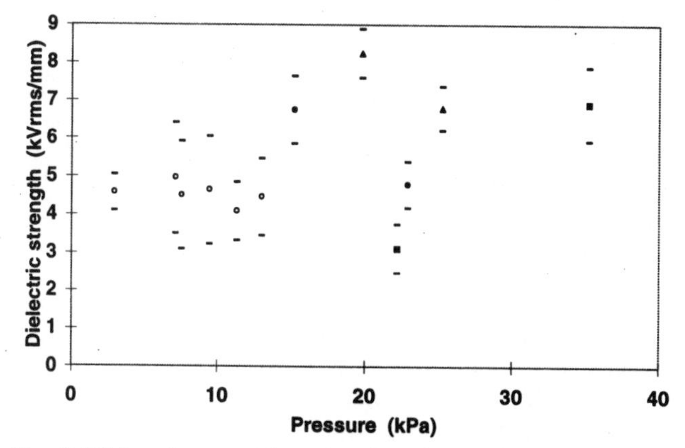

Fig. 2. Dielectric strength against interfacial pressure of greased sample (T-connector with "A" adapter, samples #1, 3 and 5) for different surface conditions (see Table 2).

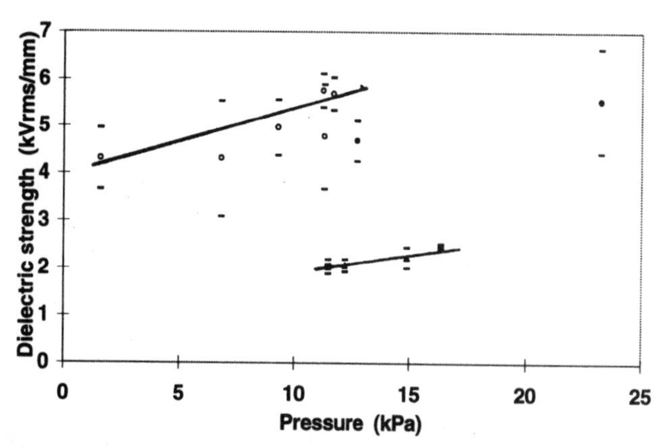

Fig. 4. Dielectric strength against interfacial pressure of ungreased sample (T-connector with "A" adapter, samples #7, 9and 11) for different surface conditions (see Table 2).

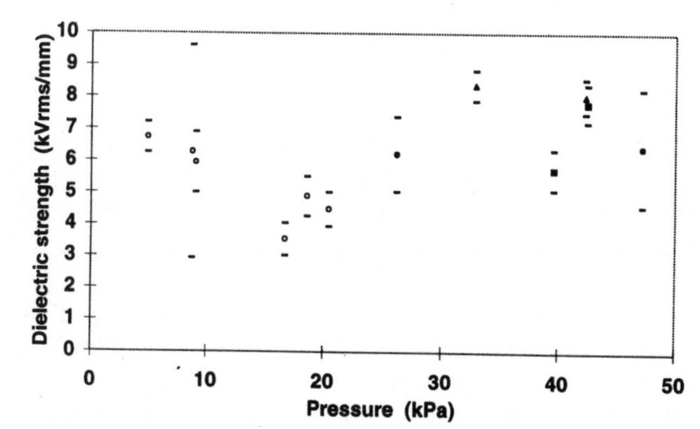

Fig. 3. Dielectric strength against interfacial pressure of greased sample (T-connector with "B" adapter, samples #2, 4 and 6) for different surface conditions (see Table 2).

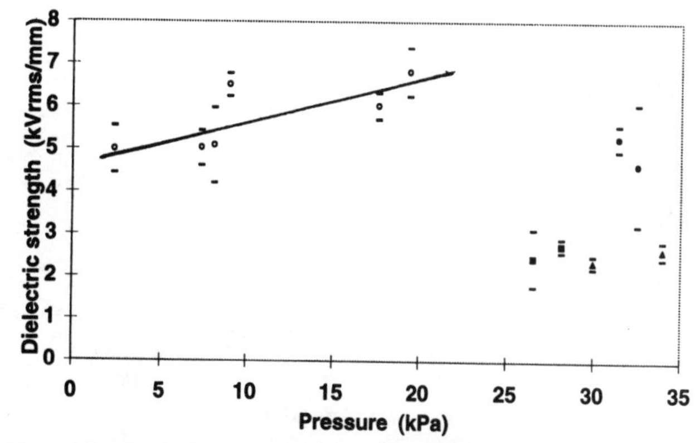

Fig. 5. Dielectric strength against interfacial pressure of ungreased sample (T-connector with "B" adapter, samples #8, 10 and 12) for different surface conditions (see Table 2).

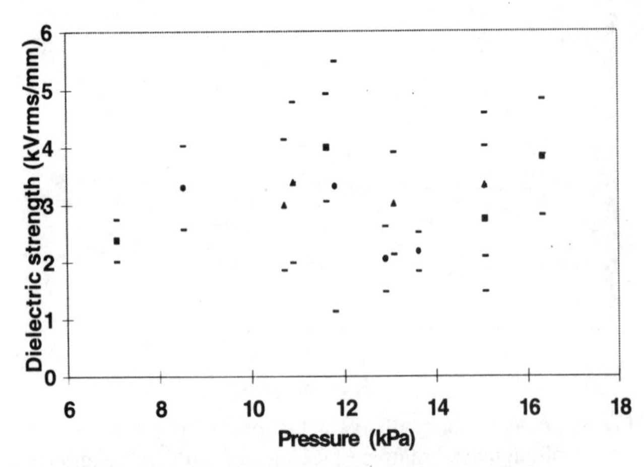

Fig. 6. Dielectric strength against interfacial pressure of greased sample (straight joint, samples #13, 14 and 15) for different surface conditions (see Table 2).

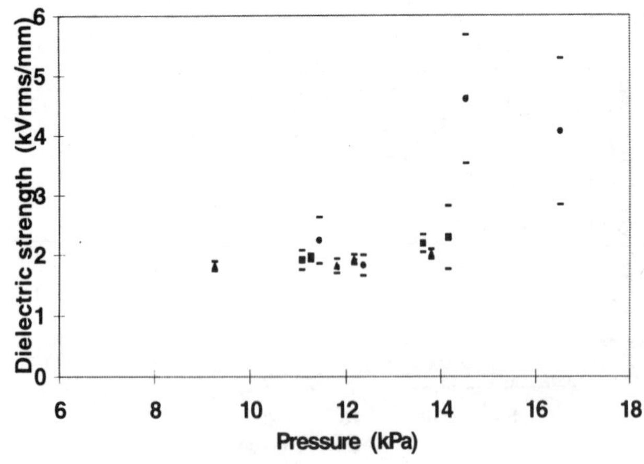

fig. 7. Dielectric strength against interfacial pressure of ungreased sample (straight joint, samples #16, 17 and 18) for different surface conditions (see Table 2).

CONCLUSIONS

Grease substantially improves the dielectric strength of EPDM/XLPE interfaces with a sanded XLPE cable surface.

Normal sanding (150-grit) of the XLPE surface is preferable, since it allows a more uniform spreading of the grease at the EPDM/XLPE interface thus improving its dielectric strength.

If no grease is used at EPDM/XLPE interfaces, sanding of the XLPE surface must be avoided to ensure a good dielectric strength there.

The dielectric strength of ungreased EPDM/EPDM interfaces is as good as that of greased interfaces. This fact supports previous work [5,6] where the EPDM/EPDM interfaces (for cable accessories aged in the field) aged better than EPDM/XLPE interfaces because it does not rely on the grease to assure a good dielectric strength at the interface.

It is preferable to choose a T-connector adapter which offers a tight fit at interfaces. This will keep interfacial pressures higher and therefore hold the dielectric strength at a good value (> 5 kV/mm), especially if degradation of the grease occurs with aging.

ACKNOWLEDGMENTS

The author wishes to thank Hydro-Québec (direction Distribution) for its financial support, and Mr. Yan Riopel (a coop student from École Polytechnique de Montréal) for carrying out all the experimental measurements.

REFERENCES

[1] L. Lamarre and C. Dang, "Characterization of Medium-Voltage Cable Splices Aged in Service," Proceedings of the Jicable 91 International Conference, Versailles, France, June 24-28, 1991, pp. 298-304.

[2] D. Fournier and L. Lamarre, "Effect of Pressure and Length on Interfacial Breakdown Between Two Dielectric Surfaces," Proceedings of the IEEE International. Symposium on Electrical Insulation, Baltimore, MD, June 1992, pp. 270-272.

[3] D. Fournier and L. Lamarre, "Interfacial Breakdown Between Two EPDM Surfaces," Proceedings of the IEE Sixth International Conference on Dielectric Materials, Measurements and Applications, Manchester, U.K., September 1992, pp. 330-333.

[4] D. Fournier and L. Lamarre, "Effect of Pressure and Temperature on Interfacial Breakdown Between Two Dielectric Surfaces, " Proceedings of the IEEE Conference on Electrical Insulation and Dielectric Phenomena, Victoria, B.C., October 1992, pp. 229-235.

[5] D. Fournier, C. Dand and L. Paquin, "Interfacial Breakdown in Cable Joints," Proceedings of the IEEE International. Symposium on Electrical Insulation, Pittsburgh, PA, June 1994, pp. 450-452.

[6] C. Dand and D. Fournier, "Dielectric Performance of Interfaces in Premolded Joints," to be published in IEEE Trans. on Power Delivery.

Conference Record of the 1996 IEEE International Symposium on Electrical Insulation, Montreal, Quebec, Canada, June 16-19, 1996

MODELING OF CABLE TERMINATIONS WITH EMBEDDED ELECTRODES

Stojan V.Nikolajevic, Member IEEE
The Electric Power Distribution Co.
Belgrade

Neda M.Pekaric-Nadj
Faculty of Technical Sciences
University of Novi Sad

Radisa M.Dimitrijevic Mirjana Djurovic
The Cable Factory
Jagodina

Yugoslavia

Abstract: **The paper describes a study of various cable termination constructions for medium voltage cross-linked polyethylene (MV XLPE) cables. A special device was used for electrical field measurements around the cable termination, which made it possible to monitor, how stress relief materials with different relative permittivity, thickness of stress relief layer and placement of isolated or grounding embedded electrodes (EE) affect electrical stress grading. The results of measurement for each construction were examined by mathematical modeling based on finite element method (FEM).**

Also, the influence of dielectric losses in the termination was considered, when relative permittivity of the stress relief material is high. Finally, the selected constructions of cable termination were tested in service conditions with load cycling.

INTRODUCTION

Modeling of cable terminations is very complex problem especially, when the analyzed system, being rotationally symmetrical, consists of more electrodes and more different dielectrics. The essential characteristic of cable termination as a part of cable line is reliability, which the higher it is, the smaller maximum electric field strength in the vicinity of the screen end is. By modeling, the service conditions of cable terminations could be simulated and the high costs of experiment, as way of investigation could be avoided.

The principal aim in designing of screened cable terminations typically above 6 kV is to decrease strong electric field at the insulation screen end, which must be, depending on circuit voltage, at appropriate so-called creepage distance from the cable conductor.

The method of stress grading in these terminations is based on application of stress relief cone or special materials in the form of pad, either of non-linear resistivity or high relative permittivity. Beside of these common way of controlling the stress distribution, the aim of this work was to investigate the influence of EE, which potential could be either zero (grounded EE) or so-called "floating" (isolated EE). The placement and form of EE were chosen on the basis of the voltage withstand tests, performed in high voltage (HV) laboratory.

The effects of position along the cable termination and grounding of EE were examined by mathematical modeling. All measurements and calculation were performed on cable Al/XLPE/PVC 1 x 120 mm^2 20 kV, manufactured in accordance with IEC publication 502 - 1983, always under the same condition. The pad of 10 cm wide typically is applied from the end of the screen over the cable insulation under the outer protection, dual wall (XLPE/EPR) heat-shrinkable tube (HST).

The cable terminations were considered as follows:

1. relative permittivity of the stress relief layer was change in range from 5.5 to 40.4 ;
2. thickness of the stress relief layer was 1,3 and 5 mm, respectively.
3. the EE was included in the cable termination construction and its position was considered.

THEORETICAL BASIS OF CABLE TERMINATION CONSTRUCTION

The first step in cable termination construction is theoretical analysis of electrical field strength in the termination. The equivalent circuit of cable termination without and with EE is presented in the Figure 1 and 2 respectively.

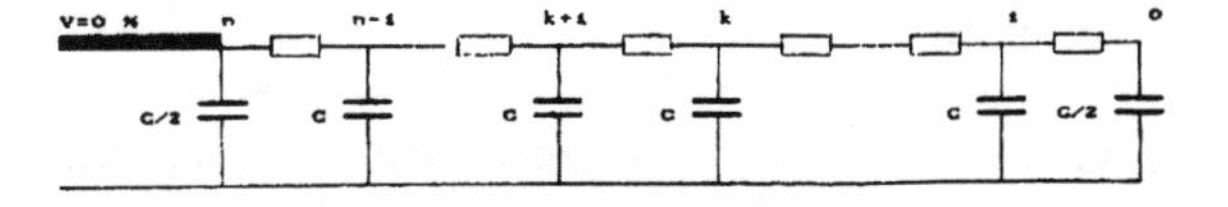

Figure 1. Equivalent circuit of cable termination without EE

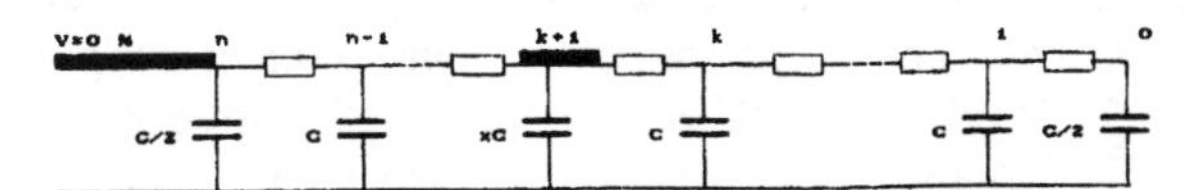

Figure 2. Figure 1. Equivalent circuit of cable termination with EE (x- width of EE)

For both cases, electric field strength can be determined by solving Laplace's equation in cylindrical coordinates:

$$\frac{\partial^2\varphi}{\partial r^2}+\frac{1}{r}\cdot\frac{\partial\varphi}{\partial r}+\frac{\partial^2\varphi}{\partial z^2}=0 \qquad (2)$$

where $\varphi = f(r,z)$ and r - radius. The boundaries were defined by the phase conductor potential, V_1=10 kV and the screen ground potential $V_2=0$. Equipotential map was generated using FEM and the electric field was calculated from the corresponding potential.

Stress control material

The properties of stress control materials, used in the experiment are shown in the Table 1.

Table 1. The properties of stress control materials

Mark of material	RZM	RZGO	No2
ρ(Ωcm)	10E+15	2E+14	3E+11
tgδ	0.0139	0.1088	0.101
ε_r	5.5	10.6	40.4
Tensile strength (MPa)	2	2	2.5
Elongation (%)	300	300	300

It can be seen, that the relative permittivity is in the range (5.5 - 40.4) which is preferred for practical application. Achieving high relative permittivity (about 10) is not problem, whereas the mixtures for electrical stress control contain, beside basic polymer (butyl rubber), such additives (SiC, ZnO), which raise total permittivity of mixtures.

However, for higher values of permittivity, as the loading of the additives is not proportional to their relative permittivity, capacitive coupling of the conductive particle platelets shall be used. Namely, the incorporation of iron and/or carbon flakes provides infinite array of conductive particles in the insulation and, thereby, raising the relative permittivity of stress control material.

Polymeric mixture i.e. stress control material is made of the following components:

- basic elastomer (BUTYL ELASTOMER) 40 %
- fillers of different chemical structure and characteristics (titanium dioxide, ferro oxide, carbon black) 40 %
- the other additives (paraffin, polyethylene vax, vinyl silane etc.) 18 %
- naphthenic oil 2%

This polymeric mixture is distinguished by good electrical and mechanical properties, good processibility and excellent self-amalgam, what is very important for its application.

Electrical stress grading with applicable materials can be explain in the following manner. Namely, using of material with adequate non-linear current - voltage characteristic (RZM, RZGO) the leakage current through the stress relief layer forms also adequate voltage drop, which should be the same per unit length as much as possible and linear function of axial distance in ideal case. That means, that characteristic of voltage distribution $U_x= f(x)$ along this layer must not have either knee point or high slope in any part.

As for using of high relative permittivity material (No2), the refraction of electric field is used at interface stress relief layer-cable insulation, so that the higher the relative permittivity, the farther the electric field refracts towards the cable end. However, for practical application a relative permittivity up to 40 is adequate to prevent excessive dielectric heating, described by formula

$$P_d = 3V_0^2\omega C_k tg\delta \quad (W) \qquad (3)$$

where capacitance of stress relief layer C_k is proportional to e_r

$$C_k = \frac{2\pi\varepsilon_0\varepsilon_r}{\ln\left(1+\frac{2t}{d}\right)} \quad (F/m) \qquad (4)$$

In this formula t is thickness of stress relief layer and d diameter over the cable insulation.

Besides, a embedded electrode is assumed to affect the equipotential map of cable termination in any way. After the voltage withstand test with several electrode shape , a ring of 10 mm width copper tape was chosen for the embedded electrode as already mentioned. This ring can be grounded or at "floating" potential and must be placed close the cable screen end at distance, which is to be optimized.

Table 2. Dielectric heating for different stress relief materials of 10 cm wide and their thickness t

ε	t (mm)	d (mm)	Ck (F)	Uo (kV)	tg δ	Pd (W)
5.5	1	25	3.97E-10	6	0.0139	0.06
10.6	1	25	7.66E-10	6	0.1088	0.94
40.4	1	25	2.92E-09	6	0.101	3.33
5.5	1	28	4.43E-10	12	0.0139	0.28
10.6	1	28	8.54E-10	12	0.1088	4.20
40.4	1	28	3.26E-09	12	0.101	14.87
5.5	1	32	5.04E-10	20	0.0139	0.88
10.6	1	32	9.72E-10	20	0.1088	13.29
40.4	1	32	3.71E-09	20	0.101	47.01
5.5	2	25	2.06E-10	6	0.0139	0.03
10.6	2	25	3.97E-10	6	0.1088	0.49
40.4	2	25	1.51E-09	6	0.101	1.73
5.5	2	28	2.29E-10	12	0.0139	0.14
10.6	2	28	4.41E-10	12	0.1088	2.17
40.4	2	28	1.68E-09	12	0.101	7.68
5.5	2	32	2.60E-10	20	0.0139	0.45
10.6	2	32	5.00E-10	20	0.1088	6.84
40.4	2	32	1.91E-09	20	0.101	24.20
5.5	3	25	1.42E-10	6	0.0139	0.02
10.6	3	25	2.74E-10	6	0.1088	0.34
40.4	3	25	1.04E-09	6	0.101	1.19
5.5	3	28	1.58E-10	12	0.0139	0.10
10.6	3	28	3.04E-10	12	0.1088	1.49
40.4	3	28	1.16E-09	12	0.101	5.28
5.5	3	32	1.78E-10	20	0.0139	0.31
10.6	3	32	3.43E-10	20	0.1088	4.69
40.4	3	32	1.31E-09	20	0.101	16.58

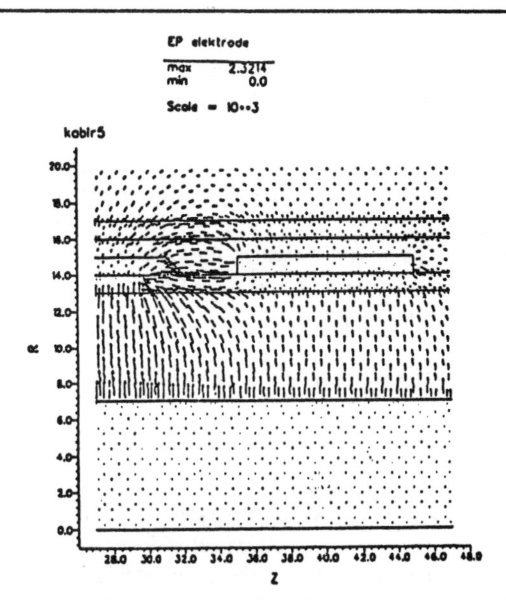

Fig. 3.1. The detail of the electric field in the vicinity of the shield edge. Cable termination formed of the RZGO material with relative permitivity εr=10.6 and the isolated embedded electrode (EE) separated 5 mm from the shield end. Thickness of the stress relief layer 1 mm.

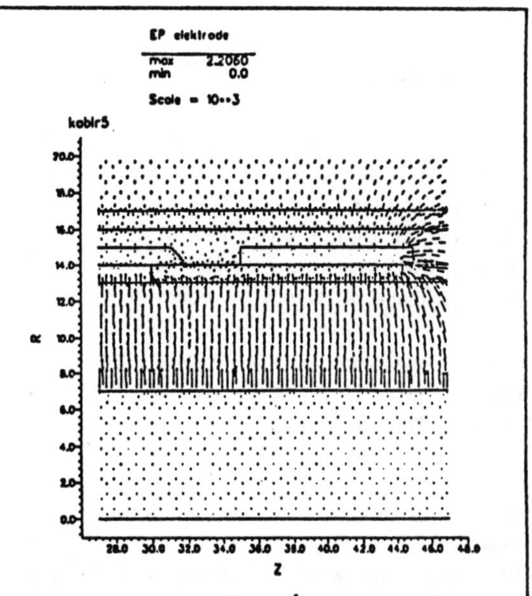

Fig. 3.2. The detail of the electric field in the vicinity of the shield edge. Cable termination formed of the RZGO material with relative permitivity εr=10.6 and the embedded electrode (EE) grounded, separated 5 mm from the shield end. Thickness of the stress relief layer 1 mm.

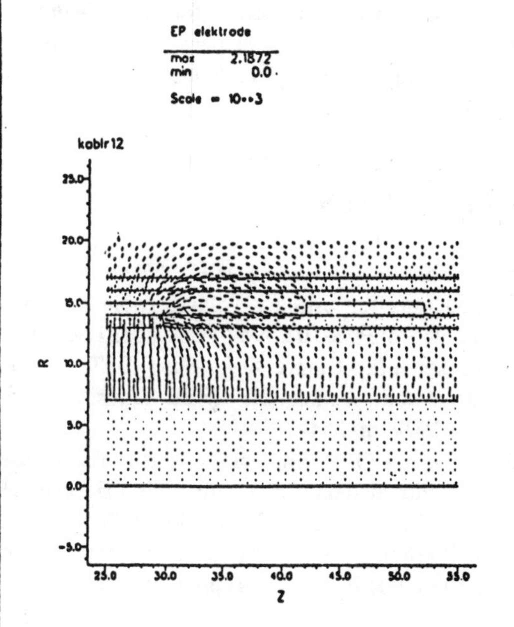

Fig. 3.3. The detail of the electric field in the vicinity of the shield edge. Cable termination formed of the RZGO material with relative permitivity εr=10.6 and the isolated embedded electrode (EE) separated 12 mm from the shield end. Thickness of the stress relief layer 1 mm.

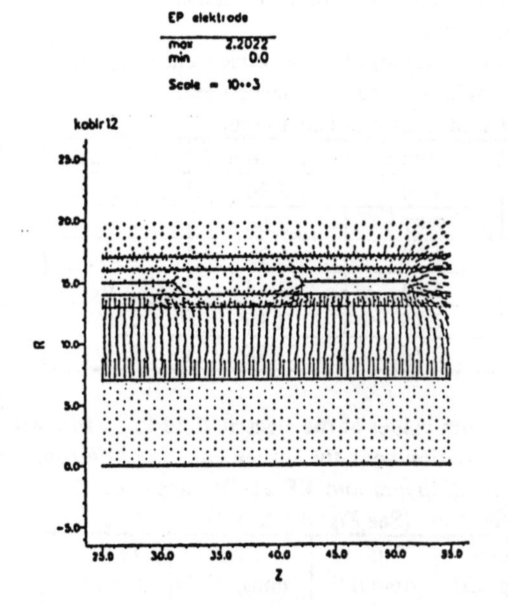

Fig. 3.4. The detail of the electric field in the vicinity of the shield edge. Cable termination formed of the RZGO material with relative permitivity εr=10.6 and the embedded electrode (EE) grounded, separated 12 mm from the shield end. Thickness of the stress relief layer 1 mm.

RESULTS OF MEASUREMENT

Measurements were performed in HV laboratory at open ended cables. The 3 axes sound was designed for axial and radial fixed movements in 1 cm steps. The readings of the sound were followed by digital voltmeter. Materials with relative permittivity ranging from 5.5 up to 40.4 were tested. There was no significant difference in effect found with increasing thickness of the layer. The layer thickness of 1 mm was found to be satisfactory for most of tested materials. Only for lower permittivity materials (ε_r=5.5) the layer thickness of 3 mm turned to be more effective than 1 mm. The thickness of 5 mm made no differences compared to 3 mm. On the other hand, higher permittivity was linked with higher dielectric loss and for that reason the increase of the permittivity was limited at value 40 approximately [Lit.1]. Addition of EE had a remarkable effect on the field. Isolated EE made the situation worse. Grounded EE in combination with the increased permittivity layer (ε_r=10.6) turned to be very effective for the stress relief.

NUMERICAL RESULTS

Numerical model for each of the designed configurations was examined. Laplace's equation (2) was solved and the electric field was calculated from the corresponding potential. Table 3 and Table 4 summarize some numerical results for different configurations.

Table 3. Maximum values of the electric field E_{max} (MV/m) in the cable insulation near the screen end for different thickness and ε_r of the stress relief layer

ε_r	t=1mm	t=3 mm	t=5 mm
1	3.30	3.30	3.30
2.3	2.87	2.66	2.80
5.5	2.39	2.25	2.30
10.6	2.13	2.14	2.14
20.5	2.15	2.16	2.16
40.4	2.16	2.18	2.17

Table 4. Maximum values of the electric field E_{max} (MV/m) in the cable insulation near the screen end for different ε of the stress relief layers and EE at distance 5 or 12 mm from the screen end (See Fig. 3.1 to 3.4)

ε_r	5 mm nongr. EE	5 mm ground.EE	12 mm nongr.EE	12 mm ground.EE
1	3.55	3.79	3.5	3.6
2.3	3.15	2.93	2.73	2.58
5.5	2.66	2.3	2.32	2.19
10.6	2.33	2.21	2.18	2.20
40.4	2.16	2.21	2.20	2.20
100	2.18	2.21	2.20	2.20

The EE placed on the top of the layer with ε_r=10.6 was examined for different distances from the screen end. The analyses was performed with EE isolated or grounded. As may be seen from Figures 3.1 and 3.3 isolated EE made the stress stronger. The situation was found to be worse when the isolated EE was closer to the screen end. It can be seen from Figures 3.2 and 3.4 that in the case when EE was grounded the stress was transferred from the screen end to the EE and surrounding material.

CONCLUSION

The results of this investigation may be summarized as follows:

- the pad made of the increased permittivity only if relative permittivity of the stress relief layer 1 mm thickness was greater then 10, has the effect on stress grading . Lower permittivity material, between 5 and 10, may be used , but thickness of material must be greater.
- addition of EE had a significant effect on the electric field. Insulated EE made the situation worse. Grounded EE in combination with the increased permittivity layer (ε_r >10) turned to be very effective for the stress relief.
- when the ε_r of the layer was between 5 and 10 the EE played and important role. Even than grounded EE was found to be a good solution for the cable termination.
- upper limit for the layer permittivity in combination with dissipation factor tgδ is dielectric loss which must not be overcome.

REFERENCES

1. Nelson P.N., Hervig H.C. "High dielectric constant materials for primary voltage cable termination", IEEE Trans.Vol.PES-103, Nov.1984, pp3211-3216.
2. Blake A.E., Clarke G.J., Starr W.T. "Improvement in stress control materials" IEEE Trans.Vol.PES-103, April 1-6,1979. pp 264-270.
3. Andersen O.W."Laplacian electrostatic field calculations by finite elements with automatic grid generation", IEEE PES Winter Meeting New York 1973.
4. McPartland, J.F.Hand book of practical electrical design, McGraw-Hill Inc., 1984, pp 9.25-9.30.

Conference Record of the 1996 IEEE International Symposium on Electrical Insulation, Montreal, Quebec, Canada, June 16-19, 1996

EFFECT OF NEW CLEANING SOLVENTS ON ELECTRICAL INSULATION MATERIALS

L. Lamarre, N. Amyot, M. Durand and C. Laverdure

Hydro-Québec (IREQ), 1800 Montée Ste-Julie, Varennes, Québec, CANADA, J3X 1S1

Abstract: **Trichloroethane which was often used to clean high voltage cables (during installation of joints or terminations), various accessories, generators and motors, now has to be replaced by friendlier solvents to protect the ozone layer. Among the new solvents considered, di-chloro-fluoro ethanes prove to be the best candidates; terpenes offered a cheaper alternative for some applications. A set of various physical, chemical and electrical tests was designed to assess compatibility and preservation of long term electrical performance of the insulation materials exposed to the cleaning solvent. Physical testing includes swelling, decoloration and elongation at break measurements. Chemical testing includes volatility, residue analysis; attention was paid to the flash point. Electrical performance was checked using tan δ, resistance to breakdown and tracking tests.**

INTRODUCTION

For the last 25 years, trichloroethane (Methylchloroform) has been used at Hydro-Québec as a cleaning (degreasing) solvent for various electrical apparatuses. Applications varied from using small volumes for cleaning extruded XLPE transport or distribution cables before jointing or terminating to using large volumes (i.e., ~300 liters) for the cleaning of hydro-generators.

However, 1,1,1-trichloroethane has been listed as an ozone depleting chemical under the Montréal Protocol that mandates the phase-out and eventual elimination of this substance [1, 4, 5]. These reasons made it necessary for utilities, solvent and accessory manufacturers to find alternative solvents for cable cleaning. The Montréal Protocol has specified a reduction of its production of 50% by January 1994, of 85% by January 1995 and of 100% by January 1996. Scientists and Engineers working in the field of electrical insulation materials have since been looking for a replacement [4-11]. For aspects regarding cables, a committee is addressing the question at ICC meetings [3].

Providing a clean insulation surface free from remaining semiconductor residues, dirt, surface oils and moisture is one of the most important preparation steps before installing a molded joint on power cable to ensure lasting integrity. 1,1,1-Trichloroethane is well suited for this application. It has no flash point, cleans oily and greasy soils extremely well, dries very quickly, leaves no residue, is inexpensive and does not affect the operation of cable's extruded components when contact is limited to short-term exposure. However, like other electrical solvents, it has been under pressure for the last several years because of the potential toxicity of its vapor.

A variety of cable cleaning solvents are currently proposed by manufacturers as candidates for the replacement of 1,1,1-trichloroethane while other solvents may also become potential candidates for use in cable splicing [6, 7]. Before using any of these available alternates, a complete evaluation of their suitability has to be done. To the present day, alternate solvents that are effective as cable cleaners have been evaluated by cable, accessory and solvent manufacturers and also by some electrical utilities [8-11] and by Hydro-Québec [12-15]. Parameters that are usually considered for the evaluation of cable cleaning solvents are reviewed in [3]. Most of the solvent evaluations include evaporation rate, flammability and cleaning effectiveness. Some evaluations also included assessment of solvent compatibility with the cable and splice materials, especially for the semiconductive shielding layers which may come in contact with the solvent during the cleaning operation.

Trichloroethylene, as well as terpenes (D-limonene) and a newly synthesized molecule (1,1-dichloro-1-fluoroethane, otherwise known as HCFC 141b), all became strong candidates to replace trichloroethane. Therefore, this paper is focusing more particularly on these three candidates for replacement of trichloroethane: trichloroethylene, a terpene based mix and pure HCFC 141b.

Among the numerous criteria to consider for a cleaning solvent, the flash point ought to be given serious consideration. Trichloroethylene has no flash point; so does the pure HCFC 141b. Hydrocarbon/Terpene blends have flash points of ~60-65°C whereas hydrocarbon/HCFC 141b blends are around 80°C.

COMPARISON OF SOLVENTS

This study includes the effect of four solvents on all materials involved in splicing: semiconductive cable shield (carbon black filled EVA), cable insulation (XLPE), splice insulation (EPDM) and splice semiconductive material. Effects on the epoxy insulations and on wedging materials of generators bars were also checked, but none was observed.

We have considered a number of physical and chemical properties that a suitable cleaning solvent or mixture should have. They fall into three categories: dielectric properties (tan δ, resistance to breakdown, resistivity), chemical properties (volatility, compatibility with materials) and others such as propensity to swell surrounding materials and practical considerations such as ease of application and cost. Most important of all, safety is the determining factor in making the final choice.

Dielectric Properties

Breakdown voltages of the solvents were found to be very satisfactory (Table I, showing breakdown voltages of the cleaning solvents considered in this study).

Table I
Breakdown strength of solvents

Solvent	Breakdown Strength(1)
1,1,1-Trichloroethane	26.8 ± 1.9
1,1,2-Trichloroethylene	N/A
Hydrocarbon/Terpene Blend	51.8 ± 4.1
HCFC 141b	18.1 ± 3.2

(1) According to ASTM D1816 (VDE electrodes, 2 mm apart).

In the unlikely event that the cleaning solvent contaminates cable oil during the jointing of an oil cable with an extruded type cable (i.e., the making of a transition joint), we have measured the changes of tan δ and resistivity of 1% mixture of solvent and cable oil (shown in Figure 1). The solvents considered were found to be good dielectrics and were not found deleterious to cable oil as shown by tan δ and resistivity measurements.

Regarding the possibility that some solvent could be left behind on the surface of cable insulation (XLPE), it was felt necessary (especially for transport cable) to check for the resistance to interfacial breakdown (i.e., internal tracking between joint and cable). A set up was built in order to verify, at least partially, if a solvent contaminated interface could withstand a certain value of voltage stress. Pieces of cable XLPE and of joint EPDM were sandwiched together and held under pressure. The interface between XLPE and EPDM was previously soaked with solvent; two strips of aluminum were laid across the interface to act as electrodes, and were separated by 4 mm. Results show that all solvents in this study did not weaken interfaces which held at least an average stress of 4 kV/mm. For a 120 kV cable, this implies that interfaces could be able to resist to ~500 kV.

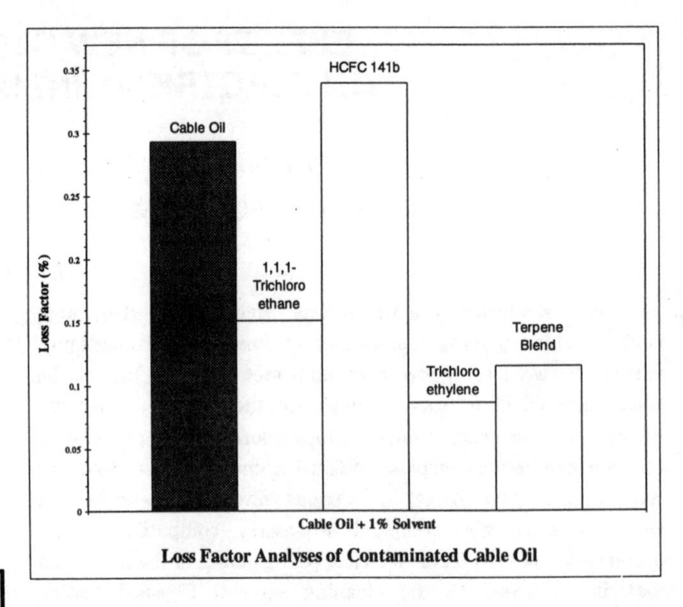

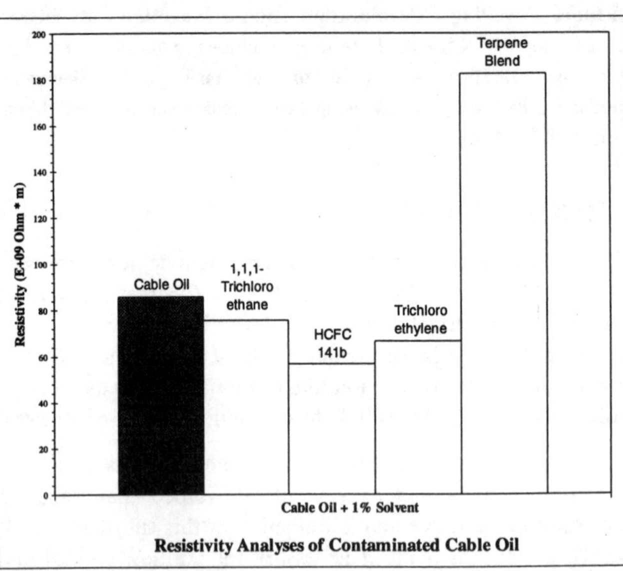

Figure 1. Effect of contaminating cable oil with the various solvents. Changes in tan δ (top) and in resistivity (bottom) are generally without major consequence. Properties shown are for 1% solvent content.

Volatility (Rate of Evaporation)

Table II gives a comparison of the volatility of the 4 solvents chosen in this study. ASTM D1901-85 was used to find relative evaporation times; although used for halogenated solvents and their admixtures, it was found suitable for the solvents in the present study. In this regard, the terpene based solvent was found the most suitable and compares well with trichloroethane.

Table II	
Relative Evaporation Rates	
Solvent	**Relative Rate**
1,1,1-Trichloroethane	1
1,1,2-Trichloroethylene	~ 1
Hydrocarbon/Terpene Blend	< 0.1
HCFC 141b	4 - 5

The evaporation rates were normalized with respect to that of trichloroethane. A relative evaporation rate of 1 means 2.7 that of ether and is obtained by taking the ratios of the respective volumes evaporated per unit of time for a given quantity of solvent.

HCFC 141b was found to be highly volatile (more so than trichloroethane); therefore, this is a limitation when used in degreasing applications such as for generators. A suitable mixture in which a slower evaporating liquid is mixed with the HCFC 141b is more desirable; the second slower evaporating ingredient picks up dirt and greases and is later removed by rinsing with pure HCFC 141b. A commercially available mixture of HCFC 141b with aliphatic hydrocarbons was used for generators in our company; however, these hydrocarbons brought down the flash point to 80°C, but nevertheless were found safe to use since no live equipment was close.

Swelling Capacity

Although the occurrence of swelling in actual applications is very unlikely, we wanted to use this criterion to help us make the optimal choice. Only accidentally a solvent will be left in contact with materials for a sufficient duration to allow any swelling to be observable.

Most solvents, including trichloroethane, have the capability to swell cable, joint or termination materials (semiconductor, EPDM rubber). Figure 2 presents increase in weight of a piece of cable XLPE cut with its insulation shield (semiconductive) still on it, when totally immersed in the 4 solvents for a very long time (144 hours). While minor swelling was noticed for insulation XLPE, it was more serious for the semiconductive layer; the terpene blend and the HCFC 141b performed better in this otherwise extreme test. Figure 3 shows the same experiments for joint materials; pieces of insulating EPDM from a 120 kV premolded cable joint were cut with some semiconductive rubber on them. As expected, swelling was important for EPDM rubber.

Using these curves to perform interpolations for short times, it is possible to get values for a 5 minute exposition of the surface to the solvents. Table III shows that swelling of the materials involved is minor. In one of our study [15], resistivity and breakdown voltage measurements on swollen pieces cut from distribution cables and joints showed that insulation materials were not significantly affected.

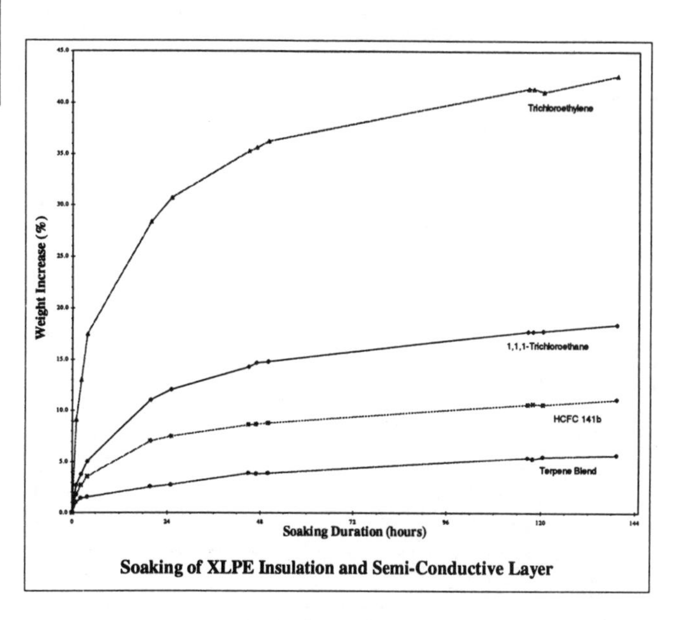

Soaking of XLPE Insulation and Semi-Conductive Layer

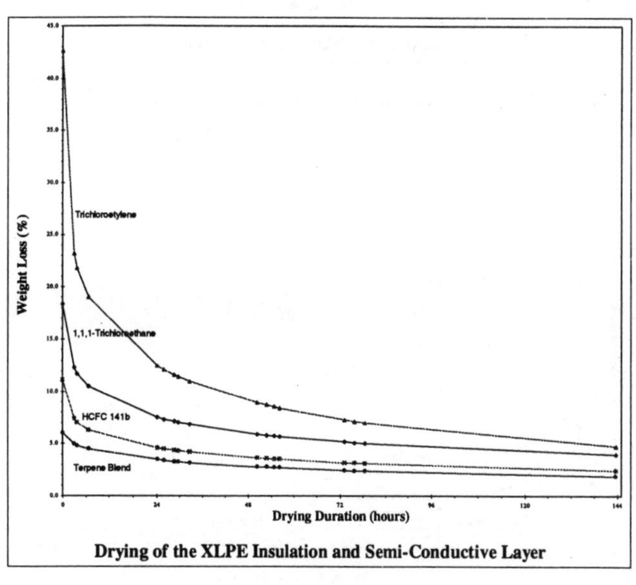

Drying of the XLPE Insulation and Semi-Conductive Layer

Figure 2. (Top) Weight increase of a piece of cable (piece of XLPE with semiconductive screen on it) during total immersion. (Bottom) Subsequent drying when piece is left in ambient air to allow solvent to desorb.

Table III Swelling for 5 minute immersions		
Solvent	% Weight Increase	
	XLPE	EPDM
1,1,1-Trichloroethane	0.22%	0.37%
1,1,2-Trichloroethylene	0.76%	0.93%
Hydrocarbon/Terpene Blend	0.09%	0.17%
HCFC 141b	0.15%	0.26%

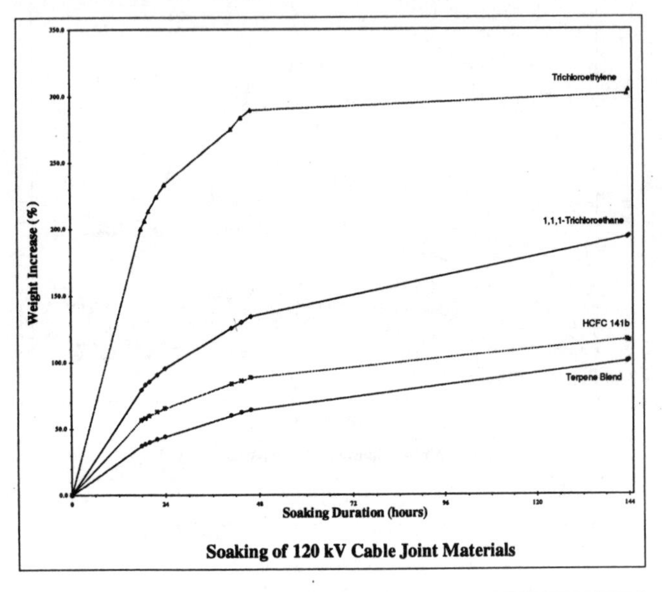

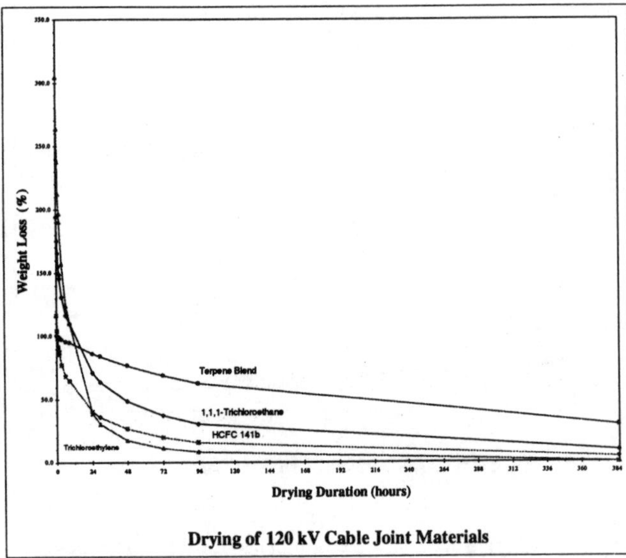

Figure 3. (Top) Weight increase of a piece of cable joint (piece of EPDM with semiconductive screen on it) during total immersion. (Bottom) Subsequent drying when piece is left in ambient air to allow solvent to desorb.

Other Considerations

Trichloroethylene was found to be a potential carcinogenic. Safety aspects gave support to the no-flash point solvent when work has to be done in the proximity of live apparatus; in those cases, pure HCFC 141b is the only one choice.

Cost comparisons are shown in Table IV. HCFC 141b is rather costly; for large volume applications, such as the cleaning of generators or motors, terpenes are an interesting alternative.

Table IV Relative Cost	
Solvent	Relative Cost
1,1,1-Trichloroethane	1
1,1,2-Trichloroethylene	~ 0.5
Hydrocarbon/Terpene Blend	~ 1.25
HCFC 141b	2 - 3

Elongation at break experiments, performed on specimens soaked in solvents and subsequently dried, show that insulation materials retain good mechanical properties [15].

CONCLUSION

Prime candidates for the replacement of trichloroethane are HCFC 141b and terpenes. The former is superior in terms of flash point but the latter is cheaper by a factor of 2 to 3. We have found that both perform very well and affect only slightly dielectric and mechanical properties. Trichloroethylene is perhaps more toxic and maybe carcinogenic. Moreover, among the 4 solvents considered in this study, it does not stand out as far superior with respect to the other ones in the comparison tests shown here. So why bother increasing the risk of health hazard?

ACKNOWLEDGMENT

The authors wish to thank Ms P.E. Beaudoin and D. Jean for help in conducting experimental work. Discussions with H. Krizou, P. Bourassa, M. Bissada and J.P. Banville also contributed to set objectives regarding specific applications (Distribution cables, Transport cables and Generators respectively) for this work.

REFERENCES

1. Montréal Protocol Conferences. Montréal 1987; London 1990; Copenhagen 1992.

2. Smith J.T.III, "Cable Cleaning Solvents: Their Use and Evaluation", IEEE Electr. Insul. Mag., Vol. 9, No. 1, pp. 18-21, 1993.

3. ICC Committee; project 10-54 "Guide for alternative Cleaning Solvents for Electrical Cables".
Chairman: Gene Weitz.

4. Walitsky P.J., "The Effect of the Clean Air Act on The Electrical/Electronics Industry", Proc. 20th Electr. Electron. Insul. Conf., #91CH2991-8, pp.78-81, 1991.

5. Richards B.P., Lodge K.J., Andrews D.C., "CFCs, Ozone Depletion and their Impact on the Electronics Industry. 1. Background", Eng. Science Educ. Jour., Vol. 1, Iss. 3, pp. 131-41, 1992.

6 Lea C., "Solvent Alternatives for the 1990s", Electron. Com. Eng. Jour., Vol. 3, Iss. 2, pp. 53-62, 1991.

7. Hayes M.E., "Semi-aqueous Cleaning: An Alternative to Halogenated Solvents", 40th Electron. Comp. Technol. Conf., #90CH2893-6, pp. 247-52, 1990.

8. Perry D.D., Bolcar J.P., "Effects of Degreasing Solvents on Conductive Separable Connector Shields and Semiconductive Cable Shields", 1989 IEEE/PES T&D Conf., 89 TD 362-5 PWRD.

9. Wilcox D.P., Hunder D.N., "Cable Cleaning Solvents: Environmental Issues and Effective Replacements", IEEE Trans Power Del., Vol. 7, Iss. 2, pp. 1023-6, 1992.

10. Yellow, J., Compatibility Testing Procedures and Results for Plastics and Elastomers Used with Alternative Cleaning Agents", Proc. 40th Electronic Comp. Technol. Conf., Vol. 1, pp.258-60, 1990.

11. Fee J., "Don't let Solvents Degrade Cable Shields", Electr. World., p. 48, 1991.

12. Amyot N., "Expertise sur les solvants de nettoyage pour câbles extrudés", IREQ Internal report #92-306, 1992.

13. Durand M., Lamarre L., "Identification de nouveaux solvants de nettoyage pour les alternateurs pour remplacer le 1,1,1-trichloroéthane", IREQ Internal report 94-350, 1994.

14. Laverdure C., Lamarre L., "Identification de nouveaux solvants de nettoyage pour les câbles haute tension pour remplacer le 1,1,1-trichloroéthane", IREQ Internal report 95-378, 1995.

15. Amyot N., "Évaluation de solvants de nettoyage pour câbles extrudés", IREQ Internal report 95-181, 1995.

Conference Record of the 1996 IEEE International Symposium on Electrical Insulation, Montreal, Quebec, Canada, June 16-19, 1996

Study on the Properties of the Wollastonite Short Fiber—PVC Cable Insulation Composition

Zhang Xianyou
Department of Electrical Marterial
Harbin Universit of Science and Technology
Harbin 150040, P.R.CHina

Zhang Xiao hong
Department of Electrical Marterial
Harbin Universit of Science and Technology
Harbin 150040, P.R.CHina

Abstract: **The surface — treated wollastonite powder is a kind of new inorganic filler on the cable insulation composition. The elemental composition, properties and main uses of the wollastonite short fiber have been described. The slenderness ratio、particle size、particle shape、content and ways of surface treatment of the wollastonite were discussed. The experiments show that the wollastonite short fiber is a kind of inorganic filler with the good reinforce and electrical insulation. By means of SEM and rheogeniometer, the distribution and fluidity of the cable insulation compositions have been observed. The balance experiments of the wollastonite powder and other inorganic fillers such as calcine clay and $CaCO_3$ were finished.** The test results showed that wollastonite is a good inorganic filler for PVC cable compound and filled polymer show better complex properties.

INTRODUCTION

Wollastonite is a kind of natural inorganic non — metallic matter. It exists as perfact aciform crystalline. Its elemental composition is calcium silicate ($CaSiO_3$). Wollastonite short fiber in different particle size and Slenderness ratio from 5/1 to 20/1 may be gotten by means of special process ways. In China, wollastonite is produced in Jilin、Liaoning and Hubei province etc. The production is not only to show high purity and good quality also to present lower price. Inorganic filler plays an important role in PVC Cable compound. Btsides reducing cost price, some properties of the composity, such as mechanical、dielectrical and heat — resistant properties etc., would be improved. At pressnt major inorganic fillers used in PVC cable compound are: calcium carbonate in light weight、superfine calcium、calcine clay and French chalke etc. Inorganic filler of calcium carbonate series is cheaper in price but not better in electrical properties. As compared with $CaCo_3$ calcine clay exhibits good electrical characteristic however its price and the process fluidity of filled PVC compound are not satisfying. Thus it is necessary to develop new inorganic fillers used in PVC cable insulation. The chemical composition of wollastonite is similar to calcium carbonate and calcine clay but wollaslonite is in short fiber crystal and good heat — proof、weather — resistant、chemical stebility and excellent electrical properties are its advantages. The study on application of wollastonite in PVC cable insulation has been made by us Since 1989.

THE STRUCTURE AND CHARACTERI STIC OF WOLLASTONITE

Wollastonite is a kind of natural inorganic non — metallic matter in white colour. It exists as aciform crystal in 5~20/1 slenderness ratio. Calcium Silicate ($CaSiO_3$) is the major element of wollastonite which is theoretically made up of calcium oxide (CaO)in 48.3% and Silica (SiO_2) in 51.7%. The chemical composition and physical characteristic of natural wollastonite are shown in Table I and II

There are three crystal shapes: ortho — wollastonite in triclinic chain structure—— TC type; Subwollastonite in monocline chain crystalline——ZM type; falsewollastonite in tricline tricyclic crystallite. Wollastonite shows good chemical stability, weak acid — proof and alkalescency — resistant but it may be solved in concentration acid. It's satistying in violet ray proof and aging resistant properties. Although wollastonite exists as aciform crystal, wollastonite short fiber in various slenderness ratio can be made by different process. The SEM photos of wollastonite in different slenderness ratio are shown is Fig. 1. and 2.

THE EFFECT OF WOLLASTONITE SHORT FIBER ON THE PROPERTIES OF PVC CABLE COMPOUND

The Effect of Wollastonite Short Fiber in Different Partide Size on PVC Cable Compound

As observed in Table III that the particle size of wollastonite short fiber does not evidently affect the bulk resistivity of PVC cable insulation hower the elongation of filled polymer lower with increasing of partiele size and filling content. The smaller is the particle size, the higher is the price of wollastonite short fiber. Thus wollastonite in 1250/cm^2 particle size was selected for PVC cable compound.

Table I
The Chemicl Composition of Wallastonite Produced in Different Countries

composition \ country / content (%)	China (lishu)	U.S.A.	Finland	India	Uzbekistan	Japan
SiO_2	50.31	48.73	43.50	52.00	37.99	50.00
CaO	46.86	37.84	47.67	47.00	40.64	46.00
Fe_2O_3	0.15	6.63	0.16	0.31	1.56	0.30
Al_2O_3	0.19	3.14	0.50	0.14	3.66	0.30
MgO	0.09	3.63	1.00	——	1.70	0.40
K_2O	0.04	——	——	——	0.42	——
TiO_2	0.25	0.00	——	——	——	0.00
P_2O_5	0.015	——	——	——	——	——
Weight loss in burning	1.58	——	6.60	0.55	13.82	2.00

Table I
The Main Physical Characteristic of Wollastonite

Characteristic	Shape of Crystalline	Density	Mohs Hardness	pH	Index of Refraction	Melt Temperature Point (℃)
Data	Aciform Crystal	2.8	4.9	9.0	1.63	1540

Fig. 1 SEM Photo of wollastonite Short Fiber in Small Slenderness Ratio

Fig. 2 SEM Photo of wollastonite Short Fiber in Large Slenderness Ratio

Property / Particle Size	Bulk Resistivity at 20℃ (Ω·m)	Tensile Strength (MPa)	elongation (%)
400/cm²	2.98×10¹²	23.53	250
800/cm²	3.28×10¹²	24.55	280
1250/cm²	2.30×10¹²	24.14	295
2500/cm²	2.21×10¹²	24.72	290

Table Ⅲ The Effect of Wollastonite Short Fiber in Different Particle Size on PVC Cable Compound*

* XS—2 PVC 100 (by weight)
Plasticizer 45
Stability 9
Filling content of wollastonite 25

The Effect of Wollastonite Short Fiber on Mechanical Properties of PVC Cable Compound

The PVC cable compound filled wollastonite short fiber treated by coupling agent is satisfying in tensile strength and elongation. With filling content increase the tensile strength of filler polymer enhances and presents the maximum value at some filling content. The elongation of the composition shows a similar variety to tensile strength. The reasons may be those: when the content of inorganic filler in polymer is lower, the particle is uniformly distributed in PVC matrix and act as a reinforcment; if the content of inorganic filler is higher, wollastonite partiele would distort original structure of PVC compound and make physical propertios of filled polymer lower. The elongation of PVC cable compound filled wollastonite untreated by coupling agent reduces with filling content increase.

The Effect of Wollastonite Short Fiber on Dielectrieal Properties of PVC Cable Insulation

As there are some low molecule weight object and polar group in PVC composition, the polymer does not exhibit good dielectrical properties. Adding inorganic filler into the matrix polymer is one of ways improving the electrical properties of PVC cable insulation. Calcine clay has been considered to be a good filler on electricl properties of PVC cable compound and $CaCO_3$ in light weight does not so. The bulk resistivity of PVC insulation filled wollastonite short fiber treated by coupling agent is middle value between that of filling calcine clay and $CaCO_3$. Wollastone is also a kind of good filler on dielectrical properties of PVC system. The reasons resulting on PVC cable insulation filled wollastonite short fiber with better electrical properties may be as follows:

- Natural wollastonite exists as aciform crystal and particle in "needle" or unregular shape after processing. The more interface between inorganic filler and polmer and structure trap can catch harmful contamination ion.
- Metal ioncontent in wollastonite produced in Lishu JiLin province China is lower.
- Si and C are elements in the same series but the charge number and atom radius of Si are Larger than that of C, then the ability of SiO_3^{-2} attractting ion is stronger than that of CO_3^{2-}.

The Compareing Study on Properties of PVC Cable Compound with Wollastonite or Calcium Carbonate ($Ca\ CO_3$) or Calcine Clay

To make a series of comparing test, PVC matrix polymer and other assistant agent were kept unchanged in different filled system. The test results are shown as Table Ⅳ and Fig. 3 and 4.

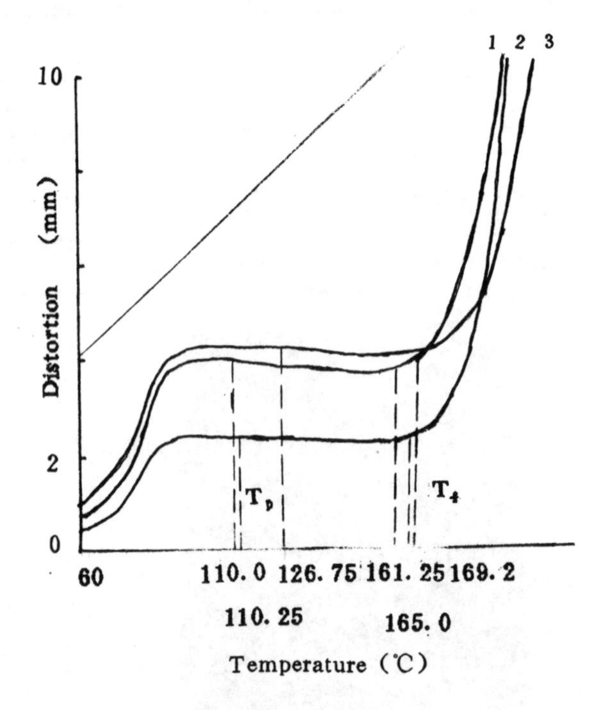

Fig. 3 The Rheological Curves of PVC Cable Compound with Different Fillers
1 — with wollostonite
2 — with calcine clay
3 — with superfine Ca

Table Ⅳ
The Compareing of Praperties of PVC Cable Compound in Different Fillers

Property / data / Composition	But Resistivity at 20℃ (Ω • m)	BUlk Rtsistivity at 70℃ (Ω • m)	Tensile Strength (MPa)	Alongation (%)	Plastified Temperature (℃)	Fluid Temperature (℃)	Plastified Time (min)	Viscosity at 165℃ (Mp)	Viscosity at 170℃ (Mp)
with wollastonite	3.28×10^{12}	7.5×10^{10}	24.55	280	110.25	161.25	10.2	0.962	0.128
with calcine clay	1.41×10^{13}	1.09×10^{11}	22.42	263	126.75	162.20	8.49	1.924	0.481
with superfine Ca	9.2×10^{11}	4.64×10^{10}	21.17	277	111.00	165.00	10.8	0.962	0.148
with Ca (light weight)	8.3×10^{11}	3.04×10^{10}	22.74	245	—	—	—	—	—

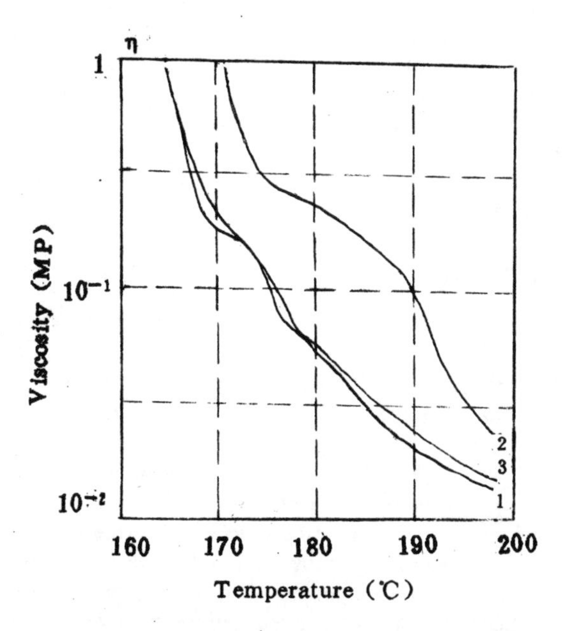

Fig. 4 Viscosity — Temperature Curves in Different System Compound with Different Fillers
1 — filled wollostonite
2 — filled calcine clay
3 — filled superfine Ca

As we observed that the tensile strength of PVC cable compound filled wollastonite short fiber is obviously higher than that of PVC composition with other fillers and its bulk resistivity is lower than that of the system filled calcine clay but higher than that of PVC compound with Ca CO_3. It was also found that the process fluidity of PVC cable composition filled wollastone is the best among the three systems by rheiogical test.

CONCLUSION

- Wollastonite short fiber is a kind of new additive for PVC cable compound. It acts as a reinfore ment and exhibits good electrical properties in PVC cable insulation
- Wollastonite short fiber used in PVC cable composity is about 10～20μm in particie diameter and 10～15/1 in slenderness ratio. The larger is the particle size of wollastonite, the lower is the rupture elongation of PVC cable compound with filler. After wollastonite was modified by coupling agent the process fluidity of filled polymer and interface binding state between PVC resin and inorganic filler would be improved.
- The advantage of wollastonite comparing with calcine clay and calciumcarbonate shows in obsvious reinfore effect, good process fluidity and excellent didelectrical properties.

REFERENCE

[1] H. S. Katz etc. *Handbook of filler and Reinforcement for Plastics.* Van Nostrand Reinhold Company, New York, 1978

[2] Zhang Xianyou etc. "The Researeh on Properties of Soft PVC Filled with Wollastonite", Journal of Harbin Institute of Electrical Technology. Vol. 13. No. 2, June 1990, PP. 147—151

[3] Zhang Xianyou etc. "Researeh of Fluidity of soft PVC — Wollostonite composites" Journal of Harbin Institute of Electrieal Technology. Vol. 13. No. 3, Sep. 1990, PP. 279—285

Conference Record of the 1996 IEEE International Symposium on Electrical Insulation, Montreal, Quebec, Canada, June 16-19, 1996

Aging of XLPE Cable Insulation under Combined Electrical and Mechanical Stresses

E. David, J.-L. Parpal, J.-P. Crine
Institut de recherche d'Hydro-Québec (IREQ)

Abstract: **Extruded crosslinked polyethylene (XLPE) insulation is widely used in high-voltage cables since it presents such attractive features as excellent dielectric properties and good thermomechanical behavior. However, its performance is affected by long-term degradation when it is subjected to the various thermal, mechanical and environmental stresses occurring in service in combination with electrical stress. The synergetic effect of superposed electrical and other stresses remains to be fully clarified. In particular, a fairly high level of mechanical stresses can be present in the insulation volume, originating from residual internal stresses created during the cooling process in the fabrication, external forces when cables are bent sharply, or thermomechanical stresses caused by differential thermal expansion between the conductor and the insulating material. In order to investigate the influence of the superposition of mechanical and electrical stresses, various measurements were conducted on XLPE and LDPE specimens in tip-plane and plane-plane geometries. Experimental data of time-to-breakdown, breakdown field and tree length are presented as a function of the magnitude of the stresses. In all cases, superposition of the mechanical stress was found to reduce the dielectric strength of the material.**

INTRODUCTION

Extruded-cable insulation is exposed to mechanical stresses from residual internal stresses created during the cooling process of the fabrication [1], external forces when cables are bent during installation, or thermomechanical stresses due to differential thermal expansion between the conductor and the insulating material. The latter subject has received considerable attention over the past decade [2, 3]. Furthermore, the internal mechanical stresses are usually nonuniformly distributed in the insulation volume, presenting a certain number of points of stress concentration [4]. It is generally accepted that mechanical strains and stresses have some influence on the electrical performance and the degradation rate of polymeric material under high electrical stress. Indeed, the last few decades have seen a considerable amount of work devoted to investigating the influence of mechanical stresses on the dielectric strength of polyethylene [5, 6] and on the propagation of electrical trees [7, 8] and water trees [9, 10] in polyethylene. However, contradictory results are often reported in different papers and a complete description of the phenomena is still out of reach.

In order to investigate the possible influence of mechanical stresses on the dielectric properties of polyethylene, measurements were conducted on samples in a plane-plane and a pin-plane geometry. For the first type of geometry, ribbons peeled from extruded cables were used. A tensile mechanical stress was then applied perpendicular to the electrical field. Measurements were also conducted on pin-plane XLPE and LDPE samples with various magnitudes of residual mechanical stresses around the embedded electrode. The time-to-inception and the growth rate of the electrical trees at different voltages are reported in this work. Specimens with the highest values of residual stress were found to have the shortest inception times and the longest trees after one hour of aging under different voltages. When the mechanical stress was allowed to relax, the measured treeing resistance improved significantly.

EXPERIMENTAL

A number of 2.5-cm-wide ribbons some 0.21 mm thick were peeled from XLPE dry-cured transmission cables. The rugosity was measured by a Sloan Dektrak II and found to be less then 3 µm. Various physico-chemical measurements were conducted to insure the homogeneity of the specimens and are reported elsewhere [5, 11]. For the measurements with the tip-plane geometry, 2-mm-thick XLPE and LDPE parallelepipeds with a steel needle electrode (10 µm tip radius) were prepared by compression moulding between highly polished copper plates in a Carver press equipped with heating platens. The insulation material was cable grade pellets with or without dicumyl peroxide for the XLPE and the LDPE specimens respectively. The pellets were moulded at 120°C and crosslinked at 180°C in the case of the XLPE specimens and then cooled to room temperature by circulating water into the platens of the press yielding a low cooling rate. The specimens were then reheated to 120°C and quenched in a cold bath of water, which yields a high cooling rate and a high magnitude of internal mechanical stresses around the needle tip. Figure 1 illustrates the isochromatic fringe patterns obtained under cross polaroids of a pin-plane specimen before and after quenching. A complete discussion on the mechanical stress field created around the electrode tip can be found elsewhere [12, 13].

The electrical tests in the plane-plane geometry were performed using a common parallel-plane geometry. The electrodes were opposing rods 6.4-mm in diameter with edges rounded to a 0.8-mm radius (ASTM D149). To avoid flashover, the tests were performed in a Teflon vessel immersed in silicone oil. The experimental setup allowing simultaneous application of electrical and mechanical stresses

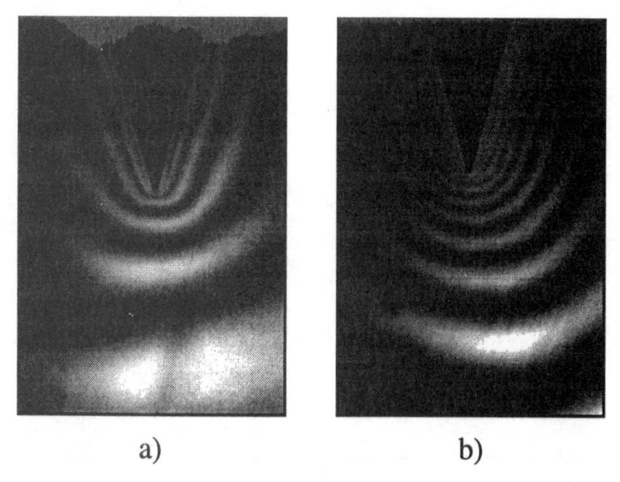

Figure 1. Photoelastic picture of XLPE steel needle-plane specimens: (a) before and (b) after quenching.

is shown in Figure 2. The electrical tests in the tip-plane geometry were performed in a pressurized SF_6 cell using an optical and electrical monitoring system described elsewhere [7]. A 60-Hz AC voltage was raised linearly at a rate of 7 kV/s and held constant at different levels. The onset of the electrical tree was estimated by both the optical monitoring system and a partial-discharge monitoring system.

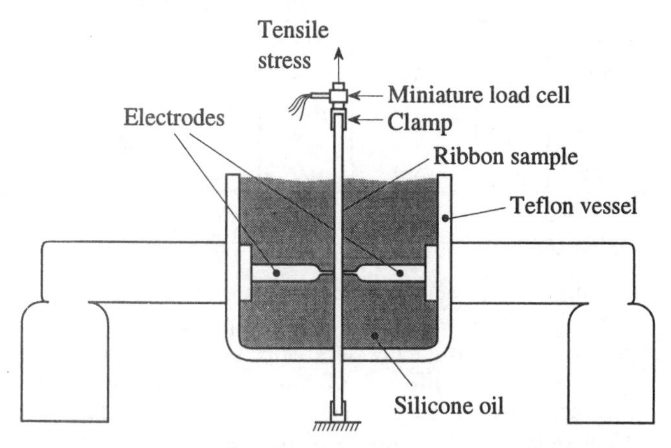

Figure 2. Experimental setup for testing XLPE ribbon specimens.

RESULTS AND DISCUSSIONS

To analyze and objectively compare the experimental data, the two-parameter Weibull distribution given by

$$P(t) = 1 - \exp\left[(t/\alpha)^{\beta}\right] \quad (1)$$

where α and β are the so-called scale and shape parameters, was used for both the breakdown times and the inception times. The estimators $\overline{\alpha}$ and $\overline{\beta}$ of the true values α and β were calculated for each sample using the maximum likelihood method. The 90% confidence bounds for the estimator $\overline{\alpha}$ were computed using the conditional method described by Lawless [14].

Plane-plane geometry

The time-to-breakdown for XLPE ribbons subjected to simultaneous electrical and mechanical stresses was measured in a constant-stress test, a 20-kV voltage (95 kV/mm) being held constant until failure occurs. The results obtained from the statistical analysis for XLPE samples under different magnitudes of mechanical stress are given in Table 1 where n is the number of specimens in the sample and $\overline{\alpha}$ and $\overline{\beta}$ are the estimators of the parameter α and β in Eq. (1). The 90% confidence bounds for $\overline{\alpha}$ are given in the last colum. XLPE A and XLPE B refer to ribbons peeled from two different transmission cables with the same insulating material (sol fraction ≈ 20%). A clear decrease in the time-to-breakdown were observed when simultaneous tensile mechanical stresses of moderate magnitude ($0 < \sigma < \sigma_y$, where σ_y is the yield stress) was applied, as shown in Figure 3. The dependency between the mechanical stress and the time-to-breakdown under these conditions is illustrated in Figure 4 in the form of a semi-log plot. A linear relationship between σ and ln t_b can be seen above a certain threshold value for σ roughly corresponding to the endurance limit of polyethylene in a fatigue test. For values of σ under the threshold (~ 3 MPa), the mechanical stress has no effect on the time-to-breakdown, which then depends only on the value of the electrical stress.

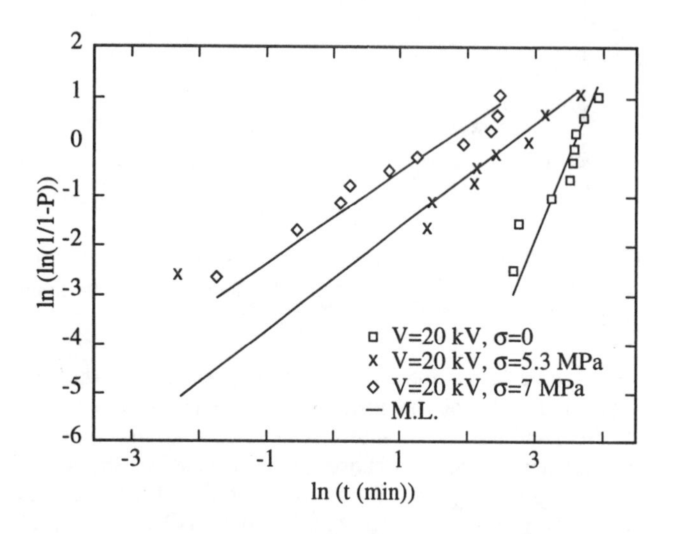

Figure 3. Two-parameter Weibull plots for specimens under combined electrical and mechanical stresses in plane-plane geometry. The solid lines are obtained from the maximum likelihood parameters.

Table 1
Results of the statistical analysis for the time-to-breakdown in plane-plane geometry.

Material	n	σ (MPa)	$\overline{\alpha}$ (min)	$\overline{\beta}$	90% conf. bounds for $\overline{\alpha}$
XLPE A	12	0	32.3	3.3	27.1 - 38.6
XLPE B	9	0	35.8	3.3	28.8 - 44.5
XLPE B	10	3.7	30.9	1.5	17 - 69.6
XLPE A	14	5	12.3	1.1	7.6 - 20
XLPE B	10	5.3	13.8	1.0	7.2 - 26.3
XLPE A	10	7	4.9	0.9	2.4 - 9.9

Tip-plane geometry

The inception time for electrical trees at the tip of a 10-μm tip radius electrode was measured in XLPE and LDPE tip-plane specimens containing various magnitudes of mechanical stress around the needle electrode. The magnitude of the mechanical stress at the tip of the needle (third column of Table 2) was calculated from the initial photoelastatic signal

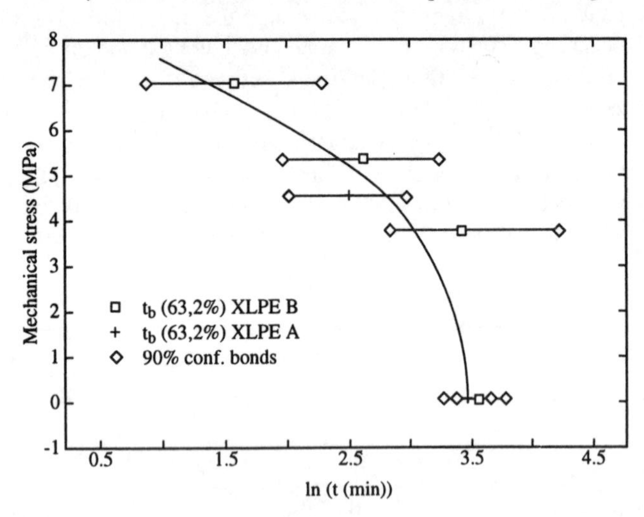

Figure 4. Semi-log plot of the mechanical stress in a constant-stress test at 20 kV vs the time-to-breakdown in plane-plane geometry.

and a stress relaxation equation to take account of the time between fabrication (moment of creation of the mechanical stresses) and the electrical test. A more detailed discussion on the calculation and distribution of the mechanical stress field in the tip-plan geometry can be found in [12, 15]. The statistical results obtained with various XLPE and LDPE samples are summarized in Table 2. The mechanical stress takes negative values since it is a compressive stress. The dependency between the inception time and the mechanical stress is illustrated in Figure 5 in the form of a semi-log plot, showing a linear relationship between σ and ln t for $6 < \sigma < 9$ Mpa.

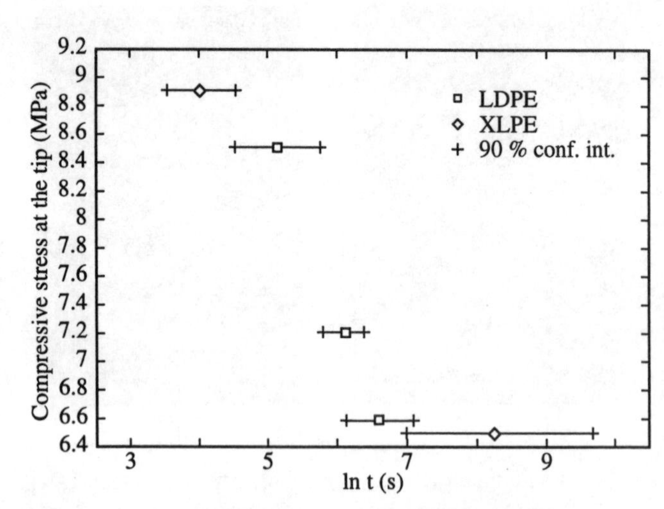

Figure 5. Semi-log plot of the mechanical stress at the tip of the needle of tip-plane specimens vs inception time in pin-plane geometry.

Table 2
Results of the statistical analysis for the electrical tree inception time in a tip-plane geometry.

Material	n	σ (MPa)	$\overline{\alpha}$ (min)	$\overline{\beta}$	90% conf. bounds for $\overline{\alpha}$
XLPE	4	-6.5	65.5	1.1	18 - 265
LDPE	9	-6.6	12.3	1.2	7.6 - 20
LDPE	11	-7.2	7.6	2.0	5.5 - 10
LDPE	9	-8.5	2.8	1.1	1.5 - 5.4
XLPE	11	-8.9	0.9	1.3	0.56 - 1.5

CONCLUSION

All the samples subjected to simultaneous mechanical and electrical stresses whether in a plane-plane or a tip-plane geometry, showed reduced dielectric properties. The dependency of the lifetime of the various samples on the mechanical stresses seemed to show a linear relationship when the logarithm of the lifetime is plotted against the mechanical stress for values of $\sigma > 3$ MPa.

ACKNOWLEDGMENTS

The authors are grateful to Lesley Régnier for the thorough revision of this manuscript. One of the authors (E.D.) thanks Hydro-Québec and the IEEE Dielectrics and Electrical Insulation Society (DEIS) for their financial support.

REFERENCES

1. Vincent, M., Agassant, J.F., De Charentenay, F.X., Oualha, A. "Calcul des contraintes résiduelles dans la gainage en polymère d'un câble de télécommunication sous-marin", J. de Méc. Th. et 1984, Appl. **3** pp. 843-859.

2. Van Aalst, R.J., Van Der Laar, A.M.F.J., Leufkens, P.P. "Contraintes thermomécaniques dans les câbles à haute tension extrudés", CIGRE 1986, paper 21-07.

3. Head, J.G., Crockett, A.E., Taylor, T., Wilson, A. "Thermo-Mechanical Characteristics of XLPE H.V. Cable Insulation", DMMA 1988, pp. 266-269.

4. Hinrichsen, P.F., Houdayer, A.J., Parpal, J.-L., David, E. "Statistical Properties of the Distribution of Bow-Tie Trees in a Field Aged HV Cable", ICPADM 1994, pp. 405-408.

5. David, E., Parpal, J.-L., Crine, J.-P. "Influence of Mechanical Strain and Stress on the Electrical Performance of XLPE Cable Insulation", ISEI 1994, pp. 170-173.

6. Mita, S., Yahagi, K. "Effect of Elongation on Dielectric Breakdown Strength in Polyethylene", Japan. J. Appl. Phys. 1975, **14**, pp. 197-201.

7. David, E., Parpal, J.-L., Crine, J.-P. "Influence of Mechanical Stress and Strain on Electrical Treeing in XLPE", CEIDP 1994, pp. 575-581.

8. Densley, R.J. "An Investigation into the Growth of Electrical Trees in XLPE Cable Insulation", IEEE Trans. on Electr. Insul. 1979, EI-14, pp. 148-158.

9. Ilstad, E., Bardsen, H., Faremo, H., Knutsen, B. "Influence of Mechanical Stess and Frequency on Water Treeing in XLPE Cable Insulation", ISEI 1990, pp. 165-168.

10. Bulinski, A.T., Bamji, S.S. "Water Treeing Degradation under Combined Mechanical and Electrical Stresses", CEIDP 1992, pp. 610-617.

11. David, E. "Influence des contraintes mécaniques sur les propriétés diélectriques du polyéthylène", Ph.D. thesis 1996, École Polytechnique de Montréal.

12. David, E., Parpal, J.-L., Crine, J.-P. "Influence of Internal Mechanical Stress and Strain on Electrical Performance of Polyethylene: Electrical Treeing Resistance", to be published in the IEEE Trans. on Dielectr. and Electr. Insul. 1996.

13. Champion, J.V., Dodd, S.J., Stevens, G.C. "Quantitative Determination of the Residual Mechanical Stress and its Relaxation in Synthetic Resin with an Embedded Electrode", J. Phys. D: Appl. Phys. 1992, **95**, pp. 1821-1824.

14. Lawless, J.F. "Confidence Interval Estimation for the Weibull and Extreme Value Distributions", Technometrics 1978, **20**, pp. 355-364.

15. David, E., Parpal, J.-L., Crine, J.-P. "Electrical Treeing in Mechanically Stressed Polyethylene", JICABLE 1995, pp. 236-241.

Conference Record of the 1996 IEEE International Symposium on Electrical Insulation, Montreal, Quebec, Canada, June 16-19, 1996

STREAMER INITIATION AND PROPAGATION IN INSULATING OIL IN WEAKLY NON-UNIFORM FIELDS UNDER IMPULSE CONDITIONS

R. Badent K. Kist A.J. Schwab
Institute of Electric Energy Systems and High-Voltage Technology
University of Karlsruhe
Federal Republic of Germany

Abstract: **This paper deals with the investigation of prebreakdown phenomena in insulating oil in weakly non-uniform fields of rod-plane geometries with gaps up to 100 mm under impulse voltages of both polarities up to 700 kV. As with the point-plane configuration, the rod-plane geometry shows a decrease of the time to breakdown with increasing voltage rate-of-rise. At a specific rate, a significantly shorter breakdown time is observed both for positive and negative polarities. Beyond this discontinuity range breakdown time decreases again but with lower rates.**

INTRODUCTION

Previous parameter studies of the predischarge behavior in insulating oil, e.g. pressure and voltage dependence (impulse voltages, AC, RF-voltages) were carried out in strongly non-uniform fields of point-plane geometries. The subject of this work is to explore the streamer initiation and propagation in *weakly non-uniform fields*. The experiments were carried out using the same set-up as described earlier [1]. However, the electrode configuration was changed in a rod-plane geometry with a hemispherical tip (radius of curvature 5 mm) and a brass plane (diameter 150 mm) with Rogowski profile.

RESULTS

Breakdown time voltage dependence

In the weakly non-uniform field of a rod-plane geometry similar results compared to the point-plane configuration (r = 50 μm) could be found. To analyze the dependence of the breakdown time and the applied voltage, and, moreover, to find a correlation with prebreakdown mechanisms and structures, it seemed to be necessary to divide the voltage spectrum in three subsequent ranges, which show very different phenomena. Decisive for this voltage range distinction is, for both polarities, the obvious characteristic, that breakdowns only appear in 2 significant time ranges, so called "upper" and "lower breakdown time bands", Figure 1.

On this background and corresponding the most typical distinguishing features we introduce the following three impulse voltage ranges.

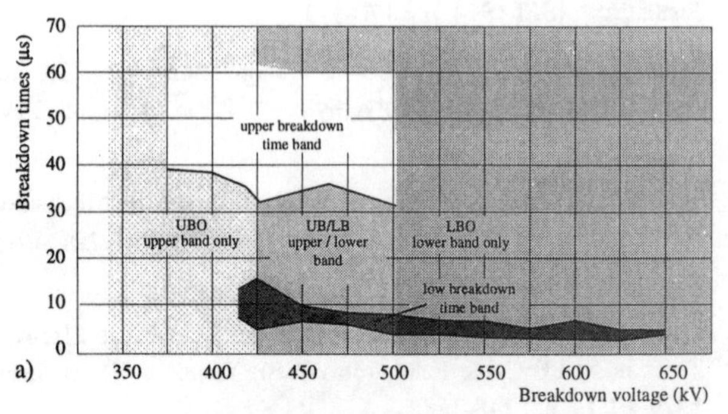

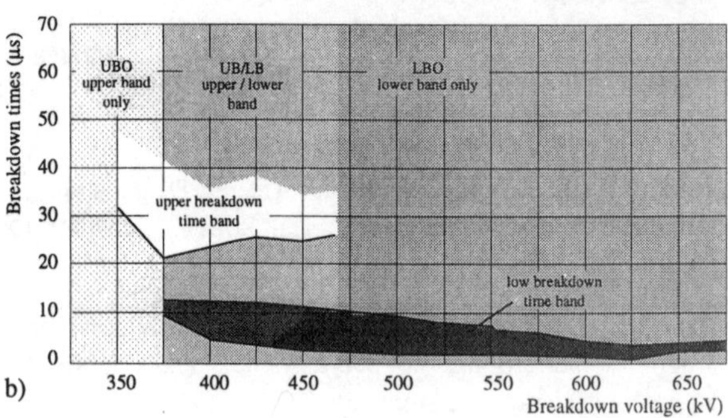

Figure 1a,b: Breakdown times for negative and positive impulse voltages for 100 mm rod-plane gap.

1) UBO *Upper Band Only* – Range of comparatively low voltage and low breakdown probability; breakdown times only in the upper band, i.e. late breakdowns initiated always by primary streamers **only**.

2) UB/LB *Upper Band / Lower Band* – Range of increasing breakdown probability; breakdown times of the upper and the lower band;
positive polarity: breakdowns in the lower band are caused by primary **and** secondary streamers;
negative polarity: no primary streamer; **only** secondary streamers appear.

3) LBO *Lower Band Only* – Range of comparatively high breakdown probability; tertiary streamer appear,
positive polarity: breakdowns in the lower band are more and more dominated by secondary, later even by tertiary streamers;
negative polarity: no primary streamer; **only** secondary and tertiary streamers appear.

Position and size of the bands and ranges are individual breakdown characteristics depending on polarity and gap distance.

Negative Predischarge

At negative polarity predischarge starts in the UBO with a leafless tree-top primary streamer (NPS). The structure consists of three or four less ramified main branches. The NPS velocity varies between 0.3 and 5.0 km/s. Moreover, it could be observed that streamers cease to grow and, therefore, no subsequent breakdown occurs, Figure 2.

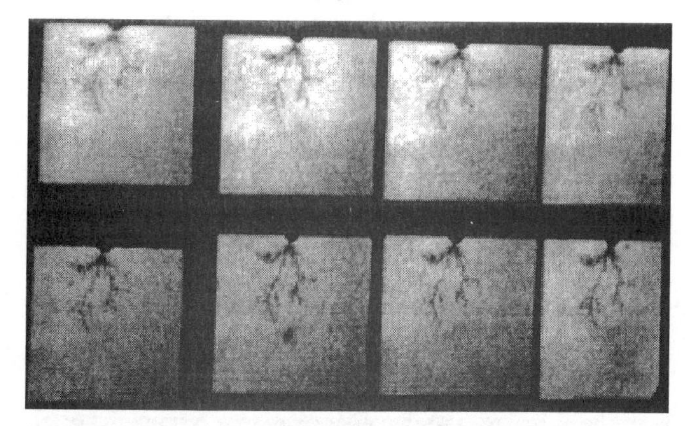

Figure 2: Negative primary streamer without subsequent breakdown

In case of long breakdown times (**upper band**) the predischarge propagation is dominated by the primary streamer bridging almost the complete gap. Figures 3 and 4 illustrate the chronological development of a NPS at 425 kV. The exposure times of both pictures were delayed by 2 µs and 46 µs, respectively. Initially, a fan-shaped structure consisting of many filigran branches is formed. The following structure consists of a few dominant branches with parallel growth. Subsequently, only 2 to 4 less branched main channels continue their propagation and stop the growth of thinner filigran side branches. In close vicinity to the anode, another very fast process overlies the propagation structure leading to breakdown.

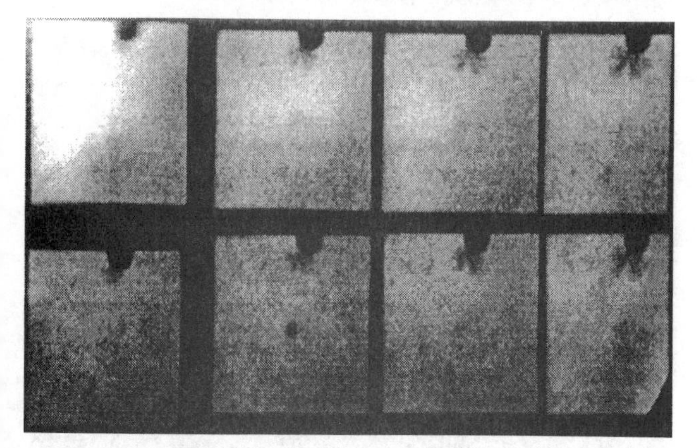

Figure 3: Negative predischarge propagation at 425 kV and 100 mm rod-plane gap; exposure time 2 µs delayed

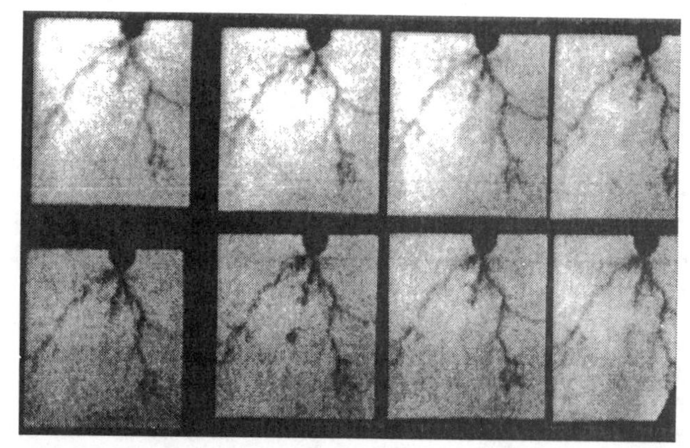

Figure 4: Negative predischarge propagation at 425 kV and 100 mm rod-planegap; exposure time 46 µs delayed

In contrast, the development at short breakdown times (**lower band**) starts with faster streamers. In view of the faster propagation velocity (v = 40 km/s) this streamer can be classified as secondary streamer. Figure 5 illustrates the propagation of a negative secondary streamer. First, in a small region, an extremely branched structure grows and subsequently, results in long, frequently ramified streamer channels. This secondary streamer propagates in batches, i.e. the velocity fluctuates during propagation (between frame 1 and 2 as well as frame 3 and 4 v > 40 km/s, between frame 2 and 3 v = 21 km/s). However, other pictures show NSS growing with constantly increased velocity. In frame 5 a further predischarge stage can be observed. A thin luminous channel, the so-called tertiary streamer, occurs. It propagates through the secondary streamer directly to the counter electrode and causes the ultimate breakdown.

Figure 5: Propagation of a negative secondary streamer and initiation of a tertiary streamer (luminous channel frame 5)

At voltages exceeding 625 kV the breakdown can even occur in less than 3.5 µs. In this case, no introductory streamer structures can be observed, i.e. the predischarge starts with a filamentary tertiary streamer growing with velocities exceeding more than 47 km/s.

Positive Predischarge

Basically, the positive polarity shows similar phenomena. The different streamer patterns have very characteristic features and allow an exact classification as primary, secondary and tertiary streamer. At low voltage levels (UBO) breakdown occurs only infrequently. The predischarge is initiated by a primary streamer appearing in its characteristic umbrella-like pattern and attains velocities up to 3 - 5 km/s, Figure 6.

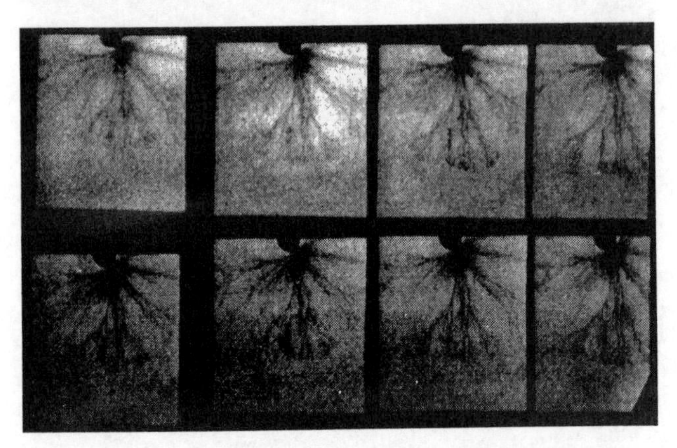

Figure 6: Primary streamer propagation of a positive predischarge

In the voltage range of UB/LB three different predischarge developments appear in the **upper band**:

1) The predischarge always starts with *primary streamer* growing *with or without* subsequent breakdown.
2) The primary streamer has bridged about 80% of the gap distance, when a secondary streamer originates and leads always to breakdown. The secondary streamer appears in its characteristic shape (coniferous branch) and with a velocity v > 40 km/s.

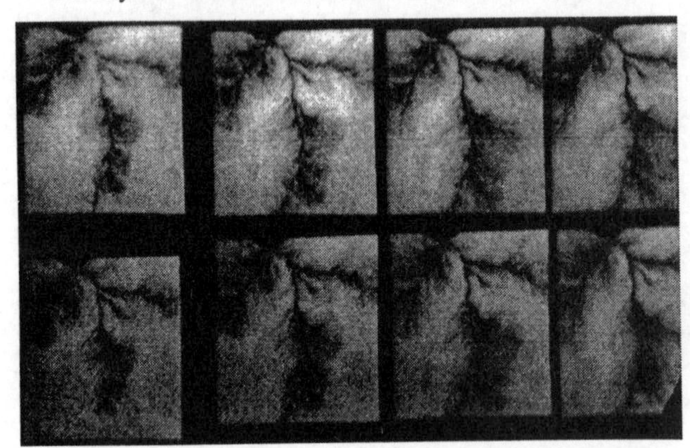

Figure 7: Degeneration of a postive secondary streamer into a primary structure

3) The *first propagation* stage is a *secondary streamer*. Two or three main branches appear in a very dark fashion and show a greater diameter compared to the other branches. After a short distance the secondary streamer degenerates and primary streamer patterns occur at its tips. Ensuing, the structure propagates with the typical primary streamer velocity of 3- 5 km/s, Figure 7. This structure sequence always leads to breakdown.

In the **lower band** primary and secondary streamer occur, but the dominating structure, which bridges most of the gap, is always the secondary streamer. The primary streamer is of secondary importance. With increasing voltage the inception probability for luminous tertiary streamer rises, Figure 8. The tertiary streamer oririginates from the anode. First it propagates through still existing primary and secondary streamer channels and bridges the remaining gap in form of a thin channel within some ten nanoseconds. The maximum tertiary streamer velocity was established to be higher than 180 km/s.

In the voltage range LBO the predischarge always starts with a less branched, straight-lined secondary streamer and no primary streamer states can be observed. During further propagation, a tertiary streamer appears and bridges the remaining gap as aforementioned.

Figure 8: Early breakdown initiated by a PSS

Figure 9: Positive tertiary streamer propagation

At very high voltages (> 675 kV) even the secondary streamer appearence is totally suppressed and only the tertiary streamer leads to breakdown with more than 200 km/s propagation velocity, Figure 9.

Inception Voltages

Although in rod-plane configurations the tip-field is at least five times lower as in point-plane configurations, the streamer inception voltages are in the same range. That means, at rod-plane configuration predischarge phenomena can still be detected at fields of approx. 60 kV/mm; at point-plane configurations 400 kV/mm. Consequently, space charges as well as electrode-liquid interface processes are the decisive factors of predischarge initiation.

DISCUSSION AND SUMMARY

Propagation coefficient π

For an easier description of the events occurring before breakdown, it is convenient to introduce formally a propagation coefficient similar to gas discharges.

In manifold single studies the influence of the parameters *ambient pressure p, electric field E, dissolved gas, voltage rate-of-rise* was investigated. The individual results allow the following deductions:

a) The electric field **E** and the channel conductivity σ have **definitevely** a great influence.
b) Inception voltage depends on ambient pressure [2].
c) The influence of dissolved gas is neglectible [2].
d) The influence of viscosity is subject of present investigations and, at this point, it will be noted as a not neglectible parameter.

So it follows for the propagation coefficient π

$$\pi = f(E; \sigma; \eta; p)$$

Due to the polarity effect of the prebreakdown phenomena in dielectric liquids, both predischarge types require separate reflections.

Postive Polarity

The predischarge of positive polarity can be divided in three stages: with sufficiently high fields, a primary streamer is generated ($\pi > \pi_{PS+}$). For $\pi < \pi_{PS+}$ the structure slows down and ceases.

The primary streamer consists of many straight thin channels growing radially from the anode and generating a hemispherical structure which resembles a fan or an umbrella. The propagation velocity is constant and ranges between 2 and 5 km/s. Since the velocity is in the range of fluid sonic speed the propagation mechanism can be assumed to be of mechanical nature.

Upon exceeding the propagation coefficient π_{SS+}, a secondary streamer originates from the PPS. The secondary streamer appears with sigificantly higher velocity (11- 32 km/s) in the pattern of a coniferous branch. The faster growth implies that secondary streamer propagation base on electronic processes. Moreover, under specific conditions, the predischarge starts immediately with a PSS.

Furthermore, after exceeding the mark π_{TS+} an ultra-high speed luminous tertiary streamer occurs. It originates from the anode and propagates towards the counter electrode using former primary and secondary streamer channels. The intense photon radiation suggests a highly conductive gaseous channel comparable to a "leader channel" in gas discharges of long range.

Negative Polarity

Similar considerations can be made for negative polarity. However, depending on the field growth rate the NPS shows three different modes: leafless tree-top, highly ramified or filamentary pattern [3]. Consequently, the propagation coefficient depends additionly on the field growth rate.

$$\pi = f(E; \frac{dE}{dt}; \sigma; \eta; p)$$

With negative predischarge a NSS or NTS occurs if the propagation coefficient exceeds π_{SS-} and π_{TS-}, respectively.

REFERENCES

[1] Badent, R.; Kist K.; Schwab A.J.;"Prebreakdown Phenomena in Insulating Oil at large Gap Distances". Proceedings of the 4th International Conference on Properties and Applications of Dielectric Materials. July 3-8, 1994, Brisbane Australia

[2] Badent, R.; Kist K.; Schwab A.J.;"The effect of hydrostatic pressure on streamer inception and propagation in insulating oil". Conference record of the 1994 IEEE International Conference on Electrical Insulation, Pittsburgh, PA USA, June 5-8, 1994

[3] Badent, R.; Kist K.; Schwab A.J.;"Voltage dependence of breakdown phenomena in insulating oil". Conference record of the 1994 IEEE International Conference on Electrical Insulation, Pittsburgh, PA USA, June 5-8, 1994

Authors´ address:

Institute of Electric Energy Systems and
High-Voltage Technology
University of Karlsruhe
Kaiserstraße 12
76128 Karlsruhe
GERMANY

Conference Record of the 1996 IEEE International Symposium on Electrical Insulation, Montreal, Quebec, Canada, June 16-19, 1996

HIGH ELECTRIC FIELD CONDUCTION IN AROMATIC HYDROCARBONS : EFFECTS OF THE POLYMERIZATION

Marwan Brouche and Jean Pierre Gosse

Laboratoire d'Electrostatique et de Matériaux Diélectriques- C.N.R.S.
25, avenue des Martyrs- BP 166 X. 38042 Grenoble Cedex
(France)

Abstract : The electrical conduction of purified toluene in the negative tip-plane electrode geometry was studied for high electric field. Mean current-voltage characteristics and current pulses (apparent charge, frequency) were measured for different tip radii.
Above a threshold electric field of 4.2 ± 0.6 MV/cm, the results depended greatly on the previous use of the tip.
For the first measurements made with a given tip, the curve I(V) showed a current jump at voltage V_S. For voltages greater than this threshold voltage, regular pulses were detected. The threshold voltage and the characteristics of the pulses were studied.
After a long use of the tip, the pulse regime faded out. Erratic pulses were detected only for voltages much greater than V_S. The current jump was no more observed at V_S. Mean current voltage characteristics and current pulses were influenced by an organic deposit at the used tip. This deposit is due to the polymerization of toluene liquid.

INTRODUCTION

Aromatic liquids are often used as liquids for impregnation of dielectric solids used in capacitors, and also added to transformer oil to improved their properties.
Discharges can appears when the potential gradient is momentary increased far away to its nominal value and then this could lead to breakdown.
The aim of this paper is thus to better understand the behaviour of an aromatic liquid under high enough electric field, and to determine under what conditions discharges appear.

Electron avalanches in saturated hydrocarbon liquids and cryogenic liquids, and consequently corona discharges, has been observed and characterised [1].
Experimental work, here presented, has been focused on the possibility of obtaining also the phenomenon of electron avalanches in aromatic liquid (such toluene) and to the new possibility of obtaining polymerization of such liquids in the region of high electric field [2].

EXPERIMENTAL PROCEDURE

The cells used for charge creation measurements were made either of pyrex* or of stainless-steel and PTFE. The point electrode was made by electrolytical etching of a tungsten wire (diameter 0.8 mm). The plane electrode was made of stainless-steel (diameter 20 mm). The electrode gap was varied from 1 to 3 mm.

Toluene was purified by passage through an actived molecular sieve and carefully degassed during three freezing-melting cycles. The purified liquid was transferred into cell by vacuum distillation.

The experimental set-up used for measuring the current and detecting discharges is shown in figure 1. It is composed of a stabilised voltage source, the tip-plane cell placed in a Faraday cage, a sensitive electrometer (Keithley 610C) or a digital discharge analyser which allows us to get the pulse apparent charge Q_i and to determine the time interval Δt between two successive pulses. This detector is associated with a microcomputer, allowing recording of Q_i and Δt. The range sensitivity of the detector was 0,05 to 200 pC, its temporal resolution was 6 µs.

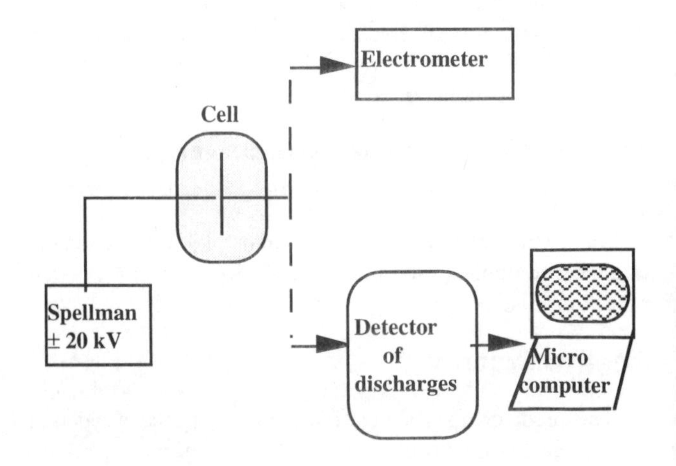

Figure 1 : Experimental arrangement.

RESULTS

Our investigation shows that the observed phenomena is influenced by the tip state. One distinguishes in this way conduction regimes for a new tip (first measurements) and for used tip (measurements after some experiments).

Numerical calculations (field distribution, carrier mobility) are obtained via the hyperbolic approximation of the point-plane electrode configuration [3,4].

1 First measurements with new tip

With the first measurements made with a given new tip, the principal characteristics of the curve I(V) were the following (figure 2) :

- for voltage below a threshold voltage V_S, the mean current is low ($I < 2.10^{-9}$A) and slightly increases with voltage. This residual current is due to impurities remaining in toluene after purification.

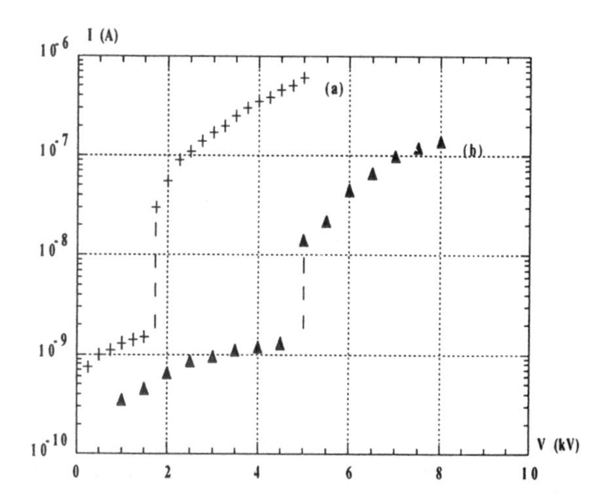

Figure 2 : Mean current vs applied voltage for negative tungsten points of radius (a) 1 μm and (b) 2.2 μm. $P = 10^5$ Pa.

- At the threshold voltage V_S a current instability occurred, and for voltage above V_S the current consisted of a continuous component on which current pulses were superposed.

For $V > V_S$, space-charge controlled current behaviour was observed, i.e. the $\sqrt{I}(V)$ curve is linear. The deduced mobilities K_- (from 1.4 to $1.8.10^{-7}$ $m^2.V^{-1}.s^{-1}$) are in good agreement with the electrohydrodynamical mobility K_{ehd} (K_{ehd} = $1.7.10^{-7}$ $m^2.V^{-1}.s^{-1}$) indicating charge convection and not electron transport ($K_e = 0.54.10^{-4}$ $m^2.V^{-1}.s^{-1}$).

V_S was proportional to r_p and this leads to a current instability occurring at a constant electric field $E_S = 4.2 \pm 0.6$ MV/cm.

The principal characteristics of the regime of current pulses for a given r_p were the following :

- the pulse apparent charge Q_i was independent of voltage V.
- The histograms of figure 3 illustrates the regular character of pulses.

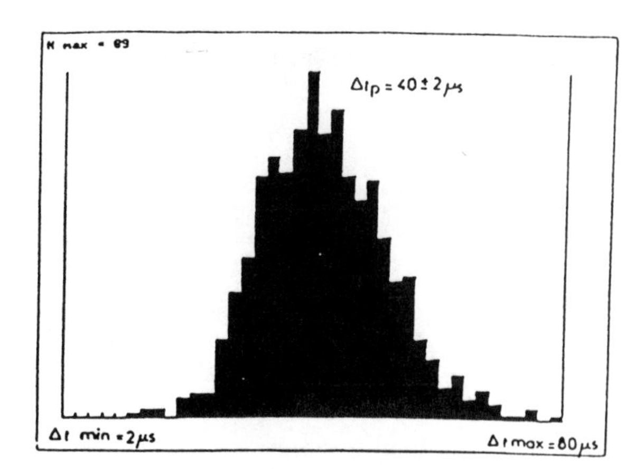

Figure 3 : Histograms of the elapsed time Δt between two successive pulses.

The repetition frequency F is proportional to the mean current (F = BI) (Figure 4).

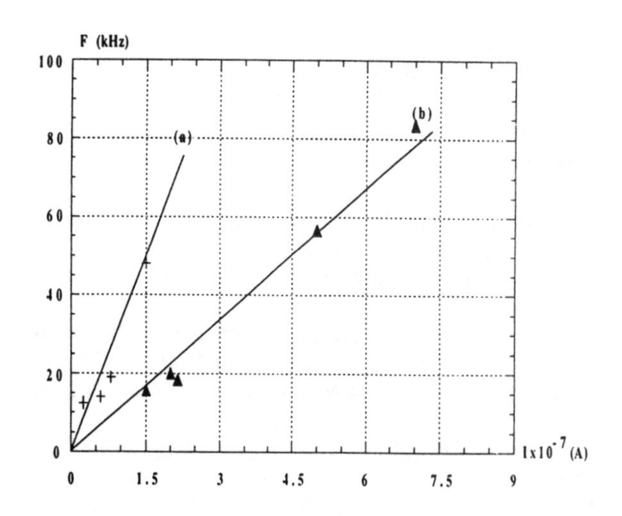

Figure 4 : Repetition frequency of the current bursts as a fonction of the mean current : negative tungsten point of r_p (a) 4.8 μm and (b) 6.2 μm. $P = 5.10^6$ Pa.

The parameter with the greatest influence on the pulses is the tip radius r_p. The decrease of B with increasing r_p is accompanied by an increase in Q_i.

2 Measurements with used tip

After some times of voltage application above V_S (5 minutes maximum), great variations of current were observed. After that, characteristics the main features of the curve I(V) (Figure 5) becomes :

-for V < Vs, the current is larger than the one measured initially with the new tip.
- The instability was no longer observed at V_S.

- For $V > V_S$, the current had nearly the same value than the one obtained in the first measurements.

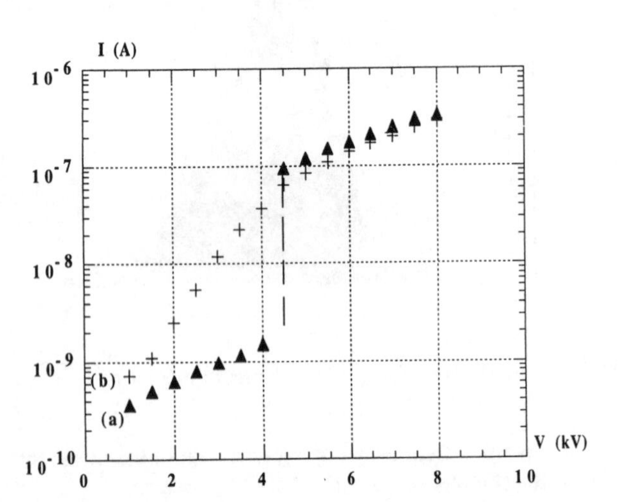

Figure 5 : Current-voltage curves for negative tungsten point of rp = 2.6 μm. Curve (a) : new tip; curve (b) : used tip.

The $\sqrt{I}(V)$ curve is also linear from voltage below V_S and the deduced mobilities are the same than for $V > V_S$.

During the period of current variation, we observed the disappearance of the regular regime of current pulses. In examining the recording of the last burst of discharges, we saw that :
- The pulse apparent charge Qi decreases with time to the minimum sensitivity of our detector (0.05 pC).
- The frequency of current pulses decreases linearly with Qi.

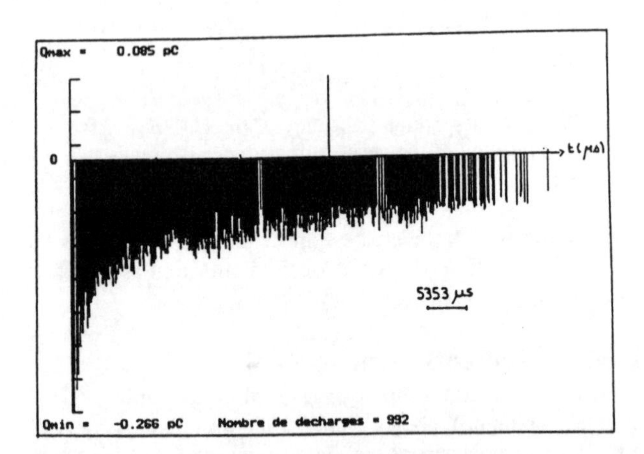

Figure 6 : Recording of the last burst of discharges

After the disappearance of the regular regime of pulses, current bursts were detected at voltage distinctly larger than V_S (Figure 7). These bursts occurred randomly and their rate increased with applied voltage.

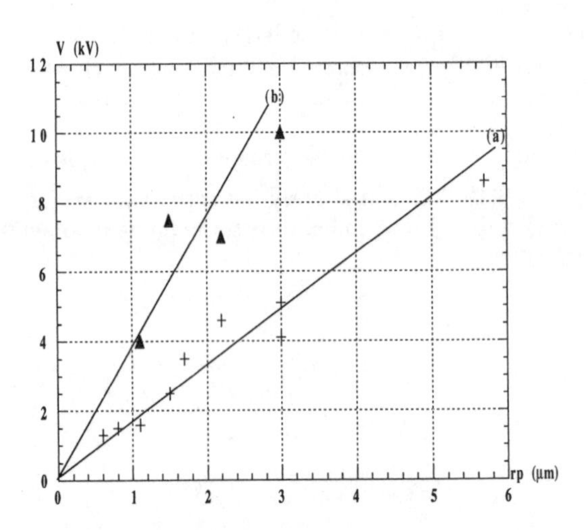

Figure 7: Threshold voltage vs tip radius. Curve (a) : new tip; curve (b) : used tip.

The modifications which occurred in the curve I(V) and in the regime of current pulses were due to a modification of the tip surface, because :
- If the cell was refilled with pure toluene, the above observations remained unchanged : no current instability observed at Vs, no regular regime of discharge detected above Vs. The only way to observe again a regular pulse regime is to replace the tip.
- We have measured the threshold voltage of discharges in a tip-plane system in air, in which the tip was the one used in toluene. We found 1.4 kV instead of 2 kV for a new tip (Figure 8).

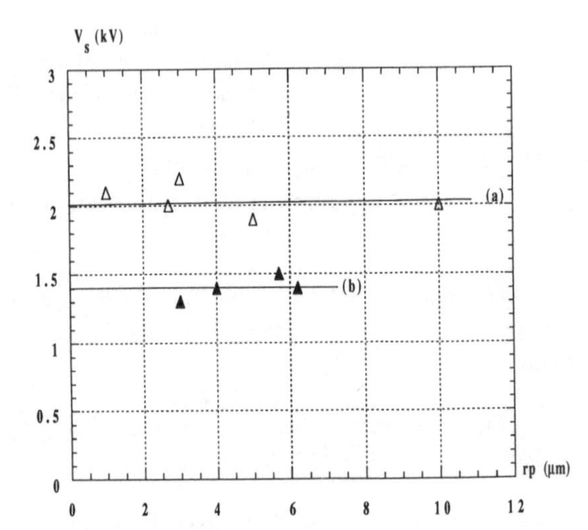

Figure 8 : Threshold voltage-vs tip radius in air. Curve (a) : new tip; curve (b) : used tip.

The electronic microscopy observation shows a deposit at the used tip. From its surface, a large number of filaments were detached. These filaments have a mean length of 0.5 μm and a mean width of 0.2 μm.

This deposit was insoluble in boiling xylene which proved its reticulate character, and has not been analysed by I.R.T.F. because of its small thickness.
The analysis by I.R.T.F. of the deposit on the plane in front of the tip, shows that it is an organic deposit which presented CH_2 groups (absorption band at 2954 cm^{-1}) and aromatics groups (absorption bands at 3027, 1611 and 730 cm^{-1}).

DISCUSION

The phenomena observed in toluene with the first measurements made with a new tip, is similar to that found in saturated hydrocarbons [5,6], and bear great similarity to those observed in electronegative gases. Above a threshold Es = 4.6 MV/cm, the pulse regime can be explained by the mechanism suggested by Loeb [7] for Trichel regime in compressed air : near the tip, electrons emitted from the cathode are accelerated by the high electrical field ; when their energy exceeds the ionisation energy of molecules. This cause an amplification in the electron and positive ion, i.e. avalanches. However, in molecular system, as all excited states with energy higher than that of the weakest chemical bond may dissociate [8]. The electronegative fragments provide deep traps for free electrons and forms the negative space charge which reduces the field strength at the tip until stopping the whole process. When the space charge has drifted away towards the anode, a another pulse can start, the phenomenon is self stabilised [9].

After having experimented for a while, substantial changes in conduction regime have appeared (i.e. I(V) curves, pulses regime). These changes have been shown to be due to a covering of the tip surface by chemical deposit.
Finding an explanation requires studies in chemical radiation for unsaturated hydrocarbons. These studies [10,11] have shown the specifically character of aromatic hydrocarbons : when an aromatic hydrocarbon, such as toluene or benzene, is irradiated, a polymer is formed. It has also been shown [2,12] that a polymerisation due to field emission in tip-plane geometry, on liquid such as styrene is also possible. Authors explained this phenomenon by a radical and cationic polymerisation near the tip.
In the case here presented, such an explanation could also be proposed : free electrons emitted at the tip for E > Es are accelerated and acquire an energy sufficient to excite and ionise toluene molecules. These molecules will after induce a radical or a cationic polymerisation. The result will be creation of a macromolecule which will constitute the deposit observed on the tip.
The phenomenon of extraction an electron from the metal toward the liquid is now far more complicated. Instead of one step mechanism, we can proposed a two steps one : the electron is first transferred to the deposit and after from it to the liquid. In the previous case (from metal directly to liquid) this requires an energy W due to the potential barrier. In the latter one, its first needs an energy W1 for electron to be transferred from metal to the deposit and a second one W2 from deposit to the liquid. This is such than W1 + W2 = W. We could however naturally supposes than W2 is positive and then that W1 is less than W.

Then it is easier for an electron to transit from metal to deposit (and then to liquid) than directly from metal to liquid. Thus for V < Vs, there are more charge carriers created but the electrical field is not sufficient to induce an avalanche process. The current I is then greater for old points than new ones, and also space charge limited but it does not appears pulses in the current.

REFERENCES

[1] A. Denat, J. P. Gosse, B. Gosse, "Electrical Conduction of Purified Cyclohexane in a Divergent Electric Field", IEEE Trans. on Elect. Ins., Vol. 23, N°4, 1988, pp. 545-554.

[2] W. Schnabel, W. F. Scmidt, "Polymerization by High Electric Fields : Field Emission and Field Ionization.", J. Polymer Sci. : Symposium N° 42, 1973, pp. 273-280.

[3] P. Sibillot, R. Coelho, "Prebreakdown Events in Liquid Nitogen", J. Phys., Vol. 35, 1974, pp. 141-148.

[4] W. L. Lama, C. F. Gallo, "Systematic Study of the Electrical Characteristics of the Trichel Current Pulses from Negative Needle to Plane Coronas", J. Appl. Phys., Vol. 45, 1974, pp. 103-113.

[5] A. Denat, J. P. Gosse, B. Gosse, "Conduction du Cyclohexane très pur en Géométrie Pointe-plan", Revue Phys. Appl. Vol. 22, 1987, pp. 1103-1111.

[6] M. Haidara, N. Bonifaci, A. Denat, "Corona Discharges in Liquid and Gaseous Hydrocarbons. Influence of Pressure.", Proc. Gaseous Dielectics VI, Knoxville USA, September 23-27, 1990.

[7] L. B. Loeb, Electrical Coronas, Chap. 4, University of California Press, Berkeley, (1965).

[8] S. Lipsky, "Ionization and Excitation in Nonpolar Organic Liquids", Journal of Chemical Education, Vol. 58 (2), 1981, pp. 93-101.

[9] M. Haidara, *Impulsions de Trichel dans le cyclohexane Liquide et les Gaz Comprimés*, Doctorat de l'Université Joseph Fourier-Grenoble I, 21 Décembre 1988.

[10] S. Gordon, A. R. Van Dyren, T. F. Doumani, "Identification of Products in the Radiolysis of Liquid benzene.", J. Phys. Chm., Vol. 62, 1958, pp. 20-24.

[11] R. R. Hentz, M. Burton, "Studies in Photochemistry and Radiation Chemistry of Toluene, Mesitylene and Ethylbenzene.", J. Am. Chm. Soc., Vol. 73, 1951, pp. 532-536,.

[12] M. Lambla, G. Sceibling, A. Banderet, "Polymérisations amorcées sous la seule influence du champ électrique.", C. R. Acad. Sc. Paris, Série C, t. 271, 1970.

Conference Record of the 1996 IEEE International Symposium on Electrical Insulation, Montreal, Quebec, Canada, June 16-19, 1996

The Effect of Particulate and Water Contamination on the Dielectric Strength of Insulating Oils

by

Kal Farooq

Pall Corporation-25 Harbor Park Drive, Port Washington, NY 11050 (Telephone - (516) 484-3600)

Abstract:

The Dielectric strength of insulating oils is influenced by particulate and water contamination. The deleterious effect of moisture on the dielectric strength is well known in the industry and is reflected in the manufacturing, commissioning and maintenance practices for oil wetted electrical equipment. The deleterious effect of solid contaminants on the dielectric strength is also well known but most often there are no specifications for the oil contamination level and it is not monitored. The subject study was undertaken to quantify the influence of naturally occurring particulate contaminants with various amounts of moisture on the dielectric breakdown voltage of serviced oil from a circuit breaker. The paper also discusses dielectric oil filtration and automatic particle counters and their use in the control of particulate contaminants in oil wetted electrical equipment.

Introduction:

The ever increasing pressure of economic competitiveness demands electrical insulation systems that can withstand higher stresses. The electrical insulation value or the dielectric strength of oil is the key factor in the viability of oil wetted electrical equipment insulation. The dielectric strength of a given insulating oil depends on the impurities contained in it. The dielectric strength of mineral based insulating oils, as measured by the breakdown voltage, is influenced mainly by their water and suspended particle content.

It is well known that both particulate and water contamination adversely affect the dielectric strength of insulating oils (Ref. 1 through 7) . However, there is lack of information regarding the combined effect of naturally occurring particulate and water contamination on the dielectric strength. Moreover, there is reluctance in the industry, especially at the user level, to establish a specification for the particulate contamination level and use it as a commissioning and maintenance tool. The objective of this study is to document the effect of naturally occurring particles and water on the dielectric strength of oil separately and together, and to determine the effectiveness with which a portable particle counter can be used to count the number of particles down to 1 μm.

The test was conducted on serviced oil obtained from a circuit breaker. The dielectric breakdown voltage tests were conducted as per ASTM 1816-84A with a 2 mm gap between the electrodes. The particle counts were obtained using an on-line, automatic particle counter. Water content was determined using a portable coulometric Karl Fischer moisture meter. A portable spinning disc vacuum dehydrator purifier (Pall model HSP-180), fitted with a 1μm ($\beta_1 \geq 1000$) filter, was used to remove the water and the particulate contaminants from the oil.

Test Procedure:

The serviced oil used for the test was obtained from a circuit breaker through a local electric utility. The primary function of the oil in a circuit breaker is arc quenching, which generates carbonaceous particulate contaminants. The oil used for the test was highly contaminated with such particles. The oil had a brownish / black tint to it initially, which disappeared following the purification and the oil turned visibly clear. Most of the particles were in the 2-5 μm size range as shown in Figure 1.

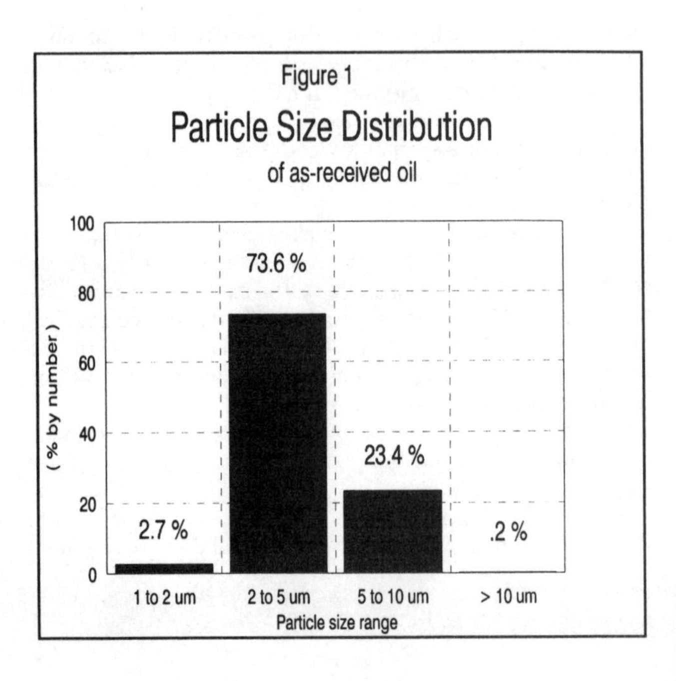

The water content of the as received oil was around 6 ppm. For the purposes of the test the oil was spiked with water. The water content of the oil was increased by first mixing a measured amount of water in a small volume of oil, ultrasonicating until homogenous and then adding to the test reservoir. A propeller type mixer was used to keep the contaminants (water and particles) in suspension during the test.

Three different test configurations were used. Initially tests were conducted with 20 gallons of the water spiked oil utilizing the purifier without the in-line discharge filter to remove water only. Later the in-line filter was re-installed to remove water and particles simultaneously. The next test was conducted with 5 gallons of oil recirculating through a 1 µm ($\beta_1 \geq 1000$) filter assembly to assess the effect of decreasing particle count while keeping the water content constant. The last part of the test was conducted by progressively cleaning up the oil contained in the ASTM 1816 dielectric test cup with a finer filter (0.2 µm absolute). The dielectric strength of the oil was measured at increasing cleanliness levels. To achieve increased water solubility, the tests were conducted with the oil temperature elevated to 140°F.

The particle counts were obtained on-line in-situ, eliminating the possibility of the introduction of environmental contaminants. During each of the tests the filter was bypassed periodically and flow diverted to the automatic particle counter to perform the on-line particle counts. Bottle samples were obtained for the dielectric breakdown voltage and Karl Fischer water analysis. For the last part of the test that utilized a 0.2 µm filter, the oil was filtered and breakdown voltage was performed without removing the oil from the dielectric test cup.

The particle counter used for the test was a bench top unit that operated on laser light interruption principle. Particles in the size ranges of greater than 1, 2, 5, 10, 15 and 25 µm were counted. The sample time was set at 15 seconds, which corresponds to a sample size of 25 ml based on the flow rate of 100 ml/minute through the sensor.

The water content and the dielectric breakdown voltage measurements were made immediately after obtaining the sample to avoid the settling of particles and temperature variation.

Test Results & Discussion:

The test results are presented in Figures 2 through 4. Figure 2 shows the effect of water content on the dielectric strength of as-received un-filtered oil samples. The decrease in the water content of the oil resulted in an increase in the dielectric strength. The dielectric strength of the oil increased from 7.9 kV to 22.4 kV (182 % increase) when the water content was reduced from 206 ppm to 6 ppm and the particle count remained constant. Figure 2 also shows that the increase in the dielectric strength with decreasing water content is exceedingly significant at lower water content values. The improvement in the dielectric strength is exponential once the water content drops below 50 ppm. It should be noted that the saturation level of the oil is ~200 ppm at the test temperature of 140 °F.

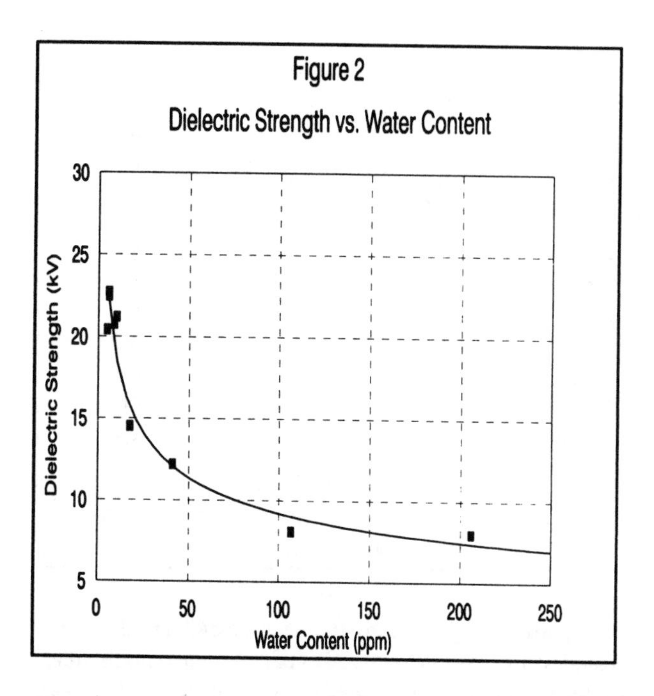

To assess the effect of particulate contaminants on the dielectric strength, tests were conducted with two grades of filters. Initially a standard hydraulic grade spin-on, 1 µm (Pall KZ grade) filter, rated at $\beta_1 \geq 1000$ to remove 1 µm and larger particles with 99.9 % efficiency was used. To further clean the oil a finer filter rated to remove 0.2 µm and larger particles with approximately 100 % efficiency was utilized.

Dielectric strength is plotted in Figure 3 as a function of the particulate contamination level for dry oil (6 ppm water) and a wet oil (150 ppm water). As can be seen from the plot the dielectric strength improved as the number of particles per unit volume of the oil decreased. The dielectric breakdown voltage was consistently higher for the drier oil, as expected. Figure 3 also shows that the dielectric strength of the wet oil did not experience any improvement as the number of 1 µm and larger particle decreased from 200/ml to less than 1/ml. This can be attributed to the high water content (150 ppm) of the oil. The dry oil (6 ppm water) showed continued improvement in the dielectric strength as the particulate contamination level decreased. It can also be concluded from the above that the particulate contamination level of the oil is more significant for its dielectric strength when the oil is well below saturation (dry oil), which is most frequently the case in actual application.

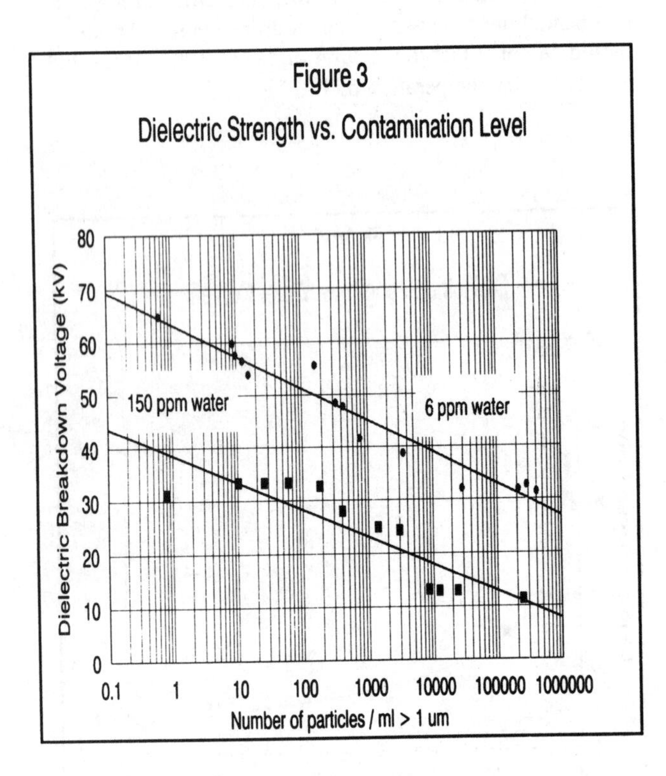

Figure 3

Dielectric Strength vs. Contamination Level

The particle count and the corresponding dielectric breakdown voltage data indicates that particles smaller than 10 μm also influence the dielectric strength of the oil. As can be seen from Table 1, the dielectric strength progressively improved towards the latter part of the test although change in the concentration of particles > 10 μm was insignificant compared to that of smaller size particles. The improvement in the dielectric strength can be contributed to the decrease in the number of smaller size (< 10 μm) particles and most likely even those particles not measured below 1μm.

The test results (see Figure3) show that the cleaner the oil the higher is the dielectric strength. Given the right filter and enough time, the dielectric oil in electrical equipment can be filtered to extremely low particle counts resulting in higher dielectric strength. OEM and end users are in the best position to determine the most appropriate oil cleanliness level for the application and its cost effectiveness. For example, a major French transformer manufacturer has determined that a particle count less than 130/ml (> 2μm) is required for some of their equipment (Ref. 7). Another transformer manufacturer has reported a typical particle count of 150/ml in the 3-150 μm size range for transformer oil based on their statistical data. Oils with a particle count above 500/ml are considered to be excessive (Ref. 8).

To determine the combined effect of the removal of water and particles, the oil was also treated with Pall HSP-180

Table 1

Results of Filtration

(oil with 6 ppm water)

Particle Count / ml >1 micron	>2	>5	>10	>15	>25	Dielectric Strength (kV)
410,000	400,000	96,000	930	75	6.4	31.3
290,000	260,000	18,000	300	52	14	32.6
210,000	150,000	4,900	120	26	5.8	31.8
29,000	13,000	710	20	4.7	0.7	32.0
3,500	2,200	170	6.6	1.6	0.3	38.7
760	460	40	1.8	0.6	0.1	41.6
430	280	30	1.7	0.6	0.1	47.7
340	210	25	1.5	0.6	0.1	48.4
160	110	12	0.7	0.2	0.1	55.5
190	130	16	1.0	0.4	0.0	55.9
15	12	3.3	1.7	1.5	1.2	53.8
12	8.2	1.8	0.4	0.3	0.1	56.4
10	7.0	2.3	0.6	0.2	0.1	57.4
8.8	6.2	1.3	0.3	0.1	0.0	59.7
0.6	0.2	0.0	0.0	0.0	0.0	64.7

purifier. There are several methods used for the dehydration of dielectric oil. A common method is the high vacuum and high heat flash distillation. The spinning disc portable fluid purifier used for the test uses moderate vacuum (26 " Hg) and low temperature (100 - 160 °F). The spinning disc purifier works on the principle of mass transfer of the moisture from the oil to a constant stream of dry air generated by the vacuum. The separation of free and dissolved water from the oil is facilitated by the generation of a large surface area of the oil. To remove particulate contaminants the purifier is fitted with a 1 μm ($\beta_1 \geq 1000$) discharge filter. Along with water, the purifier also removes gases and volatile fluids from the oil.

A 20 gallon volume of the contaminated circuit breaker oil spiked with water was treated with the Pall purifier. The results of the purification (dehydration and particle removal) are presented as Figure 4. The plot shows the improvement in the dielectric strength as the oil is dehydrated and filtered by the purifier. The number of passes refer to the equivalent number of times the test oil volume (20 gallons) has passed through the purifier. The particle count and water content of the oil decreased from 462,000 particles / ml (> 1 μm) and 183 ppm to 12 particles / ml (> 1 μm) and 4.6 ppm respectively. Consequently, the dielectric strength increased from 12.5 to 55 kV; a 4.4 fold increase.

Filter Selection:

The selection of a filter for the efficient, expedient and cost effective filtration of dielectric oil requires evaluation of several key parameters. The selection of the right filter should take into consideration the filter material compatibility with the oil and the contaminants in it, filter particle removal capability (the Beta ratio) (Ref. 9) and the dirt holding and retention capacity of the filter.

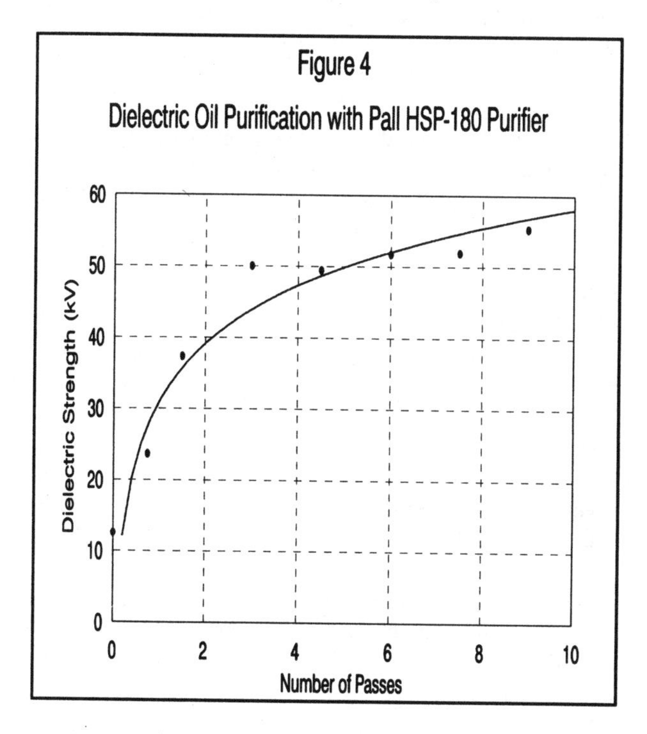

Filters constructed with organic materials (paper, cotton, etc.) are susceptible to being weakened, hence deteriorate in performance, by the acids that exist in oxidized oil and water. The type of bonding material used in the construction of the filter cartridge (end caps, side seam, medium resins) should also be compatible with the contaminated oil. Filters constructed with organic materials, because of their chemical instability, deteriorate as they age thus have limited shelf life.

Cellulose filters are sometimes promoted as water absorbing filters because of the cellulose's affinity for moisture. It should be noted though that the cellulose will absorb water from oil to the extent that it reaches equilibrium with oil under the operating conditions. As the temperature increases the cellulose's affinity for moisture decreases and a saturated cellulose filter can actually release water back into the oil. Mechanical stresses due to pulsations and increased differential pressure can also cause a cellulose filter that has absorbed water from the oil to release it back into the oil.

Contamination Analysis:

Automatic particle counters are becoming less expensive and easier to use as the technology progresses. Many of these are suited for on-line counting. On-line particle counting provides the benefit of fast real time monitoring of the oil contamination level. Automatic particle counters provide information about the number of particles per unit volume in multiple size ranges selected by the user. An alternate to on-line counting is to obtain samples in glass bottles that can be counted one at a time with an automatic particle counter or with a microscope by transferring the contaminants onto an analysis membrane. Potential for large inaccuracies exist with bottle sampling due to contaminants from the sampling process and bottle. Also particles in dielectric oil tend to settle rapidly, and, unless these particles are re-suspended, the particle count will not be representative (Ref. 10). Ultra sonication of the oil sample was found to adequately re-suspend the particles in the insulating oil (Ref. 11).

The laser light interruption automatic particle sensor / counter used for the test performed flawlessly. It was convenient and user friendly and did not require any special training for the operator. Particles in the size ranges of greater than 1, 2, 5, 10, 15 and 25 µm were selected. The sensor was calibrated as per the ISO 4402 (Ref. 12) procedure with ACFTD in MIL H-5606 oil.

Particle count specification for an in-service electrical equipment will depend on its type and the application and needs to be determined by the user. The typical contamination level and size distribution, once established for a piece or group of equipment, can serve as a reference for improvement, preventive maintenance and trouble shooting. If periodic sampling shows a significant increase in the particle count or a shift in the particle size distribution, the sources should be investigated. Arcing or pump wear, for example, can cause increased particle count levels and a shift in the particle size distribution (Ref. 8).

Conclusions:

The results of this study show that both moisture and particulate contaminants have a detrimental effect on the dielectric strength of insulating oil. The improvement in the dielectric strength is significantly higher as the water content dropped below 50 ppm; well below its saturation level (200 ppm).

The lower the particle count the higher is the dielectric strength. Given the right filter and enough time, the dielectric oil in electrical equipment can be filtered to extremely low particle count resulting in higher dielectric strength.

The particle count and the corresponding dielectric breakdown voltage data indicates that fine particles in the size range of 1 to 10 µm influence the dielectric strength.

Because of the importance of particles greater than 1 µm on the dielectric strength as shown by the test results, it advisable that the oil cleanliness specification be based on a particle count at 1 µm.

The selection of filter for the efficient, expedient and cost effective filtration of dielectric oil requires evaluation of the filter material compatibility with the oil and the contaminants in it, filter particle removal characteristic (Beta ratio) and the dirt holding and retention capacity of the filter.

The typical contamination level and size distribution, once established for a piece or group of equipment, can serve as a guideline for preventive maintenance and trouble shooting. A significant increase in the particle count or a shift in the size distribution indicates an abnormality and should be investigated.

Automatic particle counters calibrated with ACFTD can effectively be used for the on-line monitoring of contaminant in dielectric oil. This technique can be used in the instantaneous monitoring of the fluid cleanliness level as it is filtered.

References:

1. Mathes, K. N. and Atkins J. M. "Influence of Minute Particles on Partial Discharge and Breakdown in Oil" IEEE Symposium on Electrical Insulation, Philadelphia, June 12-14 1978.

2. Flanagan, P. E., et al. "The Measurement of Particulate Contaminants in Transformer Oil and Correlation with 60 Hz Dielectric Strength" Minutes of the Forty-Third Annual International Conference of Doble Clients, 1976.

3. Miyao, H., et al. "Influence of Naturally Existing Micro Particles on AC and Impulse Breakdown Strengths of Industrially Purified Transformer Oil", IEEE Conference on Electrical Insulation and Dielectric Phenomena, Paper No: VI-II October 1982.

4. Kako, Y., et al. " Influence of Minute Particles on Breakdown Characteristics of Insulating Oil' CIGRE 15-03, Aug. 27- Sept. 4 1986.

5. Oommen, T. V., "Particle Analysis on Transformer Oil for Diagnostic and Quality Control Purposes" Minutes of the Fifty-First annual International Conference of Doble Clients, 1984. Sec 10-701.

6. Miners, K., "Particles and Moisture Effect on Dielectric Strength of transformer Oil Using VDE Electrodes" IEEE Trans. PAS, Vol. PAS-101, No: 3, March 1982, pp 751-6.

7. Samat, J., Lacaze, D., "Micro Particles in Transformer Oil and Dielectric Withstand Effects" Alsthom Review No: 11-1988.

8. Oommen, T. V., "Particle Analysis of Insulating Fluids" ASTM Symposium, Committee D-27, Fort Worth, Nov. 16, 1993.

9. "Hydraulic Fluid Power - Filters - Multi-pass method for evaluating filtration performance", ISO 4572 1981(E).

10. Danikas, M., "Particles in Transformer Oil" Technical Report, IEEE Electrical Insulation Magazine. March/April 1991-Vol 7., No. 2.

11. Vincent, C., et al., "Ultrasonic Homogenization of Particles in Transformer Oil" Minutes of the Fifty-Fourth annual International Conference of Doble Clients, 1987.

12. "Hydraulic Fluid Power- Calibration of automatic-count instruments for particles suspended in liquids- Method using Air Cleaner Fine Test Dust contaminant" ISO 4402: 1977.

Conference Record of the 1996 IEEE International Symposium on Electrical Insulation, Montreal, Quebec, Canada, June 16-19, 1996

Prebreakdown and breakdown phenomena in large oil gaps under AC

A. Saker, P. Gournay, O. Lesaint, R. Tobazéon
Laboratoire d'Electrostatique et de Matériaux Diélectriques, CNRS, BP 166, 38042 Grenoble, FRANCE

N. Giao Trinh
Institut de Recherche d'Hydro-Québec (IREQ), 1800 Boulevard Lionel Boulet, Varennes, Québec 1S1J3X CANADA

C. Boisdon
Jeumont-Schneider Transformateurs, groupe Schneider, 84 Avenue Paul Santy, 69371 Lyon, FRANCE

Abstract: **This paper presents a study of prebreakdown and breakdown phenomena under AC voltage in mineral oil in large gaps to 60 cm. The investigations presented concern the study of streamers and the measurement of breakdown voltages in rod-plane and sphere-plane gaps. Also, the influence of a contamination by solid particles in the oil has been considered.**
A specific breakdown mode under AC voltage is evidenced, where "bursts" of streamers lead to the lowest breakdown fields recorded. Numerical values of the mean field in oil required for "direct" or "burst" breakdown modes are derived from the experiments. As a consequence, the great sensitivity to the presence of particles of EHV transformers insulation with large oil gaps is pointed out.

INTRODUCTION

The breakdown mechanisms in large oil volumes under AC stress for transformer applications are not well known, especially the conditions required for the inception and propagation of breakdown precursors. Investigations so far have consisted mainly of breakdown voltage measurements (up to gaps of 1 m) [1] and studies of streamer propagation under impulse voltages (up to d= 35 cm [2,3]) or AC voltages (up to d= 20 cm [4]).

The work reported here was undertaken to explain the high failure rate of bushing shields of EHV transformers, where large volumes of oil are subjected to a low electric field (around 10 kV/cm).

For easier comparison with streamer inception voltages Vi, the applied voltage V is systematically expressed as a peak value in this paper.

EXPERIMENTAL TECHNIQUES

Two test cells were build: the first for measurements up to V= 400 kV (volume: 150 L) and the second for full-scale tests (voltages up to 1200 kV) with a much larger volume of $11 m^3$. Both test cells were equipped with a grounded plane and two propellers to stir the oil.

The experimental work described here was performed with bare electrodes to simplify test preparation. The HV electrodes were either stainless steel rods with a hemispherical end of radius r, or spheres. In some experiments, a triggering tungsten wire (length l, diameter a) was attached on the sphere surface. The experiments in sphere-plane configurations were designed to simulate typical field distributions present around transformer bushing shields for 400 kV (gap distance d= 30 cm, V= 325 kV, field at the electrode: 40 kV/cm, sphere diameter: 14 cm), and for 800 kV (d= 60 cm, V= 625 kV, field at the electrode: 34 kV/cm, sphere diameter: 28 cm).

Two voltage sources with a maximum amplitude of 420 and 1400 kV were used. A device in the transformer primary provided a means of interrupting the voltage with an approximately one-cycle delay. This system, which can be triggered by the appearance of either a breakdown or a streamer, was designed to avoid the occurrence of a large number of streamers or breakdowns, which would cause rapid degradation of the oil.

Streamer propagation was observed with a streak camera and the streamer light emission intensity was recorded using a photomultiplier.

The oil under study was filtered, degassed and dried prior to experiments, and pollutants could be added to the oil (particles of cellulose, carbon and copper, or water). All measurements presented in the following sections, unless specifically mentioned, were obtained with the oil continuously stirred to keep it homogenous.

STREAMER INCEPTION FREQUENCY

Figure 1 illustrates the influence of the length of the triggering wire fixed on a 28 cm sphere on the streamer rate of occurrence. These measurements were obtained by counting the number of cumulative negative and positive streamers generated by a 10-min voltage application. The maximum voltage applied was limited by the breakdown event.

The streamer occurrence increases with the wire length l, which is logical because the wire produces a field reinforcement at the sphere surface which increases with l.

The appearance of streamers is not regular but tends to be more of a random phenomenon, taking the form of bursts alternating with periods of inactivity.

This electrode configuration with its selected triggering wire size thus offers a means of controlling the rate of occurrence of streamers under AC voltage conditions. As will be seen later, the breakdown voltage measured in these conditions increases substantially with shorter wire lengths because of the reduced discharge occurrence. Thus, by modifying the wire length it is possible to induce breakdown at different voltage levels, keeping the other parameters (distance, oil conditioning) constant. Depending on the breakdown voltage level, two main types of prebreakdown events were observed.

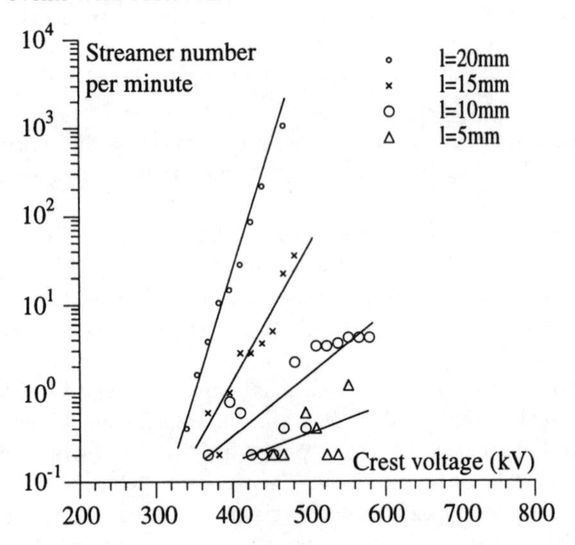

Figure 1: Influence of the triggering wire length (l) on the streamer rate of occurrence (sphere diameter 28 cm, d= 60 cm, a= 0.5 mm, l= 20 mm).

BREAKDOWN MODES

"Streamer burst" breakdown mode

Figure 2 presents typical luminous signatures of these events detected by the photomultiplier prior to breakdown. Each impulse on the photomultiplier signals corresponds to the propagation of a streamer. In Figure 2a, corresponding to a low breakdown voltage measured when many discharges are generated (with l= 10 mm), breakdown is preceded by a burst of streamers generated at each half-cycle. This last burst was preceded by many others that did not lead to breakdown.

In Figure 2b, at a higher breakdown voltage with a shorter wire (l= 5 mm) the number of streamers in the burst is lower, whereas in Figure 2c, with no triggering wire and at a still higher voltage, we see an example of direct breakdown in which a single streamer leads to breakdown.

In bursts, streamers of opposite polarity are systematically generated at each half-cycle at approximately maximum voltage, as illustrated in Figure 2a.

The maximum duration of a burst is usually less than 20 cycles before breakdown, and this, as well as the number of streamers in the burst, decreases as direct breakdown conditions are approached. The large majority of bursts begin with a positive streamer and end with positive or negative streamers.

In this paper the term "direct" is used to describe breakdown for which no streamer appeared in the last 20 cycles preceding breakdown. These two breakdown modes under AC voltage were observed in the rod-plane and sphere-plane configurations (with or without a triggering wire).

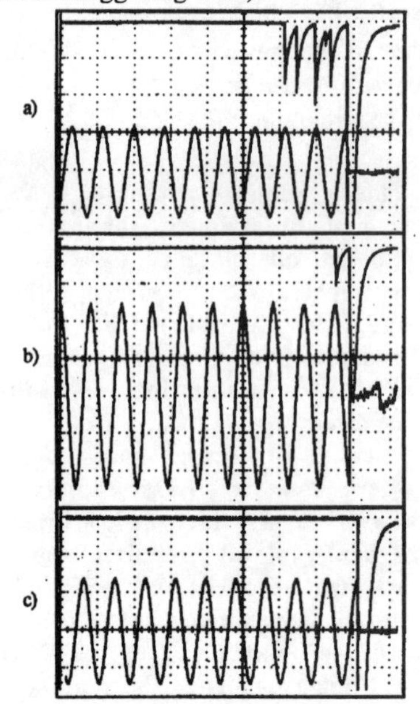

Figure 2: Examples of events leading to breakdown. (Sphere-plane configuration with triggering wire, d= 60 cm, a= 0.5 mm). a): long burst, l= 10 mm, V= 552 kV; b): short burst, l= 5 mm, V= 565 kV; c): direct breakdown, l= 0 mm, V= 890 kV.

"Direct" breakdown mode

The propagation of single streamers (i.e. those not preceded by others) was studied from recordings of the first streamer generated following voltage application. A typical recording is shown in Figure 3. All the characteristics observed at shorter gaps may be found in these measurements [2,3], namely:

- during propagation, periodic re-illuminations of the main channel of the streamer are observed;
- these re-illuminations are superimposed on a weak luminous phenomenon propagating steadily at the head of the streamer;
- the propagation velocity is practically uniform from the tip to the plane ($\approx$2 km/s);
- breakdown usually occurs when the streamer touches the opposite electrode.

BREAKDOWN VOLTAGE MEASUREMENTS

Breakdown voltages were measured by raising the applied voltage in 10-kV steps every 10 min. In figure 4, values of average breakdown fields (E= V/d) measured in sphere-plane configurations with different wire sizes are plotted versus the occurrence frequency of discharges measured at breakdown voltage in all these experiments.

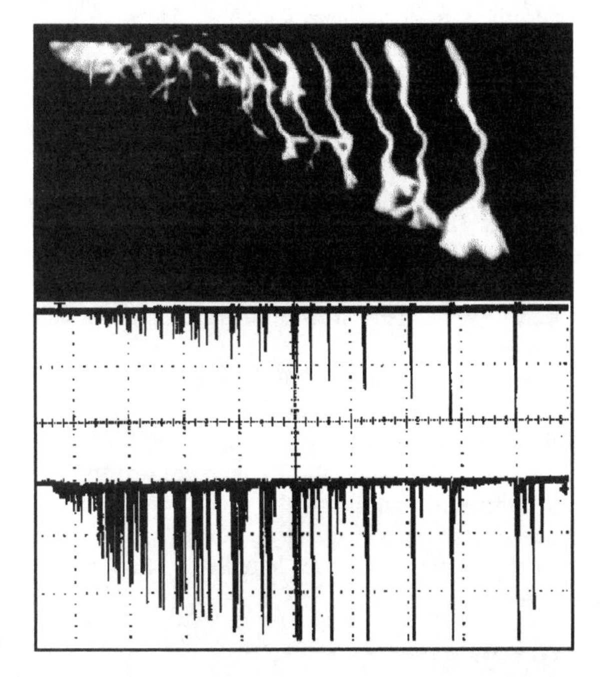

Figure 3: Streak camera image of a positive streamer (x axis: 13.6 μs/cm, y axis: 6.4 cm/cm), and corresponding oscillogram (10μs/cm, upper trace: current 2A/div., lower: emitted light intensity). Rod-plane geometry, Vi= 224 kV, r= 4 mm, d= 35 cm.

From figure 4 it may be concluded that:

- an obvious correlation exists between the average breakdown field in burst mode and the occurrence of streamers for the two gaps considered;
- the breakdown field for d= 60 cm is approximately 25% less than that measured for d= 30 cm. This reflects the relative degradation of the oil performance as the gap increases;
- the measurement at d= 30 cm in filtered oil or oil polluted with cellulose show that, for the same discharge occurrence, the presence of cellulose has no effect on the breakdown field in burst mode.

An estimate of direct breakdown conditions under AC voltage may be obtained from figure 4 by extrapolating the curves to frequencies for which a single streamer is generated when the voltage is applied (0.1 discharge/min for 10-min steps). The extrapolation yields the following values: 12.5 kV/cm at d= 30 cm; 10.3 kV/cm at d= 60 cm.

It is also possible to deduce the "minimum" values of breakdown fields measured with a very high discharge frequency (for a frequency arbitrarily set at 1000 discharges/min): 10.5 kV/cm at d= 30 cm; 7.5 kV/cm at d= 60 cm.

These values show clearly the degradation of breakdown fields due to the streamer burst mechanism under AC. At d= 30 cm, this discharge mode led to a 15% reduction in the breakdown field compared to the value for direct breakdown. The reduction doubled when d= 60 cm. This indicates that the influence of burst discharges increases significantly with voltage and distance.

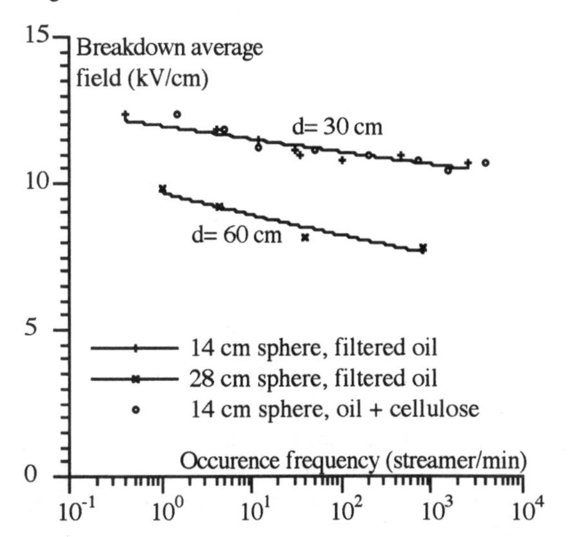

Figure 4: Correlation between the average breakdown field and the occurrence of streamers in the sphere-plane configuration with a triggering wire.

These values for the transition (direct - burst mode) and the "minimum" breakdown field values are in very good agreement with breakdown measurements performed under a variety of conditions (rod or sphere-plane configurations, addition of various pollutants), in which the mode has been determined. In particular, it is amazing that identical results are obtained in the rod-plane and sphere-plane configurations, which shows that once streamers have been generated they propagate singly or in bursts in exactly the same way in both configurations. This supports the use of the average electric field as the parameter characterizing streamer propagation.

The results in figure 4 emphasize the close relationship between the generation mechanisms (which determine the occurrence of streamers and are greatly affected by particles) and the average field in the burst mode of breakdown.

STREAMER GENERATION, INFLUENCE OF PARTICLES

In sphere-plane geometry without triggering wire, it was impossible to measure the streamer rate of occurrence, but the influence of particles on breakdown is revealed by the effect of stirring the oil on the breakdown voltages (figure 5).

The measurements emphasize the effect that stirring the oil has on the discharge triggering process: the average breakdown voltage, which is more than 1100 kV after leaving the oil to rest for four days, drops to 660 kV (a 40% reduction) when stirring is resumed.

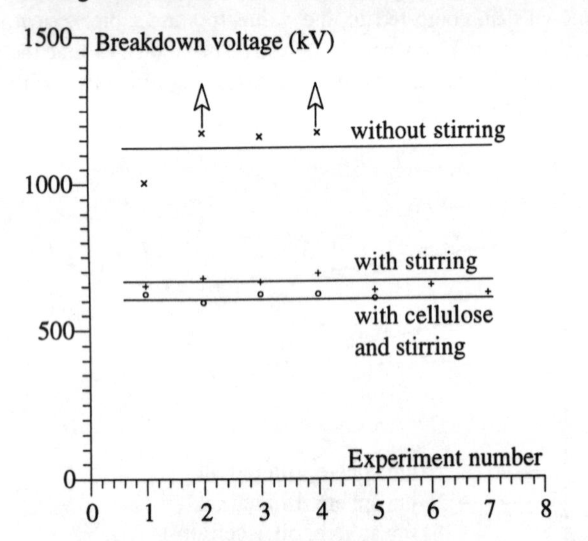

Figure 5: Evidence of the effects of stirring the oil and adding cellulose on the breakdown voltage with a sphere diameter of 28 cm (d= 60 cm). Arrows: no breakdown occured at V= 1180 kV.

This is most probably due to decantation of the heaviest particles. Just a few susceptible particles are enough to trigger breakdown under the effect of stirring, because the movement of oil carries them toward the high-voltage electrode. The probability of detecting these particles by counting is practically zero considering the volume of oil sampled (a few hundred cm^3) compared to the cell volume (11 m^3).

CONCLUSIONS

In the case of single streamers, no great differences in the propagation process were noted; in fact it is very similar to that observed in smaller oil volumes. It is the positive streamers that cause breakdown in most instances. In large gaps subjected to relatively low average fields (<10 kV/cm), the breakdown was observed to occur in the streamer burst mode.

The characteristic values of breakdown fields determined above offer an understanding of the relative fragility of mineral oil insulation in extra-high-voltage transformer applications. A 400-kV bushing shield insulation with an average field of 10.8 kV/cm (gap= 30 cm) is close to the extreme value for which breakdown can only occur at a very high discharge frequency, i.e. in the presence of a low-probability permanent flaw. On the other hand, 800-kV insulation with a field of 10.4 kV/cm for a gap of 60 cm represents a range where direct breakdown can occur as soon as a discharge is generated. Breakdowns under conditions very similar to these were indeed observed, especially in the case of oil stirring. Controlling the conditions of streamer inception is therefore a determining factor in these circumstances.

With regard to the generation of discharges in practical cases and correlation with the presence of particles, electrode coatings represent an essential parameter in the discharge triggering process. Since this aspect has not been covered by the study reported here, it is difficult to draw a definitive conclusion.

Acceptance tests on EHV transformers are usually performed with carefully filtered oil and no stirring. For power frequency tests, on large-oil-volume insulating structures, the findings reported here show that the presence of particles could have a substantial impact on the breakdown probability. It therefore seems to be a suitable time to take a critical look at some of the existing procedures in order to better take account of real operating conditions.

AKNOWLEDGMENTS

The authors wishes to express their sincere thanks to P. Kieffer for his technical assistance in the experiments conducted at IREQ, J. Aubin, C. Vincent and F. Rizk for their valuable advices throughout this project, IREQ and Jeumont Schneider Transformers that both supported this study.

REFERENCES

1. Kamata Y., Kako Y., "Flashover characteristics of extremely long gaps in transformer oil under nonuniform field conditions", IEEE Trans. EI, Vol. EI-15, No. 1, 1980, p. 18.
2. Rain P., Lesaint O., "Prebreakdown phenomena in mineral oil under step and ac voltage in large-gap divergent fields", IEEE Trans. DEI, ITDIES Vol. 1, No. 4, 1994, p. 692.
3. Saker A., Lesaint O., Tobazéon R., "Propagation of streamers in mineral oil at large distance", Conf. on Elec. Insul. and Diel. Phen., IEEE No. 94CH3456-1, October 1994, p. 889.
4. Rain P., Boisdon C., Lesaint O., Tobazéon R., "Behaviour of streamers under divergent ac fields in transformer oils at large gaps", IEEE Trans. EI, Vol. 26, No. 4, 1991, p. 715.

Conference Record of the 1996 IEEE International Symposium on Electrical Insulation, Montreal, Quebec, Canada, June 16-19, 1996

Transition to fast streamers in mineral oil in the presence of insulating solids

O. Lesaint and G. Massala

Laboratoire d'Electrostatique et de Matériaux Diélectriques
CNRS, BP 166, 38042 Grenoble Cedex 9
FRANCE

Abstract: **This paper presents a study of the transition to fast streamers in mineral oil in overvolted point-plane gaps. Streamer propagation is studied in 5 cm and 10 cm gaps, either in the liquid alone or in the presence of solid surfaces parallel or perpendicular to the gap axis. The experiments presented concern the visualization of streamers, the measurement of breakdown and acceleration voltages. In the liquid alone, above some critical voltage, a large increase in propagation velocity is recorded from 2 km/s up to about 100 km/s. The presence of insulating solids parallel to the electric field is greatly in favor of the inception and propagation of fast streamers (velocities up to 300 km/s have been measured). Conversely it is observed that a solid perpendicular to the electric field stops more efficiently fast streamers than slow ones, which easely creep and get round such obstacles.**

INTRODUCTION

Previous studies of the propagation of positive streamers in mineral oil in gaps up to 60 cm have shown that, at the breakdown voltage Vb, the propagation velocity of streamers is nearly constant and close to 2 km/s, under AC [1] and with "step like" 0.4/1400 µs impulses [2]. In overvolted point-plane gaps, it was observed that above some critical voltage Va, a sudden transition to fast streamers occurs, which velocity grows very quickly up to more than 100 km/s [3]. For gaps to 10 cm, Va is about twice as large as Vb, and the sudden increase in velocity is correlated to specific changes in the shape and charge of streamers.

In this paper, we investigate the influence of solid insulating surfaces on the propagation and acceleration of streamers, and the consequences on the measured breakdown voltages Vb. Solid surfaces either parallel or perpendicular to the electric field have been considered.

EXPERIMENTAL TECHNIQUES

The experimental set-up was essentially the same as that reported in [2]. A 13 liters test cell with a grounded plane 20 cm in diameter has been used. The HV electrode consisted of a steel point (tip radius: 100 µm) supported by a holder containing an optically coupled current or charge measurement device. The mineral transformer oil was filtered, degassed and periodically renewed after about 20 impulses. Positive-polarity impulses, 0.4/1400 µs in shape, were applied to the point electrode from a 500 kV Marx generator via a 320 Ω resistor. With this impulse shape, the applied voltage decreases of about 10% from its maximum value V during 100 µs. Thus, the voltage was practically constant during the propagation of streamers. Photographs of the light emitted by streamers were obtained with a gated image intensifier. The light emission was recorded by a photomultiplier.

Experiments with insulating solids were made with transformer board, 2 mm in thickness. Two configurations A and B were studied (figure 1). A: solid parallel to the point-plane axis, B: insulating disc (15 cm in diameter) perpendicular to the gap axis, located at a distance x from the point electrode.

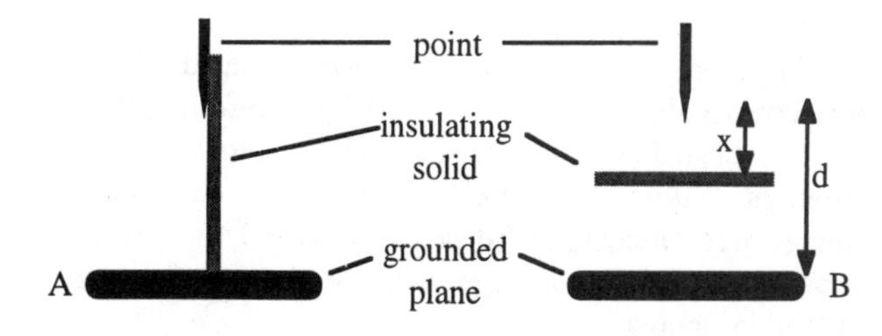

Figure 1: Electrode configurations A and B used.

SURFACES PARALLEL TO THE FIELD

Breakdown voltage

Breakdown voltages were determined for gaps d = 5 and 10 cm by measuring the breakdown probability versus voltage (figure 2). This was done by counting the number of breakdown produced during the application of series of 25 impulses at a fixed voltage level. The applied voltage was incremented by 10 kV between each series. The breakdown voltage noted Vb later in the text refers to the 50% breakdown probability.

Experiments were made either in the liquid alone, or with a solid placed in position A (figure 1). In this latter case, the solid was systematically replaced after breakdown has occured.

There is no significant influence of the solid surface on the breakdown probability. Since one streamer is generated at each shot, the breakdown probability represents the probability for a streamer to reach the plane and produce breakdown. Thus we can conclude that the presence of this interface has no significant influence on the propagation of streamers generated at the breakdown voltage, i.e. streamers with a velocity ≈ 2 km/s. A similar result was obtained at d = 5 cm. It was observed that streamers always propagate along the solid/liquid interface.

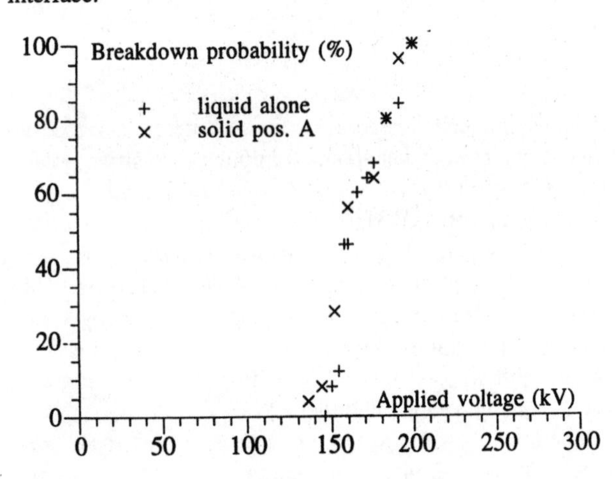

Figure 2: Breakdown probability versus voltage (d = 10 cm).

Effect of overvoltages

The average velocity of streamers initiated above the breakdown voltage was investigated by measuring the time delay to breakdown (Td) versus applied voltage for several distances. Within the 450 kV limit of our equipment, the voltage was raised until Td decreased to ≈ 1 µs. A typical measurement of the average velocity deduced from Td is showed in figures 3 and 4.

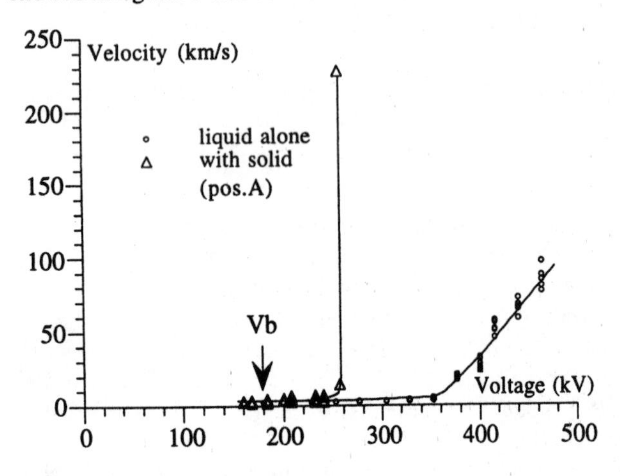

Figure 3: Average streamer velocity versus voltage in the liquid alone and with a solid parallel to the field (d = 10 cm).

Without a solid interface, within the range Vb to ≈ 2Vb, the average velocity grows slowly from 2 to 3 km/s, and a large increase of the streamer charge and number of branches is recorded [3]. When the applied voltage exceeds some critical value Va, the average velocity starts to grow strongly with voltage. This is correlated to a sudden decrease of the streamer charge and to a change in streamer shape (it tends to become monofilamentary) [3].

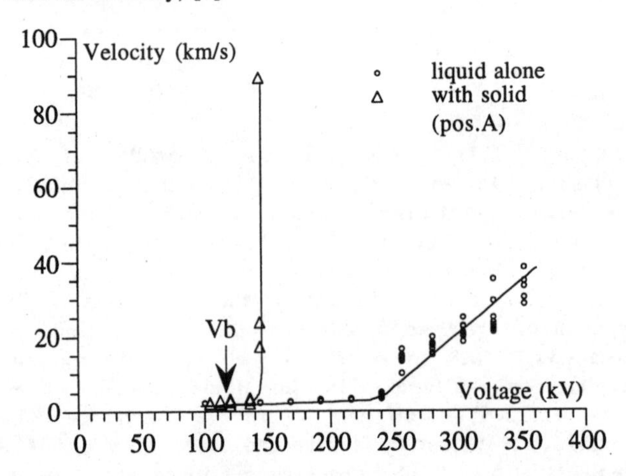

Figure 4: Average streamer velocity versus voltage in the liquid alone and with a solid parallel to the field (d = 5 cm).

With an interface parallel to the electric field, a similar behaviour is recorded, but at a much lower voltage: the speeding up of streamers is very steep and occurs at 250 kV instead of Va = 380 kV in the liquid alone with d = 10 cm (resp. 140 kV and 256 kV for d = 5 cm).

SURFACES PERPENDICULAR TO THE FIELD

Experiments were made with an insulating solid placed as a "barrier" perpendicular to the field (pos. B on figure 1).

Breakdown voltage

Figure 5 shows the measured breakdown voltage with a 15 cm insulating disc placed at different distances x from the point. These measurements were obtained by measuring the breakdown probability versus voltage such as in figure 2. Ten shots were applied at each voltage level, and the solid was replaced after each breakdown. Values reported on figure 5 correspond to the 50 % breakdown probability.

When the solid is close to the point tip, the breakdown voltage is significantly higher (≈ 30 %) than in the liquid alone. The visualisation of prebreakdown events with a gated image intensifier shows that streamers iniated at the point reach the solid, creep on this solid up to its side, and then propagate in the liquid toward the plane electrode to induce breakdown (figure 6).

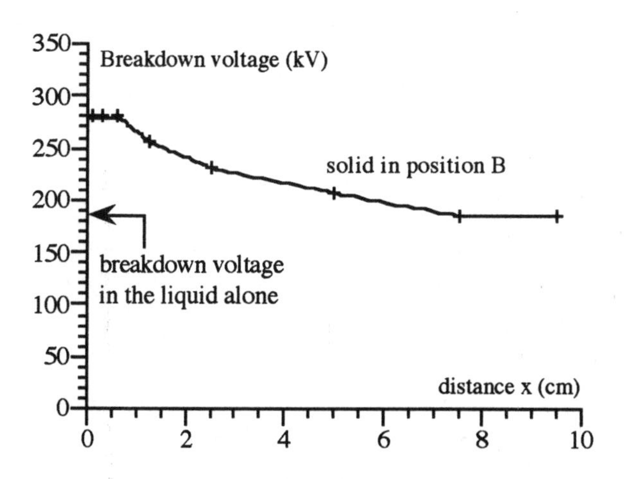

Figure 5: Breakdown voltage versus point-barrier distance. (15 cm insulating disk, position B in figure 1, d = 10 cm).

To obtain these photographs, the intensifier's gate was opened synchronously with the voltage rise, and closed before the streamer reaches the plane to avoid image saturation when breakdown occurs.

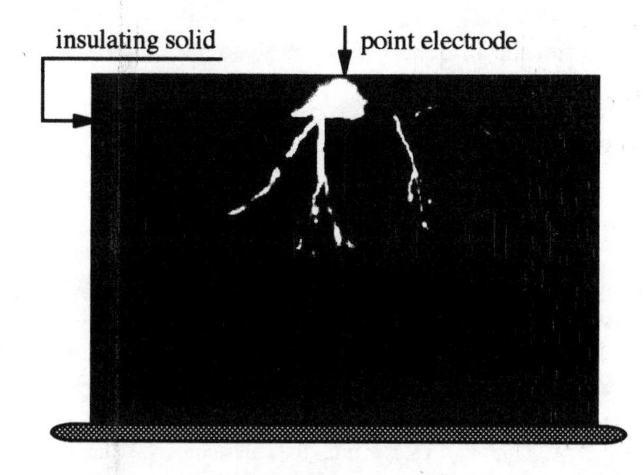

Figure 6: Photograph of a streamer getting round a solid barrier (d = 10 cm, V = 304 kV, x = 1.2 cm, barrier diameter: 15 cm).

When the solid is far from the point (x > 7 cm), breakdown results from the propagation of lateral branches of the streamer that reach directly the plane electrode without creeping at the liquid/solid interface (figure 7). The measured breakdown voltage is then the same as that measured without barrier (figure 5). This is due to the fact that streamer filaments in the liquid naturally expands within a roughly conical outer volume. Then, a barrier located at a distance x has no effect if its size is smallest than the cone cross section at this distance.

Notice that within the voltage range investigated on figure 5, no perforation of the solid was observed.

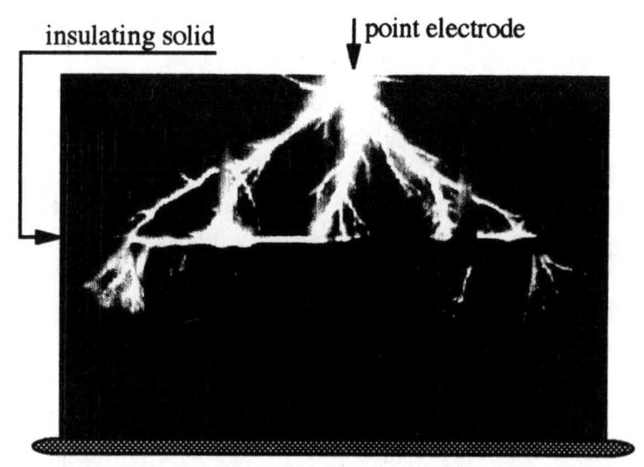

Figure 7: Photograph of a streamer getting round a solid barrier (d = 10 cm, V = 256 kV, x = 5 cm, barrier diameter: 15 cm).

Effect of overvoltages

When overvoltages are applied with the configuration B (solid perpendicular to the field), the plot of the average streamer velocity versus voltage (figure 8) is very different from that recorded with a solid in position A (figures 3 and 4). In this situation, breakdown also results from streamers creeping and getting round the surface, but the propagation velocity remains low (2 to 3 km/s) up to the acceleration voltage Va in the liquid alone (380 kV). Above Va, the speeding up of streamers is much lower than that recorded in the liquid alone, and decreases with x (at x = 1.2 cm, the recorded acceleration of streamers is negligible).

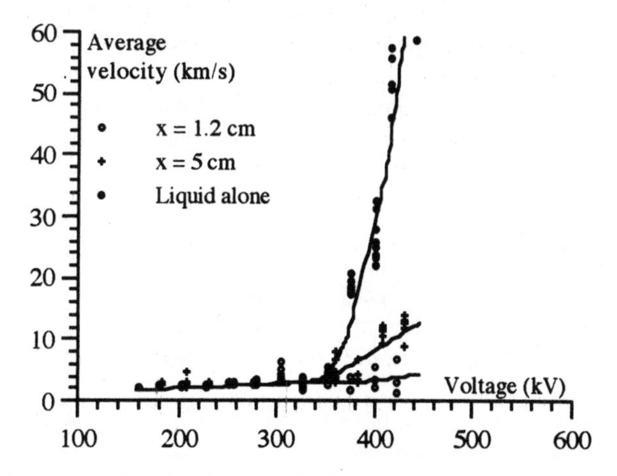

Figure 8: Average streamer propagation velocity versus voltage with a 15 cm barrier in position B (d = 10 cm).

With a barrier in position B, the average velocity is calculated taking into account the total streamer path (propagation up to the solid, creepage, propagation toward the plane electrode). Thus in figure 8, a path of 17.5 cm was considered.

With a solid in position B at d = 5 or 10 cm, and above the acceleration voltage Va, streamers propagate with a high velocity up to the barrier, and then slow down to 2 -3 km/s until breakdown. When the barrier is close to the point (at x = 1.2 cm), that streamers propagate rapidly to the barrier (with a velocity > 10 km/s) has a negligible effect on the calculated average velocity. With a larger distance x, the effect is more significant, and it is enhanced by the fact that lateral branches of streamers can reach the plane more quickly, since their path with a low velocity is shorter.

DISCUSSION

The main result evidenced in this study is the effect of the position of solid/liquid interfaces on the transition to fast streamers in mineral oil. With an interface parallel to the electrode axis, the inception and propagation of fast streamers is greatly enhanced. Conversely, a perpendicular interface makes the propagation of such streamers more difficult: in these investigations, the creepage of fast streamers has not been observed. When fast streamers impinge on a solid perpendicular to the field, they slow down to creep over its surface.

It was previously hypothetised that for slower streamers (velocity 2-3 km/s), the branching contributes to keep the field surrounding the streamer nearly constant when the voltage is increased. This can explain the constancy of the velocity over a large voltage range (from Vb to Va). When this effect of "electrostatic" field regulation is not any more possible (i.e. when the streamer gets a roughly spherical shape), the transition to a faster streamer mode is observed [3].

In the presence of a solid interface, the natural expansion of the streamer is hindered and its geometry is modified. In both cases of parallel or perpendicular interfaces, the field acting on the streamer is enhanced at the interface, compared to the field thay would be present on the streamer in the liquid alone. Then with the preceding hypothesis, the presence of an interface should favor the inception of fast streamers, what is effectively observed with an interface parallel to the electrode axis. In this case, the field on the streamer head at the interface is mainly tangential.

With a perpendicular interface, a high field at the interface also exists, but it is not oriented in the possible propagation direction of streamers (i.e. tangentially).

Thus the propagation of fast creeping streamers would require a higher applied voltage.This may explain why creeping fast streamers have not been observed in our experiments limited to 450 kV.

At the breakdown voltage, it has been observed that streamer propagation is not sensitive to the presence of a solid surface parallel to the gap axis (figure 2). At this voltage in the liquid alone, streamers are constituted typically with 1-2 branches, and the modification of their overall shape by a surface is much more limited than with overvoltages.

With a perpendicular interface, the increase in breakdown voltage seems to be related to the lengthening of the streamer path. Below the breakdown voltage in the liquid alone, streamer stop at some length ls (stopping length) that increases on average with the voltage V [2]. At d = 10 cm, a rate of increase $\Delta ls/\Delta V \approx 0.7$ mm/kV has been measured. With a 15 cm insulating disk, the lengthening of the streamer path is 7.5 cm. Thus if streamers behave as in the liquid alone, an excess voltage of ≈ 100 kV should be necessary for streamers to overcome this increase in distance. In fact, a maximum breakdown voltage increase of 100 kV has been measured with the insulating solid close to the point (figure 5).

ACKNOWLEDGMENT

The authors wish to thank M. Hilaire for his help in the construction of the experimental device, and the Jeumont-Schneider Transformers society that provided the materials used in this work.

REFERENCES

1. Saker A, P. Gournay, O. Lesaint, R. Tobazéon, N. Giao Trinh and C. Boisdon, "Prebreakdown and breakdown phenomena in large oil gaps under AC", this conference.
2. Saker, A, O. Lesaint and R. Tobazéon, "Propagation of streamers in mineral oil at large distances", Annual Report of the 1994 Conference on Electrical Insulation and Dielectric Phenomena. IEEE Publication 94CH3456-1, pp. 889-894.
3. Massala, G, A. Saker and O. Lesaint, "Study of streamer propagation in mineral oil in overvolted gaps under impulse voltage", Annual Report of the 1995 Conference on Electrical Insulation and Dielectric Phenomena. IEEE Publication 95CH35842, pp. 592-595.

Conference Record of the 1996 IEEE International Symposium on Electrical Insulation, Montreal, Quebec, Canada, June 16-19, 1996

Anomalous Oil Electrification Phenomena in Couette Flow Between Bare Stainless Steel Electrodes

D. Schlicker and M. Zahn
Laboratory for Electromagnetic and Electronic Systems
Department of Electrical Engineering and Computer Science
Massachusetts Institute of Technology
Cambridge, MA 02139

Abstract: **Electrification, which occurs in transformers due to the entrainment of electrical double layer charge from the recirculatory flow of the oil over internal insulating pressboard surfaces, is simulated in an apparatus consisting of two coaxial cylinders filled with oil. The inner cylinder is rotated, resulting in a turbulent flow in the fluid region between the inner and outer cylinder. Measurements of oil charge density, open circuit voltage and short circuit current are made. Experiments using bare stainless steel cylinders demonstrate transient behavior having time constants which are much greater than known time constants of the oil dielectric. Anomalous and non-reproducible system responses occurring due to changes in rotational speed, as a result of electrical prestressing, and for no known reasons are observed.**

INTRODUCTION

Flow electrification is a phenomena which has been a problem in many industries due to the transport of highly insulating fluids. In some cases, such as the use of jet fuel, conductivity enhancers can be used to decrease the relaxation time of the fluid such that charge can not accumulate. However in other applications of insulating fluids, as in the use of oil dielectrics in large power transformers, increasing the conductivity has the unsatisfactory results of increasing losses and decreasing the dielectric breakdown strength.

The source of fluid electrification is the separation of charge from the electrical double layer at the interface between the insulating fluid and generally a solid material. With no fluid flow, this charge extends into the fluid in an exponentially decaying manner. When the fluid near the surface starts to flow, the charge is also transported along with the fluid. If the dielectric relaxation time is short as compared to the transport time, this entrained charge will quickly dissipate from the fluid. However if the relaxation time of the fluid is large compared to the transport time, the fluid can maintain a charge density. The determining factors of the fluid charge density depend on both the flow characteristics and the material properties.

In power transformers, the flow of the oil dielectric is used as a means to convect thermal energy out of the transformer in order to limit the temperature. The insulating nature of both the solid pressboard and fluid oil, allow large quantities of charge to be redistributed within the transformer, both in the volume and on insulating or isolated surfaces, under appropriate conditions. This redistribution of charge results in the development of quasistatic potentials, which can lead to discharges and possible catastrophic failure of the unit.

The complexity of the fluid flow and total charge distribution in a full scale transformer make it difficult to study the electrification phenomena of the materials. Therefore, electrification experiments are carried out in a less complex system in which flow patterns and charge distributions can be better determined.

Because flow electrification induced failures of transformers often occur immediately after recommissioning and are a relatively rare phenomena, this work focuses on anomalous behavior. This behavior usually occurs immediately after initial system start up or when the system experiences an abrupt change in operating condition.

APPARATUS

The experimental apparatus used for observing the electrification phenomena is shown in Figure 1. This unit is referred to as the Couette Charger (CC) because of the Couette fluid flow characteristics and the resulting charge density in the fluid. The inner cylinder and outer cylinder of the CC are constructed of stainless steel, while the top and bottom are made of polycarbonate plastic providing electrical isolation between cylinders. A motor connected through a rotating seal delivers the mechanical torque to rotate the inner cylinder at variable rotation rates. The outside of the CC is encircled with heating tape and cooling coils (not shown) to maintain the systems temperature. The gap between the inner and outer cylinder is filled with transformer oil. As a result of the dimensions of the CC and properties of the oil, rotation at rates exceeding 62 RPM result in turbulent flow keeping the bulk of the oil well mixed [1]. This is significant in that the charge density in the well mixed region can be taken as uniform.

Many measurements are made while the CC is in operation. Basic oil properties such as moisture content, conductivity, dielectric constant, and temperature are monitored. The charge density in the well mixed core is also measured using an Absolute Charge Sensor (ACS) [2]. The inner and outer cylinder are utilized as electrical terminals from which an electrometer measures open circuit voltage or short circuit current. These terminals can also be used to apply an external voltage to the CC.

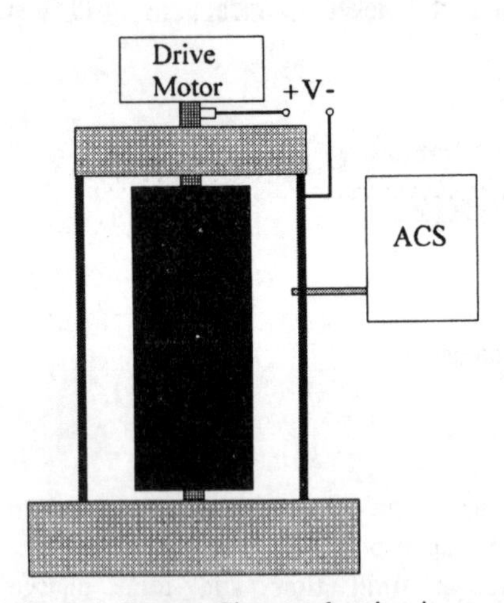

Figure 1. Couette Charger showing inner and outer cylinders along with polycarbonate top and bottom. The drive motor rotates the inner cylinder. Transformer oil fills the gap between inner and outer cylinders. The ACS unit is used to measure the charge density of the oil, while moisture, temperature, and dielectric sensors (not shown) measure other fluid properties. The inner and outer cylinder also act as electrodes for the measurement of terminal voltages and currents.

MEASUREMENTS

Initial System Transients

The first set of measurements to be discussed are those taken immediately after the cleaning and assembly of the system with fresh Shell Diala A transformer oil. The cleaning procedure of the system involves complete disassembly of all components which contact the oil. The surfaces of all components are then cleaned with methanol or a combination of acetone and methanol, with methanol used as the final solvent because of its low residue. The components are next treated with fresh Shell Diala A to remove any possible residue and the system is assembled. After assembly the system was evacuated to less than 15 mT for 24 hours. Fresh oil which was also dried at 15 mT was loaded into the CC under vacuum. The temperature of the system was then set to 15°C and approximately 20 hours was allowed for thermal equilibrium to be achieved.

The measurement in Figure 2 was then started. The terminal voltage reaches a positive peak within ten minutes after the start of rotation with a negative charge density. After a sudden dip in voltage and simultaneous rise in charge density with polarity reversal, a constant decline in terminal voltage and increase in charge density is experienced for approximately 1.75 hours. At this point the terminal voltage inverts over the next ten minutes and becomes largely negative saturating the electrometer measurement. At the same time an increase in charge density is observed.

The measurement is continued in Figure 3 after rotation was stopped for several minutes to reset the data acquisition. Even though rotation was stopped the charge density continues near the ending value in Figure 2, and the terminal voltage measurement remains saturated. After an additional 0.5 hours, the terminal voltage and charge density appear to rapidly stabilize to a relatively constant value. Continued measurements show no extreme jumps in charge density and terminal voltage.

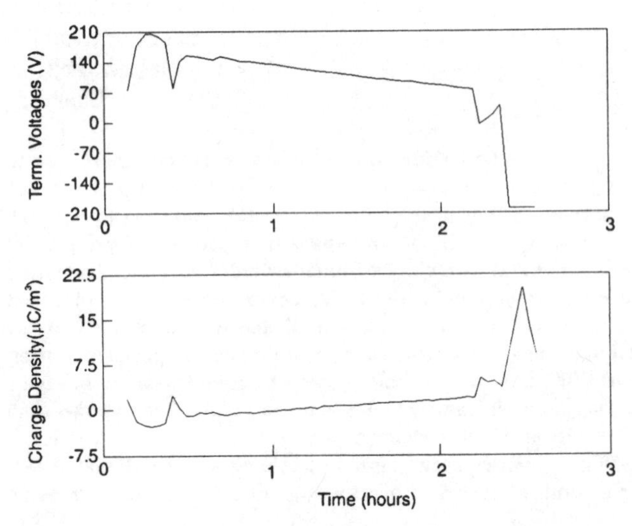

Figure 2. First set of terminal voltage and charge density measurements taken after cleaning of CC. The rotational rate is a constant 400 RPM. The oil in this measurement was at 15°C, 2.5 ppm moisture content, 0.3 pS/m conductivity and had a relative dielectric constant ε_r=2.19.

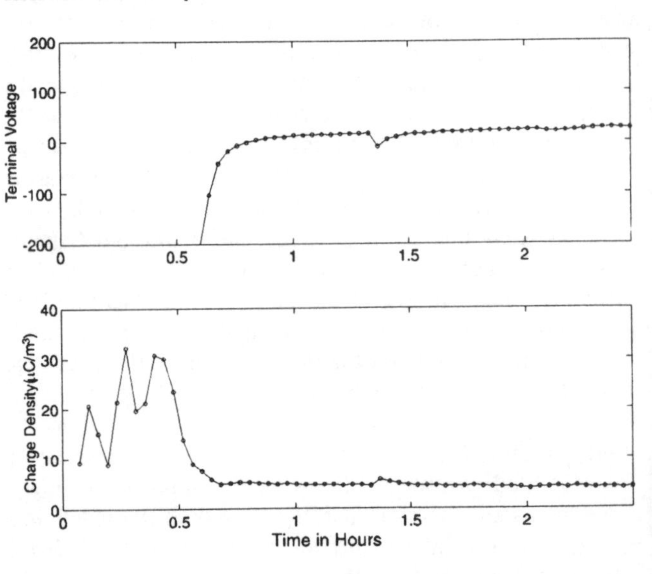

Figure 3. Second set of terminal voltage and charge density measurements taken after cleaning of CC. The start of this data follows the end of that in Figure 2 by several minutes during which rotation was stopped and data acquisition was restarted. The rotation was then continued at a constant 400 RPM. The terminal voltage in the measurement is initially saturated at -200V DC due to limitations of the electrometer.

Explanations for this phenomena have yet to be tested, but in light of the measured behavior, one hypothesis suggests a surface conditioning at each interface. In this system, with only bare stainless steel electrodes and Shell Diala A transformer oil, one may expect the behavior to depend only on the fluid flow and dielectric properties. In the case of the fluid dynamics, the time constant for turbulent development is less than a few seconds [1]. Furthermore, the relaxation time of this oil, based on its measured conductivity and dielectric constant, is about 64 s. In the measured data, events taking significantly longer than this appear. Also, the two sharp transients occur with no external change to the system, indicating two internal events. Once a conditioned state was reached this observed anomalous phenomena did not reoccur.

Rotation Induced Transients

The plots in Figure 4 show the response of the terminal voltage and charge density to step increases in the rotational rate of the CC. These measurements were taken after the initial transient behavior of the previous section had ceased. The measured oil parameters have not changed from this earlier experiment except for an increase in moisture to 7 ppm.

At about 4.4 hours into the experiment, the rotational rate was stepped from 700 to 800 RPM. This resulted in a large negative transient in voltage, which was accompanied by a transient increase in charge density. The data points in this experiment are spaced at around 2.3 minutes. The time constant of the decay of this transient is approximately 6 minutes, which is again greater than the turbulent development time or relaxation time. The event is very nonlinear compared to the response of the system to other steps, and repetition of the same experiment does not result in the large observed transient indicating the time variance of the system. However, measurements show much greater repeatability in the steady-state values of voltage and charge density for a set of experiments.

Electrical Prestress Transients

The next set of measurements is a comparison of data taken before and after the application of a prestressing voltage source to the terminals of the CC. Figure 5 shows the temporal behavior without prestressing after rotation at 400 RPM was started. The terminal voltage reaches a steady-state after several minutes, while the charge density remains steady from nearly the start of the measurement.

Once rotation was stopped, 500V DC was applied to the terminals of the CC for one hour. The terminals were then open circuited and the system was allowed to relax for ten minutes down to a terminal voltage of approximately 0.4V. The conductivity of this oil was 2.5 pS/m and the relative dielectric constant was approximately 2.2, resulting in a dielectric relaxation time of less than ten seconds. The plot in Figure 6 shows the temporal response of the system once

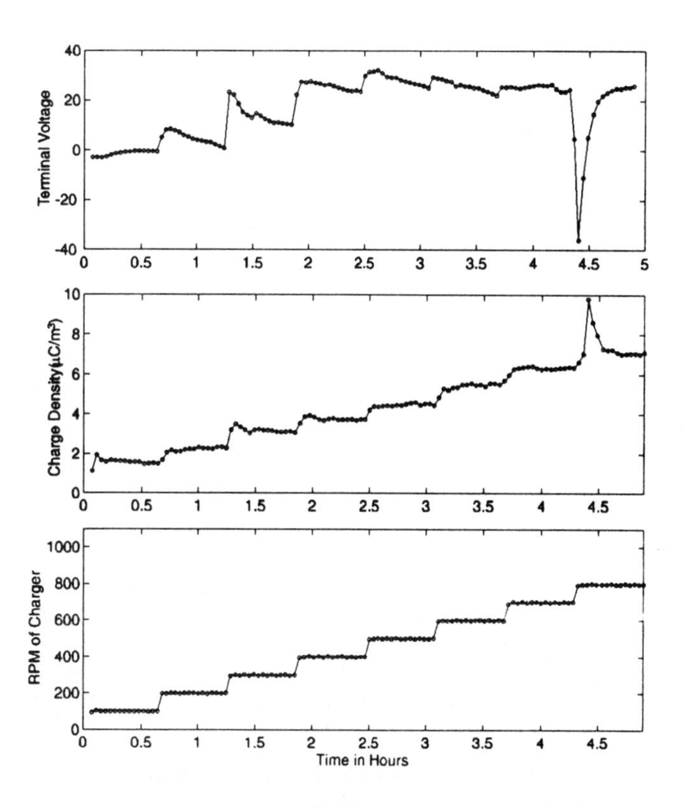

Figure 4. Temporal data of CC terminal voltage and charge density as a result of the applied rotation at rates shown in lower plot. The oil in this measurement was at 15°C, 7 ppm moisture content, 0.3 pS/m conductivity and had a relative dielectric constant ε_r=2.19.

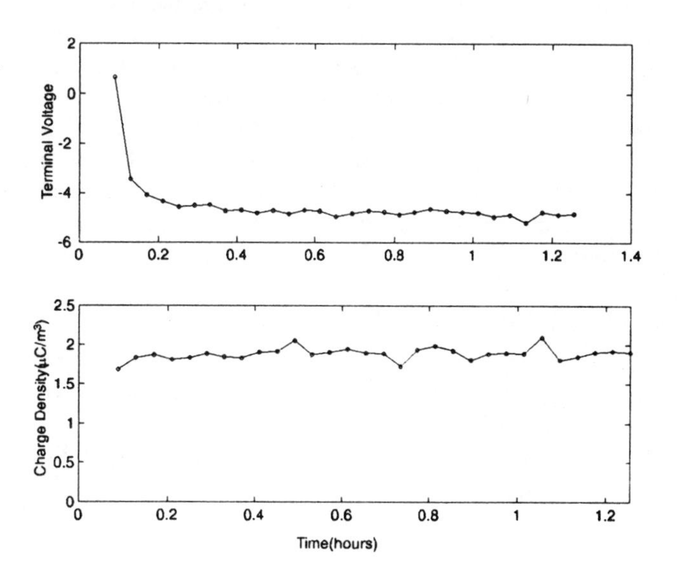

Figure 5. Baseline data at a constant 400 RPM, showing typical start up transient. The oil in this measurement was at 30°C, 25 ppm moisture content, 2.5 pS/m conductivity and had a relative dielectric constant $\varepsilon_r \approx 2.2$.

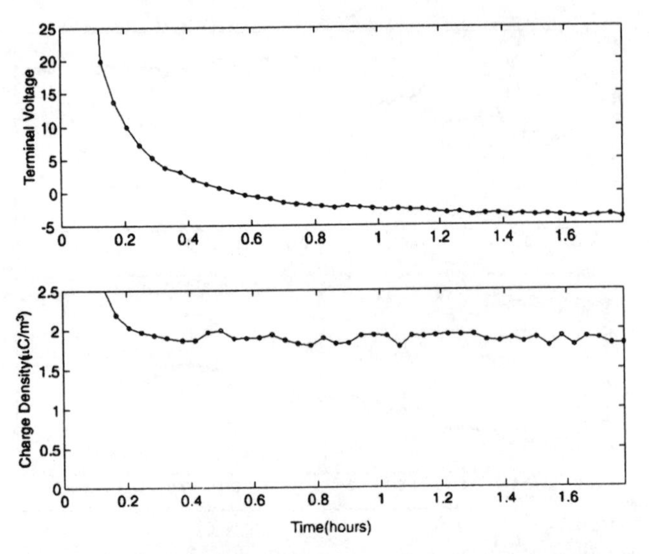

Figure 6. Once rotation had stopped from the measurement in Figure 5, the CC was stressed with 500 V at its terminals for one hour and then allowed to relax for ten minutes. This open circuit voltage and charge density data was taken immediately after the relaxation period with a rotation rate of 400 RPM. Oil parameters are the same as Figure 5.

rotation was restarted at 400 RPM. The terminal voltage immediately rises to greater than 25V, which is much greater than the voltage in the non-prestressed case in Figure 5, and then decays with a time constant which is almost two orders of magnitude longer than the relaxation time of the oil. The charge density also immediately rises to greater than $2.5\mu C/m^3$ after rotation begins and then returns to a steady state value of approximately $1.8\mu C/m^3$, in agreement with the steady state charge density value in Figure 5. However, the charge density decays with a time constant much greater than the relaxation time, but shorter than the open circuit voltage time constant. It appears that the 500V DC prestress has initially increased the electrical double layer wall charge density so that at the start of rotation, the values of terminal voltage and charge density are larger than without prestressing..

CONCLUSIONS

The data presented in this paper demonstrate several phenomena, which are inconsistent with time constants based on only dielectric relaxation. The first phenomena was observed after initial system cleaning and assembly, when the system experienced two sudden simultaneous changes in both voltage and charge density, with no known external stimulus. These two events may indicate a sudden change in interfacial properties at the cylinder surfaces. A second phenomena was observed when the rotation rate was increased in steps. In the step from 700 to 800 RPM, a very non-linear response was observed with a comparatively long time constant. Interestingly, repetition of the measurement did not yield such a phenomena, indicating some unmeasured parameter of the system had changed. In an attempt to alter the system while it was at rest, a DC voltage source was applied to the CC terminals. After allowing the system to relax for nearly ten minutes (≈60 relaxation time constants), rotation at constant RPM demonstrated that a change in the system had occurred. This change resulted in a large transient in terminal voltage and charge density with a time constant much longer than the relaxation time. Further investigation of interfacial parameters will be done to help better explain these observations.

ACKNOWLEDGMENT

This work was supported by the Electric Power Research Institute under Research Project WO 3334-1 under the management of S. R. Lindgren. Special thanks go to R. Albano and M. Grossman for the excellent machining of the Couette Charger and associated parts, and W. Ryan for assembly and construction of both mechanical and electrical components.

REFERENCES

1. Morin, A.J. II, M. Zahn, and J.R. Melcher. "Fluid Electrification Measurements of Transformer Pressboard/Oil Insulation in a Couette Charger". IEEE Trans EI. Vol. 26 No. 5, October 1991, pp. 870-901.

2. Morin, A.J. II, M. Zahn, J.R. Melcher and D.M. Otten. "An Absolute Charge Sensor for Fluid Electrification Measurements". IEEE Trans EI. Vol. 26 No. 2, April 1991, pp. 181-199.

Conference Record of the 1996 IEEE International Symposium on Electrical Insulation, Montreal, Quebec, Canada, June 16-19, 1996

Static Electrification Induced by Oil Flow in Power Transformers

L. Peyraque
Jeumont Schneider Transformateurs
84, ave Paul Santy, 69371-Lyon,France

A. Beroual and F. Buret
CEGELY-Ecole Centrale de Lyon
BP 163, 69131-Ecully Cedex, France

***Abstract*: This work concerns the study of the static electrification phenomena on a real transformer equiped with an exceptional set of oil pumps allowing to carry out different tests with an oil flow reaching nearly five times the rated value. The transformer was filled with two kinds of oils, aged and new, in order to compare their Electrostatic Charging Tendency (E.C.T.). The static electrification levels are analyzed through the measurements of the leakage currents in unenergized and energized conditions, and by varying the temperature. Dielectric contraints and temperature are adjusted to the values inducing the maximum electrification level. The consequences of static electrification on the insulating reliability are analyzed basing on the electrification level and acoustically measured partial discharges.**

INTRODUCTION

During the last twenty years, several cases of dielectric breakdown in large power transformers, attributed to static electrification induced by flowing oil through insulating structures, have been reported. These failures concern both technologies of transformers (ie, shell and core types). Numerous works have been undertaken by makers and users to understand this phenomenon for preventing and avoiding its annoying consequences. These studies have been carried out on various types of laboratory cells or physical models to reproduce oil flowing cooling ducts [1-3]. However, because of the complexity of the transformer structure, the study of the parameters influencing the static electrification on reduced scale models in laboratory is insufficient to reproduce accurately the different phenomena occuring in real transformers. Thus, it is essential to make studies on that way.

This paper is devoted to the static electrification phenomena on a real transformer equiped with an exceptional set of oil pumps allowing to carry out different tests with an oil flow reaching nearly five times the rated value. The static electrification phenomena is studied through the leakage currents and partial discharges measured on unenergized and energized transformer filled with two types of oils, and by varying the temperature.

EXPERIMENTAL TECHNIQUE

Tested Transformer

The tested apparatus was a single phase 100 MVA, 500/230/35 kV autotransformer of shell type technology. The design and technology of the internal structure was strictly representative of the standard shell type. The rated value of oil flow for this design was 350 m^3/h. For the test, it is necessary to have the capability of adjusting an unusual parameter in such type of structures, namely the oil flow rate in addition of the temperature and the supply voltage. For this purpose, a battery of 12 pumps with their associated valves has been installed. The rated characteristic values of each pump were respectively 225 m^3/h and 9 meters water column for the nominal oil flow and the pressure.
To reduce the oil pressure drops, the heat exchangers were replaced by straight pipes. Oil flow adjustments were achieved by reducing the number of pumps in operation. Figure 1 shows an external view of such an equipment.

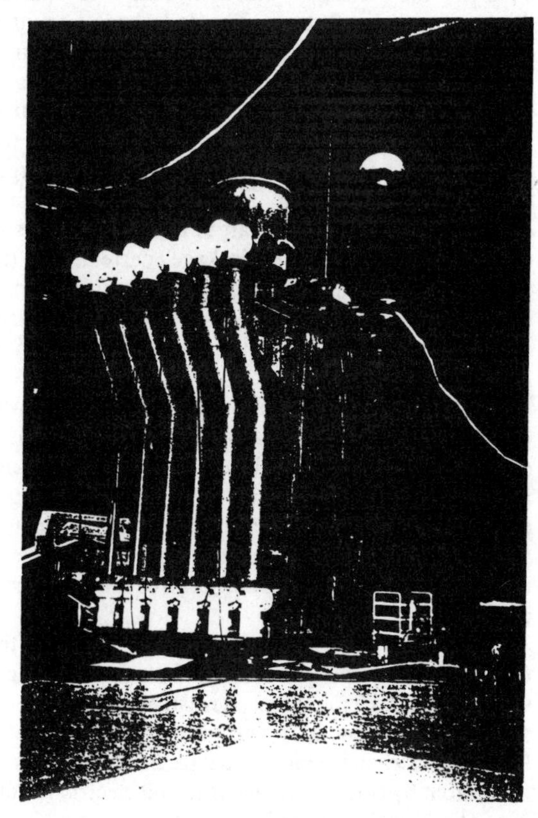

Figure 1: Tested transformer with its uprated pumps

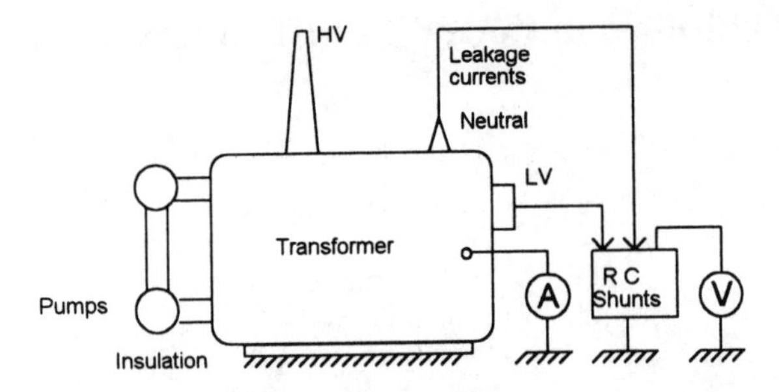

Figure 2a: Transformer test circuit under unenergized conditions

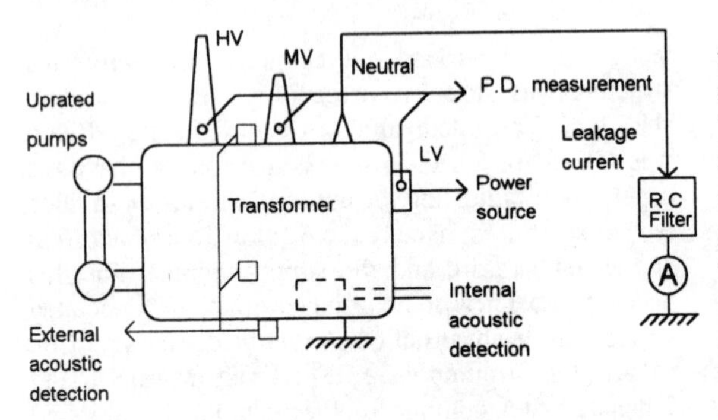

Figure 2b: Transformer test circuit under energized conditions

Leakage Current and Partial Discharges

The measurement that we have undertaken concern the leakage currents and the partial discharges in transformers. The dc-leakage current is the best indicator of the global static electrification in transformers. Generally, the oil flow produces a negative leakage current on the windings resulting from the dominant separation process. Indeed, on the windings, this current is due to the negative charges created on the surface of insulating materials. But a small amount of positive charges in the oil relaxes to them. Consequently, the leakage current on the windings is the superposition of a negative current caused by the separation of charges on press boards and a small positive current due to the relaxation of the charged oil.

The level of the partial discharges is a common way of measurement and evaluation of the incipient dielectric phenomena. Dielectric strength and insulation reliability are correlated with the partial discharge level. Figures 2a and 2b present the test circuits used for the measurement of the leakage currents on transformer in unenergized and energized conditions.

To study the effect of the applied voltage, the transformer is energized from LV (low voltage) winding. In this case, the alternative component of the leakage current is by passed to the ground by an RC filter. The detection of the partial discharges is carried out by two means: acoustic and electric (figure 2b). The electrical detection is achieved through the classic test circuit connected to the test taps of bushings. The measurer is connected alternatively to the HV, MV and LV windings. The acoustic method uses a hydrophone, sensitive to high level partial discharges which is implanted in the lower plenum, near the entrance of insulating ducts. This place is a preferential region where audible electrostatic discharges could occur. In the second serial of tests external ultrasonics sensors have been added to the internal detection by hydrophone.

Charging Tendency of Oil

To complete the above measurements, the transformer oil is analyzed with a mini-static tester, allowing a quatification of its Electrostatic Charging Tendency (E.C.T.). This latter indicates the quantity of charges susceptible to be created in oil [4]. For this, oil supposed electrically neutral is forced through a cellulose filter where a charge separation takes place. In results, the oil is positively charged and the filter negatively. So, we obtain the quantity of charges created by the flow, for a certain volume of the oil. An electrometer measures this charge generation on the filter (figure 3). If Q is the charge, V the volume of the oil flowing during a time t and I the mean current, the charging tendency ρ is given by the ratio of mean current to flow rate:

$$\rho = Q / V = I\,t / V = I / (V/t)$$

The measurement carried out with this technique minimizes the efffect of the relaxation time τ of the oil, τ being the ratio of the permittivity (assumed to be constant) to the conductivity. As the level of the charge accumulation is determined by the competition between the charge generation and the relaxation, other characteristics, such the conductivity must be regarded to analyze static electrification in the transformer[3].

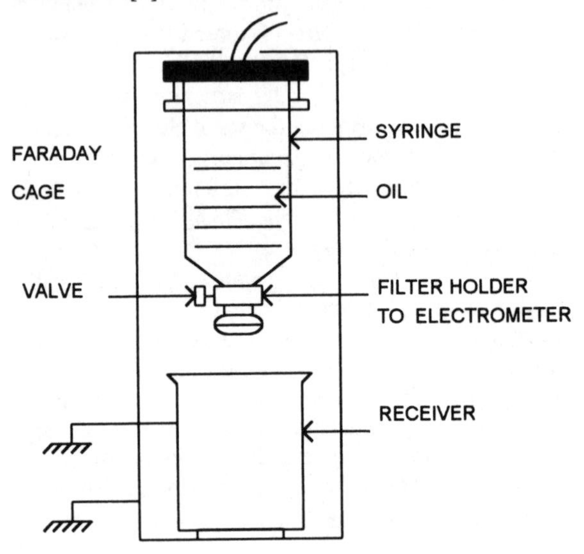

Figure 3: E.C.T. measuring apparatus

The variation of E.C.T. with the temperature and the conductivity have been also considered to forecast the maximum electrification level.

Conditioning of transformer and oil

The transformer oil is dried and filtered before the beginning of the tests. Standard procedures have been used. Since the oil nature is an important parameter of static electrification, we choose two kinds of nonstandardized oils. The first one (oil A) is an aged oil with a high level of electrostatic charging tendency allowing to increase the level of electrification in the transformer. The second one (oil B) is a new oil (virgin) with high resistivity and low charging tendency. The characteristics of both oils are summarized in table 1.

	T (°C)	OIL A AGED	OIL B NEW
Dissipation factor (Tan δ)	25 60 90	$5.6\ 10^{-4}$ $50.5\ 10^{-4}$ $136\ 10^{-4}$	$1.6\ 10^{-4}$ $15.4\ 10^{-4}$ $41.1\ 10^{-4}$
Conductivity (K : S/m)	25 60	$3.4\ 10^{-12}$ $29\ 10^{-12}$	$1.0\ 10^{-12}$ $9.5\ 10^{-12}$
Relaxation time (τ = ε/K : sec)	25 60	2.6 0.3	8.8 0.9
E.C.T .(μC/m³)	25	370	48
Interfacial tension (T.I.F. : mN/m)	25	27	37
Water content (ppm)	25	6	5
Breakdown voltage (kV)	25	94	91

Table 1

EXPERIMENTAL RESULTS

Unenergized transformer

a) Influence of flow rate: The experiments performed on transformer at room temperature (21 - 25 °C) show that the leakage currents strongly depend on the flow rate (figures 4a and 4b). In the tests with the aged oil A, the currents are low for the nominal rate (350 m^3/h): - 19 nA on HV winding and + 8 nA on LV winding, for a flow rate of 390 m^3/h. But these currents increase with the power of 2 to 3 of the flow rate. They reach respectively - 2045 nA and - 615 nA at the flow rate equal to 4.8 times the rated value.

On the characteristics of transformer oil A (table 1), we note that both E.C.T. and the conductivity are rather high. These parameters play a significant role in the increase of the leakage currents. With the new oil B, the currents are still lower (at 25°C) in a ratio of 1 to 3 comparing to oil A. At the same temperature, the leakage current varies in the same way as the charging tendency.

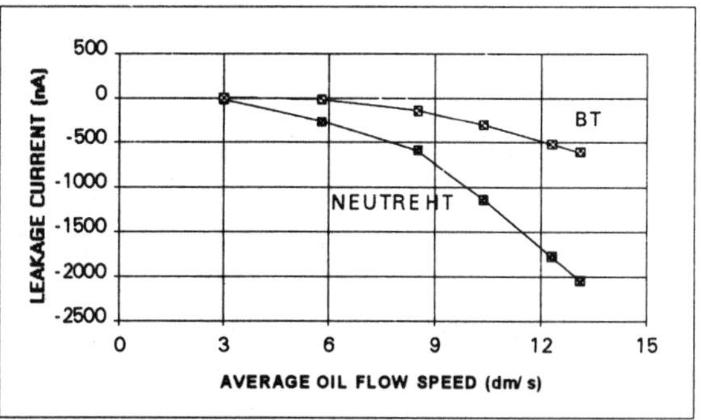

Figure 4a: Leakage current on unergized transformer with aged oil A

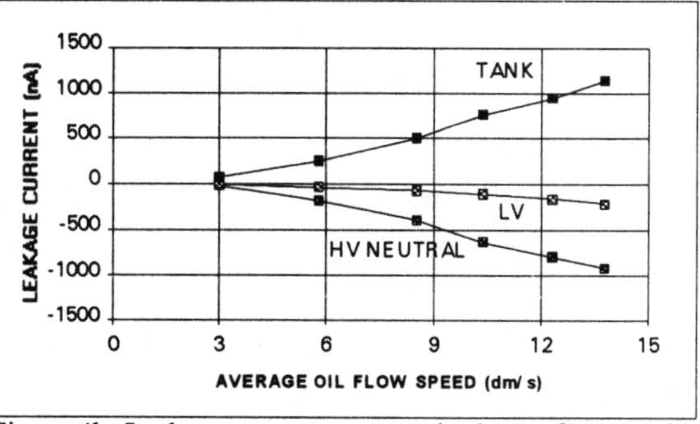

Figure 4b: Leakage current on unergized transformer with new oil B

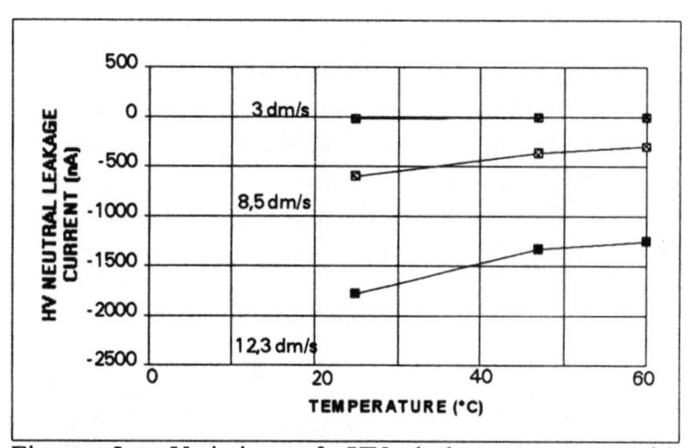

Figure 5a: Variation of HV leakage current with temperature on aged oil A

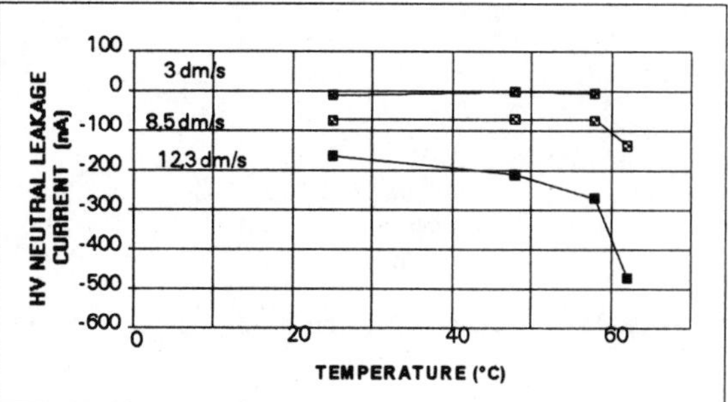

Figure 5b: Variation of HV leakage current with temperature on new oil B

***b) Temperature dependence*:** The measurements have been carried out during the natural cooling of the transformer (duration ~ 72 hours) after a short-circuit heating period. Figures 5a and 5b illustrate the temperature dependence of the leakage currents in transformer. As previously reported, the temperature level for which the leakage currents reach a maximum value depends strongly on the oil characteristics (charging tendency and conductivity). For the aged oil A, the maximum of the leakage current occurs at room temperature in this experiment. As the conductivity of the oil decreases strongly with the temperature, the maximum value of this current could occur for a lower conductivity of the oil that we used here i.e., for a higher relaxation time and then for a lower temperature. With the new oil, the maximum leakage current is observed at 62°C, let be - 2200 nA. This is 20% higher than the maximum current obtained with the aged oil A.

Energized Transformer

The transformer is energized at 200 Hz and a voltage varying from 0.5 to 1.5 times the nominal value. The flow rate is comprised between 1.7 and 4.5 times the nominal one. For each level of voltage and flow rate, the leakage currents are measured only on the neutral of HV winding for a minimum period of 15 minutes. The results, reported in figure 6a confirm that the static electrification increases gradually with the applied voltage. The leakage currents reach - 6000 nA for 4.5 times the nominal flow rate and 1.5 times the rated voltage for the aged oil A.

At 25°C, the leakage current for the new oil is lower in a ratio of 3 (figure 6b) than with the aged one, but this value increases, and for 65°C the maximum value (-10000 nA) exceeds that one obtained with an aged oil (figure 7).

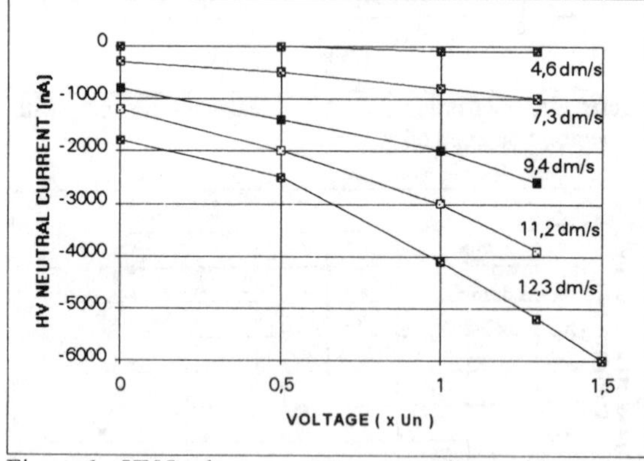

Figure 6a: HV Leakage current versus voltage with aged oil

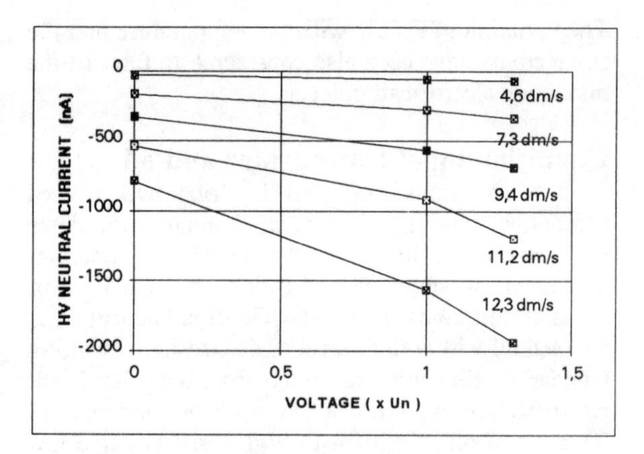

Figure 6b: HV Leakage current versus voltage with new oil

However, no partial discharges had been heard or detected by the hydrophone, or by the ultrasonic sensors. The P.D. maximum level measured electrically on the bushing taps seems to be more dependent on the voltage than on the flow rate and then on the electrification level. All the PD are lower than the usual aceptance level (500 pC) with an applied voltage 1.3 times the nominal one (tables 2, 3 and 4).

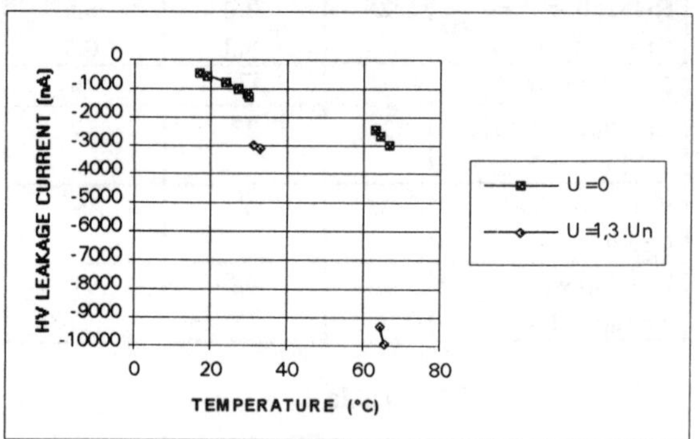

Figure 7: HV Leakage current versus temparature with two level of induced voltages (new oil B)

Temperature - 25 °C		MAXIMUM LEVEL OF PARTIAL DISCHARGES. (pC)			
Each point - 15 minutes		$1.0 \times U_n$		$1.3 \times U_n$	
FLOW RATE (m³/h)	OIL SPEED (dm/sec)	MV	HV	MV	HV
0	0	22	44	178	158
600	4.6	80	71	251	126
940	7.3	141	50	398	159
1220	9.4	141	50	50	100
1455	11.2	32	63	158	282
1600	12.3	28	56	178	398

Table 2: Partial discharge level with aged oil A

Temperature 25°C Each point 15 minutes		MAXIMUM OF PARTIAL DISCHARGES. (pC)			
		U_n		1.3 U_n	
FLOW RATE (m^3/h)	OIL SPEED (dm/s)	HV	MV	HV	MV
0	0	34	195	38	210
600	4.6	32	34	22	44
940	7.3	8	31	30	192
1220	9.4	12	26	36	117
1455	11.2	9	26	16	44
1600	12.3	24	158	49	153

Table 3: Partial discharge level with new oil B

The leakage currents measured in this work are in the same range as those obtained with transformers of higher sizes (300 MVA) under nominal voltage. In these experiments, partial discharges occured only for a very high value of oil flow: 3 to 4 times the nominal value [5,6]. Under nominal voltage and when partial discharges occur, the leakage currents measured in a core transformer is of 5 to 8 μA [7]. Electrification and partial discharges phenomena could be observed from 15 to 50°C.

OIL SPEED = 12.3 dm/sec		MAXIMUM OF LEVEL OF D.P.			
		U_n		1.3 U_n	
Test time (mn.)	T (°C)	HV	MV	HV	MV
15	25	24	158	49	153
90	65	/	/	210	444

Table 4: Partial discharge level with new oil B, for different temperatures

CONCLUSION

This work shows a very large safety margin on oil flow without occurence of any dielectric phenomenon due to static electrification (more than 4 times the nominal oil flow) for a shell type transformer. The fixed design parameters in this type of transformers (ie., cooling ducts geometry, blocks inclination, linear oil speed, type of boards,...) play undubitably an important role in the electrification phenomenon. It is noticeable that at nominal oil flow, the winding leakage current value is more than forty times lower than the maximum observed value where no partial discharges occured. Consequently, in this standard design, the shell type transformer does not present particular sensitiveness to electrification phenomena but a large safety margin on the known parameters of static electrification in normal operation.

REFERENCES

[1] A.J. Morin, M. Zahn, J.R. Melcher, "Fluid Electrification Measurements of Transformers Pressboard/Oil, Insulation in a Couette Charger", I.E.E.E. Transactions on Electrical Insulation, Vol. 26, N°5, October 1991, pp.870-901.

[2] H.P. Moser, CH. Krause, G. Praxl, G. Spandonis, R. Stonitsch, "Electrostatic Charging in Large Size Models of Power Transformer Cooling Ducts", GIGRE, Session 1992, J.W.G. 12/15.13, T.F. 01-T.F. 02.

[3] L. Peyraque, A. Beroual, C. Boisdon, F. Buret, "Phenomènes d'Electrisation des Matériaux Isolants pour Transformateurs de Puissance", Journal de Physique III, Vol.4, n°7, 1994, pp.1295-1304.

[4] T.V. Oommen, E.M. Petrie, "Electrostatic Charging Tendency of Transformer Oils", I.E.E.E.Transactions on Power Apparatus & Systems, Vol. PAS-103, N°7, 1984, pp. 1923-1931.

[5] R. Tamura, Y. Miura, T. Watanabe, T. Ishi, N. Yamada, T. Nitta, "Static Electrification by Forced Oil Flow in Large Power Transformer", I.E.E.E. Transactions on Power Apparatus & Systems, Vol. PAS-99, N°1, 1980, pp. 335-343.

[6] M. Higaki, Y. Kako, M. Moriyama, H. Hirano, K. Hiraishi, K. Kurita, "Static electrification and partial discharges caused by oil flow in forced cooled core type transformers", I.E.E.E. Transactions on Power Apparatus & Systems, Vol. PAS-98, N°4, 1979, pp. 1259-1267.

[7] S. Shimizu, H. Murata, M. Honda, " Electrostatic in power transformers", I.E.E.E. Transactions on Power Apparatus & Systems, Vol. PAS-98, N°4, 1979, pp. 1244-1250.

Conference Record of the 1996 IEEE International Symposium on Electrical Insulation, Montreal, Quebec, Canada, June 16-19, 1996

A Suppression Method of Oil Flow Electrification on Surface of Insulating Paper

Yang Jia-Xiang Chi Xiao-Chun Liu Yong

Department of Electrical Materials
Harbin University of Science and Technology
Harbin , 150040 , P.R.China

ABSTRACT

In this paper, the charge accumulation caused by oil flow electrification on the surface of insulating paper was observed, using an oil-paper pipe model. In order to eliminate this accumulation of charges on the surface of insulating paper, a charge generator consisted of polymer films was made in experiment. Experimental results indicates that the negative charges on the surface of the insulating paper was eliminated by the charge generator. Thus, partial discharge due to oil streaming electrification can be suppressed.

INTRODUCTION

The tendency of power transformers towards increased scale, in both capacity and voltage, has led to in increasing amount of oil flow for cooling purposes. Thus, the amount of streaming electrification both in oil and surface of pressboard can no longer be ignored. Actually, the electrostatic discharge caused by the charges accumulation in large power transformers has been reported by many efforts[1][2][3]. It has been clarified that adding BTA in oil is an effective method to prevent flow electrification in transformer oil [4][5].

In this paper, the charges accumulation on surface of insulating paper has been simulated experimentally using an oil-paper insulating pipe model and a charge generator. The charge density on the surface of the insulating paper can be controlled by generated the charge generator. Thus, this method is useful for eliminating the partial discharge due to oil flow electrification in large power transformer.

EXPERIMENTAL SYSTEM

The experimental system of oil flow electrification is shown in Figure 1. This system is made up of oil reservoir, relaxation tank, oil circle unit, temperature control unit, charge generator, measurement unit and pipe model.

Oil reservoir (volume : 200 liter) is used to reserve transformer oil in circle system. The temperature control unit is made up of electric heater (power 15 kW) in the oil reservoir, contacting thermometer and temperature control instrument. The highest temperature in this system reaches up to 85 ℃ . The relaxation tank (volume : 100 liter) is used to neutralize the static charges in the oil, it is grounded generally.

In oil circuit system, transformer oil driven by pump (the flow rate : 100 l/min) flows along with pipe-line. The oil flow rate can be adjusted by the throttle valve and overflow valve. Its range of oil flow rate adjusted is between zero and 5 m/s. The oil flow rate is measured by turbine-flow-meter. Every part in the whole system is connected with steel pipes to close circuit of system.

The structure of oil-insulating paper pipe model is shown in Figure 2. This model is made up of inner and external electrodes connected with lead wire. The external electrode is steel pipe, (internal diameter : 13 mm, thickness : 5 mm) and inner electrode is copper pipe, winded by three layers of insulating paper(thickness : 0.7 mm). The spacers made of polytetrafluoroethylene(PTFE) is fixed between inner and external electrode .

Configuration of the charge generator is shown in Figure.3. It is consisted of polymer films (40 layers). Adjusting the valve A and valve B in this system, the oil charge density can be controlled. . The maximum oil charge density generated by this charge generator is 53 $\mu c/m^3$ (16 ℃), 140 $\mu c/m^3$ (40 ℃) and 208 $\mu c/m^3$ (60 ℃) at oil flow rate of 2.25 m/s.

Measurement unit consists of the oil charge density instrument; the surface potential detector and the microgalvanometer. The oil charge density instrument is mainly composed of the absolute oil following:

1. The measurement of the charge density in oil generated by charge generator

In the oil flow electrification system, oil-paper insulating model was taken out and the charge density ρ in the oil generated by the charge generator was measured only. The Figure 5 presents the dependence curve of the charge density ρ in the oil to the oil flow rate when the oil temperature is 16 ℃, 37 ℃ and 60 ℃ respectively.

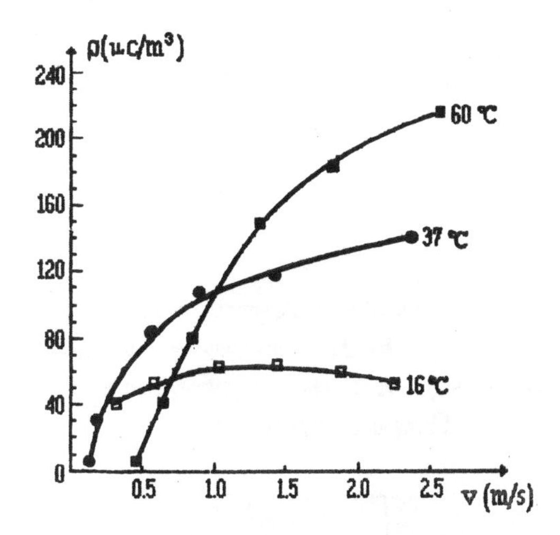

Figure 5: Dependence curve of charge density generated by the charge generator at oil flow rate

It is clear from measurement results above that the charge generator has the following characteristics:

(1) The polarity of the charge generated by the charge generator is positive.

(2) Regardless of low flow rate or high flow rate, the charge density generated in oil by the charge generator is much more than that in the oil-paper model at the same conditions. In the case of the flow rate of 2.25 m/s, when the temperature is 16 ℃, 37 ℃ and 60 ℃, the charge density in oil is 53 $\mu c/m^3$, 140 $\mu c/m^3$ and 208 $\mu c/m^3$ respectively.

(3) The charge density in the oil increases with the increase of oil temperature; and in the range of $v < 1.0$ m/s and $v < 0.7$ m/s, the charge density in oil at higher temperature is smaller than that at low temperature at the same flow rate.

2. Influence of charge density generated by charge generator on surface potential of insulating paper

When the temperature is 30 ℃ and the oil flow rate is 1.4 m/s and the charge ensity ρ generated by the charge generator is 0 $\mu c/m^3$, 10 $\mu c/m^3$, 20 $\mu c/m^3$, and 40 $\mu c/m^3$ respectively, the dependence curve of surface potential caused by the charge on insulating paper on time is shown in Figure 6.

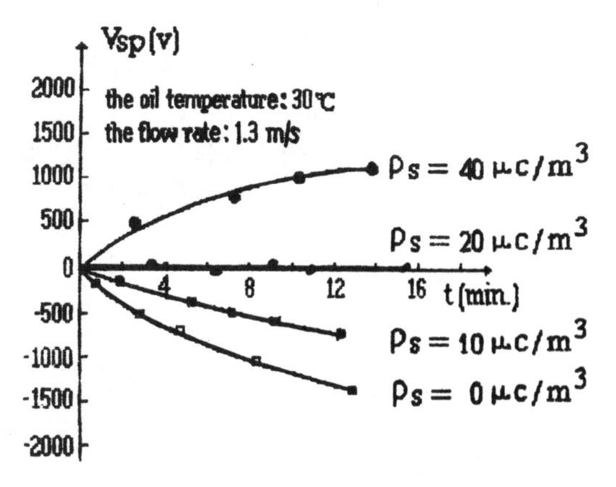

Figure 6: Dependence curve of surface potential on the insulating paper on time.

In Figure 6, when the charge generator is switch off, that is $\rho_s = 0$, the negative surface potential on the insulating paper (V_{sp}) increases along with the time, and tends to be constant eventually. When charge density generated ρ_s y charge generator is 10$\mu c/m^3$, the negative surface potential of the insulating paper goes down along with the time. When ρ_s is 20$\mu c/m^3$, V_{sp} becomes 0, in other words, all negative charges on the surface of insulating paper are neutralized by positive charges generated by the charge generator. When ρ_s= 40$\mu c/m^3$, the polarity of V_{sp} is reversed, that is, the polarity of V_{sp} changes from negative to positive. The changing of polarity of V_{sp} with the change of the charges generated by the charge generator is related to the phenomena of oil-paper interface.

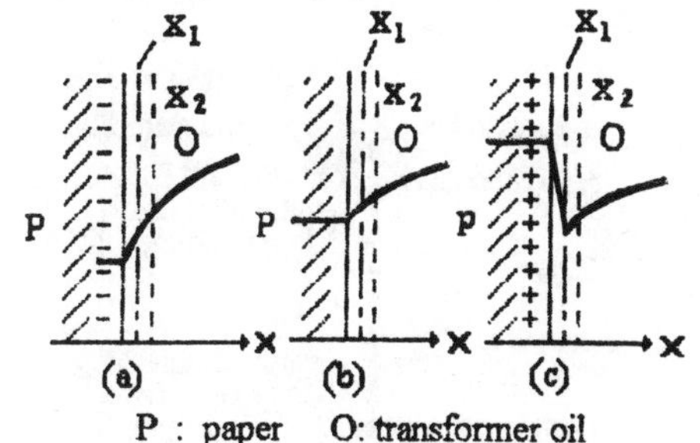

P : paper O: transformer oil

x_1: Inner Holmholtz plane

x_2 : External Holmholtz (slipping plane)

Figure 7: Distribution of electrical potential in electrical double layer of oil-paper interface.

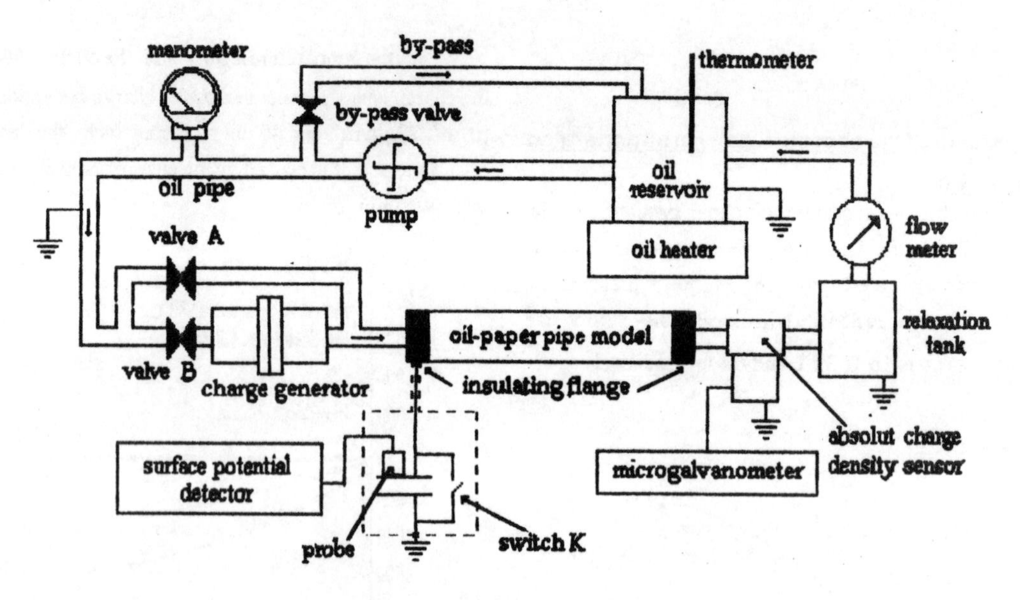

Figure 1: Experimental system.

charge density sensor made of Faraday Cup[6]. When the oil was drawn into Faraday Cup from pipeline and returned back to pipeline, the charge in the oil through Faraday cup and leakage current caused by it can be measured (Fig.4). The minimum absolute charge density in the oil measured by this instrument is $1 \mu c/m^3$ and its accuracy of measurement is 5%. The surface potential detector can measure

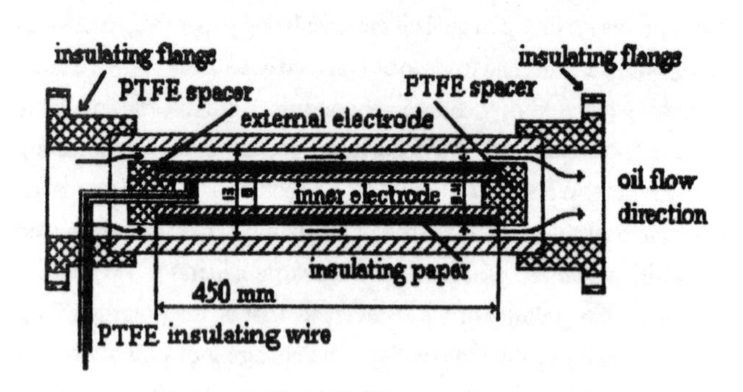

Figure 2: The structure of oil-paper model.

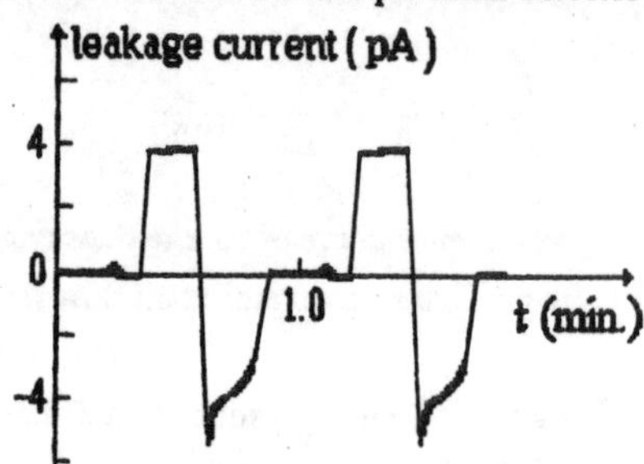

Figure 4: Measured current as a function of time from the absolute charge density sensor.

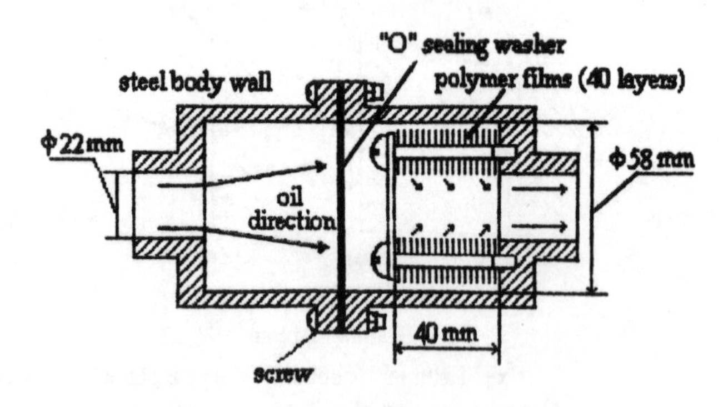

Figure 3: The structure of charge generator.

the surface potential formed by the accumulated charge on the surface of the insulating paper through a probe located on the capacitor (C=20 pF) shown in Fig.1. Maximum value of the surface potential measured is 3000V. The microgalvanometer can measure the leakage current formed by the charge on the surface of the insulating paper. The range of leakage current measured is 10^{-1} ~10^{-14} A.

EXPERIMENTAL RESULTS AND DISCUSSION

The accumulation of the charges on the surface of insulating paper was observed, and the effect of the charge generator in oil circuit system was investigated. The experimental results were obtained as

Figure 7(a), and 7(b), 7(c) show the distribution of electrical potential in electrical double layer of oil-paper interface in the case of ρ_s=0 and ρ_s > 0 respectively. When ρ_s=0, there are negative charges caused by the oil flow on the surface of insulating paper only, the electrokinetic potential ζ decided by the slipping plane is negative; as shown in Figure 7(a). When ρ_s > 0 , the negative charges on the surface of insulating paper is neutralized completely by the charge generator, the electrokinetic potential ζ is zero; as shown in Figure 7(b). When ρ_s>> 0, because the positive charge density generated by the charge generator is more much than the negative charges in the oil, the polarity of charges on the surface is reversed, electrokinetic potential ζ changes into positive, as shown in Figure 7(c). This three models of electrical double layer can prove the changing of the surface potential along with the charge generated by the charge generator.

3 Influence of charges generated by the charge generator on the charge density in the oil

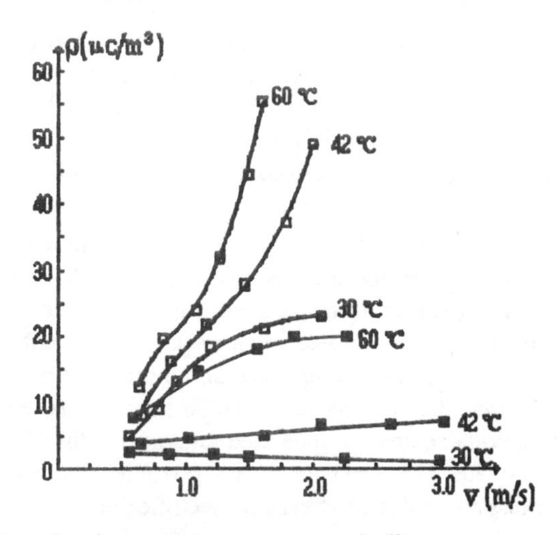

■ : the charge generator was turned off.

□ : the charge generator was turned on and kept 0 v of potential on the insulating paper.

Figure 8: Relationship between charge density in the oil and oil flow rate.

Since the charge generated by the charge generator can change the surface potential of the insulating paper, it may be influence the charge density in the oil. When temperature is 30 ℃, 47 ℃ and 60 ℃ separately, the charge density in the oil is measured in the case of on/off charge generator, and the surface voltage keeps zero always until charge generator turn off. Fig. 8 shows the relationship between the charge density in the oil and oil flow rate under the two conditions mentioned can be seen from Figure 8 that in the case of turning on the charge generator the charges in the oil increase along with increasing oil flow rate under certain temperature, especially in the high temperature this changing of the charge in the oil is faster than that the in the low temperature. It is attributed to that charges due to streaming electrification and the charges generated by the charge generator are positive all.

CONCLUSIONS

According to the experimental results and analysis mentioned above, the conclusions are obtained as following:

The positive charges generated by the charge generator can eliminate the negative surface potential due to the charges accumulation on the surface of the insulating paper, at the same time, the charge density in the oil increases, especially at the higher temperature, these effects are obvious. Therefore, the failure due to oil flow electrification can be suppressed using this charge generator. However, it is necessary that further investigation is made in large power transformer.

REFERENCES

1. D.W.Crofts, "Failure of 450 MVA, 345/138 KV Autotransformer form Static Electrification", EPRI AC/DC Transmission Substations Task Force Meeting, New Orleans, LA, May 1984.
2. M.M.Dixon, "Failure and Repair of a Westing-house 345/138 KV, 560 MVA Autotransformer", EEI Electrical Systems and Equipment Committee Meeting, Paper P5, Pensacola, FL, February 1985.
3. D.A.Ditzler, "Failure and Repair of a Westing-house 345/138 KV, Autotransformer", Minutes of the Fifty-Second Annual International Conference of Doble Clients, Doble Engineering, Watertown, MA, 1985.
4. M.Yasuda, K.Goto, H.Okubo, T.Ishii, E.Mori and M.Masunga, "Suppression of Static Electrification of Insulating Oil for Large Power Transformers", IEEE Trans. on Electr. Ins., Vol.23, No.1, PP. 153-157, 1988.
5. Chi Xiao-Chun, "The Effect of BTA Content on the Oil Streaming Electrification in Power Transformer", Jornal of Power Transformer (in chinese), No.3, 1990.
6. A.J.Morin, M.Zahn, J.R.Melcher and D.M.otten, "An Absolute Charge Sensor for Fluid Electrification Measurements", IEEE Trans. on Electr. Ins., Vol. 26, NO. 2, PP. 181-199,1991.

Conference Record of the 1996 IEEE International Symposium on Electrical Insulation, Montreal, Quebec, Canada, June 16-19, 1996

A Preliminary Study of Oil-Polypropylene Insulating System

Chi Xiao-Chun Ding Li-Jian Yang Jia-Xiang

Department of Electrical Materials Engineering

Harbin Institute of Electrical Technology

Harbin, 150040, P. R. CHINA

Abstract: **In this paper, creeping discharge of oil-polypropylene insulating system have been studied using an experimental model with the flowing oil under ac voltage. The ac creeping discharge inception voltage of the oil-polypropylene system has been tested under varied conditions. According to the experimental results, all of the oil temperature, the oil flow rate and the position of polypropylene film can influence the ac creeping discharge inception voltage of oil-polypropylene system. The ac inception voltage of this system is increased along with the increase of oil temperature and oil flow rate. Moreover, the ac inception voltage is higher at the position near to oil entrance than that at the position far to oil entrance. Considering the experimental results, a new composite insulating system is put forward for reducing the negative effects of streaming electrification in larger power devices with forced-oil cooling systems.**

INTRODUCTION

Nowadays polymers become one of the great majority insulating materials in use because of their excellent electrical and mechanical properties. Polypropylene, as one of the superior polymers with high dielectric strengthen and low dissipation factor characteristics, has found extensive use as the main dielectric in many high voltage devices, especially in power capacitors. In terms of the performances of oil-impregnated polypropylene and composite polypropylene insulating system, many investigations have been done [1-4]. However, these investigations are under the condition that the insulating fluid in the composite polypropylene insulating system is motionless. In order to know more about the performances of the composite polypropylene insulating system, a study on the performances of oil-polypropylene insulating system with the flowing fluid has been carried out experimentally.

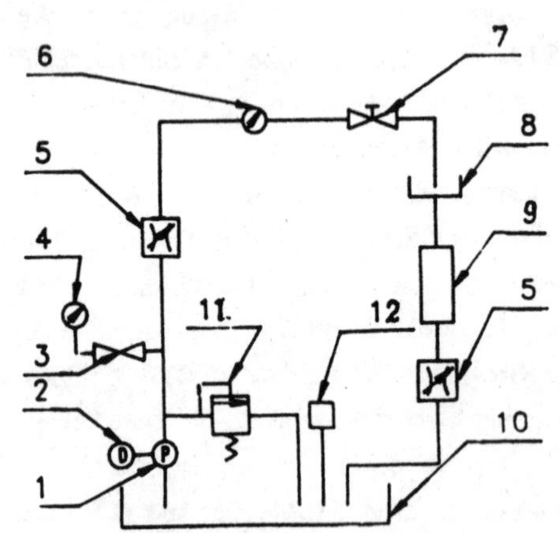

1. Oil pump 2. Electric motor 3. Damper 4. Oil manometer 5. Throttle valve 6. Oil flow meter 7. Stop valve 8. Oil relaxation tank 9. Oil-polypropylene insulating experimental model 10. Oil storage tank 11. Overflow valve 12. Temperature controller

Figure 1 Schematic diagram of oil circuit system

In this paper, the influences of insulating oil temperature, oil flow rate and polypropylene film's position in composite polypropylene insulating system on the creeping discharge of oil-polypropylene insulating system have been studied using an experimental model. The experimental results are presented and discussed.

DESCRIPTION OF TEST MODEL

In order to study the performances of oil-polypropylene system with the flowing oil, an oil circuit system is used in the experiment. Figure 1 is the schematic diagram of the oil circuit system. This system mainly includes an oil storage tank, a relaxation tank, a temperature control unit, a unit of oil flow rate regulation and an experimental model.

In this system, the temperature control unit can adjust the oil temperature from room temperature up to 350K. It severs to investigate the role of temperature in the experiment. The flow rate regulation unit consists of an oil pump; an overflow valve; two throttle valves; a stop valve; an oil manometer and an oil flow-meter. It can easily to set the oil flow rate as required by the experiment, and it can be used for studying the effect of the oil flow rate. The relaxation tank is grounded, and the charges in insulating oil due to streaming electrification can be neutralized in it.

Figure 2 shows the inside diagram of the experimental model for creeping discharge of oil-polypropylene insulating system. It is mainly composed of a needle-plate electrode system and a set of fastening screw. The set of fastening screw is used for fastening polypropylene film on the surface of plate electrode. This unit is set into a plexiglass container, and then joins into

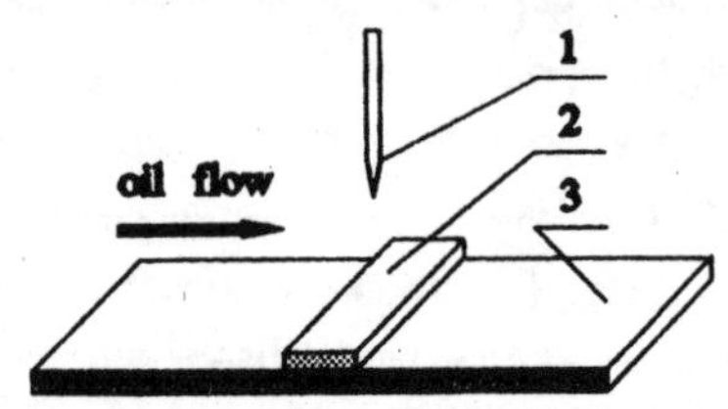

1. Needle electrode 2. Polypropylene film 3. Plate electrode

Figure 2 Creeping discharge of oil-polypropylene model

the oil circuit system.

Figure 3 shows the position of polypropylene film in the creeping discharge model for investigating the influence of the polypropylene film position in oil-polypropylene insulating system. Three pieces of polypropylene films were fastened at different places on the surface of plate electrode. The distances of three polypropylene films' centers from oil entrance are 140 mm, 190 mm and 240 mm respectively, and the width between two polypropylene films is 20 mm.

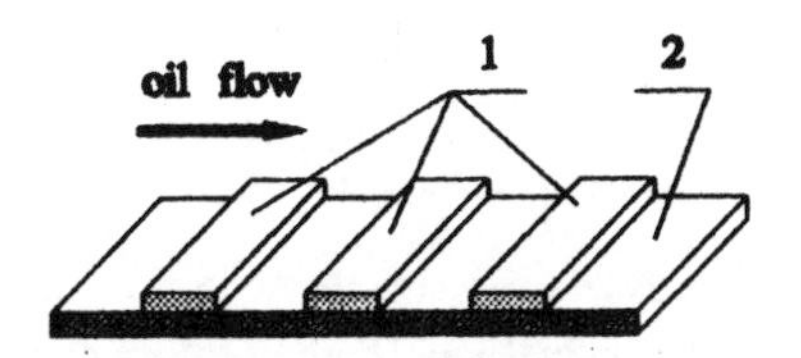

1. Polypropylene Film 2. Plate electrode

Figure 3 position of polypropylene film in model

MEASUREMENT AND RESULTS

In this investigation, type 45# transformer oil is selected as the insulating oil in oil-polypropylene insulating system. Its moisture content is 34.9 PPM, air content is 8.88%, dielectric permittivity is 2.3, and its $\tan\delta$ is 0.05. The thickness of polypropylene film is 20 μm; before impregnated, its $\tan\delta$ is 0.026, and its dielectric permittivity ε is 1.86; after impregnated, the $\tan\delta$ changes to 0.004, the dielectric permittivity ε becomes 2.21. In the experiment four layers of polypropylene film was fastened on the surface of plate electrode, and the needle electrode was kept above the center of polypropylene film. The plated electrode is grounded, and the needle electrode is connected with high ac voltage. The test of creeping discharge of oil-polypropylene insulating system starts after the oil circuit system has run half an hour, and the speed of increasing voltage is around 1 kv/s.

The experimental results are presented in the following.

The influence of oil temperature on ac creeping discharge inception voltage of oil-polypropylene system is given in figure 4. From this curve, it can be seen that the inception voltage is increased along with the increase of oil temperature.

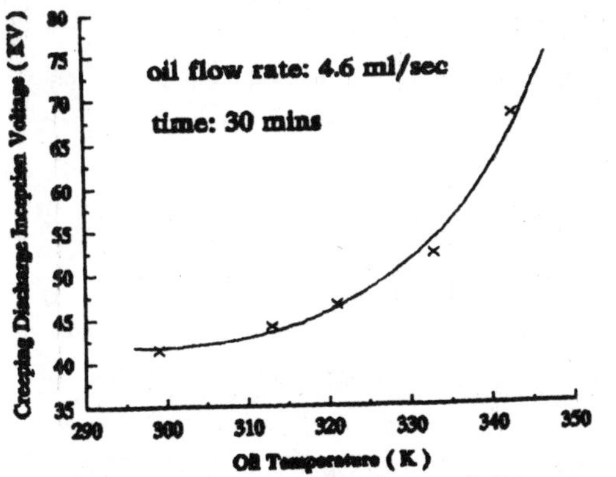

Figure 4 Relation between inception voltage and oil temperature

Figure 5 shows the relation between the ac creeping discharge inception voltage and the oil flow rate. It can be found that the inception voltage is increased when the oil flow rate is increased.

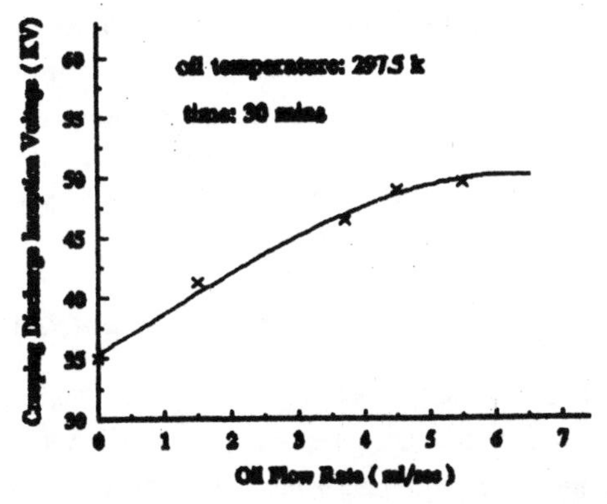

Figure 5 Dependence curve between inception voltage and oil flow rate

Figure 6 gives the experimental result of different polypropylene film position. The creeping discharge inception voltages of oil-polypropylene system at different place were tested (refer to figure 3). Clearly the position of polypropylene film in the experimental model can influence the inception voltage, and the inception voltage at the place near to oil entrance is higher than that at far place.

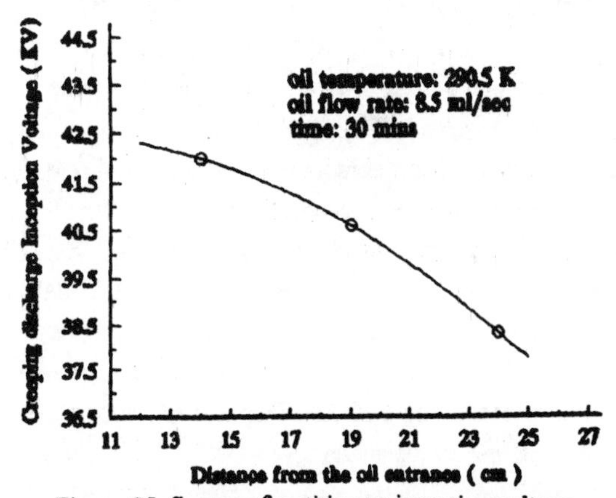

Figure 6 Influence of position on inception voltage

Moreover, the creeping discharge inception voltages of oil-polypropylene system under different polarity dc voltages have been tested. It is found that there is a lower inception voltage when the dc voltage is negative, in other words, it is easier to generate the creeping discharge with negative dc voltage. In addition, the surface potential of polypropylene film and the charge density in the oil have been measured in the experiment. It reveals that the polypropylene film is tend to be charged negatively, and there are positive charges in the oil.

It is worth noting that the following phenomena are observed. During the test of creeping discharge, the corona generates around the needle electrode firstly, and then a glimmer light following the oil flow can be watched on the surface of polypropylene film. It seems that the creeping discharge of oil-polypropylene system tends to develop towards downstream area.

DISCUSSION

According to the experimental results mentioned above, it is obvious that the creeping discharge of oil-polypropylene insulating system is influenced by the oil temperature, oil flow rate and the polypropylene film position under the flowing oil condition. It can be attributed to the streaming electrification in the oil-polypropylene system

The influences of oil flow rate and oil temperature

It is well known that streaming electrification is generated by friction between flowing insulating liquid and a fixed solid. Usually, the higher flow rate will bring more streaming electrification. So does the insulating oil temperature [5]. In this experiment, polypropylene film is the fixed solid, and when the insulating oil flows along with the surface of polypropylene film, there is streaming electrification in this system. It is found in the experiment that there are positive charges in the oil and there are negative charges on the surface of polypropylene film. So the higher oil flow rate, the more negative charges on the surface of polypropylene film. On the other side, it is also found that the negative dc creeping discharge inception voltage is lower. In other words, it is easier to generate creeping discharge in the negative half cycle when the applied voltage is ac voltage. Consequent to the facts that mentioned above, the real voltage between the needle electrode and the plate electrode is smaller than no streaming electrification during the negative half cycle under ac voltage. Therefore, the electric field in the oil-polypropylene system is relatively low, and so the creeping discharge inception voltage of the oil-polypropylene system will be increased along with the increase of the oil flow rate. Figure 5 confirms above analysis.

Considering the same reason mentioned above, increasing oil temperature can increase the ac creeping discharge inception voltage too. Moreover, the moisture content in oil can also affect the inception voltage of oil-polypropylene system. As we know, when the oil temperature is increased, the dissolvability of water in insulating oil is improved. It means that there is less water in insulating oil, and so the electrical strengthen of insulating oil is improved. On the other hand, it is insulating oil that is the weakness of the oil-polypropylene insulating system. Therefore, the creeping discharge inception voltage of oil-polypropylene system is increased. It is proved by figure 4.

The influence of polypropylene film position

Figure 6 shows that the creeping discharge inception voltage at different position is different, and also the position close to the oil entrance, the inception voltage is higher than that far to the oil entrance. This can also be attributed to streaming electrification in the oil-polypropylene insulating system with the flowing oil.

According to the theory of streaming electrification [6], the leakage current $J_{(l)}$ from insulated pipe at a distance l from oil entrance can be expressed as

$$J_{(l)} = J_{(0)} \exp\left(-\frac{l}{v\tau}\right).$$

Here, the $J_{(0)}$ is the leakage current density at the oil entrance (it depends on oil conductivity, shape of flow path, solid material, oil flow rate and so on). The v is the average oil flow rate. τ is the relaxation time of oil ($\tau = \varepsilon/k$, ε is the oil dielectric permittivity, and k is the oil conductivity).

It is clear from above expression that the leakage current is the largest at oil entrance and decreases in the direction of flowing oil. Considering the relation between the leakage current and the charge density, it can be deduced that the charge density on the insulated pipe have the same relation as the leakage current. In other words, the charge density on insulated pipe is the highest at the oil entrance and decreases along with the oil flow direction. Therefore, the charge density on polypropylene film is different at different places. As a result, the creeping discharge inception voltage is influenced. Because there are more negative charges on polypropylene film at the place near to oil entrance, the creeping discharge inception voltage of oil-polypropylene system is higher at this place.

In addition, the experimental phenomena that are observed can be explained using the same analysis mentioned above. Because there are less negative charges on the polypropylene film placed far to oil entrance, the electric field in the downstream area is stronger than that in the upstream area. Therefore, the creeping discharge develops towards downstream area.

Comparing with the experimental results of oil-paper insulating system under the flowing oil condition [7], the effects of streaming electrification on oil-polypropylene insulating system with the flowing oil are just reverse. In the experiments of oil-paper insulating system, the creeping discharge inception voltage of oil-paper system is decreased because of the streaming electrification. However, the creeping discharge inception voltage of oil-polypropylene insulating system is increased owing to the streaming electrification. Therefore, if the two types of composite insulating structures are combined properly in the flowing insulating fluid system, such as in large power devices with the forced oil cooling system, the negative effects of streaming electrification would be reduced.

CONCLUSIONS

The performances of oil-polypropylene insulating system have been studied experimentally under the flowing oil condition. In the oil-polypropylene insulating system with the flowing oil, streaming electrification play an important role. Considering the experimental results and analysis mentioned above, the following conclusions can be obtained.

1. The ac creeping discharge inception voltage of the oil-polypropylene system increases with the increase of insulating oil temperature under the flowing oil condition.

2. The oil flow rate in the oil-polypropylene system affects the ac creeping discharge inception voltage of this system. The inception voltage is higher when the oil flow rate is increased.

3. The ac creeping discharge inception voltage of the oil-polypropylene system is higher at the place near to oil entrance than that at the place far to oil entrance.

4. The creeping discharge of oil-polypropylene system generates towards the oil flow direction under the flowing oil condition.

This study is helpful to develop new composite insulating system for reducing the danger owing to streaming electrification in large power devices with forced-oil cooling system.

REFERENCES

1. A. N. Hammoud, J. R. Laghari and B. Krishnakumar, "Characterization of Electron-Irradiated Biaxially-Oriented Polypropylene Films", IEEE Trans. on Nuclear Science, Vol. 35, No. 3, June 1988, pp. 1026-1029
2. D. G. Shaw, S. W. Cichanowski and A. Yializis, "A Changing Capacitor Technology--Failure Mechanisms and Design Innovations", IEEE Trans. on Elec. Insul., Vol. EI-16, No. 5, October 1981, pp. 399-413.
3. Akio Tomago, Tokohiko Shimizu, Yasuo Iijima and Ichiro Yamauchi, "Development of Oil-Impregnated, All-Polypropylene-Film Power Capacitor", IEEE Trans. on Electr. Insul., Vol. EI-12, No. 4, August 1977, pp. 293-300.
4. S. Yasufuku, T.Umemura and Y. Yasuda, "Dielectric Properties of Oil-Impregnated All Polypropylene Film Power Capacitor Insulation System", IEEE Trans. Electr. Insul., Vol. EI-13 No. 6, December 1978, pp. 403-410.
5. D. W. Crofts, "The Static Electrification Phenomena in Power Transformer", IEEE Trans. on Electr. Insul., Vol. 23, No. 1, 1988, pp. 137-146.
6. S. Shimizu, H. Murata and M. Honda, "Electrification in Power Transformer", IEEE Trans. on Power Apparatus & System, Vol. 98, July/August 1979, pp. 1244-1250.
7. Yang Jia-Xiang, Chi Xiao-Chun, Ding Li-Jian, Li Zhong-Hua and Jin Chong-Jun, "Influence of Streaming Electrification on Creeping Discharge of Oil-Paper Insulation System", Conference Record of the 1994 IEEE International Symposium on Electrical Insulation, Pittsburgh, PA USA, June 5-8, 1994, pp. 538-540.

Conference Record of the 1996 IEEE International Symposium on Electrical Insulation, Montreal, Quebec, Canada, June 16-19, 1996

Electrical Double Layer at Dielectric Liquid/Solid Interface. Space Charge measurements using the Thermal Step methods

Alain Toureille
Laboratoire d'Electrotechnique
Université de Montpellier
Place Eugène Bataillon
34095 Montpellier Cedex 5
FRANCE

Gérard Touchard
L.E.A. - Laboratoire de Physique et Mécanique des Fluides
40 Avenue du Recteur Pineau
86022 Poitiers Cedex
FRANCE

Tony Richardson
Engineering Mathermatics
University of Bristol
Queen's Walk
Bristol BS8 1TR
U.K.

INTRODUCTION

The experimental evidence of the Electrical Double Layer (E D L) at the Dielectric Liquid-Solid Interface is an important question :first, it is a fundamental problem which has given several physical models and then numerous industrial applications concerns the electrical ingeneering (transformers , wet cables and capacitors ..)
The Thermal Step Technique , recently , has shown its great sensitivity to measure space charge in several geometries in particular in cables [1-3].
So this method has been tried here to verify the assumptions made in models established few years ago [4]

Experimental Method-Overview

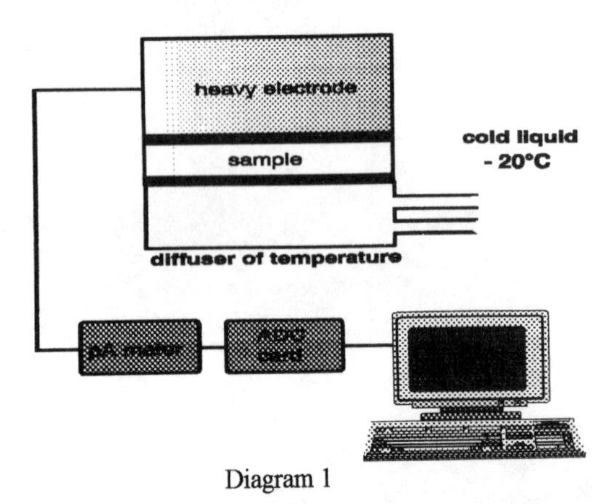

Diagram 1

The sudden application of a significant temperature difference across a solid dielectric polymeric material causes a transmission of a thermal wave which inherently induces an expansion or a contraction of the material. Any intrinsic volumetric space charge trapped in the bulk and any surface charge is then subjected to a small spatial displacement that will contribute to a transient electrical current with a peak value usually of the order of pico-ampères. This is detected by measuring the motion of image charges induced on suitably placed electrically conducting electrodes amplified and stored on a computer for later processing. Diagram 1 illustrates schematically a simplified version of the process for a slab. The resulting transient current $I(t)$ can be related theoretically to the resident local electrical field caused by the presence of trapped electrical space charge. The process of numerical resolution involves, in a cylindrical geometry, the use of Fourier-Bessel expansions for the electrical field and the space-charge density. The number of terms retained obviously determines the accuracy of the method and consequently mathematical calculations are performed before and after the experiments to ensure both self-consistency and that the accuracy is within experimental tolerance and acceptability levels [1-3].

The transient current $I(t)$ that arises in the external circuit joining both probe electrodes, on application of a thermal step, has the form (Xe: insulating thickness) :

$$I(t) = -\alpha C \int_{Xo}^{Xe} E(x) \frac{\delta T}{\delta t} dx \qquad (1)$$

where $T(x,t)$ is the temperature field at position x at time t, C is the system capacitance and α is correlated with coefficient of thermal variation of C.

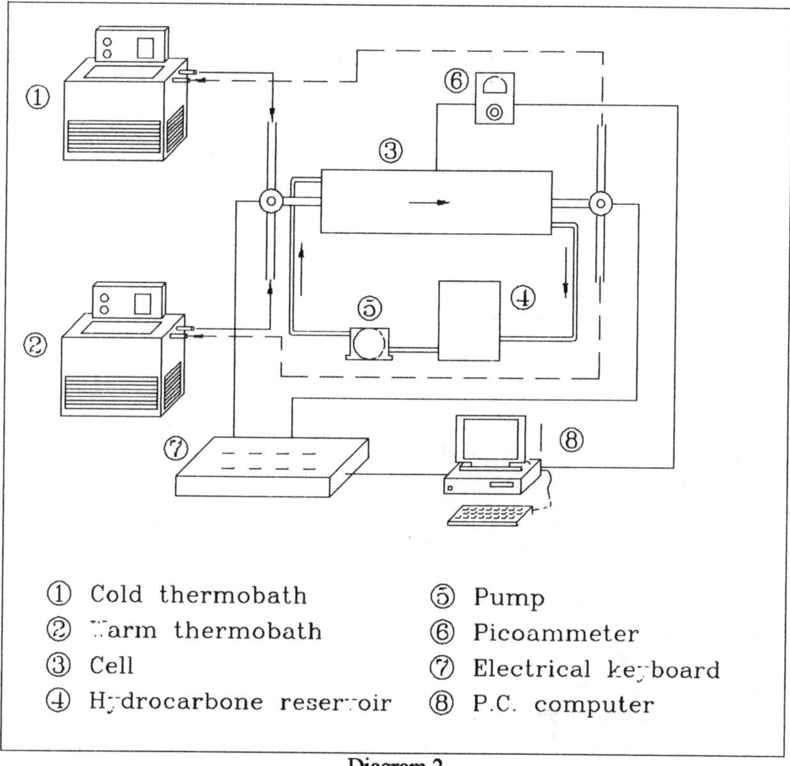

Diagram 2

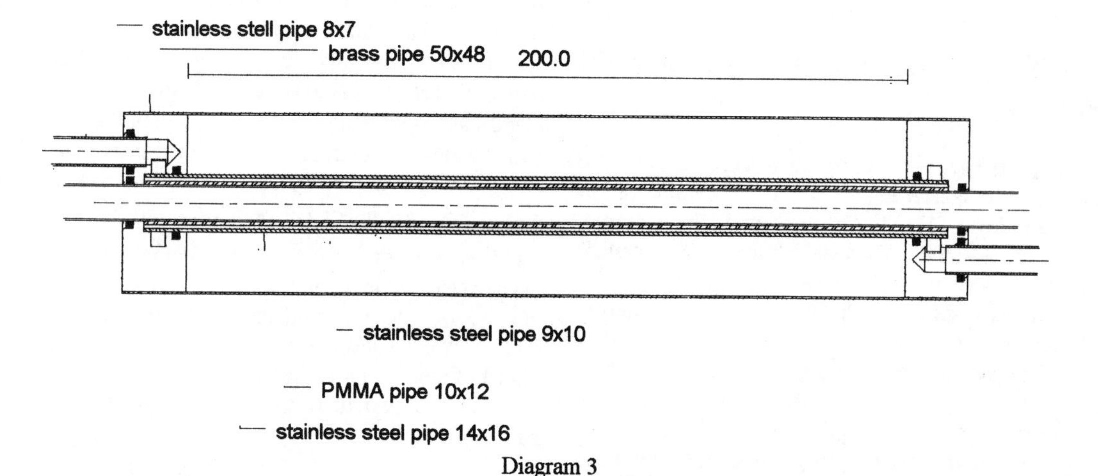

Diagram 3

Experimental Configuration

The experimental equipment is shown Diagram 2 and details on the cell in Diagram 3. An axi-symmetrical concentric cylindrical geometry was chosen as it appeared the most appropriate to the cable industry. A straight test section of earthed stainless steel tube of length 21 cms with inner and outer diameters 9 mm and 10 mm respectively was chosen to confine the temperature controlled liquid refrigerant. Placed in a circuit and using a sequence of control valves, the liquid refrigerant from a thermostatically controlled bath could be passed at a precise moment through the tube and so provide a thermal step of -40°C to the remaining part of the system otherwise at ambient temperature. An attached thermocouple was used to measure the tube temperature. This whole tube section was surrounded by a second annular tube, with inner and outer diameters 10 mm and 12 mm respectively, of PMMA (Polymethylmetacrylate) as this is a material often found in practical electrical insulation situations.

Enclosing this latter was a second stainsteel tube of inner and outer diameters 14 mm and 16mm respectively. This was effectively the second electrode and was therefore used as the electrical probe via which measurements of transient current were made. The annular region between this and the solid polymeric PMMA contained liquid heptane which is connected to a second and separate fluid circuit. Finally an earthed exterior brass tube of inner and outer diameters 48 mm and 50 mm was used as an electrical shield.

Experimental Verification Procedure

The electrical capacitance of the combined solid-liquid test section of the system was measured using a Hewlett Packard Bridge Device. At 20°C the capacitance was 118pF and by circulating the proprietary refrigerant at -13°C for between 2 and 3 hours the instantaneous fractional growth rate with temperature was found to be $\alpha \cong -2.1\times10^{-3}$. Temperature measurements were monitored by thermocouple probes inner stainless steel tube. Electrical current measurements were effected by a Keithley Current Amplifier No.428. The reliability and reproducibility of the experiments, in which current transients were measured over a period of about 10 sec, were tested by conducting a series of six identical experiments performed every three minutes. The test section was returned to ambient temperature each time by an automatically controlled forced flow of refrigerant through the innermost copper tube.

Space Charge Density at Solid/Liquid Interface

At the contact between a liquid and a solid the specific adsorption phenomenon leads to the development of an electrical double layer [4]. Thus in the solid, for example, negative ions are adsorbed when in the liquid the positive one in excess remain. It is generally assumed that in the liquid phase the part of the double layer is in fact composed of two different layers : one very close to the solid, with a thickness in the order of the ions size, is called the compact layer, the other one is called the diffuse layer and has a thickness proportional to the sqare root of the electrical resistivity of the liquid. In our configuration, for which the radius of the pipe is much greater than the diffuse layer thickness, a good representation of the space charge density distribution is

$$\rho(x) = \rho_i \exp(-x / \delta_o) \qquad (2)$$

where ρ_i is the space charge density at the interface, x the distance to the wall and δ_0 the diffuse layer thickness. Thus an approximate solution, of the charge exponentially decreasing in the liquid, can be chosen for a numerical verification procedure.

Numerical Verification Procedure

An approximate but physically plausible distribution of space charge density $\rho_{pos}(x)$ in a solid-liquid cylindrical system was postulated in order to verify the self-consistency of the numerical computations. This was chosen to have the form with the radius x :

$$\rho_{pos}(x) = 0 \text{ with } 4.0mm \le x \le x_i \qquad (3)$$

$$\rho_{pos}(x) = a(x - x_i)(x - x_d)\exp[-2x/(x_d - x_i)] \qquad (4)$$

for $x_i \le x \le x_e$

where $a=160000$, x_i, x_d and x_e are constants respectively equal to 5.2mm, 5.6mm, 7.0mm giving the position of the solid-liquid interface, the position of sign reversal in the electrical double layer and the effective external liquid boundary respectively. The corresponding postulated electrical field $E_{pos}(x)$ was then determined by using the fact that there is no net voltage is applied across the system.

Using the known analytical form for $T(x,t)$ and the postulated distribution of electrical field $E_{pos}(x)$ it was possible to calculate the corresponding postulated transient $I_{pos}(t)$. Then followed the numerical resolution algorithm using series expansions to evaluate the corresponding electrical field $E(x)$ and space charge density $\rho(x)$ distributions. Adequate agreement between $E(x)$ and $E_{pos}(x)$ was achieved by using only ten term series but for equivalent agreement between $\rho(x)$ and $\rho_{pos}(x)$ somewhat more terms in the case of a good signal/noise.

The result appears on the fig 1 where we have the given electric field *Epos(x)*and calculated one and on the fig 2 where we show the corresponding space charge distributions . This check permits to achieve the difference between the two distributions (given and calculated)in space and modulus

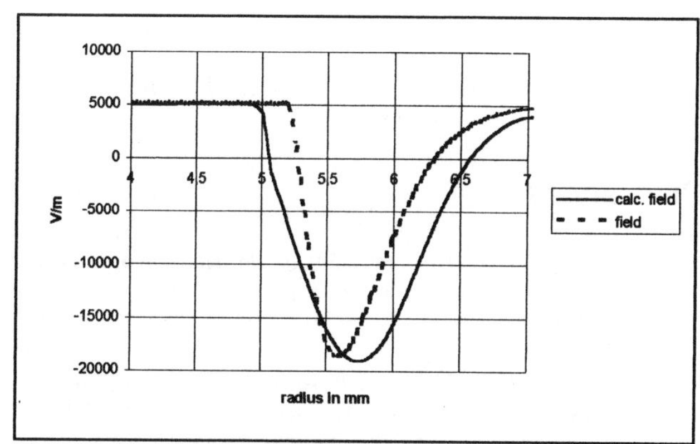

fig 1 Electric Field (given and calculated)

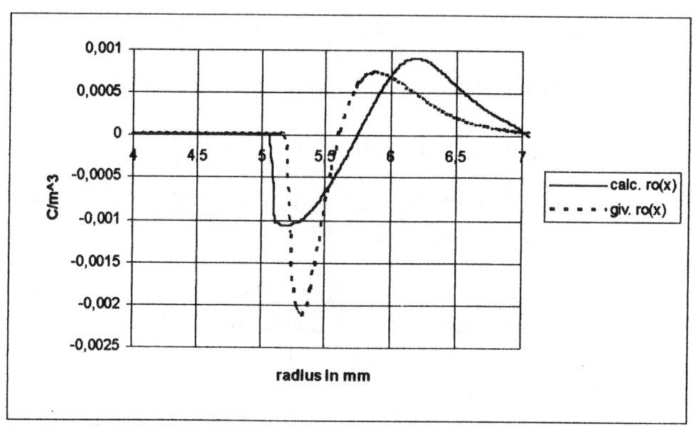

fig 2 Space Charge (given and calculated)

Preliminary Experimental Procedures

Before starting any of the experimental measurements the whole system was discharged. A thermal step was applied to the system in which there was no liquid heptane in the outer fluid circuit. The resulting transient current was found to have magnitude less than 0.1 pA. The heptane was then introduced in the annular section and then remains in situ for all experiments. A thermal step of -40°C was then applied and $|I(t)| \cong 1.5$ pA measured. The thermal step method thus revealed the existence of some remaining space charge at the interface between the solid and liquid phases experimental evidence of the existence of the electrical double layer.

Experimental Results

These thermal step experiments were the first to be performed on such a solid-liquid system. The resulting electrical field and space charge distributions are shown in Figures 3 and 4, where the region (x radius) 5mm < x< 6mm represents the solid PMMA domain and the region 6mm ≤ x ≤ 7mm the liquid heptane surrounding it. These confirm the presence of negative charge in the bulk of the solid insulating material, as was to be expected from previous analyses, but also positive charge to be found in the liquid. The important result is the clear evidence of sign reversal of the space charge density at the interface between the solid and liquid dielectrics. Furthermore, the magnitude of this positive charge in the liquid is of the order of 10^{-3} C/m^3. However a further sign reversal in the liquid region needs to be looked at more carefully to establish whether it has real physical significance or is an artefact of the method of analysis.

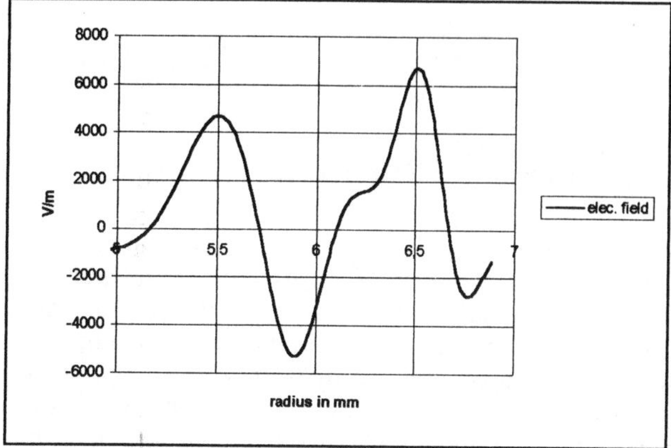

fig 3 Experimental Electric Field

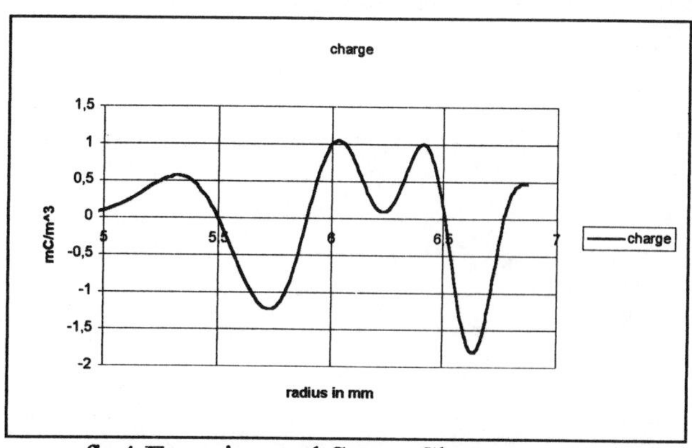

fig4 Experimental Space Charge

The measured and numerically calculated transient currents $I(t)$ are presented in Figure 5. Using ten terms in the deconvoluting algorithim still produces some discrepancy and this needs to be addressed.

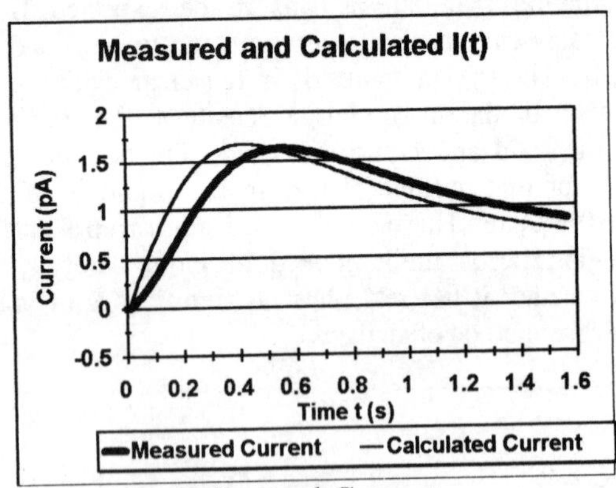

Fig5 Experimental Current

Conclusion and Prospects

There could be complex interactions occurring within the liquid region arising from the passage of the thermal wave. The significance and possible complications arising from thermally induced and electrohydrodynamic convection need to be assessed.[5] Mathematical instability analyses will be used to determine over what length and time scales instabilities are likely to contribute to experimental complications. The numerical algorithm will also be reassessed to establish its potential for further economisation of representation.

References

1. Toureille, A. "Sur une méthode de détermination de la densité de charges d'espace dans le PE", 1987, JICABLE 87, pp. 89-103.

2. Toureille, A., Reboul, J.P. & Merle, P. "Détermination des densités de charges d'espace dans les isolants solides par la méthode de l'onde thermique", 1991, J. Phys. III, **1**, pp. 111-123.

3. Cherifi, A., Abou Dakka, M. & Toureille, A. "The validation of the thermal step method", 1992, IEEE Trans. on Elec. Insul., **27**, N°. 6, pp. 1152-1158.

4. Touchard, G., Patzek, T.W. and Radke, C.J. "A physicochemical explanation for flow electrification in low conductivity liquids", Proc. IEEE-IAS Annual Conf., 1994, **3**, pp. 1669-1675.

5. Richardson, A. T. "Electrohydrodynamics : An Interdisciplinary Subject Come of Age", Inst. Phys. Conf. Ser., 1995, 143, 2, pp. 41-46.

Conference Record of the 1996 IEEE International Symposium on Electrical Insulation, Montreal, Quebec, Canada, June 16-19, 1996

AC Breakdown Strength of N2, SF6 and a Mixture of N2+SF6 Containing a Small Amount of SF6

Kevork Mardikyan, Özcan Kalenderli
Istanbul Technical University
Electric - Electronics Faculty
80626 Maslak - Istanbul - TURKEY

Orhan Ersen, Ergun Canarslan
Schneider Electric Co.
81080 Goztepe - Istanbul - TURKEY

Abstract: AC breakdown strengths of N_2, SF_6 and a mixture of N_2+SF_6 containing 1% of SF_6 were experimentally studied. For this purpose, 50 Hz AC breakdown voltages in both uniform and non-uniform fields up to a pressure of 400 kPa from 50 kPa were measured. Test results show that the addition of 1% of SF_6 to nitrogen increases the breakdown voltage up to 250 kPa in a non-uniform field. The relative breakdown strength of the mixture with respect to components was also calculated in 50 Hz AC voltage. The analysis of Paschen curves reveals an improvement of 40% in the breakdown strength of mixture in uniform field but in non-uniform fields, the maximum AC breakdown voltage of the mixture is 24% lower than that of pure SF_6.

INTRODUCTION

In recent years, sulphur hexafluoride (SF_6) has been of considerable technological interest as an insulation medium in high-voltage apparatus because of its superior insulating properties, high dielectric strength at relatively low pressure and its thermal and chemical stability.

In uniform field, the dielectric strength of pure SF_6 is approximately three times that of nitrogen in the same conditions. However in non-uniform fields it depends on many factors such electrode geometry, voltage wave shape, polarity, gas pressure, etc. Although in gas insulated systems (GIS) a highly non-uniform field would be generally avoided, such divergent fields can occasionally exist due to electrode surface roughness or dust and conducting particles between electrodes [1, 2]. As a result of this, breakdown could occur due to local field enhancement.

Several investigations have been reported in the literature on the breakdown behaviour of SF_6, N_2 and SF_6+N_2 mixtures [3-5]. Most of the published data refer to power frequency breakdown in uniform or nearly uniform field gaps [6, 7]. As part of an extensively study of the dielectric behaviour, we are investigating these gases in both uniform as well as non-uniform field gaps.

Many researchers have studied the behaviour of air mixed with a small percentage of SF_6 as an additive. Kurimoto found that addition of a small amount (1%) of SF_6 to dry air produced a considerably increase in the breakdown voltage of a 30 cm rod-plane gap raising the breakdown value above that of pure SF_6 over wide pressure range [8]. Qiu and Kuffel reported an impulse strength for 30 mm point-sphere gap in 1% SF_6 + air mixture, higher than that of pure SF_6 at a pressure of 400 kPa and ac strength nearly the same as that for pure SF_6 [9].

Previous investigations show that the addition of a small amount of SF_6 to buffer gases like He, N_2, CO and CO_2 can considerably improve the breakdown strength depending on the pressure and reduces insulation cost of the system.

The main purpose of this work is therefore to study the breakdown strength of N_2, SF_6 and a mixture of N_2 + 1% SF_6 and to compare the results with previous investigations.

Experimental Set-up

Experiments were carried out first using sphere-sphere electrodes of 50 mm radius and 15 mm spacing and rod-sphere electrodes with a rod tip radius of 1 mm and sphere radius of 50 mm (Fig. 1). In this case gap spacing was kept at 30 mm in order to obtain a highly divergent field. The rod electrode was made of stainless steel. Electrodes were mounted in an aluminium tank of 550 mm diameter and 883 mm length. In sphere-sphere arrangement one of the spheres was connected to the high-voltage supply while the second was earthed, but in rod-sphere arrangement the rod was connected to the high-voltage supply while the sphere was earthed. Before each series of tests, the electrodes were polished and cleaned thoroughly.

The test vessel was first evacuated for at least two hours and then filled with the desired gas up to a relative pressure of 400 kPa. The gas mixture was left for at least 2 hours before test, for the purpose of obtaining a uniform mixture. For the 50 Hz AC tests with voltages up to 300 kV rms a cascade high-voltage transformer was employed. AC breakdown voltage was measured by means of a capacitive divider. The mean value of breakdown voltage and standard deviation were calculated by means of twenty voltage applications.

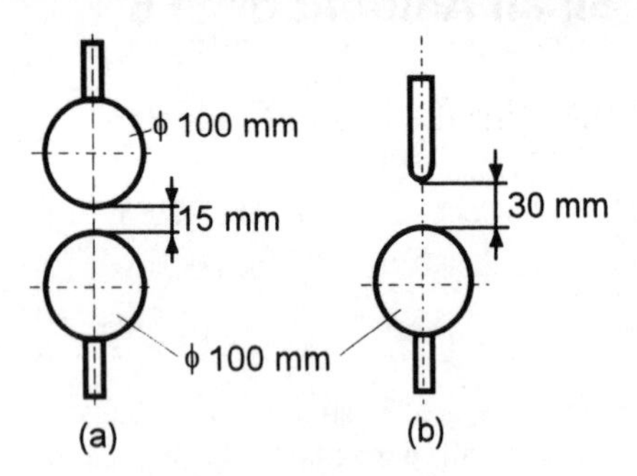

Figure 1. Electrode configurations.
(a) Sphere-sphere gap;
(b) Rod-sphere gap.

Breakdown voltage measurements were carried out with SF_6, N_2 and $N_2 + 1\%\ SF_6$ in the p·d range of $75 \leq p \cdot d \leq 600$ kPa.cm in the sphere-sphere arrangement with p varying in the interval $50 \leq p \leq 400$ kPa. The purity of gases in experiments were of commercial purity.

Test Results

The breakdown voltages of nitrogen, sulphur hexafluoride and a mixture of $N_2 + 1\%\ SF_6$ were measured up to a pressure of 400 kPa both in uniform and non-uniform fields. The analysis of Paschen curves reveals that AC breakdown voltage of the mixture is approximately 47% lower and 45% higher than that of pure SF_6 and N_2 at 150 kPa, respectively.

In uniform field, the addition of 1% of SF_6 to nitrogen increases considerably the breakdown strength of N_2 alone (Fig. 2). In non-uniform field, however measurements have shown that AC breakdown voltage for nitrogen increases with pressure up to 350 kPa and beyond that pressure, begins to decrease slowly (Fig. 3).

The breakdown voltage of pure SF_6 increases with pressure up to 125 kPa where it reaches the maximum value then decreases according to a negative slope up to 300 kPa where it reaches the minimum value and beyond that pressure, decreases again.

The breakdown curve of the mixture decreases slowly with pressure. The comparison of this three curves show that in non-uniform fields the maximum AC breakdown strength of the mixture of $N_2 + 1\%\ SF_6$ reaches 76% than that of pure SF_6 and is approximately 50% higher than N_2 alone.

One may see that up to 250 kPa the breakdown voltage of nitrogen is lower than that of the mixture but beyond that pressure, the breakdown voltage of nitrogen became higher than that of the mixture. In fact the breakdown voltage of pure SF_6 is higher than that of nitrogen and the mixture of $N_2 + 1\%\ SF_6$ in the whole pressure range.

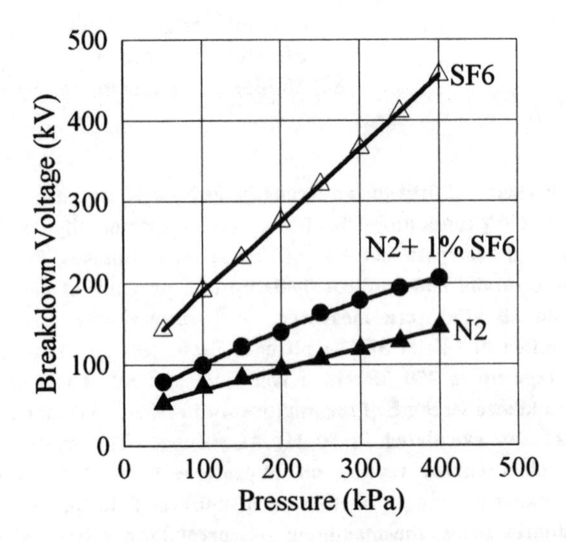

Figure 2. Variation of breakdown voltage with pressure in N_2, SF_6, N_2+ 1% SF_6 for sphere-sphere gap. Paschen Curves.

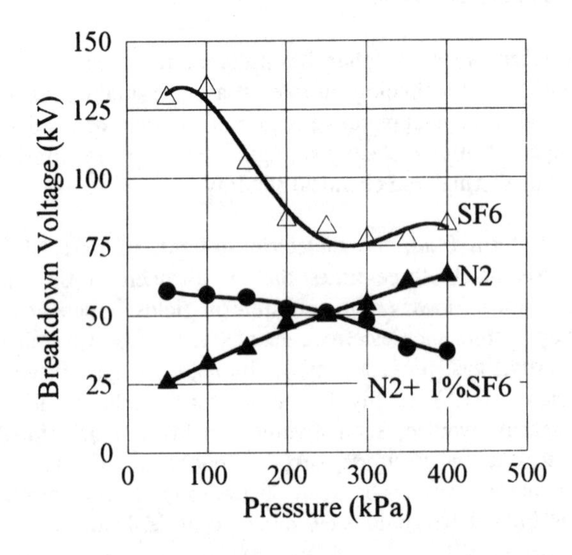

Figure 3. Variation of breakdown voltage with pressure in N_2, SF_6, N_2+ 1% SF_6 for rod-sphere gap.

DISCUSSION

From Paschen curves it is seen that in uniform field, pure SF_6 ranks first, the mixture of $N_2 + 1\%\ SF_6$ ranks second and pure N_2 ranks third as a dielectric irrespective of the gas pressure. In non-uniform fields, pure SF_6 was found to have highest dielectric strength in the whole pressure range. In fact for pressures under 250 kPa the mixture of N_2 +1% SF_6 was better than pure N_2 alone while beyond that pressure breakdown strength of pure N_2 was found to be higher than the mixture.

CONCLUSION

The present work shows that addition of 1% SF_6 to N_2 can significantly increase the AC breakdown voltage with respect to N_2 but the improvement in divergent fields depends on gas pressure. At pressures lower than 250 kPa, addition of 1% SF_6 to N_2 increase AC breakdown strength of N_2. For higher pressures however dielectric strength of the mixture was found to be smaller than pure nitrogen alone.

REFERENCES

[1] Ryan, H. M., Watson, W. L., Dale, S. J., Tedford, D. J., Kurimoto, A., Banford H. M. and Hampton B. F. "Factors Affecting the Insulating Strength of SF_6 Filled Systems". CIGRE Paper 15-02, 1976.

[2] Wooton, R. E. and Cookson, A. H. "AC Particle Initiated Breakdown in Compressed Gas Mixtures of SF_6 with He, N_2 and CO_2". 5th International Conference on Gas Discharges and Their Applications, London, 1978.

[3] Li, Z., Kuffel, R. and Kuffel, E. "Volt-Time Characteristics in Air, SF_6/Air Mixture and N_2 for Coaxial Cylinder and Rod-Sphere Gaps". IEEE Trans. on Electrical Insulation, Vol.21, No. 2, April 1986, pp.151-155.

[4] Watanabe, T. and Takuma, T. "The Breakdown Voltage and Discharge Extension of Long Gaps in Nitrogen-SF_6 and Air-SF_6 Gas Mixtures". Journal of Applied Physics, Vol.48, No.8, 1977, pp. 3281-3287.

[5] Ermel, M. "Das N_2 - SF_6 Gasgemisch als Isoliermittel der Hochspannungstechnik". ETZ-A Bd. 96, H-5, 1975, pp.231-235.

[6] Bouvièr, B. and Fallou, B. "Caractéristiques diélectriques de l'hexafluorure de soufre et des mélanges d'azote et d'hexafluorure de soufre". RGE, Tome 75, No. 2, 1966, pp. 174-175.

[7] Takuma, T., Watanabe, T. and Kita, K. "Breakdown Characteristics of Compressed-Gas Mixtures in Nearly Uniform Fields. Proc. IEE, Vol.119, No.7, July 1972, pp. 927-928.

[8] Kurimoto, A. "Prebreakdown Phenomena in Long Non-Uniform Field Gaps in SF_6/N_2 and SF_6/Air Mixtures". 7th International Conference on Gas Discharges and Their Applications, 31 August - 3 September, 1982, pp. 211-213.

[9] Qiu, Y. and Kuffel, E. "The Breakdown strengths of Gas Mixtures Containing a Small Amount of Electronegative Gas in Non-uniform Field Gaps". 7th International Conference on Gas Discharges and Their Applications, 31 August - 3 September, 1982, pp. 215-218.

Conference Record of the 1996 IEEE International Symposium on Electrical Insulation, Montreal, Quebec, Canada, June 16-19, 1996

Investigation of SF_6-N_2, SF_6-CO_2 and SF_6-Air as Substitutes for SF_6 Insulation

Y. Qiu and Y. P. Feng
High voltage Division
Xi'an Jiaotong University
Xi'an, China 710049

Abstract: The dielectric strength of SF_6-N_2 is slightly higher than that of SF_6-CO_2 in slightly non-uniform fields encountered in gas-insulated switchgear (GIS) and gas-insulated cables (GIC), but the opposite is true in the cases of highly non-uniform fields and gas/film insulation existing in gas-insulated transformers (GIT). Experiments also show that the SF_6-air mixture with the SF_6 concentration being 0.8 can be used in cubicle type GIS (C-GIS), where the impulse dielectric strength of the gas mixture is apparently higher than that of SF_6 gas.

INTRODUCTION

From a practical point of view only very few SF_6 gas mixtures have prospects for industrial applications. SF_6-N_2 is often considered to be the best substitute for SF_6 in GIS and GIC because nitrogen is a cheap inert gas and its dielectric strength in a uniform field is higher than that of gas mixtures of SF_6 with most common gases[1]. According to the authors' research work, SF_6-CO_2 is also quite promising, especially in highly non-uniform fields and in a gas-impregnated film insulation system encountered in the case of GIT[2]. Our research work also shows that the SF_6-air mixture has a higher impulse ratio than SF_6 gas in highly non-uniform fields[3], and therefore it is an ideal substitute for SF_6 in C-GIS. The present paper compares the dielectric strength of SF_6-N_2 and SF_6-CO_2 in different electric fields and compares the dielectric strength of SF_6-Air and SF_6 in highly non-uniform fields.

UNIFORM FIELD BREAKDOWN

The uniform field breakdown strength $(E/p)_0$ can be estimated by measuring the ionization coefficient α and the attachment coefficient η in the gas mixture. The $(E/p)_0$ value of SF_6-N_2 and SF_6-CO_2 can also be estimated if the ionization and attachment coefficients in the component gases are known because a simple approximation based on an assumption that there is no interaction between different component gas molecules is applicable to both gas mixtures, i. e. the coefficient in the gas mixture can be calculated by the sum of the partial-pressure-weighted coefficients in the component gases.

$$\alpha_m = F\alpha_1 + (1-F)\alpha_2 \quad (1)$$

$$\eta_m = F\eta_1 + (1-F)\eta_2 \quad (2)$$

where the subscripts 1, 2, and m denote SF_6, N_2 (or CO_2), and the mixture, respectively; and F is the fractional concentration of SF_6 in the mixture.

Equations (1) and (2) are linear functions of F, and therefore whether the measured α_m and η_m are linear functions of F can be a simple criterion for determining the validity of the approach of linear addition of the partial-pressure-weighted coefficients[4].

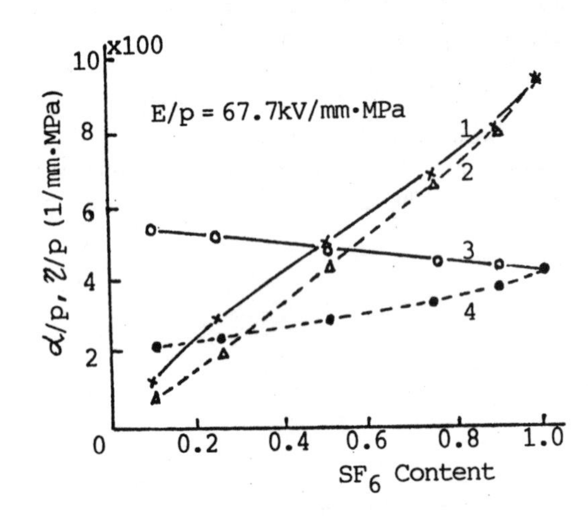

Fig. 1 α/p and η/p as a function of F
1. η/p in SF_6-CO_2 2. η/p in SF_6-N_2
3. α/p in SF_6-CO_2, 4. α/p in SF_6-N_2

Figure 1 gives the measured α_m/p and η_m/p as a function of F in SF_6-N_2 and SF_6-CO_2 at an E/p value of 67.7 kV/mm · MPa using the steady-state Townsend method, and it indicates that Eqs. (1) and (2) apply to both gas mixtures.

Figure 2 shows the relative electric strength

(RES) of SF_6-N_2 and SF_6-CO_2 in the uniform field. For engineering application purposes, the RES of SF_6-N_2 and SF_6- CO_2 can be expressed by the following expressions when F is greater than 0.1[2].

$$(RES)_{SF_6\text{-}N_2} = F^{0.18} \quad (3)$$

$$(RES)_{SF_6\text{-}CO_2} = F^{0.32} \quad (4)$$

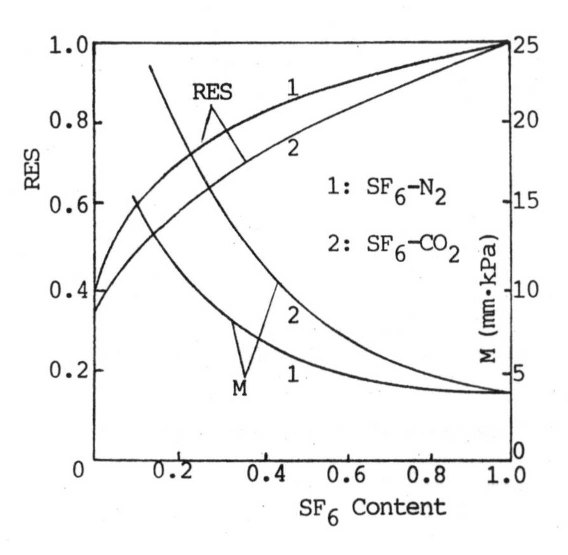

Fig. 2 RES and M values of SF_6-N_2 and SF_6-CO_2

SLIGHTLY NON-UNIFORM FIELDS

The breakdown voltage of any gas mixture in a slightly non-uniform field can be computed if the figure-of-merit (M) value and the electric field distribution function are known in addition to the theoretical breakdown threshold $(E/p)_0$. The M value is defined as[5]

$$M = k/c(E/p)_0 \quad (5)$$

Where k is the constant in the streamer breakdown criterion; c and $(E/p)_0$ are given in the following equation.

$$(\alpha - \eta)/p = c[E/p - (E/p)_0] \quad (6)$$

The advantage of introducing M which can be determined experimentally is to avoid the unknown constant k for mixtures, and the breakdown criterion can thus be re-written as

$$\int_0^{x_c} [E - p(E/p)_0]dx = M(E/p)_0 \quad (7)$$

The M values measured for SF_6-N_2 and SF_6-CO_2 using a simplified method[6] are given in Fig. 2.

For a coaxial cylindrical electrode system and a concentric spherical electrode system, it is more convenient to use an electrode surface curvature factor h in calculation of the breakdown voltage V_b[2].

The factor h is defined as

$$h = E_m/p(E/p)_0 \quad (8)$$

where E_m is the maximum electric field in the gap. And therefore the following equation can be given for a gap length of d and a field non-uniformity factor of f.

$$V_b = (E/p)_0 pdh/f \quad (9)$$

The surface curvature factors for a coaxial electrode system (h_c) and for a concentric spherical electrode system (h_s) can be expressed as

$$h_c = 1 + \sqrt{2M/pr} = 1 + A/\sqrt{pr} \quad (10)$$

$$h_s = 1 + \sqrt{4M/pr} = 1 + B/\sqrt{pr} \quad (11)$$

The values of A and B for SF_6, SF_6-N_2 and SF_6-CO_2 are shown in Table 1. It can be seen that when $pr \geqslant 50$MPa · mm, the curvature factors for SF_6, SF_6-N_2 and SF_6-CO_2 are quite close to each other though the h value for SF_6-CO_2 is still the highest.

Table 1 A and B for SF_6, SF_6-N_2 and SF_6-CO_2

	A	B
SF_6	0.092	0.13
SF_6-N_2(F=0.5)	0.103	0.146
SF_6-CO_2(F=0.5)	0.13	0.185

SURFACE ROUGHNESS EFFECT

Electrode surface roughness leads to a microscopically non-uniform field problem, which can also be calculated if the electric field distribution is known. Two roughness models were used, i. e. the multi-ridge model simulating surface profile formed in the machining process, and a single protrusion model simulating a fixed conducting particle or any similar surface defect[7].

Figure 3 compares the dielectric strength of SF_6, SF_6-N_2 and SF_6-CO_2(F=0.5) in the case of a rough electrode surface (R = 110μm). It can be seen that the difference in dielectric strength between these mixtures and SF_6 becomes smaller when gas pressure increases. Figure 4 gives a comparison of these two gas mixtures and SF_6 in a plane-parallel gap (d=15mm) with a needle- tip protrusion (h=2. 5mm). In this case the dielectric strength of SF_6-CO_2 exceeds that of SF_6-N_2 and SF_6 gas.

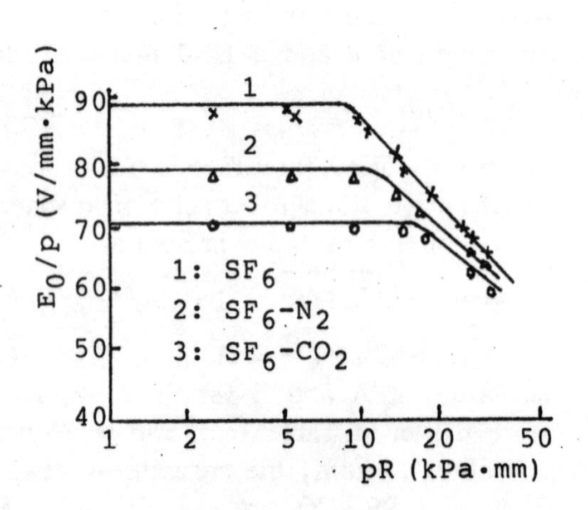

Fig. 3 Electrode surface roughness effect (R=110μm)

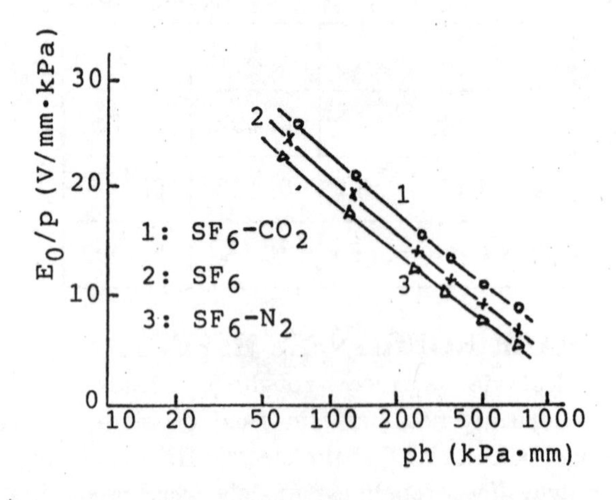

Fig. 4 Effect of a needle-tip protrusion (h=2. 5mm)

HIGHLY NON-UNIFORM FIELDS

Experiments show that addition of some gases to SF_6 can improve its breakdown characteristics in highly non- uniform fields, which is inevitable in GIT and C-GIS.

Figure 5 shows that adding nitrogen to SF_6 (F =0. 75) increases the maximum breakdown voltage by 12%. Figure 6 shows that the impulse breakdown strength of SF_6-CO_2 is 40% higher than that of SF_6, which looks very promising because it is the impulse withstand level that determines the insulation dimensions.

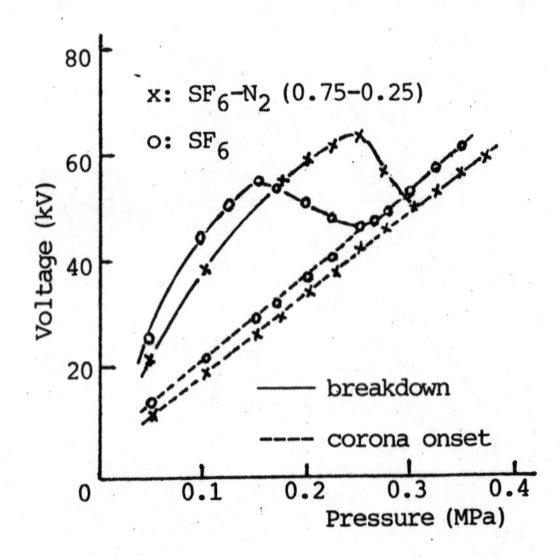

Fig. 5 Comparison of SF_6 and SF_6-N_2 in a point-plane gap

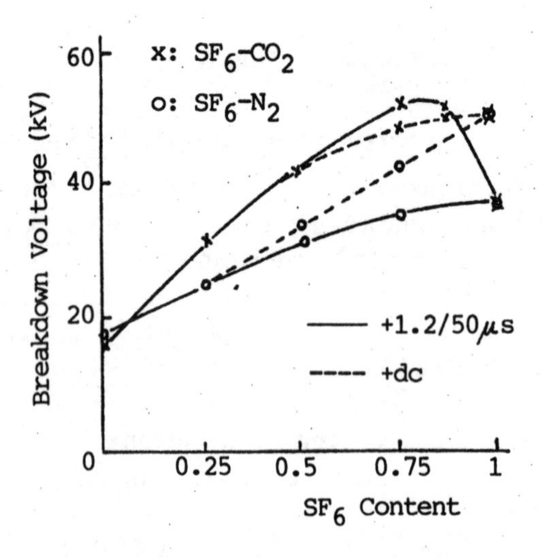

Fig. 6 Comparison of SF_6-N_2 and SF_6-CO_2 in a point-plane gap (p=0. 125MPa)

Experiments show that SF_6-air is quite similar to SF_6-CO_2 in respect of impulse breadown strength. Figure 7 shows adding air to SF_6 can increase the impulse ratio, which is 0. 7 for pure SF_6 and approximately 1 for the mixture with an optimum mixing ratio(F=0. 8)

GAS/FILM INSULATION

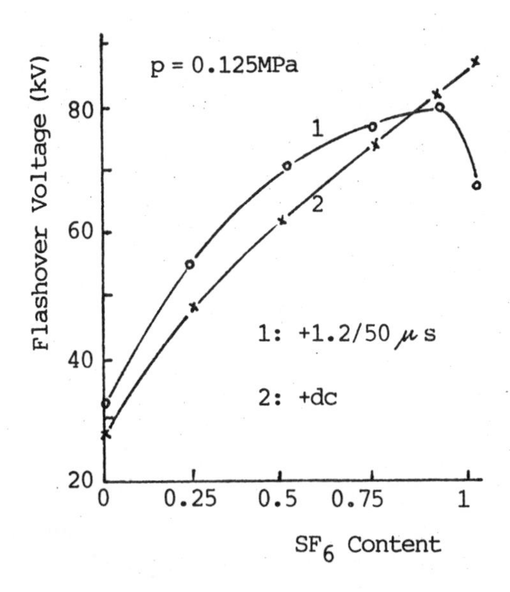

Fig. 7 Flashover voltages of SF_6-Air as a function of F

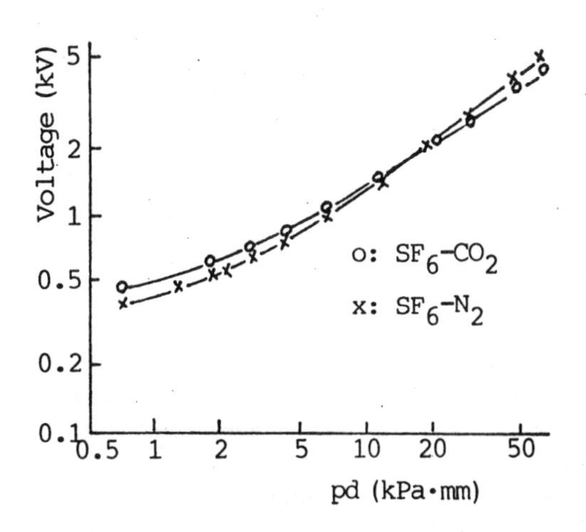

Fig. 8 Comparison of SF_6-N_2 and SF_6-CO_2 at low pd values

Partial discharge in a gas/ film insulation system is actually a problem of breakdown in gaseous dielectrics at very low pd values. If the pd value is within the linear part of the Paschen curve, then the breakdown voltage of the gas-filled void can be expressed as[2]

$$V_b = (E/p)_0[pd + M] \qquad (12)$$

And therefore it can be expected that the linear part of the Paschen curve for SF_6-CO_2 intersects that of SF_6-N_2 at a pd value of 21mm • kPa, corresponding to a breakdown voltage of 2kV, which is in good agreement with the experimental results (Fig. 8).

CONCLUSIONS

While SF_6-N_2 is probably the best gas mixture to be used in GIS and GIC, it can also be used in GIT to improve the impulse breakdown characteristics. In the latter case SF_6- CO_2 might have some advantages over SF_6-N_2 because gas/film insulation and highly non-uniform field problems are encountered. Though some manufacturers specify that SF_6 concentration in their C-GIS is not less than 0. 95, it is actually better to use the SF_6-Air mixture with SF_6 concentration being around 0. 8.

ACKNOWLEDGMENT

The present work is supported by the International Development Research Centre of Canada.

REFERENCES

1. Qiu, Y. and Y. P. Feng. "Calculation of Dielectric strength of the SF_6/N_2 Gas Mixture in Macroscopically and Microscopically Non- Uniform Fields". Proc. 4th Int. Conf. on Properties and Applications of Dieletric Materials. Vol. 1, 1994, pp. 87-90.
2. Qiu, Y. and D. M. Xiao. "Dielectric strength of the SF_6/CO_2 Gas Mixture in Different Electric Fields". Proc. 9th Int. Symp. on High Voltage Engineering. Vol. S2, 1995, paper 2255.
3. Qiu, Y. et al. "Breakdown of SF_6/Air Mixtures in Non-Uniform Field Gaps under Lightning Impulse Voltages". Proc. 10th Int. Conf. on Gas Discharges and Their Applications. Vol. 1, 1992, pp398-401.
4. Qiu, Y. and D. M. Xiao. "Calculation of Dielectric Strength of Some SF_6 Gas Mixtures". Gaseous Dielectrics VII (Edited by L. G. Christophorou and D. R. James). Plenum Press, 1994, pp. 161-167.
5. Pedersen, A. "On the Assessment of New Gaseous Dielectrics for GIS", IEEE Trans. Power Apparatus and Systems. Vol. PAS-104, 1985, pp2233-2237.
6. Qiu, Y. and Y. F. Liu. "A New Approach to Measurement of the Figure- of- Merit for Strongly Electronegative Gases and Gas Mixtures". IEEE Trans. Electrical Insulation. Vol. EI-22, 1987, pp 831-834.
7. Qiu, Y. and I. D. Chalmers. "Effect of Electrode Surface Roughness on Breakdown in SF_6- N_2 and SF_6-CO_2 Gas Mixtures". J. Phys. D: Appl. Phys. Vol. 26, 1993, pp. 1928-1932.

Conference Record of the 1996 IEEE International Symposium on Electrical Insulation, Montreal, Quebec, Canada, June 16-19, 1996

Dielectric Characteristics of SF_6 under the Steep-Fronted Impulses

Qiaogen Zhang, Yuchang Qiu, Pei Wang, Wenguo Gu

School of Electrical Engineering

Xi'an Jiaotong University

Xi'an (710049), China

Abstract: The influence of the wavefront duration on impulse breakdown voltage of the nonuniform field gap in compressed SF_6 gas has been studied using a point-plane electrode arrangement. Voltage-time curves are also given in this paper. For comparison purposes, dielectric characteristics of SF_6 in a sphere-plane gap have been measured. The wavefronts used in this work are 0.05, 0.3, 0.6, 0.9, 1.2 and 1.5μs. The experimental results show that the wavefront duration has significant influence on the dielectric behaviour of SF_6 in the nonuniform field. It is found that the breakdown voltage-wavefront relationships and voltage-time curves of nonuniform field tend to be U-shaped.

INTRODUCTION

Disconnector switching operation in GIS often initiates a multitude of transient phenomena, especially the very fast transient overvoltage (VFTO) with frequencies in the range from 1 to 100MHz [1, 2]. The fast transient overvoltage represents an increased stress on the insulation of GIS, particularly for higher voltage level, and is therefore attracting much attention[2,3,4].

In the GIS without imperfection, the electrical field is slightly non-uniform, and VFTOs should not affect the insulation strength of GIS [2, 5]. In practical systems, local inhomogeneous field caused by protrusion or particle contamination may lead to a reduction of the dielectric strength of SF_6. Some investigations under different wavefront impulses have shown the down trend of the breakdown voltage with increasing the steepness of impulses in the strongly inhomogeneous field, but few data are available in the shorter time range, especially under very fast transient overvoltages[2].

In the following, main attention will be paid to the dielectric characteristics of SF_6 under the steep-fronted impulses (SFI) with the aim to indicate the influence of the wavefront duration on the breakdown behaviour. Also, some dielectric characteristics of SF_6 in the case of lightning impulse (LI) are presented in this paper.

EXPERIMENTAL APPARATUS

A coaxial steepening circuit was used to generate the non-oscillating steep-fronted impulses of different wavefronts for simulating VFTOs. In the authors' opinion, the non-oscillating steep-fronted impulse has advantages over oscillating waveshapes in the standardization of the testing waveshape and the comparison of the different experimental results although very fast transient overvoltages are oscillating by nature. In this paper, the dielectric characteristics of SF_6 were obtained under the non-oscillating waveshapes. The wavefront durations used in this work were 0.05, 0.3, 0.6, 0.9, 1.2 and 1.5μs while the wavetail was kept about 50μs in all cases.

The measurements were carried out in a point-plane gap and a sphere-plane gap simulating respectively the local field inhomogeneity and the slightly non-uniform field in GIS. The point-electrode was a 6mm dia. rod tapered with a 30° cone to a point tip (r = 0.5mm). The field nonuniformity factor in the sphere-plane gap (d = 8mm) was 1.26. The measuring system used included a 500MHz storage oscilloscope and a compact, low resistance divider with response time less than 4ns.

EXPERIMENTAL RESULTS AND DISCUSSION

The 50% breakdown voltages of SF_6 under the 0.05/50μs impulse were measured as a function of gas pressure in the range of 0.1 and 0.5MPa. Figure 1 shows the breakdown voltage-gas pressure characteristics with positive and negative polarities. In the case of positive polarity, the breakdown voltage at high gas pressure ($P > 0.15$MPa) is nearly independent of the gas pressure. This is because rapidly rising impulses produce a smaller corona than slowly rising impulses. In the case of negative polarity, the saturation characteristic against gas pressure is less pronounced. This shows that with negative polarity, even when the steep-fronted impulse voltage is applied, corona stabilization is effective.

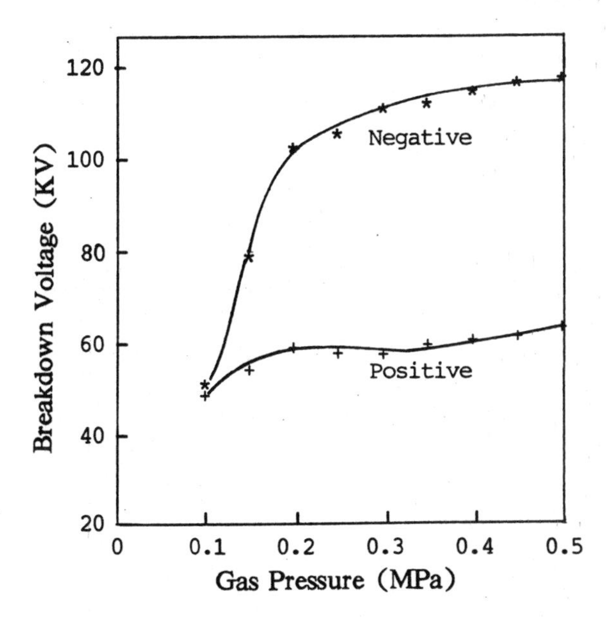

Figure 1 Breakdown voltage in a 10mm point-plane gap under the 0.05/50μs impulse

Experiments were conducted under the 0.05/50μs impulse to study the influence of the point-plane gap distances on the breakdown voltage. The results for three gap distances (10mm, 20mm and 30mm) are compared in figure 2. A similar dependence of breakdown voltage on pressure is found with different gap distances. The only difference, although not very obvious, is that the shorter the gap distance, the smaller is the corona stabilization.

Figure 3 shows the voltage-time characteristics in a point-plane SF_6 gap for gas pressure of 0.1MPa under the different wavefront impulses. The reduction of the insulation strength in the case of the steep-fronted impulse is obvious, and the breakdown level of the 0.05/50μs impulse is lower than the lightning impulse level. In Figure 3 there is a distinct minimum breakdown voltage at about 1μs with positive polarity and slightly shifting to a longer time domain with negative polarity, which is in general agreement with the findings reported elsewhere[6]. Also the remarkable similarity for both positive and negative polarity is that the V-t curves tend to be U-shaped.

For comparison purpose, the V-t curves in a sphere-plane gap are shown in Figure 4. The results in Figure 4 are in agreement with those reported in [5, 7]. It is evident that the breakdown levels especially in the point- plane gap are strongly depending on the applied impulse waveshapes, and V-t curves for different impulse waveshapes should not be mixed. Extrapolation to shorter times is often misleading.

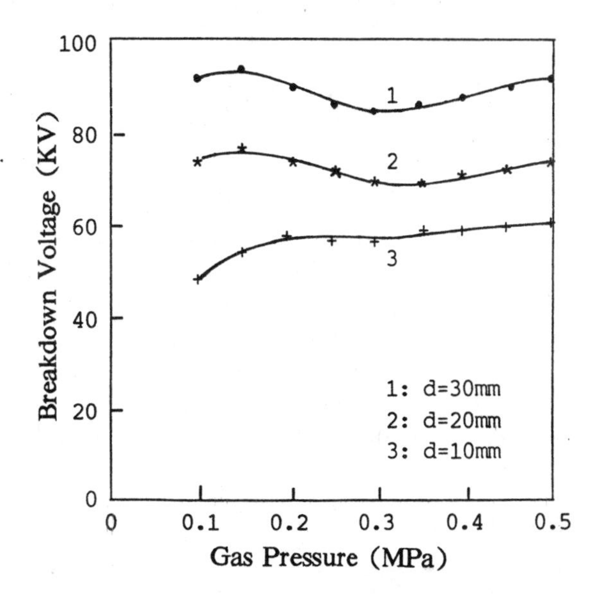

Figure 2 V-p characteristics for different point-plane gaps under the 0.05/50μs impulse

Figure 5 gives the 50% breakdown voltage-wavefront relationships in the 10mm point- plane gap. From Figure 5, we concluded the following:

1) Breakdown voltages are higher with negative polarity than with positive polarity. This trend is more distinct in higher pressure.

2) The breakdown voltage-wavefront relationships is U-shaped. The time domain where the minimum breakdown voltage appears is different and shifts to a shorter time domain for both polarities with lower pressures.

It is interesting to note that in Figure 5 for the posivive polarity the voltage-wavefront relationship curves of the pressures of 0.1MPa and 0.4MPa intersect. This intersection is closely related to the electric field distortion by space charges. For understanding of the U-shapes, the statistical time lag, the corona formation time, the leader stepping time and the corona stabilization have to be considered.

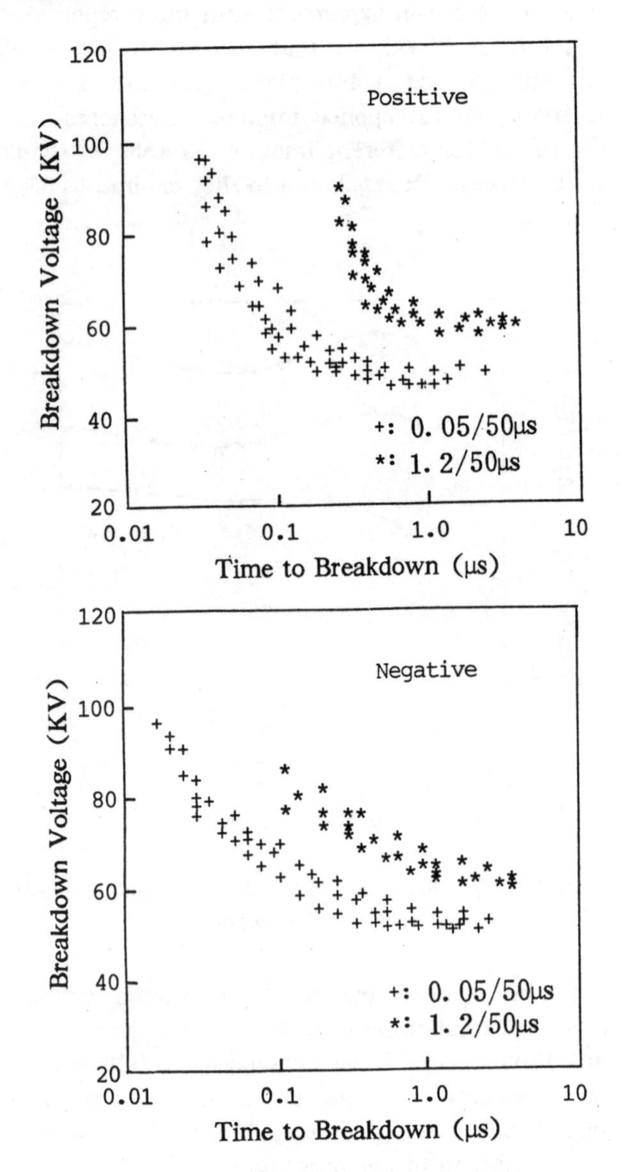

Figure 3 V-t characteristics for LI and SFI in a 10mm point-plane gap (gas pressure 0.1MPa)

Figure 6 compares the breakdown characteristics of a 10mm point-plane gap and a sphere-plane gap (d = 10mm) at the pressure of 0.1MPa when the wavefront durations of impulses were decreased. The data indicate that, while the wavefront duration has an insignificant effect on the breakdown voltage of the sphere-plane gap, it drastically influences the breakdown voltage of the point-plane gap.

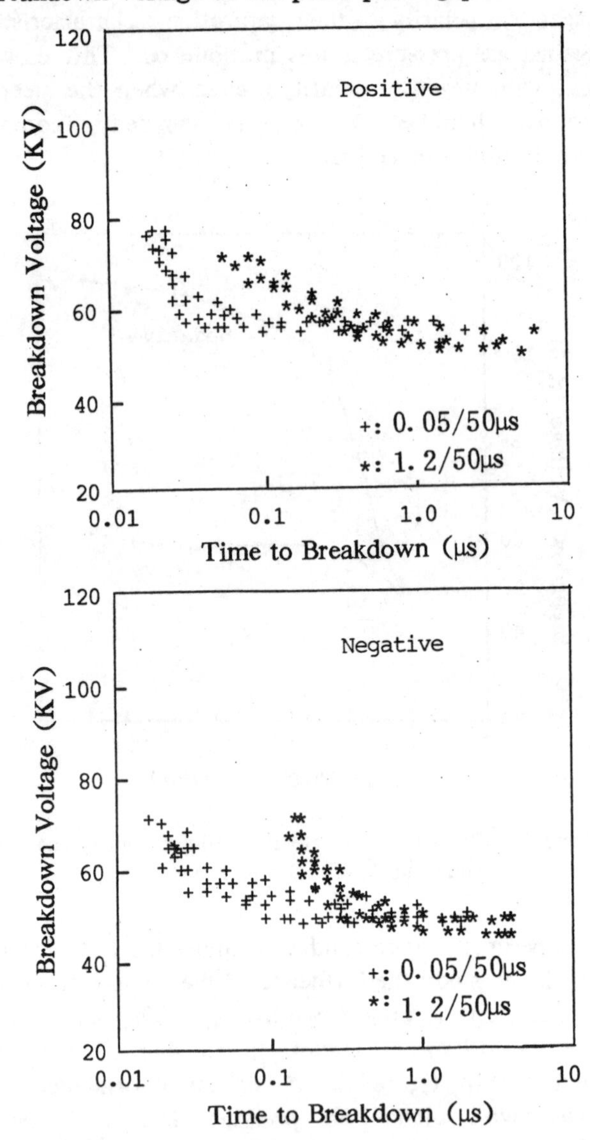

Figure 4 V-t characteristics for LI and SFI in a 8mm sphere-plane gap (gas pressure 0.1MPa)

CONCLUSIONS

The breakdown characteristics of SF_6 gas under the nonoscillating steep-fronted impulses were investigated in the strongly inhomogeneous field simulated by a point-plane gap and the results were compared with those obtained in a sphere-plane gap. As a result, the following were concluded:

1) V-t curves under the steep-fronted impulses obtained in this paper extend to longer time lags and are more complete because of the use of the non-oscillating waveshapes.

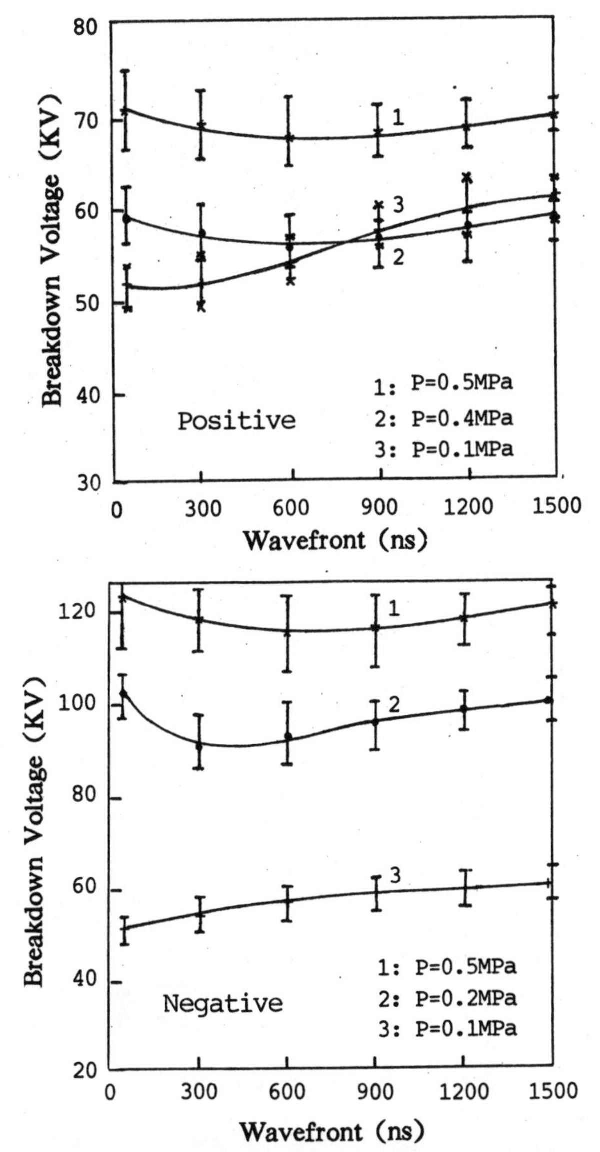

Figure 5 Impulse breakdown voltage- wavefront relationships for a 10mm point-plane gap

2) V-t characteristics reveal that the insulation strength is lower under the steep-fronted impulse than under the lightning impulse in the strongly inhomogeneous field. Contrary, the insulation level for steep-fronted impulse is higher than that for lightning impulse in the slightly nonuniform field.

3) The breakdown characteristics in the strongly inhomogeneous field are wavefront-dependent for both the positive and negative polarities. Both the breakdown voltage-wavefront relationship and the V-t curve are somewhat U-shaped.

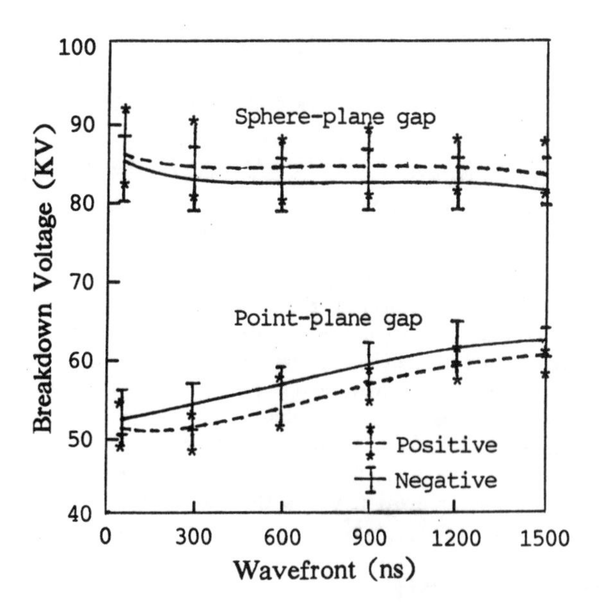

Figure 6 Influence of wavefront on breakdown voltage (gas pressure 0.1MPa, gap length 10mm)

REFERENCES

1. Working group 33/13-09. "Very Fast Transient Phenomena Associated with Gas Insulated Substations". CIGRE, 33-13, 1988.
2. W. Boeck et al. "Insulating Behaviour of SF_6 with and without solid insulation in case of fast transients". CIGRE, 15-07, 1988.
3. S. Kobayash et al. "Particle-initiated Flashover Caused by Disconnector Restriking Surges in GIS". 5th International Symposium of High Voltage Engineering, Braunschweig, Paper No. 12-13, 1987.
4. S. Matsumoto, H. Okubo, H. Aoyagi and S. Yanabu. "Nonuniform Flashover Mechanism in SF_6 Gas under Fast Oscillating and Non-oscillating Impulse Voltages". 6th International Symposium on High Voltage Engineering, New Orleans, Paper No. 32. 16, 1989.
5. Y. Qiu, C. Y. Lu and M. Zhang. "Voltage-time Characteristics of SF_6 and SF_6-N_2 under Lightning and Steep-fronted Impulses". 1992 IEEE Int. Symp. on Electrical Insulation, Baltimore, June 7-10, 1992, PP. 310-313.
6. Lee, B. H et al. "Dielectric Characteristics of SF_6 Gap Stressed by the Steep-Fronted Oscillating Transient Overvoltages". 8th International Symposium on High Voltage Engineering, Yokonama, August 23-27, 1993, Paper No. 33. 03.
7. G. Luxa et al. "Recent Reseach Activity on the Dielectric Performance of SF_6 with Special Reference to Very Fast Transients". CIGRE, 15-06, 1988.

Conference Record of the 1996 IEEE International Symposium on Electrical Insulation, Montreal, Quebec, Canada, June 16-19, 1996

The Effect of Fast Transient Overvoltages on 550 kV SF_6/Oil Transformer Bushings

Helvio J. A. Martins   Alexandre Neves
Cepel - Centro de Pesquisas de Energia Elétrica
Rio de Janeiro - Brazil

Ivan B. Amorim   Fernando Maranhão
Itaipu Binacional
Paraná - Brazil

Abstract: **This paper presents the results of the studies in a GIS - Gas-insulated substation about disconnector operations and short-circuit and their dielectric stress in the form of steep front wave on the main power transformer bushings of Itaipu hydroelectric power plant. As a singular point is presented also the results of some tests specially developed to confirm and determine safe operational level and some modifications required in the equipment in order to enable them to live with these kind of stresses.**

INTRODUCTION

Some bushings from the main power transformers of the generation groups have showed abnormal level of acetylene. This fact was detected after some occurrences, possibly associated to overvoltages in the GIS. Therefore, a study was established in order to find out the causes and define corrective and predictive actions necessary to assure the reliability of the plant.

The investigations carried out showed that the high level of acetylene was probably related to damage on the paper-oil insulation due to high voltage variations with a risetime of some nanoseconds. These kind of stresses are related to disconnectors switching or short-circuit-to-ground, associated or not to overvoltages.

Although the bushings passed in all the conventional tests at the time of manufacture; in service, they failed when submitted to fast transient waves, typical of a GIS.

Besides computer simulations, a group of tests were performed in order to assess dielectric suportability of the equipment under fast transients, with the monitoring of dissolved gases, dielectric power loss factor, capacitances and partial discharge levels.

GIS OVERVOLTAGES

The aim of the study was to identify/characterize the overvoltages generated inside the GIS (Figure 1), potentially damaging to the bushings, in order to support the discussions between utility and manufacturer as well as to define a methodology for laboratory tests. As a digital tool, ATP (Alternative Transient Program) was used, taking into account two types of events: Disconnector switchings and Phase-to-ground fault.

The models adopted for the GIS components, as circuit-breakers, busbars, voltage transformers, surge arresters, etc., were the same of a previous study [1]. Field measurements performed in disconnectors were taken to compare with digital simulations.

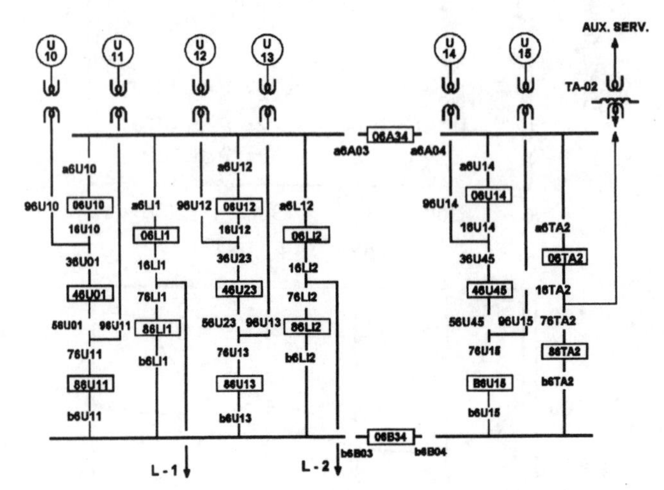

Figure 1 - Single line diagram of a power plant section

Disconnector Switchings

Preliminary simulations with different configurations and disconnector switched positions were performed with the following results:

-For constant configuration, the disconnector that caused the worst overvoltages was located near the bushings (generator disconnector).

-For the constant position of the disconnector switched, the worst configuration was the simplest, which involves a minimum length of busbar and a maximum trapped charge.

The most critical values of trapped charges, measured in December of 1986 [2], were adopted for the simulations.

Table 1 presents the results of simulations based on the most critical conditions of configurations and trapped charges.

Cases "d" and "e" were obtained with switchings of disconnectors far from the bushings and with great lengths of busbar. One can see that, the amplitudes for these configurations are lower than for "a", "b" and "c" situations.

Disconnector Switched	Steep front amplitude (kVpp)	Overvoltage (pu)	Oscillations Frequency (MHz)
76U-case "a"	515	1.58*	3.6 / 12.0
56U-case "b"	667	1.63	3.7 / 8.3
96U-case "c"	685	2.00	5.0
B6LI-case "d"	385	1.41	2.0 / 7.5
A6LI-case "e"	249**	1.28**	5.0/7.5

* - amplitude after 1.3μs   ** - amplitude after 1.7μs

Table 1 - Results of GIS overvoltages study (first oscillation peak)

GIS Internal Faults

Internal faults in the GIS, generally due to overvoltages, constitute a different class of occurrence that can impose critical stresses for the bushings. A typical condition was chosen, with a standard lightning impulse (1,2 x 50 μs) being injected at the first tower not protected by the guard cable, with an amplitude of 900 kV which corresponds to the surge arrester protection level of line bay entrance with simultaneous phase-to-ground fault. For this condition, the maximum overvoltage at the bushing was 680 kV.

Three distinct faults due to lightning and their consequent overvoltages in the bushings were analyzed. The amplitude of steep front wave obtained were: 615 kV for faults in bushing entrance, 880 kV in generator surge arrester and 1000 kV in disconnector.

All the faults were chosen to occur at the maximum amplitude (crest) of the lightning impulse wave.

Commentary

The worst amplitudes due to disconnector switchings (cases "a" and "b") present repeated steep fronts with amplitudes about 660 kVpp associated to voltages of 1.58 up to 1.63 pu. The frequency of oscillations of major amplitude, in the more critical configuration is about 3 MHz.

The steep fronts as a result of internal faults caused by a lightning impulse overvoltage presented values between 600 and 1000 kV in function of the failure point.

The steep fronts caused by the failures at the disconnector and in the surge arrester are followed by oscillations. In the case of surge arrester the second peak amplitude is higher than the first and also shows a lower temporal gradient.

According to the results, the most severe occurrences are caused by the internal faults in the GIS.

Electrical stresses, similar to that calculated, and of the same order of amplitude can be generated by simultaneous faults to temporary overvoltages (1,5 pu) or by switchings out of GIS (1.7 pu) which have a reasonable probability to occur [1].

The occurrence of disconnector switchings is much more frequent (some switchings per day); however their severity is lower than the internal faults.

Since 1986, disconnector 96U together with all generator's disconnectors were blocked to operate with voltage. So, these values must not be considered as a cause for the appearing of acetylene in the bushings.

Considering the digital simulations we had gone to the development/planning and performing the representative tests of the most critical situation (internal faults in the GIS).

TESTS

The digital simulations indicated that the most severe dielectric stress on the paper-oil condenser bushing was caused by internal faults in the GIS. Thus, the laboratory tests were developed to represent this condition. The critical transients originated by this kind of faults are characterized by a first steep front, usually higher than the observed with disconnector switchings, followed by lower level oscillations.

The laboratory set-up imposed some simplifications on the original model used in the digital simulations. The most remarkable was the chopping of the voltage wave on the top of the bushing, instead of doing it at the extremity of a 40 m busbar.

Procedure

The tests were performed according to the cycle presented in Figure 2. The procedure was a result of discussions among Utility, Manufacturer and Laboratory.

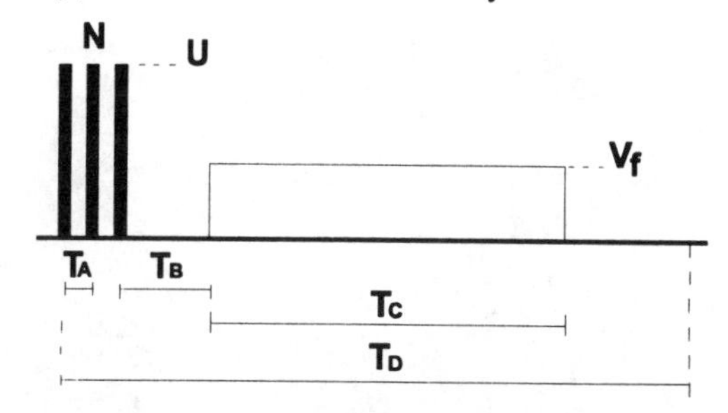

Figure 2 - Typical test cycle

where:

N = Number of chopped wave impulse applied at each cycle test (in this case 3)

U = Chopped wave impulse amplitude

T_A = Interval between successive impulses

T_B = Interval between the last applied impulse and the starting of power frequency voltage application (~ 5 min). The circuit was mounted in order to allow a prompt switching from the impulse generator to the sinusoidal source.

T_C = Time interval of power frequency voltage application (10 min). During this period, with applied voltage V_f, measurements of partial discharge level, capacitance and dielectric power loss factor were taken every 2 min.

V_f = Power frequency voltage amplitude
$1.3 \times (\sqrt{2}/\sqrt{3}) \times 500 = 412.8\ kV_{rms}$

T_D = Total time of the cycle (not fixed at first).

At the end of the application of Power frequency voltage, the impulse generator was connected again to the circuit in order to start a new cycle.

The first test cycle was started with 450 kV. For the successive cycles the chopped impulse wave was increased in steps of 100 kV up to the level of 1050 kV. From this point, the voltage step was 50 kV. V_f (power frequency voltage) was 413 kV for all cycles.

During the cycles, the chopped impulse wave, the current and the time-to-breakdown were registered. Also the pressure of the bushings and SF_6 compartment were monitored and a sample of oil was collected at the end of each cycle for gascromatography.

Test set

The test set was mounted in order to reproduce as closely as possible the real situation on-site. The bushing under test was partially immersed in a tank with mineral oil, simulating the insulation medium of the transformer. The upper part of the bushing was mounted in a GIS enclosure, with 420 kPa of SF_6. The application of voltage was applied via a SF_6/Air bushing. Figure 3 shows the test set-up for Chopped impulse wave and Power frequency applied voltage.

Figure 3 - Test set-up for chopped impulse (left) and power frequency voltage application (right)

Voltage step generation

It is well known that during a phase-to-ground short-circuit the voltage at the fault point goes down from a value U to zero in a short interval of time. In the case of a GIS, the distance between conductor and enclosure is very short, and the voltage rise time stays in the order of some nanoseconds. The consequence of the voltage collapse is the injection of travel voltage steps in all the busbars directly connected to the fault point. The point acts as a source, generating two step waves in opposite directions.

In laboratory, this situation is reproduced with the chopping of an impulse wave (2.3 x 365 μs) in a SF_6 medium, with a pre-established voltage via a device especially constructed and placed at the top of the bushing.

RESULTS ANALYSIS

Three bushings were tested using the procedure described above [3]. During the tests some parameters were monitored, to provide data for the diagnostic of some abnormality related to the dielectric integrity of the equipment.

Every parameter will be follow commented showing or not its capability as a tool for diagnostic purpose.

Bushing pressure

The pressure was permanently monitored during the tests. It is not an effective parameter though, for any significant variation would indicate irreversible damage to the bushing.

Capacitance, Dielectric power loss factor and Partial discharge level

Table 2 shows a typical example of the evolution of these parameters for each cycle of the test. Tap T_x (more external) was chosen for the measurements, based on inspections of an equipment that failed in a previous test [4].

Cycle (kV)	Capacitance (pF)		Diel. Power Loss Factor x 10^{-3}		Partial Discharge (pC)	
	C_{TX}	C_{BPD}	tanδ TX	tanδ BPD	Tap TX	Tap BPD
1-450	348.07	368.70	1.866	0.452	0.4	0.6
2-750	347.92	369.35	1.865	1.964	0.4	0.6
3-900	347.48	368.85	1.954	2.016	1.0	0.8
4-1050	347.33	368.76	1.950	2.030	0.5	0.8
5-1100	347.18	368.68	1.964	2.054	0.4	0.8
6-1150	347.35	368.78	1.960	2.056	0.4	1.0
7-1200	348.13	378.90	1.952	2.064	0.3	0.7
8-1250	347.00	368.64	1.956	2.098	0.3	0.3
9-1300	355.55	369.26 **	5.020 *	2.090 **	-	$>10^3$

* - Measurement with 5kV ** - Measurements with 200 kV

Table 2 - Measurements of capacitance, tanδ and partial discharge level

Until the 8th cycle (1250 kV) these parameters did not present significant variation, although, as shown later, there was evidence of discharges in the oil. In the 9th cycle the measurements made in both taps (B_{PD} and T_X) showed expressive changes.

The measurements made on tap B_{PD} (more internal) showed significant values of partial discharge, but capacitance and dielectric power loss factor have remained invariable. Measurements on tap T_X (with only 5 kV applied to the bushing) presented very expressive variations of power loss factor, but it was not possible to measure partial discharge.

From 1300 kV level, the measurements via taps pointed out with a good precision some deterioration activity in paper-oil dielectric.

Gascromatography

The gascromatographic analysis stands out as the most adequate technique to take a decision about the continuation of the test and the most sensitive indication about the dielectric integrity of the insulation system.

Every test has at least one collected sample for each step of applied voltage. Table 3 shows a typical example of the evolution of dissolved gas in oil.

n	H_2	O_2 x10^3	N_2 x10^3	CH_4	CO	CO_2	C_2H_4	C_2H_6	C_2H_2
0	55	3.4	15.0	6	54	57	1	3	nd
1	62	2.3	10.0	9	58	53	1	1	1
2	35	11.0	6.0	5	24	19	1	nd	1
3	130	5.6	22.0	17	76	72	5	1	7
4	130	0.9	6.9	21	63	39	7	1	10
5	190	3.8	15.0	29	71	66	8	1	17
6	190	2.8	12.0	35	67	60	13	1	19
7	230	15.0	8.8	45	61	49	16	1	25
8	200	1.6	11.0	34	56	43	13	1	19
9	660	1.3	8.4	140	780	230	27	4	160

nd - not detected cycle "0" was before energization

Table 3 - Gascromatography results of collected samples after 2 to 3 hours of the test end.

The acetylene (C_2H_2) content presents a monotonic increasing up to the 8th cycle and a remarkable variation at the 9th cycle (1300 kV). The generation of acetylene was noted since the first cycle (450 kV), due to discharges in the top of the bushings, between the copper conductor and the aluminum tube (zero foil- first capacitive layer, over which the next layers are wound).

The ratio CO/CO_2 is practically constant until the 8th cycle, suffering a variation from this cycle to the next, indicating at this point a deterioration of the cellulose. It is not recommended to go on with the tests after this point, for the risk of a crack in the porcelain (top bushing) and the breakdown of the paper layers, mainly in the region of "50% butt" paper joint (Figure 4).

Figure 4 - Damaged bushing

CONCLUSIONS

a-The study of internal overvoltages in a GIS, due to switching or short-circuit played an important role to identify critical points of the installation, in this case, bushings with a conventional insulation.

b-The tests performed are representative and stress the equipment realistically.

c-There is a clear correlation between points of failure of the equipment out of service and the equipment tested in laboratory.

d-Fast transient waves, with risetime shorter than the traveling time along the equipment, impose a longitudinal electric stress to the bushing insulation. This stress is not contemplated by standard tests despite its critical importance. The treeing depicted in Figure 4 is a consequence of this longitudinal stress.

e-Gascromatography points out as the best, among the techniques used to assess the dielectric integrity during the tests.

f-Laboratory test results in conjunction with on-site inspections show an inadequate design for the stresses imposed by a GIS.

g-These tests were considered significant and were approved by the bushing manufacturer for the development of improvements in the design of the equipment.

h-Some improvements were made on the equipment like the use of "Moving butt" paper joint, reinforcement of paper layers between the taps. The connection between the copper conductor and the "zero foil" was also moved to the side of GIS bushing terminal.

i-An equipment comprising all the modifications suggested was successful tested up to 1600 kV level.

j-Nowadays, one chopped wave impulse is enough for performing a typical test cycle (Figure 2)

ACKNOWLEDGMENT

The authors wish to thank Mrs. Rita de Cássia Barúqui (Light) and Mr. Helcio Gleizer (Cepel) for their technical discussions and assistance.

REFERENCES

1. Faria, F. M. M. et alli. "Evaluation of surges produced by disconnectors in the 60 Hz sector of Itaipu GIS". IX SNPTEE, Belo Horizonte, 1987 (in Portuguese)
2. Faria, F. M. M. et alli. "Measurements and digital simulations of high frequency surges due to disconnector switchings in Itaipu GIS". Itaipu Report, 1987 (in Portuguese)
3. Martins, H. J. A; Silva A. C. F; Neves A. "Chopped wave in transformer bushings, class 550 kV", Cepel Reports n°s 253 and 315/1991 (in Portuguese)
4. Martins, H. J. A. "Inspection in bushing transformers", Cepel Report n° 316/1991 (in Portuguese)

Conference Record of the 1996 IEEE International Symposium on Electrical Insulation, Montreal, Quebec, Canada, June 16-19, 1996

COMPARATIVE EVALUATION OF BREAKDOWN STRENGTH OF THIN INSULATING MATERIALS UNDER FAST TRANSIENT VOLTAGES

J. Sundara Rajan and K. Dwarakanath
Central Research and Testing Laboratory,
Central Power Research Institute, Bangalore, INDIA 560 094

Abstract: **Laboratory methods of generating fast transient overvoltages are discussed in this paper. The results of the breakdown studies carried out with the Transmission line model Generator using plastic materials is presented. The breakdown is observed to be a fast front phenomenon and the electrode effects are critical. The magnitude of the breakdown voltage is one order higher than the corresponding ac and dc values.**

INTRODUCTION:

Fast transient voltages are generated in Gas Insulated Systems under switching operations or when line to enclosure faults occur within the GIS. The transients generated are characterized by (i) front time of few nano seconds, (ii) duration of few micro seconds and (iii) magnitudes of the order of 2 to 3 pu. The entire Insulation System consisting of the SF_6 gas, Insulation of the Transformer, Bushings, CT's etc., are also subjected to these transients. Therefore it is essential to study the behaviour of different Insulating materials and Insulation Systems used in GIS under Fast Transient Voltages.

Many studies have been carried out in recent years on the performance of GIS under fast transient voltages [1,2] and a few studies are also reported on paper oil Insulation Systems[3,4]. However, plastic materials like Polypropylene, Polyester, polyamide etc., are also used in Gas Insulated Apparatus like Transformers and Devices like Capacitors etc. Hence there is a need to study the performance of plastic materials under Fast Transient Voltages. This paper presents the results of the Breakdown measurements carried out with Fast Transient Voltages on thin Insulating materials.

METHODS OF GENERATING FAST TRANSIENT VOLTAGES

The two methods generally employed in the Laboratory for generating fast Transient Voltages are : (i) By simulating a transmission line with finite number of stages of capacitances and inductances, having a series spark gap and a terminating impedance. When the line is charged with a dc voltage sufficient to cause the breakdown of the spark gap, a rectangular pulse with initial oscillations is generated. The circuit diagram is shown in Figure 1.

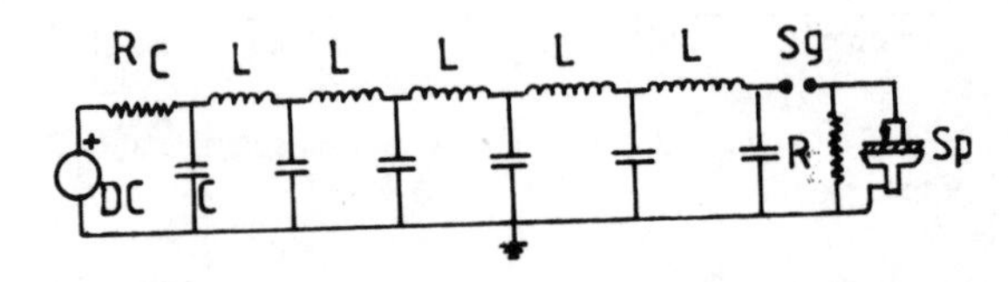

Figure 1. Fast Transient Generator (Transmission Line Model)

(ii) In the other method, a spark gap pressurized with SF_6 gas is charged with a 1.2/50 micro second Lightning Impulse voltage. The pulse shape gets modified into a fast transient with a very sharp rise time of few nano seconds. The Generator is schematically shown in figure 2.

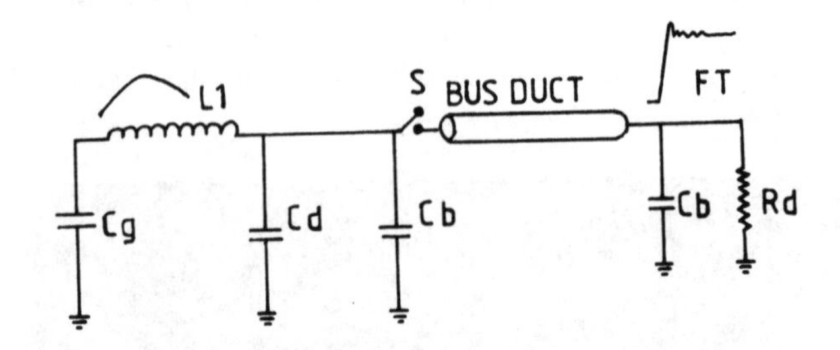

Figure 2. Fast Transient Generator (Spark Gap Method)

The basic difference in the two methods lies in the wave shape of the resulting pulse. In the first method, a rectangular pulse with initial oscillations is obtained due to the mismatch of the impedance, whereas in the second case, the rise time of the charging voltage gets sharply reduced to generate a fast transient voltage. The first method is very useful for Dielectric material testing where the magnitude of the dc charging voltage is about 50 kV. But the capacitors are expensive and the cost becomes a major constraint.

RESULTS AND DISCUSSIONS

A typical voltage withstand pulse characteristic in case of polypropylene is shown in figure 3. The pulse is characterized by a first peak of magnitude 70.6 kV, having a rise time of about 10 nano seconds, with a constant portion of 7.3 kV and a pulse duration of 5 micro seconds.

In figure 4, a comparison of the voltage withstand and breakdown of the specimen is made in case of kraft paper. The two waveforms are different only after the 1st peak and this indicates that breakdown is initiated within about 120 nano seconds. Similarly in case of polyamide, the differences are seen only after the 3rd peak which on time scale corresponds to about 100 nano seconds. These results and their comparison with the corresponding ac and dc values are summarized in Table I.

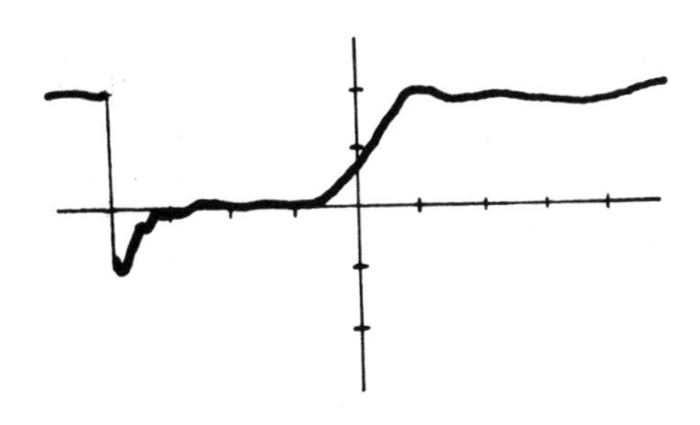

Figure 4. Oscillogram of a Fast Transient with polypropylene (Voltage Withstand)

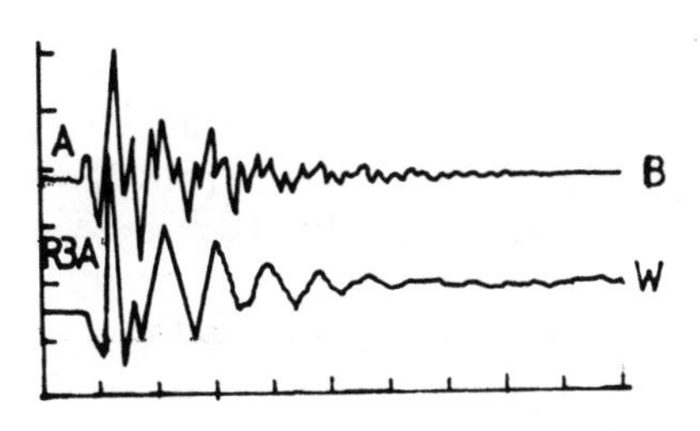

(W: Withstand; B: Breakdown) { 250 nano Sec/div}

Figure 5. Oscillogram showing the fast transient breakdown and withstand voltage of kraft paper

TABLE I COMPARISON OF THE ac, dc AND THE FAST TRANSIENT BREAKDOWN VOLTAGES

MATERIAL	BREAKDOWN STRENGTH kV/mm		
	ac	dc	Fast transient
POLYPROPYLENE	233	460	1900
POLYAMIDE	396	400	2700
POLYESTER	283	340	2000
KRAFT PAPER	26	20	135

In the above table, the ac breakdown strength refers to the peak ac value and the fast transient breakdown strength is based on the magnitude at breakdown and does not necessarily correspond to the peak value. It can be seen from the above table that polyamide has the highest breakdown strength both under ac and fast transient voltages, and polypropylene has the highest value under dc voltages. The fast transient breakdown strength values are nearly 8 times the ac value in case of polypropylene, 8 times in case of polyamide, 7 times in case of polyester and 14 times in case of kraft paper. Further, the breakdown in case of the plastic materials was in the form of a pin hole whereas in case of kraft paper it resulted in rupture of the material which extended to several millimeters. Hence it clearly indicates that the mechanisms of breakdown involved are different in different materials.

The average time period in which the breakdown is initiated in case of polypropylene is 60 nano sec, it is 90 nano sec in case of polyamide, 100 nano sec in case of polyester and 120 nano sec in case of kraft paper. These values clearly indicate that the breakdown is caused by the fast transient and hence the breakdown is essentially a fast front phenomenon.

The electrode roughness plays an important role in the breakdown phenomenon in case of the fast transient voltages. For example, repeated breakdown measurements carried out with the same set of electrodes resulted in a reduction of the breakdown strength of the material by more than 50 % whereas similar measurements with either ac or dc resulted in 15 to 20 % reduction in the breakdown voltage values. Thus the influence of the surface roughness is predominantly seen in case of fast transient voltages.

When very high magnitudes of transient voltages are required, the spark gap method is convenient and less expensive. The spark gap distance, pressure of SF_6 gas and the magnitude of the charging Lightning Voltage are the variable parameters which provide flexibility to the Generator. This method is very useful for Insulation system studies.

In this paper, the results of the studies on breakdown of thin Insulating materials carried out with the Transmission Line model of the Fast Transient voltage generator is presented and discussed.

EXPERIMENTAL DETAILS

The Fast Transient Generator essentially consists of six stages of capacitances (30 nF/ 50 kV) and inductances (each of 5 micro Henry) and a series spark. The line is terminated by an impedance (18 ohms). The circuit simulates a transmission line of few kms length. When the line is charged with dc and discharged through the impedance, rectangular pulse of 5 micro seconds duration are generated. Since the impedance is not matched with the characteristic impedance of the line, there is an overshoot in the front portion of the pulse. The oscillations have a rise time of 30 nano seconds and the magnitude is 3 to 4 times the charging voltage or even higher. The amplitude of the current is given by the expression:

$$V/[(2/\sqrt{(Ln/Cn)}] \ldots (1)$$

and the duration of the pulse is given by the expression :

$$2/\sqrt{Ln\,Cn} \ldots\ldots. (2)$$

where Ln and Cn are the distributed inductances and capacitances of the line. The pulse amplitude can be increased by increasing the spark gap distance, which in turn necessitates a higher charging voltage. A typical pulse shape for a perfect and imperfect matching of the load impedance is shown in figure 3.

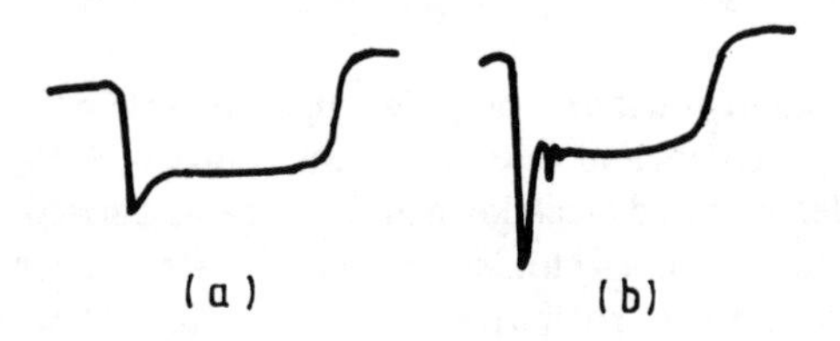

Figure 3. Pulse shape of a Fast Transient with perfect (a) and imperfect (b) matching impedance.

MEASUREMENT OF TRANSIENT VOLTAGES

A liquid column resistive divider was designed and fabricated for measurement of the fast Transient voltages. The resistances were closely monitored before and after each measurement. A 100 M Hz digital storage oscilloscope (Phillips model 3375) was used in conjunction with a recorder for measurements. The divider was connected across the the terminating impedance.

METHOD OF BREAKDOWN VOLTAGE MEASUREMENTS

A pair of cylindrical electrodes were used for the measurement of the breakdown voltages. The high voltage electrode had a diameter of 25 mm and height of 30 mm while the earthed electrode had a diameter of 75 mm and height of 10 mm. The electrode assembly with the test specimen in-between was connected across the terminating impedance, in parallel with the measuring system. By varying the spark gap distance, the dc charging voltage for the gap breakdown was varied. By closely monitoring the oscillogram, a clear distinction between the withstand voltage and the breakdown voltage was achieved. The voltage is increased until the specimen breaks down consistently (i.e., until 100 % breakdown is reached). The magnitude of the charging voltage is reduced in small steps to a value at which the breakdown probability is less than 100 %. The breakdown voltage is defined as the lowest value of the voltage which results in consistent breakdown of the material. At each voltage level, a minimum of 20 measurements were taken.

In order to investigate whether the breakdown phenomenon is due to the fast transient or not, and in order to get an insight into the breakdown events on the time scale, two oscillograms, one representing the breakdown voltage and the other, though very close to it (less by about 5 %) not resulting in a breakdown, representing the withstand voltage were recorded and compared. The point of deviation in the oscillogram was considered to give a measure of the time periods in which the fast transients were causing the breakdown events.

Using the same electrode system, the breakdown voltages were measured with 50 Hz ac and dc voltages for a comparative evaluation of the materials.

INSULATING MATERIALS FOR THE STUDY

Four different insulating materials used are polypropylene (0.018 mm), polyester (0.030 mm), polyamide (0.025 mm) and kraft paper (0.035 mm).

CONCLUSIONS

[1] The transmission line method isa convenient method of producing transient voltages at low voltage levels for dielectric studies.

[2] The breakdown strength of materials under fast transient voltages are one order higher than the corresponding ac or dc values.

[3] The breakdown with fast transient voltages occurs within 100 nano seconds thereby confirming that breakdown is a fast front phenomenon.

[4] Electrode surface roughness significantly affects the breakdown voltages of materials under fast transient voltages.

ACKNOWLEDGEMENTS

The authors wish to thank the authorities of the Central Power Research Institute, Bangalore, for sponsoring this work and also for their permission to publish this paper.

REFERENCES

[1] WG 33/13-09, "Very fast transient phenomena associated with gas insulated substations", CIGRE 1988 session.

[2] Luxa G. et al., "Recent research activity on the dielectric performance of SF_6 with special reference to very fast transients", Paper no. 15-06, CIGRE 1988 session.

[3] Vandermaar A.J., M.G. Wang, J.B Neilson and K.D. Srivastava," The characteristics of oil-paper insulation under fast front impulsevoltages", 7th ISH, Yokohoma, Japan, 1993, Paper no 25.03 .

[4] Dzikowsski D., " Electrical strength of paper-oil insulation at steep front switching impulses", Proc. 6th ISH, 1989, Paper 25.08 .

Conference Record of the 1996 IEEE International Symposium on Electrical Insulation, Montreal, Quebec, Canada, June 16-19, 1996

Surface Charging and Charge Decay in Solid Dielectrics

Alexandre Neves Helvio J.A. Martins
Centro de Pesquisas de Energia Elétrica -- CEPEL
Rio de Janeiro -- BRAZIL

Abstract: This paper presents the results of the investigation on surface charging and charge decay in solid dielectrics such as Teflon (P.T.F.E.), Epoxy (filled with alumina) and Polyethylene. The experiments were carried out in atmospheric air and SF_6 environment. Impulse and ac voltage were used to generate charges, by partial discharges, in a point / insulation sample / plane electrode arrangement. The results showed that surface charging and charge decay are influenced by several parameters such as ambient gas, surface resistivity, voltage level and polarity and subsequent voltage application.

INTRODUCTION

The intrinsic quality of insulating materials is of great importance to the dielectric behavior of high voltage systems. However, for solid dielectrics, their surfaces play an important role, particularly in gas insulated equipment under higher pressures.

It is well know and supported by a great number of evidences that surface charge accumulation on insulator distorts the electric field distribution and may increase or decrease the flashover voltage of the insulating system [1, 2, 3].

The mechanisms and factors governing the magnitude and distribution of surface charging, charge decay and how they influence flashover levels, are not fully understood. However, a better understanding of these phenomena can contribute to some improvements and solutions such as the use of low-loss insulating materials and the control or inhibition of abnormal faults caused by surface charge accumulation.

The main purpose of this paper is to show some characteristics of surface charging and charge decay in solid dielectrics and how they can influence the electrical behavior of high voltage equipment.

TEST SET UP

The experimental equipment, Figure 1, consists mainly of an aluminum chamber, charge generation system and residual electric field measurement system.

The charges were generated by partially discharging a point / insulation sample / plane electrode arrangement. The point electrode had a tip radius of approximately 1mm and was conected to the hv sources through hv bushing. The samples had a form of a plane disc with diameter of 90mm and thickness of 3mm, and they were fixed on a sample holder in the earthed plane electrode with a circular metal clamp. The bottom side of the samples was painted with a conducting paint, to provide a good electrical contact with the plane electrode. The samples and their holders were mounted in a aluminum circular plane disc, with radius of 300mm and thickness of 4mm. Up to four samples can be dealt with simultaneously with this arrangement.

The residual electric field, normal to the solid dielectric surface, was measured using an induction probe [4]. The main probe components are the probe head and a buffer amplifier of unity gain and high input impedance. The signal from the buffer amplifier was stored in a digital oscilloscope. The gap between the sample and point electrode and between the sample and probe, was adjusted using a step motor system.

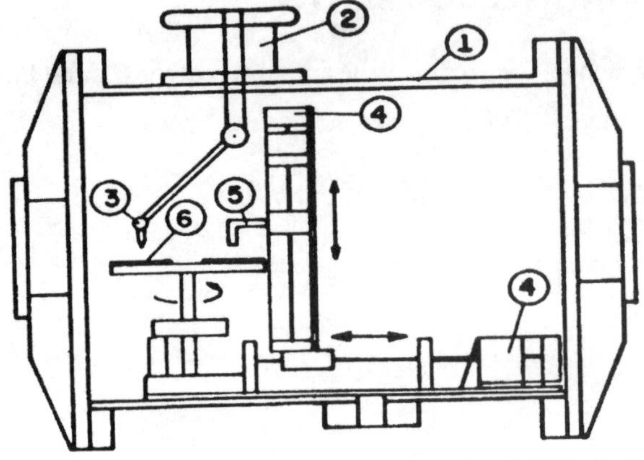

Figure 1. Test chamber - 1-Aluminum cylinder, 2-H.V. Bushing, 3-Point electrode, 4-Step motors, 5-Induction Probe, 6-Plane electrode and solid dielectric.

EXPERIMENTAL TECHNIQUES

Before any test, the insulation samples were cleaned with ethanol and dried so that they could be considered free of charges. They were fixed on the plane electrode and centralized under the point electrode.

After adjusting the gap distance, the voltage was applied to the test arrangement.

Immediately after voltage application the point electrode was lifted and the earthed plane disc was rotated so that the insulation sample passed under the induction probe head. Therefore, a scan of the residual electric field intensity was obtained along an arc line on the charged sample surface. After this probe scan procedure, the sample was removed from the test chamber and dust figure technique was used to obtain the surface charge figure. This technique consists of dusting a charged surface with a powder mixture and gently blowing the surplus from the surface. Therefore, only the attracted particles remain on the surface. The powder used in this work is the same used in photocopier machines.

To study the charge decay, the residual electric field intensity was periodically monitored by scanning along an arc line passing through the center of the sample over a long period of time.

The experiments were carried out at room temperature in atmospheric air with 50% R.H. and SF_6 with commercial purity. In both cases the gas pressure was 0.1MPa.

To investigate the influence of surface charges on the flashover level, an insulation barrier was held vertically and centrally between two spheres with radius of 5mm. The Teflon and Bakelite insulation barrier had dimensions of 300mm x 80mm and thickness of 5mm.

EXPERIMENTAL RESULTS

Surface Charging

Initially some characteristics of surface charging were investigated with impulse voltage in both polarities. An impulse generator was adjusted to produce 30/350μs, $10kV_{peak}$ impulses.

The point-sample gap and probe-sample gap were adjusted respectively with 10mm and 3mm.

After voltage application the charged sample surface was scanned by induction probe along an arc line passing through the center of the sample. The sample was then removed from the test chamber and dusted in order to obtain the surface charge figure. Figure 2 shows the results obtained for Teflon in atmospheric air for negative and positive impulse voltages.

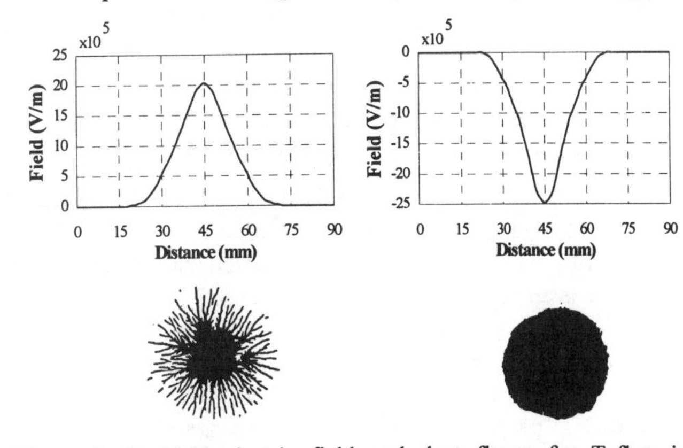

Figure 2. Residual electric field and dust figure for Teflon in atmospheric air. Point-sample gap=10mm, probe-sample gap=3mm, impulse voltage=10kV. Positive (Left) and Negative (Right) Impulses.

According to the test results it was observed that, for the electrodes configuration used, the surface charge has the same polarity of the applied impulse.

Although the residual electric field profiles are very similar for both polarities, the dust figure revealed that positive surface charge distribution consists of a large number of filamentary streamers. On the other hand, the negative surface charges are distributed in a well defined circular area.

Similar results were obtained for Epoxy and Polyethylene samples under identical test condition.

It is of great importance, particularly for high voltage tests, how the deposited surface charge varies with repeated impulse applications of increasing amplitude. This aspect was investigated by performing surface charging with impulses of increasing peak values. The point-sample gap was adjusted to 5mm and positive and negative impulse voltages were applied to the arrangement. After each impulse application the surface was scanned through its center by the induction probe. Typical results are shown in Figure 3.

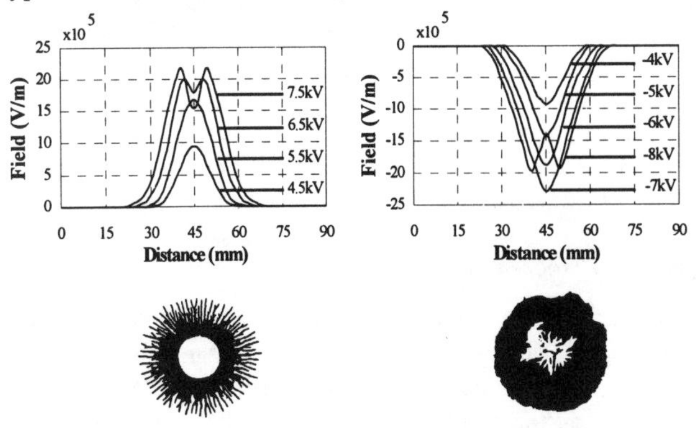

Figure 3. Effect of repeated impulse of increasing peak value on surface charging for Teflon sample. Point-sample gap=5mm, probe sample gap=3mm,Teflon sample. Residual electric field and dust figure for positive (Left) and negative (Right) impulses.

Two main characteristics can be identified from these figures. First, with increasing peak value of the applied impulse a significant increasing of the residual electric field intensity and surface charged area were observed. This effect was attributed to discharges occurred during the wavefront of both positive and negative impulses. This type of discharges carried a large amount of charges to the sample surface and it is called direct discharges.

Second, above a given voltage level the field profile was substantially modified and it was observed that the surface area directly under the point electrode becomes almost free of charge, as can be seen by the dust figures. The large amount of charges accumulated on the surface, due to direct discharges, significantly reduce the field strength at the point electrode. After the impulse crest, at the beginning of the wavetail, the resultant electric field is predominantly influenced by the stored surface charges. The field at the point electrode reaches an intense value such that reverse discharge occurs. This type of discharges controls the maximum charge density in a particular area on the dielectric surface.

Similar experiments were performed to investigate the characteristics of surface charging with ac. 60Hz voltage. The point-sample gap was adjusted with 5mm and the voltage was applied to the arrangement during 15 minutes for each voltage level. Figure 4 shows the results of the residual electric field , in atmospheric air, for a Teflon sample.

No surface charging was detected with a peak voltage up to 3.5kV. Above this level the charging process was initiated.

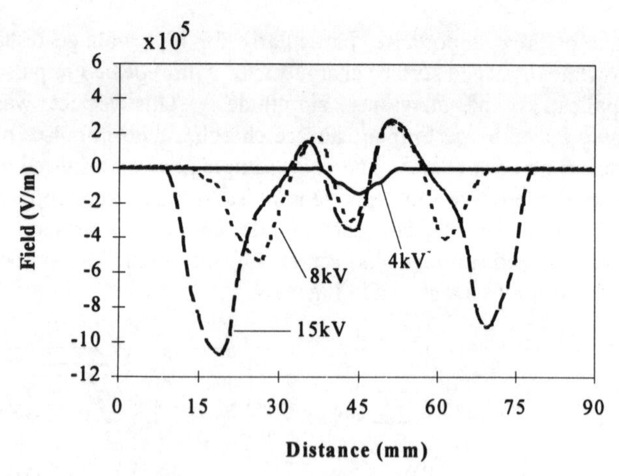

Figure 4. Characteristics of surface charging with a.c voltage application in atmospheric air for Teflon sample. Point-sample gap=5mm, probe-sample gap=3mm.

According to the results only negative charges were detected with a peak voltage of 4kV. As the voltage amplitude was increased bipolar charges and a significant increasing of the charged surface area were observed. An interesting characteristic for the electrode configuration used was the predominance of negative surface charges.
Comparing surface charging for impulse and ac voltages, it was observed that the residual electric field intensity is much higher for impulse. On the other hand, the charged surface area is greater for a.c voltage.
Similar results were observed for Polyethylene and Epoxy samples.

Influence of Ambient Gas

Tests similar to those just described were also carried out using SF_6 as the ambient gas in the test chamber. Figure 5 shows some results obtained for Epoxy sample.
With repeated impulse application of increasing peak value it was noted that residual electric field and surface charged area increase significantly. Reverse discharges were also observed. For ac voltage application it was observed the predominance of negative surface charges and no significant increasing in surface charges was observed for peak voltages between 10kV and 20kV. Although the results with SF_6 were very similar to those obtained in air, Figures 3, 4 and 5, it was observed that for SF_6 the charges were stored in a smaller surface area for both positive and negative impulse and ac voltage. Another aspect is that for SF_6 the inception voltage for direct and reverse discharges are higher compared to those obtained in atmospheric air. Similar results were obtained for Teflon and Polyethylene samples.

Charge Decay

Solid insulating materials may hold surface charges for a long time. This behavior is of great importance in the design of high voltage equipment. Decay process starts immediately

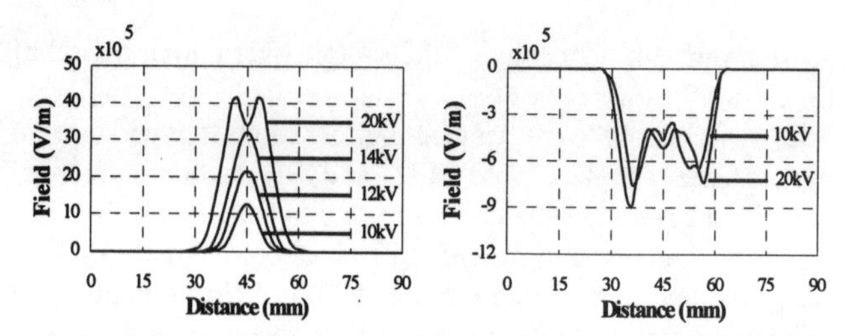

Figure 5. Influence of ambient gas on surface charging for Epoxy sample and SF_6 ambient gas. Positive impulse (Left) and ac (Right) voltages. Point-sample gap=5mm,.probe-sample gap=3mm.

after surface charging, due to bulk conduction, diffusion along the surface or neutralization by ion created by background radiation. The decay rate depends also on the nature of the material, temperature, condition and type of ambient gas and surface condition.
In order to examine the charge decay characteristics a single impulse wave was applied to the arrangement. The experiments were performed in atmospheric air and SF_6 gas for Teflon, Epoxy (filled with alumina) and Polyethylene. Typical results are shown in Figure 6 .

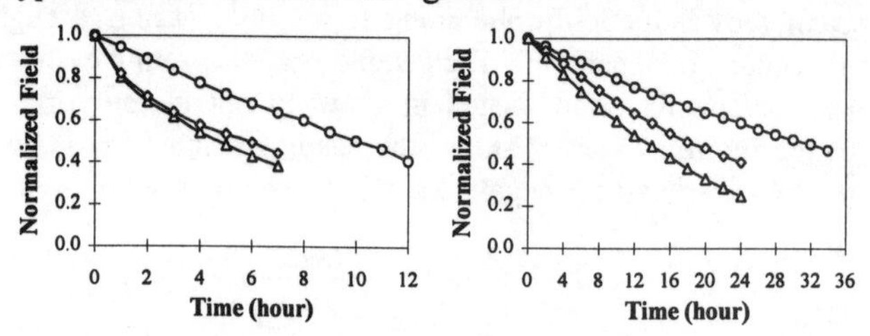

Figure 6. Surface charge decay. Teflon (O), Polyethylene (◊), Epoxy (Δ). Atmospheric air (Left) and SF_6 (Right).

For all the insulating materials used here it is clearly indicated that surface charge decay in atmospheric air is much faster than in SF_6. This characteristic can be attributed to the low humidity level in SF_6.
Teflon exhibits slower charge decay time compared to other materials. This fact is due to its higher surface and bulk resistivity. Another relevant aspect of charge decay is show in Figure 7. The field at the center of the sample decays at a faster rate than that at the periphery. It can be explained by the neutralization of the surface charges by charge carriers present in the gas volume above the sample surface. In a high intensity field region (center of the sample) the charge carriers are easily attracted so it is expected that the charges decay much faster in this region. It was verified that this effect is totally independent of the insulating material, ambient gas and voltage polarity.

Insulation Flashover Characteristics

Two test arrangements were used to investigate the influence of surface charge and charge decay on flashover level. The first was a Teflon or a Bakelite insulation barrier held vertically between two spheres as previously described.

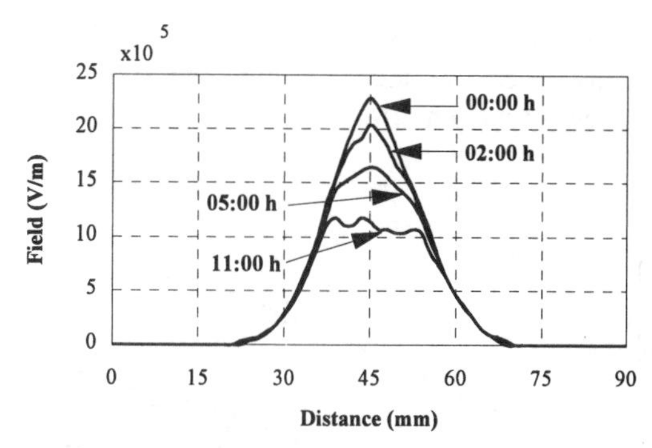

Figure 7. Surface charge decay for Teflon sample in atmospheric air

Up and down test method with standard lightning impulse voltage (1.2/50µs) was used to determine the 50% flashover probability. The experiments were performed under laboratory atmospheric condition with 22°C and 50% relative humidity. The second arrangement consisted of a SF_6 circuit breaker model with arc contact and an insulation Teflon nozzle similar to a real circuit breaker as described in [5]. The same test method was used to determine 50 % flashover level with non-standard impulse voltage (50/1000µs). The results are shown in Figure 8.

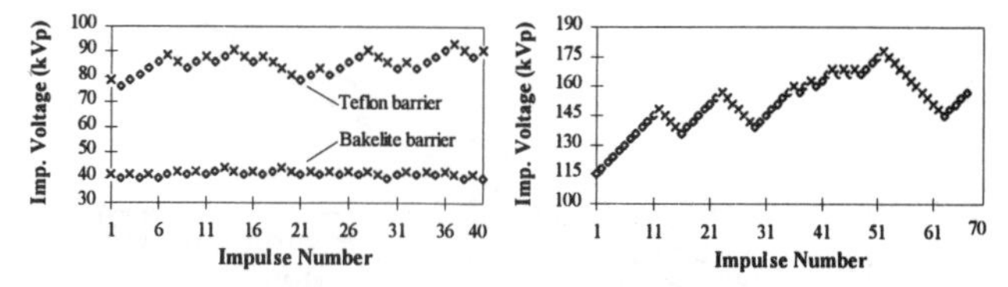

Figure 8. Flashover characteristics. Withstand (◊), Flashover (x). Sphere gap arrangement (Left) and SF_6 Circuit -Breaker "cold-characteristic" test (Right).

For Bakelite barrier the result of 50% flashover probability was quite consistent. Flashover and withstand voltages were very close over the 40 impulse applications. On the other hand for Teflon the results indicate a large dispersion on flashover level. From surface charging and charge decay investigations it was observed that Bakelite dissipates the charges very fast (few seconds) due to its low surface and volumetric resistivity. The wide difference on the flashover characteristics for these two materials can be explained in terms of stored surface charge and time decay.

The tests result for SF_6 circuit-breaker shows a large dispersion on the flashover level. Although it was not possible to measure the residual electric field on nozzle surface, due to the more complex geometry, dust figure revealed traces of surface charges. A visual inspection of the material shows a large number of surface discharges.

CONCLUSIONS

A series of tests was carried out to investigate some characteristics of surface charging and charge decay in solid dieletrics.

For impulse voltage tests it was observed that, for the electrode configuration used here, monopolar surface charges, with the same polarity of the applied impulse, were produced. Under both positive and negative impulse voltages reverse discharges were observed for all interface gas / insulating materials used here. Reverse discharges had the effect of limiting the maximum surface charge density in a particular area of the dielectric surface. When repeated impulses of increasing amplitude were applied to the system a significant increasing in residual field intensity and surface charged area were observed for both impulse polarities and both gas used. This characteristic contributes to increase the flashover level to a value well above that for an uncharged system. However, it is evident that under other condition, such as polarity reversal of the applied voltage, the stored surface charge will cause a reduction on the flashover level.

For ac voltage it was observed a predominance of negative surface charge for all gas / insulating material interface used. Surface charged area is greater than for impulse voltages but the residual electric field intensity is lower.

Surface charge decay is mainly governed by the ambient gas, neutralization of surface charge by charge carriers in the gas and surface and volumetric resistivity of the insulating materials. Teflon / SF_6 interfaces exhibit slower time decay than the others used.

It was found that 50% flashover probability can be misinterpreted when insulating materials, that hold surface charge for a long period of time, is used as insulation barriers.

ACKNOWLEDGMENT

The authors would like to acknowledge Messrs. Fernando A.Chagas, Walter R.Cerqueira, Enio Alvarenga and Orsino Oliveira for their technical assistance and discussions.

REFERENCES

1. Fujinamy, H; Takuma, T; Yashima, M; Kawamoto,T. "Mechanism and effect of DC charge accumulation on SF_6 gas insulated spacers". IEEE Trans. Power Delivery, PWRD-4 No.3, 1989.
2. Stoop, Th; Tom, J; Verhaart, H.F.A.; Verhage,A.J.L. "The influence of surface charges on flashover of insulator in gas insulated systems". 4th International Symposium on H.V. Engineering. Athens-Greece, 1983, Paper No. 32-10
3. Al-Baway, I; Farish,O. "Insulator Flashover in SF_6 under impulse voltage conditions". IEE Proceedings-A, Vol. 138, No 1 January 1991, pp. 89-97.
4. Connoly, P; Farish, O. Surface Charge measurements in air and SF_6. Gaseous Dielectric IV. Pergamon Press, 1985, pp. 405-413.
5. Carvalho, A.C.; Junqueira, A.J.S.; Lacorte, M.; Neves,A.; Mpalantinos, A. "Alternative test method for the cold caharacteristic determination of high voltage circuit-breakers". XI SNPTEE, Group XIII-G.E.M., Rio de Janeiro-Brazil, 1991(in Portuguese).

air gap and open air without UV irradiation are also compared in Figure 2.

The difference between the two sets of data can be explained by the fact that, in the enclosed gap, due to shielding from cosmic radiation and stray UV irradiation, there will be something of a scarcity of free electrons or negative ions capable of yielding free electrons. In the open gap UV irradiation from the impulse generator[8] as well as cosmic radiation(which can produce approximately 40 ion pair/s.cm^3 MPa[7]) will more or less guarantee a relative abundance of free electrons which cannot be significantly increased by artificial irradiation.

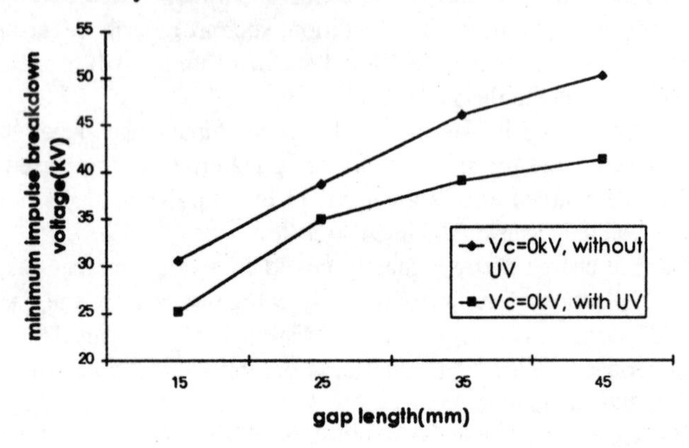

Figure 1 Minimum impulse breakdown voltage versus gap length, enclosed air gaps, p=0.1MPa

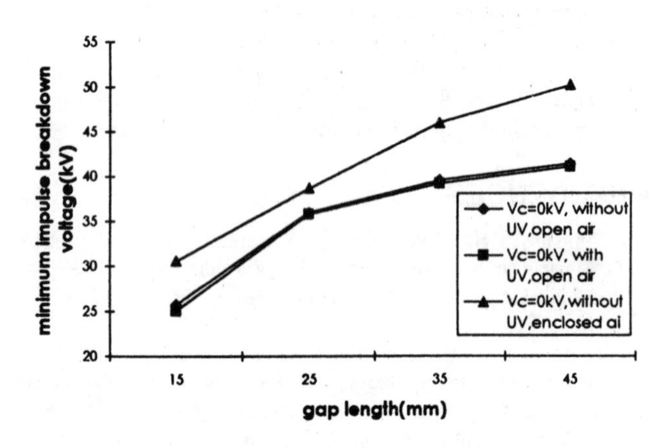

Figure 2 Minimum impulse breakdown voltage versus gap length, open air and enclosed air gaps, p=0.1MPa

The above observations strongly suggest that initiatory electrons make an important contribution to the breakdown process. In order to further investigate this, the corona pin arrangement was used to inject negative ions into the point /plane gap volume. Figure 3 shows a typical result when a 12kV negative bias voltage was applied to the pins. It can be observed that for an enclosed gap, negative ions injection has the same effect in decreasing impulse breakdown voltage as UV irradiation in enclosed gaps. Also it is noticed that in Figure 3 with negative ion injection there is little further effect on breakdown voltage when UV irradiation was also applied. This seems to imply that there are already sufficient initiatory electrons resulting from detachment of plentiful negative ions injected by means of applying -12kV dc bias voltage on the corona pin arrangement.

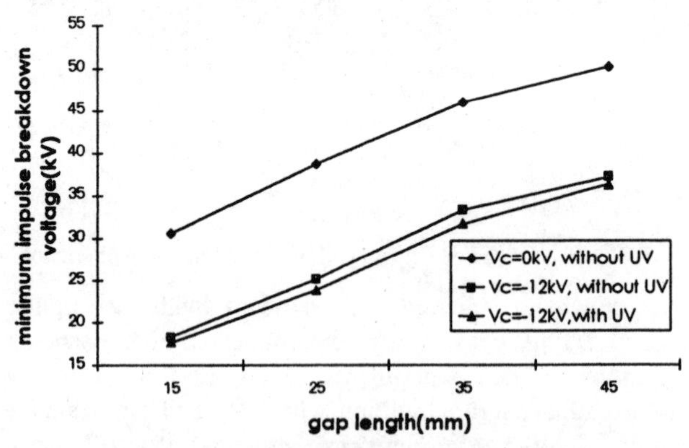

Figure 3 Minimum impulse breakdown voltage versus gap length, enclosed air gaps, p=0.1MPa

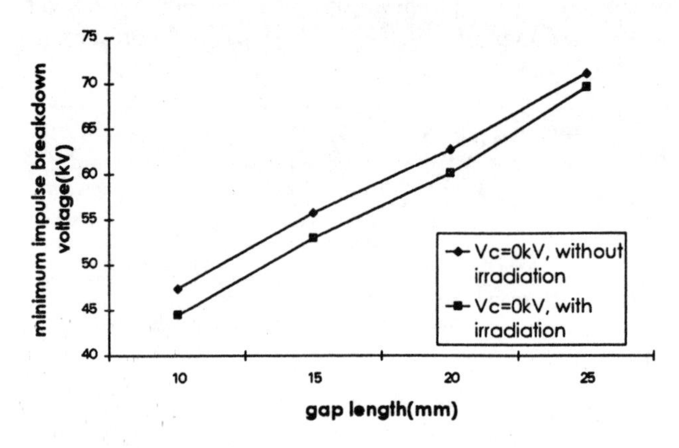

Figure 4 Minimum impulse breakdown voltage versus gap length, SF_6, p=0.1MPa

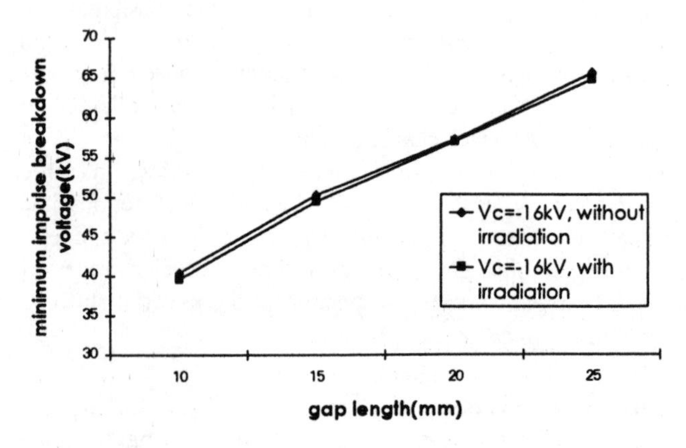

Figure 5 Minimum impulse breakdown voltage versus gap length, SF_6, p=0.1MPa

Figure 4 shows the impulse breakdown voltage versus gap length with and without ^{137}Cs irradiation in SF_6. The same test was repeated with a -16kV dc bias voltage applied to the

Conference Record of the 1996 IEEE International Symposium on Electrical Insulation, Montreal, Quebec, Canada, June 16-19, 1996

THE EFFECT OF INITIATORY ELECTRONS ON BREAKDOWN CHARACTERISTICS IN HIGHLY NONUNIFORM GAP

X.Q.Qiu and I.D.Chalmers
Centre for Electrical Power Engineering
University of Strathclyde
204 George Street
Glasgow G1 1XW, U.K.

Abstract: **Previous work[1,2] has shown that the positive impulse breakdown voltage of SF_6, N_2 and SF_6/N_2 mixture in a point-plane gap is affected by the presence of space charge. With negative dc bias voltage on a corona needle arrangement external to the test gap, the impulse breakdown strength decreases significantly. This phenomenon has little to do with the electronegative properties of gases, since the same result is found in both electron-attaching and non-attaching gases. The reason was supposed to be the abundance of initiatory electrons resulting from detachment of plentiful negative ions injected.**

The aim of the present work is to confirm the above explanation. Experiments were carried out in air, SF_6 and an SF_6/N_2 mixture both with and without irradiation. The results show that with either artificial irradiation or a negative dc bias voltage, the impulse breakdown strength decreases. The present experimental results confirmed to a large extend that the breakdown strength of a non-uniform field gap is greatly affected by the availability of initiatory electrons within the gap volume.

INTRODUCTION

In a non uniform field, the space charges can affect the breakdown strength greatly[3]. Some work relating to the effect on breakdown in N_2, SF_6 and its mixtures has been carried out[1,2,3] and it is believed that the abundance of initiatory electrons plays an important role in breakdown strength[1,2]. Previous work [4] pointed out that the production of initiating electrons in the critical volume of the gap is quite important for the development of breakdown. It has also been found that in an enclosed highly non uniform field gap in air[4,5] and SF_6[6] the impulse breakdown voltage decreases under the condition of strong irradiation. From the industrial viewpoint it was concluded in [7] that the probability of failure of SF_6 insulated switchgear(GIS) under positive surge waveforms depends on the presence of free electrons. Thus comparison of the effect on breakdown characteristics of irradiation with that of negative ions injection in a highly non uniform field gap would be beneficial to a better understanding of the breakdown mechanism with particular reference to the role of negative ions.

EXPERIMENTAL SETUP AND PROCEDURE

Positive lightning impulse voltage with a waveform of 1.2/50 μs was applied to a point-plane gap in which charges could be injected by means of four auxiliary needle electrodes surrounding the gap with the needles being approximately in line with the point electrode tip. Negative polarity dc voltage was applied to the needle electrodes for injection of negative ions into the main gap.

With the experiments carried out in air, UV irradiation was produced by a 125W quartz-mercury arc lamp. Because UV irradiation is often insufficient for gaps due to photon absorption in pressurised SF_6 and its mixtures, so a 5mCi capsule ^{137}Cs was placed close to the point electrode when experiments were carried out in SF_6 and its mixtures.

All tests were carried out in a pressure vessel which has four orthogonally arranged windows for observation. The vessel was first evacuated by a rotary vacuum pump and then flushed several times with appropriate gas before filling. After filling, the vessel was left for at least 24 hours before testing in order to achieved a uniform gas distribution.

Tests were carried out to determine the minimum impulse breakdown voltage as a function of negative D.C. bias voltage applied to the corona pin under different conditions, i.e., artificial irradiation, negative ion injection or both. The significance of the minimum impulse breakdown level has been discussed previously[10]. The procedure for determining its value was as follows. Using the normal up-and-down method, the V_{50} was first established. From the V_{50} value, the voltage was then reduced by 0.5kV and 40 impulses applied at this voltage. The number of breakdown in 40 applications was noted and the procedure was repeated until the voltage was reached at which the gap just failed to breakdown in 40 applications. This value was taken as the minimum impulse level.

RESULTS AND DISCUSSION

Figure 1 compares the minimum impulse breakdown voltage versus gap length in enclosed air gap with and without UV irradiation. The test was repeated using the same point-plane geometry in open air and the results are shown in Figure 2. It is very clear from both figures that in an enclosed air gap there is a significant decrease in minimum impulse breakdown voltage with UV irradiation but there is almost no decrease under the same condition if the gaps are in open air. The minimum impulse breakdown voltages in enclosed corona pin arrangement with and without ^{137}Cs irradiation in SF_6(Figure 5).

Figure 6 and 7 show both results in SF_6/N_2(50/50), respectively. It is quite interesting to note that, from Figure 4 to 7, either in SF_6 or SF_6/N_2 there is a significant decrease in breakdown voltage when the ^{137}Cs irradiation was applied with no dc bias voltage. As the -16kV dc bias voltage was applied, there is very little, if any, decrease in breakdown voltage when the irradiation source was added.

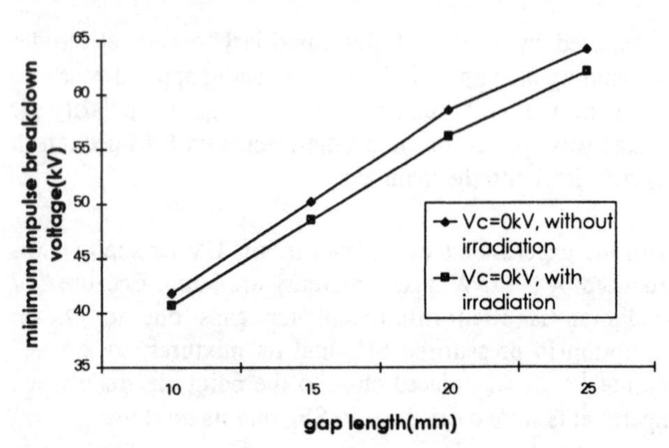

Figure 6 Minimum impulse breakdown voltage versus gap length, SF_6/N_2(50/50), p=0.1MPa

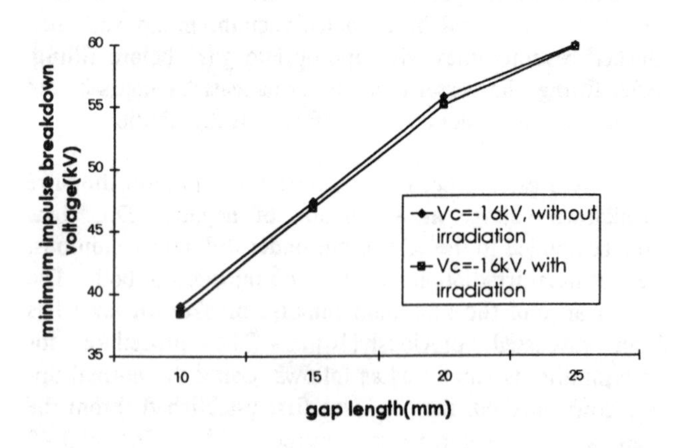

Figure 7 Minimum impulse breakdown voltage versus gap length, SF_6/N_2(50/50), p=0.1MPa

From the test results obtained in Air, SF_6 and SF_6/N_2, it is quite clear that artificial irradiation has exactly the same effect on breakdown voltage as that of negative ions injected by the corona pin arrangement in producing abundance of initiatory electrons. Availability of seed electrons within the point/plane gap volume has been discussed previously by the authors[10] in terms of its effect upon propagation of a stepped discharge from the point. As the availability of electrons increases then the probability of a corona discharge at the end of each leader step also increases and this leads to an associated increase in breakdown probability and consequent decrease in breakdown strength. It would appear from the present work, however, that there is a limit beyond which the breakdown strength cannot be further decreased by increasing the electron or negative-ion population.

CONCLUSIONS

In nonuniform fields, the initiatory electrons can affect breakdown characteristics greatly. It has also been proved that the effect of initiatory electrons is not only true in electron-attaching gases but also in non-attaching gases.

Negative ions injection has the same effect as that of irradiation on impulse breakdown strength which implies detachment of negative ions is probably the main source of initiatory electrons in an enclosed gap.

ACKNOWLEDGEMENT

One of the authors (XQQ) wishes to thank the Committee of Vice-Chancellors and Principals of the Universities (UK) in providing an Overseas Research Students Award to carry out a PhD study in the Centre for Electrical Power Engineering at Strathclyde University.

REFERENCES

[1]Qiu,Y., Chalmers, I.D. and Li, H.M. "Effect of injected space charges on the positive impulse breakdown of SF_6 in a point-plane gap", J.Phys.D: Appl.Phys., 25, 1992, pp.326-328

[2]Qiu,Y., Chalmers,I.D.,Li,H.M.,Feng, Y.P.,Z.Wu and L.M.Zhou, "Space charge effect on impulse breakdown of SF_6, N_2 and SF_6/N_2 in a highly non-uniform field gap", 8th International Symposium on High Voltage Engineering, Yokohama, Japan, August 23-27, 1993, pp.53-56

[3]Berg, D. and Works, C.N. "Effect of space charge on electric Breakdown of sulfur hexafluoride in nonuniform fields", AIEE Trans, 1958,77, pp.820-822

[4]Chalmers, I.D., Farish, O and Macgregor, S.J. "The effect of impulse waveshape on point/plane breakdown in SF_6", Gaseous Dielectrics IV, Knoxville: 1984, pp.344-353

[5]Guenfoud, O., Jordan, I.B. and Saint-Arnand, R. "Atmospheric ions and positive point corona", Proc.4th ISH, Athens:1983, paper 41.03

[6]Somerville, I.C., Farish, O. and Tedford, D.J. "The influence of atmospheric negative ions on the statistical time lag to spark breakdown, Gaseous Dielectrics IV, Knoxville:1984,pp.137-144

[7]Boggs,S.A. and Wiegart, N. "Influence of experimental conditions on dielectric properties of SF_6 insulated systems -- theoretical considerations", Gaseous Dielectrics IV, Knoxville:1984, pp.531-539

[8]Meek, J.M. "The influence of irradiation on the measurement of impulse voltage with sphere gaps", IEE Journal, Vol 93, pt.II, 1946, pp.97-115

[9]Qiu, Y.,Kuffel, E. and Raghuveer, M.R. "Irradiation effect on breakdown voltages of gaps enclosed in test chambers", IEEE MEXICON-81, Vol 3,1981,pp.4-8

[10]Chalmers, I.D., Farish, O and Shelton, R., "Insulation Performance of SF6 mixtures", Gaseous Dielectrics VII, pp 601-608, Plenum Press,NY, 1994

Conference Record of the 1996 IEEE International Symposium on Electrical Insulation, Montreal, Quebec, Canada, June 16-19, 1996

Characteristics of Abnormal Glow Discharges

S.V. Kulkarni, D. Panchali, V. Kumari and D. Bora

Institute for Plasma Research
Bhat, Gandhinagar 382 424, Gujarat, INDIA

Abstract: Here we report the work done in case of abnormal glow discharge in air at low pressures. The breakdown voltage measurements and I-V characteristics are obtained to find the region and the pd conditions corresponding to abnormal glow discharge which also tries to occupy the maximum possible volume of the chamber. The charge density, electron and ion temperature and floating potential measurements are carried out using Langmur probe along with breakdown current pulses for understanding different regions of glow discharges. The floating potential measurements show fluctuations upto 7% level near the anode and the level decreases as one goes away from the anode in positive column of the discharge. These fluctuations are related to the ion acoustic oscillations in positive glow region. We could obtain the maximum charge density of 10^{13} cm^{-3} with ion temperature in the range of 7-9 eV. It is expected that these results can help in understanding the behavior of abnormal glow discharge which shows different characteristics under different experimental conditions and has various applications.

INTRODUCTION

Glow discharge and its different versions like normal, subnormal and abnormal glow is being under investigation for a long time. A good amount of experimental as well as theoretical work is available in literature [1-3]. Even now, some of the regions of glow discharge like negative glow are not understood fully. Sometimes glow discharge shows some kind of unusual characteristics like striations, localized or propagating oscillations or waves etc. under different experimental conditions. By taking into consideration all above points here we report the preliminary work done regarding I-V characteristics, density and temperature measurements and the floating potential measurements which show low frequency oscillations of few kHz which most probably seem to be ion acoustic waves which are generated near anode dark region and the fluctuation level decreases as the probe is moved away from the anode in the positive column of the discharge.

EXPERIMENTAL DETAILS

The experiments are performed in a stainless steel high vacuum chamber of volume more than 1 m^3 and the vacuum of the order of 10^{-6} Torr is produced before introducing a gas. Two parallel plate electrodes (10 cm dia.) are used for producing a discharge. Although the distance between electrodes was 8 cm, it was not possible to isolate the chamber from the ground and hence the field distribution gets modified. In fact, it was observed that the discharge was mainly existing between the high voltage electrode and the chamber. The connecting rods of the electrodes were well insulated with PTFE tapes and it was possible to vary the distance between electrodes, position of the Langmur probe and the pressure in the chamber. The experiments are conducted at various distances as well as at different pressures in the range of 10^{-2} Torr to 1.0 Torr. Langmur probes are used for the measurements of density, temperature and floating potential measurements are carried out at different positions in the discharge between electrodes as well as at different applied voltages for a fixed position of the probe near hv electrode. I-V characteristics are obtained at different pressures using 20 kV, 50 Hz high voltage ac power supply and by measuring the voltage across the electrodes with dividers and the current passing through the gap which is controlled by current limiting resistance. Breakdown current pulses are recorded on oscilloscope of 300 MHz bandwidth using a 50 Ω resistance in series with low voltage electrode and by using a 50 Ω transmission line to obtain exact current pulses.

RESULTS AND DISCUSSIONS

In first set, the breakdown voltage measurements are conducted at different pressures to verify the complete set- up and measurement techniques and it is observed that the obtained Paschen curve matches with the one available in literature [4].

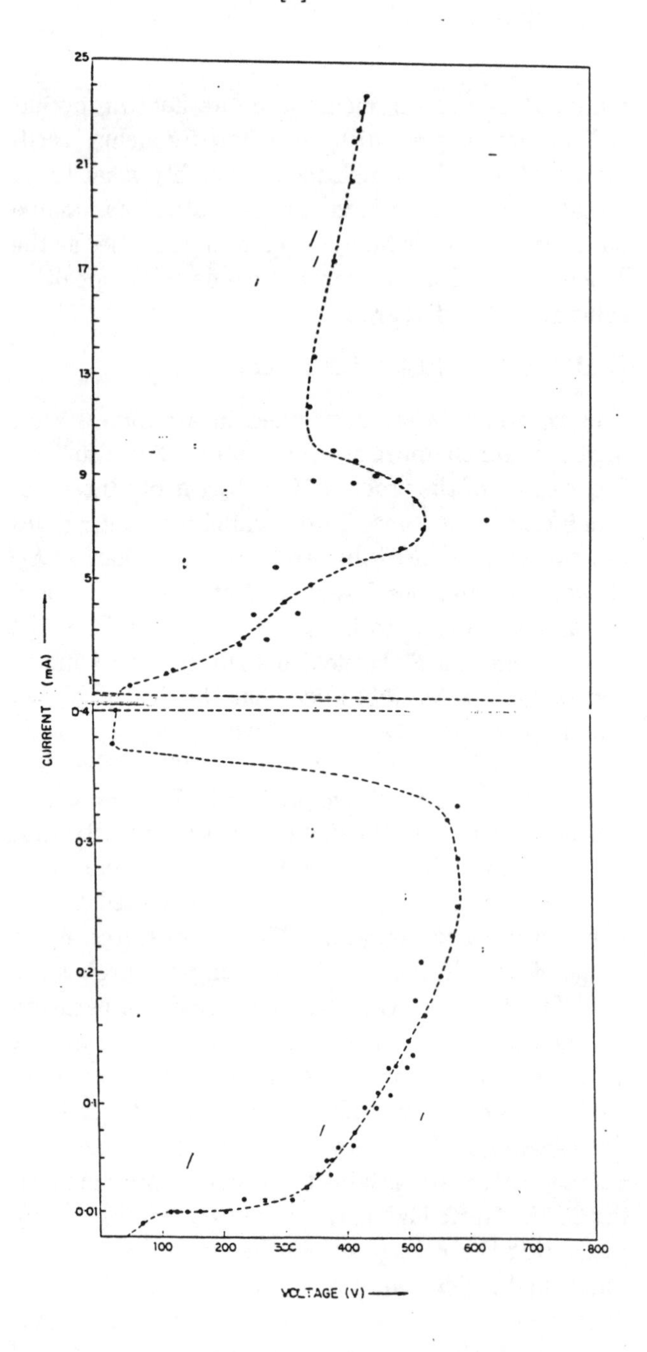

Fig.1 Measured I-V characteristics of the glow discharge in air

Figure 1 shows I-V characteristics obtained by measuring the current at different applied voltages across electrodes. It was not possible to measure the initial part of the current growth due to limitations of the current measuring device. However, one can clearly see the subnormal, normal and abnormal glow regions. In order to study the characteristics of abnormal glow region, the applied voltage is increased gradually. It is observed that the cathode glow which exists in a small area of the electrode increases with increase in voltage, then spreads all over the electrode even to occupy the back portion of the electrode and then goes to the leads and the supports. In case of abnormal glow region, the whole volume of the chamber is occupied by the intense glow discharge. Since, the applied voltage is alternating voltage, the polarity of the electrodes changes in every cycle and hence to the naked eye it is a mixture of the different regions existing from the cathode such as Aston dark space, cathode glow, Crooks dark space, negative glow, Faraday dark space, positive column the anode dark space and anode glow region. However, one can clearly distinguish the behavior in two different portions of the alternating voltage by looking at the oscilloscope waveforms of breakdown current pulses.

The breakdown current pulses are recorded for different applied voltages and currents using a 50 Ω resistance in the ground path and using 50 Ω transmission line with matched impedance to get exact waveshapes of breakdown current pulses on oscilloscope. It can be seen from the Fig. 2 that as the applied voltage is increased, although the total current increases, there is little increase in the magnitude of the current pulses which indicate that the walls of the chamber play an active role in a discharge which occupies the available volume. It can also be concluded that the current pulses do not follow the applied electric field and certainly this is due to existence of strong space charge field which can dominate the applied field for the discharge to take its own path.

The floating potential measurements are done at two conditions. In one case, the potential is measured at different positions (1.0 cm and 4.0 cm) for applied voltage of 360 Volts and for a current of 11.8 mA (Fig. 3). It can be seen that the potential fluctuations exist only in positive half cycle of the applied alternating voltage indicating that they exist near the anode after the anode dark region but in positive column of the discharge which has charge neutrality to satisfy the plasma properties. It can also be seen that as the probe is moved away from the anode, the fluctuation level decreases from 7% to almost 1% of the floating potential. The magnitude of the potential is almost close to the applied voltage indicating that the positive column of the discharge has negligible voltage drop and most of the drop exists in the region near the cathode. The frequency of fluctuations observed is in the range of 1- 2 kHz. Since, in case of a 4.0 cm position of the probe, which is almost at the center of the discharge, there are no oscillations in negative half cycle, one can conclude that these fluctuations originate from the anode dark region of the discharge and are not due to nonuniform electric field of the chamber and electrodes.

In another set, the measurements are done at different voltages and currents for a fixed position of probe at 1.0 cm from the high voltage electrode (Fig. 4).

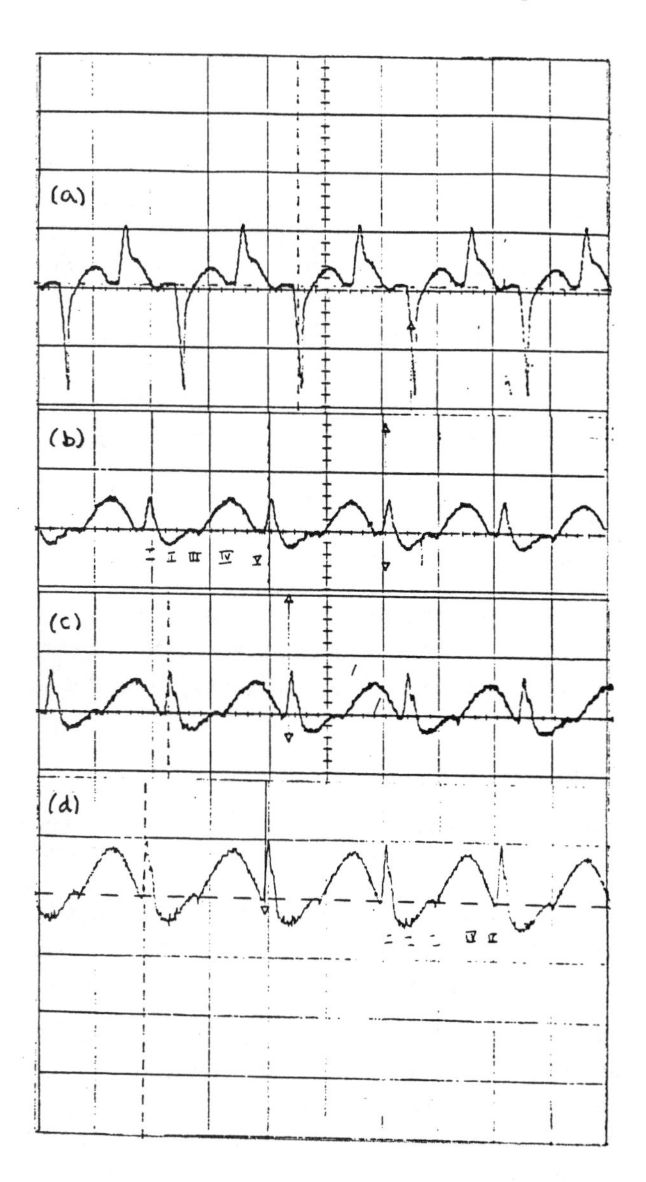

Fig.2 The brekadown current pulses at 422 Volts and 30.17 mA (a), at 451.0 Volts and 40 mA (b), 526.0 Volts and 45 mA (c) and 620.0 Volts and 60 mA (d). Scale: x axis- 10 ms/div and y axis- 10 mV/div.

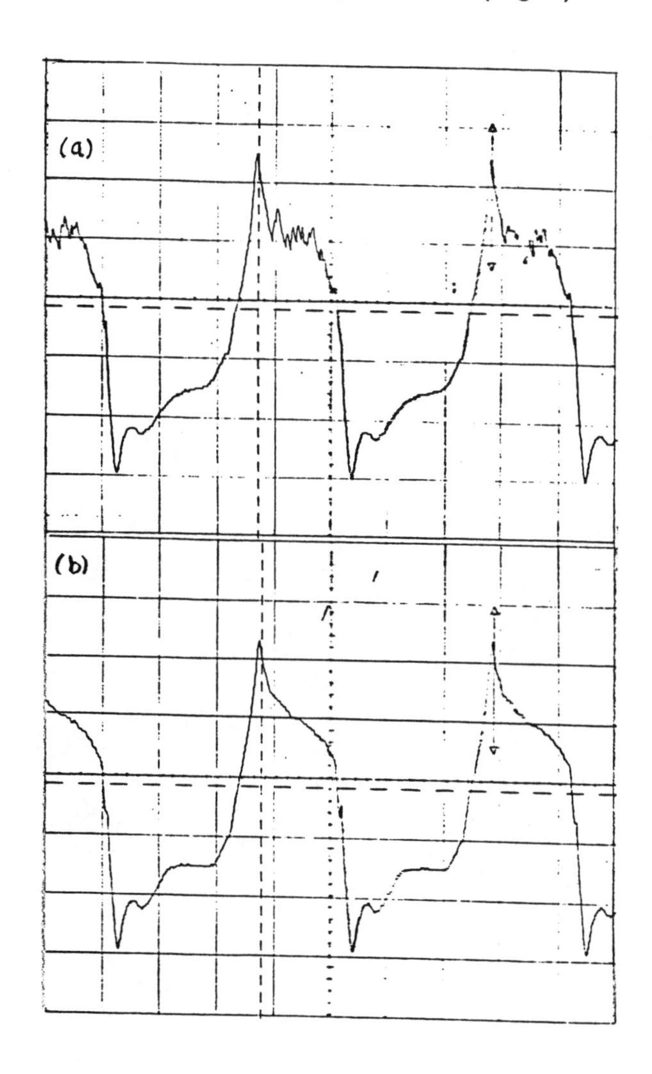

Fig. 3: Floating potential measurements at different positions of 1.0 cm and 4.0 cm from anode in a discharge gap of 8.0 cm for Vapp= 360.0 Volts. x: 5 ms/div, y: 100 V/div

Here also we observe the fluctuations only in positive half cycle and the fluctuation level decreases with increase in applied voltage. In general, in case of abnormal glow, as the voltage is increased the plasma tries to occupy the maximum possible volume of the chamber. However, since we observe the decrease in fluctuation level at higher voltages, it is quite possible that the anode dark region may get modified with increase in voltage to influence the fluctuations in the positive column of the discharge.

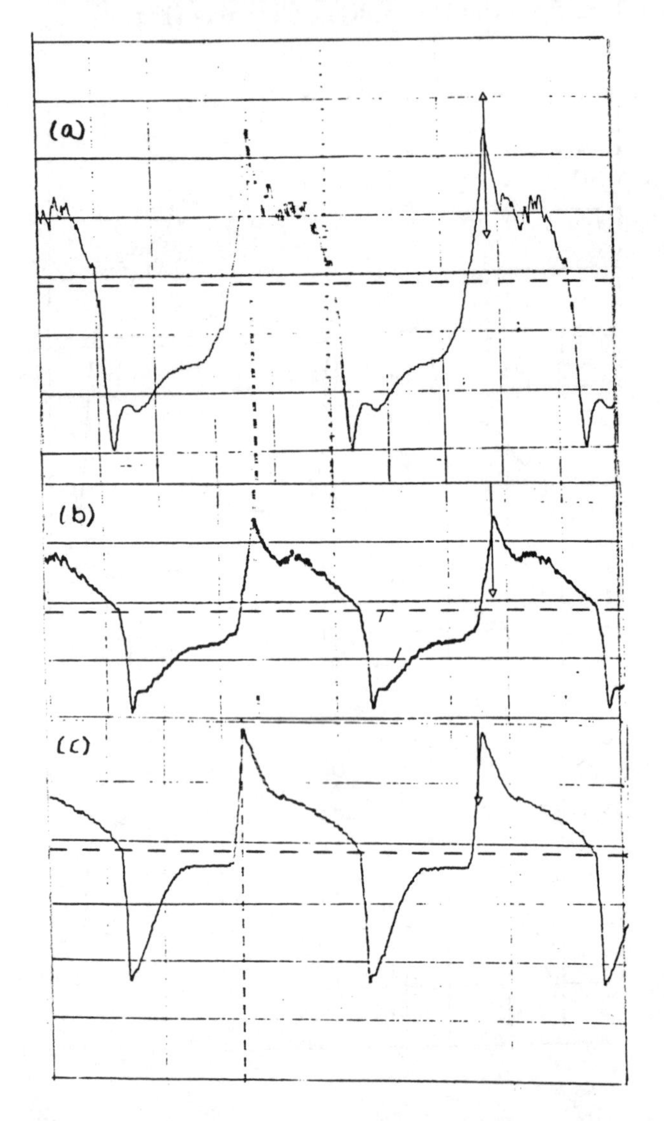

Fig.4: Floating potential measurements at different applied voltages and currents: i.e. 360 Volts and 11.89 mA (Fig.a), 441.0 Volts and 26.0 mA (Fig. b) and 534.0 Volts and 47.0 mA (Fig.c). x: 5 ms/div, y: 100 V/div

Since, the present measurements are done using a single probe, one cannot say whether the fluctuations detected are localized oscillations or propagating instabilities. It is however most probable that these instabilities are generated due to anode dark region and propagate inside the positive column of the plasma. We did the calculations of ion acoustic oscillations and the frequency lies in the same range of observed fluctuations.

In order to confirm these results some more work is required which is under progress.

CONCLUSIONS

The preliminary study done of glow discharge shows that it can have different unusual behavior like ion acoustic oscillations or fluctuations in potential which generate from the anode dark region and propagate inside the positive column of the discharge and get attenuated with distance. The I-V characteristics show that the glow discharge exist only if the pd is greater than pd minimum of Paschen curve. These are however preliminary results and further experimentation is required to confirm the results.

ACKNOWLEDGEMENT

The authors are thankful to Prof. Kaw, Director of the Institute for useful discussions and encouragement regarding this work and Mr. Raghuraj Singh for the technical help.

REFERENCES

1. Nasser E., Fundamentals of Gaseous Ionization and Plasma Electronics, Wiley Interscience, pp.397-425.

2. Marode E., The Glow-Arc Transition, in Electrical Breakdown and Discharges in Gases, Editors: E. Kunhardt and L.H. Luessen, Plenum Press, pp.119-166.

3. Ferreira C.M., Current Research Topics in Low Pressure Glow Discharges in Rare Gases and in Pure Nitrogen, in Electrical Breakdown and Discharges in Gases, Editors: E. Kunhardt and L.H. Luessen, Plenum Press, pp.395-417.

4. Meek J.D. and Craggs J.M., Electrical Breakdown in Gases, John Wiley and Sons, 1978.

Conference Record of the 1996 IEEE International Symposium on Electrical Insulation, Montreal, Quebec, Canada, June 16-19, 1996

A TIME-RESOLVED OPTICAL STUDY OF THE AVALANCHE AND STREAMER FORMATION IN ATMOSPHERIC NITROGEN

E.H.R. Gaxiola and J.M. Wetzer

High Voltage and EMC Group, Department of Electrical Engineering
Eindhoven University of Technology
PO Box 513, 5600 MB Eindhoven, The Netherlands
phone +31 40 2473993 fax +31 40 2450735

ABSTRACT

This work deals with a time-resolved optical study of the avalanche and streamer formation phases leading to breakdown in atmospheric nitrogen. We present the results obtained for nitrogen, from experiments and two-dimensional model simulations. This model is used to obtain a better insight in the relevant mechanisms and processes by a comparison of measurements and simulation data. The trends of externally measured quantities correspond with those predicted by our model.

1. Introduction

In the design of electrical gas insulated systems data on breakdown fieldstrengths is extensively used. Breakdown voltages for gases are usually given by Paschen curves, which serve as an engineering tool in high voltage engineering. In many practical situations non-uniform fields require a more sophisticated approach (corona, void discharges, ESD). This work is a continuation of earlier work [1,2], and deals with a time-resolved optical study of the avalanche and streamer formation in atmospheric nitrogen. The goal of this work is to develop and verify discharge models for practical insulating geometries on the basis of physical discharge processes. In this paper we present time-resolved optical and electrical measurements of the avalanche and streamer formation phases leading to breakdown. Through these diagnostics we obtain a better understanding of the processes fundamental to gas-breakdown. Here we will discuss results obtained for nitrogen, from experiments and two-dimensional (2-D) model simulations. A sophisticated simulation program describes the physical discharge processes in the gap. This model is used to obtain a better insight in the relevant mechanisms and processes by a comparison of measurements and simulation data.

2. Modelling of prebreakdown phenomena

To obtain a better understanding of the processes fundamental to gas-breakdown we have developed 1-D and 2-D models to describe the prebreakdown phenomena and thereby reproduce and get an understanding of the externally measurable quantities. Several approaches are possible which basically lead to a description by means of a Monte-Carlo simulation or by means of a model based on continuity equations (hydrodynamic model) [1,3,4,5 and 6]. The last one loses its validity at non-equilibrium conditions, like might exist within a streamerhead of a streamer discharge. At the electron energies occuring at our conditions it has been shown that a 2-D hydrodynamic model correctly describes the phenomena until the moment at which the cathode directed streamer (or the anode-directed streamer) reaches the cathode (anode), i.e. hundreds of picoseconds before actual breakdown occurs [1]. The description of prebreakdown phenomena in gases for homogeneous laplacian fields (e.g. parallel plate electrode geometry) with superimposed space charge field necessitates a 2-D discharge model with a 3-D electrical field calculation routine. For inhomogeneous laplacian fields (e.g. point to plane or wire to cylinder electrode geometries) a 3-D model is necessary. Closed form analytical solutions only exist for the case without space charge effects [7]. We here discuss results from a 2-D, rotationally symmetric, discharge model with a 3-D field calculation routine. The modelling techniques are presented in detail in [1]. The 2-D model describes the spatio-temporal development of the charged species, i.e. for nitrogen electrons and positive ions. The incorporated processes are drift, diffusion, ionization, recombination, gas-phase photo-ionization and secondary photo-electron emission. For nitrogen attachment and detachment as well as conversion processes don't have to be taken into consideration. The calculation mesh consists of 120 linearly spaced axial and 22 linearly spaced radial elements. The simulation is started by an initial gaussian electron distribution at the cathode, which represents the initial electrons in the experimental setup induced by cathode photo-emission from a pulsed N_2-laser (pulse duration *0.6 ns* (full width at half maximum)).

3. Experiments

A fast experimental setup with a *250 MHz* bandwidth is used to measure the discharge current. An ICCD camera (Intensified Charge Coupled Device) registeres the optical discharge activity. The time resolved swarm method is used for the avalanche current measurements [1,7]. The ICCD camera used here has a minimum shutter time of *5 ns*. The ICCD camera is triggered via an antenna near the N_2-laser which initiates the discharge through the release of a sufficient number of electrons from the cathode surface. A programmable delay enables the choice of any opening moment with reference to the start of the avalanche. We first evacuate the vessel to a pressure of less than $2*10^{-5}$ *Torr*. The vessel is filled to atmospheric pressure with (technically pure) nitrogen. We sandpaper the cathode surface to ease the release of large electron number densities through photo-emission.

The *0.6%* magnesium content in the aluminium electrode material is responsible for photo-emission at photon-energies as low as *3.68 eV*. The software controlled automated setup starts measurements by triggering the N_2-laser, which triggers the pulsed ICCD camera and an HP 54542A digital oscilloscope, to register the spatio-temporal light activity and the discharge current waveform. The type of gas under study is nitrogen; pressure = *768 Torr* = $1{,}02*10^5$ *Pa*; voltage = *30 kV*; electrode gap = *1 cm*.

4. Results and discussion

We present results for a uniform applied field with space charge field distortion. Figure 1 shows results of previous work in which we used *20 ns* ICCD camera gate-shutter times [1]. Characteristic is the cloudy character of the optical activity. Images *a* through *d* show the primary avalanche's arrival at the anode and the subsequent cathode directed streamer (CDS) formation in the midgap enhanced field region, initiated by delayed electrons residing in this region. In nitrogen, gas-phase photo-ionization is the predominant (non-local) source for these delayed electrons. On the left hand side in Fig.2 the measured current waveform is shown both on a linear- and log-scale. Different electron avalanches are clearly distinguishable. The reproducibility of the waveforms is quite good. Here only one (of a sequence of 52) is shown. Also shown in Fig.2 on the right hand side are the corresponding simulated avalanche current waveform, the total net charge (positive ions minus electrons) and the total number of electrons. On the horizontal axes in Fig.2 the time is normalized to the electron transit time in a uniform field (T_e). According to data in [7] T_e is calculated to be *83 ns*. This agrees with the result of the 2-D model simulation. From our experimental data we found a primary electron transit time of *65 ns* for the considered conditions. The difference is due to the space charge enhanced electron motion. The initial electron number is derived from measurements at *26 kV* because deviations from exponential growth at higher voltages hinder an accurate estimation. The illuminated area here is *0.03* cm^2. In earlier work [1] the current peak was

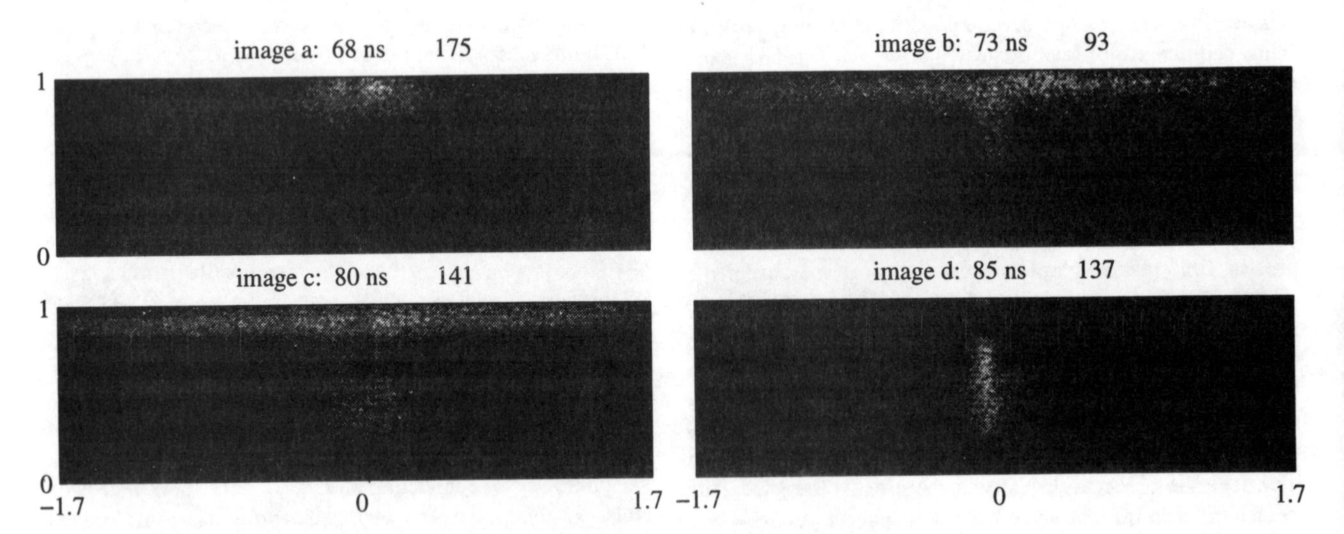

Fig.1 Sequence of ICCD camera images describing a laser-induced avalanche and it's transition to a streamer. In the figure the corresponding times on the measured current waveforms are indicated. The cathode (bottom) and anode (top) surfaces are located at *0* and *1 cm* respectively on the vertical axis. The horizontal scaling in centimetres is also indicated. The initial electrons are released from the cathode (bottom) at $t=0$ *s*. The maximum photon count per pixel is denoted in the upper right corner of each image. *20 ns* ICCD camera gate-shutter time.

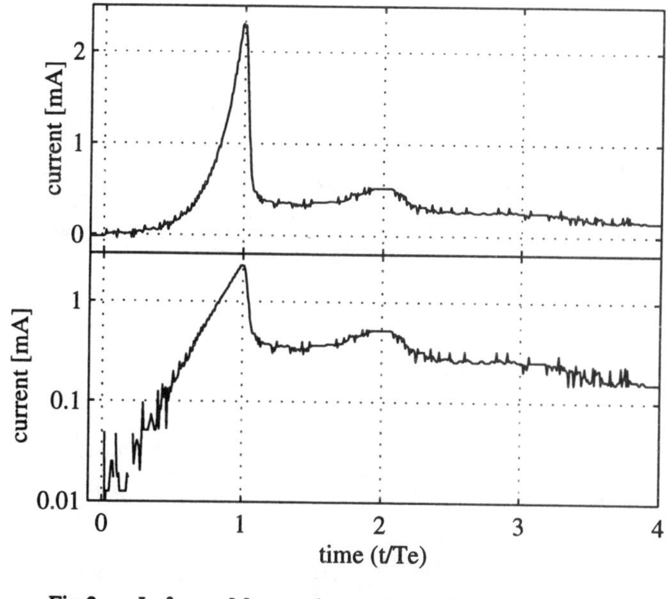

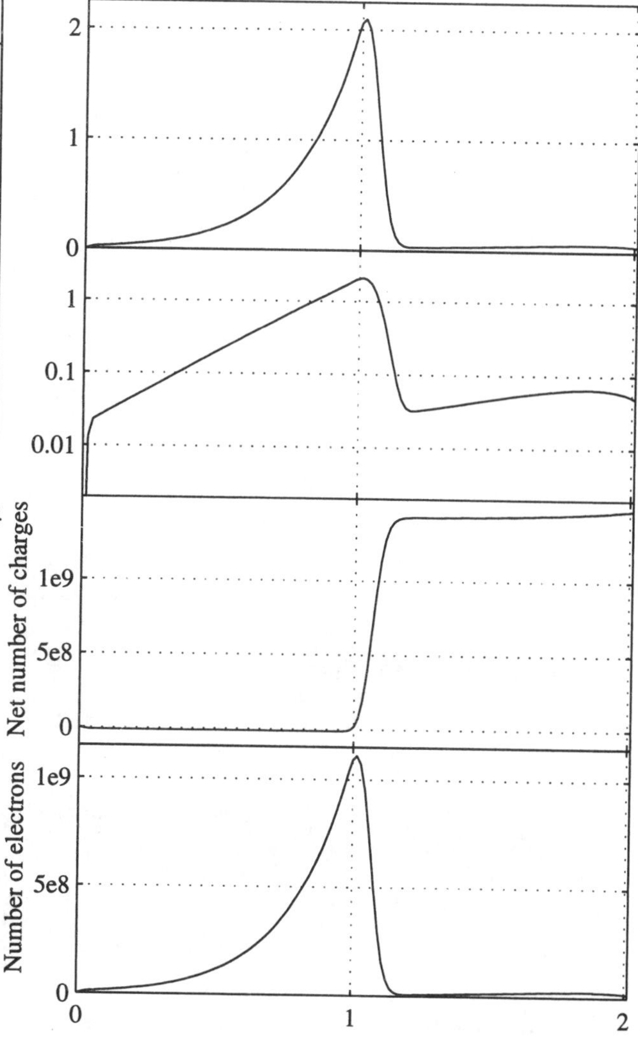

Fig.2 Left: Measured current waveforms ($N_0=4*10^7$, $T_e=65$ *ns*)
Right: Simulated current waveform ($N_0=2*10^7$, $T_e=83$ *ns*).
Conditions: atmospheric nitrogen (*p=768Torr*) *E=30kV/cm, gap=1cm.*

observed to broaden due to space charge effects. Such "peak broadening" is not oberved here. As was shown earlier [1] the current waveform strongly depends on the initial conditions. Figure 3 shows ICCD-images with *5 ns* gate-shutter times, recorded earlier in the avalanche development, prior to streamer onset. It is clear that space charge effects are present: the distribution is scattered in radial direction and we observe a considerable radial expansion during the primary avalanche's transit. Without a significant radial field the discharge would be more strongly pinched. Diffusion alone cannot account for the relatively large radial expansion of the optical discharge activity. In earlier work [1] we showed that the measurements with a *20 ns* shutter time at later times (see also Fig.1) were in good agreement with our model. When comparing the ICCD-images with different exposure times, we conclude that the "clouds" of optical activity obeserved at "long" (*20 ns*) exposure times actually consist of a number of small active regions which show a large radial expansion. It is at present not clear whether this is an inherent property of avalanches, or whether it is due to non-uniform initial conditions.
Two generations of optical activity crossing the gap from cathode to anode are observed, and in the same time span two electron avalanches are distinguished in the measured current. From the value of the maximum photon count per pixel we see that the optical activity is rather intens at first and decreases upon the radial expansion with progressing time, despite the exponential charge growth and increase in number of ionizing collisions. Eventually the gap broke down after several series of measurements since we are just slightly below the d.c. breakdown voltage. Also in the simulations only a slight change in the chosen initial electron distribution at these conditions leads to a complete breakdown. The 2-D hydrodynamic model provides a good description of the external current.

5. Conclusions

From the above the following can be concluded:

1. A strongly non-uniform avalanche structure observed at very short camera exposure times.
2. A strong radial expansion is observed which is not yet well understood.
3. The 2-D model correctly predicts the externally measured current, and the discharge structure in the streamer phase, but the model does not show the non-uniform avalanche structure nor the strong radial expansion in the early avalanche phase.

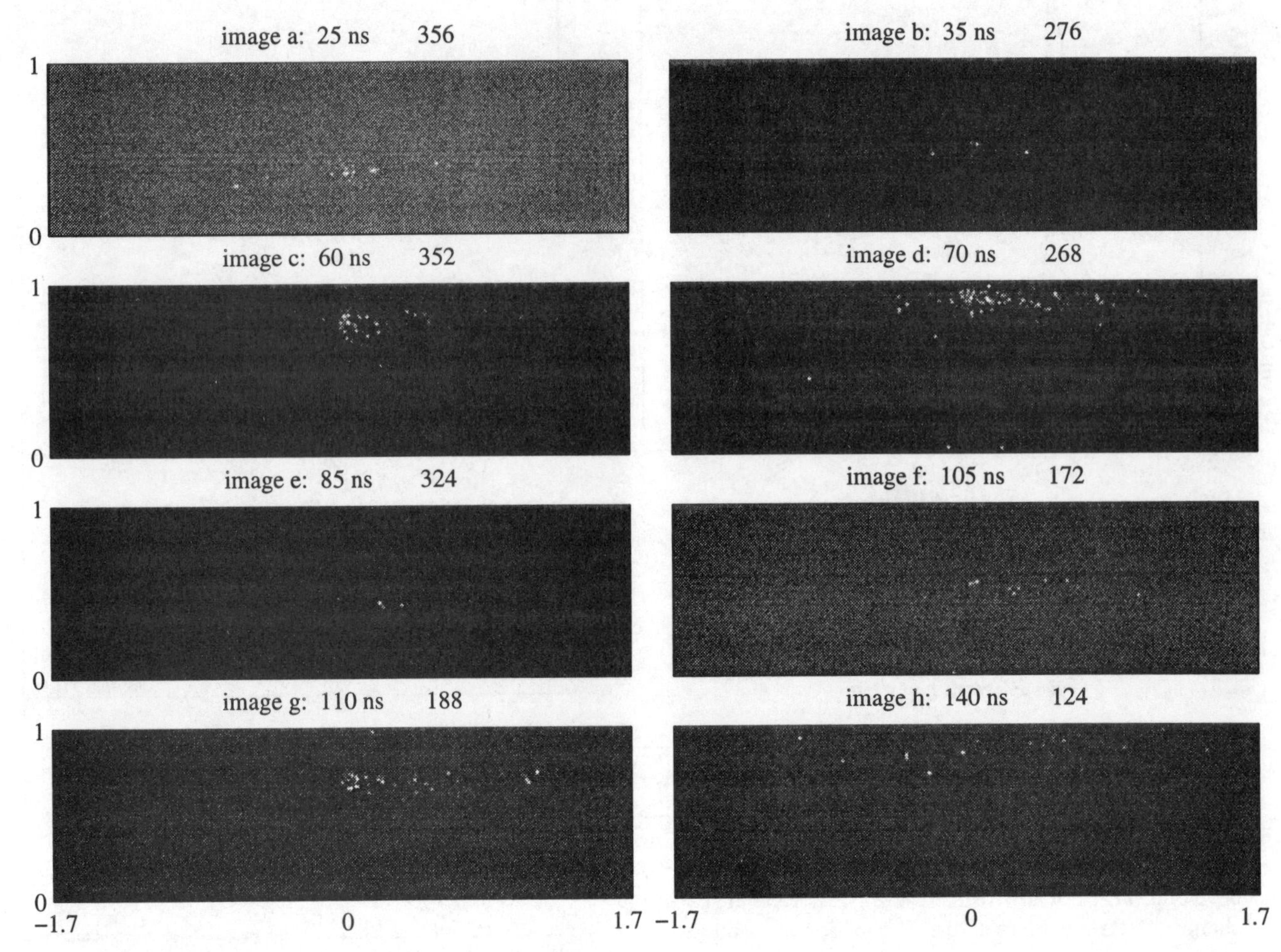

Fig.3 (Legends as in Fig.1.) *5 ns* ICCD camera gate-shutter time.

Literature

1. Kennedy, J.T. "Study of the avalanche to streamer transition in insulating gases". Ph.D. thesis, Eindhoven Univ. of Techn., The Netherlands, 1995.
2. Kennedy, J.T. and J.M. Wetzer. "Air Breakdown in Under-Volted Gaps Time-Resolved Electrical & Optical Experiments". 9th Int. Symp. on High Voltage Eng., Graz, Austria, August 28 to September 1, 1995, Paper No.2153, pp.1-4.
3. Djemoune, D. and S. Samson, E. Marode, P. Ségur. "A time resolved two dimensional modelling of the electrical behaviour and the chemical yield of streamer induced discharge". 11th Int. Conf. on Gas Discharges and Their Applications, Tokyo, Japan, Vol.2, September 11 to September 15, 1995, pp.484-487.
4. Babaeva, N. Yu. and G.V. Naidis. "2D Model of Streamer Propagation in Nonuniform Electric Fields". 11th Int. Conf. on Gas Discharges and Their Applications, Tokyo, Japan, Vol.2, September 11 to September 15, 1995, pp.488-491.
5. Loiseau, J.F. and F. Grangé, N. Spyrou, B. Held. "Laser-Induced Space Charge Initiating a Corona Discharge: Numerical Simulation and Comparison with Experiment". 11th Int. Conf. on Gas Discharges and Their Applications, Tokyo, Japan, Vol.2, September 11 to September 15, 1995, pp.492-295.
6. Held, B. "Coronas and Their Applications". 11th Int. Conf. on Gas Discharges and Their Applications, Tokyo, Japan, Vol.2, September 11 to September 15, 1995, pp.514-526.
7. Wen, C. "Time-Resolved Swarm Studies in Gases with Emphasis on Electron Detachment and Ion Conversion". Ph.D. thesis, Eindhoven Univ. of Techn., The Netherlands, 1989.

Conference Record of the 1996 IEEE International Symposium on Electrical Insulation, Montreal, Quebec, Canada, June 16-19, 1996

Pulsative Corona From Free Spherical Conducting Particles in SF_6 / Gas Mixtures

Anwar H. Mufti
High Voltage Laboratory
Elcectrical and Computer Eng. Dept.
King Abdulaziz University
Jeddah , Saudi Arabia

Nazar H . Malik
High Voltage Laboratory
Electriccal Engineering Dept.
King Saud Uinversity
Riyadh , Saudi Arabia

***Abstract* : Pulsative corona discharges from free spherical shaped conducting particles are investigated experimentally using SF_6 and its mixtures with nitrogen N_2 , perfluorocarbon $C_8F_{16}O$ triethylamine (C_2H_5)$_3$N and freon $C_2Cl_3F_3$ gases . Corona inception , particle lift - off and breakdown voltages as well as charge - voltage (q -v) characteristics were determined in these mixtures .The results show that corona characteristics are affected by particle diameter and gas mixtures . Generally , small percentage of these additive gases results in reduced corona charge levels . The particle movement is also observed and reported in this paper .**

INTRODUCTION

SF_6 insulated high voltage apparatus such as gas insulated substation (GIS) and gas insulated transmission line (GITL) need a high degree of insulation reliability . The insulation strength of such systems can be greatly reduced by the presence of contamination in the form of conducting particles [1] .This behavior has lead to a considerable research interest in the corona and breakdown characteristics of SF_6 and SF_6 gas mixtures containing fixed and free conducting particles [2,3]. These mixtures are reported to exhibit reduced sensitivity to microscopic filed nonuniformities and higher corona stabilized breakdown levels as compared to SF_6 gas alone [4,5] . The presence of N_2 in SF_6 tends to reduce the undesireable influnce of particles or surface asperities [6,7] .Similarly , SF_6 mixtures with C_8F_{16} O , $(C_2H_5)_3N$ and $C_2Cl_3F_3$ show improved insulation strength in nonuniform fields [8] .

In a previous report [3] , the authors investigated the corona from wire shaped free conducting particles of different parameters and discussed the q-v characteristics for SF_6/gas mixtures . In this paper, the investigations are extended by using spherical particles and SF_6 mixtures with N_2 , $C_8F_{16}O$, $(C_2H_5)_3N$ and $C_2Cl_3F_3$ gases . Corona onset , particles lift-off and breakdown voltages as well as pulsative corona charge levels are measured under different experimental conditions . The particle motoin is also observed. The results show that the q-v characteristics are affected by gas composition. Generally , the adddition of a small percentage of nitrogen, perfluorocarbon , triethylamine and freon to SF_6 reduces the pulsative corona charge levels and increases the particle-initiated breakdown voltage level .

EXPERIMENTAL SYSTEM

The measurements were carried out using a transparent, Plexiglas cylindrical pressure vessel of 130 mm internal diameter and 520 mm height . Concave electrodes of 60 mm diameter with an interelectrode gap separation of 13.3 mm were used . Sphere particles having diameters ∅ of 3.2 mm and 4.5 mm were employed .The high voltage source was a 60 Hz , 100 kV test transformer . A current limiting water resistor was connected in series with the high voltage supply . A partial discharge detector consisting of a coupling capacitor, measuring impedance and a partial discharge meter was used to measure the pulsative corona charge levels . Without the test geometry in the circuit , the system had a partial discharge level of < 5 pC for voltages of 100 kV.

The SF_6 gas used was of commercial purity i.e 99.8% pure. The total gas pressure was kept constant at 0.3 MPa throughout the investigations . The percentage of the additive gas indicated is by volume and was chosen arbitrarily .

Charge-Voltage (q-v) characteristics were measured for electrodes in the absence of particles .Then these measurements were repeated with particles untill a corona onset was detected . Subsequently , corona charge levels were measured for different values of the applied voltage . Particle motion was observed by using a telescope . In addition to q-v characteristics , particle lift-off as well as breakdown voltages were also recorded .

RESULTS

Pulsative corona charge levels associated with positive (+pC) and negative (-pC) half cycles were measured for SF_6 gas without any particles and with a single spherical particle of ∅=3.2 mm. Figure 1 shows that without particles corona started at 40 kV with a charge level of 1.5 pC which increases with applied voltage V .With 3.2 mm diameter particle the corona initiated at 33 kV with considerably higher pulse charge level .As expected q increases with V and positive corona charge levels (+pC) are in an order of a magnitude higher than the negative corona charge levels (-pC) . Moreover , positive corona charge levels measured using a particles of ∅ =4.8 mm is also shown in Figure 1 . It shows that as particle diameter is increased , the onset voltage increases . Furthermore for a given V , the charge level associated with a bigger diameter particle

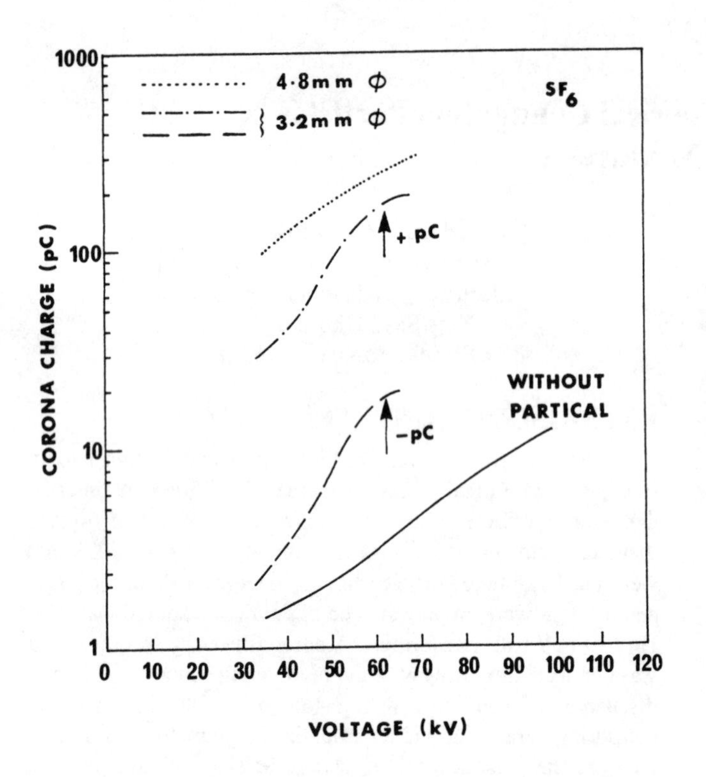

Figure 1 Corona characteristics for positive (+pC) and negative (-pC) half cycles for two spherical particles in SF_6.

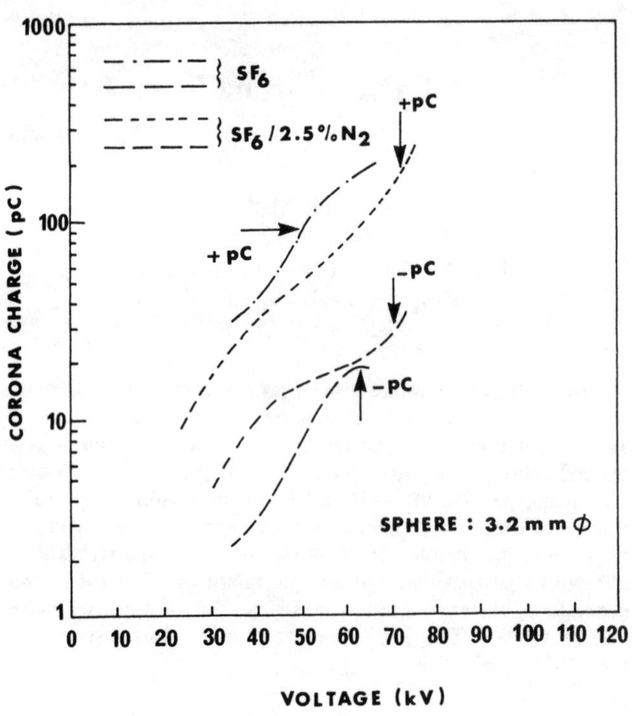

Figure 2 Corona characteristics for positive (+pC) and negative (-pC) half cycles for one spherical particles in $SF_6/2.5\%N_2$ mixtures.

is more than the corresponding value for smaller particle. At the corona onset, the 3.2 mm particle excited a circular motion on the bottom electrode. At V=45 kV, this particle started to bounce during its circular motion. At 65 kV, the bounce height increased significantly. 4.8 mm particle started bouncing on the bottom electrode at the onset voltage of V≈ 35 kV with pulses of 100 pC on the positive half cycle.

With $SF_6/2.5\%$ N_2 mixture and 3.2 mm diameter particle, corona starts with q = 9 pC on the positive half cycle at V= 25 kV as shown in Figure 2. In this mixture + pC values are lower whereas - pC values are higher than the corresponding values in SF_6 gas. Thus the addition of a small N_2 percentage to SF_6 modifies the positive as well as negative corona pulses. For this mixture, particle started bouncing at 35 kV at the bottom electrode. The bouncing height increased with voltage. However, for this mixture, the circular motion of particle on the bottom electrode which was noticed in SF_6 gas was not observed.

Figure 3 shows the q-v characteristics for SF_6 mixture containing 2.5% and 1.5% of perfluorocarbon vapour for a single spherical particle of 3.2 mm diameter. With these mixtures corona pulses were detected at 30 kV (1.5% $C_8F_{16}O$) and 25 kV (2.5% $C_8F_{16}O$) with positive corona charge levels of 1.5 and 11 pC respectively. Generally + pC values for these mixtures are lower and - pC values are higher than the corresponding values in SF_6 gas. Furthermore, the corona evels for the mixture containing 2.5% $C_8F_{16}O$ are higher than

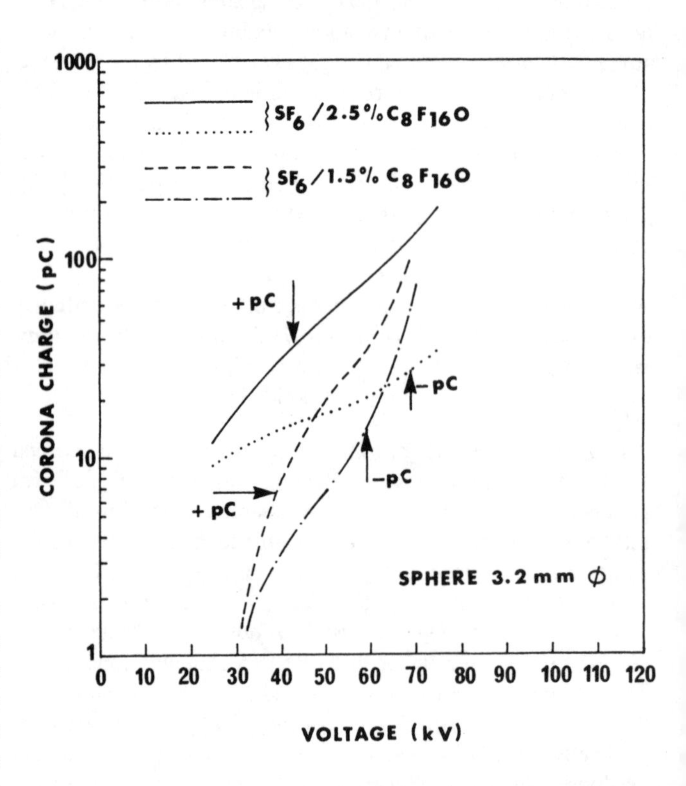

Figure 3 Corona characteristics for positive (+pC) and negative (-pC) half cycles for one spherical particles in $SF_6/2.5\%$ $C_8F_{16}O$ mixtures.

the corresponding values for 1.5% $C_8F_{16}O$ - SF_6 mixtures . In the SF_6 /2.5% $C_8F_{16}O$ mixture , the particle exhibited irregular movement at the bottom electrode at a threshold voltage of 30kV. The particle motion increased and had a bouncing pattern at a higher voltage . In SF_6 /1.5% $C_8F_{16}O$ mixture , the particle motion was noticed at a higher threshold voltage of 50 kV.

Figure 4 shows the q-v characteristics with a single 3.2 mm diameter particle present in SF_6 mixtures with a triethylamine $(C_2H_5)_3N$ and freon $C_2Cl_3F_3$ additives . Each of these additives had 2.5% content in the mixture . With triethylamine additive, the corona charge at the inception voltage of 25 kV was as low as 3 pC during the positive half cycle. Similar corona charge value is observed for the $C_2Cl_3F_3$ additive but at a higher onset voltage of 30 kV. For both of these mixtures , corona charge increases rapidly with voltage. Also the difference between - pC and + pC values are significantly smaller in these mixtures as compared to SF_6 gas . In SF_6 /$(C_2H_5)_3N$ mixture, the particle exhibited circular motion at the bottom electrode starting around 30kV which later on leads to bouncing motion at a higher voltage. In SF_6/ freon mixture only the bouncing motion was noticed .

DISCUSSION

When a conducting particle is in contact with one of the electrodes , it gets charge under applied stress . The particle motion is controlled by electrostatic and gravitational forces.

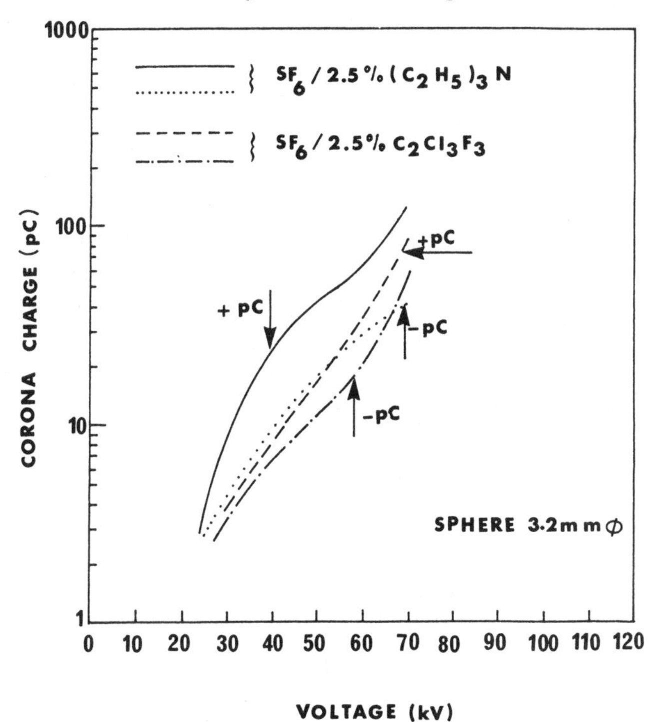

Figure 4 Corona characteristics for positive (+pC) and negative (-pC) half cycles for one spherical particles in SF_6 /2.5 % $(C_2H_5)_3N$ and SF_6/2.5%$C_2Cl_3F_3$ mixtures.

In addition , if corona discharges occur at the particle surface, corona wind becomes an additional force .Generally , under ac stress, the particle lifts off the bottom electrode and assumes a bouncing motion , reaching a height determined by the applied voltage , particle parameters and gas composition . As the applied voltage is increased , the bouncing height as well as corona charge levels increase till a breakdown is triggered when the particle approaches the opposite polarity electrode [9]. Small spark discharges also occur at the instant of a particle- electrode impact [10] .

The modes of the particle motion as well as the required stresses are different in different gases [1] . These results indicate that even a small percentage of an additive can make significant differences in a particle motion as well as a corona charge characteristics . Generally , the pulsative charge in the positive half cycle is higher than its negative half cycle counterpart . In SF_6 gas , this difference in the q values between the two half cycles is very large with + pC values in an order of a magnitude higher than the - pC values. Generally, breakdown takes place in a positive half cycle in SF_6 gas. Therefore, this large difference in q values may explain the sensitivity of SF_6 to contamination at a high gas pressures [1,9]. For the SF_6 /gas mixtures investigated in this work , the differences between + pC and - pC values are relatively small. All such mixtures suppress the positive corona pulse levels while increasing the negative corona pulse levels . Consequently , these additives reduce the sensitivity of SF_6 gas to microscopic field nonuniformities and increase the corona stabilized breakdown voltage levels. For the mixtures investigated , particle-initiated breakdown voltages are shown in Table 1 . This table further shows the maximum corona charge levels and the applied voltage levels when the particle started motion (mobility mode) and when the particle started lift-off and bouncing motion (bouncing mode).

Table 1 : Particle-initiated breakdown voltage and motion modes

SF_6/ gas mixtures	Mobility mode		Bouncing mode		Breakdown
	kV	pC	kV	pC	kV
SF_6	33	30	45	45	75
SF_6/2.5% N_2	25	9	35	25	92
SF_6/2.5% $C_8F_{16}O$	30	18	40	32	89
SF_6/2.5% $(C_2H_5)_3N$	30	9	40	25	75
SF_6/2.5% $C_2Cl_3F_3$	30	4	45	12.5	83

In SF_6 mixtures, the mobility mode starts with a low q value at a lower threshold voltage as compared to SF_6 gas. A similar feature is noticed in the bouncing mode. The higher breakdown voltage of SF_6/gas mixtures shows the improvement of the corona stabilization process .Hence , the additives investigated here exhibit improved particle - initiated breakdown voltages and are good candidates for large scale testing and possible industrial uses [4,6].

CONCLUSION

Small percentages of N_2, $C_8F_{16}O$, (C_2H_5)$_3$N and $C_2Cl_3F_3$ additives to SF_6 modify the pulsative corona from free conducting spherical particles . Generally, these additives reduce the corona charge levels , and increases the particle - initiated breakdown voltages , thereby making such mixtures less sensitive than SF_6 to the effects of microscopic field nonuniformities . These mixtures also show different particle motion characteristics . Extensive research effort is required to fully understand the dielectric behavior of these mixtures .

REFERENCES

1. Laghari,J.R. and A.H.Qureshi. "A Review of Particle-Contaminated Gas Breakdown". IEEE Trans EI Vol.16. No.5, October 1991, pp. 388-398.
2. Yunping,F., O. Farish, I.D. Chalmers and A. Aked. "Particle-Initiated Breakdown in Coaxial Systems in SF_6 and SF_6/Air Mixtures". Proc. Eight International Conference on gas discharges and their applictions, 1995, pp. 235-238.
3. Mufti,A.H., A.A. Arafa and N.H. Malik. "Corona Characteristics for Free Conducting Particles in Various SF_6-Gas Mixtures". IEEE Trans EI. Vol.1.No.3, June 1994, pp . 509-519.
4. Devins,J.C."Replacement Gases for SF_6".IEEE Trans EI.Vol.15.No.2, April 1980, pp. 81-86.
5. Mufti,A..H and A.A.O.Arafa. "N_2 Synergism with SF_6/Gas Mixtures in Nonuniform Field Breakdown". Conference Record of the 1994 IEEE International Symposium on Electrical Insulation, IEEE Publication 94 CH3445-4, pp. 512-514.
6. Christophorou, L.G. and R.J. Van Brunt. "SF_6/N_2 Mixtures Basic and HV Insulation Properties". IEEE Trans EI. Vol.2.No.5, October 1995, pp. 952-1003.
7. Malik, N.H., A.H. Qureshi and G.D. Theophilas."Static Field Breakdown of SF_6-N_2 Mixtures in Rod-Plane Gaps".IEEE Trans EI. Vol.14, 1979, pp. 61-69.
8. Mufti,A.H and A.A.O.Arafa. "Nonuniform Field Breakdown Characteristics of SF_6/Gas Mixtures Containing Small Amounts of N_2". Proc. Tenth International Conference on gas discharges and their applications, 1992, Vol.1., pp. 394-397.
9. Cookson, A.H., O. Farish and G.M.L. Sommerman. "Effect of Conducting Particles on a.c. Corona and Breakdown in Compressed SF_6". IEEE Trans PAS,Paper No71-TP-508-PWR, 1971, pp. 1329-1338.
10. Cookson, A.H and O.Farish. "Particle-Initiated Breakdown Between Coaxial Electrodes in Compressed SF_6". IEEE Trans PAS-92, 1973, pp. 871-877.

Conference Record of the 1996 IEEE International Symposium on Electrical Insulation, Montreal, Quebec, Canada, June 16-19, 1996

Negative Corona Discharge in Air with Small SF_6 Content

M. F. Fréchette, S. Kamel,[1] R. Bartnikas,[2] and R.Y. Larocque

Câbles et Isolants, Technologie et IREQ, Hydro-Québec, Varennes, Canada
[1]Lignes aériennes, Technologie et IREQ, Hydro-Québec, Varennes, Canada
[2]DTAE, Technologie et IREQ, Hydro-Québec, Varennes, Canada

Abstract: **The influence of small amounts of SF_6 in air was examined over the corona pulse regime, using a plane-point-plane geometry at a pressure of 800 Torr. With pure air, normal Trichel-type pulse behavior prevailed. Introduction of 1% by volume of SF_6 into the intervening air-gap space gave rise to drastic changes in the discharge-regime characteristics. Narrower and more intense pulses resulted, and the pulse amplitudes as well as the pulse intervals in the discharge were more broadly distributed: the corona pulse regime exhibited characteristics similar to those found in pure SF_6. Data were statistically treated, and analyzed in terms of the most recurrent values of the current pulse amplitude and associated pulse interval. The variation of these parameters, when taken one by one or in a combination, was found to be sufficient to detect and characterize the observed changes in comparison to the discharge regime in air.**

INTRODUCTION

Implementation of a large number of SF_6-insulated equipment in electrical networks has brought us to the point in time where research efforts are moving towards the assessment of the condition of the equipment, and in particular, of its insulation. In this context, carrying on continuous monitoring or punctual diagnostics on the insulation state necessitates the selection of a set of sensitive and revealing parameters, which would be selective towards the processes involved. To qualify, the established limit of these parameters or behavioral signatures should eventually result in the prediction of maintenance and/or identification of the nature of the developing fault.

The presence of partial discharge is recognized to be detrimental to the well being of SF_6-insulated HV equipment. This field of study covers a wide spectrum and shows much effervescence [1]. The present research program focuses rather on the parameters and their limit to be selected. In a recent work [2], the idea of transitional behavior in terms of a more integral aspect, *viz.* changes in discharge-regime characteristics, was contemplated. To this effect, Sellars *et al.* [3] had already documented the change in discharge regime due to a transition from streamer to leader mechanism. Our contended approach was later tested [4], involving the effect of water-vapor contamination on the corona discharge regimes in SF_6. Distinctive changes could be observed due to the presence of moisture.

The present study offers a different perspective. A reasonably tractable corona discharge, i.e. that observed in atmospheric air, is modified by the adjunct of a small quantity of SF_6. The changes in discharge regime that result are quantified in reference to the observed characteristics in air. The experiment aims at the determination of meaningful traits characterizing the changes. Additionally, it documents the partial-discharge behavior of a gas mixture with a low SF_6 content when alternatives for pure SF_6 are sought due to its potentially detrimental contribution to the global warming effect [5].

INVESTIGATION

Setup and Procedure

The properties of the partial-discharge generator [2,4,6] are mainly defined by the gas-insulated gap particularities. The electrode arrangement consists of two 90° Rogowski-profiled electrodes forming a parallel uniform-field assembly. The plane electrodes are 8.35 cm in diameter. At the center of the earthed electrode, a small rod with a diameter of 1 mm extends into the inter-plane space, the rod length being 10 mm. The end of the rod is beveled at 45° to form a conical tip. The gap length, defined as the space between the point and facing plane, can be varied. For the present, it was fixed at a constant value of 5 cm (± 0.005 cm). Figure 1 offers a schematic view of the experimental setup, including the associated Laplacian electric field distribution along the discharge axis, the calculated field nonuniformity factor being ~50. This gap geometry favors the axial development of the discharge.

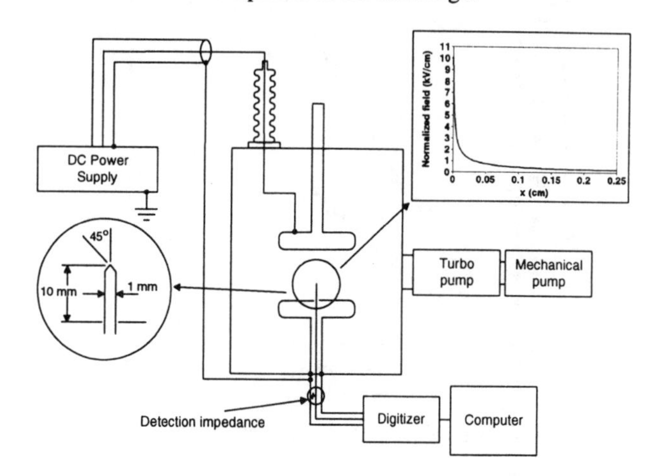

Figure 1. Schematic of the experimental setup; insets: the point electrode and field distribution on the discharge axis.

The gas pressure (p), fixed at 1.07 bar (800 Torr), was kept constant throughout. Positive DC voltage was applied to the plane opposite to the rod. Operated under these conditions, the generator produces a rather well-defined negative discharge of true-corona type. A digital electrometer coaxially connected to the point electrode was used to record the currents. The maximum value of the time-averaged current tolerated in the course of the experiment was in the range of 50 μA. Pulse activity was studied as a function of applied voltage over the current-voltage characteristic. Replacing the electrometer, the signal from the point electrode, the point being isolated from its associated plane, was measured across a 50 Ω detection impedance and fed across the 50 Ω input of a fast digitizer (time resolution of 0.5 ns). Sampling by sequence (of a duration of 0.5 ms) of a large number of pulses was used to establish statistical distributions.

The electrode arrangement is contained in a cylindrical stainless steel chamber. The system vacuum is ensured by the use of a turbomolecular pump backed mechanically. A vacuum of the order of 10^{-7} Torr was currently attained following discharge. The system was pumped down immediately after each experiment. Gas mixtures were prepared directly within the test chamber, by firstly introducing the low-pressure SF_6 followed by topping with air. After reaching equilibrium, representing an elapse of 30 min as verified using gas chromatography, the gas mixture containing 1% (± 0.1%) SF_6 in volume was subjected to discharge. The gap length was adjusted after filling the cell at the operating gas pressure. The SF_6 was CP grade, and the dry air, UHP grade.

Measurements

Much effort has been devoted to obtain reproducibility and stability of the data. A conditioning protocol was arrived at which allowed the I-V curves to be displayed as in Figure 2. The I-V characteristics feature current data taken over many experiments. That corresponding to air exhibits a slightly growing standard deviation occurring with the applied voltage; on average, it amounts to about 7%. The standard deviation associated with the mixture was observed to be rather constant (i.e. ± 0.4 μA). Overall, the nature of the currents in the mixture is more reproducible. Other major differences can be noticed. As compared to the situation with air, the corona threshold has slightly increased (by ~ 10%), and the slope of the I-V curve has been reduced indicating a lower mobility of the ionic carriers (see for instance [7]). Sequences of pulses, typically containing 100 events recorded in a one-shot, were acquired for conditions corresponding to several voltage-values taken on the I-V curves. During the acquisition, the corona current remained stable within ± 3%.

Typical recorded waveforms are displayed in Figure 3. In pure air (curve 1), the pulses have a rise time varying from 8 to 10 ns and a total duration between 300 to 400 ns. This shape is dominantly recurrent, although there exists a significant number of measurements, on average 40%, exhibiting a double-peak structure (see Fig. 3, inset). The SF_6 adjunct greatly affects the pulse characteristics. The maximum amplitude is typically enhanced by a factor ranging from 2 to 5, and the rise time, steepened. The rapid rate of change induced by the discharge produces a somewhat faulty behavior of the signal (curve 2, under- and overshoot; never exceeding 15% of the peak-value), indicating the limit of our experimental arrangement. Overall, the total duration was found to be reduced considerably, not exceeding 40 ns. In both cases, no continuous current component was detected.

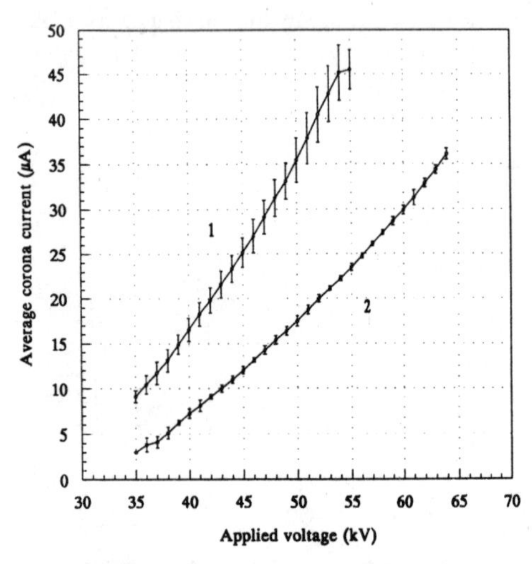

Figure 2. I-V Characteristics; 1- air, 2- air/SF_6, $[SF_6]$=1%.

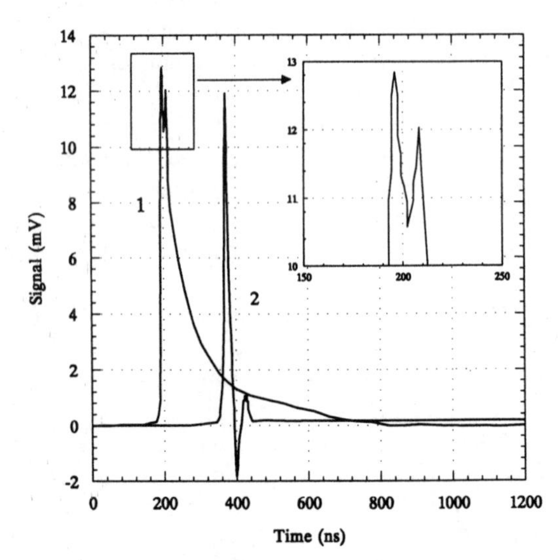

Figure 3. Typical pulse shape in the discharge; 1- air, 2- air/SF_6, $[SF_6]$=1%.; inset: observed variants for air. Scales are as indicated for curve 1; curve 2: multiplication factors for the signal amplitude and time scales are 0.2 and 4, respectively.

PULSE REGIME IN AIR

In order to define a framework for comparison, the discharge regime in air was investigated in detail. Through the performed statistical analysis of the experimental data, it was sought to firmly establish parameters or parametric patterns with a potential to discriminate and quantify eventual changes in the discharge regime. Overall, the observed regime exhibits the recognized traits generally associated with the present conditions [8,9], the so-called Trichel regime.

A total of 30 pulse sequences, totaling over 3000 discrete pulses, were recorded in various conditions. A single experiment consisted in recording the pulses at the fixed voltages of 35, 40 and 50 kV. The data taken at each voltage was analyzed in terms of the peak-values of the waveform, here defined as the amplitude (A), and the pulse interval (Δt). The statistical analysis showed that the measured amplitudes were distributed evenly and very narrowly (on average $\pm 3.5\%$) around a most recurrent or probable value (A_p) that remained independent (within 9%) of the applied voltage. Normal distributions were found to fit quite well these amplitude variations, as shown in Figure 4. Although the I-V curve is quite reproducible, the pulse characteristics may differ or evolve in time. The time associated with these changes would usually exceed that taken to run one experiment, e.g. 60 min. In any case, the same distribution in amplitude is then obtained, implying a generic signature even if A_p is altered. This point is exemplified in Figure 4, where two amplitude distributions corresponding to the same applied voltage, yet obtained in different experiments, are compared. The thus obtained distributions are characterized with small standard deviations ($\sigma \sim$ 4%), within which an average of 68% of the data is contained.

The inverse of the time interval between two consecutive pulses, $1/\Delta t$, was defined as the frequency (F). With the rise of the applied voltage, it was confirmed that the frequency increased. For instance, increasing the voltage from 35 kV to 50 kV resulted in the rise of the frequency by a factor of almost 4. Figure 5 summarizes the measurements for air, where A_p and F_p, the most probable amplitude and frequency, were arbitrarily normalized, and F_p gathered in terms of a fixed applied voltage. It is worth mentioning that when the most recurrent amplitude taken at fixed voltage was found to differ, the associated frequency had varied accordingly to keep the product A_p times F_p almost constant (e.g. within 12%).

EFFECT DUE TO ADJUNCT OF SF6

The 1-% SF_6 adjunct to air brought about major changes in the discharge regime. Overall, the pulse behavior was still Trichel-like, characterized by an occasional appearance of a high intensity pulse which would then momentarily interrupt the Trichel pulse sequence. This would translate into a longer time lapse before the occurrence of the next pulse. At higher voltage, it was also observed that sporadic increases and subsequent decreases in pulse amplitude may take place. The perturbed discharge regime exhibits some of the usual characteristics observed in pure SF_6 [10,11].

Pulse sequences were recorded for conditions corresponding to the I-V curve. Some voltage-values for which pulse acquisition was achieved were selected to match the currents observed during measurements in pure air. Over 10^4 pulses were recorded and statistically processed. The pulse amplitudes are now more broadly distributed as shown in Figure 6a.

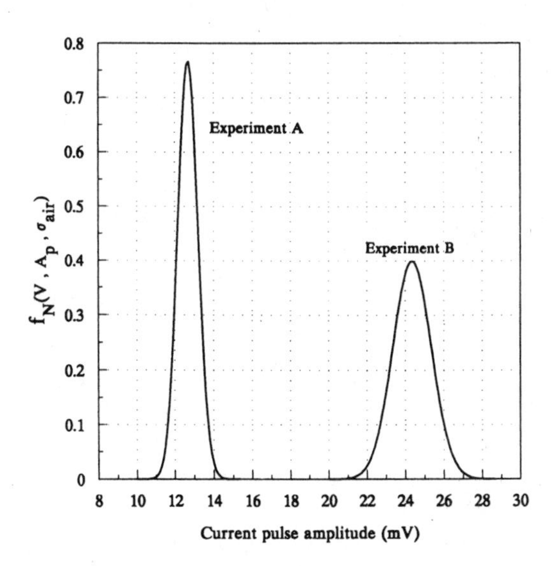

Figure 4. Amplitude distributions demonstrating a generic signature; data taken at V=40 kV, I=16 µA.

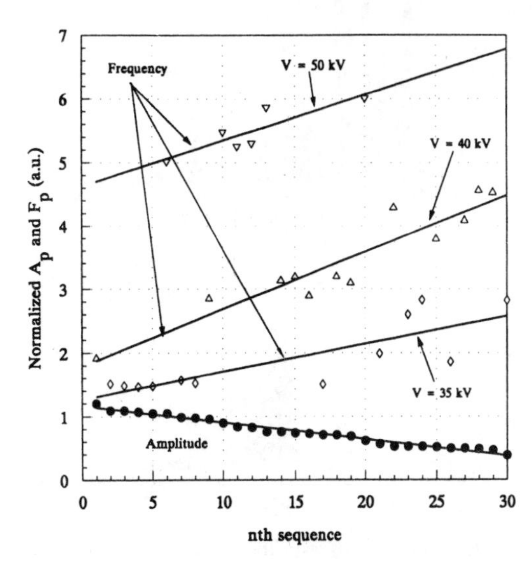

Figure 5. Normalized parameters A_p and F_p for air; all measured A_p confounded and displayed in a decreasing order.

The analysis of the amplitude data led to the conclusion that a normal distribution was not statistically representative, mainly because of the asymmetry displayed by the data towards higher values. The 50-% cumulative probability was used to obtain the most recurrent amplitude. To this value was ascribed σ_{air} obtained from the analysis of the air data, in order to evaluate the portion (S) of the data distributed around A_p and to establish the corresponding F_p. Analyses showed that on average not more than 40% of all pulses fell within $A_p \pm \sigma_{air}$. This result was taken as giving way to the detection of a broadly distributed and skewed function of the occurring amplitudes. To a less degree, similar statistical analysis using a variable σ-value could be helpful in quantifying the skewness of the distribution. Several pulse sequences were taken at a fixed voltage. In some, S was found to be rather constant, but for the majority of cases, S was quite dispersed. For the complete data set, S varied from 10 to 80%. Over the voltage span ranging from 40 to 63 kV, the large dispersion displayed by S, and the fact that the maximum ratio between the A_p obtained at fixed applied voltage is typically large (between 1.5 and 3), precluded any conclusion relative to a definite trend potentially exhibited by A_p as a function of applied voltage.

Figure 6b presents a typical histogram relative to the measured parameter Δt from which the frequency is determined. In all cases, analysis showed that the most recurrent amplitude was associated with the most probable frequency F_p. This, and the observation in air that a variation in A_p is coupled with a counter adjustment in F_p, allowed to consider the product A_p times F_p as a denominating parameter for comparison with the regime in air.

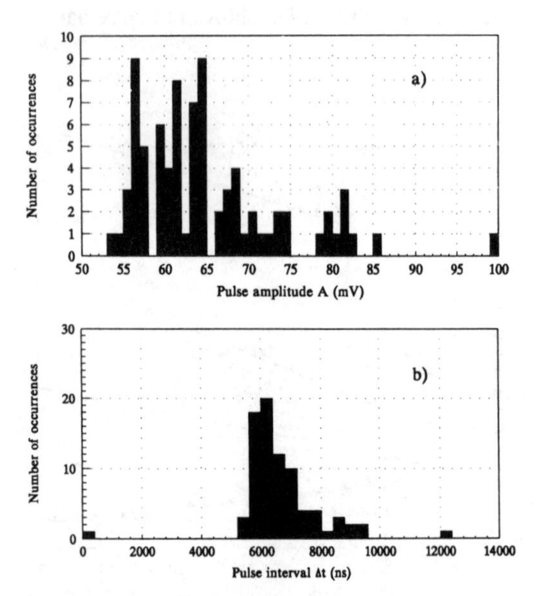

Figure 6. Histograms relative to the current pulse amplitude A and time interval Δt; air/ SF_6 , $[SF_6]$=1%, V= 40 kV, I= 6.5 μA.

Figure 7 presents the results of such an analysis. The product A_p F_p is displayed in terms of the corona current at which numerous pulse sequences were recorded. As in air, the product A_p F_p was found to be a quasi-invariant at a fixed voltage. In both air and the mixture, A_p F_p is observed to increase with the corona current. In air, the rate of increase is much less than that of the mixture, and it tends to diminish with increasing corona current. Since A_p is more or less constant, the rate of increase is due to the variation of F_p as a function of the applied voltage (see Fig. 5). The upper limit on F_p is determined by the capacity to clear the injected space charge. Once this limit is attained, the regime will transit to glow-type. For the mixture, the case is more complex, since A_p and F_p are both variables. Measurements taken at 63 kV (34 μA) indicate the simultaneous coexistence of two distinct regimes: low A_p values with high frequencies, and the reverse situation. Due to the limit on the clearing time, the A_p F_p function will divide as the corona current is increased leading to glow-type or corona streamers. For the mixture, the rate of rise is seen to be comparatively higher than for air and the variation with the corona current, rather linear. In the present context, the product A_p F_p represents a function inversely proportional to most of the injected charge into the gap. Once the typical pulse shape is known, the ratio of the slopes could serve in the comparison of the injected charge.

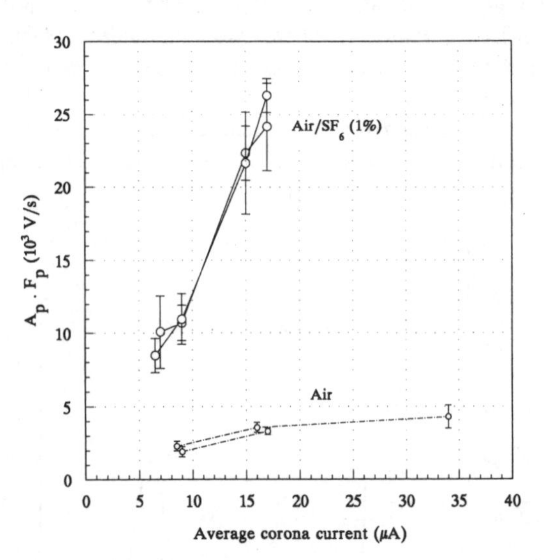

Figure 7. Comparison of the product A_p F_p for air and its mixture with SF_6.

CONCLUSION

Perturbation of the discharge regime in air due to a 1-% admixture of SF_6 was investigated. The observed Trichel regime in air was found to be strongly affected by the small SF_6 adjunct. Statistical analysis involving the current pulse characteristics, namely the most recurrent amplitudes (A_p) and frequencies (F_p), was carried out in an attempt to unravel global parameters capable of distinguishing the changes in the discharge regime. The results have demonstrated that the product A_pF_p constitutes an effective means of characterizing the differences in the Trichel pulse behavior in air and air containing small admixtures of SF_6 gas. Extending this work into mixtures containing a wide range of SF_6 admixtures could help in determining the applicability of the A_pF_p approach in measurements relating to SF_6-insulated high-voltage apparatus.

ACKNOWLEDGMENT

This work was sponsored in part by Vice-présidence Technologie et IREQ, Hydro-Québec. Thanks are due to M. Côté (student at Ecole Polytechnique) for performing some preliminary measurements.

REFERENCES

1. Proc. Int. Conf. on Partial Discharges, IEE Conf. Proc. No. 378, Canterbury, 1993.

2. Fréchette, M.F., M. Côté, N.G. Trinh and R.Y. Larocque. "Partial-Discharge Activity for a Negative DC Corona in Atmospheric SF_6". Int. Symp. on Electrical Insulation, Pittsburgh Penn., USA, June 5-8 1994.

3. Sellars, A.G., O. Farish and M.M. Peterson. "UHF Detection of Leader Discharges in SF_6". IEEE Trans. on Diel. and Electr. Insul., Vol. 2, No. 1, pp. 143-154, 1995.

4. Côté, M., M.F. Fréchette and R.Y. Larocque. "Negative DC Corona Characteristics in SF_6 under Moisture Contamination". Conf. on Electr. Insul. and Dielec. Phen., Arlington TX, USA, Oct. 23-26 1994.

5. Final Conf. Proc. Electrical Transmission and Distribution Systems, Sulfur Hexafluoride, and Atmospheric Effects of Greenhouse Gas Emissions Conference, U.S. Environmental Protection Agency, Washington DC, USA, August 9-10, 1995.

6. Fréchette, M.F., M. Côté and R.Y. Larocque. "Discharge Characteristic in an Asymmetrically-Perturbed SF_6 Gap". Conf. on Electr. Insul. and Dielec. Phen., Pocono Manor Penn., USA, Oct. 17-20, 1993.

7. Fréchette, M.F. and S. Kamel. "DC Corona Discharge in SF_6: Evidence of an Electronic Component for the Negative Polarity". Proc. IEEE Int. Symp. on Electrical Insulation, Boston, USA, 1988.

8. Loeb, L.B. *Electrical Coronas: Their Basic Physical Mechanisms*. University of California Press, Berkeley, 1965.

9. Lama, W.L. and C.F. Gallo. "Systematic Study of the Electrical Characteristics of the Trichel Current Pulses from Negative Needle-to-Plane Coronas". J. Appl. Phys., Vol. 45, No. 1, pp. 103-113, 1974.

10. Van Brunt, R.J. and D. Leep. "Characterization of Point-Plane Corona Pulses in SF_6". J. Appl. Phys., Vol. 52, No. 11, pp. 6588-6600, 1981.

11. Fréchette, M.F. and S. Kamel. "Negative DC Corona at Low Pressure and Short Gap Length". Proc. IX Int. Symp. on Gas Discharges and their Applications, Venice Italy, Sept. 19-23, 1988.

Conference Record of the 1996 IEEE International Symposium on Electrical Insulation, Montreal, Quebec, Canada, June 16-19, 1996

Avalanche Development in SF_6 Contaminated with Water Vapor

N. Bouchelouh,[1] M. F. Fréchette, S. Kamel,[2] and R.Y. Larocque

Câbles et Isolants, Technologie et IREQ, Hydro-Québec, Varennes, Canada
[1]Dept. of Engineering Physics, Ecole Polytechnique, Montréal, Canada
[2]Lignes aériennes, Technologie et IREQ, Hydro-Québec, Varennes, Canada

Abstract: **In the present study, the growth of an externally triggered avalanche in SF_6 was voluntarily modified using a substantial amount of moisture content. Results show that contamination did not increase the number of initially released photoelectrons but enhanced the avalanche growth rate. The avalanche current amplitude was found to increase with the contamination level, which permitted measurements at higher pressures compared to pure SF_6. However, even with 15% water contamination, it was possible to extend the measurement pressure limit only from about 30 to 50 Torr.**

INTRODUCTION

There is a definite interest in developing the ability to produce rapid space-charge growth in a gaseous dielectric. To this effect, several basic discharge principles were recently combined with a view to designing a *space-charge generator* [1-5] that would be useful, for instance, for the study of recovery mechanisms and space-charge relaxation in the occurrence of mixed voltages. This concept has already been demonstrated but the discharge was found to be totally quenched when initiated in atmospheric SF_6. The idea put forward to tentatively resolve the situation was to contaminate the SF_6 with water vapor. Depending on the conditions, the presence of water may increase the number of initiatory electrons and enhance avalanche growth [6-7] or reduce the breakdown strength [8]. This approach was implemented in an aim to extend the operating-pressure range of the space-charge generator.

The present paper reports on avalanche growth in SF_6 contaminated with water vapor. The gas pressure and the contamination level were both varied to make the avalanche growth detectable. Interpretation of the experimental data is supported by calculations obtained by numerical modeling.

INVESTIGATION

Measurements

The experimental setup is schematized in Figure 1. The plane-plane electrode arrangement is housed in a stainless-steel cylindrical cell to which turbomolecular and mechanical pumps are connected. The anode is Bruce-profiled with a 0.5-cm-diameter circular hole at its centre. The hole through which the beam passes was designed so that it produces depressed-field conditions towards the anode. The gap length is kept constant at 1 cm, resulting in a Laplacian distribution that has an almost flat-field value over approximately half the gap length followed by a strongly decreasing field near the anode. The UV pulse (FWHM 4 ns, ~2 mJ) produced by a Nitrogen laser (337 nm) enters the gap through the hole and falls onto the cathode for the release of initial photoelectrons. A converging lens with a 1-m focal length is used to control the laser beam focus. A fragmented-plane cathode [9] is used to improve the system time response. The signal from the central part is fed through a 50-Ω coaxial cable to a digital scope (single shot bandwidth: 400 MHz).

The applied voltage, gas pressure and moisture content were varied while the UV beam intensity was kept constant and unfocused. The experiments started by evacuating the test cell using the turbomolecular pump and filling it first with pure SF_6 at the desired pressure. Many avalanche current pulses are recorded at different voltage levels. The cell is pumped out again for a relatively short time (typically 30 to 60 min) to avoid a change in the experimental parameters, then distilled water is injected through a septum inlet using a syringe. When equilibrium is attained, pure SF_6 is added until the mixture pressure reaches the same level as that of the corresponding experiment with pure SF_6. The avalanche current waveforms are then recorded for several voltage levels and, if possible, at the same values as those for pure SF_6. The effect of water vapor contamination is deduced by comparing the avalanche growth to that observed in pure SF_6.

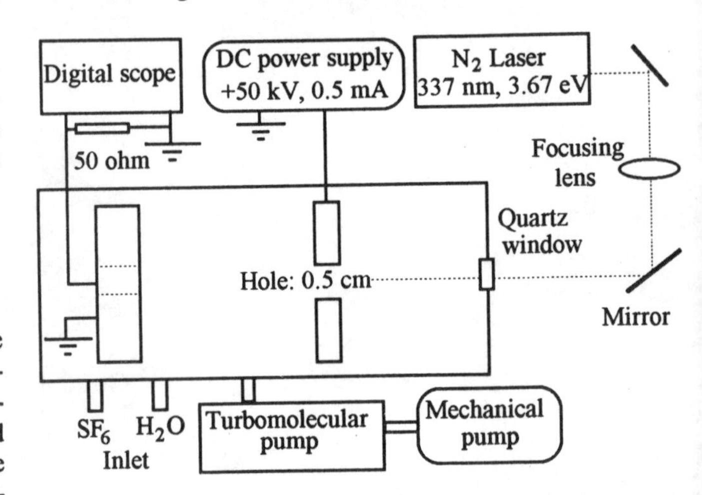

Figure 1. Schematic of the experimental setup.

Calculations

The theoretical model, introduced in an earlier publication [5], is based on the one-dimension continuity equations for electron, positive and negative ions, which include drift, ionization, attachment and secondary-electron production by photons (γ^p) and positive ions (γ^i) at the cathode. The equations are nonlinear and coupled by the one-dimension Poisson's equation. The space-charge field is computed using the disk charge method of Davies [10] in order to account for the finite radial extent of the discharge. Reported swarm data for SF_6 [11] and H_2O [12] shown in Figure 2 are used. For the SF_6/H_2O mixture, the swarm parameters are calculated using a weighted linear addition of the parameters of the separate components. The Flux Corrected Transport (FCT) algorithm of Boris and Book [13-15] is used to solve the continuity equations.

RESULTS AND DISCUSSION

Test results show that, up to a few thousand ppm_v of water in the SF_6, no observable effect on the avalanche current waveform could be detected. Only when the water contamination level increased to few percent did a clear difference emerge between the waveforms in pure SF_6 and those in the contaminated gas. Figure 3 shows typical experimental results for three pressure values, 5, 15 and 50 Torr, with a contamination level of ~15%; the E/p values are the equivalent uniform reduced-field values. It can be seen that both the amplitude and the rate of growth of the avalanche currents measured in the water-contaminated SF_6 are higher than the corresponding ones in pure SF_6. The figure also shows that when avalanche growth takes place, the current maximum in the contaminated gas is reached earlier than in pure SF_6 (for both 5 and 15 Torr).

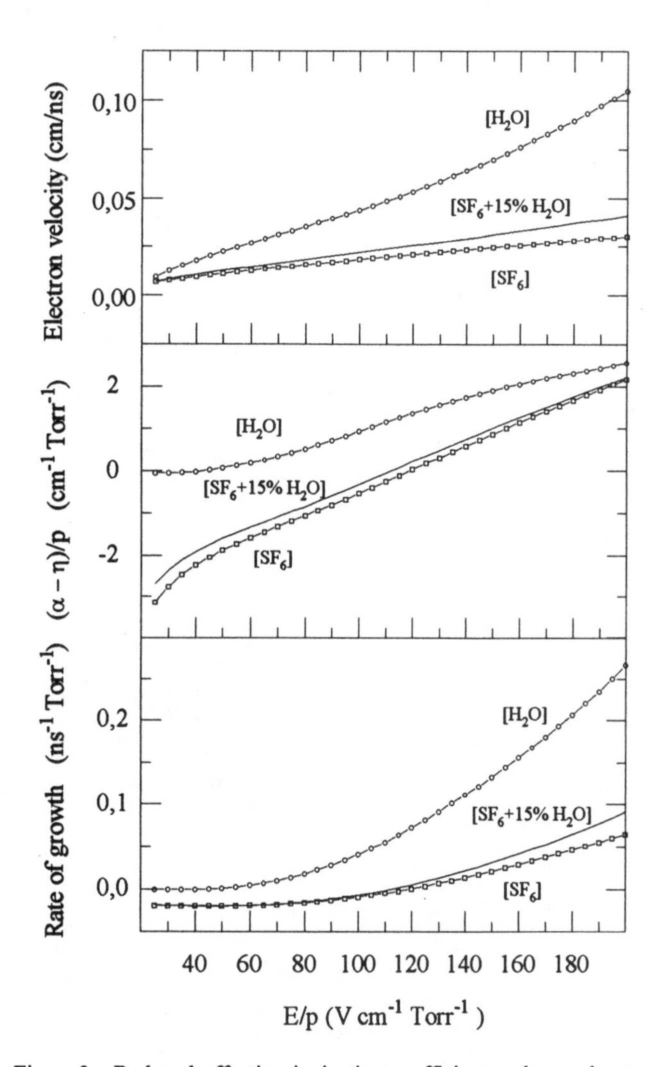

Figure 2. Reduced effective ionization coefficient and growth rate for SF_6 [11], H_2O vapor [12] and their mixture (calculated using a weighted linear sum).

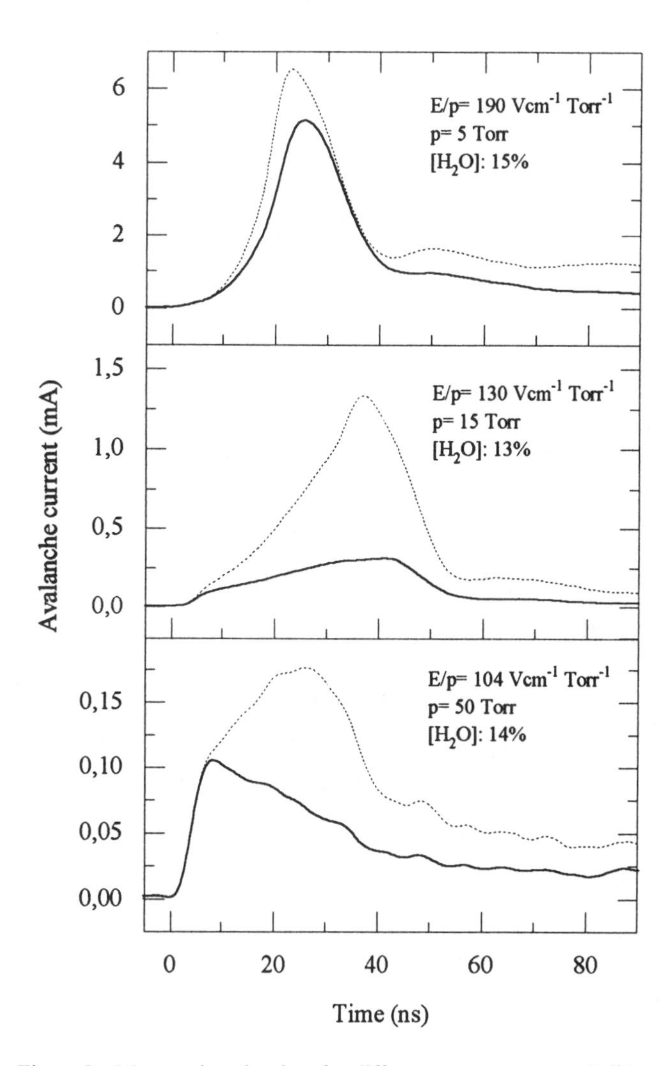

Figure 3. Measured avalanches for different gas pressures. Solid line: pure SF_6; dotted line: water-contaminated SF_6.

The effect of water contamination on the avalanche amplitude is more pronounced at 15 Torr than at 5 Torr. No avalanche growth is measured in pure SF_6 at 50 Torr and it was impossible to create positive avalanche growth by raising the applied voltage because of the breakdown limit of the gap. It is only with water contamination that positive growth is observed for the same conditions. At 100 Torr, however, even with this contamination level no avalanche growth was measured.

The fundamental swarm parameters in SF_6 and water vapor (Figure 2) may help to understand the above observations. The measured avalanche current is induced by movement of the charge carriers (mainly electrons) in the gap. If the ions contribution is neglected, the current becomes proportional to the product of the electron density and velocity and the Laplacian electric field. For a given reduced field (E/p), the effective ionization coefficient (α-η) and the electron velocity are greater in H_2O than in SF_6 (Figure 2). Therefore, water contamination is expected to increase the values of these two parameters compared to those in pure SF_6. The rise in the net ionization results in a higher electron density and, consequently, a larger avalanche current amplitude, while the rise in the electron velocity contributes to a further increase in the avalanche current amplitude. While the effective ionization and electron velocity have the same effect on the avalanche maximum amplitude, they have an opposite effect on the time to reach the maximum: in the present experimental setup, the electron cloud evolves in a decreasing Laplacian field and the maximum avalanche current is reached when the electrons arrive at a position in the gap where the ionization process is balanced by the decreasing electric field and velocity. Increasing the effective ionization (due to water contamination) pushes the balance point further from the cathode and consequently increases the time of the avalanche current peak compared to pure SF_6. On the other hand, when the velocity increases, the time to reach the balance point is shorter. From the experimental results of Figure 3, it can be deduced that the velocity effect is more dominant regarding the time to reach the avalanche maximum.

All the foregoing qualitative explanations are confirmed by numerical simulation. Figure 4 shows the results of calculations using the following parameters: the discharge radius, 0.2 cm; the total number of initial photo-released electrons (N_o), 1.25×10^5 electrons; $\gamma^p = 0.01$; $\gamma^i = 0.1$. At the pressure value of 5 Torr, the avalanche maxima are more distinct in time than in Figure 3. This is due to the fact that the initial time (t=0) for the measured avalanches, is adjusted rather arbitrarily because of the lack of the signature of the first photoreleased electrons caused by the noise and the measuring scale of the scope. Comparison of Figures 3 and 4 shows that, at the same pressure, the relative increase in the maximum current due to contamination is lower in measurements than in the calculations. This may be due to water affecting the cathode surface in such a way that the number of the initial photoelectrons is lower than in pure SF_6, while in the calculations the initial number (N_o) is taken to be the same. This hypothesis is sustained by the observation that the higher the water vapor partial pressure, the greater the discrepancy between the calculated and the measured relative amplitude ratio of the avalanches. As in the measurement, the effect of water contamination is more pronounced at 15 than at 5 Torr. For the case of 50 Torr and with E/p= 104 $Vcm^{-1}Torr^{-1}$, we were unable to reproduce the avalanche growth under water contamination using the data of Figure 2. A reduced field value of 114 $Vcm^{-1}Torr^{-1}$ had to be used to ensure a similar growth to that of the measurement. The discrepancy between these field values does not lie within the experimental error. One possible reason for this is the use of a weighted sum when calculating the swarm parameter of the mixture. On the other hand, due to the scarcity of the work on H_2O in the literature, the precision of the data given in Figure 2 was not ascertained.

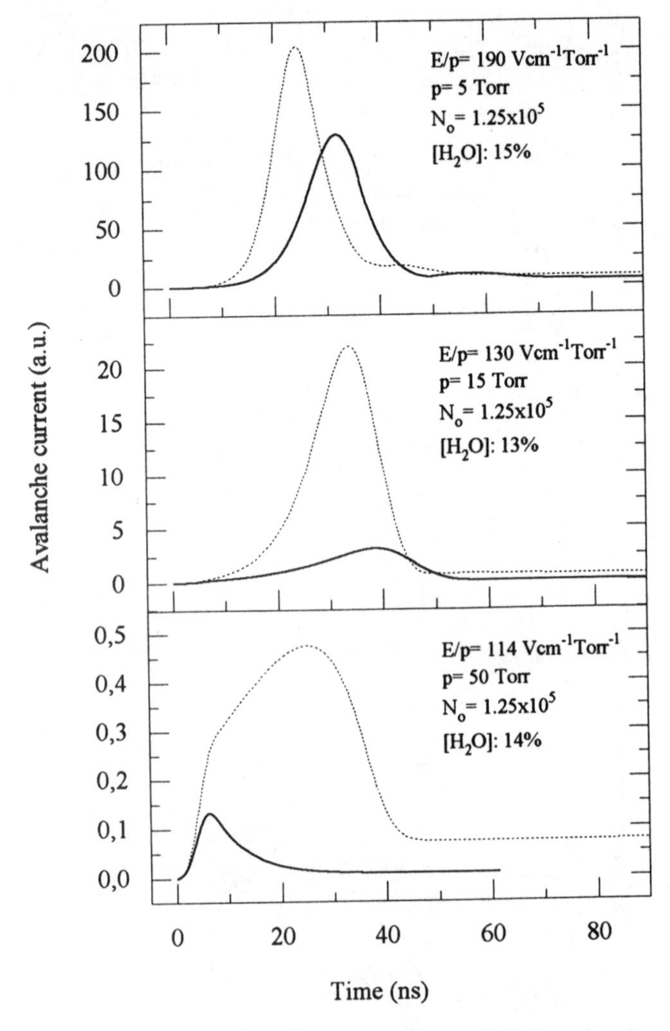

Figure 4. Calculated avalanche waveforms in SF_6 illustrating the effect of water vapor contamination. Solid line: pure SF_6; dotted line: contaminated SF_6.

Figure 5 shows a calculation-based comparison between the separate, then combined, effects of increasing the effective ionization and electron velocity due to water contamination on the shape of the avalanche current waveform as compared to the case of pure SF_6.

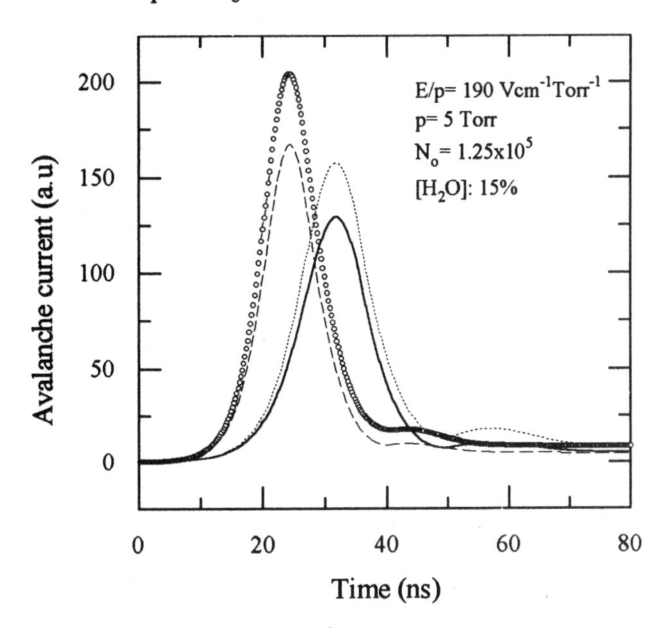

Figure 5. Effects of increasing the effective ionization and electron velocity due to the water contamination. Solid line: pure SF_6; dotted line: net ionization effect; dashed line: velocity effect; line with symbols: cumulative effect.

SUMMARY

The present contribution represents an attempt to overcome quenching of the avalanche growth in our experimental arrangement when operated at higher SF_6 pressure. Avalanche current waveforms were measured and numerically simulated in pure and water-contaminated SF_6. The effect of water vapor contamination was deduced by comparing the avalanche growth to that observed in pure SF_6. Results showed that water contamination increased the avalanche growth rate and current amplitude, which allowed measurement at higher pressures than in pure SF_6. However, even with 15% water contamination, it was only possible to extend the measurement pressure limit from about 30 to 50 Torr.

ACKNOWLEDGMENT

This work was sponsored in part by Direction Distribution, Hydro-Québec. N. Bouchelouh is a graduate student, beneficiary of a Hydro-Québec fellowship.

REFERENCES

1. Fréchette, M.F., N. Bouchelouh and R.Y. Larocque. "Time-Resolved Avalanches in SF_6 under Slightly Non-uniform Field". 46th Annual Gaseous Electronics Conf., Montréal, Can, Oct. 1993.

2. Fréchette, M.F., N. Bouchelouh and R.Y. Larocque. "Time-resolved Avalanches in SF_6 under Geometrically-perturbed Uniform Field". Seventh Int. Symp. on Gaseous Dielectrics, Knoxville, USA, April 24-28, 1994.

3. Fréchette, M.F., N. Bouchelouh and R.Y. Larocque. "Laser-Induced SF_6 Undervoltage-Breakdown Characteristics in a Geometrically Decreasing Field". Ninth Int. Symp. on High-Voltage Engineering, Graz Austria, Aug. 28- Sept. 1, 1995.

4. Bouchelouh, N., M.F. Fréchette and R.Y. Larocque. "Avalanche Study in Low-Pressure SF_6 under Geometrically Decreasing Field". Eleventh Int. Conf. on Gas Discharges and their Applications, Tokyo, Japan, Sept. 11-15, 1995.

5. Bouchelouh, N., M.F. Fréchette and R.Y. Larocque. "Large Avalanche Buildup in Low-Pressure SF_6 under Nonuniform-field Conditions". Conf. on Electrical Insulation and Dielectric Phenomena, Virginia Beach, Virginia, USA, Oct. 22-25, 1995.

6. Van Brunt, R. J. "Water Vapor-enhanced Electron-avalanche Growth in SF_6 for Nonuniform Fields". J. Appl. Phys., Vol. 59. No.7, 1986, pp. 2314.

7. Côté, M., M.F. Fréchette and R.Y. Larocque. "Negative DC Corona Characteristics in SF_6 under Moisture Contamination". Conf. on Electrical Insulation and Dielectric Phenomena, Arlington TX, Oct. 23-26, 1994.

8. Berger, G. and B. Senouci. "The Role of Impurities on the Deviation from Paschen's Law of SF_6". J. Phys. D: Appl. Phys., Vol. 19, 1986, pp. 2337-2342.

9. Verhaart, H.F.A. and P.C.T. Van der Laan. "Fast Current Measurements for Avalanche Studies". J. Appl. Phys., Vol. 53, 1982, pp. 1430.

10. Davies, A.J., C.J. Evans and F. Llewellyn-Jones. "Electrical Breakdown of Gases: The Spatio-Temporal Growth of Ionization in Fields Distorted by Space Charge" Proc. Roy. Soc. London Ser. A, Vol. 281, p. 164, 1964.

11. Aschwanden, Th. "Die Ermittlung physikalischer Entladungs-Parameter in Isoliergasen und Isoliergasgemischen mit einer verbesserten Swarm-Methode". Doctorale dissertation, ETH, Zurich, Switzerland, unpublished, 1985.

12. Ness, K. F. and R.E. Robson. "Transport Properties of Electrons in Water Vapor". Phys. Rev. A, Vol. 38. No.3, 1988, pp. 1446.

13. Boris, J.P. and D.L. Book. "Flux-corrected Transport I: SHASTA, a Fluid Transport Algorithm that Works". J. Comp. Phys., Vol. 11, 1973, pp. 38-69.

14. Book, D.L. and J.P. Boris. "Flux-Corrected Transport II: Generalization of the Method". J. Comp. Phys., Vol. 18, 1975, pp. 248.

15. Boris, J.P. and D.L. Book. "Flux-Corrected Transport III: Minimal Error FCT Algorithm", J. Comp. Phys., Vol. 20, 1976, pp. 397.

Conference Record of the 1996 IEEE International Symposium on Electrical Insulation, Montreal, Quebec, Canada, June 16-19, 1996

Simulation of Positive Corona Discharge in SF_6 Gas by Multichannel Model

Shozo Okabe and Tsunenari Hara
Dept. of Electrical Eng.
Tokai University
Kanagawa, Japan 259-12

Furuhata Takaaki*
Numazu Works,
Meidensha Corp.
Shizuoka, Japan 410

Abstract: **R. Morrow has simulated the positive corona discharge of SF_6 gas by using the Flux-Corrected Transport (FCT) method for one-dimensional continuity equations, in which a single channel model was applied and only the longitudinal drift of charged particles and diffusion of electrons was considered. In this work, we simulate the discharge phenomena of the SF_6 gas using a model that considers both of the longitudinal and the transverse diffusion of electrons and the bidirectional drift of charged particles. We adopt a multichannel model: A double channel model and a triple channel model. We expand the continuity equations by considering the conditions mentioned above and also take account of the photoionization process in the equations. Then we apply the FCT method to the multichannel model. The result of simulation based on the triple channel model shows that charged particles move actively within a certain radius. The simulation results using the multichannel model are a better approximation to the actually measured waveform of corona pulse current than those of the single channel model.**

INTRODUCTION

Computer simulation has been widely used as a method for precisely investigating the properties of gas discharges. R. Morrow has simulated the positive corona discharge of SF_6 gas [1] and the negative corona discharge of O_2 gas [2] by using the Flux-Corrected Transport (FCT) method [3] that is stable for numerically computing one-dimensional continuity equations. In his analysis, he set one cylindrical channel to a plane electrode, similar to the model of Kato et al. [4], and kept the density of charged particles at a given position constant. In the FCT method, the number of points that have to be computed over the discharge region is set finely where the density of charged particles peaks near the top of the streamer or the tip of the point electrode and the number of calculating points is set coarsely at other points. In this way, the method can reduce the number of calculation points and computing time, and calculate correctly the density of charged particles. We have also used this FCT method previously, and reported the results of the simulation of gas mixtures (SF_6, N_2, c-C_4F_8) [5]. However, a question arises as to whether the actual behavior of charged particles in gases is well described by the single channel model. Only the longitudinal drift of charged particles and diffusion of electrons is considered in the single channel model. Therefore, we have simulated the positive corona discharge of SF_6 gas using a model that takes account of both the longitudinal and the transverse diffusion of electrons and the bidirectional drift of charged particles.

In this paper, we describe the simulation results obtained under the above conditions. First, the simulated waveforms of the corona pulse current are considered, then the charged particle density distribution and electric field distribution are shown, and finally the differences between simulated results are described. The distance that the streamer progressed is also discussed.

ANALYSIS MODEL AND BASIC EQUATIONS

The electrode system that is the subject of this work is a point to plane electrode. We adopt a multichannel model (MCM) that is a model arranged coaxial-cylindrically forming a plurality of channels outside of a single channel. The single channel model (SCM) was shown by Morrow, in which one channel seemed to be formed between electrodes which consisted of a point electrode and a plane electrode. There are two models in the multichannel model: A double channel model (DCM) and a triple channel model (TCM) which consists of two and three channels, respectively.

The radius of the streamer of SF_6 is set as 0.01 cm [1], and in the other case, it is set as 0.00666 cm. Figure 1 shows the triple channel model. The radius of each channel is set as shown in Table 1. The volume ratio of each channel is also shown in Table 1. In this analysis, we expand the continuity equations by considering the conditions mentioned above and also take account of the photoionization process in the equations, because the process becomes important around the top of the positive corona which is

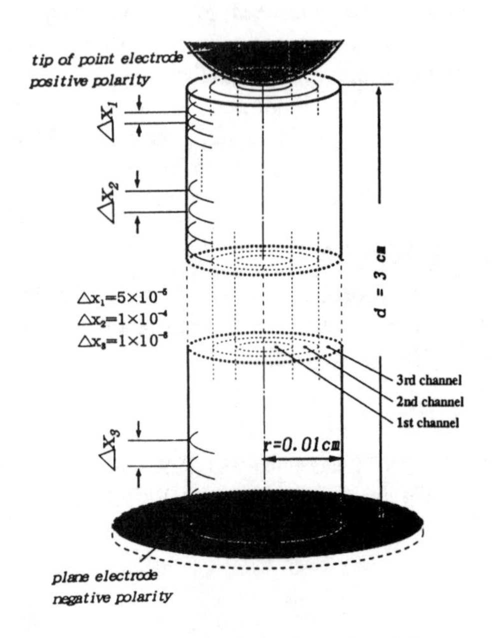

Figure 1. Triple channel model

Table 1 Radius and Volume Ratio in Each Channel

Radius r=0.01 [cm]		r1 [cm]	r2 [cm]	r3 [cm]
Single Channel		0.01	--------	--------
	Volume Ratio			
Double Channel	1:1	0.00707	0.01	--------
	1:3	0.005	0.01	--------
	1:8	0.00333	0.01	--------
Triple Channel	1:3:5	0.00333	0.00666	0.01

Radius r=0.00666 [cm]		r1 [cm]	r2 [cm]	r3 [cm]
Single Channel		0.00666	--------	--------
	Volume Ratio			
Double Channel	1:1	0.00471	0.00666	--------
	1:3	0.00333	0.00666	--------
	1:8	0.00222	0.00666	--------
Triple Channel	1:3:5	0.00222	0.00444	0.00666

developing. Then we apply the FCT method to the multichannel model. The continuity equations for electrons, positive ions, and negative ions are as follows:

$$\frac{\partial N_e}{\partial t} = S(x,r) + N_e\alpha_a|W_{ea}| - N_e\eta_a|W_{ea}| - N_eN_p\rho - \frac{\partial(N_eW_{ea})}{\partial x} + \frac{\partial}{\partial x}\left\{D_a\frac{\partial N_e}{\partial x}\right\} - \frac{\partial(N_eW_{er})}{\partial r} + \frac{\partial}{\partial r}\left\{Dr\frac{\partial N_e}{\partial r}\right\}, \quad (1)$$

$$\frac{\partial N_p}{\partial t} = S(x,r) + N_e\alpha_a|W_{ea}| - N_eN_p\rho - N_nN_p\rho - \frac{\partial(N_pW_{pa})}{\partial x} - \frac{\partial(N_pW_{pr})}{\partial r}, \quad (2)$$

$$\frac{\partial N_n}{\partial t} = N_e\eta_a|W_{ea}| - N_nN_p\rho - \frac{\partial(N_nW_{na})}{\partial x} - \frac{\partial(N_nW_{nr})}{\partial r}, \quad (3)$$

where t is time, x is the longitudinal distance from the anode, r is the transverse distance from the central axis, N_e, N_p, and N_n are the densities of the electron, positive ion, and negative ion, and W_{ea}, W_{pa}, and W_{na} and W_{er}, W_{pr}, and W_{nr} are drift velocities of electrons, positive ions, and negative ions due to longitudinal electric fields and transverse electric fields, respectively. The material functions α, η, ρ, and D_a & D_r are the coefficients of ionization, electron attachment, recombination, and the coefficients of longitudinal electron diffusion and transverse electron diffusion, respectively [1]. The term S(x, r) is the charged particles density per unit second at discharge point (x, r) due to photoionization. The term S(x, r) is given by as follows [1]:

$$S(x,r) = \sum_i \beta_i \int_0^d \{N_e(x',r)|W_{ea}(x',r)| \cdot \alpha_a(x',r)G|x-x'| \cdot \exp(-\mu_i|x-x'|)\}dx', \quad (4)$$

where μ_i, α_i^*, and β_i are the coefficients of absorption, excitation, and secondary ionization for i-th excited state, and d is the gap spacing. x' is the distance from the anode when the particles densities or the various coefficients at x' affect to the point x. G is the solid angle subtended at x' by the disc of charged at x defined as follows:

$$G = 0.5\left(\frac{r_1}{\sqrt{r_2+(x-x')^2}} - \frac{r_3}{\sqrt{r_2+(x-x')^2}}\right), \quad (5)$$

where r is the discharge radius.

An electric field strength can be calculated from a superposition of the field strength due to space charges on the applied field strength when space charges are formed by discharging. The applied field strength is computed by a charge stimulated method, and an electric image charge method is used for calculating the space charge fields. When the space charge field is computed, image charges are assumed inside of the plane electrode, as the surface of the electrode is a boundary. Calculation of the current is carried out by using Sato's expression [6]. As an example, the expression for the current of the 1st channel is as follows:

$$I_{1st} = \frac{\pi r_1^2 e}{V_A}\int_0^d \left[N_p|W_{pa}(x,0)| - |N_n|W_{na}(x,0)| - N_e|W_{ea}(x,0)| + D_a(x,0)\frac{\partial N_e}{\partial x}\right] \cdot E_L(x,0)dx, \quad (6)$$

where, r_1 is the radius of the 1st channel, V_A is applied voltage, e is electric charge of electron, and E_L is electric field strength right after voltage applied.

ANALYSIS PARAMETERS

For the computation, three kinds of mesh were used to indicate the calculating points. Fine, coarse, and medium fine meshes were used at the regions of the anode and the streamer head, around the cathode, and at the other regions. The total number of calculating points on an axis connecting the center of the anode and cathode was 400 for simplicity. We used a point-plane electrode system. The diameter of the point electrode was 1 cm, the point angle 30°, and its edge curvature radius 0.05 cm, and the opposite electrode was a disc 8.6 cm in diameter. The gap spacing was set at 3 cm in order to neglect electron emissions from the cathode. The applied voltage was a rectangular wave of amplitude 30 kV, the gas pressure was 101.3 kPa, and the temperature was 293 K. Initial seed electrons were set as Gaussian distribution at a point of 0.01 cm in the 1st channel apart from the tip of the anode, where the number of electrons is approximately 500.

The data required for computing Eqs. (1) to (6) was as follows: Electron swarm parameters α, η, W_e and D are given as the functions of the reduced electric field E/p_{20} that were calculated from a Boltzmann Equation Analysis by the SST method [7]. Also we used the following values for ρ, W_p, and W_n referring to the data reported by Morrow [1]. That is, $\rho = 1.0 \times 10^{-6}$ cm^3s^{-1} at 133.3 Pa and 293 K, and the values of W_p and W_n were calculated from the functional equation of the mobility for the electric field strength. The computation time was set as 3 ns under the conditions stated above. The computations were carried out by a supercomputer (NEC SX-3).

RESULTS AND DISCUSSION

Figure 2 shows waveforms of the positive corona pulse current in the single channel model (SCM) and the double channel model (DCM) with volume ratios of 1:1, 1:3, and 1:8 at the discharge

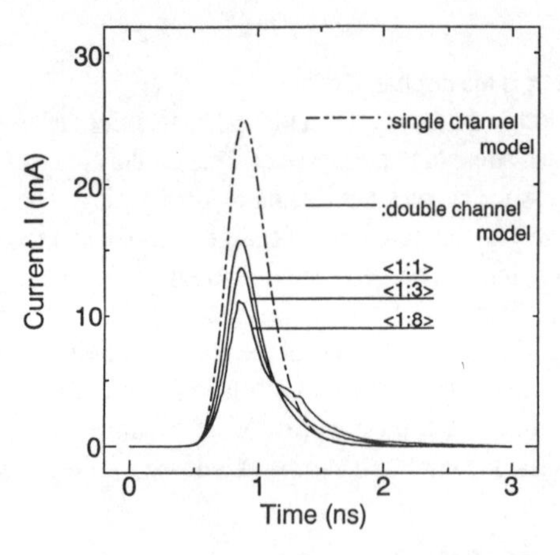

Figure 2. Waveforms of the corona pulse current in single channel model and double channel model

radius of 0.05 cm. The peak values of the corona pulse current for DCM are 63~46% of the value for SCM and the pulse width of the corona pulse current in DCM becomes about 80% for SCM. The fall times of the corona pulse current in DCM become shorter than those in the SCM. In DCM, the peak value of the corona pulse current decreases as the volume (radius) of the 1st channel decreases.

Figure 3 (a), (b) show the distribution of net charge per unit length in each double channel of discharge radius of 0.05 cm. The net charge is defined as the value of the difference between the number of positive ions and the number of negative ions and electrons. The discharge region is occupied by positive charge because of the positive polarity discharge, and many positive space charges exist around the top of the streamer in each channel at each time. The density of net charge in the 1st channel is much higher than that of the 2nd channel. There are many electrons around the streamer head because of the high electric field strength near the top of it, and thus electron collision ionization and photoionization occur actively in the front region of the streamer head.

Figure 4 (a) shows electric fields of an axial direction at the center of the 1st channel in each double channel model at discharge radius of 0.05 cm. The electric field intensity at the streamer head is higher than the electric field strength at the other positions due to the positive space charge existing around the streamer head region. At t=2.0ns, the field intensity around the streamer head in volume ratio 1:8 is still high. As a result, the corona is still developing due to active collision ionization and photoionization in this region.

Figure 4 (b) shows electric fields of a radial direction at each radius of the 1st channel. The electric field strength of the radial direction is much lower than that of the axial direction. Positive ions which have a low mobility seem to show less movement toward the 2nd channel from the 1st channel. However, some electrons do move to the 2nd channel due to the electric field in the inverse direction just behind the streamer head.

Figure 5 shows the waveform of the corona pulse current in TCM at the volume ratio of 1:3:5, and shows the waveforms of each current component in each channel. The peak value of the current of the 2nd channel is about half of the 1st channel. The current component of the 3rd channel is much less than the others. In this case, it is understood that the movement of the charged particles and electron diffusion are more active inside a certain radius for SF_6 gas.

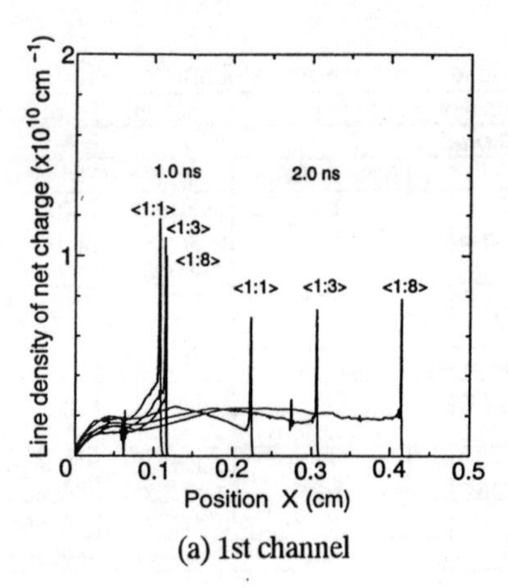

(a) 1st channel

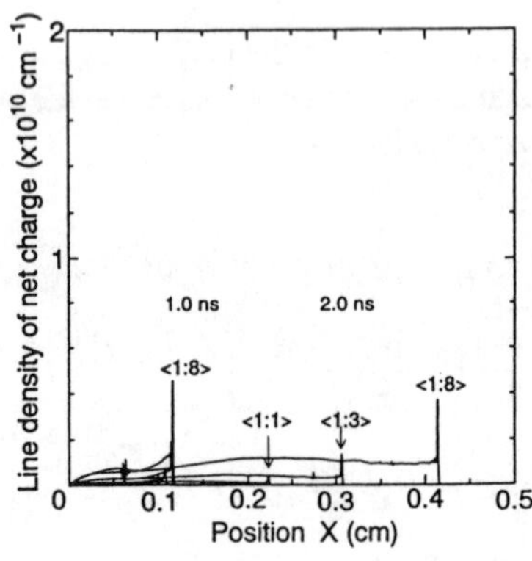

(b) 2nd channel

Figure 3. Distribution of line density of net charge

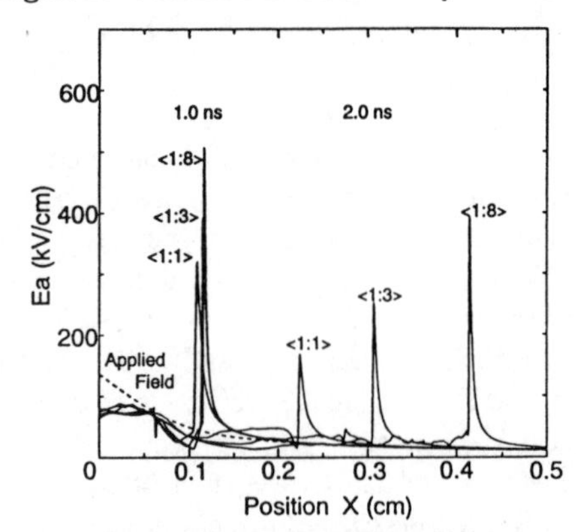

(a) electric fields of axial direction at the center of first channel in each double channel

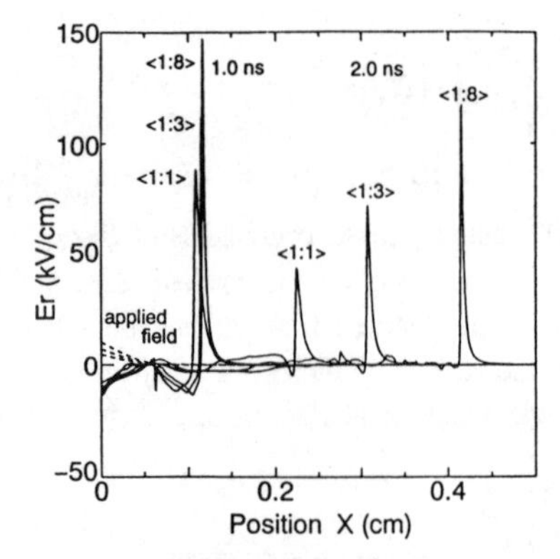

(b) electric fields of radial direction at each radius of 1st channel

Figure 4. Electric fields

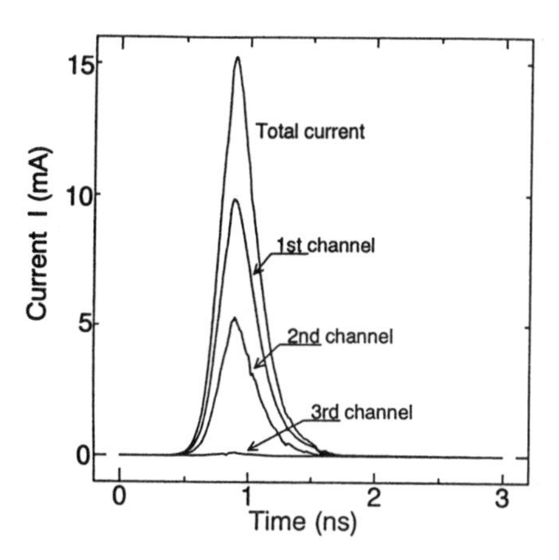

Figure 5. Corona pulse current at each channel in a triple channel model (r=0.01cm)

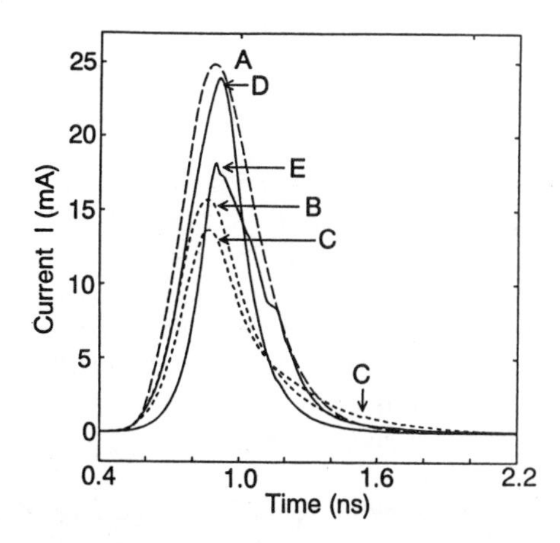

Type I [r=0.01cm]
A : SCM
B : DCM 1:1
C : DCM 1:3

Type II [r=0.00666cm]
D : DCM 1:1
E : DCM 1:3

Figure 6. Waveforms of corona current in single channel model and double channel model

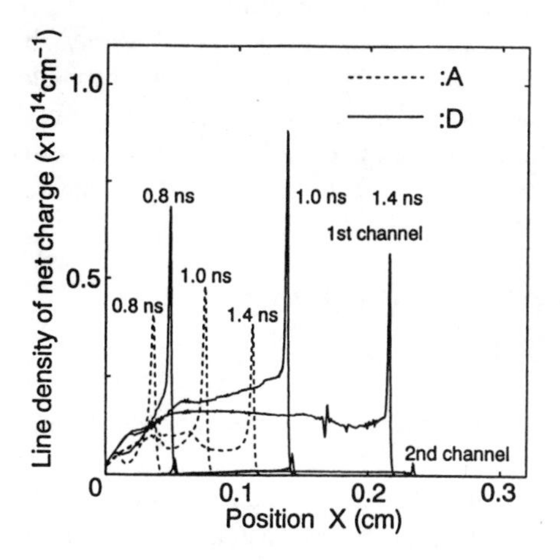

A : SCM (r=0.01cm)
D : DCM (r=0.00666cm) 1:1

Figure 7. Distribution of line density of net charge

Figure 6 shows the waveforms of the corona pulse current as the radius of each channel is changed. Type I in Fig. 6 indicates that the radius is 0.01 cm, and the radius of type II is 0.00666 cm. In SMC, the peak value of the corona pulse current decreases when the radius of the channel decreases. However, the peak value of the case of 1:1 in DCM (r=0.0066 cm) is almost the same as the peak value of SCM (r=0.01 cm). Figure 7 shows the distribution of line density of the net charge in SCM and DCM. The density of the 1st channel of DCM is higher than the density of SCM. On the other hand, the density of the 2nd channel of DCM is lower than the SCM. The line density is affected by the radius of the channel. The smaller the radius of the 1st channel, the higher the density of the net charge and the longer the distance of corona development.
Thus, the waveform of the corona pulse current in MCM is different from that of SCM. An actually measured waveform is complex by itself, but however, the waveform computed from SCM seems to be simple and we can easily show changes of waveforms of the corona pulse current using MCM.

CONCLUSIONS

The electric field intensity of radial direction is considerably lower than that of axial direction, however, a few electrons move toward the 2nd channel due to weak radial electric fields of inverse direction and some electrons diffuse with the passage of time. Thus, discharges in the 2nd channel also start for a short while. Electrons are multiplied by collision ionization around the streamer head in the 2nd channel and then these electrons move toward the streamer head in the 1st channel by strong radial electric fields near the streamer head of the 1st channel. Therefore, ionization actions occur actively in the 1st channel whose volume is small, and as a result, the positive corona streamer develops. The simulation results by a multichannel model are a better approximation to the actually measured waveform of the corona pulse current than that of a single channel model.

REFERENCES

1. Morrow,R. "Theory of Positive Onset Corona Pulses in SF_6".IEEE Trans EI. Vol. 26. No.3 , June 1991,pp.97-106.
2. Morrow,R. "A Survey of the Electron and Ion Transport Properties of SF_6". IEEE Trans P.S. Vol. PS-14, No .3, June 1986, pp. 234-239.
3. Boris,J.P and Book,D.L. "Flux-Corrected Transport". J. Comput.Phys. Vol. 11, 1973, pp.38-69.
4. Kato,S. "Computer Simulation of Positive Streamer in Air Between Rod-Plane Electrodes Gap by Two Region Model". IEEJ A-107, No.4, April 1987, pp.169-176
5. Okabe,S and Furuhata,T et al. "A Simulation of Positive Corona Pulse Current in the Gas Mixtures". IEEE. Int. Symp. EI, June 1994, pp.508-511.
6. Sato,N. "Discharge Current Induced by the Motion of Charged Particles". J.Phys.D. Vol. 13, 1980, pp.L3-L6.
7. Okabe,S and Ienaga,E et al. "A Simulation of the Positive Corona Pulse Current in SF6". Proceedings of the School of Engineering Tokai University. Vol. 33, No. 1, 1994, pp.1-7.

Conference Record of the 1996 IEEE International Symposium on Electrical Insulation, Montreal, Quebec, Canada, June 16-19, 1996

ABOUT HIGH-VOLTAGE INSULATING OF 750kV CIRCUIT-BREAKER

Petre Tusaliu
University of Craiova
Faculty of Electrotechnics
Craiova 1100, Romania

Marian Ciontu
University of Craiova
Faculty of Electrotechnics
Craiova 1100, Romania

Virginia Tusaliu
National College of Craiova
Romania

ABSTRACT

The study analyses the non-uniform repartition of high voltage on the blow-out chambers of an 750kV circuit breaker. We achieve the 750kV circuit breaker model, on which base we establish the repartition of an large spectrum of impulse waves.

Keeping account of the behaviour of the isolation at over-voltage waves and of self-capacity and stray capacity of the apparatus construction, we elaborate the circuit breaker model. On this base is analysed the repartition of impulse waves of voltage, which are altering in a large spectrum of frequency, from the which moulds the thunder, to the wave which moulds the commutation over-voltage.

The mathematic calculus is maiden with the aid of "Theory of Systems", using the functions "in-state-out" of the model. The study present also an adequate program on computer, which praises the non-uniform repartition of the electric stresses on the blow-out chambers of the circuit breakers: with many break places and the influence of this repartition of the slope of voltage impulse wave.

Certainly that at once with the non-uniform repartition of the circuit breaker, the parameters of electric field are changing, the amplitudes out it's variations presenting interest for the whole ensemble circuit-breaker-network environment.

In the study there are calculations concerning the bettering of this situation, finding the best solution concerning the connecting in parallel on the blow-out-chambers of the circuit breaker, with resistor or with condensers.

At the same time, the study present proposal and recommendations of construction, function, exploitation, in the purposals to decrease the stresses resulted from the perturbation created by the non-uniform repartition of the over-voltage on the blow-out having in view the rise of security in function of the ensemble circuit breaker-network.

KEYWORDS: High-voltages insulating, systems, circuit-breakers.

1. INTRODUCTION

The aspects referring to electromagnetic disturbances produced within the framework of National Power Systems, receives a special importance, many researches being recently preoccupied about this [1..9]. These have specific features at high and very high voltages [7..9].

Potentially, there is a large number of possible sources of disturbances in electrical networks of high voltages and very high voltages like: the fast transitory phenomena as a result of the switching in normal and damage conditions; the effect of short-circuits and unbalanced loading current ; the electromagnetic impulses of nuclear origin; the un-uniform repartitions of overvoltages on the insulations parts of high voltages .

Thus, at high and very high voltages, electrical requirements repartitions is strongly influenced by the existence of stray capacities and by the ununiformity of electrical fields [1..5].

As the normal voltages of the transport of electric power are in a highly increasing[3], for switching apparatus the dimensioning and reliable operation of insulation, receives a major importance[9].

Thus, the behaviour of the insulation of circuit breaker in transitory conditions generated either by switching processes or by electrical atmospheric discharge , is determinative in the ensurance of high reliability within the frame of Electromagnetic System. To realise a deeper analyse, it is necessary to make a convenient model as well as possible[2].

2. DETERMINATIONS OF REQUIREMENTS

For determinations of requirements produced by un-uniform repartition of high voltage on the blow-out chambers of the 750 kV circuit breaker was elaborated the equivalent electrical scheme of insulation depending on number of blow-out chambers and number of support insulatings [8..10]. Thus, in fig.1 and fig.2 were presented the equivalent schemes for circuit breaker with compressed air and respectively circuit breaker with oil, in which:

C_1 - represents the electric capacity of a extinction chamber (the circuit breaker with opened contacts);

C_2 - represent the equivalent capacity of a insulating from the support insulating.

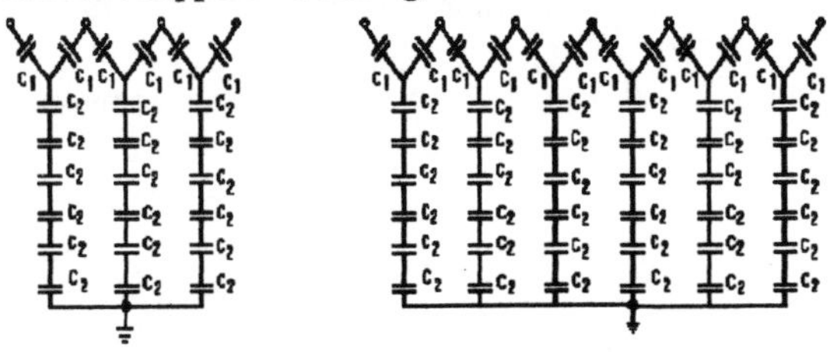

Fig.1 The insulation model of pneumatic circuite breaker

Fig. 2 The insulation model of circuite breaker with oil

For the standardising the highvoltages repartition on the inserted extinction chambers of the circuit breaker, two solution are recommended.[1..4]:

- the connection, in parallel, on the extinction chambers of some resistors (R), solution whose utilisation implies certain technological difficulties[9];
- the connection, in parallel, on the extinction chambers of some condenser h(C_d) which is frequently used in the construction and exploiting of electric apparatus[1,8].

Based on the scheme from fig. 1 and fig.2, there were elaborated the equivalent electric schemes of these sorts of 750 kV circuit breaker (fig.3 and fig.4) as well equivalents schemes with the capacities of the chambers shunting (fig.5 and fig.6):

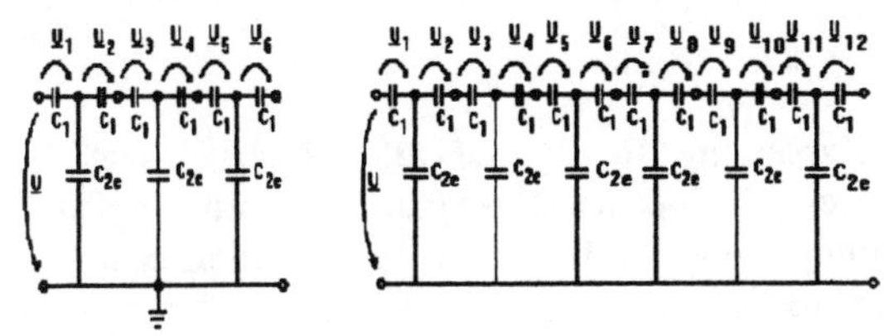

Fig. 3 The electric scheme of 750 kV pneumatic circuite breaker

Fig. 4 The electric scheme of 750 kV circuite breaker with oil

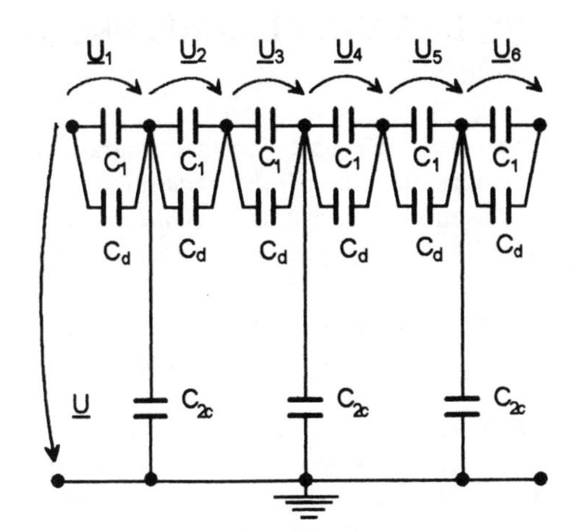

Fig5. Equivalent electrical scheme of 750kV pneumatic circuite breaker, with shunting capacities Cd

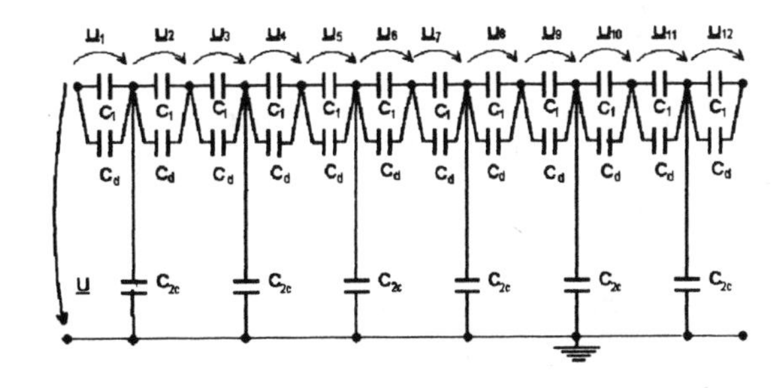

Fig.6 The equivalent electric scheme of the 750 kV circuit breaker with oil, with shunting capacities C_d

Based on the equivalent schemes, the functions of transfer were calculated for circuit breaker with compressed air (1) and for circuit breaker with oil (2).

a) 6 chambers

$$H_1(p)=\frac{U_1(p)}{U(p)}=\frac{Y_1^3+9Y_1^2Y_2+12Y_1Y_2^2+4Y_2^3}{(3Y_1+2Y_2)(2Y_1^2+5Y_1Y_2+2Y_2^2)}$$

$$H_2(p)=\frac{U_2(p)}{U(p)}=\frac{Y_1(Y_1^2+4Y_1Y_2+2Y_2^2)}{N_1(p)}=H_3(p)=\frac{U_3(p)}{U(p)}$$

$$H_4(p)=\frac{U_4(p)}{U(p)}=\frac{Y_1^2(Y_1+Y_2)}{N_1(p)}=H_5(p)=\frac{U_5(p)}{U(p)}$$

$$H_6(p)=\frac{U_6(p)}{U(p)}=\frac{Y_1^3}{N_1(p)} \qquad (1)$$

b) 12 chambers

$$H_1(p)=\frac{U_1(p)}{U(p)}=\frac{Y_1^6+27Y_1^5Y_2+420Y_1^4Y_2^3+896Y_1^3Y_2^3+864Y_1^2Y_2^4+384Y_1Y_2^5+64Y_2^6}{23Y_1^6+292Y_1^5Y_2+1036Y_1^4Y_2^2+1600Y_1^3Y_2^3+1216Y_1^2Y_2^4+448Y_1Y_2^5+64Y_2^6}$$

$$H_2(p)=\frac{U_2(p)}{U(p)}=\frac{Y_1(32Y_2^5+160Y_1Y_2^4+280Y_1^2Y_2^3+200Y_1^3Y_2^2+50Y_1^4Y_2+2Y_1^5)}{N_2(p)}=$$
$$=H_3(p)=\frac{U_3(p)}{U(p)} \qquad (2)$$

$$H_4(p)=\frac{U_4(p)}{U(p)}=\frac{2Y_1^2\left(Y_1^4+16Y_1^3Y_2+40Y_1^2Y_2^2+32Y_1Y_2^3+8Y_2^4\right)}{N_2(p)}=H_5(p)=\frac{U_5(p)}{U(p)}$$

$$H_6(p)=\frac{U_6(p)}{U(p)}=\frac{2Y_1^3\left(2Y_1^3+9Y_1^2Y_2+12Y_1Y_2^2+4Y_2^3\right)}{N_2(p)}=H_7(p)=\frac{U_7(p)}{U(p)}$$

$$H_8(p)==\frac{U_8(p)}{U(p)}=\frac{2Y_1^4\left(Y_1^2+4Y_1Y_2+2Y_2^2\right)}{N_2(p)}=H_9(p)=\frac{U_9(p)}{U(p)}$$

$$H_{10}(p)=\frac{U_{10}(p)}{U(p)}=\frac{2Y_1^5(Y_1+Y_2)}{N_2(p)}=H_{11}(p)=\frac{U_{11}(p)}{U(p)}$$

$$H_{12}(p)=\frac{U_{12}(p)}{U(p)}=\frac{2Y_1^6}{N_2(p)}$$

in which : $Y_1(p)=p(C_1+C_2);\ Y_2(p)=p\frac{C_2}{6}$

In the case of resistors use for the standardisation of overvoltages (fig.7 and fig.8) the following functions of transfer were obtained, for the circuit breaker with compressed air (3) and respectively for the circuit breaker with oil (4) [4,8,9].

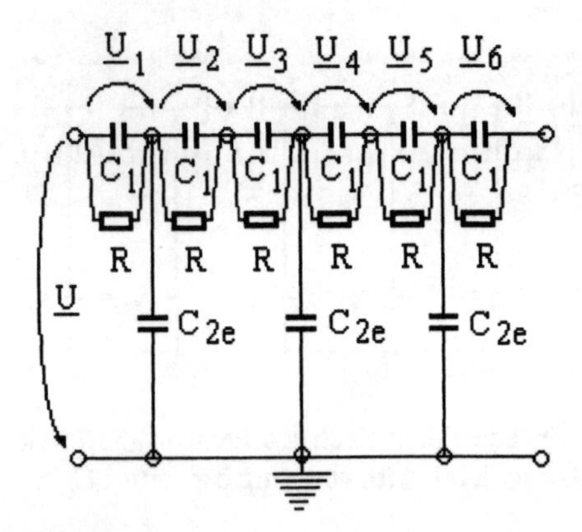

Fig.7 The equivalent electric scheme of the 750 kV pneumatic circuit breaker, with shunting resistors R

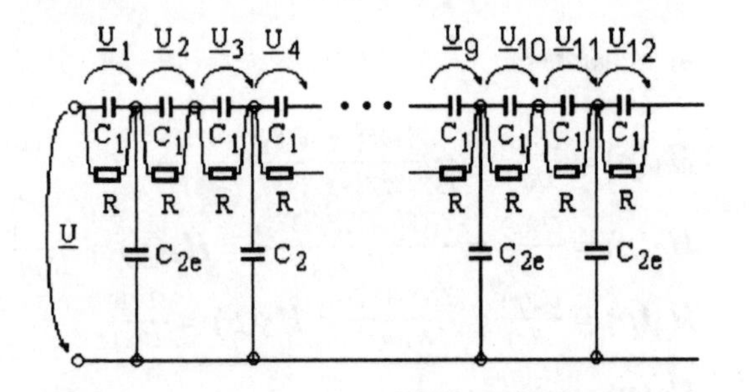

Fig.8 The equivalent electric scheme of the 750 kV circuit breaker with oil, with shunting resistors R

$$H_1(p) = \frac{U_1(p)}{U(p)} = \frac{p^3\left(27T_1^3+81T_1^2T_8+36T_1T_8^2+4T_8^3\right)}{H(p)} +$$

$$+\frac{p^2\left(81T_1^2+162T_1T_8+36T_8^2\right)+81(T_1T_8)}{H(p)}$$

$$H_2(p) = \frac{U_2(p)}{U(p)} = \frac{(1+pT_1)(1+pT_2)(1+pT_3)}{H(p)} = H_3(p) = \frac{U_3(P)}{U(p)}$$

$$H_6(p) = \frac{U_6(p)}{U(p)} = \frac{(1+pT_1)^3}{H(p)} = H_6(p) = \frac{U_6(p)}{U(p)} \qquad (3)$$

in which : $T_1 \div T_8$ are time constants (RC or RC equivalent)

$$H_1(p)$$
$$\cdot$$
$$\cdot \qquad (4)$$
$$\cdot$$
$$H_{12}(p)$$

Writing the first relations on the form :

$$H(p) = \frac{K'(p+p_1)^3}{(p+p_2)(p+p_3)(p+p_4)}$$

with : $K' = \frac{KT_1^3}{T_2T_3T_4}$; $p_i = \frac{1}{T_i}$; for i = 1, 2, 3, 4

and using the method of series programmations, it was elaborated the block scheme of the circuit breaker (fig.9)

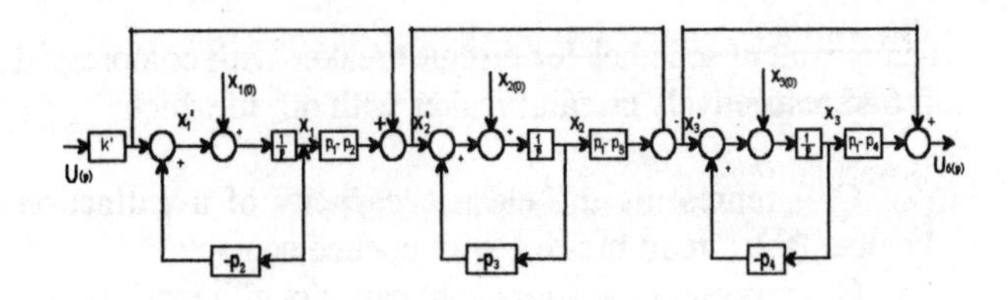

Fig.9 The block scheme of the circuit breaker

On these based, there were written the input-output equations [5,8]

$$H_6(p) = \frac{U_6(p)}{U(p)} = H_1(p)\, H_1'(p)\, H_1''(p) \qquad (6)$$

in which :

$$H_1(p) = K'(1+\frac{p_1-p_2}{p+p_2});\ H_1'(p) = 1+\frac{p_1-p_3}{p+p_3};\ H_1''(p) = 1+\frac{p_1-p_4}{p+p_4}$$

$$\dot{x}_1 = -p_2x_1 + K'u$$
$$\dot{x}_2 = (p_1-p_2)x_1 - p_3x_2 + K'u$$
$$\dot{x}_3 = (p_1-p_2)x_1 - (p_1-p_3)x_2 - p_4x_1 - K'u$$

which put under the matrix - vectorial model, they become :

$$x = A\cdot\dot{x} + bu \qquad (7)$$
$$u_6 = C^Tx + du$$

when :

$$A = \left\|\begin{matrix} -p_2 & 0 & 0 \\ p_1-p_2 & -p_3 & 0 \\ p_1-p_2 & p_1-p_3 & -p_4 \end{matrix}\right\| \quad ; \quad b = \left\|\begin{matrix} K' \\ K' \\ K' \end{matrix}\right\| \quad ;$$

$$c = \left\|\begin{matrix} p_1-p_2 \\ p_1-p_3 \\ p_1-p_4 \end{matrix}\right\| \quad ; \quad d = \|K'\|$$

Explaining the amplification-throb characteristics, there were obtained resistors values R, in order to distribute uniformly the overvoltages on the extinction chambers of circuit breaker.

3. THE NUMERICAL MOULDING OF REQUIREMENTS

It was realised a adequate computer program, on this based, there were determined the repartitions and the electric requirements on the extinction chambers of 750 kV circuit breaker, with compressed air and respectively with oil.

There were obtained interesting results concerning at influence of self frequency of re-establishment oscillatory voltage and, of number of chambers compled in series about stresses repartition.

There were also obtained variation in time of high voltages on the first extinction chamber when at the entrance there are applied impulse voltage waves with a very big slope (the step wave).

The voltage repartition on the extinction chamber.

I. 6 chambers

a) normal regime

U_1	U_2	U_3	U_4	U_5	U_6
0.281U	0.178U	0.178U	0.126U	0.126U	0.11U

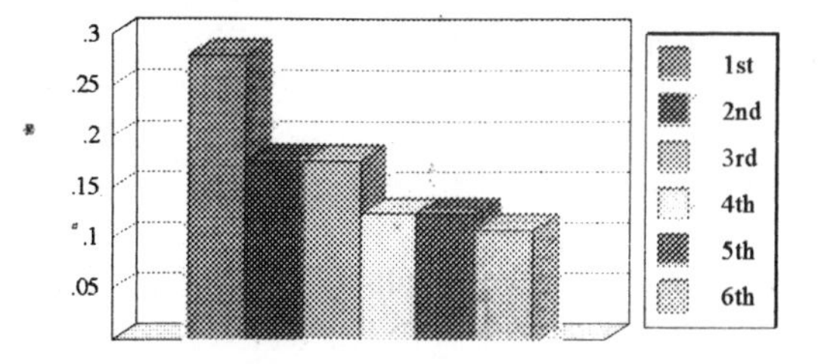

b) using condensers

U_1	U_2	U_3	U_4	U_5	U_6
0.164U	0.178U	0.178U	0.169U	0.169U	0.142U

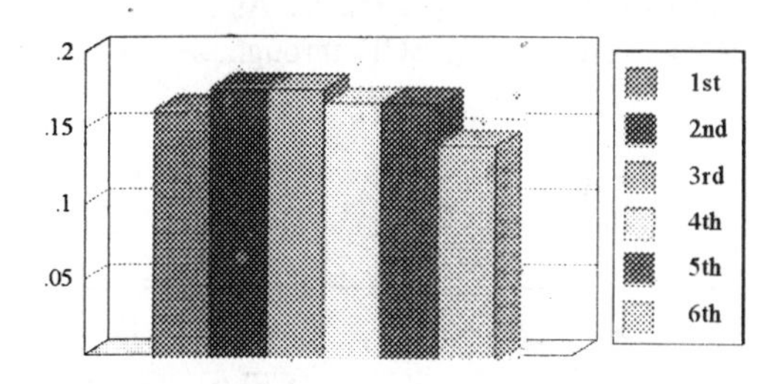

c) using resistors, R=2MΩ

5kHz	0.1668U	0.1668U	0.1668U	0.1664U	0.1664U	0.1663U
100kHz	0.3349U	0.1878U	0.1878U	0.1132U	0.1132U	0.0894U

II. 12 chambers

a) normal regime

U_1	U_2	U_3	U_4	U_5	U_6
0.144U	0.133U	0.133U	0.091U	0.091U	0.075U

U_7	U_8	U_9	U_{10}	U_{11}	U_{12}
0.075U	0.064U	0.064U	0.058U	0.058U	0.072U

b) using condensers

U_1	U_2	U_3	U_4	U_5	U_6
0.107U	0.096U	0.096U	0.087U	0.087U	0.080U

U_7	U_8	U_9	U_{10}	U_{11}	U_{12}
0.080U	0.075U	0.075U	0.073U	0.073U	0.072U

c) using resistors, R=100k

5kHz	0.103U	0.091U	0.091U	0.085U	0.085U	0.083U
100kHz	0.257U	0.121U	0.121U	0.093U	0.093U	0.071U

0.083U	0.079U	0.079U	0.075U	0.075U	0.071U
0.071U	0.037U	0.037U	0.035U	0.035U	0.029U

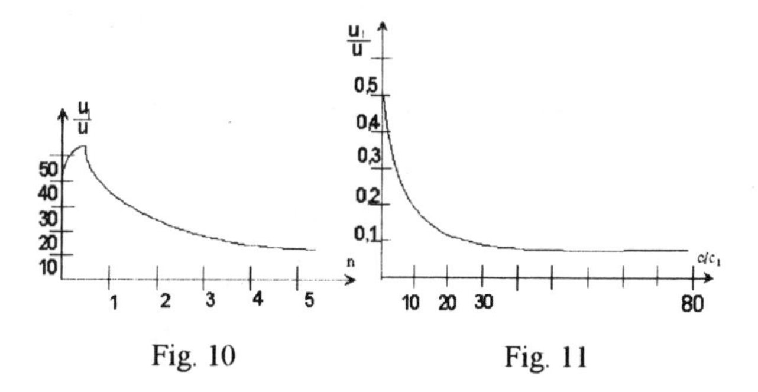

Fig. 10 Fig. 11

4. CONCLUSIONS

The analytical, numerical and graphical results demonstrates un-uniformity of high voltages and overvoltages repartitions on the extinction chambers of the 750 kV circuit breakers with compressed air and with oil.

It was resulted, the un-uniformity increases in the same time with the increasing number of extinction chambers, with the slope of overvoltage wave and respectively with the increasing of the self frequency of re-establishment oscillatory voltage oscillation.

The un-uniform repartition of overvoltage to high and very high voltage can produce complex electromagnetical perturbations, extremely damaging for the whole assembly circuit breaker-network-environment.

To limit this effects it is necessary to continue the researches in order to perfectionate the model of circuit breaker insulation and the approach of these phenomena in dynamic system.

5. REFERENCES

[1]. IEC-1988 - High voltage altenativ-courent circuit breakers.

[2]. Tuşaliu,P.,others - Portcontact assembly and arc extinction chamber for IUP-M-27,5kV/1250A switch. Invention patent no 628/01.07.1987 Bucharest, Romania.

[3]. Tuşaliu, P. - Research in domain of high voltage engineering. "Research, Invention, Creation', Dialogue technical-scintifical, Băniei House, Craiova, 11 nov., 1994.

[4]. Tuşaliu, P, Ciontu, M. - High voltage repartition over the extinction chamber of switch, analyse with helps of Systems's Theory International AMSE Conferences, London, (UK), sept. 13, 1993.

[5]. Tuşaliu,P, Ciontu, M. - Aspects concerning the determination of high voltages distribution on very high voltages insulating strings, "The International Conference on applied an theoretical electrotechnics" Proceeding Session A,B, 21-23 nov., Craiova, Romania, 1991.

[6]. Tuşaliu, P., Ciontu, M. - Aspects concerning re-estabishment oscillatory voltage as disconecting kilometric defect. Seventh International Conference on switching arc phenomena, Lodz, Poland, 1993.

[7]. Tuşaliu, P., Ciontu, M. - Perturbation generees par l'interaction appareil-reseau au debranchement des batteries multiples de condensateurs. Symposium"La compatibilté électromagnetique et les resseau d'énergie électrique",Laussane, Suisse, 18-20 oct., 1993.

[8]. Tuşaliu,P, Ciontu, M.,Tuşaliu,V - Interfernces of electric field from non-uniform repartition of overvoltage on extinction chambers of 750 kV circuite breaker. International Symposim on Electromagnetic Compatibility, EMC'94, Roma,13-16, sept. 1994.

[9]. Tuşaliu, P. - About repartition of high voltage on extinction chambers of the circuite breaker. Romanian Academy, Bucharest, 21-22 dec., 1994.

[10]. Tuşaliu, P. - Aspects concerning at repartition overvoltages on the extinction chambers of 750 kV circuit breaker. International Symposium, Very High Voltage Networks, May 31, 1995, Sibiu, Romania.

Conference Record of the 1996 IEEE International Symposium on Electrical Insulation, Montreal, Quebec, Canada, June 16-19, 1996

Absorption of SF_6, SF_4, SOF_2 and SO_2F_2 in the 115-220 nm Region

Catherine Pradayrol Anne-Marie Casanovas Joseph Casanovas
CPAT, URA n° 277, Université Paul Sabatier
118, route de Narbonne, 31062 Toulouse cedex, France

Abstract **: Absorption coefficients k_o ($m^{-1}.100\ kPa^{-1}$) of SF_6 and of its main gaseous by-products SF_4, SOF_2 and SO_2F_2 were measured in the 115-220 nm region. The experiments were carried out at a temperature of 298 K and a spectral resolution of 0.1 nm over the wavelength range 115-118 nm for SF_6, 115-220 nm for SF_4, 120-195 nm for SOF_2 and 120-210 nm for SO_2F_2. The highest absorption coefficient values were obtained for SF_4 and the lowest for SF_6.**

INTRODUCTION

Radiative transfer in SF_6 arcs is an important mechanism of energy transport which has several consequences in SF_6 circuit breakers. The absorption of radiation by the gas surrounding the plasma heats this gas [1] leading to a pressure increase in the circuit breaker and this may have a strong influence on the apparatus behaviour. The main aim of our work was to determine the spectral dependence of the absorption coefficient of cold gaseous SF_6 (T = 298 K) in the wavelength range 115-220 nm. Concurrently we measured, over the same wavelength region, the absorption coefficients of the main gaseous by-products of SF_6 under arcs : SF_4, SOF_2 and SO_2F_2 [2-5].

EXPERIMENT

The experimental set-up used for measuring the absorption spectra of SF_6, SF_4, SOF_2 and SO_2F_2 is shown in figure 1.

The absorption cell was attached to the exit slit of a vacuum-UV spectrophotometer (Spex model 1500) which was evacuated to a pressure of 10^{-2} Pa and a low pressure microwave powered hydrogen-argon lamp equipped with a MgF_2 window was located in front of the entrance slit of this apparatus. This lamp emits a complex spectrum between ≅ 115 nm and 230 nm. As a very similar system was already used for photoconductivity studies in non-polar liquids (hydrocarbons, silicone oils), further information can be found in some of our papers dealing with this subject [7, 8]. All absorption measurements were made at a resolution of 0.1 nm.

The different gases were fed into the absorption cell via a stainless steel gas handling system and the pressure in the cell measured using two MKS Baratron type 122 B absolute pressure gauges (accuracy : ± 0.15%) one for the range 5-10^3 Pa and the other for the range 10^3-10^4 Pa and with a Membranovac 1 VS Leybold-Heraeus device connected to a Cl absolute pressure sensor head (accuracy : ± 0.5 %) for the range 10^4-10^5 Pa.

As we know from literature results [9-11] that gases like O_2 and H_2O, which may be present as impurities in SF_6, exhibit strong absorption in the region 110-200 nm, we chose to measure the absorption of three SF_6 samples differing from one another in : initial purity, storage time, volume of the storage containers, containers that had been opened or left sealed. Commercially available SO_2F_2 (> 99.5 %) and SF_4 (> 94 %) were used in our experiments. As SF_4 readily reacts with water traces and leads to SOF_2 through the reaction [12]

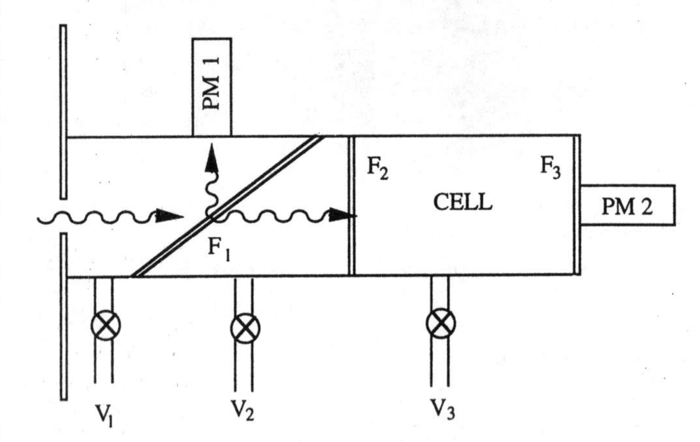

Figure 1. Schematic representation of the absorption cell. ∿∿► : Monochromatized VUV photons. F_1, F_2 and F_3 : MgF_2 plates (thickness 1.3 mm ; cut-off wavelength 115 nm). F_2 constitutes the entrance window of the absorption cell. F_3 is coated on its outside surface with sodium salicylate (fluorescent material whose maximum intensity of fluorescence is located at 420 nm [6] and which coincides with the maximum sensitivity of the cathode of the photomultipliers PM_1 and PM_2). PM_1 and PM_2 : Hamamatsu model R268 photomultipliers. PM_1 : reference photomultiplier ; its photocathode is coated with sodium salicylate. It measures the photons reflected by F_1. PM_2 : photomultiplier measuring the photon flux transmitted through the cell. The absorption cell (inside diameter 40 mm ; optical length d = 136 mm) is made of stainless steel and can be filled up to a gas pressure of 100 kPa. V_1 and V_2 : connecting valves to primary and secondary pumps. V_3 : connecting valve of the cell to the pump and to the gas filling system. It is also connected to a vacuum gauge and to three pressure gauges (5 Pa ≤P≤100 kPa).

$$SF_4 + H_2O \longrightarrow SOF_2 + 2\,HF \quad (1)$$

special precautions were taken in order to avoid this reaction or at least to minimize it. Before filling the cell with the SF_4 sample to be studied, the cell and the gas filling circuit were, after being evacuated, flushed several times with SF_4. This gas was then pumped out and the cell filled to the required pressure with new SF_4. It should also be noted that the experiments with the lower SF_4 pressures were carried out after those at higher pressures and that new SF_4 samples were used every time. SOF_2 was prepared through reaction (1) by putting SF_4 in contact with water vapor. This procedure only allowed us to obtain low pressure samples of SOF_2 and this explains why the absorption of this compound was restricted to wavelengths lower than 196 nm.

The absorption spectrum of each gas was divided into between two and four overlapping regions of about 30 nm width. The absorption spectrum over these different wavelength intervals was measured by determining the photon flux intensity when the cell was evacuated, followed by the intensity when the cell was filled with the gas. For each region, absorption measurements were made at five or more values of the gas pressure. The output signals of the photomultipliers were graphically recorded synchronously as the monochromator scanned at a rate of 0.06 nm per second. The uncertainty on the wavelength values from data set to data set was typically ± 0.05 nm.

The absorption coefficient k_o was calculated from Beer-Lambert's law

$$I = I_0 \exp(-k_0 \frac{P}{P_0} d) \quad (2)$$

where I_o and I are the transmitted intensities through vacuum and gas respectively at a wavelength λ and a temperautre T, d is the path length, $P_o = 10^5$ Pa and P is the absolute pressure in the absorption cell in Pa. For all our experiments d was equal to 13.6 cm (see fig. 1) and T to 298 ± 2K. From reproducibility tests carried out over the total wavelength range considered we asses the uncertainty on the absorption coefficient to be about ± 10 % for the highest and the lowest values and slightly better (± 6 %) for the others.

RESULTS

SF_6

Figure 2 shows the results obtained for the three samples of SF_6 studied. For $\lambda < 135$ nm, the three curves are practically superimposed indicating that in this zone, the absorption measured corresponds only to SF_6. The differences that appear between curve A and curves B and C over 135 nm and between curves B and C over 150 nm confirm the presence of variable quantities of O_2 (maximum absorption at around 142 nm [9-11]) and of H_2O (maximum absorption at around 165 nm [9-11]) in the samples of SF_6 studied. In particular, the shape of curve B between 160 and 180 nm is almost identical to that of the variation of H_2O absorption over the same wavelength range [9-11]. It is therefore curve C that best represents the absorption of pure SF_6.

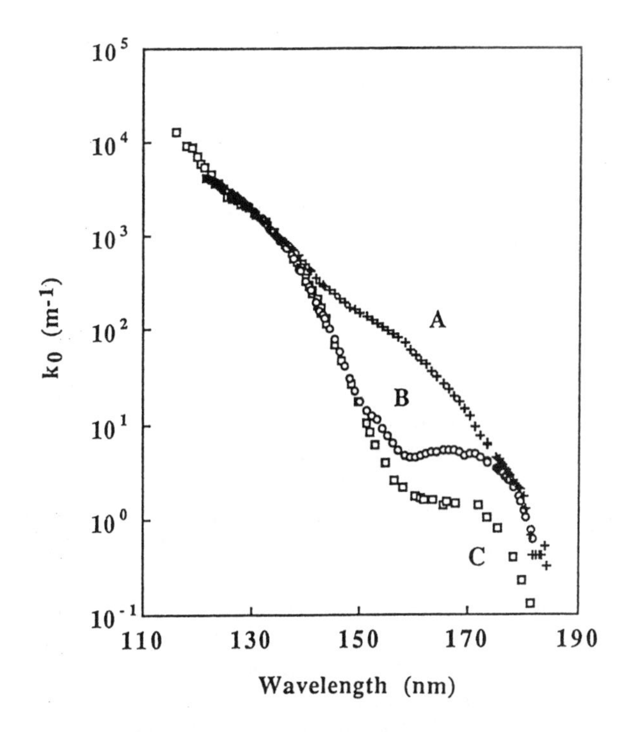

Figure 2. Measured absorption coefficient (k_o) for three gaseous SF_6 samples. T = 298 K. The k_o values are given for a gas pressure of 100 kPa.

From these curves and those of the absorption of O_2 and H_2O published in the literature [9-11] we determined the quantities of O_2 and H_2O contained in the three samples of SF_6 from the relationship :

$$(\frac{k_0}{\rho})_{mixture} = (\frac{k_0}{\rho})_{SF_6} \cdot \alpha_{SF_6} + (\frac{k_0}{\rho})_{O_2} \cdot \alpha_{O_2} + (\frac{k_0}{\rho})_{H_2O} \cdot \alpha_{H_2O} \quad (3)$$

where $(\frac{k_0}{\rho})_{SF_6}$, $(\frac{k_0}{\rho})_{O_2}$ and $(\frac{k_0}{\rho})_{H_2O}$ are the mass absorption coefficients and α_{SF_6}, α_{O_2} and α_{H_2O} the percentages in mass of pure SF_6, of O_2 and of H_2O respectively for a given wavelength.

The cuve $(k_0 / \rho)_{SF_6}$ versus λ was obtained from curve C of figure 2 by considering that in the absence of water vapour, the absorption coefficient of SF_6 continues to decrease between 155 nm and 185 nm like it did from 140 nm to 155 nm. This hypothesis is in agreement with the results of previous tests performed at a pressure of 300 kPa and an optical path of 8 cm which showed that the absorption of SF_6 is practically nil for $\lambda \geq 190$ nm.

Fitting the theoretical curves, $k_o = f(\lambda)$, calculated from equation 3, for variable concentrations of O_2 and H_2O to the experimental curves A, B and C of figure 2 enabled us to determine the quantities of these two gases present in the various samples of SF_6 studied and led to the following results :

- the sample of SF_6 used to plot curve A contained 600 ppm_v H_2O and 4 500 ppm_v O_2 ;
- the sample of SF_6 giving curve B contained 500 ppm_v H_2O and 60 ppm_v O_2 ;
- the sample of SF_6 corresponding to curve C contained 125 ppm_v H_2O and less than 5 ppm_v O_2.

Figure 3 enables a comparison to be made between our results (curve C of figure 2) and those obtained by Bastien et al [13]. The corresponding curve A was obtained by digitisation of the curve published by these authors. It can be noted that the two curves of figure 3 are practically superimposable between 130 about 145 nm whereas between 120 and 130 nm, the results of Bastien et al. [13] lead to values of k_o that are higher than the ones we obtained independently of the quality of the SF_6. This could be due to a difference of accuracy (and/or to calibration errors) of the gauges used to measure the absolute pressure of SF_6 (note that for this wavelength range, the pressures of SF_6 that must be used to determine k_o are very low) and also possibly to the presence, in their SF_6, of impurities other than water and oxygen. The difference that appears for $\lambda > 145$ nm can be put down to O_2 and H_2O : applying equation 3 to the results of Bastien et al. [13] indicated that the SF_6 they used contained about 200 ppm_v of H_2O and 1 200 ppm_v O_2.

Our results, like those of the literature [13, 14] show that SF_6 presents a continuous absorption between 180 and 115 nm. This absorption is not related to the ionisation of the gas, which is known to have an ionisation potential at about 16 eV [14-16]. According to Sasanuma et al.[15] it can be assigned to the transition $5t_{1u}$ - $6a_{1g}$ with an energy of 8.7 eV. It therefore most probably gives rise to dissociation processes that, as emphasized by Herzberg [17], are difficult to identify owing to the existence, in each electronic state, of several dissociation limits corresponding to different dissociation products. A molecular photoelimination process leading to the production of SF_4 was indeed demonstrated for values of λ of 121.6 nm, 106.7 nm and 104.8 nm by Smardzewski and Fox [18].

SF_4

The values of k_o obtained for SF_4, SOF_2 and SO_2F_2 are reported in figure 4. Moreover, for the sake of comparison, the curve of the absorption of our purest SF_6 sample (curve C of figure 2) has also been added to this figure. Between 118 and 220 nm, SF_4 presents a much greater absorption than SF_6 with values of k_o of between 8.9 10^4 m^{-1} and 5 m^{-1}. Evidence of structure is also observed with clear peaks occuring at 123,

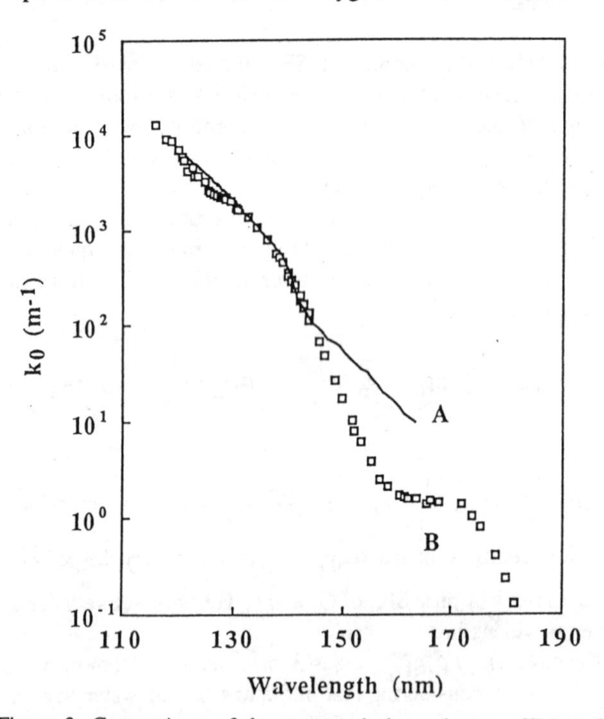

Figure 3. Comparison of the measured absorption coefficient k_o for our purest gaseous SF_6 sample (curve B ; this curve corresponds to curve C of figure 2) with that published by Bastien et al. [13] (curve A). The k_o values are given for a gas pressure of 100 kPa.

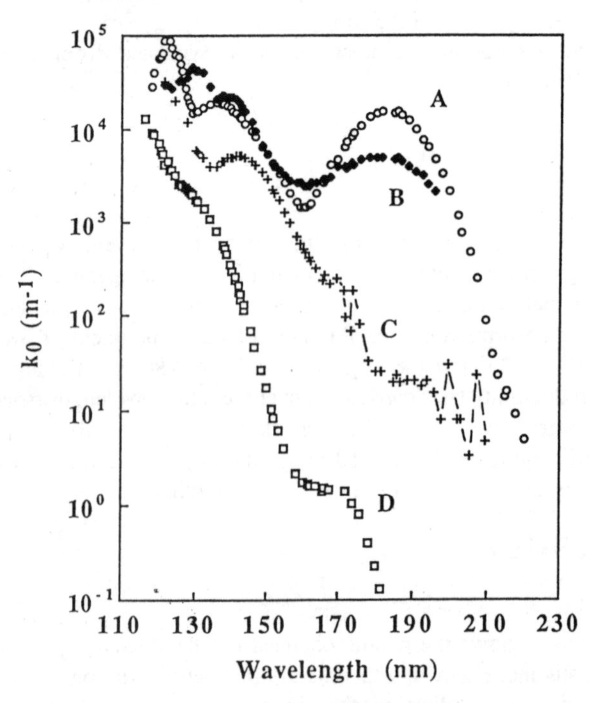

Figure 4. Measured absorption coefficient k_o for SF_4 (curve A), SOF_2 (curve B), SO_2F_2 (cure C) and SF_6 (curve D ; this curve corresponds to curve C of figure 2). T = 298 K. The k_o values are given for a gas pressure of 100 kPa.

135.9 and 182.5 nm.

SOF_2

This compound also presents very strong absorption which, between 135 and 200 nm, resembles that of SF_4 with two maxima in this zone, one situated at 138 nm and the other at 180.6 nm i.e. slightly shifted with respect to those of SF_4. Between 120 nm and 135 nm, SOF_2 presents a third absorption peak ($k_o = 4.5\ 10^4\ m^{-1}$) for $\lambda = 129.4$ nm whereas for the same wavelength, the absorption of SF_4 goes through a minimum. On the other hand, SOF_2 presents a minimum of absorption at around 123.6 nm whereas for this value of λ, SF_4 presents its maximum absorption ($k_o = 8.9\ 10^4\ m^{-1}$).

SO_2F_2

The absorption curve of SO_2F_2 occurs between that of SF_6 and those of SF_4 and of SOF_2. It presents a structure which is much more complex than that of the others with a strongly attenuated maximum at 141.7 nm ($k_o = 5.3\ 10^3\ m^{-1}$) and a series of narrow bands of variable amplitude superimposed on a continuum and situated at 166, 169.5, 173.6, 193.7, 199.7 and finally 207.4 nm. It should be noted that the absorption coefficients on these bands were independent of pressure (factor of about 10) ; thus, our spectral resolution was probably adequate.

CONCLUSION

Our results on SF_6 agree quite well with those previously published in the literature [13] and illustrate the importance of even low concentrations of impurities like oxygen and/or water on « SF_6 » absorption.

The results obtained for SF_4, SOF_2 and SO_2F_2 are new. They show that these gases present considerable absorption in the wavelength range 115-220 nm. So, as these compounds, at least the first two, are formed in large amounts when SF_6 is submitted to arcs they would also contribute to the heating of the gas surrounding the plasma and therefore to the pressure increase inside the circuit-breaker.

ACKNOWLEDGEMENT

We gratefully acknowledge the GEC Alsthom Company for its financial support.

REFERENCES

1. Gleizes, A., M. Bouaziz, C. Pradayrol, G.Raynal and M. Razafinimanana. « Absorption of SF_6 Arc Plasma Radiation by Cold SF_6 Gas ».11th International Conference on Gas Discharges and their Applications, September 11-15, 1995, Tokyo, pp. I-38-I-41.

2. Rüegsegger, W., R. Meier, F.K. Kneubuhl and H.J. Schotzau. « Mass Spectrometry of Arcs in SF_6 Circuit Breakers ». Appl. Phys. B. Vol. 37, 1985, pp. 115-135.

3. Chu, F.Y.. « SF_6 Decomposition in Gas-Insulated Equipment ». IEEE Trans. EI. Vol. 21. No.5, October 1986, pp. 693-725.

4. Belmadani, B., J. Casanovas, A.M. Casanovas, R.Grob and J. Mathieu. « SF_6 Decomposition under Power Arcs. Physical Aspects ». IEEE Trans. EI. Vol. 26. No. 6, December 1991, pp. 1163-1176.

5. Belmadani, B., J. Casanovas and A.M. Casanovas. « SF_6 Decomposition under Power Arcs. Chemical Aspects ». IEEE Trans. EI. Vol. 26. No. 6, December 1991, pp. 1177-1182.

6. Samson, J.A.R. *Techniques of Vacuum Ultraviolet Spectroscopy*. Wiley, New-York, 1967, pp. 212-216.

7. Casanovas, J., R. Grob, R. Sabattier, J.P. Guelfucci and D. Blanc. « Photoconductivity of Liquid 2,2-Dimethylbutane and Liquid 2,2,4-Trimethylpentane ». Radiat. Phys. Chem. Vol. 15, 1980, pp. 293-296.

8. Baron, P.L., J. Casanovas, J.P. Guelfucci and R. Laou Sio Hoi. « Photoconductivity Induced by VUV Photons in Polydimethylsiloxane and Polymethylphenylsiloxane Oils ». IEEE Trans. EI. Vol. 23. No. 4, August 1988, pp. 563-570.

9. Watanabe, K. *Advances in Geophysics*. Academic Press, New-York, 1958, pp. 153-221.

10. Thompson, B.A., P. Harteck and R.R. Reeves, Jr. « Ultraviolet Absorption Coefficients of CO_2, CO, O_2, H_2O, N_2O, NH_3, NO, SO_2 and CH_4 between 1850 and 4000 A » J. Geophys. Res. Vol. 68. No. 24, December 1963, pp. 6431-6436.

11. Hudson, R.D. « Critical Review of Ultraviolet Photoabsorption Cross Sections for Molecules of Astrophysical and Aeronomic Interest ». Rev. Geophys. and Space Phys. Vol. 9. No. 2, May 1971, pp. 305-406.

12. Sauers, I., J.L. Adcock, L.G. Christophorou and H. W. Ellis. « Gas Phase Hydrolysis of Sulfur Tetrafluoride : A Comparison of the Gaseous and Liquid Phase Rate Constants ». J. Chem. Phys. Vol. 83. No. 5, September 1985, pp. 2618-2619.

13. Bastien, F., P.A. Chatterton, E. Marode and J.L. Moruzzi. « Photoabsorption Measurements in SF_6, SF_6 Mixtures and some Fluorocarbon Gases ». J. Phys. D : Appl. Phys. Vol. 18, 1985, pp. 1327-1337.

14. Nostrand, E.D. and A.B.F. Duncan. « Effect of Pressure on Intensity of Some Electronic Transitions in SF_6, C_2H_2 and C_2D_2 Vapors in the Vacuum Ultraviolet Region ». J. Am. Chem. Soc. Vol. 76, 1954, pp. 3377-3379.

15. Sasanuma, M., E. Ishiguro, H. Masuko, Y. Morioka and M. Nakamura. « Absorption Structures of SF_6 in the VUV Region ». J. Phys. B : Atom. Molec. Phys. Vol. 11. No. 21, 1978, pp. 3655-3665.

16. Hitchcock, A.P. and M.J. Van Wiel. « Absolute Oscillator Strengths (5-63 eV) for Photoabsorption and Ionic Fragmentation of SF_6 ». J. Phys. B : Atom. Molec. Phys. Vol. 12. No. 13, 1979, pp. 2153-2169.

17. Hergberg, G. *Molecular Spectra and Molecular Structure, III. Electronic Spectra and Electronic Structure of Polyatomic Molecules* ». Van Nostrand, Princeton, 1967.

18. Smardzewski, R.R. and W.B. Fox. « Vacuum Ultraviolet Photolysis of Sulfur Hexafluoride and its Derivatives in Argon Matrices. The Infrared Spectrum of the SF_5 Radical ». J. Chem. Phys. Vol. 67. No. 5, September 1977, 2309-2316.

Conference Record of the 1996 IEEE International Symposium on Electrical Insulation, Montreal, Quebec, Canada, June 16-19, 1996

Influence of O_2 and H_2O on the Spark Decomposition of SF_6 and SF_6 + 50 % CF_4 Mixtures

Catherine Pradayrol Anne-Marie Casanovas Joseph Casanovas

CPAT, URA n° 277, Université Paul Sabatier

118, route de Narbonne, 31062 Toulouse cedex, France

Abstract **: The spark decomposition of SF_6 and SF_6 + 50 % CF_4 mixtures was studied at a gas pressure of 200 kPa. The sparks were generated between a point and a plane either under 50 Hz ac voltage (0.09 J per spark) or by discharging a capacitor (3.59 J par spark). Our attention was only focused on the gaseous SF_6 by-products : SOF_2 + SF_4, SOF_4, SO_2F_2, S_2F_{10}, S_2OF_{10}, $S_2O_2F_{10}$ and $S_2O_3F_6$ which were assayed by gas chromatography. Their yields were studied varying the cell preparation technique, the metal constituting the plane electrode (aluminum, copper, stainless-steel) and the concentrations of two additives, O_2 (between O and 1 %) and H_2O (between O and 0.2 %). The cell preparation procedure had a strong effect on the formation of all products except SOF_2 + SF_4. The aluminum led, whatever the procedure used, to the highest levels of products. With SF_6 as well as with the SF_6 + 50 % CF_4 mixtures and whatever the spark energy, an increase of the oxygen content led to a decrease of SOF_2 + SF_4 and S_2F_{10} production rates and to an increase of that of SOF_4 + SO_2F_2. Moreover, an increase of the H_2O content induced a lowering of the S_2F_{10} production rate and an increase of those of SOF_2 + SF_4 and SOF_4 + SO_2F_2. With the SF_6 + 50 % CF_4 mixture the by-products production rates were lower (10 to 50 % depending on the by-product considered and on the O_2 or H_2O concentration) than those measured with pure SF_6.**

INTRODUCTION

Today, sulfur hexafluoride (SF_6) is widely used as insulating medium in high voltage systems and its decomposition, under for example sparks, has been the object of various studies over the last years [1-11]. However, at very low temperatures, pure SF_6 can no be longer used and has to be mixed with other gases. One of the most realistic possibilities is a mixture constituted of SF_6 and 50 % of perfluoromethane (CF_4). It therefore seemed interesting to compare the spark decomposition of pure SF_6 and that of SF_6 + 50 % CF_4 mixtures under the same experimental conditions. In this paper we present the results obtained for different spark energies, sparking cell preparations, types of metal constituting the electrode and additives and their concentrations (O_2 : O to 1 % ; H_2O : 0 to 0.2 %).

EXPERIMENT

The sparks were generated in cells made of stainless steel or Monel 400 (volume : 340 cm^3) between a steel point (radius of curvature : 10 µm or 0.5 m) and a plane (aluminum, copper or stainless steel) either under 50 Hz ac voltage (0.09 J per spark) or by discharging a capacitor (3.59 J per spark) [4]. The distance between the two electrodes was 0.9 mm. Owing to the deterioration of the point and the plane during the experiments, a new set of electrodes was used for each test.

Depending on the initial preparation of the cell, the experiments (with or without O_2 or H_2O addition) were called « very clean » when the cell was submitted to several heating (60°C)-pumping cycles, evacuated to a final pressure lower than 1 Pa before filling with the gas to be studied and « clean » when the cell was only evacuated to about 1 Pa before filling.

The gas mixtures were prepared in the previously evacuated cell by setting up the required partial pressure of O_2 (corresponding to 0.1, 0.25, 0.5 or 1 %) or H_2O (corresponding to 600 ppm_v, 1000 ppm_v or 2000 ppm_v) and then introducing SF_6 (99.995 %) or a mixture of 50 % SF_6 + 50 % CF_4 (99.995 %) to 200 kPa total pressure. We called « dry » the ones prepared with the above procedure without water addition and « wet » the ones with different H_2O concentrations. For the « dry » experiments the cells were left to stand for 19 hours before sparking in order to allow an equilibrium to be reached between the residual water vapor on the walls and the main gas volume.

We checked that the yields of the various by-products rose linearly with the number of sparks (at least up to 1500 for the more energetic of them) and we expressed the corresponding production rates in nmoles per joule.

After each series of sparks, the cell content was analysed by gas chromatography. The analytical conditions used to measure the yields of the different SF_6 by-products observed : SOF_2 + SF_4, SOF_4, SO_2F_2, S_2F_{10}, S_2OF_{10}, $S_2O_2F_{10}$ and $S_2O_3F_6$ were described in detail in [12]. Under these analysis conditions the detection limits for SOF_4 + SO_2F_2 and SOF_4 + SF_4 were respectively 20 ppm_v and 10 ppm_v, while for the other compounds the detection limit was about 1 ppm_v.

It should be noted that $S_2O_3F_6$ was identified thanks to the work of Castonguay and Dionne [8, 13] and that, as the response of our detector was practically the same for S_2F_{10}, S_2OF_{10} and $S_2O_2F_{10}$, we considered that it most likely responded in the same way to $S_2O_3F_6$.

Reproducibility tests carried out under all the different experimental conditions used showed that the mean uncertainty of our data was about ± 10 %.

RESULTS AND DISCUSSION

As said in the experimental part, two methods of cell preparation were tested for both types of sparks. We observed that the initial « cleanliness » of the cell had a greater influence on the SF_6 by-products yields for the high-energy sparks, related to a greater heating of the electrodes under these sparks. The use of a « very clean » cell enabled us to observe the formation of compounds like $S_2O_2F_{10}$ and $S_2O_3F_6$ and with an aluminum electrode, an increase of the S_2F_{10} production rate by a factor of ≅ 17 (see table 1). It also allowed us to show that, of the three metals used for the plane electrode, aluminum was that leading to the greater production of SOF_2 + SF_4, S_2F_{10} and $S_2O_3F_6$. The lower production rates being obtained with copper. The « very clean » procedure and aluminum were therefore used in all later experiments.

Table 1. Production rates, in nmoles/J, of SOF_2 + SF_4, S_2F_{10}, S_2OF_{10}, $S_2O_2F_{10}$, $S_2O_3F_6$ and SO_2F_2 + SOF_4 in purest SF_6 and in SF_6 + 50 % CF_4 mixtures sparked through capacitor discharge. P = 200 kPa. Aluminum plane.

Compound	Type and % of added impurities	SF_6 « Clean » cell	SF_6 « Very clean » cell	SF_6 + 50% CF_4 «Very clean »cell
SOF_2+SF_4	0	49.8	44.7	47.35
	1 % O_2	17.4	19.9	15.6
	0.2 % H_2O	41	40.6	44.8
S_2F_{10}	0	0.11	1.89	1.75
	1 % O_2	0.14	0.14	0.11
	0.2 % H_2O	0.075	0.17	0.42
S_2OF_{10}	0	1.1	0.02	0.02
	1 % O_2	O.61	0.65	0.64
	0.2 % H_2O	O.96	0.39	0.19
$S_2O_2F_{10}$	0	0	0.09	0.11
	1 % O_2	0.66	0.66	-
	0.2 % H_2O	0	0	0
$S_2O_3F_6$	0	0.24	0.48	0.54
	1 % O_2	0.36	0.36	0.44
	0.2 % H_2O	0	0.28	0.5
SO_2F_2 + SOF_4	0	3.84	0.19	0.51
	1 % O_2	10.8	9.7	11.3
	0.2 % H_2O	4.43	2.8	2.9

Figure 1 displays the effect of increasing concentrations of

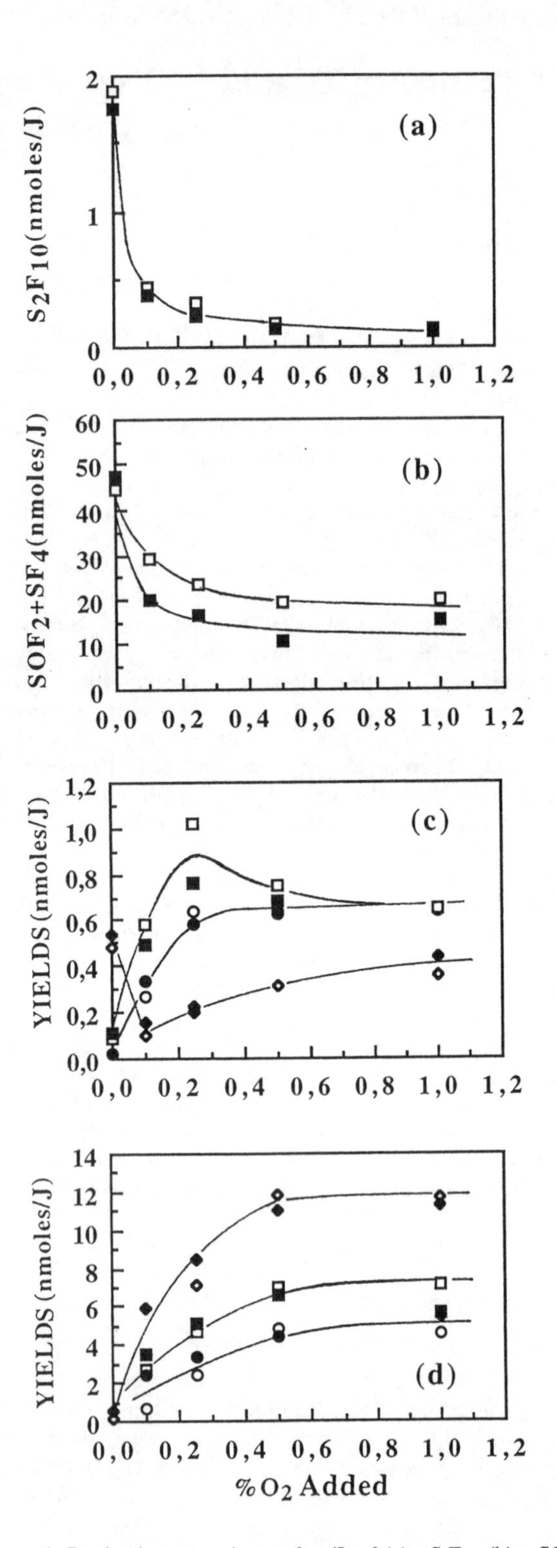

Figure 1. Production rates, in nmoles /J, of (a) : S_2F_{10}, (b) : SOF_2 + SF_4, (c) : S_2OF_{10} (○), $S_2O_2F_{10}$ (□) and $S_2O_3F_6$ (◇), (d) : SOF_4 (□), SO_2F_2 (○) and SO_2F_2 + SOF_4 (◇) versus added oxygen percentage. Purest SF_6 (open symbols) and SF_6 + 50 % CF_4 mixtures (closed symbols) sparked through capacitor discharge. P = 200 kPa. Aluminum plane. « Very clean » cell.

O_2 (up to 1 %) on the SOF_2 + SF_4, S_2F_{10}, S_2OF_{10}, $S_2O_2F_{10}$, $S_2O_3F_6$, SOF_4, SO_2F_2 and SO_2F_2 + SOF_4 production rates in SF_6 and SF_6 + 50 % CF_4 mixtures submitted to high-energy sparks. It can be seen that in accordance with the reactions of formation of these compounds proposed in the literature [3, 10, 11], oxygen leads to a decrease of SOF_2 + SF_4, S_2F_{10} and to a lesser extent of $S_2O_3F_6$ and to an increase of the other above mentioned compounds. It should also be noted that very similar trends are observed in pure SF_6 and in the SF_6 + 50 % CF_4 mixtures. The small differences that appear in the levels of compounds formation in the pure gas and in the 50-50 mixture are perfectly explainable by the effect of partial pressures of oxygen and/or of SF_6 [14] and thus do not necessarily imply an active role for CF_4 in the degradation of SF_6.

For its part, whatever the gas studied, water has a very little influence on the SOF_2 + SF_4 production, leads to a decrease of S_2F_{10}, $S_2O_2F_{10}$ and $S_2O_3F_6$ and to an increase of S_2OF_{10} and SO_2F_2 + SOF_4. Table 1 summarizes the effect of the cell preparation procedure and of the addition or not of 1 % of O_2 and 0.2 % of H_2O on the production rates of the different by-products. Our results for a « clean » cell agree quite well, except for $S_2O_2F_{10}$ and $S_2O_3F_6$ whose formation was not previously reported under high-energy sparks, with those obtained, under similar experimental conditions, by Piémontési and Zaengl [9], but are about one order of magnitude higher than those of Sauers [3, 11], Sauers et al. [5, 7] and Van Brunt et al. [10]. This difference can be due to the method used by these authors to determine the energy dissipated in the spark.

For low-energy sparks the plane electrode material has a much lesser effect than for the high-energy ones due to the fact that the vaporisation of the metal constituting the plane electrode is very unimportant under 50 Hz ac voltage sparks. Figure 2 shows the effect of oxygen (concentration up to 1 %) on the SOF_2 + SF_4, S_2F_{10}, S_2OF_{10}, $S_2O_2F_{10}$ and SO_2F_2 + SOF_4 production rates in SF_6 and SF_6 + 50 % CF_4 mixtures submitted to low-energy sparks. As for high-energy sparks oxygen leads to a lowering of the SOF_2 + SF_4 and S_2F_{10} yields and to an increase of those of S_2OF_{10}, $S_2O_2F_{10}$ and SO_2F_2 + SOF_4. In the presence of water, SOF_2 + SF_4 and SO_2F_2 + SOF_4 increase while S_2F_{10} decreases. It should be noted (see table 2) that S_2OF_{10}, $S_2O_2F_{10}$ and $S_2O_3F_6$ are only observed when O_2 is present in the sparked gas and that the production rates of SOF_2 + SF_4 and S_2F_{10} are, whatever the impurity (O_2 or H_2O) or its concentration, lower in the SF_6 + 50 % CF_4 mixtures than in SF_6. Once again, these differences can be accounted for SF_6 pressures effects [14]. Concerning the 50 Hz ac voltage sparks, the only results that have been published in the literature are those of Castonguay and Dionne [8] who observed the production of very small quantities of $S_2O_2F_{10}$ and $S_2O_3F_6$ in addition to SOF_2, S_2F_{10} and SO_2F_2 + SOF_4. The levels of formation that they obtained for the three latter compounds are comparable to those we report (see table 2) for damp SF_6 (2000 ppm$_v$ H_2O).

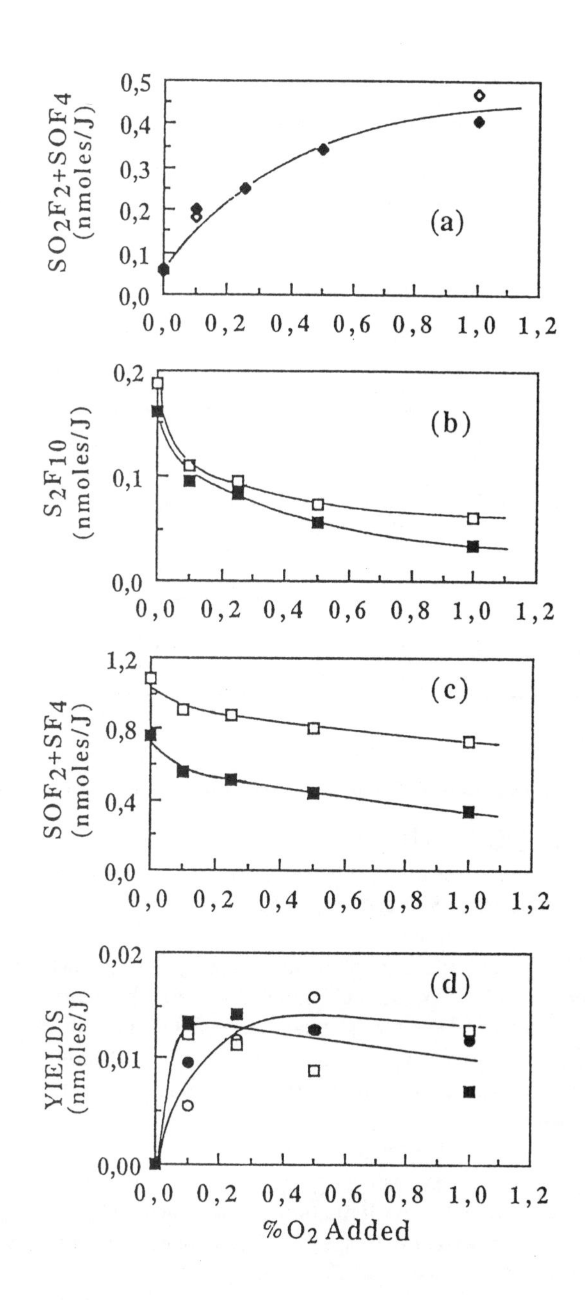

Figure 2. Production rates, in nmoles /J, of (a) : SO_2F_2 + SOF_4, (b) : S_2F_{10}, (c) : SOF_2 + SF_4, (d) : S_2OF_{10} (o) and $S_2O_2F_{10}$ (□) versus added oxygen concentration. Purest SF_6 (open symbols) and SF_6 + 50 % CF_4 mixtures (closed symbols) sparked through 50 Hz ac voltage. P = 200 kPa. Aluminum plane. « Very clean » cell.

Table 2 : Production rates, in nmoles/J, of $SOF_2 + SF_4$, S_2F_{10}, S_2OF_{10}, $S_2O_2F_{10}$, $S_2O_3F_6$ and $SO_2F_2 + SOF_4$ in purest SF_6 and in SF_6 + 50 % CF_4 mixtures sparked through 50 Hz ac voltage. P = 200 kPa. Aluminum plane. « Very clean « cell.

Compound	Type and % of added impurities	SF_6	SF_6 + 50 % CF_4
$SOF_2 + SF_4$	0	1.08	0.76
	1 % O_2	0.73	0.33
	0.2 % H_2O	1.78	1.32
S_2F_{10}	0	0.19	0.16
	1 % O_2	0.062	0.033
	0.2 % H_2O	0.1	0.088
S_2OF_{10}	0	0	0
	1 % O_2	0.013	0.012
	0.2 % H_2O	0	0
$S_2O_2F_{10}$	0	0	0
	1 % O_2	0.013	0.007
	0.2 % H_2O	0	0
$S_2O_3F_6$	0	0	0
	1 % O_2	0.017	0
	0.2 % H_2O	0	0
$SO_2F_2 + SOF_4$	0	0.054	0.059
	1 % O_2	0.47	0.41
	0.2 % H_2O	0.94	0.97

CONCLUSION

The main results obtained on the spark-induced decomposition of SF_6 and SF_6 + 50 % CF_4 mixtures can be summarized as follows :

- The greater the energy of the sparks, the greater the importance of the initial « cleanliness » of the discharge cell. The desorption of varying quantities of O_2 and of H_2O, alongside the greater vaporisation of the metal constituting the electrodes, explain, at least in part, the dispersion of the results in the literature.
- The formation of compounds like S_2OF_{10}, $S_2O_2F_{10}$ and $S_2O_3F_6$ has been demonstrated.
- Oxygen inhibits the formation of S_2F_{10} and $SOF_2 + SF_4$ and to a lesser extent that of $S_2O_3F_6$ but strongly favours that of S_2OF_{10}, $S_2O_2F_{10}$ and $SO_2F_2 + SOF_4$.
- Water inhibits, even more than oxygen, the formation of S_2F_{10} and of $S_2O_2F_{10}$ and $S_2O_3F_6$, and encourages that of S_2OF_{10} (at least under high-energy sparks) and $SO_2F_2 + SOF_4$.
- Finally, the addition of 50 % CF_4 to SF_6 either without other additives or containing O_2 or H_2O hardly modifies the levels of formation of the main gaseous decomposition products of SF_6 during high-energy sparking and even brings about a large decrease of the levels of $SOF_2 + SF_4$ and S_2F_{10} on low-intensity sparking. These results can, it seems, be adequately explained by the effects of the partial pressure of O_2, H_2O and/or SF_6. In other words, CF_4 seems to have little or no effect on the yields of the gaseous products resulting from the spark-decomposition of SF_6.

ACKNOWLEDGEMENTS

Financial support from GEC Alsthom is gratefully acknowledged.

REFERENCES

1. Sauers, I., W. Ellis and L.G. Christophorou. « Neutral Decomposition Products in Spark Breakdown of SF_6 ». IEEE Trans EI. Vol. 21, 1986, pp. 111-120.

2. Chu, F.Y. « SF_6 Decomposition in Gas-Insulated Equipment ». IEEE Trans EI. Vol. 21. No. 5, October 1986, pp. 693-726.

3. Sauers, I. « By-Product Formation in Spark Breakdown of SF_6/O_2 Mixtures ». Plasma Chem. Plasma Proc. Vol. 8. No. 2, June 1988, pp. 247-262.

4. Derdouri, A., J. Casanovas, R. Grob and J. Mathieu. « Spark Decomposition of SF_6/H_2O Mixtures ». IEEE Trans EI. Vol. 24, n° 6, December 1989, pp. 1147-1157.

5. Sauers, I., G. Harman, J.K. Olthoff and R.J. Van Brunt. « S_2F_{10} Formation by Electrical Discharges in SF_6 : Comparison of Spark and Corona ». Gaseous Dielectrics VI, Plenum, New-York, 1991, pp. 553-562.

6. Sauers, I. and S.M. Mahajan. « Detection of S_2F_{10} Produced by a Single-Spark Discharge in SF_6 ». J. Appl. Phys. Vol. 74. No. 3, August 1993, pp. 2103-2105.

7. Sauers, I., S.M. Mahajan and R.A. Cacheiro. « Production of S_2F_{10}, S_2OF_{10} and $S_2O_2F_{10}$ by Spark Discharges in SF_6 ». Gaseous Dielectrics VII, Plenum, New-York, 1994, pp. 423-431.

8. Castonguay, J. and I. Dionne. « S_2F_{10} and other Heavy Gaseous Decomposition Byproducts Formed in SF_6 and SF_6-Gas mixtures Exposed to Electrical Discharges ». 7th International Symposium on Gaseous Dielectrics, April 24-28, 1994, Knoxville, paper 61.

9. Piemontesi, M. and W. Zaengl. « Analysis of Decomposition Products of Sulfur Hexafluoride by Spark Discharges at Different Spark Energies ». 9th International Symposium on High Voltage Engineering, August 28 - September 1, 1995, Graz, pp. 2283-1 - 2283-4.

10. Van Brunt, R.J., J.K. Olthoff, S.L. Firebaugh and I. Sauers « Measurement of S_2F_{10}, S_2OF_{10} and $S_2O_2F_{10}$ Production Rates from Spark and Negative Glow Corona Discharge in SF_6/O_2 Gas Mixtures ». 11th International Conference on

GasDischarges and Their Applications, September 11-15, 1995, Tokyo, pp. I-316 - I-319.

11. Sauers, I. « Effect of Water Vapor on the Production of S_2F_{10} and S_2OF_{10} by Spark Discharges in SF_6 ». 11th International Conference on Gas Discharges and Their Applications, September 11-15, 1995, Tokyo, pp. I-320 - I-323.

12. Belarbi, A., C. Pradayrol, J. Casanovas and A.M. Casanovas. « Influence of Discharge Production Conditions, as Pressure, Current Intensity and Voltage Type, on SF_6 Dissociation under Point-Plane Corona Discharges ». J. Appl. Phys. Vol. 77, no. 4, February 1995, pp. 1398-1406.

13. Castonguay, J. and I. Dionne. « Analysis of S_2F_{10} in Decomposed SF_6 by Gas Chromatography : Various Aspects ». 7th International Symposium on Gaseous Dielectrics, April 24-28, 1994, Knoxville, paper 41.

14. Pradayrol, C., A.M. Casanovas, A. Hernoune and J. Casanovas. « Spark Decomposition of SF_6 and SF_6 + 50 % CF_4 Mixtures ». Proposed for publication to J. Phys. D : Appl. Phys.

Conference Record of the 1996 IEEE International Symposium on Electrical Insulation, Montreal, Quebec, Canada, June 16-19, 1996

Sorption of SF_6 and SF_6 decomposition products by activated alumina and molecular sieve 13X

M. Piemontesi[#], L. Niemeyer[*]

[#]High Voltage Laboratory, Swiss Federal Institute of Technology, 8092 Zürich, Switzerland

[*]ABB Corporate Research, 5405 Baden-Dättwil, Switzerland

***Abstract*: This paper presents measurements of the adsorption characteristics of various SF_6 decomposition products on activated alumina and on a molecular sieve type 13X (synthetic zeolite). The adsorption process is analysed by measuring the concentrations of the gas component to be adsorbed as well as its desorbed reaction products by IR absorption in a 2.5 litre stainless steel cell. The adsorption characteristics (adsorption isotherms) and the chemical reactions in the adsorber were determined for the pure gases SF_4, SOF_2, SO_2, SOF_4, SO_2F_2, S_2F_{10}, $S_2O_2F_{10}$ and SF_6. "Precharging" the adsorber with humidity has been shown to lower the adsorption. The importance of the chemical reaction energy associated with chemisorption processes was established by measuring the temperature of the adsorber substances during the adsorption process. The transport mechanisms responsible for the transfer of the adsorbable gas from the volume into the adsorber (adsorption kinetics, diffusion, convection) were identified by comparing the measured decay times of pure SO_2 as well as SO_2 in SF_6 with and without forced convection. A preliminary attempt is made to understand the physical chemistry of the observed phenomena.**

1. INTRODUCTION

SF_6 insulated electric power equipment is normally equipped with adsorbers which have the task to remove humidity and reactive/corrosive/toxic gaseous decomposition products. Humidity originates mainly from desorption from polymeric insulation materials and from surfaces previously exposed to atmospheric humidity. SF_6 decomposition products may be generated in insulation compartments by partial discharges when insulation defects are present. They are inevitably generated in compartments which contain switchgear such as disconnector switches, grounding switches, load break switches, and circuit breakers.

The adsorbers are provided for controlling the humidity level and for keeping the concentrations of reactive decomposition products below a level at which they would start to impair the function of the equipment or constitute a health risk for personnel. The latter issue has been extensively discussed in the past with respect to highly toxic SF_6 decomposition products such as S_2F_{10} [1], [2], [3] and $S_2O_2F_{10}$ [1], [4].

In addition to the above functions, adsorbers also play an important role in an environmental context. Like all complex and chemically stable fluorine compounds, SF_6 is a greenhouse gas [5] and as such should not be released into the atmosphere but recycled as much as possible [6]. Recycling is a natural part of SF_6 handling due to the contained character of its application in electrical power equipment [7]. An important part of SF_6 recycling equipment are adsorbers which remove chemically reactive decomposition products from the gas to be recycled. A thorough knowledge of these materials is thus essential for the design of efficient recycling equipment.

In the above context, various research activities on adsorbers for SF_6 equipment have been triggered. Early work on adsorbers includes [8], [9], [10] and a literature review [11]. In a recent publication T. Lussier, M.F. Frechette and R. Y. Laroque [12] presented a procedure for the study of gas/adsorbent dynamics.

In the present paper, an attempt is made to approximately quantify some of the technically most relevant aspects of the adsorption process, namely

- the equilibrium relation between adsorbate partial pressure and adsorber charging (adsorption isotherms)
- the time history of the adsorption process
- the interaction between competing adsorbates

There is a wide variety of adsorber materials used in SF_6 equipment which will not be comprehensively treated in this paper. Instead, two typical materials will be chosen to illustrate the main issues, namely, activated alumina as the "traditional" adsorber material and the molecular sieve 13X (synthetic zeolite) as an example of a "designed" adsorber material with extremely high internal surface area.

2. EXPERIMENTAL METHODS

The gas adsorption on activated alumina or on molecular sieve type 13X at room temperature is directly analysed by measuring the partial pressure decay of the adsorbate by IR absorption in a 2.5 litre stainless steel gas cell (see figure 1). The adsorbers are placed in a 0.03 litre glass chamber, initially isolated by a butterfly valve of high conductance when open. The gas cell and the adsorber chamber can be evacuated separately. The molecular sieve 13X, an alkali alumino-silicate, with an effective pore opening of about 10 Å is used in form of spherules with an average diameter of 2-3 mm. The activated alumina, an activated aluminium oxide, is also used in form of spherules with an average diameter of 2-5 mm. The adsorbers are prepared by a 10-hour vacuum bake at 120 °C and stored in a vacuum-tight container.

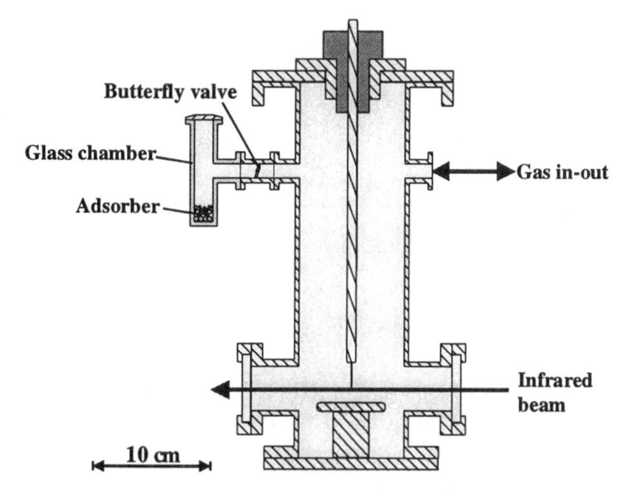

Figure 1. 2,5 litre stainless steel gas cell connected with glass chamber for adsorption experiments

After measuring the weight at ambient temperature, a quantity of 1 to 6 grams of the adsorber substance was introduced into the glass chamber and then the chamber was evacuated down to about 1 Pa. The stainless steel cell was separately evacuated (butterfly valve closed) to less than 1 Pa and the gaseous adsorbates were introduced either in pure form or diluted with SF_6 at 100 kPa as a background gas.
We call the procedure discussed above **"standard"** adsorber preparation. In this case the estimated residual water content is (2.2 +/- 0.8) mol/kg for 13X and (1.2 +/- 0.4) mol/kg for activated alumina. The water content was estimated by heating the adsorber to a temperature higher than 400 °C with a gas heater and measuring the amount of desorbed water. The humidity "precharge" obtained with the "standard" preparation procedure comes close to that encountered with the normal handling practice with SF_6 insulated equipment.
To analyse the effect of "preadsorbed" water we have the possibility to "dry" the adsorber at a temperature higher than 400 °C under vacuum at a pressure of about 0.3 Pa. The residual water content for a **"dry"** adsorber is lower than 0.2 mol/kg (estimated by water adsorption-desorption experiments).
Subsequently the **"wet"** adsorbers are prepared by "charging" "dry" ones with water, 2.5 mol/kg for activated alumina and 4 mol/kg for 13X.
The gas losses on the cell surface and by gas phase hydrolysis (for very reactive gases) have been assessed by checking the stability of the gases of interest in the 2.5 l cell before commencing the adsorption experiment (butterfly valve closed). The losses are very low in comparison to the total quantity of gas adsorbed during the experiment.
After the measurement of the initial gas phase concentration of the different adsorbates, the experiment was started by opening the butterfly valve. By comparing the initial gas concentrations of the different adsorbates with the concentrations in the course of the experiment, the adsorbed quantities can be determine in dependence on time. With this method we can also determine the adsorption equilibrium.
The IR absorption measurements were carried out directly in the gas cell (path length 0.175 m) by means of a FT-IR spectrometer equipped with a DTGS detector for the spectral range 400 cm^{-1} to 4000 cm^{-1} with a resolution of 0.5 cm^{-1}.
Quantitative analysis was made by simple IR-spectra subtraction. The calibration of the FT-IR spectrometer was carried out with pure gas samples of SF_4, SOF_2, SO_2, SOF_4, SO_2F_2, S_2F_{10}, $S_2O_2F_{10}$ and H_2O and with these pure gases diluted in argon or SF_6 with a background pressure of 100 kPa. The sensitivities obtained varied from some Pa to below 1 Pa partial pressure. A quantitative calibration procedure for the IR absorption of H_2O results in an uncertainty of the water concentration of about +/- 15%.

3. ADSORPTION EQUILIBRIUM RELATIONS

For the interaction of SF_6 decomposition products with adsorber materials two limiting cases can be distinguished. If the adsorption occurs reversibly and without chemical reaction, it will be referred to as **"physisorption"** in this paper. If the adsorbate reacts chemically and the adsorber material is irreversibly transformed, it will be referred to as **"chemisorption"**. In reality adsorption processes generally lie between these two extremes but in the case of SF_6 decomposition products they can be taken as good approximations.
It has to be noted that the concepts of physi- or chemisorption only apply to a specific combination of adsorbate and adsorber and are not general properties of adsorbate or adsorber materials. In particular, a specific adsorber material may be a "chemisorber" for one adsorbate and a "physisorber" for another one according to the respective reactivities.

3.1. Physisorption

3.1.1 General considerations

If an adsorber is exposed to an adsorbate gas, the latter is adsorbed until an equilibrium situation is reached in which the adsorber contains a specific "charge" (measured as the molar quantity M [mol] of adsorbate gas divided by the adsorber mass m [kg]) and a residual gas pressure p remains in the gas volume. The equilibrium relation $\mu = M/m = \mu(p)$ between the adsorber charge μ and the residual adsorbate pressure p is referred to as the adsorption isotherm [13]. The latter generally depends on the temperature T, the adsorbate gas, the adsorber material, and on precharges μ_i of the adsorber with other competing adsorbates:

$$M/m = \mu = f(p, T, \text{adsorbate, adsorber}, \mu_i) \quad (1)$$

The form of the adsorption isotherm is determined by details of the molecular interaction between adsorbate and adsorber.

A general classification of adsorption isotherms has been developed by Brunauer et al. reported in [13]. The physisorption type gases studied in this work can be classified as Type I or II following the classification cited above. As an example, figure 2 shows the adsorption isotherms of water vapour for activated alumina and for molecular sieve 13X.

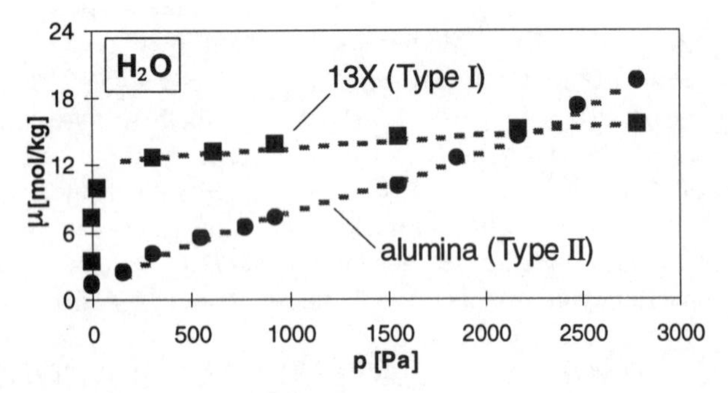

Figure 2. Adsorption isotherms of water on activated alumina (●) and on molecular sieve 13X (■) [14].

Adsorption isotherms of the Types I and II have the following general features:

(a) μ increases monotonously with the residual pressure p. For a specific application it is required that the pressure of a specific adsorbate does not exceed a specific maximal value p_1. A simple measure of the efficiency of adsorber for this adsorbate is the "charge" pertaining to this pressure:

$$\mu_1 = \mu(p_1) \qquad (2)$$

In this paper, μ_1 will be referred to as the **"adsorption efficiency"**. Note that the adsorption efficiency thus defined is not a general property of the adsorbate or adsorber but only reflects the adsorption efficiency for the particular application.

For the specific case of SF_6-insulated power equipment, the values of p_1 for the various SF_6 contaminants are those up to which the function of the equipment is not degraded or *above* which toxic risk may occur. Values for these limit pressures are presently being discussed in the CIGRE Working Group 23-10 and will be published [7]. In this paper we use p_1 = 100 Pa for all reactive and toxic SF_6 decomposition products to provide a sufficient safety margin. For water p_1 = 500 Pa and for SF_6 p_1 = 100 kPa are chosen. Preliminary measured adsorption efficiencies μ_1 are given in table 1.

(b) The slope ($d\mu/dp$) of the adsorption isotherm $\mu(p)$ is normally high at low pressure p and tends to decrease with increasing pressure (only in the low pressure region for Type II).

(c) At very high pressures μ tends to saturate (with Type I).

(d) The adsorption isotherms are shifted to lower μ-values if the adsorber is "precharged" by one or more competing adsorbates. This competition is not, in general, an additve phenomenon because the adsorption sites in the adsorber may be different for different adsorbates. This means that one adsorbate only partly blocks sites for another adsorbate so that more adsorbate molecules can be accommodated than would be expected on the basis of a simple addition. The degree to which this is the case has to be explored experimentally for all adsorbate combinations of interest.

(e) With increasing temperature T the isotherms are shifted to lower μ-values, i.e. the adsorber becomes less efficient. When it has been "charged" at low temperature it releases adsorbate when the temperature is increased.

In SF_6-insulated power equipment the features (a), (d), and (e) are of highest interest with respect to the SF_6 contaminants. The feature (c) is only relevant for SF_6 itself as it is present at very high pressure so that its adsorption isotherm is in the saturation range. The feature (d) is particularly important because the adsorbers are usually exposed, in addition to the background gas SF_6, to a mixture of different contaminants. The two most important "precharges" are the background gas SF_6 and humidity. The latter is partly introduced with the adsorber material which has inevitably been exposed to atmospheric humidity during assembly and partly originates from the desorption of humidity from enclosure walls and polymeric insulators after installation. The feature (e) is important up to about 100 °C which is the highest operating temperature of the equipment.

3.1.2 Experimental results

For both adsorber materials studied here, namely, activated alumina and molecular sieve 13X, it is found that the gases SF_6, SO_2, SO_2F_2, S_2F_{10}, $S_2O_2F_{10}$ and H_2O are physisorbed. Adsorbers prepared by the "standard" procedure are used to measure the adsorption isotherms of SF_6, SO_2, SO_2F_2 and S_2F_{10} at ambient temperature. The resulting isotherms are shown in figures 3 to 6. It is seen that the isotherms are not linear in the pressure range up to p_1.

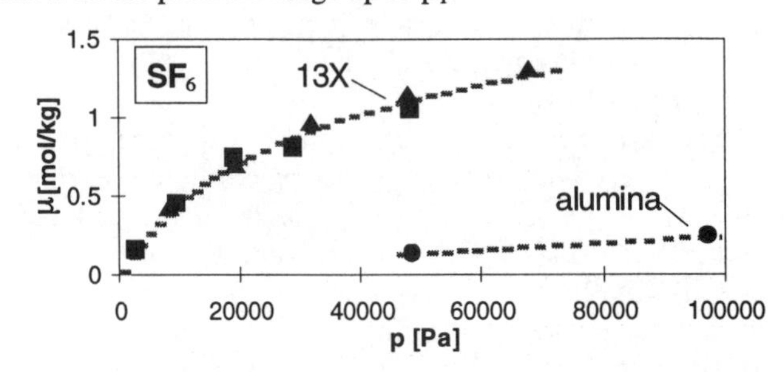

Figure 3. Adsorption isotherms of SF_6 on activated alumina (●) and on molecular sieve 13X (■) (our experiment) and data given for 13X by D. Berg and W. M. Hickam [15](▲).

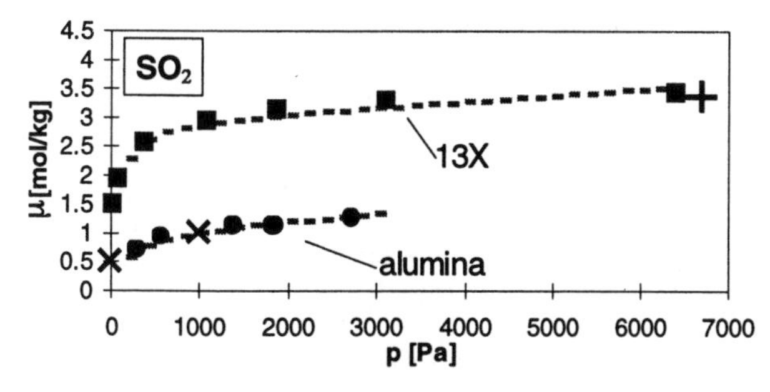

Figure 4. Adsorption isotherms of SO_2 on activated alumina (●) and on molecular sieve 13X (■) and of SO_2 diluted in SF_6 at 100 kPa on activated alumina (✖) and molecular sieve 13X (✚).

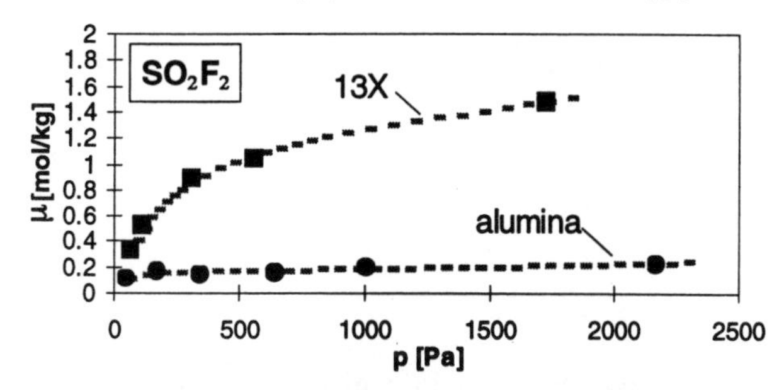

Figure 5. Adsorption isotherms of SO_2F_2 on activated alumina (●) and on molecular sieve 13X (■).

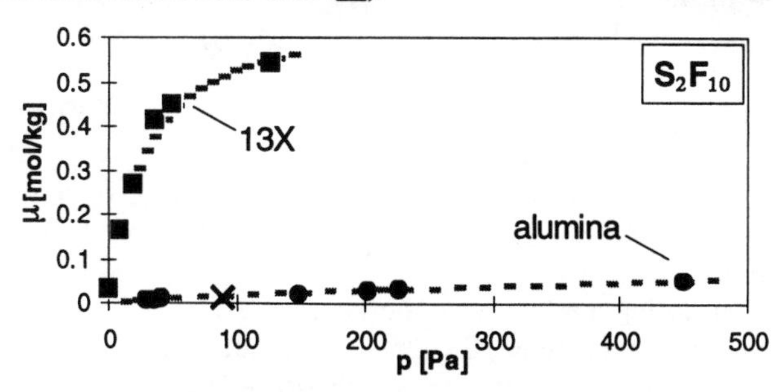

Figure 6. Adsorption isotherms of pure S_2F_{10} on activated alumina (●) and on molecular sieve 13X (■) and of a S_2F_{10} diluted in SF_6 at 100 kPa on activated alumina (✖).

The adsorption efficiencies derived from these data are listed in the upper section of table 1. It is seen that:

(1) The adsorption efficiencies of the two adsorber materials are quite different being higher for 13X than for activated alumina. This is probably due to the higher specific internal surface of the molecular sieve 13X [13].

(2) For both adsorber materials the SF_6 decomposition products can be ranked in order of increasing adsorbtion efficiencies from the decafluorides (S_2F_{10} and $S_2O_2F_{10}$) via SO_2F_2 to SO_2. The fact that the adsorption efficiency is low for SF_6 and high for humidity (see figures 2 and 3) is advantageous for SF_6 insulated equipment as the SF_6 is required to stay in the gas volume whereas the humidity has to be removed from it.

(3) For typical SF_6 operating pressures of 120 to 500 kPa, the SF_6 "precharge" can be extrapolated to be less than 1 mol/kg for alumina and a few mol/kg for 13X. This is lower than the respective humidity charges of 5 mol/kg and 13 mol/kg at 500 Pa, respectively. The charge for SO_2F_2 and SO_2 are between 0.15 and 0.7 mol/kg for alumina and 0.52 and 2 mol/kg for 13X at 100 Pa. For S_2F_{10} and $S_2O_2F_{10}$ the charges are between 0.015 and 0.05 mol/kg for alumina and 0.5 mol/kg for 13X, at 100 Pa (see table 1).

(4) Charging the adsorber with SF_6 at high pressure does not cause a significant effect on the adsorption efficiency for SO_2 and S_2F_{10} as can be seen from figures 4 and 6. As the SF_6 charge is much higher than the S_2F_{10} charge this means that the adsorbed SF_6 molecules do not block the adsorption sites for the S_2F_{10} molecules. Preliminary measurements of the interaction of the adsorbates H_2O, SO_2F_2, and $S_2O_2F_{10}$ with SF_6 indicate a similar behaviour. This non-interference of SF_6 in the adsorption of SF_6 contaminants is another advantageous feature for SF_6-insulated equipment.

(5) The effect of a humidity precharge was explored for the adsorbates SO_2 and S_2F_{10}. Figure 7 shows the change of the SO_2 adsorption isotherms for activated alumina and 13X with a different humidity precharge. At an SO_2 equilibrium pressure of about 1000 Pa a humidity precharge of 2.5 mol/kg (alumina "wet") reduces the adsorption efficiency for activated alumina by about 25 %. At about the same pressure, a humidity precharge of 4 mol/kg (13X "wet") reduces the adsorption efficiency for 13X by about 40%. At lower SO_2 equilibrium pressure the reduction effect is lower for both adsorbers. A similar tendency was observed for S_2F_{10} in preliminary experiments. This shows that humidity only partly blocks the adsorption sites for SO_2 and S_2F_{10}. More detailed measurements remain to be made for the other adsorbates. An important practical consequence of the above findings is that the humidity inevitably introduced into the adsorber material during equipment assembly causes a non-negligible reduction of the adsorption efficiencies for SF_6 decomposition products .

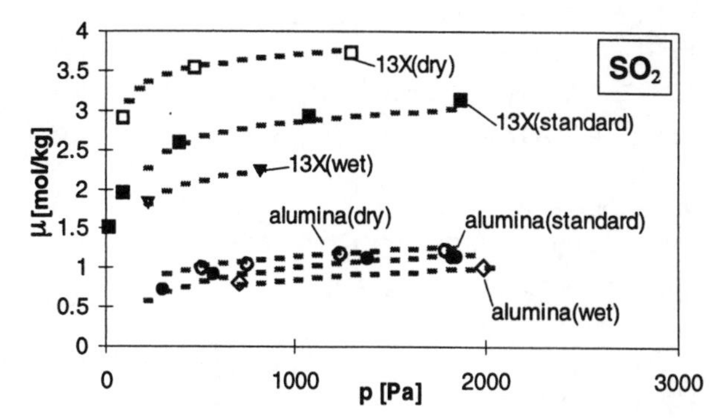

Figure 7. Adsorption isotherms of SO_2 on "dry" (○),"standard"(●) and "wet" activated alumina (◇) and on "dry" (□),"standard" (■) and "wet" molecular sieve 13X (▼).

(6) The effect of a temperature increase on the adsorption efficiency has been explored for SO_2. Figure 8 shows the SO_2 adsorption isotherms for activated alumina and 13X at 25°C and at 100°C, the latter being the highest operating temperature of power equipment. It is seen that at a pressure of about 5 kPa the temperature increase reduces the adsorption capability of alumina and 13X by about 30% and 15%, respectively. At lower SO_2 equilibrium pressure the adsorption reduction is lower.

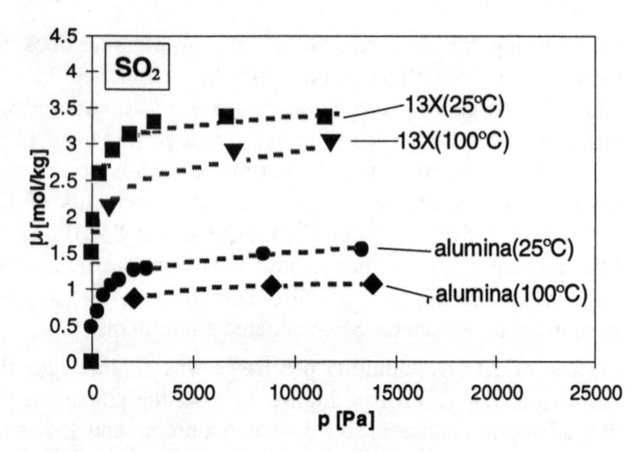

Figure 8. Adsorption isotherms of SO_2 on alumina at 25 °C (●) and 100 °C (◆) and on 13X at 25 °C (■) and 100 °C (▼)

3.2. Chemisorption

3.2.1 General considerations

The limiting case of chemisorption occurs when adsorbate and adsorber react irreversibly with one another. The adsorber materials studied here have the chemical compositions Al_2O_3 (activated alumina) and $Na_2O{\bullet}Al_2O_3{\bullet}2.5SiO_2{\bullet}nH_2O$ (molecular sieve 13X). From a thermodynamic point of view (minimal Gibbs enthalpy), these substances could react irreversibly with adsorbates like SF_4, SOF_2, and SOF_4 via direct fluorination reactions of the following type :

$$3\,SF_4 + 2\,Al_2O_3 \rightarrow 4\,AlF_3 + 3\,SO_2 \qquad (3)$$
$$3\,SF_4 + Al_2O_3 \rightarrow 2\,AlF_3 + 3\,SOF_2 \qquad (4)$$
$$3\,SOF_2 + Al_2O_3 \rightarrow 2\,AlF_3 + 3\,SO_2 \qquad (5)$$
$$3\,SOF_4 + Al_2O_3 \rightarrow 2\,AlF_3 + 3\,SO_2F_2 \qquad (6)$$

$$SF_4 + SiO_2 \rightarrow SiF_4 + SO_2 \qquad (7)$$
$$2\,SOF_2 + SiO_2 \rightarrow SiF_4 + 2\,SO_2 \qquad (8)$$
$$SF_4 + 3Na_2O \rightarrow 4\,NaF + Na_2SO_3 \qquad (9)$$

The first 4 reactions refer to both alumina and 13X and the last three ones to 13X only. It is seen that in all reactions but the last one each adsorbate molecule is transformed into one SO_2 or SO_2F_2 molecule so that the molar sum of all adsorbate destroyed is equal to the molar quantity of SO_2 or SO_2F_2 produced.

As, in practice, the adsorber is inevitably precharged with some moisture due to handling in atmospheric air, an additional set of reactions is possible in which H_2O hydrolyses the adsorbates and is subsequently recycled:

$$3\,SF_4 + 3\,H_2O \rightarrow 3\,SOF_2 + 6\,HF \qquad (10a)$$
$$3\,SOF_2 + 3\,H_2O \rightarrow 3\,SO_2 + 6\,HF \qquad (10b)$$
$$12\,HF + 2\,Al_2O_3 \rightarrow 6\,H_2O + 4\,AlF_3 \qquad (10c)$$
$$3\,SF_4 + 2\,Al_2O_3 \rightarrow 3\,SOF_2 + 4\,AlF_3 \qquad (10d)$$

$$3\,SOF_2 + 3\,H_2O \rightarrow 3\,SO_2 + 6\,HF \qquad (11a)$$
$$6\,HF + Al_2O_3 \rightarrow 3\,H_2O + 2\,AlF_3 \qquad (11b)$$
$$3\,SOF_2 + Al_2O_3 \rightarrow 3\,SO_2 + 2\,AlF_3 \qquad (11c)$$

$$3\,SOF_4 + 3\,H_2O \rightarrow 3\,SO_2F_2 + 6\,HF \qquad (12a)$$
$$6\,HF + Al_2O_3 \rightarrow 3\,H_2O + 2\,AlF_3 \qquad (12b)$$
$$3\,SOF_4 + Al_2O_3 \rightarrow 3\,SO_2F_2 + 2\,AlF_3 \qquad (12c)$$

It is seen that these reactions are catalysed by H_2O and lead to the same end products as the direct fluorination reactions with the same molar SO_2 or SO_2F_2 yield. Similar reaction cycles might also exist for silica (SiO_2) but will not be discussed here.

For all the above reactions the damage to the filter by fluorination can be determined from the number of AlF_3 molecules generated per adsorbed reactive gas molecule. This is seen to be twice as strong for SF_4 as for SOF_4 and SOF_2.

Concerning the relative importance of the above reactions at ambient temperature, the direct fluorination reactions (7) and (8) are known to be very much slower than corresponding hydrolytic reactions [16]. Evidence for reaction (11) is shown in figure 14 which gives a comparison of the time history of adsorption of SOF_2 in "dry" and humidity "precharged" alumina. This indicates that the velocity of the chemisorption is much higher for the humidity precharged adsorber than for the "dry" adsorber. It is therefore expected (though still to be proven for the remaining reactions) that the humidity-catalysed reactions are much faster than the direct fluorination reactions and therefore determine the kinetics of the chemisorption of SF_6 decomposition products under practical conditions.

The transformation of chemisorbed SOF_2 to SO_2 in activated alumina had already been observed by Boudene [9].

Chemisorption of reactive SF_6 decomposition products is thus seen to lead to a fluorination of the adsorber material and to the formation of the gaseous end products SO_2 and SO_2F_2. These are stable in the SF_6 environment and are subject to physisorption in the adsorber in which they are generated. Their release is therefore controlled by their physisorption isotherms. However, these isotherms are those of the chemically transformed adsorber material. For a precise quantification of the chemisorption process it would therefore be necessary to determine, for each gaseous end product, a family of adsorption isotherms at various degrees of chemical degradation.

As a first approximation we use a simplified approach by defining a **"virgin chemisorption curve"** which gives the partial pressure p of the released gaseous reaction products (SO_2 or SO_2F_2) with increasing adsorber charge μ when the adsorber is exposed to a reactive adsorbate for the first time.

This chemisorption curve can be viewed as a continuous transition through the family of physisorption isotherms associated with the progressive degradation of the material.

For practical purposes it is useful to characterise the virgin chemisorption curve by an adsorption efficiency in the same way as the physisorption curves (see equation (2)).

3.2.2. Experimental data

Figure 9 shows the measured virgin chemisorption curves of SOF_2 on activated alumina after "standard" preparation with SO_2 as reaction product. It is seen that μ may recede again as the damage increases and reduces the number of adsorption sites for SO_2.

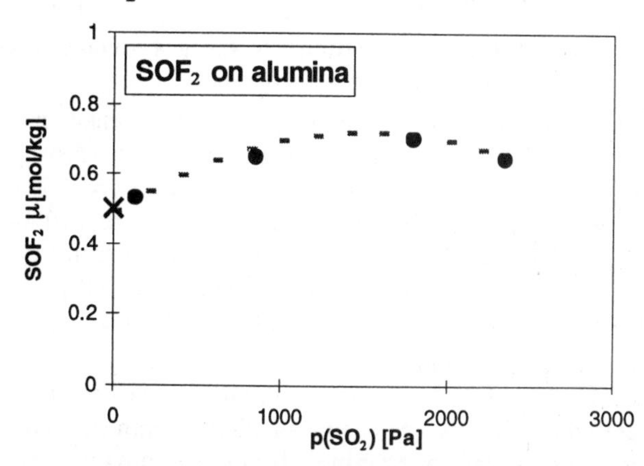

Figure 9. Chemisorption curve of SOF_2 on "standard" alumina (●) and of SOF_2 diluted in SF_6 at 100 kPa on "standard" alumina (✖).

In order to explore the degradation of the physisorption efficiency directly, the physisorption curves of alumina and 13X for SO_2 were measured in the virgin state and in the absence of a humidity precharge ("dry" adsorber preparation), see figures 10 and 11. The same adsorber materials were subjected to fluorination by SF_4 up to a charge of about 0.7 mol/kg for activated alumina and 1 mol/kg for 13X. The adsorbers thus degraded were then "emptied" by evacuation and heating to 400 °C and their physisorption curves were once more measured with SO_2, see figures 10 and 11.

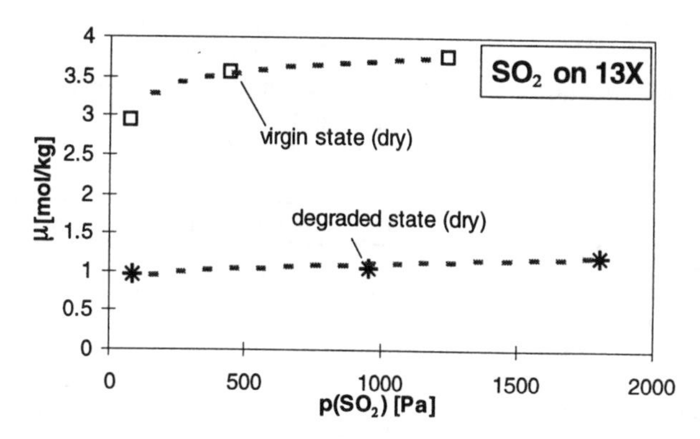

Figure 10. Adsorption isotherms of SO_2 for "dry" 13X (□), and after a 1 mol/kg SF_4 "precharge" and a subsequent "dry" preparation for 13X (✱).

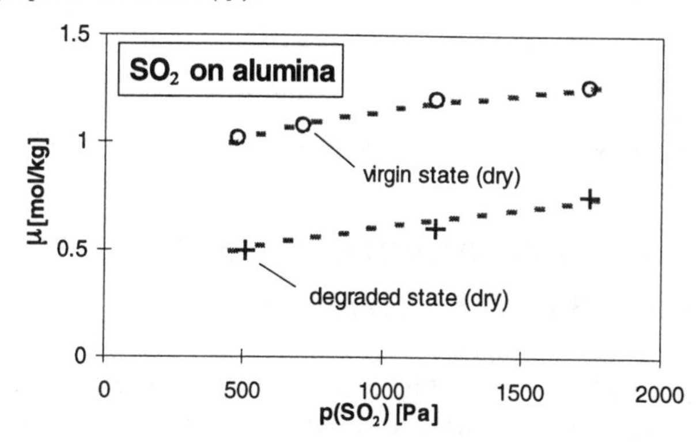

Figure 11. Adsorption isotherms of SO_2 for "dry" alumina (O) and after a 0.7 mol/kg SF_4 "precharge" and a subsequent "dry" preparation for alumina (+)

It is seen that the damage due to chemisorption degrades the physisorption efficiency of both adsorbers substantially. The chemisorption efficiencies measured hitherto are given in the lower part of table 1.

The following conclusions can be drawn from the experimental data:

(1) Like physisorption, chemisorption also occurs more efficiently in the 13X material than in alumina.

(2) An SF_6 "precharge" does not measurably affect the chemisorption efficiency for SOF_2.

(3) The chemisorption efficiencies rank in increasing order from SOF_4 via SF_4 to SOF_2 and are in the range of 0.08 to 0.55 mol/kg for alumina and 0.3 to 1.3 mol/kg for 13X.

(4) The degradation of the physisorption characteristics by chemisorption can be substantial. With an SF_4 precharge the physisorption-capability for SO_2 is here reduced by about 50 % for alumina and by about 70 % for 13X (see figures 10 and 11). A chemically degraded adsorber has thus a substantially reduced physisorption capability.

Table 1: Adsorption efficiencies μ_1 of SF_6, H_2O and various toxic and corrosive SF_6 decomposition products on activated alumina and molecular sieve 13X ("standard" preparation procedure).

Gas Type	Pressure p_1	Adsorber efficiencies μ_1[mol/kg]	
		13 X	Alumina
SF_6	100 kPa	1.5	0.3
H_2O	500 Pa	13	5
SO_2	100 Pa	2	0.7
SO_2F_2	100 Pa	0.52	0.15
S_2F_{10}	100 Pa	0.5	0.015
$S_2O_2F_{10}$	100 Pa	0.5	0.05
SOF_2	100 Pa	1.3	0.55
SF_4	100 Pa	1	0.5
SOF_4	100 Pa	0.3	0.08

4 ADSORPTION DYNAMICS

4.1 Basic considerations

In practical applications it is important to know the time required to remove SF_6 decomposition products from the equipment. In general, three characteristic time scales are involved in this process:

(1) The time t_{trans} required to transport the decomposition products from the location where they have been generated to the location of the adsorber.

(2) The time t_{pen} required for the decomposition products to penetrate into the interior of the adsorber package.

(3) The kinetic time constant t_{kin} of the adsorption process itsself.

The **kinetic time scale** can be determined experimentally by exposing a thin layer of the adsorber material to the "pure" adsorbate gas in the absence of any background gas. The full adsorbate concentration is then always present at the surface of the adsorber spherules.

To characterise the adsorption kinetics quantitatively, we define an initial kinetic adsorption time constant t_{kin}^0 which can be evaluated from the initial adsorbate pressure p_0 at the beginning of the adsorption process and the initial pressure decay rate $(dp/dt)_0$ as

$$t_{kin}^0 = [(dp/dt)_0/p_0]^{-1} \qquad (13)$$

This time constant can be interpreted by the simplest possible adsorption model of Langmuir [13]. It assumes that every adsorption site is equivalent and the probability of adsorption at this site is independent of whether or not nearby sites are occupied.

The number N of molecules lost from the gas volume is equal to the number N_a of adsorbed molecules and the latter is, with the above assumptions, proportional to the partial pressure p of the adsorbate gas, to the fraction of adsorption sites (1-Q) still vacant, and to the mass m of the adsorber where Q is the fraction of adsorption sites already occupied:

$$d N_a /dt = k_{ads} \cdot p \cdot m \cdot (1-Q) \qquad (14)$$

The proportionality factor k_{ads} is the adsorption velocity constant which, in general, depends on temperature T, adsorber-adsorbate combination, and on already existing precharges μ_i due to other adsorbates:

$$k_{ads} = f(T, \text{adsorbate/adsorber}, \mu_i, ...) \qquad (15)$$

This approximate relationship is valid both for physisorption and chemisorption.

At the beginning of the adsorption process, i.e. for $Q \rightarrow 0$, eqn. (15) yields, with $p = n \cdot k \cdot T = (N/V) \cdot k \cdot T$ the relation

$$k_{ads} = [(dp/dt)_0/p_0] \cdot (k \cdot T)^{-1} \cdot (V/m) \qquad (16)$$

where $n = N/V$ is the molecule density of the adsorbate in the gas and V the gas volume. The adsorption velocity constant k_{ads} can thus be determined from the experimental data p_0, $(dp/dt)_0$, V, T, and the adsorber mass m.

Once k_{ads} has been determined in this way from a specific experiment, the kinetic decay time constant can be determined for arbitrary conditions as

$$t_{kin}^0 = (k \cdot T \cdot k_{ads})^{-1} \cdot (V/m) \qquad (17)$$

As expected, t_{kin}^0 is proportional to the gas volume V and inversely proportional to the adsorber mass m.

The **penetration time t_{pen}** into a packaged adsorber depends on the configuration of the adsorber package, the density of the spherules in it, the surface of the package accessible to the gas, and other factors. It cannot be determined by a simple estimate but requires a solution of a usually three-dimensional transient reaction-diffusion problem [17]. In this paper, no attempt will be made to estimate this time t_{pen}.

Following a reasoning originally proposed by T. Ushio, I. Shimura and S. Tominaga [8], the **transport time t_{trans}** is determined by the geometry of the equipment and the disposition of the adsorber in it. To a first approximation this configuration can be represented by the schematic shown in figure 12.

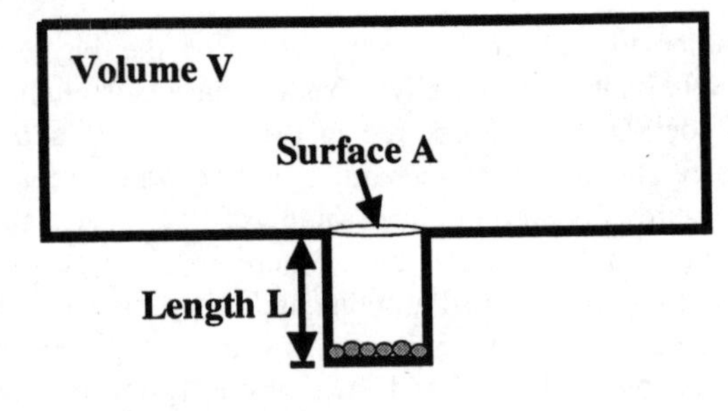

Figure 12. Schematic of gas diffusion model.

The decomposition products are generated in a generation volume V (e.g. the switching chamber of a circuit breaker) which is connected via some duct to the adsorber. The duct is represented as a channel of length L and cross sectional area A and the depth of the adsorber is ignored. In the volume V convective mixing normally occurs either by natural convection or by the flow associated with electric discharges (e.g. electric corona wind or turbulent gas flow generated by sparks and arcs). As a consequence the decomposition products are convectively mixed in the volume V so that their average particle density n becomes

$$n = N/V \qquad (18)$$

when N is the total number of adsorbate molecules.
At the surface of the adsorber, the concentration of the decomposition products is assumed to be approximately zero due to the proximity of the adsorber. The particle density gradient along the connection duct is thus approximately n/L. If we assume further, as a limiting case, that there is no convection in the duct, the adsorbate is transported to the adsorber by diffusion. In this case, the particle flux from the volume V to the adsorber results as

$$F = D \cdot A \cdot n/L \qquad (19)$$

where D is the diffusion coefficient of the adsorbate in the background gas. With eqs. (18) and (19) we can set up the particle balance in the volume V as

$$dN/dt = - F = - D \cdot A \cdot N/(L \cdot V) \qquad (20)$$

which yields the diffusive transport time constant

$$t_{trans} = [(dN/dt)/N]^{-1} = [(dp/dt)_0/p_0]^{-1}$$

$$t_{trans} = L \cdot V/(D \cdot A) \qquad (21)$$

The diffusion coefficients D for adsorbate molecules in SF_6 are of the same order of magnitude as those for SF_6 molecules. For the latter one has $(D \cdot p_{SF6}) = 1\ Pa \cdot m^2/s$ where p_{SF6} is the SF_6 background pressure.

4.2 Experimental data

4.2.1 Adsorption kinetics of "pure" gases

Figure 13 shows measured pressure decay curves for the adsorption of SO_2 on activated alumina and 13X. In the two cases the adsorbers were prepared by the "standard" procedure described in chapter 2. The kinetic time constants are different for the different adsorbates, but always in the same order of magnitude. The data show that the kinetic time constants t_{kin} are of the order of

$$t_{kin} = \text{several minutes} \qquad (22)$$

and that the adsorption occurs faster for 13X than for alumina. It is thus seen that the adsorption itself is a very fast process once the adsorbate has been transported to the adsorber.

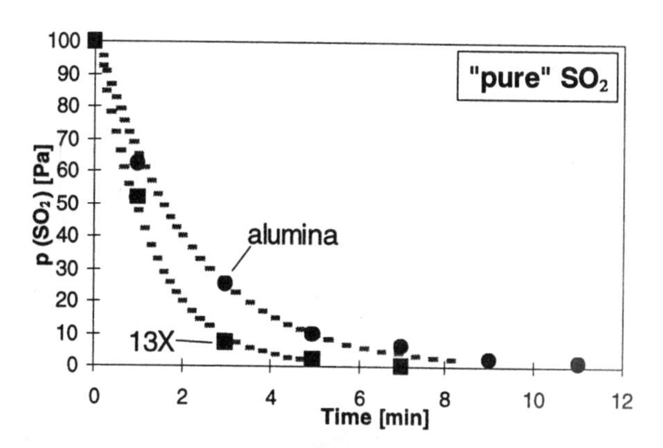

Figure 13. Adsorption kinetics of "pure" SO_2 on 1 g of activated alumina (●) and on 1 g of molecular sieve 13X (■)

The influence of a humidity "precharge" on the chemisorption process is illustrated by figure 14, which shows the chemisorption kinetics of SOF_2 for the case of a "dry" adsorber and for an adsorber prepared by "standard" procedure (residual water content of (1.2+/-0.4) mol/kg) .

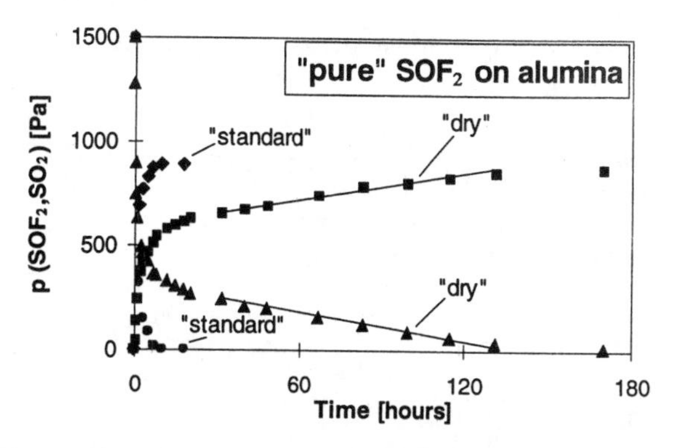

Figure 14. Adsorption kinetics of SOF_2 on 1 g "standard" alumina (● for adsoption of SOF_2 and ◆ for desorbed SO_2) and on 1 g "dry" alumina (▲ for adsoption of SOF_2 and ■ for desorbed SO_2)

One can recognise that the presence of humidity substantially accelerates the chemisorption process. This is interpreted as being due to the higher velocity of the hydrolysis-controlled reactions eqn. (11) as compared to the direct fluorination reaction eqn. (5) as discussed in chapter 3.2.1.
As the adsorption process is exothermic and rather fast, the released heat causes a heating-up of the adsorber material which was monitored roughly by a temperature sensor inserted into the adsorber material. An example of the measurements is given in figure 15 for the chemisorption of SF_4 on 13X. The temperature is seen to rise to 185 °C in the initial phase of the adsorption process. As the reaction kinetics are strongly temperature dependent this means that kinetic measurements have to be corrected for this effect.

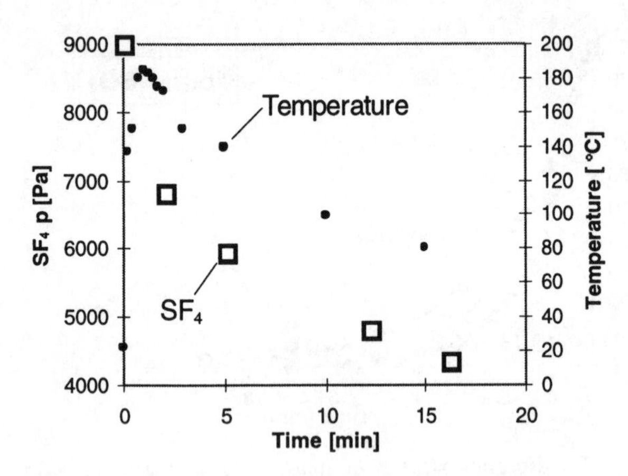

Figure 15. Measurement of the adsorber temperature (•)during the adsorption of SF_4 (□) on 6 g molecular sieve 13x after "standard" preparation procedure.

4.2.2 Adsorption kinetics in the presence of background SF_6

The role of the background gas for the adsorption dynamics is illustrated by figure 16 which shows the adsorption of SO_2 on alumina in the absence and in the presence of the background gas. It is seen that the adsorption is significantly delayed by the presence of the background gas. It is experimentally found that this delay can be eliminated by providing forced circulation between the gas volume and the adsorber. This confirms that in the absence of gas circulation the transport of the adsorbate to the adsorber is diffusion controlled in our measurement set up.

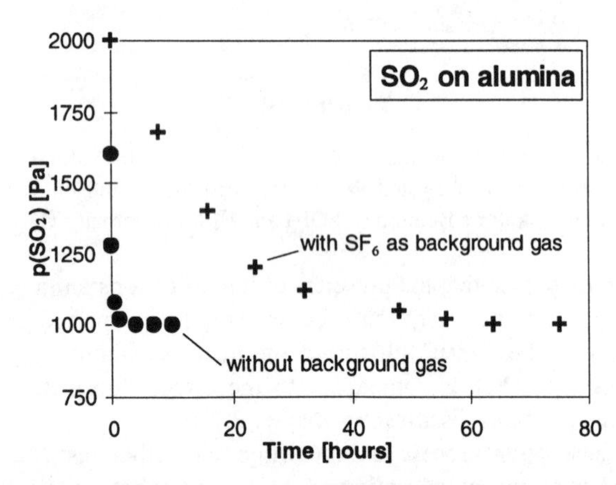

Figure 16. Adsorption kinetics of 2000 Pa SO_2 (●) and of 2000 Pa SO_2 diluted in SF_6 at 100 kPa (✚) on 1 g activated alumina.

Figure 17 shows normalised pressure decay plots for SOF_2 and SO_2 in the presence of a background pressure of 100 kPa SF_6 in the configuration of figure 1. It is seen that all adsorbates decay with the same time constant of about 70 h. Inserting the geometry data of the arrangement of figure 1 ($L \approx 0.15$ m, $A \approx 1.5 \cdot 10^{-4}$ m^2, $V = 2.5 \cdot 10^{-3}$ m^3), the SF_6 pressure $p_{SF6} = 100$ kPa, and the diffusion coefficient $D = 10^{-5}$ m^2/s into eqn. (21) one obtains a decay time constant of the same order of magnitude.

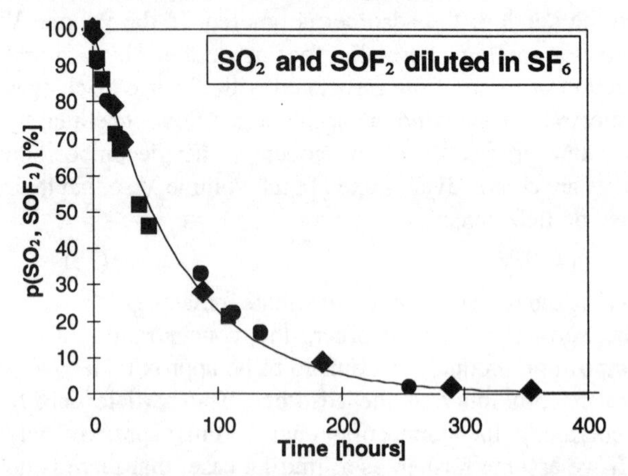

Figure 17. Adsorption kinetics for mixtures of 1500 Pa of SOF_2 on 1 g alumina (●), 1500 Pa of SOF_2 on 1 g molecular sieve 13x (■) and 500 Pa SO_2 on 1 g alumina (◆)diluted in SF_6 at 100 kPa.

An estimate of the diffusion-controlled time constant in practical SF_6 equipment by eqn. (21) yields, with typical values of $L = 0{,}05$ m, $V = 0{,}1$ m^3, $A = 10^{-2}$ m^2, $p_{SF6} = 500$ kPa, and $D = 2 \cdot 10^{-6}$ m^2/s

$$t_{trans} = 2.5 \cdot 10^5 \text{ s} \approx 3 \text{ days} \qquad (23)$$

which is in order of magnitude agreement with practical experience. It can thus be seen that

$$t_{trans} >> t_{kin} \qquad (24)$$

i.e. that the diffusion limited transport of the adsorbate from the volume in which it is generated to the adsorber determines the time required to remove SF_6 decomposition products from power equipment.

5. COMPOSITION OF CONTAMINANTS IN SF_6 POWER EQUIPMENT

The quantities of contaminants in SF_6 insulated power equipment which have to be handled by adsorbers depend on various factors like the equipment design, the projected lifetime operation cycles and the materials employed.
Humidity is normally introduced into the SF_6 enclosure by desorption from internal surfaces, particularly from rough metal surfaces, and from polymeric materials which can adsorb substantial quantities in the bulk. The quantity of humidity to cope with in a specific system depends on the assembly procedure and on material processing and cannot be specified in general.
The quantities of SF_6 decomposition products to be expected over the lifetime of the equipment depend on the discharge activity in the system.

In gas compartments which have only an insulation function, SF_6 decomposition products are not formed under normal operating conditions. They can only be expected in the case of insulation defects which cause partial discharge (PD) activity. In this case the size and position of the defects control the discharge intensity and with it the production rate of decomposition products. An outline of how the associated quantities can be estimated is given in [18].

In disconnector switches the spark current is usually very low and operation is infrequent so that the quantities of decomposition products generated over the lifetime of the equipment are quite low. Field sampling in old equipment not equipped with adsorbers indicates that the total concentrations of all decomposition products are definitely below 100 ppm_v (e.g. [19]). At an SF_6 filling pressure of some 500 kPa and typical compartment volumes of 0.2 m^3 this corresponds to a few mmol.

The only equipment components which, under normal operating conditions, produce sizeable quantities of decomposition products are load break switches and circuit breakers. In these, SF_6 decomposition is essentially controlled by the chemical reaction of thermally fragmented SF_6 with arc-evaporated contact and insulator materials. The primary gaseous decomposition product is SF_4. As it is highly reactive, it is transformed into secondary gaseous products like SOF_2, SO_2, HF and in metal fluorides by reactions with humidity and enclosure materials, and, last but not least, by chemisorption in adsorbers.

Again the quantities of decomposition products cannot be specified in general because they strongly depend on design, on the materials employed, on the switched currents, and on the operation sequence during the lifetime of the switchgear.

6. ADSORBER DESIGN

Adsorbers for gas compartments which have to fulfil only an insulating function have to cope mainly with humidity and with the decomposition products generated by partial discharges, such as SF_4, SOF_4, SOF_2 and SO_2F_2. SF_4 is hydrolysed, SOF_4 and SOF_2 are adsorbed, and their reaction products SO_2F_2 and SO_2 are physisorbed. The highly toxic decafluorides S_2F_{10} and $S_2O_2F_{10}$ are produced in much lower quantities [1], [2], [3], [4] and are subject to decomposition at the enclosure surface [1], [4], [18]. They need therefore not be taken into account when dimensioning an adsorber.

Adsorbers for switchgear compartments have to cope mainly with SF_4, SOF_2, SO_2, HF and with humidity. Their dimensioning has to be based on the specific design features, materials employed in the arc zone, currents to be interrupted, and the projected lifetime operating cycle. To a first approximation, the estimated quantity of the primary product SF_4 can serve as a guideline for adsorber dimensioning (see table 1).

7. SUMMARY AND CONCLUSIONS

(1) An attempt has been made to explore the major physico-chemical processes underlying the removal of humidity and SF_6 decomposition products from SF_6-insulated power equipment by adsorbers.

(2) For the adsorber materials usually applied, two types of adsorbates can be distinguished between, namely, "physisorbates" which do not react with the adsorber (SF_6, SO_2, SO_2F_2, S_2F_{10} and $S_2O_2F_{10}$,) and "chemisorbates" which react with it (SF_4, SOF_4, and SOF_2).

(3) Each adsorbate-adsorber combination can be quantitatively characterised by an adsorption efficiency for the specific application in SF_6 insulated power equipment.

(4) Two basic types of *interaction* between different adsorbates have to be accounted for:

- the physisorption of one adsorbate may partly block the adsorption of another one
- the chemisorption of reactive adsorbate impairs the physisorption efficiency for non-reactive adsorbates.

(5) The chemisorptive interactions tend to be stronger than the physisorptive interactions.

(6) The temporal dynamics with which contaminants are removed from SF_6-insulated power equipment seem to be controlled by diffusion.

(7) The measurements presented in this paper provide a first overview of the complex physico-chemical processes involved in the removal of contaminants from SF_6 equipment. They are by no means comprehensive.

(8) The following particular issues require further study:

- adsorbate interactions in multi-component adsorption
- adsorption characteristic of gases like HF, SiF_4, and WF_6 which have not been studied in this paper
- dimensioning methods which take multicomponent interaction and compound adsorber designs into account
- methods to quantify the adsorption dynamics in equipment accounting for the combined effects of convection, diffusion and adsorption kinetics.

ACKNOWLEDGEMENTS

The authors gratefully acknowledge helpful discussions with B. Brühl, H.J. Knab, A. Maggi, K. Seppelt, M. Textor and W. Zaengl. Special thanks also to Th. Heizmann and T.H. Teich for critical revision of the manuscript.

REFERENCES

[1] CRADA "Workshop on investigation of S_2F_{10} production and mitigation in compressed SF_6-insulated power systems", Pittsburgh, 6 June 1994

[2] M. Piemontesi, R. Pietsch and W. Zaengl "Analysis of decomposition products of sulphur hexafluoride in negative DC corona with special emphasis on content of H_2O and O_2", IEEE Int. Symp. on Electr. Insulation, Pittsburgh, 1994, pp. 499-503

[3] M. Piemontesi and W. Zaengl "Analysis of decomposition products of sulphur hexafluoride by spark discharges at different spark energies", 9th Int. Symp. on High Voltage Engineering, Graz, 1995, Paper No. 2283

[4] M. Piemontesi, L. Niemeyer "Generation and decay of $S_2O_2F_{10}$ in SF_6 insulation", 9th Int. Symp. on High Voltage Engineering, Graz, 1995, Paper No. 2284

[5] "Radiative Forcing of Climate Change" *The 1994 Report of the Scientific Assessment Working group of IPCC*, Summary for Policymakers WMO/UNEP

[6] CIGRE WG 23-10 "SF_6 and the Global Atmosphere", ELECTRA, December 1995

[7] CIGRE WG 23-10 TF 01 "SF_6 Recycling Guide", to be published in ELECTRA

[8] T. Ushio, I. Shimura and S. Tominaga "Practical problems on SF_6 gas circuit breakers", IEEE Winter Power Meeting, 1971, pp. 2166-2174

[9] C. Boudene, J-L. Cluet, G. Keib and G. Wind "Identification and study of some properties of compounds resulting from the decomposition of SF_6 under effect of electrical arcing in circuit-breakers" Revue Generale de l'Electricite, Special Issue, June 1974, pp. 45-78

[10] EPRI report "Study of arc by-products in gas-insulated equipment", EL-1646 Project 1204-1, Greensburg, 1980

[11] F. Y. Chu "SF_6 decomposition in gas-insulated equipment", IEEE Transactions on Electrical Insulation, Vol. 21 (5), 1986 , pp. 693-725

[12] T. Lussier, M.F. Frechette and R. Y. Laroque "Gas/adsorbent interaction dynamics: an experimental approach", IEEE Conference on Electrical Insulation and Dielectric Phenomena, Virginia Beach, 1995, pp.153-158

[13] W. Gerhartz "Ullmann's Encyclopedia of Industrial Chemistry", 5th edition, VCH, Weinheim, 1988, Vol. B 3, pp. 9/1-9/51

[14] A. Maggi (CU Chemie Uetikon AG), private communication

[15] D. Berg and W. M. Hickam "Sorption of sulfur hexafluoride by artificial zeolites", Journal of physical chemistry, Vol. 65, 1961, pp.1911-1913

[16] I. Sauers, H. W. Ellis and L.G. Christophorou "Neutral decomposition products in spark breakdown of SF_6", IEEE Transactions on Electrical Insulation, Vol. 21 (2) , 1986, pp. 111-120

[17] E. Dietrich, G. Sorescu and G. Eigenberger "Numerische Methoden zur Simulation verfahrenstechnischer Prozesse", Chemie Ingenieur Technik, Vol.64 (2), 1992, pp. 136-147

[18] L. Niemeyer "S_2F_{10} in SF_6 insulated equipment", Gaseous Dielectrics VII, L.G. Cristophorou and D. R. James (Eds.), Plenum Press, New York, 1994, pp. 239-245

[19] V.H. Tahiliani, K.B. Miners and W.J. Lannes "Diagnostic techniques to direct abnormal operating conditions in gas insulated substations", CIGRE, Paris, 1984, paper 23-08

Conference Record of the 1996 IEEE International Symposium on Electrical Insulation, Montreal, Quebec, Canada, June 16-19, 1996

Gaseous Contaminants in SF_6 Interacting with Alumina

T. Lussier, M.F. Fréchette and R.Y. Larocque
Câbles et Isolants, Technologie et IREQ,
Hydro-Québec, Varennes, Québec, Canada J3X 1S1

Abstract: **The interaction between CO_2 or SOF_2 (10^3 to 10^4 ppm$_v$) and alumina (10 g) in SF_6 was investigated. The adsorption kinetics was monitored by taking on-line gas samples. A gas chromatograph coupled to a thermal-conductivity detector served to assess the concentration of each species in the mixtures. Under the chosen experimental conditions, no matrix effect was detected, since the sorptive kinetics involving CO_2/SF_6 proved similar to that observed with CO_2/He. The adsorption of SOF_2 was found to be stronger than that observed for CO_2, although part was transformed into SO_2. When the alumina sample was exposed to CO_2 then to SOF_2, some of the CO_2 adsorbed was released, demonstrating a preferential adsorption on alumina.**

INTRODUCTION

The use of SF_6-insulated equipment in the field of electricity is growing in popularity due to its performance, reduced space requirements and low maintenance needs [1]. However, for a long-term high performance of the insulation and for safety, all gaseous impurities and/or by-products that appear under normal operating conditions (i.e. chemical or electrical stress) must be removed from the gas phase by adding an adsorbing material to the medium [2]. In this context, it is essential that the dynamics of the interactions be well understood and detailed.

In a previous work [3], an experimental methodology using a simple parametric system was developed and tested: trace amounts of CO_2 and CF_4 in a helium matrix interacting with alumina. Results showed that both the sorption capacity of alumina and the kinetic process depend strongly on the nature of the gaseous contaminants involved and their initial concentration, as well as on the exposure history of the adsorbing material. The present contribution will focus on trace amounts of either CO_2 or SOF_2 in an SF_6 matrix interacting with alumina. The former was chosen for comparison and to detect any matrix effect on the sorption kinetics, while the latter constitutes the by-product most often detected in SF_6-insulated equipment in common use [4-6].

SF_6 by-products are generally known to be strongly acid gases which should interact with the alumina and common sieves through basic or basic-like sites. However, they also have ligands containing free electron pairs which could react with acid-base sites, as expected for CO_2. Comparison of the sorption kinetics of both species should therefore lead to a better knowledge of the processes involved.

EXPERIMENTAL

Apparatus

The analytical arrangement in Figure 1 is described in a previous work [3]. The system consists mainly of a double-chamber test cell connected to a gas chromatograph (C) by a 2-m long tube (F). Part A (150 mL) of the cell contains the adsorbent tested (10 g) while part B (9.5 L) is filled at a pressure of 2 bar abs with different SF_6 gas mixtures. Opening the gate valve (G) between the two sections initiates the gas-adsorbent interaction. The concentration of all gaseous constituents can then be followed as a function of time by on-line sampling. The analytical measuring system comprises a chromatograph equipped with a rotary valve and a 100-µL sample loop in lieu of the injector, a Poraplot Q column (25 m x 0.32 mm with a 10-µm adsorbent thickness) connected to a mass spectrometer (D) and a Poraplot Q column (25 m x 0.53 mm with a 20-µm adsorbent thickness) connected to a thermal conductivity detector (E). The system operating conditions used are summarized in Table 1. An HPChem data station served to computerize the results.

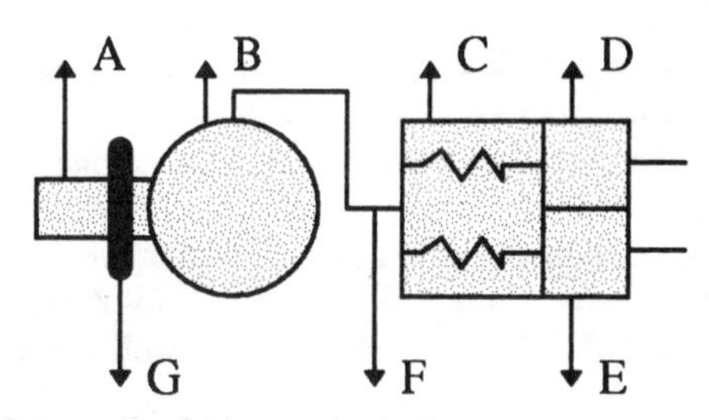

Figure 1. Schematic of the experimental arrangement.

Table 1 Analytical conditions

MSD carrier gas:	2.03 mL/min
TCD carrier + makeup gas:	18.95 mL/min
Capillary reference gas:	15.60 mL/min
Injector block temperature:	100°C
TCD block temperature:	145°C
Oven temperature program:	50°C for 1.5 min
	20°C/min for 3.5 min

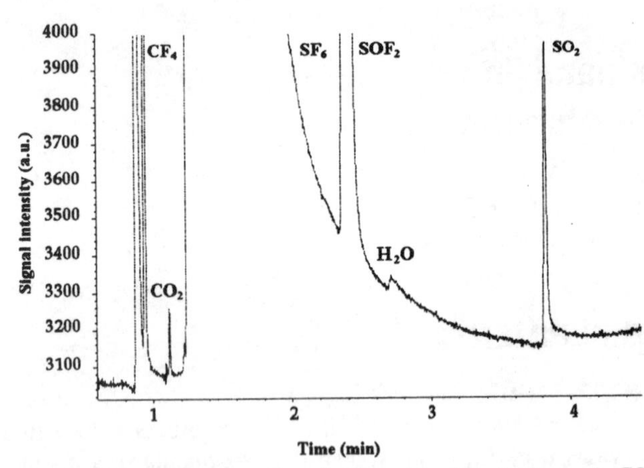

Figure 2. Chromatogram of a mixed sample: CF_4 (218 ppm$_v$), CO_2 (19 ppm$_v$), SOF_2 (9070 ppm$_v$), H_2O (80 ppm$_v$) and SO_2 (257 ppm$_v$).

Table 2 Analytical performance

Species	Response factor (a.u. / ppm$_v$)	Precision** at low level* (± ppm$_v$)	Precision** at high level (± %)
CF_4	7.85	4	0.9
CO_2	6.31	2	0.6
SF_6	9.06	-	1.8
SOF_2	6.61	7	0.4
SO_2	6.57	2	0.7

** For 10 replicates of a given sample; * < 1,000 ppm$_v$.

Chemicals

Both certified and freshly prepared standards were involved in the GC-TCD calibration process. The former comprised stock gas mixtures containing CF_4, CO_2, SF_6 and SO_2F_2 species at levels ranging from 100 to 10,000 ppm$_v$ in helium, while in the latter case the level of SOF_2 or SO_2 was adjusted to cover the same range in SF_6. These two matrices can be used indiscriminately in the present context without any effect on the calibration parameters, since the chromatographic technique allowed sequential extraction of all analytes in a carrier gas consisting of helium.

All other gas mixtures involved in the course of this work were constituted of SF_6 to which low levels of CF_4, CO_2, SOF_2 or SO_2 were added. Apart from mixtures containing species at levels higher than 4,000 ppm$_v$ which can be prepared with precision at a very low contamination level directly in chamber B of the experimental test cell, they were premixed in a well vacuumed and purged cylinder. In all cases, the final composition of the gas mixture was fixed by adjusting the partial pressure of each constituent to the desired value. The high-purity grade of the raw gas, the inertness of the containers and the appropriate configuration of the manifold helped to keep all impurities minimal. Both the materials and the detailed protocol are well described elsewhere [3].

The absorbing material was alumina, mainly in the form of χ-Al_2O_3. The molecular water content of this material was reduced by heating the samples at 200°C under vacuum for about 12 h prior to use.

Experimental procedure

As in our previous work, the test cell was pumped down for several hours prior to each experiment. The gas under study was added to chamber B at room temperature and a fixed pressure of 2 bar abs, with the gate valve G (see Figure 1) closed. The pre-treated adsorbent (10 g) contained in a cylindrical metallic net was placed in chamber A using nitrogen or helium to keep the sorbent in a dry environment during the operation. Chamber A was pumped down to vacuum before the gate valve was opened, which determined the starting point of the interaction dynamics.

Prior to each set of measurements, 100 µL of a concentrated SO_2 gas mixture (10^5 ppm$_v$) was injected into the analytical system in an aim to favor the detection of this species when present at the trace level in the next sample to be analyzed. All samples were circulated through the on-line sampling interface and the sample loop for at least 5 min (flow rate = 3 mL/min) to remove most of the impurities from the inner walls of the instrumentation prior to analysis.

Chromatography

As shown in Figure 2, the analytical conditions given in Table 1 yield a chromatogram where the peaks of interest are well resolved. When the peak areas associated with these species were compared over a concentration range of 0-10,000 ppm$_v$, a linear relationship passing through zero was obtained. Table 2 reports the regression coefficient established and the typical precision obtained.

RESULTS AND DISCUSSION

In order to discriminate between and interpret the sorption processes of the species of interest, the SF_6 gas used as the raw material in preparing the samples was investigated. It was tested first as received from the supplier to determine the nature and potential effect of the contaminants and then in diluted form (10^5 ppm$_v$/He) to monitor the SF_6 alone in the presence of alumina. In both cases, the test cell was filled with the gas and left to equilibrate for several hours. The gate valve was then opened, exposing the adsorbent to the gas. After a few days, no detectable interaction of the SF_6 with the adsorbent, nor with the CF_4 naturally present at a low level in this particular gas, was observed within the experimental

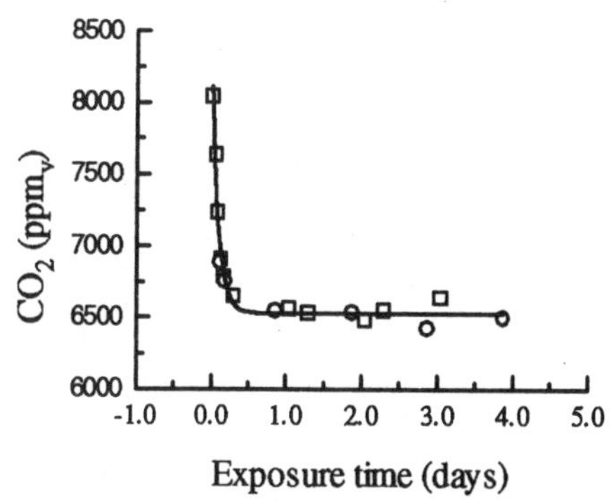

Figure 3. Concentration versus exposure time. Conditions for a one-shot exposure of CO_2 (8,000 ppm$_v$) to alumina: (o) He matrix; (□) SF_6 matrix.

Table 3: Reproducibility of SOF_2 mixtures

N*	CF_4 (ppm$_v$)	CO_2 (ppm$_v$)	SOF_2 (ppm$_v$)
1	87	20	10149
2	90	21	10336
3	89	21	9739
4	94	20	10122
5	84	20	10103
6	89	19	10012
7	83	21	10005

* Number associated to a given preparation.

error. Although the water and CO_2 levels were both found to be low (<20 ppm$_v$), particular attention should be paid to these impurities in further experiments. The moisture content will play a role in the conversion of SOF_2 into SO_2. The possibility of CO_2 interference in the SOF_2 sorption kinetics will depend on the relative affinity of the sorbent for each species and the sorption processes involved.

The inertness of the SF_6 matrix with respect to the sorption kinetics of interest was evidenced using a CO_2/SF_6 mixture with a CO_2 concentration of 8,000 ppm$_v$ and a trace amount of CF_4 (90 ppm$_v$) to perform a single exposure to alumina. As seen in Figure 3, the decaying concentration curve obtained for CO_2 matches exactly those obtained under similar conditions in helium. The data collected during the sorption process was normalized on the basis of the mean value obtained when the gate valve was still closed and the gas phase was at equilibrium. The solid line depicts the average value of the two sets of measurements.

When a single sorbent sample was subjected to three consecutive exposures of a CO_2/SF_6 mixture with a CO_2 concentration of 845 ppm$_v$ and a trace amount of CF_4 (90 ppm$_v$), the same reproducible phenomenon was repeated. The curve associated with the first exposure had an initial decay rate comparable to those expected in helium at this level. Despite an exposure of several days, a residual CO_2 concentration amounting to 283 ppm$_v$, i.e. 67% of the initial level, remained unabsorbed. With subsequent exposures, the initial decay rate was found to decrease and the residual CO_2 concentration to increase, showing the limited sorbing action of the adsorbent. Consequently, the assumption of a gas interaction driven by concentration gradients was maintained, and the estimated sorbent capacity for CO_2 was found as in helium: approximately 0.5% g [CO_2]/g alumina, in the case where low-level CO_2 contamination diffuses freely in a gas matrix.

A first test trial for the SOF_2 species consisted in preparing reproducible stable gas mixtures containing different concentration levels of the constituent. As seen in Table 3, gas mixtures of a given composition can be prepared with precision, accuracy and reproducibility, provided an appropriate material and a well defined procedure are used. Otherwise, unexpected results are produced. For example, two SOF_2 gas mixtures were freshly prepared and left to equilibrate for a couple of days. When analyzed for the first time, the SOF_2 content was estimated to be 550 ppm$_v$ in premix #1 and around 1,225 ppm$_v$ in premix #2. Part of the SOF_2 continued to disappear with time, as illustrated in Figure 4, while all the other constituents remained constant. No detailed explanation for this instability was attempted but it was suspected that the inside wall of the containers, which was made of hardened steel, probably had a catalytic effect on the conversion of the SOF_2 species into SO_2 according to the reaction:

$$H_2O + SOF_2 \rightarrow SO_2 + 2\,HF$$

The reaction rate of this last process was previously studied in gas phase by Van Brunt, who mentioned that sorbents could catalyze the reaction [7-8]. All mixtures prepared in a well-conditioned stainless steel container remained stable much longer than the time required for the experiments.

To study the interaction between SOF_2 and alumina, a mixture with a SOF_2 concentration of 845 ppm$_v$ was used to perform a multi-exposure sequence (series of three consecutive exposures) to a single sorbent, followed by exposure of a

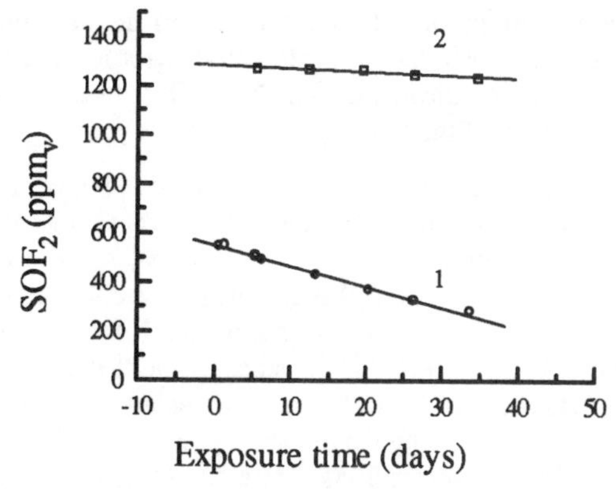

Figure 4. Concentration versus exposure time. Initial SOF_2 content: curve 1 - 550 ppm$_v$; curve 2 - 1,225 ppm$_v$.

single virgin sample of alumina to a gas mixture with a SOF_2 concentration of 10,000 ppm_v. As summarized in Figure 5, the SOF_2 species completely disappears after a few days, independently of both the number of exposures and the initial concentration of the analyte over the range tested. The steep concentration decrease characterizing the beginning of curve 2 reflects the tendency of the alumina to adsorb the SOF_2 molecule. Compared to the case involving CO_2, SOF_2 seems to be more reactive toward free active sites on the alumina surface. The leveling off that follows was attributed to a gradient concentration effect. Overall, the curves illustrating the sorption kinetics were found to be remarkably reproducible, even if a slight degradation of the SOF_2 species was detected at this stage. The CF_4 traces detected as a contaminant and used as a tracer agent were unaffected throughout, remaining constant within less than 2 ppm_v.

In an attempt to distinguish the sorption process from the degradation of the SOF_2, two gas mixtures with the same initial concentration of the species were compared; one was exposed to alumina for a couple of days while the other was left alone over the same period. It can be deduced from curve 1 in Figure 6 that less than 4% of the initial amount of analyte was converted into SO_2 during the time required to sorb more than 90% (see Figure 5) of the initial SOF_2. Considering the lapse of time from the beginning of the experiment to a total disappearance of the SOF_2 species, it was very convenient to take this percentage as a maximum in the present context.

The high stability of curve 2 also suggests a negligible amount of SOF_2 degradation but the possibility of a catalytic effect for the alumina cannot be discarded. In fact, SO_2 species were detected in the medium with a maximum concentra tion (90 ppm_v) at around 1.5 days of exposure. Almost all of it was later removed by the sorbent. The initally determined CO_2 traces were also totally adsorbed, demonstrating the presence of enough sites (type or number) for all species.

When an additional amount of SOF_2 (10,000 ppm_v) was put in the presence of the same sorbent, the CO_2 species sorbed previously was released and the level of SO_2 rose significantly without any further decrease in the amount detected. This demonstrated the preferential adsorption of SOF_2 on alumina, and confirmed the catalytic effect of the sorbent on the conversion of the SOF_2 species to SO_2.

Figure 7 compares the interaction kinetics for two different SOF_2 concentrations, corresponding to a first-exposure condition. The ordinate (DEL) is a dimensionless ratio given by (N-No) / (Nr-No), where N is the measured concentration, and No and Nr are the initial and residual concentrations respectively; DEL is shown as the square root of the exposure time. As was the case for CO_2, these curves are accompanied by an initial steeply rising curve which gradually flattens out. The initial linear shape reflects a rate-controlling flux of the sorbate in the matrix proportional to a concentration gradient (see [9], for instance, and references therein). As the initial

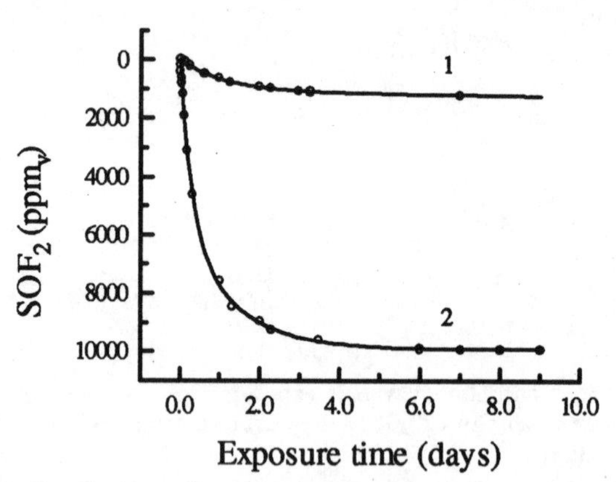

Figure 5. Concentration adsorbed versus exposure time. Initial conditions for SOF_2: curve 1 - 845 ppm_v; curve 2 - 10,000 ppm_v.

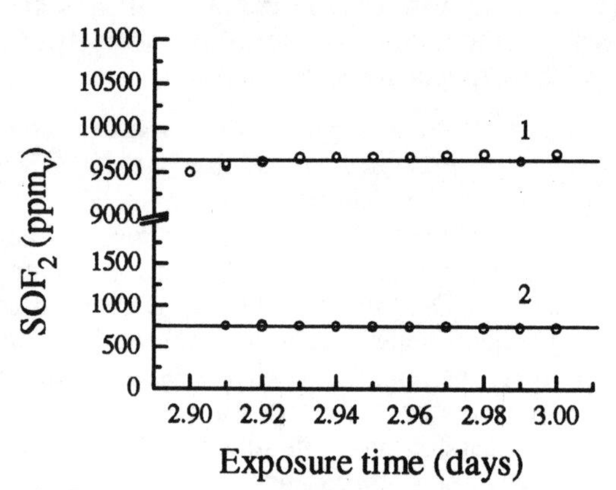

Figure 6. Concentration versus exposure time. Initial conditions for SOF_2: curve 1 - 10,125 ppm_v at equilibrium without adsorbent; curve 2 - 10,103 ppm_v with adsorbent.

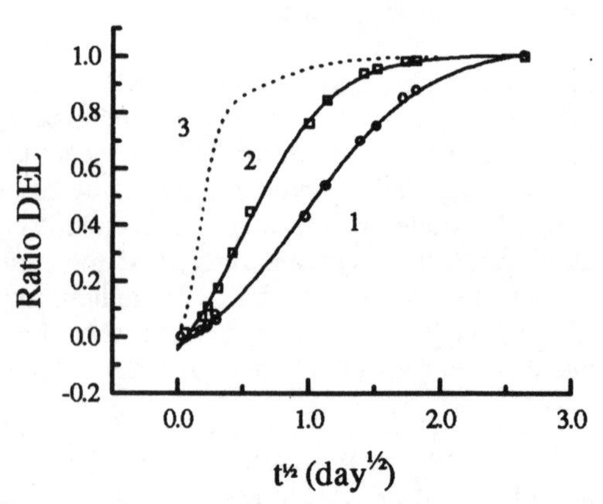

Figure 7. Ratio DEL as a function of $t^{½}$ for a first-exposure condition: curve 1 - SOF_2, No = 1,225 ppm_v, curve 2 - SOF_2, No = 10,000 ppm_v; curve 3 - CO_2, No = 8,750 ppm_v.

SOF_2 concentration increases, the slope of the ratio DEL is found to get steeper. Compared to what was observed [3] in the case of CO_2, the equilibrium point (DEL = 1) here is reached later, after 6 days instead of 2.25 days. Under comparable conditions, the total interaction time and, especially the removal rate in the last stretch, are determined mainly by the acidity and capacity of the adsorbent for a particular compound ($Nr \neq 0$ at equilibrium). Thus, experimental observations confirm that the reactivity with alumina is stronger with the SOF_2 than with CO_2. On the other hand, the rate of rise of DEL is primarily more dependent on the diffusion process involved. Differences in molecular weight will favor a higher collisional frequency in the case of CO_2 interacting with alumina, although this process is not independent of the number of available sites.

SUMMARY

A low concentration of gaseous compounds in a SF_6 matrix, namely SOF_2 and CO_2, interacting with alumina was investigated. Results demonstrated that the alumina can effectively sorb the SOF_2 species, which is usually one of the dominant by-products in the SF_6-insulated equipment in use. Experimental observations support a stronger capacity and reactivity of the sorbent with respect to this compound, compared to CO_2. The observed kinetics provides supporting evidence for a catalytic role played by the alumina in the conversion of the SOF_2 species into SO_2. Under the selected experimental conditions, the sorptive interaction was found to remain unaffected by the gas matrix used.

ACKNOWLEDGMENT

The authors thank the Distribution and Environment departments of Hydro-Québec for their financial support.

REFERENCE

1. Fréchette, M.F. "Colloque SF_6 - Recueil des conférences". Hydro-Québec report on SF_6 technologies in electric power systems. IREQ-4790, April 1991.

2. Fréchette, M.F. "Mitigation Studies". Proc. Workshop on Investigation of S_2F_{10} Production and Mitigation in Compressed SF_6-insulated Power Systems, Pittsburgh, USA, Ch. 6, June 1994.

3. Lussier, T., Fréchette, M.F. and Larocque, R.Y. "Gas/Adsorbent Interaction Dynamics: An Experimental Approach". Annual Report of the 1995 Conference on Electrical Insulation and Dielectric Phenomena. IEEE Publication 95CH35842, pp. 153-158.

4. Boudène, C., Cluet, J.-L., Keig, G., Wind, G. "Identification and Study of some Properties of Compounds Resulting from the Decomposition of SF_6 under the Effect of Electrical Arcing in Circuit Breakers". Revue Générale de l'électricité, Special issue, June 1974, pp. 45-78.

5. Van Brunt, R.J. "Production Rates for Oxyfluorides SOF_2, SO_2F_2, and SOF_4 in SF_6 Corona Discharges". Journal of the National Bureau of Standards, Vol. 90 No. 3, May-June 1985, pp. 229-253.

6. Chu, F.Y. Stuckless, H.A. Braun J.M. "Generation and Effects of Low Level Contamination in SF_6-insulated Equipment". Gaseous Dielectrics IV, 1984, pp. 462-472.

7. Van Brunt, R.J. and Sauers, I. "Gas-phase hydrolysis of SOF_2 and SOF_4". J. Chem. Phys., Vol. 85 No. 8, 1986, pp. 4377-4380.

8. Van Brunt, R.J. and Herron, J.T. "Fundamental Processes of SF_6 Decomposition and Oxidation in Glow and Corona Discharges". IEEE Trans. Elect. Insul., Vol. 25 No. 1, 1990, pp. 75-94.

9. Doelle, H.J. and Riekert, L. "Kinetics of Sorption, Desorption, and Diffusion in Zeolites". Angew. Chem. Int. Ed. Engl., Vol. 18, 1979, pp. 266-272.

Conference Record of the 1996 IEEE International Symposium on Electrical Insulation, Montreal, Quebec, Canada, June 16-19, 1996

SPARKS IN SF_6 AND SF_6/CF_4 ATMOSPHERES: IMPACT ON METALLIC AND INSULATING SURFACES

M.L. Trudeau, M. F. Fréchette[1] and R.Y. Larocque[1]

Technologie des Matériaux, Technologie et IREQ, Hydro-Québec, Varennes, Québec, Canada
[1]Câbles et Isolants, Technologie et IREQ, Hydro-Québec, Varennes, Québec, Canada

Abstract: **Solid materials (copper and Teflon) used in HV circuit breakers were exposed to a closely spark discharge in SF_6 and SF_6/CF_4 mixtures. The surface of the different samples was studied using x-ray photoelectron spectroscopy to reveal possible interactions with the by-products. Results show that discharging in SF_6 and SF_6/CF_4 mixtures produces active species, mainly F-groups, which exhibit strong interactions with metallic surfaces. The surface modification due to the discharge processes was found to be less apparent in the case of insulating materials.**

INTRODUCTION

New gas mixtures are now finding applications such as interrupting media for high-voltage circuit breakers. These mixtures consist of a high content of carbon tetrafluoride (CF_4) added to sulfur hexafluoride (SF_6). Despite the fact that CF_4 is only slightly electronegative [1], its addition to SF_6 enhances the rate of rise of the recovery voltage (RRRV) capability compared to SF_6/N_2 mixture [2]. This performance makes it a suitable candidate to supplant SF_6/N_2 mixtures for low-temperature environmental applications.

The aim of the present study was to unravel any differences in material compatibility due to the new mixture. Copper and Teflon/ceramic bulk samples were exposed to gaseous by-products originating from the close presence of a spark discharge. Discharges were run in different gaseous media to provide a comparative scheme. The surface of the different samples was subsequently studied using x-ray photoelectron spectroscopy to reveal possible interaction with the by-products. Also, the gas composition following the discharge was analyzed. Details on the on-going experiments will be published elsewhere [3-5].

EXPERIMENTAL DETAILS

The discharge experiments were done in a reaction chamber (~ 22 L in volume) connected in-situ to an x-ray photoelectron spectrometer. The chamber can be evacuated down to 10^{-8} Torr. Two thorium-stabilized tungsten electrodes with a diameter of 0.3 cm were used; the sparks occurred between the end-points. The rods were placed at a 60-degree angle resulting in a fixed horizontal gap of 1 cm. The bulk sample was located below the discharge gap, parallel to it. For copper, the surface of the virgin sample was sputtered clean. In the case of the ceramic, between each discharge experiment, the sample used was removed from the chamber and its surface was scraped. XPS analysis was done on the new surface to ensure it was comparable to the original material. For the copper sample, the surface was placed at a distance of 6 to 10 mm below the two electrodes. For the insulating sample, this distance was about 1 mm. Because of the high surface reactivity of copper, the conditions were less severe than for the insulating sample.

All experiments were done using DC voltage from a stabilized source (80 kV, 7.5 mA max). The resulting conditions correspond to negative spark discharges. For Cu, the typical values for the voltage and current were 7 kV and 150 μA respectively, with a spark frequency of about 0.5 Hz. For the ceramic sample, the voltage was set at 14 kV and the resulting average current was about 250 μA. The spark frequency was about 2 Hz. Three different atmospheres were investigated: N_2, SF_6 and a 50:50% mixture of SF_6 and CF_4 at a pressure of 200 Torr. Following the spark discharge, the sample and chamber were left undisturbed for about 10 min to allow reaction with active species.

Following the discharge, the gas pressure in the chamber was increased with helium (to a total of 900 Torr) to allow the sparked gas to be sampled. The gas samples were analyzed using gas chromatography (GC/TCD, see for instance [6]). In the case of N_2, no by-products were detected, irrespective of the time duration of the discharge. For SF_6 and the mixture, detection of by-products was negative for sparking duration less than 3 hr. The by-product concentration of SOF_2 and SO_2F_2, after sparking for 63 hr, was non-negligible. For the discharge in pure SF_6, the concentration of SOF_2 was found to exceed that of SO_2F_2: 110 ppmv compared to 45 ppmv. The discharge in the mixture led to an inversion in the proportion of the these two compounds. The following concentrations $[SO_2F_2]$ = 45 ppmv and $[SOF_2]$ = 20 ppmv were estimated.

The XPS data was acquired using a PHI-5500 ESCA system from Perkin-Elmer using 300W Al K_α monochromatic radiation. High resolution acquisitions of the elemental regions were performed using a pass energy of 11.75 eV with a 0.1 eV step size. The peaks were referenced to the C-(C,H) components of the C 1s band at 284.8 eV. Depth profiling was done using Ar ions at 2 keV and with a 4x4 mm rastered region, which gives a sputtering rate for SiO_2 of about 1 nm/min.

RESULTS

Copper

N_2 atmosphere

Sparks in nitrogen were found to substantially modify the surface of copper. While only metallic Cu was present prior to the discharge, Cu, O, C and N were found after the discharge. In addition to the deposition of surface contaminants, the chemical state of Cu was also modified. Prior to the sparks, the peak position of the Cu $2p_{3/2}$ emission was found at 932.9 eV (binding), while the Auger emission had its maximum at 918.3 eV (kinetic). These two values indicate that mainly metallic Cu is present at the surface. After the discharge, the major emission of the Cu $2p_{3/2}$ emission is found at a binding energy of about 933.8 eV. Two smaller contributions are present at binding energies of 932.8 and 936.0 eV respectively. For its part, the LMM Auger emission has its maximum at 917.7 eV with a shoulder contribution at a lower kinetic energy, probably corresponding to the emission at about 916.7 eV found in the middle of the reacted layer. Analysis of these results suggests that three different Cu species are present at the surface. The peak position at 933.8 eV, in association with the Auger emission at 917.7 eV, has been associated in the past to the presence of CuO [4]. The emission at 936.3 eV can be related more to the presence of hydroxide species such as $Cu(OH)_2$. Lastly, the emission at 932.8 eV is indicative of the presence of Cu_2O. It is possible to calculate, on the basis of the different spectral area, that the +1 (Cu_2O) emission corresponds to about 15% of the surface composition.

The analysis of the oxygen x-ray emission is also in agreement with the copper data. In the case of carbons, two species were identified. The position of the major emission, at about 284.7 eV, suggests the presence of C-(C,H) groups at the surface, while the second at 288.3 eV is normally associated more with O-C-O groups. Because of the very low signal to noise ratio, the nitrogen emission is difficult to analyze. Three individual contributions seem to be resolvable at about 406.6, 403.8 and 397.5 eV, which according to published data could be associated with nitrate (NO_3), nitrite (NO_2) and nitride (N) groups respectively.

Finally, depth profiling reveals that the reacted layer is of the order of 2 to 4 nm. Also, it was found that the Cu^{+2} species are present only at the very near surface. Once these species have been removed, Cu is present mainly in the form of Cu_2O. The C and N contaminants found at the surface were not observed after the first sputtering cycle.

SF_6 atmosphere

A series of similar spark discharges were done in 200 Torr of SF_6. The duration of the discharge experiments was varied between 7 to 30 min. Figure-1 presents the XPS survey of the Cu surface after 30 min of sparks in SF_6, showing that the surface was strongly affected by this process. Compared to the pure metallic surface, a number of contaminants can be seen at the surface. Apart from the Cu, F and O, which are found in large quantities, traces of S, C, W and Th are also present at the surface, the latter two coming from the erosion of the electrodes. A semi-quantitative evaluation of the surface composition gives a concentration for the Cu, F, O, C, S and W of 28, 38, 26, 5, 1, and 2 at.% respectively. It was found that, as the sparking time increased, the Cu and F concentration decreased to the detriment of the O concentration. The amount of C and S was found to remain independent of the discharge time. Finally, the W and Th content at the surface increased continuously with the sparking time, in agreement with the erosion process of the electrodes. Depth profiling reveals that Cu, O and F are the major constituents of the reacted layer, with the O being mainly present at the very near surface. Once the O was removed, the composition of the reacted layer was relatively constant at a F:Cu ratio of about 60:40 up to about 25 min of sputtering (25 nm). After that, the F was found to decrease rapidly and only metallic Cu was then observed.

As in the case of the nitrogen experiment, the surface and sub-surface chemistry was studied by looking at the different elemental spectra. Figure-2 presents the Cu $2p_{3/2}$ emission for the reacted surface layer and after different sputtering cycles. At the surface, six different individual contributions were used in order to best fit the total emission, with three of these contributions associated with a shake-up structure found at a binding energy above 940 eV. The main emission can be decomposed into three contributions: a major one at 937.2 eV and two smaller ones at 935.7 and 933.3 eV. Based on these values, it is possible to determine that the major state for Cu at the surface takes the form of CuF_2. The emission at 935.7 eV, for its

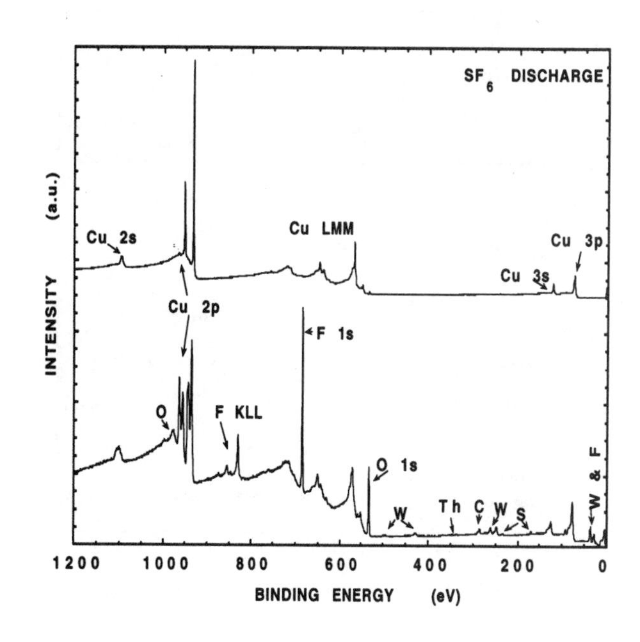

Figure 1. XPS survey of the Cu surface before and after 30 min of sparks in SF_6

part, can be identified with Cu^{+2} states such as $Cu(OH)_2$ or $CuCO_3$. Finally, the small emission at 933.3 eV (4% of the total emission), as found in the case of the discharge in nitrogen, can be associated with Cu_2O.

The variation in this Cu 2p emission as a function of the sputtering time reveals that the emission associated with this Cu^{+1} state increases at the beginning of the sputtering, followed by a decrease and a slight shift to lower binding energies. Based on the relative emission area, the contribution from this +1 state increases to 23% of the total emission after 1.5 min of sputtering. After 4 min of sputtering, however, this area drops to about 16%, a value that remains stable until a rapid increase at the end of the depth profile, corresponding to an increased presence of metallic Cu (since Cu_2O and metallic Cu have the same peak position). The identification of Cu^0 was confirmed by the Auger peak at 918.3 eV. The second feature worth noting is the rapid disappearance of the emission at about 935.7 eV, which was identified with the presence of some hydroxyl or carbonate groups. Finally, the emission related to CuF_2 which, after the first few minutes of sputtering, remains stable up to its rapid decrease at the interface of the metallic Cu.

The 1s emission spectra for the oxygen indicates that two different species are present at the surface with a peak emission at 532.6 and 531.0 eV. The major emission, at 532.6 eV at the very near surface, can be associated with adsorbed water [4], while the second, at 531.0 eV, can be related more to the presence of hydroxyl or carbonate groups (e.g.: $Cu(OH)_2$ or $CuCO_3$). The small emission at low binding energy remains at the same position and gradually disappears after about 20 min of sputtering.

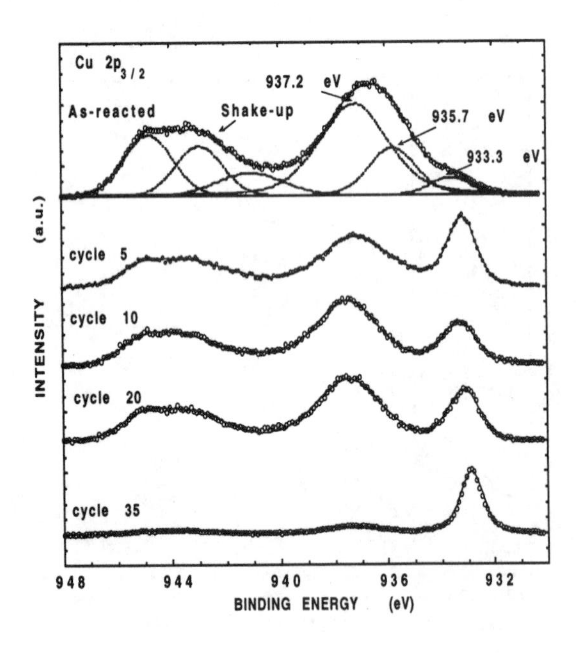

Figure 2 Cu $2p_{3/2}$ emission for the reacted layer and after different sputtering cycles (30 min sparks in SF_6)

As shown in Figure-1, some minor adsorbed contaminants can be observed, which disappear readily with the first sputtering cycle. The spectra for the C 1s emission, which is dominated by an emission at 284.9 eV, is characteristic of C-(H,C) bonds. Also, two other contributions at higher binding energies are observed. The first, at about 286.7 eV, could possibly be assigned to some C-O(H) bonds, the other, at about 289.3 eV, to some O-C-O or CO_3 bonds. These two emissions could also be due to the presence of some CF bonding. The 1s emission spectra for sulfur reveals a peak position at about 170.4 eV, which could be associated with compounds such as SOF_2. Finally, the $4f_{7/2}$ emission for tungsten, with a peak position at about 36.4 eV, if compared to the metallic position at 31.4 eV, is indicative of the presence of a significant ionization state possibly due to the presence of compounds such as WO_3, WF_5 or WF_6.

SF_6/CF_4 atmosphere

The same arcing experiments were done in a 50:50 SF_6/CF_4 atmosphere. As in the pure SF_6 case, the surface chemistry was greatly changed. The relative elemental surface concentration after 30 min of arcing indicates the presence of Cu, F, O, C, S and W in a concentration of 28, 38, 25, 5, 1 and 2 at.% respectively. No major differences compared to the SF_6 experiments are observed. The only small difference is a small increase in the C content at the surface. As in the previous case, the different elemental emissions were deconvoluted to determine the chemical species present. For all the elements present, the same peak emissions, as found with SF_6 were observed, suggesting that the same active species are formed in this SF_6/CF_4 mixture. At the surface, Cu is mainly in the form of CuF_2, with an amount of hydroxide species. The presence, also, of a large peak at 532.4 for the oxygen suggests the presence of adsorbed surface water. Depth profiling indicates that hydroxide species are removed rapidly and analysis of the Cu and O emission suggests the presence of Cu_2O only near the surface and, once removed, only CuF_2 is found, with traces of metallic Cu.

Teflon/ceramic

The surface reactivity of insulating material is normally greatly reduced compared to a metallic surface. XPS analysis of this material reveals that the surface is composed mainly of F and C in a relative concentration of 70 and 29 at.% (based on PHI sensitivity factor), which is not far from the expected value for Teflon C_2F_4. The only other noticeable constituent of this material was Ca, at a concentration of about 1 at%. In all the analyses, all peaks are referenced to the F 1s emission for n-CF_2 at 689.0 eV. In the as-received material, only one fluorine contribution could be found. The C 1s emission had its peak at 292 eV, in agreement with studies on $p(CF_2{=}CF_2)$ [7]. Finally, the Ca $2p_{3/2}$ had a peak position at 349.4 eV indicative of its presence in the form of CaF_2.

N_2 atmosphere

It is well known that polymer surfaces are susceptible to low pressure plasma or glow discharge [7], which can result from etching, surface chemical modification or thin film deposition. As in the Cu case, spark discharges were produced in N_2 for 3 h. Elemental analysis of the surface shows a concentration for the F, C and Ca at 66.5, 30.6 and 1.5 at.% respectively. Traces of N and O at around 0.5 to 1 at.% could also be detected. The peak analysis indicates that the original chemical structure was present for the three original elements, but in all three cases, smaller contributions were found, indicative of the presence of new chemical species. For the F, a smaller contribution at 686.8 eV was detected, which suggests the formation of lower-bond compounds (CF group, for example). The case of C was more complex, since three new contributions, at 293.8, 290.5 and 288.2 eV, were necessary to deconvolute the total 1s emission. The presence of these emissions suggests some kind of modification of the polymeric chain. Finally, a small contribution at about 347.4 eV was present in the case of the Ca $2p_{3/2}$ contribution. For the N and O, the peak position was found at 401.8 and 535.4 respectively. For the N, this could be the result of the presence of surface species such as nitrides, while in the case of O, it could be due to the presence of adsorb water.

SF_6 atmosphere

Compared to the previous experiments and the as-prepared material, the surface of the ceramic after sparks in SF_6 or SF_6/CF_4 was found to be less stable against x-ray or electron emission. For sparks in the SF_6 case, some similar types of surface changes were observed but no drastic modification of the surface composition. While the overall surface concentration of F and C was found to slowly decrease with the sparking time, due to the deposition of surface contaminants, the F:C ratio remains at about the same value as the original material (2.3 compared to 2.4). However, even if this ratio remains stable, the surface chemistry of these two elements was found to be modified by the discharge process. Figure-3a and b presents the F and C 1s emission for the as-prepared material and after 60 hr of sparks in SF_6. The emission spectra of these two elements can be seen to become more complex, indicating that different carbon and fluorine environments appear as a function of the sparking time. In the case of the fluorine emission, two different contributions are necessary to deconvolute the 1s emission: the major one remains at 689 eV but a second one is now present at 685.9 eV. For the C emission, after 60 h of sparks, three different species are found: a major emission at 291.5 eV with two smaller contributions at 288.4 and 284.9 eV, the latter probably being due to the presence of some hydrocarbon contamination. No changes were observed in the surface concentration and peak position for the Ca.

Three other elements were found at the surface, in varying amounts depending on the arcing time. After 60 h, up to 2 at.% of O, with a peak position of 532.5 eV, indicated adsorbed water. The other two contaminants come from degradation of the electrodes: thorium, which was found to be present at a level of about 1 at.% and in the form of ThF_4, and tungsten, which could be found at about 0.5 at.% in the form of WF_6 WF_5 or WO_3. Compared to the Cu case, no traces of sulfur were found at the surface.

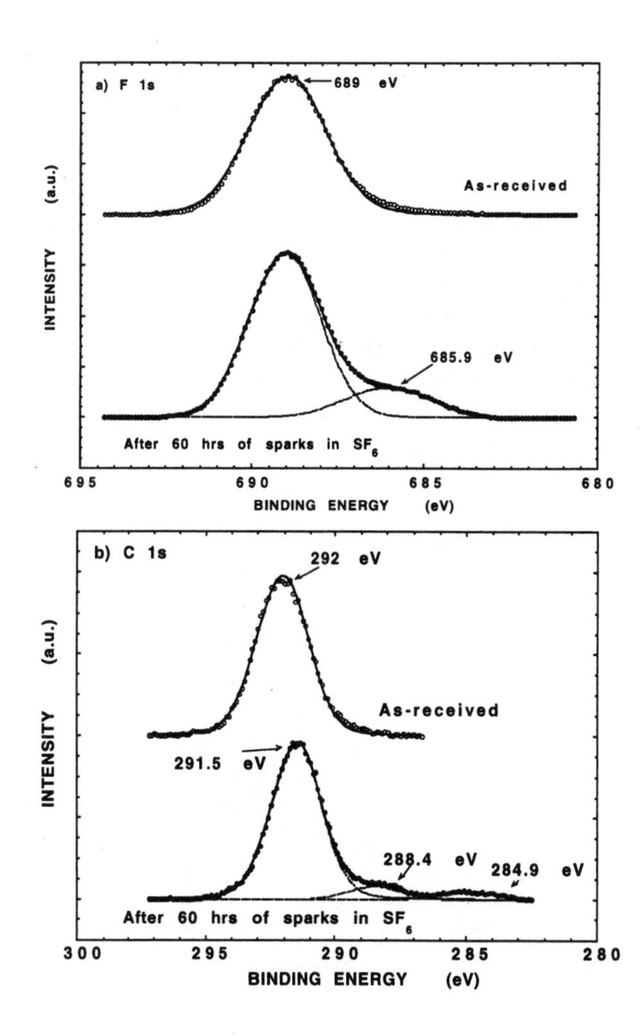

Figure 3 XPS 1s emission for a) F and b) C before and after 60 hrs of sparks in SF_6

SF_6/CF_4 atmosphere

Similar surface modifications were observed in the case of arcing in the SF_6/CF_4 atmosphere. The fluorine 1s emission was found to be composed of two different contributions, one remaining at the normal value for the Teflon (689.0 eV) and a second at 686.1 eV. The carbon 1s level, for its part, shows a more complex structure than for the SF_6 case but not too different the N_2 case. Figure-4 presents the C 1s emission after 63 h of arcing. At least five different contributions can be distinguished: a major one remaining at 292 eV and four smaller ones at 294.1, 290.4, 288.2 and 285.7 eV.

As for the SF_6 case, there was no marked variation for the Ca emission. Also, the surface was found to be contaminated by the same element: O at about 1 at.%, W at a very low value of about 0.1 at.% and Th at a higher value of 3.5 at.%. The peak positions for these three contaminants were nearly identical to those in the SF_6 case.

DISCUSSION

Sparks in N_2, SF_6 or SF_6/CF_4 atmospheres have a strong impacts on metallic surfaces, such as copper. In the case of nitrogen, most of the effects observed are due to interaction between the sparks and air or water molecules present in the gas or adsorbed on the electrodes. The probable dissociation of these molecules will produce active oxygen species that will react readily with metallic surfaces to produce some kind of oxide layer. On the other hand, the by-products produced by the discharge process in SF_6 and SF_6/CF_4, which are mainly fluorine group, have a more damaging effect on the metallic surface, resulting in the formation of a rapidly growing CuF_2 layer. For similar spark durations, this layer was found to be 5 to 10 times thicker for fluorine-based gas than for the nitrogen experiments. It was also found that the surface of this reacted layer, following the sparks, will adsorb and react readily with water molecules present in the reaction chamber, to form some surface oxide or hydroxide compounds.

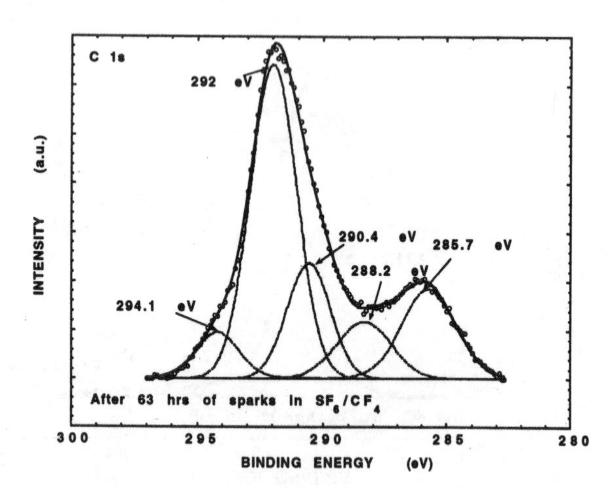

Figure 4 C 1s emission after 63 hrs of sparks in a 50:50 SF_6/CF_4 mixture.

In the case of the Teflon/ceramic sample, much more energy was needed to modify the surface chemistry of the material. After a long period of sparks, only small changes in the chemical species could be observed apart from the presence of small amount of surface contaminants. Because of the greater shift in the binding energy of the C 1s emission, compared to the F 1s case, this modification of the surface chemistry could be more easily observed in the carbon case. The effect of fluorine substituents on carbon has been studied in the past [7]. The effect of replacing a C or a H bond by a fluorine has been known to increase by 2.7 eV (per fluorine substituent) the binding energy of the C 1s core level. Moreover, replacing a C or H bond on a next carbon atom along the chain by a fluorine atom has been shown to produce a 0.4 eV increase in binding energy. This means that going from a CH_2-CH_2 chain to a CHF-CHF or to a CF_2 -CF_2 chain will result in an increase in the binding energy of the C 1s level from about 285 eV to 288.5 eV or 292 eV respectively. From the observed C 1s emission, obtained after the sparks, it can be seen that this discharging process has the effect of modifying the polymeric chain, probably through the incorporation of some H bond. Finally, the formation of a peak at about 294. eV is probably a result of the break-up of the chain owing to the formation of a CF_3 group at some points. Finally, the fact that the surface of the reacted material was found to react to x-ray or electron flux more marked after the discharge in SF_6 or SF_6/CF_4, as compared to the as-received material or even the material exposed that was submitted to sparks in nitrogen, suggests that some surface instabilities were created by these discharges.

CONCLUSION

Spark discharges were produced in N_2, SF_6 and a 50:50 mixture of SF_6/CF_4, and the interaction of the by-products produced by these sparks with copper and Teflon surfaces was investigated using XPS. As expected, spark discharges, especially in fluorine-based gas, have a strong impact on the surface chemistry of adjacent materials, especially metallic elements. The formation of a rapidly growing CuF_2 layer and breakdown of the polymeric CF_2 chain of Teflon were shown to occur readily due to sparks in SF_6 or SF_6/CF_4.

However, these experiments did not unravel any major differences in behavior between the effects of sparks in SF_6 or SF_6/CF_4, except for an increase in the C contamination of the copper surface or a more complex change in the C 1s level environment for the Teflon sample for sparks produced in the SF_6/CF_4 as compared to the pure SF_6 case.

ACKNOWLEDGMENTS

This work was sponsored in part by Service Appareillage électrique 2, Gestion de la technologie and Vice-présidence Technologie et IREQ, all from Hydro-Québec.

REFERENCES

1. Fréchette, M.F. and C. Laverdure. "Performances des mélanges SF6/CF4 pour applications électrotechniques", Internal Report, No IREQ-92-153C, 1992.

2. Frind, G., R.E. Kinsinger, R.D. Miller, H.T. Nagamatsu and H.O. Noeske. "Fundamental Investigation of Arc Interruption in Gas Flow". EPRI Report EL-284, 1977.

3. Fréchette, M.F., M. Trudeau, R. Schulz, I. Dignard-Bailey, R.Y. Larocque, and R. Dubuc. "Gaseous SF6/CF4 By-product Interactions with a Metallic Cu Surface". Int. Symp. on High Voltage Engineering, Graz, Austria, Aug. 28/Sept. 1, 1995.

4. Trudeau, M.L., M.F. Fréchette and R.Y. Larocque. "Surface Modification of Cu due to Arc Discharge in SF6 and SF6/CF4 Atmospheres". submitted to Surf. & Interf. Anal.

5. Trudeau, M.L., M.F. Fréchette and R.Y. Larocque. " Surface Modification by Electrodischarge in SF6 and SF6/CF4 of a Teflon/ceramic insulating material", to be published.

6 Lussier, T., M.F. Fréchette and R.Y. Larocque. "Gas/Adsorbent Interaction Dynamics: An Experimental Approach". Conf. on Electr. Insul. and Diel. Phen, Virginia, USA, Oct-22-25 1995.

7. Dilks, A., "X-ray Photoelectron Spectroscopy for the Investigation of Polymer Surfaces", in Photon, Electron and Ion Probes of Polymer Structure and Properties, Ed: D.W. Dwight, T.J. Fabish and H.R. Thomas, ACS Symposium Series 162, p. 293, 1981.

Conference Record of the 1996 IEEE International Symposium on Electrical Insulation, Montreal, Quebec, Canada, June 16-19, 1996

DIELECTRIC AND MICROSTRUCTURE PROPERTIES OF POLYMER CARBON-BLACK COMPOSITES.

C. Brosseau(a,*), C. Bourbigot(a,*),
P. Queffelec(b,*Y. Le Mest(c,*), J. Loaec(a,*)

A. Beroual(d)

(a) Laboratoire d'Étude des Matériaux.
Département de Physique
(b) LEST, Département d'Electronique
(c) Département de Chimie
(*) Université de Bretagne Occidentale.
6 avenue Le Gorgeu, B.P. 809,
29285 Brest Cedex, France.

(d) CEGELY, Ecole Centrale de Lyon.
36 avenue de Collongue, B. P. 163,
69131 Ecully Cedex, France.

Abstract: Dielectric and physicochemical properties of a composite material prepared by incorporating carbon black particles into a polymer matrix were investigated. The analysis of the carbon black concentration dependence of the amplitudes and linewidths of the ESR signals indicates that ESR is an important experimental probe of the structure of the elasticity network. Sorption kinetics experiments with toluene indicate that the uptake of solvent exhibits a square root dependence in time as a consequence of Fick's law and permit to evaluate the effective diffusion coefficient in the range 10^{-10}-10^{-11} cm^2s^{-1} depending on the mass fraction of the carbon black in the sample.

INTRODUCTION

The pace of research on dielectric properties of heterogeneous materials has accelerated in recent years. Industries such as the aerospace, electronics and others, have continuously provided the impetus pushing the development of new materials in a fascinating and rich variety of applications, e. g. shielding enclosures, capacitive video disk units, antistatic devices and electromagnetic absorbing materials [1-2]. At the same time, these materials hold general interest since they provide fundamental problems which are not completely answered, e.g. stochastic transport in disordered media and metal-insulation transition [3].

Several reports in the past decade have been aimed at determining the dielectric response of carbon-black filled polymer composites [2,4]. Transmission electron microscopy (TEM) studies have evidenced that carbon black particles are organized into clusters of average size of a few microns (primary aggregates). These studies have also shown that a secondary structure built by the agglomeration of these primary aggregates by van der Waals interactions may coexist. In these materials, at a critical concentration, the sharp insulator-conductor transition which is detected from the dc resistivity behavior, is commonly regarded as a signature for a threshold percolation and suggests that the topology of disorder is a key feature for understanding their electric behavior.

An added dimension is afforded to experimenters by performing electron spin resonance (ESR) which has the potential to provide also some insight into the interactions which occur between the polymer chains and carbon black particles [2,4]. Some of the aspects concerning the nature of the charge carriers involved in the process are now better understood. Specifically, a mechanism of charge exchange through electron transfer between carbon black particles has been advanced to explain the mechanism of the aggregation.

The objective of this paper is to explore the correlation existing between the internal structure of carbon-black filled polymer composites and their dielectric properties. For that purpose, the technique of ESR combined with swelling effects by a good solvent are important probes of the semilocal structural properties of the chain-carbon black network. In Section II, a brief descriptive summary of the context of the problem is given. Next in Section III the specifications of the experimental conditions are given. Experimental results and their physical interpretation are discussed in Section IV. Finally Section V presents a few concluding remarks.

BACKGROUND

Elasticity network and swelling

A defining physical picture of the microstructure of polymer carbon-black mixtures is nicely embodied through the concept of elasticity network. The mixture can be modeled by a three-dimensional lattice of particles interconnected by polymer chains (Fig.1). The resulting networks is elastic with properties similar to a three-dimensional gel. Thus polymer carbon-black composites can be studied in much the same way as gels. At least two main properties are of importance for the characterization of these networks: elasticity under deformation and swelling by a compatible penetrant.

When a cross-linked elastomer or a gel is placed in contact with a compatible penetrant, the liquid enters holes and microvoids of these structures. This gel swells to some equilibrium state at which the elastic response of the swollen network balances the osmotic pressure [5]. Swelling results from the diffusion of solvent into the polymer with a characteristic time which is proportional to the square of a linear dimension of the gel and also inversely proportional to the diffusion coefficient D of the gel network. Although these porous structures are random with an unknown void volume distribution the solvent penetration in these structures is generally accounted for by a Fickian diffusion mechanism.

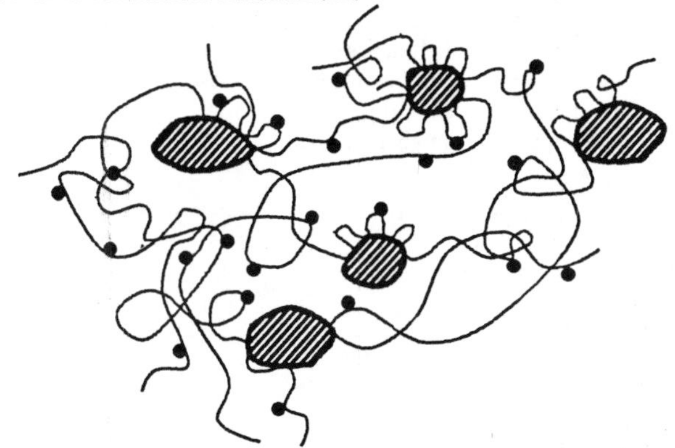

Figure 1. A schematic picture of our polymer-carbon black network. The polymer chains are connected to carbon black clusters. Two types of clusters may coexist simultaneously: primary aggregates (shadowed area), secondary aggregates (solid circle). The former describes the network properties of Raven 7000 while the latter is relevant to Y50A.

Relation to percolation

It is generally recognized that the electric behavior in these materials reflects a mechanism of percolation, including in particular an insulator-conductor transition at a critical concentration f_c of the conductive phase. The dc electrical conductivity is an important means to probe the geometry of disorder. The connectivity and its physical implications in continuum systems of objects is manifested by the critical behaviors of conductivity and permittivity: in the vicinity of the percolation threshold, the conductivity σ scales as $\sigma \sim (f - f_c)^{-\alpha}$ and the permittivity varies as $\varepsilon \sim |f - f_c|^{-\beta}$ where α is in the range 1.5-2 and $\beta \approx 0.8$-0.9 in three dimensional systems respectively. This is a significant result since it suggests that the observed $\sigma(f)$ characteristics should be described completely by one parameter f_c which depends on the topological arrangement of the system since the critical exponents α and β depend only on the dimensionality of the system [3].

One must also keep in mind that polymer carbon-black composites form a complicated class of materials since tunneling or hopping mechanisms make the system to become conductor before a physical contact between the particle has been realized [2].

EXPERIMENTAL

Materials

The polymer matrix material used in this investigation was prepared by polycondensation of Diglycidyl ether of Bisphenol F (Araldite XPY 306) obtained from Ciba Geigy with 1,12 Diamino 4,9 dioxadodecan purchased from Aldrich. Two kinds of carbon blacks were blended into this polymer: Raven 7000 (of high surface area) procured from Columbian Carbon International and Y50A (of low surface area) procured from SN2A. Some information concerning the two types of carbon black powders is reported in Table 1.

type of carbon black	Raven 7000	Y50A
mean size of carbon black powder (nm)	20-30	40-60
density ($10^{-3}kgm^{-3}$)	1.87	1.87
surface area, BET (m^2g^{-1})	625	70
f_c (% in mass), from [6]	4.5-5	0.5-0.75

Table 1. Properties of the two kinds of carbon blacks investigated in this study. The values for surface area (obtained by BET), mean size of carbon black powder and density are those given by the manufacturer as typical. Fractions (% mass) at the percolation threshold concentrations are taken from [6].

Relatively monodisperse distributions of carbon black powders were used; all powders were roughly spherical with particle sizes ranging from 20 to 60 nm [6]. After weighing all samples to the proper mass fractions the batches were mixed for 5 min with a propeller blade at 200 rpm and 30°C as previously described. It is important to appreciate here that the good quality of the dispersion in the samples was checked by TEM [6]. Specifically, it was shown by Meraoumia that the two kinds of carbon blacks produce very different structures of carbon black aggregates. On the one hand, TEM micrographs of mixtures containing Raven 7000 show primary aggregates of typical mean size of about of a few µm and secondary aggregates of typical mean size of about of 0.2 µm. On the other hand, mixtures with Y50A contain only secondary aggregates of submicronic size. The mass fractions of the carbon black in the composite ranged from 0.2% to 0.8% for Y50A and from 2% to 7% for Raven 7000, i.e. below and above the percolation threshold concentration. The specimens were further compression molded into cylinders that were 12 mm in diameter and 40 mm in height. The samples for swelling measurements were further cut into 3x6x10 mm rectangular parallelepipeds.

Methods

The data presented in this study are based on two sets of experiments. First, ESR spectra were recorded at room temperature with a JEOL FE 3X spectrometer. Second, we have studied the diffusion of small molecules in polymer carbon-black composites by the method of sorption kinetics. In this technique, a sample is immersed in a bath of the liquid penetrant. The sample is removed periodically and after blotting the surfaces to remove excess liquid, the change of mass is recorded as a function of time by using a conventional microbalance. All experiments were carried out at room temperature.

RESULTS AND DISCUSSION

ESR spectra

The ESR spectra of the composite samples were first characterized as a function of the mass fraction of the carbon black. Figure 2 shows a typical ESR signal, displayed as the first derivative of the absorption band, for a sample containing 2% in mass of Raven 7000.

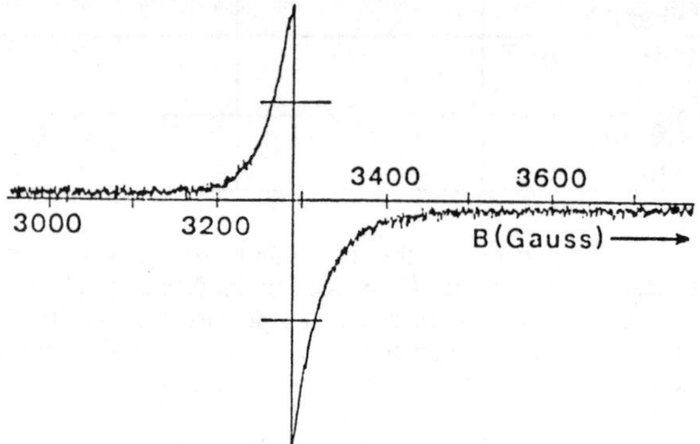

Figure 2. Typical room temperature ESR spectrum of carbon black (Raven 7000) composite materials investigated in this study, at a concentration of 2% in mass. Spectrometer settings: microwave power=1 mW, frequency=9.2 GHz.

In our experiments, we find that the g-factor (spectroscopic splitting factor) takes a value between 2.0032 and 2.0057: thus the magnetic flux density within our samples can be attributed to the spins of unpaired electrons (for the free electron, the value of g is known to be 2.0023). An interesting observation is the dependence of the amplitude A of the absorption band on the concentration f of carbon black. Our results are summarized in Fig.3 for Raven 7000 and Fig.4 for Y50A where we indicate with a bar the region corresponding to the percolation threshold f_c observed from the measurements of dc conductivity.

An increase of carbon black concentration produces a break in the slope of the amplitude plot that corresponds to the percolation threshold f_c. It also leads to a significant increase in the value of the amplitude (typically by a factor of 3).

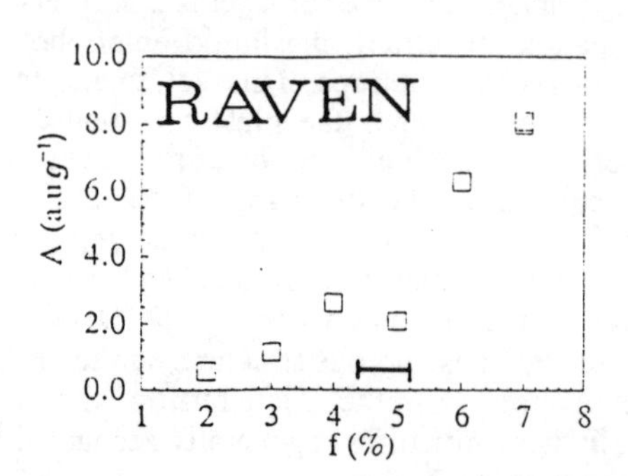

Figure 3. Amplitude A of the ESR signal of Raven 7000 in the polymer composite as a function of the mass fraction of carbon black. The amplitude has been normalized to the sample mass. The bar indicates the region corresponding to the dc conductivity percolation threshold f_c.

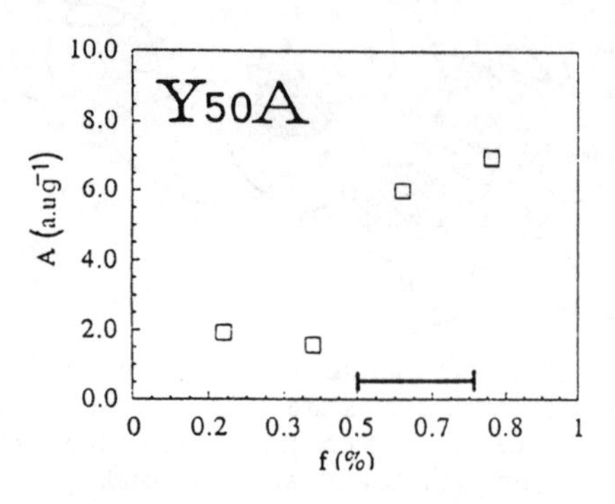

Figure 4. Amplitude A of the ESR signal of Y50A in the polymer composite as a function of the mass fraction of carbon black. The amplitude has been normalized to the sample mass. The bar indicates the region corresponding to the dc conductivity percolation threshold f_c.

Composites with low surface area carbon black (Y50A) have an identical behavior under the same experimental conditions. We find interesting in light of prior discussion of the amplitude to plot the linewidth LW of the absorption band vs the concentration of carbon black. The points on Fig. 5 corresponding to samples containing Raven 7000 show a discontinuity at about the percolation threshold f_c. In contrast with the case of Raven 7000, it can be seen in Fig.6 that the linewidth is practically constant for Y50A over the range of concentrations studied, i.e. below and above f_c. This behavior is consistent with the observation of the existence of two types of carbon black networks. We conclude that the elastic properties of the network of samples containing Raven 7000, and their dc conductivity, will be mainly driven by primary aggregates. The behavior of the linewidth, which is sensitive to the distribution of the aggregates, for samples with Y50A is consistent with the TEM studies indicating that these samples contain only one type of monodisperse small aggregates.

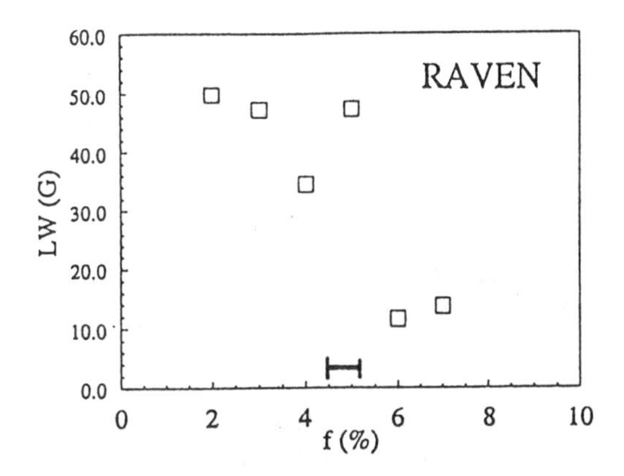

Figure 5. Linewidth LW of the ESR signal of Raven 7000 in the polymer composite as a function of the mass fraction of carbon black. The bar indicates the region corresponding to the dc conductivity percolation threshold f_c.

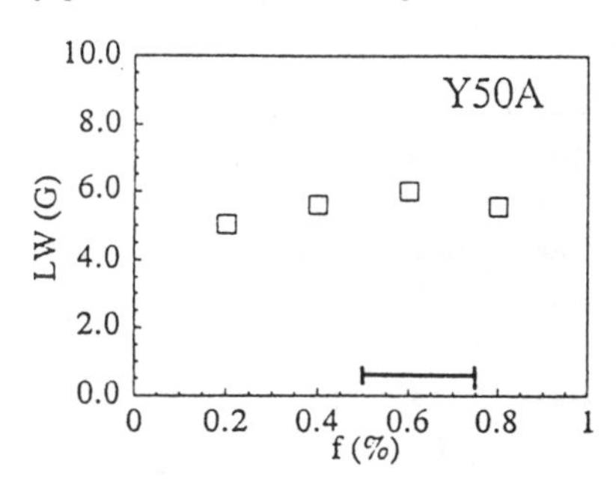

Figure 6. Linewidth LW of the ESR signal of Raven 7000 in the polymer composite as a function of the mass fraction of carbon black. The bar indicates the region corresponding to the dc conductivity percolation threshold f_c.

Sorption kinetics

We have studied the mass of sorbed penetrant normalized to the initial mass $\Delta m = \frac{m(t) - m(0)}{m(0)}$, where m(t) denotes the mass of the sample (containing Raven 7000) at time t, as a function of time and for toluene (ε(293 K)=2.38). This is an important information since the diffusion of small molecules is an important probe of the porosity of these structures. We observe that the plots in this figure are increasing functions of time and then saturate to a plateau value. The data shown in Fig.7 clearly show that the mass of sorbed penetrant $\Delta m \sim t^{0.5}$: thus diffusion is purely Fickian and we expect that the diffusion coefficient will depend only on temperature. Referring to Fig.7, it is interesting to note that the equilibrium uptake of toluene is small, i.e. $\leq 10\%$ in mass. The values of the effective diffusion coefficient of toluene for two different polymer-carbon black networks studied in this paper can be calculated from the linear slope of Δm vs $t^{0.5}$: we find that $D = 5\ 10^{-11} cm^2 s^{-1}$ for 2%

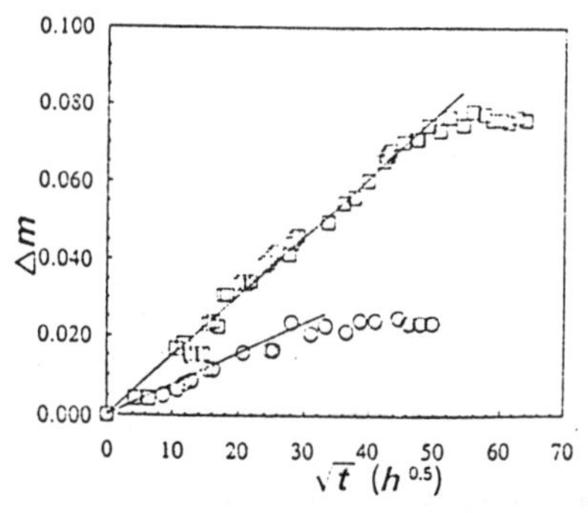

Figure 7. Comparison of experimental data and fits (full lines) to the data of sorption kinetics. This plot demonstrates that Δm scales as $t^{0.5}$ dependence in agreement with a model of swelling based upon a Fickian diffusion. Squares and circles denote respectively the sample composites with Raven 7000 (2%), Raven 7000 (6%). Conditions include: room temperature, solvent=toluene.

of Raven 7000 (i.e. $f<f_c$) and $D=10^{-11} cm^2 s^{-1}$ for 6% of Raven 7000 (i.e. $f>f_c$). It must be noted that the value of D is higher for carbon black concentrations below the dc percolation threshold. Since the solvent molecules can more easily penetrate into the more porous structure, this can be interpreted to mean that the pore structure is more connected for this range of carbon black concentrations.

CONCLUSIONS

Based on the study presented in this paper, we close with the following brief remarks. We want to stress again that a useful feature of our work is that we are able to correlate the ESR signals (amplitude and linewidth), originating from the electron acceptance by the carbon black, to the static conductivity threshold of dry and swollen composite samples. For future research it would be of interest to perform dynamic mechanical measurements of carbon black filled polymers to investigate the role of agglomerates within the polymer matrix.

ACKNOWLEDGMENTS

We are indebted to Dr. T. Méraoumia for providing us with the samples that allowed us to make this study.

REFERENCES

1. Priou,A. *Progress in Electromagnetics Research: Dielectric Properties of Heterogenous Materials*. Elsevier, 1992.
2. Donnet, J. B., Bansal R. C. and Wang M. J. *Carbon-Black: Science and Technology* . Dekker, 1994.
3. Stauffer, D. and Aharony, A.*Introduction to Percolation Theory* .Taylor and Francis, 1992.
4. Sichel, E. K. *Carbon-Black Polymer Composites*. Dekker 1982.
5. Flory, P. J. *Principles of Polymer Chemistry*. Cornell University 1953.
6. Meraoumia, T. Ph. D. Dissertation, University of Bordeaux, France, 1995.

Conference Record of the 1996 IEEE International Symposium on Electrical Insulation, Montreal, Quebec, Canada, June 16-19, 1996

Electronic Flexible Hybrid Composite Films PVDF/SiO_2*

D. Yang and W. Bei
Dept. of Materials Science & Engineering, EPRL
Univ. of Electronic Science and Technology of China
Chengdu 610054, Sichuan, CHINA

Abstract: **A new method for electronic flexible hybrid composite film PVDF/SiO_2 by Sol-Gel process are described. The process of preparing the hybrid films and dielectric, piezo-& pyroelectric properties, charge storage & electet effects have studied in detial from the theory and experiment.**

INTRODUCTION

Electronic Functional Composite (EFC) Materials and their applications are one of the important directions in the world — wide development of modern electronic High & New Tech materials[1]. In recent years, the new development directions of modern Electronic Functional Composite (EFC) have been two tendecy: (1) flexible composite; (2) fine composite. The electronic flexible composite piezo- and ferroelectric films have brought up wide research interest and been a focus of international ferro- and piezoelectric study[2]. The concepts of structure. mechanism and characteristics of flexible composite functional films have been investigated and described in detail[3]. Within the framework of fine composite characteristics signify into the following ways in organic/ inorganic 2 phase system:

(1) organic flexible phase structure: flexibility, thin film, and incresed thermostability.

(2) inorganic functional and treatment processing at low temperature.

(3) fine composite phase structure: micro (μm), nono (nm), and atom-molecular level combination.

Table 1 summarizes the fine composite principles and characteristic types of modern fine composites.

Table I
The Fine Composite Principles

materials Type	Composite Characterization
Composite	μm~mm, Micro—mixture
Composite—factor	1~100nm, Nano—dispersion
Hybrid	<10nm, Atom—Molecular combination

Recently, electronic ceramic manufacturing by Sol-Gel process have brought up wide research interest and been a main way of inorganic fine particle for fine composite treatment at low temperature.

The author studies the most typical Electronic Polymers PVDF, PI and PVDF — bosed flexible fine composite from a new method for hybrid films PVDF/ SiO_2, PT, PZT etc by sol-Gel process.

As well as other kinds of flexible composite piezo-, ferroelectric, electret films of different organic/inorganic system, these flexible hybrid composite films contain remarkable good scientific and technological values in modern electronic sensitivity & transducing, electret & storage[4], pyroelectric detectivty etc.

EXPERIMENT

Preparation of SiO_2 by Sol-Gel Process

Sol-Gel process is one of the important way on modern new inorganic and ceramic materials by chemical preparation at low temperature and microstructure control (molecular level about 50—1000Å).

The SiO_2 solution by Sol-Gel process was prepared from the silicon alkoxide by a hydrolysic — polycondensation reaction[5]. In the reation process, C_2H_5OH ethanol (ErOH) was used as the solvent of the reaction, and tetraethoxysilance (TEOS) Si $(OC_2H_5)_4$ as monomeric oxide precursor. The Sol-Gel process of silicon dioxide are described as follow:

$$\equiv Si-(OC_2H_5)+H_2O \rightleftharpoons \equiv Si-OH+C_2H_5OH \quad (1)$$

$$\equiv Si-(OC_2H_5)+HO-Si\equiv \rightleftharpoons \equiv Si-O-Si\equiv +C_2H_5OH \quad (2)$$

$$\equiv Si-OH+HO-Si\equiv \rightleftharpoons \equiv Si-O-Si\equiv +H_2O \quad (3)$$

Here, $\equiv Si-O-Si\equiv$ is the final reaction product. A small amount of hydrogen chloride, HCl (37%, extrapure) was used as an acid-catalyst to promote a hydrolysic-polycondensation reaction.

In order to speed the gel for gel forming, the higher reaction temperatures is necessary for preparing gel. Reaction rate is a sensitive function of the reaction temperature. Fig. 1 show that the relation amongthe solution viscosty, reaction temperature and time.

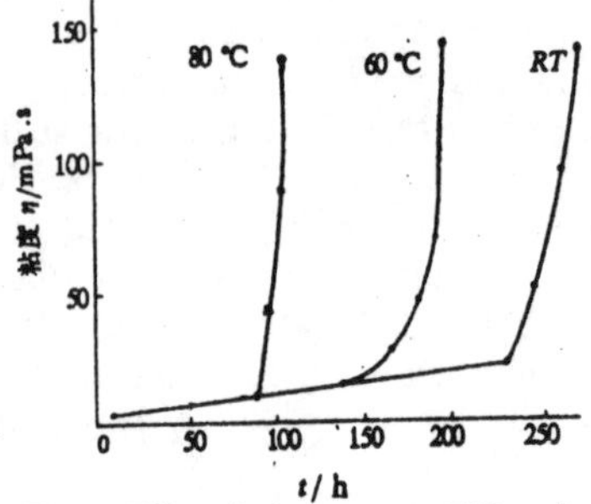

Fig. 1 The relation among SiO_2 solution

* Supported by the National Natural Science Fundation of China

Preparation of flexible hybrid composite films PVDF/ SiO_2

As an newimportant brance of modern flexible functional composite materials, author studies two kinds of flexible fine 2 — phase organic/inorganic by Sol — Gel process composite: Type I polymer/ceramic powder by Sol—Gel fine composite: Type II polymer/ceramic phase by Sol—Gel hybrid composite.

Here, the PVDF (Sorlvay) for two types composite was dissolved all DNF (Dunethyl) formanmide HCON $(CH_3)_2$). The concentration of the PVDF solution depended on the amount of DMF. In the dissolution process, the neating temperature is necessary for speeding the PVDF dissolution.

1) Polymer/ceramic powder by Sol — Gel fine composite

The type I composite with ceramic powder by Sol—Gel processing. Herefirst, the ceramic were prepared by Sol — Gel method into gelling state. Then, the ceramic powder were obtained by calcining the aried solid gels at rarious temperatures for some time into crystalline state.

2) Polymer/ceramic phase by Sol — Gel hybrid composite

The type II composite is polymer with ceramic phase by Sol—Gel processing. Here, the ceramic were prepared by Sol—Gel method also into gelling state, but were no calcined and direct form gelling state to composite.

Here e. g., PVDF/SiO_2 composite process; SiO_2 Solution Closed to gelling point as mixed with PVDF solution in the wright rate 1:1 of SiO_2 and PVDF content, It is very important to control the viscosity of SiO_2 and PVDF solution in forming the high quality hybrid composite film. Also the hybrid film was formed with flow extension compound method in thickness 70~150μm.

Polarization treatment of flexible hybrid composite films.

One or two side of hybrid film sample was vacuum evaporated with Al electrode. Polarization treatment have been two process: Thermopolarization or corona polarization.

Electronic properties measurement

The dielectric, piezoelectric, flexible coupling of PVDF/SiO_2 flexible composite films have been measured.

The open circuit TSD (Thermally Stimulated Discharge) experiments were performed with a heating rate of 3℃/min from RT to 200℃. The surface potential was measured by compensation method.

Microstructure analysis

The microstructure of flexble hybrid composite PVDF/SiO_2 films are examined by SEM、XRD、IR、XPS etc comparison with the hypothetical model and explanations are given.

REULTS AND DISCUSSIONS

Microstructure of flexible hybrid Composite film

Many microstructure analyses SEM、IR、XPS etc show that the P/C flexible fine composite have been two kind of composite connectivity types for 2 phase composite systems: type I Polymer/Ceramic by Sol — Gel with calcining, it is 3—0 type; type II Polymer/Ceramic phase by Sol—Gel with noncalcining, it is hybrid type.

1)PVER/SiO_2 3—0 type composite

2)PVDF/SiO_2 hybrid composite

Here, PVDF/SiO_2 by Sol—Gel with non—Calcining is a new hybrid composite that been different form 3—0 type. Fig. 2 shows the SEM of PVDF/SiO_2, the two phase solutions were mixed in the molecular levels by Sol—Gel process. The goals of material structure design and process control (water H_2O content and solution viscosity η) have been achieved by the Chemical method.

The microstructure XRD、IR、XPS analysis (Fig. 3 Fig. 4) show: PVDF and SiO_2 in hybrid composite film from two phase solutions were mixed in the molecule level & interface hybrid by Sol—Gel process, i. e. 3D networt and take together. The interface of two phase PVDF/SiO_2 is continue and has some hybrid ferm, the SiO_2 network structure is a network and the PVDF/SiO_2 hybrid composite process is also from point to network.

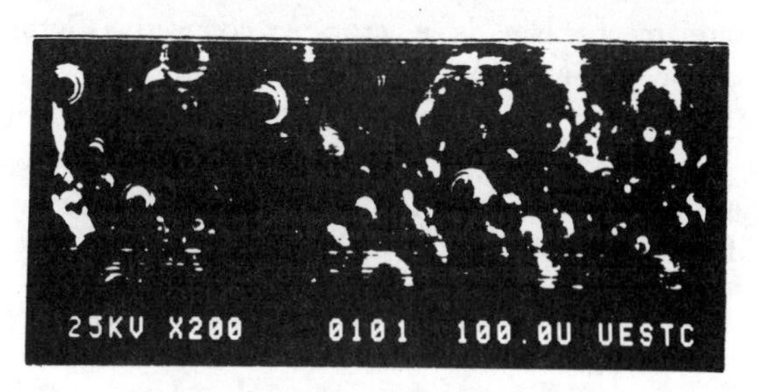

Fig. 2 SEM of PVDF/SiO_2 film

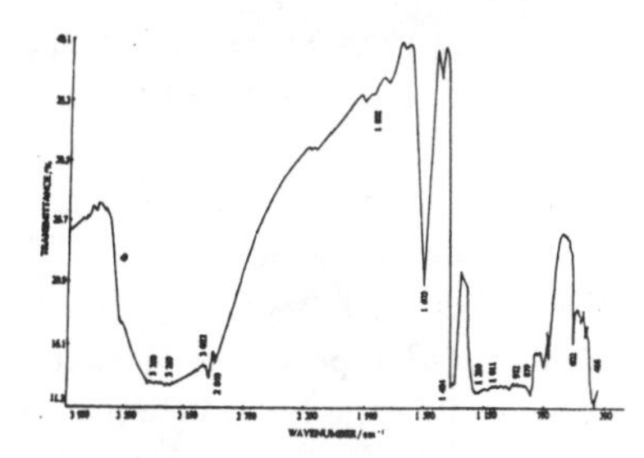

Fig. 3 IR of PVDF/SiO_2 film

From a comparative study of the XRD and IR (Fig. 3) shows that the PVDF phase of PVDF/SiO_2 composite film X—ray diffraction peaks for β & α—crystal form. But SiO_2 phase is amorphous networkstate. Besides the XRD and IR absorption spectrum of β-crystal form of PVDF phase in composite film have a reversed strength charge before and after polarization, i. e. under high temperature and field, some α form transfer into β form of PVDF place in PVDF/SiO_2 film and correspondingly, the IR absorption peak of β form have shifted from 472 cm^{-1} to 486cm^{-1}. As to SiO_2 phase at 1000 ~ 1404 cm^{-1}, IR sectrum, IR absorption band for Si—O—Si bond LO form is at ~ 1210cm^{-1}, the Si — O — Si bond TO form is at ~ 1802cm^{-1}; the IR absorption band for SiO_2 network tetrahedron bond is at ~1011cm^{-1}.

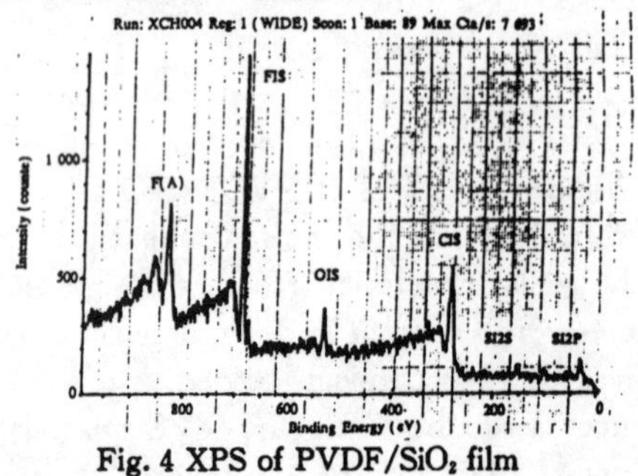

Fig. 4 XPS of PVDF/SiO_2 film

The binding energy of different atom in 2—phase interface is shown in Fig. 4. Out microstructure shows that the interface atoms of 2—phase are only PVDF— and SiO_2—atoms, but no other impurity atom. Therefore, out IR and XPS experimental tests were demonstrated convincingly that the 2 — phase PVDF/SiO_2 flexible composite mechanism is a cross — connective hybrid composite in the interface.

According to microstructure analysis, the percolation model of two — phase PVDF/SiO_2 hybrid composite continue medium by preparatin of Sol — Gel process is derived.

$$\begin{cases} \dfrac{\varnothing(\varepsilon_i^n - \varepsilon^n)}{\varepsilon_i^n + [\varnothing(1-\varnothing_c)]\varepsilon^n} + \dfrac{(1-\varnothing)(\varepsilon_h^n - \varepsilon^n)}{\varepsilon_h^n + [\varnothing_c/(1-\varnothing_c)]\varepsilon^n} = 0 \\ \varnothing_c = (d-1)/d \end{cases} \quad (4)$$

here, ε_i, ε_h, ε—dielectric coefficient of doping phase, base—mass phase and composite phase, respectively; $\varnothing$—volume percent of doping phase; d—space dimension 1, 2, 3; n—mixed factor of series connection and parallel connection.

The formula (4) is the percolation equation of weophase continue medium for any formal distribution doping phase. Therefore, the two — phase compostioe priciple is a hybrid mixed type of series connection and parallel connection; i. e. it is the hybrid composite principle of mixed form of series connection and parallel connection.

Dielectric and Charge Storage Effects

A great number of repeated experiments show that the dielectric and different electronic functional properties of the PVDF/SiO_2 by Sol—Gel flexible hybrid composite films are defermined by the best Sol—Gel technological process parameters and flexible hybrid composite content and treatment process etc. In addition, they are continuously enhanced when technological factors are improved. The level reached to date is shown in Table 2.

Table 2
The Electronic Properties of PVDF/SiO_2 by sol—Gel

Property	PVDF	PVDF/SiO_2
ε'	9—15	4.2—11.7
$\varepsilon'' \cdot 10^{-3}$	15~24	0.8~9.1
d_{33}PC/N	10~25	11—13
$g_{33}10^{-3}$Vm/N	75~313	106—349
p nc/℃ · cm^2	3.8~4.0	3.1—3.7

The results of dielectric experiment coincide with the results of the percolation model of two—phase continue medium. Therefore, the composite dielectric coefficient ε'' will obtained by the calculation formula as follow:

$$\begin{cases} \varepsilon^n = (1-\phi)\varepsilon_1^n + \phi\varepsilon_2^n \\ \varepsilon = (1-\delta)\varepsilon_1 + \delta\varepsilon_2 \\ 1/\varepsilon = (1-\delta)1/\varepsilon_1 + 1/\delta\varepsilon_2 \end{cases} \quad (5)$$

Where, ε_1, ε_2, ε—dielectric coefficient of PVDF-phase, SiO_2-phase and composite film, respectivity; ϕ—Volume percent of SiO_2; δ, φ—Composite series, parallel coefficient.

The relative curve of PVDF/SiO_2 composite coefficient ε' with temperature τ and frequency f are shown in curve of $\varepsilon'(T, f)$. The relaxation phenomenonare related to orietation polarization of electric dipole moment and space charge polarization of shallow energy trap, and H_2O content and distribution in PVDF/SiO_2 composite.

The PVDF/SiO_2 flexible hybrid composite film has very excellent polarization and electret charge storage effects. The current — voltage relations of single plane thermo—charging in composite film have been studied. The charging space charges are exponent distribution in energy and space. The results of charging show that surface potentical and sample thickness greatly influence the electrical properties of samples under corona charging.

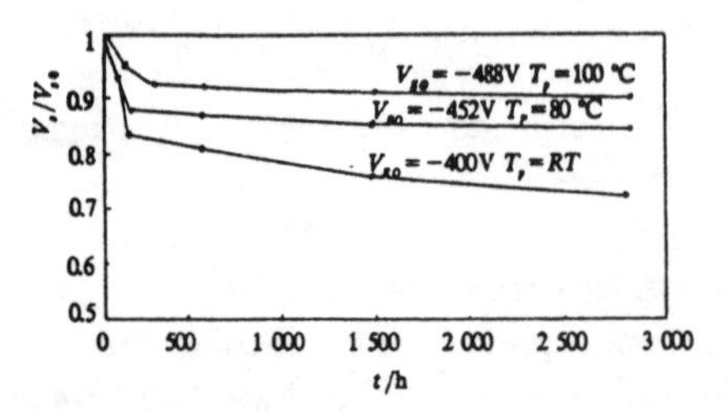

Fig. 5 The surface potential curves of PVDF/SiO_2 film

The charge storage effect and stability of PVDF/SiO_2 hybrid composite film related to charge corona charging (poling) temperature (T_p), time (T_p), gate voltage (V_g) and discharging voltage (V_p). Fig. 5 shows the surface potential decay curves of negatively charged samples under the different temperature T_p of hybrid film $PVDF/SiO_2$.

The decary of the charge storage effect of hybrid composite films is examined over a period of time. The aging rate (τ) of $PVDF/SiO_2$ film is about ~160/years at 20℃. Howere, an increase of there storage environment temperature will decay the aging time (τ). Obviously the $PVDF/SiO_2$ hybrid film is excellent charge storage electriet materials.

Besides, the charge storage stability of composite film is improved by chemical surface modification (using HMDS).

Piezo—& pyrroelectric properties

The good piezo — and pyroelectric properties of $PVDF/SiO_2$ are shown in Table 2. The piezoelectric constants d_{ij}, g_{ij}, e_{ij}. etc. are great relation to the composite process and polarization treatment processing. The curve of $d_{33}-E_p$ shows the influence of polarizing filed intensity E_p on the piezoelectric strain constant d_{33} of $PVDF/SiO_2$ film.

The formula of thermally stimulated discharge (TSD), and the active energy of detrapped charges and dipole in composite electret film $PVDF/SiOs_2$ have been derived. There calculation formulas of the piezoelectric properties constant d, e and pyroelectric property constant P of 2— phase hybrid composite films as follow:

$$\begin{cases} d = d/G \\ e = \frac{\varepsilon-1}{\varepsilon_1-1}(1-\varphi)[\Phi(\frac{3\varepsilon_1}{2\varepsilon_1+\varepsilon_c})e_{cg} - \frac{\partial \ln L}{\partial s}P_s] + \frac{\varepsilon-1}{\varepsilon_2-1}\varphi(P_{s1}+P_0)\frac{\partial \ln L}{\partial s} \\ p = \frac{\varepsilon-1}{\varepsilon_1-1}(1-\varphi)[\frac{\Phi}{V}(\frac{3\varepsilon_1}{2\varepsilon_1-\varepsilon_c})\frac{\partial(vP_{sc}}{\partial T} - \beta P_s] + \frac{\varepsilon-1}{\varepsilon_2-1}\varphi[(\frac{\partial \ln L}{\partial T} - \frac{1}{T})P_0 + P_{s1}\frac{\partial \ln L}{\partial T}] \end{cases} \quad (6)$$

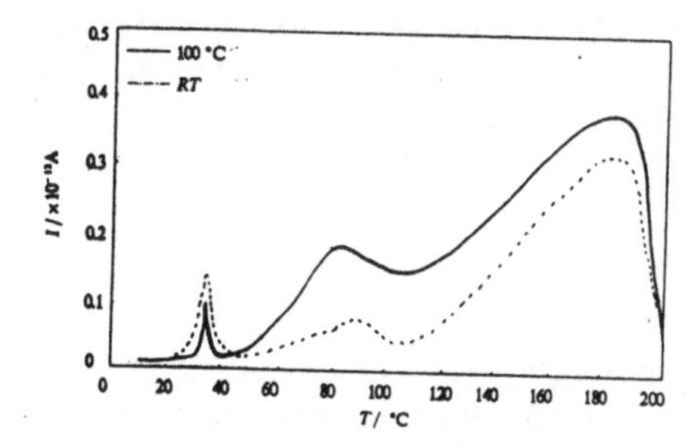

Fig. 6 Open—curcuit TSD spectra of $PVDF/SiO_2$ film

Fig. 6 shows the open—circuit TSD current spectra of $PVDF/SiO_2$ by Sol—Gel hybrid composite film after negrative corona chaging a room and elevated temperature.

The theoretical and experimental results show: sharp TSD peak of low temperature about 35℃ was the detrapped space charge peak of shallow levels. The transport process of the detrapped space charge of shollow levels in film body belong to fast retrapped effect. The TSD peak about 80℃ is dipolar charge current peaks in PVDF phase. The TSD peak about 180℃ includes the real space charge and dipolar charge current peaks. The transport process of the detrapped space charge in film body about 180℃. Belong to slow retrapped effect. The experimental results show that mian TSD peak in pure PVDF film about 120℃ shifts to about 180℃ TSD peak in composite film. It indicates that composite film $PVDF/SiO_2$ has better charge storage ability and stability. The work temperature areas of composite film $PVDF/SiO_2$ are widened.

The conntributions of real space charge for piezo and pyroelectricity in composite film $PVDF/SiO_2$ were first obtained in the researce.

The piezo — and pyroelectric experimental results show that real space charge not noly provides for piezo— and pyroelectricity in composite electret $PVDf/SiO_2$, but also forms the self — field of charge layer in order to sustain dipole order in the film body.

CONCLUSION

The flexible composite films of Electronic Polymers base PVDF and ferroelectric ceramics by Sol—Gel process $PVDF/SiO_2$ are one of the new important method way for making to flexible fine and hybrid 2—phase composite functional films system.

The concept of "hybrid composite film" shows the composite film from two phase solutions were mixed in the molecular leve & interface hybrid by Sol—Gel process, and i. e. 3 D network and take together.

The resëarced flexible fine hybrid films $PVDF/SiO_2$ have excellent dielectric, piezo — & pyroelectric charge storage, transport & electret effects and functional properties, as new electronic flexible composite materials have extensive theoretical and applied value in the widely field of electronic sonsors, transducers and electret charge storage & transport devices & systems.

REFERENCE

1. R. E. Newnham et al., J. Am. Cersm, Soc. 8, 1990, p. 73.

2. Daben YANG, ISAF'92, 1992.

3. Daben YANG, ICECM'92, 1992, pp. 245—248.

4. G. M. Sessler, *Electrets*, Springe Verlag, 1987.

5. Bai Wei, Yang Daben, 1994 IEEE ISEI, 1994, p. 590.

6. R. A. C. Altafim, et al., IEEE trans, Electr. Insulaton, Vol. 27, No. 4, 1992, pp. 739—743.

7. Yang Daben, et al, ICECM'95, 1995, p. 154/1—4.

Conference Record of the 1996 IEEE International Symposium on Electrical Insulation, Montreal, Quebec, Canada, June 16-19, 1996

Space Charge Near Microbes During Pulsed Electric Field Pasteurization of Liquid Foods

R. E. Bruhn , P. D. Pedrow, and R. G. Olsen
School of Electrical Engineering and Computer Science
B. G. Swanson
Department of Food Science and Human Nutrition

G. V. Barbosa-Canovas
Department of Biological Systems Engineering
Washington State University
Pullman, WA 99164

***Abstract*: Inactivation of microbes by the application of pulsed electric fields could result in low temperature pasteurization of liquid foods. Advantages over conventional heat pasteurization include longer shelf life, better flavor, and less enzyme damage. In this work, fields as high as 40kV/cm have been applied to milk, apple juice, and electrolyte that was inoculated with microorganisms. Modeling of the microbes during exposure to these intense electric fields is described. Suspension solution and liquid protoplasm are modeled with a relative pemittivity of 81 and each contains two species of ionic charge carriers (one species plus and one species minus). The microbe membrane is modeled with a relative permittivity of 2 and zero conductivity. The continuity equation has been solved numerically in 1 dimension for low ion concentration to investigate the transient behavior of space charge sheaths near the microbes. Free surface charge density, which accumulates on both sides of the cell membrane is also described by this model. Mesh size and simulation time step were adjusted to resolve space charge sheath dynamics near the microbes.**

INTRODUCTION

Pasteurization of liquid foods by pulsed electric fields has been studied as an alternate to conventional thermal pasteurization [1,2,3,4,5]. Microbe mortality may be caused by electroporation which is the creation of pores in the cell membrane when voltage drop across the membrane exceeds about 1 volt [6]. Electric fields on the order of 40 kV/cm have been applied to a variety of microbes that have included *Escherichia coli, Staphylococcus aureus, Bacillus subtilis, Saccharomyces cerevisiae, Yersinia enterocolitica, Listeria monocytogenes, and Candida albicans.* Liquid suspensions have included milk, apple juice, NaCl solution, and a simulated milk solution (described in [5]). Depending on parameters, these pulsed electric fields have produced survival fractions smaller than 10^{-8}. Since this inactivation of microorganisms takes place at temperatures substantially lower than for conventional heat pasteurization, improved flavor, longer shelf life, and reduced enzyme damage are possible improvements in the final product [7,8].

Previous work, which was cited above, focused on microbial and high voltage engineering issues such as culture techniques and dielectric breakdown of the test chamber. Electrical modeling of the space charge sheaths that form within the liquid suspension near the liquid/electrode interfaces has been described [9]. Response of microorganisms to electric fields has been investigated by others who were studying electroporation [6] and cell fusion [10,11]. In that work, cell inactivation usually was not the objective as it is in the present work.

In its simplest form, the cell membrane is composed of a lipid bilayer with the hydrophobic ends of the molecules being shielded from the suspending liquid (and from the protoplasm of the microbe) by the hydrophilic ends of the molecules [6]. In reality, cell membranes are quite complex [12,13].

In this work, we have used numerical techniques to simulate the cell membrane of a microorganism exposed to large electric fields. During early model development, the membrane has been represented by a shell of lossless dielectric encasing a region of lossy dielectric (the protoplasm) with the entire system immersed in a third lossy dielectric (the suspension liquid). Each of these three regions have been assumed linear, homogeneous, and isotropic.

We have assumed planar geometry and that the space charge sheaths are too weak to have significant influence upon the externally applied electric field. More realistic assumptions (to be used in future work) are described below in the Discussion section.

PLANAR MODEL

Assumptions

Figure 1 shows the five regions being modeled and Table 1 describes parameters used to characterize each region. Initially all interfaces are assumed to have zero free surface charge density but free surface charge accumulates at the interfaces as the simulation progresses.

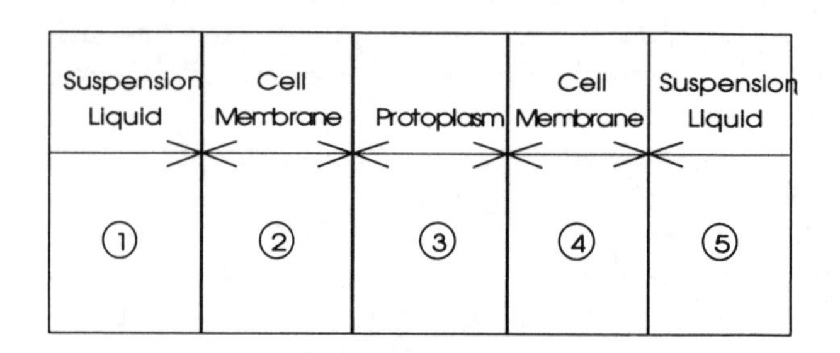

Figure 1. Five regions being considered in the planar model.

Table 1.
Parameters used to characterize the various regions shown in Figure 1.

Region of Simulation	Initial Uniform Density of Pos. Ions	Initial Uniform Density of Neg. Ions	Relative Permittivity	Mobility of Pos. Ions	Mobility of Neg. Ions	Thickness
Suspension Liquid (1 & 5)	$(n_0)_L$	$(n_0)_L$	$(\varepsilon_r)_L$	$(\mu_+)_L$	$(\mu_-)_L$	Regions 1 & 5 are Semi-infinite
Cell Membrane (2 & 4)	0	0	$(\varepsilon_r)_M$	0	0	d_M
Protoplasm (3)	$(n_0)_P$	$(n_0)_P$	$(\varepsilon_r)_P$	$(\mu_+)_P$	$(\mu_-)_P$	d_P

For each ion species α we can write the continuity equation as

$$\frac{\partial n_\alpha}{\partial t} + \nabla \cdot (n_\alpha \bar{u}_\alpha) = 0 \quad (1)$$

where n_α and $\bar{u}_\alpha$ are concentration and fluid velocity, respectively, for ion species α. We assume there are no volume sources nor volume sinks of ions; however, ions can impinge on and exist as free surface charge density at interfaces. Ion mobility can be used to write

$$\bar{u}_\alpha = \mu_\alpha \bar{E} \quad (2)$$

where μ_α is mobility of species α and $\bar{E}$ is electric field. Ion diffusion and motion of the suspension liquid have been ignored. Using a vector identity, Eqs. (1) and (2) can be combined to give

$$\frac{\partial n_\alpha}{\partial t} + \mu_\alpha \bar{E} \cdot \nabla n_\alpha + \mu_\alpha n_\alpha \nabla \cdot \bar{E} = 0 \quad (3)$$

The third term is assumed small enough to be neglected (this will be valid only for small volume charge density) giving

$$\frac{\partial n_\alpha}{\partial t} + \mu_\alpha \bar{E} \cdot \nabla n_\alpha = 0 \quad (4)$$

Numerical Algorithm

The numerical simulation begins at t=0 by assuming a step increase in the D field (throughout all 5 regions) from zero to some value D_0. Evolving volume charge density in sheaths and surface charge density at interfaces are assumed small enough that D is not significantly modified from its externally applied value, D_0. Thus we have

$$D_1 = D_2 = D_3 = D_4 = D_5 = D_0 \quad (5)$$

and

$$(\varepsilon_r)_L E_1 = (\varepsilon_r)_M E_2 = (\varepsilon_r)_P E_3 = (\varepsilon_r)_M E_4 = (\varepsilon_r)_L E_5 = D_0 \quad (6)$$

Since D is uniform throughout the five regions, E will be uniform within each region and so Eq. (4) applies. It can be represented [14] by the numerical algorithm

$$\frac{\{n_\alpha\}_i^j - \{n_\alpha\}_i^{j-1}}{\Delta t} + \mu_\alpha E \frac{\{n_\alpha\}_{i+1}^j - \{n_\alpha\}_{i-1}^j + \{n_\alpha\}_{i+1}^{j-1} - \{n_\alpha\}_{i-1}^{j-1}}{4\Delta x} = 0 \quad (7)$$

where i and j are indices that represent the ith spatial grid point and the jth time step, respectively. This is a time centered algorithm, characterized by a high order of accuracy [14]. For convenience, we form the dimensionless quantity

$$\xi_\alpha = \frac{1}{4}\left(\frac{\mu_\alpha E \Delta t}{\Delta x}\right) \quad (8)$$

which is proportional to the fraction of a grid spacing traversed during a time step by an element of the ion fluid traveling at the speed $\mu_\alpha E$. Substituting Eq. (8) into (7) and placing known densities on the right hand side and unknown densities on the left hand side gives

$$-\{n_\alpha\}_{i-1}^j + \frac{1}{\xi}\{n_\alpha\}_i^j + \{n_\alpha\}_{i+1}^j = \{n_\alpha\}_{i-1}^{j-1} + \frac{1}{\xi}\{n_\alpha\}_i^{j-1} - \{n_\alpha\}_{i+1}^{j-1} \quad (9)$$

This yields a set of simultaneous linear equations at each time step for each ion species. Solution to these is obtained by inverting a tridiagonal matrix. Volume space charge density is found from $(1.6 \times 10^{-19})(n_+ - n_-)$ and surface charge density at interfaces is found from the net flux of ions to the interface. Results are shown in the next section.

Table 2. Numerical Values Used in Simulation.													
$(n_0)_L$ (m^{-3})	$(n_0)_P$ (m^{-3})	$(\varepsilon_r)_L$ (-)	$(\varepsilon_r)_M$ (-)	$(\varepsilon_r)_P$ (-)	$(\mu_+)_L$ ($\frac{m^2}{Vs}$)	$(\mu_+)_P$ ($\frac{m^2}{Vs}$)	$(\mu_-)_L$ ($\frac{m^2}{Vs}$)	$(\mu_-)_P$ ($\frac{m^2}{Vs}$)	d_M (μm)	d_P (μm)	D_0 (C/m^2)	Δt (μs)	Δx (μm)
6.23 $\times 10^{15}$	6.23 $\times 10^{15}$	81	2	81	7 $\times 10^{-8}$	7 $\times 10^{-8}$	7 $\times 10^{-8}$	7 $\times 10^{-8}$	1.0	4.0	3.59 $\times 10^{-3}$	0.14	0.1

Results

For the parameters shown in Table 2 the numerical simulation showed the development of volume space charge sheaths on both sides of the cell membrane as shown in Figure2.

Space charge sheaths result from uncompensated charge that is born when charge of the opposite sign vacates a region in response to the applied electric field. In addition to the volume charge density shown in Figure 2, the magnitude of the free surface charge density on the four membrane surfaces reached about 1.7×10^{-9} C/m^2 at t = 5μs. While the size of the protoplasm (4 μm) is consistent with real microbes, membrane thicknesses are known to be [12,13] in the range 5 to 20 nm rather than the 1 μm used here. A thicker membrane was used for illustrative purposes.

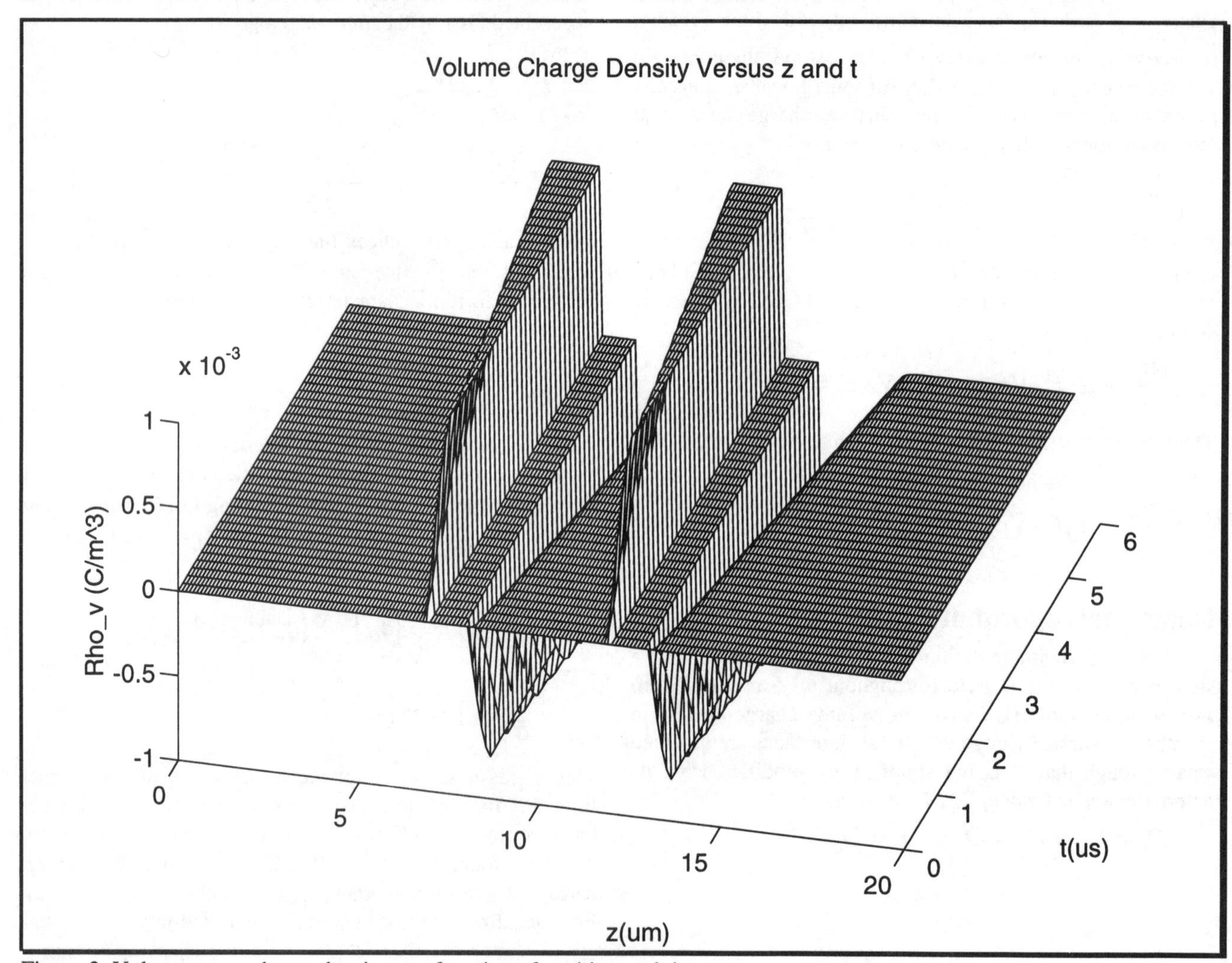

Figure 2. Volume space charge density as a function of position and time.

DISCUSSION

For more dense ion fluids, volume charge density in the sheaths and free surface charge density at interfaces will significantly modify the D field so that it will no longer be equal to the externally imposed value D_0. For that condition, the more complicated model described in Eq. (3) will be used. Following that, the model will be extended to two dimensions.

ACKNOWLEDGMENT

This work was supported in part by U. S. Army Research Office Grant #DAAH04-94-G-0113.

REFERENCES

[1] B. L.Qin, G. V. Barbosa-Canovas, B. G. Swanson, P. D. Pedrow, and R. G. Olsen, "A continuous Treatment System for Inactivating Microorganisms with Pulsed Electric Fields," Paper #EP-4-4-196, Presented at the 1995 IEEE Industry Applications Society Annual Conference, Orlando, Florida, October 8-12, 1995.

[2] B. Qin, Q. Zhang, G. V. Barbosa-Canovas, B. G. Swanson, and P. D. Pedrow, "Pulsed Electric Field Treatment Chamber Design for Liquid Food Pasteurization Using a Finite Element Method," Transactions of the American Society of Agricultural Engineers, Vol. 38, No. 2, pp. 557-565, 1995.

[3] B. Mazurek, P. Lubicki, and Z. Staroniewicz, "Effect of Short HV Pulses on Bacteria and Fungi", IEEE Transactions on Dielectrics and Electrical Insulation, Vol. 2, No. 3, pp. 418-425, June 1995.

[4] B. L. Qin, Q. Zhang, G. V. Barbosa-Canovas, B. G. Swanson, and P. D. Pedrow, "Inactivation of Microorganisms by Pulsed Electric Fields of Different Voltage Waveforms," IEEE Transactions on Dielectrics and Electrical Insulation, Vol. 1, No. 6, pp. 1047-1057, December 1994.

[5] Q. Zhang, A. Monsalve-Gonzalez, G. V. Barbosa-Canovas, B. G. Swanson, "Inactivation of E. Coli and S. Cerevisiae by Pulsed Electric Fields Under Controlled Temperature Conditions," Transactions of the American Society of Agricultural Engineers, Vol. 37, No. 2, pp. 581-587, 1994.

[6] J.C. Weaver, "Electroporation: A General Phenomenon for Manipulating Cells and Tissues", Journal of Cellular Biochemistry, Vol 51, pp 426-435, 1993.

[7] D. Knorr, M Geulen, T. Grahl, and W. Sitzmann, "Food Application of High Electric Field Pulses," Trends in Food Science & Technology, Vol. 5, pp. 71-75, March 1994,

[8] B. Mertens and D. Knorr, "Developments of Nonthermal Processes for Food Preservation," Food Technology, pp. 124-133, May 1992.

[9] D. B. Spencer, "Space Charge Evolution in Dilute Binary Electrolytes Exposed to High Voltage Transients," Masters Thesis, Washington State University, May 1994.

[10] L.Y. Song, et al., "Divalent Cations, Phospholipid Asymmetry and Osmotic Swelling in Electrically-Induced Lysis, Cell Fusion and Giant Cell Formation with Human Erythrocytes", Biochimica et Biophysica Acta, Vol. 1148, pp 30-38, 1993.

[11] M. Kiyomota and H. Shirai, "Reconstruction of Starfish Eggs by Electric Cell Fusion: A New Method of Detecting the Cytoplasmic Determinant for Archenteron Formation", Develop. Growth & Differ., Vol. 35 , pp 107-114, 1993.

[12] T. S. Tenforde, "Mechanisms of Interaction and Biological Effects of Extremely Low Frequency Electromagnetic Fields," Proceedings of the 1994 Japan-U. S. Science Seminar on Electromagnetic Field Effects Caused By High Voltage Systems, pp. 311-321, June 28-July 1, 1994, Sapporo, Japan.

[13] F. S. Barnes and C. J. Hu, "Nonlinear Interactions of Electromagnetic Waves with Biological Materials," in **Nonlinear Electromagnetics**, edited by P. Uslenghi, NY: Academic Press, pp. 391-426, 1980.

[14] E. S. Oran and J. P. Boris, **Numerical Simulation of Reactive Flow**, NY: Elsevier, 1987, pp. 110-111.

Conference Record of the 1996 IEEE International Symposium on Electrical Insulation, Montreal, Quebec, Canada, June 16-19, 1996

Influence of Negative Corona-charged Surface Protective Material on Surface Depletion Region Width

Zhang Feng Xu ChuanXiang Zhang ShaoYun
Department of Electrical Engineering
Xi'an Jiaotong University, China

Abstract: **The change of the surface depletion region width (SDRW) of pn junction for high voltage silicon rectifiers caused by negative corona-charged surface protective material has been studied. Negative corona charge makes the surface charge density of a SiO_2-insulator interface negative and thus, cause the SDRW to expand. The photocurrent method is used to measure the SDRW of reverse-biased pn junction with positive bevel angle mesa structure. The change SDRW can be caused by the decay of the surface charge density. The expanse of the SDRW tend to saturate as the voltage of the corona is increased. The decay process is different with different surface protection materials. Numerical calculation were performed and the surface charge density of negative corona-charged insulators was found to be approximately -10^{11} charges/cm^2.**

INTRODUCTION

Polymers have been widely used throughout the electrical and electronic industry, such as electrical devices, electrical wire, cable, etc. Besides conductivity, dielectric loss, potential discharge and breakdown, there has been considerable interests in the characteristics of the corona-charged insulators. Many experiments and theoretical works have been carried out to determined the processes by which a corona-charged insulator becomes discharged when resting on an earth plane[1]. Many methods have been developed, such as the constant-current mode and constant-voltage mode[2], TSC and many others. The surface potential measurement is one of the more widely used method[3] and the work done by Ieda[4] is a typical example.

To improve the junction-breakdown characteristics at the surface, the high voltage p-n junction of rectifier is usually designed bevel angle mesa structure. The surface is coated with a appropriate passivation and protective material. The change of the SDRW of rectifiers caused by negative corona-charged surface protective material is studied in the present paper. Negative corona discharge makes the equivalent interface charge density of Si-SiO2-insulator negative and thus, cause the SDRW to expend.

MODEL

The model used in our calculation is shown schematically in Fig. 1. Based on doping and compensating, the surface depletion region of a positive bevel angle rectifier bends upward. When there is actual surface charge, the depletion region changes with the different surface charge density. The positive surface charge narrows the surface depletion region, while negative ones widens it.

For a definite bevel angle, θ is a constant. There is

$$W_0 = W_D / Sin\theta = \sqrt{\frac{2\varepsilon_S U_R}{qN_D}} / Sin\theta \quad (1)$$

$$W_s = W_D' + \Delta W_D + \Delta W_C \quad (2)$$

ΔW_D is the degree of depletion region bends upward due to bevel angle. ΔW_C is the change due to surface charge. Ws is actual depletion region width, which can be measured by photocurrent method[5,6].

Standard numerical methods were used in calculating of the potential distribution in positively beveled p-n junctions. The ΔW_D and ΔW_C can be calculated by solving Poison's equation with suitable boundary conditions. The inner potential of depletion region is given by Possion's equation:

$$\frac{\partial^2 V(x,y)}{\partial x^2} + \frac{\partial^2 V(x,y)}{\partial y^2} = -\frac{\rho(x.y)}{\varepsilon_0 \varepsilon_s} \quad (3)$$

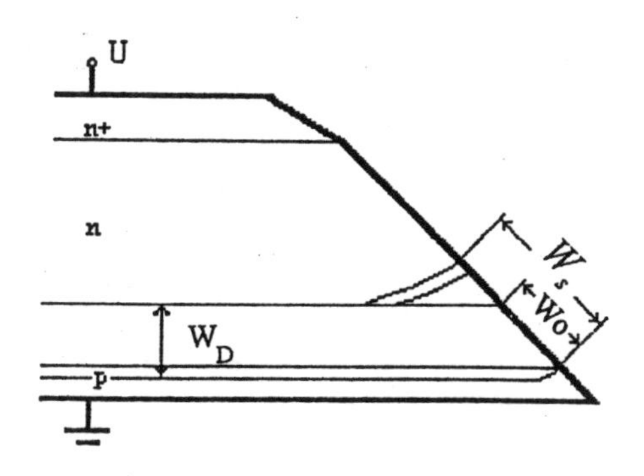

Figure 1. The surface depletion region width of reversed-biased high voltage p-n-n+ junction of silicon rectifier with positive bevel angle mesa structure.
θ =25° , θ 1=5°

If the space charges of dielectric coating and location out of depletion region are equivalent to interface charges, the potential distribution can determined by Laplace's equation:

$$\frac{\partial^2 V(x,y)}{\partial x^2}+\frac{\partial^2 V(x,y)}{\partial y^2}=0 \qquad (4)$$

for the interface of Si-insulator, neglected the SiO_2 of tennis millimicrons, the equivalent interface charge density is σ, we obtain the boundary conditions due to Gauss's law:

$$\varepsilon_d E_{dn} - \varepsilon_s E_{sn} = \sigma \qquad (5)$$

ε_s ,ε_d is the dielectric constant of Si and of the dielectric respectively. E_{sn} ,E_{dn} is the normal field in the semiconductor and in the dielectric.

Then, we use difference equation, through calculating of SLOR (successive line over relaxation)[5], the depletion region width of certain charge density under actual voltage is obtained.

RESULTS

For a bevel angle rectifier p-n junction, when the doping of n-side is certain, ΔW_D is definite with a reserved-biased voltage and bevel angle θ. Under this condition, the change of Ws is determined by the change of Δ Wc, which is caused by the interface charge. There is a certain interface charge corresponding with SDRW one-to-one. Therefore, if we measure the SDRW of p-n junction with the surface protective material being corona-charged, the interface charge density could be obtained by calculation. Most authors report the surface-potential decay of insulator charged with a corona discharge, and the charge-decay plots show the same general features. In our experiment, the charge-decay of corona discharged protective material is obtained by measuring the change of the SDRW of p-n junction with photocurrent method.

An n-type silicon crystal of thickness 600 μ m p-n junction produced by diffusion method is used, and its actual structure is n+np, with ρ =120 ~ 125 Ω · cm, θ 1=5 ° θ =25 ° . For the uncovered p-n junction, which has been angle lapped and corroded, then, covered with organopolysiloxane gel directly as surface protection. Its SDRW is 444 μ m.

The gel, protecting the mesa structure of rectifier, is corona discharged by many negative needles, then ground contacted for 10 seconds. The SDRW is measured immediately after ground contacted. From the measurement of the SDRW, the corona discharged interface charge density of gel is computed. With the different corona discharge voltage, the maximum SDRW is various, as in Fig. 2. From the Fig. 2, it is observed that the widen of SDRW of p-n junction, caused by corona discharge, occurs when corona discharge voltage goes beyond 3.2 kV. With the increasing of corona discharge voltage, the widen of SDRW rises quickly and tends to saturate.

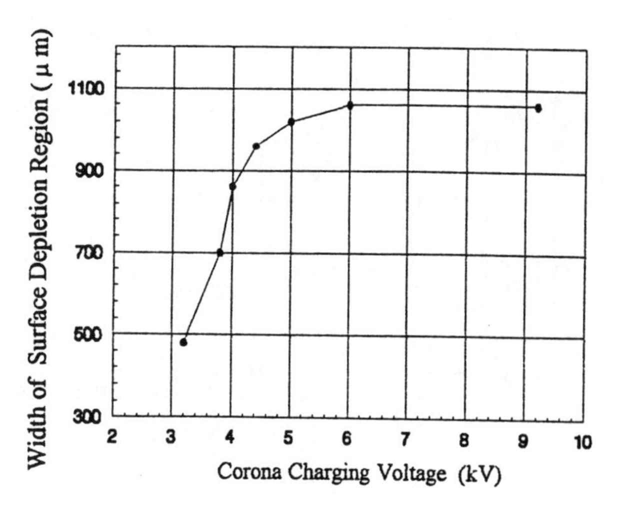

Figure 2. The change of surface depletion region width of reversed-biased p-n junction with the surface protection material was corona charged with needles .

When the material is corona discharged, a charge-decay process causes the change of the interface charge density of rectifier .Correspondingly, there is a decay process of SDRW, and it finally stabilized on the condition of which it uncharged, as in Fig. 3.

The another material, which is chosen for surface protection on silicon rectifiers, is SP(silicone resin modified polyester). The influence of negative corona discharge on rectifiers is studied. We also found that the SDRW of rectifiers expand with the increasing corona discharge, but its decay process takes long time. It takes more than 2 hours for the SDRW to back its uncharged condition after being corona discharged with needles.

DISCUSSION

Using the model mentioned in section 1, the equivalent interface charge density of positive angle mesa structure rectifier was calculated theoretically. Within general technical condition, the interface charge density of Si-SiO2 is $\sigma = 1.01 \times 10^{11}$ charges/cm², for the SiO2 layer of several tennis millimicrons is invaded. The interface charge density of organopolysiloxane gel is $\sigma = -6.8 \times 10^{10}$ charges/cm². Although the gel has the negative interface charge density, the total effect of the interface shows positive charge density. The interface charge density of SP is $\sigma = 5.4 \times 10^{10}$charges/cm².

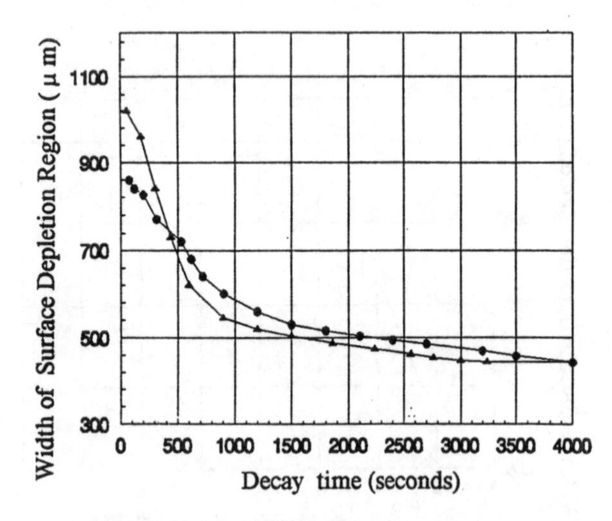

Figure 3. The change of SDRW with time after the surface protection material being corona discharged, (a) ● discharged voltage is 4.0 kV. (b) ▲ discharged voltage is 6.0 kV

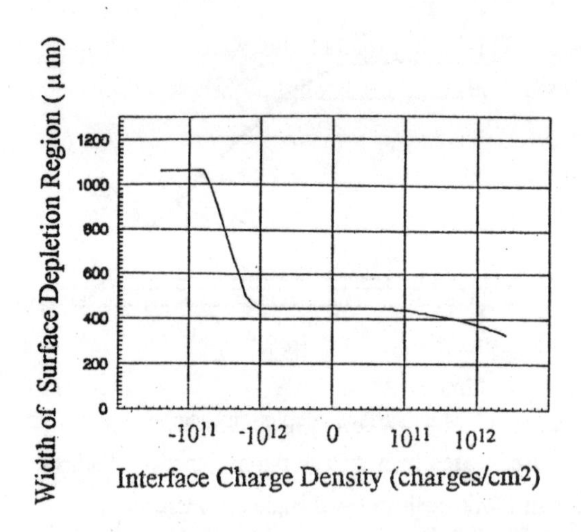

Figure 4. The change of SDRW with the various equivalent surface charge density, obtained from the calculation of SLOR

With the negative corona discharge, the charges have been trapped on their way through the material, causing the change of dielectric charges. If the charges of the dielectric considered as interface charges, the interface charge density increases with the corona discharge voltage rising and more charges being trapped on deeper traps. Thus , it changes the interface charge density. When the negative interface charge density goes up to a certain degree, it causes the widen of SDRW.

The calculation results are shown in Fig. 4 and are discussed else where[5]. When corona discharge voltage goes up to 3.2 kV, the equivalent interface charge density of the corona-charged material is approximately little above 10^{11}charges/cm² and shows negative characteristics, causing the SDRW to widen. With the increasing corona discharge voltage, the equivalent interface charge density rises, and the SDRW is much more expanded. When corona discharge voltage goes beyond 6.0 kV ,the expansion of SDRW is halted, as shown in Fig. 2. This means that the charge trapped in material is more than -8×10^{11} charges/cm², and no expansion of SDRW can be observed, as shown in Fig. 4.

When the SDRW measured by photocurrent method with a spot light, a reserved-bias voltage is applied (U=200V in experiment). This causes a high surface electric field, about 2 ~ 6 kV/mm for the maximum, which is measure by a metal probe. This will promote the charge-decay process, and organopolysiloxane gel is slightly cross linking polymer, so that the corona-charged charges decays rapidly within 1000 seconds, and then it decays slowly to the uncharged condition, as in fig. 3. The degree of cross linking of gel is 19.8%, there is more linear molecular, and its structure allows its linear molecular to free vibration, and complete the charges exchange. While SP solidified films, its degree of cross linking is about 73%, the more network structure restricted the vibration of high molecular, so it is difficult to exchange charges, causing the decay process to take longer time.

For different corona discharge voltages, the rate of SDRW decay caused by charge-decay process is different. We observe the cross phenomenon of various corona discharge voltages, which is also obtained by Ieda[4].

CONCLUSION

Negative needles corona discharge of organopolysiloxane gel, which is protective material of positive angle mesa structure rectifiers, causes the negative charge to be trapped. The corona-charged negative interface charge density of various corona discharge voltage causes the change of SDRW. With the corona discharge voltage from 3.2 kV to 6.0 kV, the equivalent interface charge density is approximately -10^{10} to -8×10^{11} charges/cm^2, and makes the great change of the SDRW. With the increasing of corona discharge voltage, the expansion of SDRW tends to saturate. The decay of SDRW caused by corona discharge is time dependent. The result of corona-charged studied by measurement of the SDRW using photocurrent method is similar to the surface-potential decay method. The decay of higher corona discharge voltage is more rapidly than that of lower voltage. For high network structure material, the decay process takes longer.

Reference

1. G. F. Leal Ferreira, "Corona Charging of Electrets". IEEE Transactions on Electrical Insulation. Vol. 27 No. 4 , 1992 pp.719-738.
2. Xia. Z. F, "Constant-current Corona Charging of Teflon PAF". IEEE Transactions on Electrical Insulation. Vol. 26 No.1, 1991 pp35-41.
3. R.A.Moreno."Measurement of Potential Buildup and Decay, Surface Charge Density, and Charging Currents of Corona-charged Polymer Foil Electrets. Journal of Applied Physics. Vol. 47 No. 8, 1976 pp3397-3402.
4. M. Ieda. "Electrical Conduction and Chemical Structure of Insulating Polymers". IEEE Transactions on Electrical Insulation. Vol. 21 No. 21, 1986, pp301-306.
5. Wan. H. *Study and Measurement of the Equivalent Interface Charges of Organic Protective Materials-Silicon p-n Junction contact.* Thesis of Doctor Degree. Xi'an Jiaotong University. 1994
6. M .Bakowski and K. Ingemar Lundstrom. "Depletion Layer Characteristics at the Surface of Beveled High-voltage p-n Junctions". IEEE Transactions on Electron Devices Vol. ED-20, No. 6, 1973, pp550-563.

Conference Record of the 1996 IEEE International Symposium on Electrical Insulation, Montreal, Quebec, Canada, June 16-19, 1996

Dielectric Response in Natural Clinoptilolite

Adolfo Delgado
Centre of Biomaterials
University of Havana
Havana 10400, Cuba.

A.Rabdel Ruiz-Salvador
IMRE, Zeolites Engineering Lab
University of Havana
Havana 10400, Cuba

Gerardo Rodríguez-Fuentes
IMRE, Zeolites Engineering Lab
University of Havana
Havana 10400, Cuba

Antonio Berazaín-Iturralde
Higher Pedagogical Institute EJV
76 and 31 Playa, Havana, Cuba.

Abstract: **Several samples of cation forms of the natural clinoptilolite from Tasajeras deposit were studied using dielectric spectrometry. The behaviour of the response was analyzed for different water contents and, for the calcium and sodium forms, it was investigated for several temperatures too. The results show the influence of cation nature and water content in the shape and magnitude of the dielectric response. Two relaxation domains appear in the frequency range from 34 Hz to 500 kHz, which were studied following the formalism of the Universal Dielectric Response in solids. The existence of correlated motions between water-induced species and cations is discussed.**

INTRODUCTION

Natural clinoptilolite figures in the seventh group of the Meier's zeolite classification [1]. It constitutes a set of bidimensional arrays of interconnected channels, forming a low-dimensional system for cationic motions.

Dielectric relaxation techniques could provide a better understanding about the properties of zeolites as ionic exchangers, ionic conductors, catalysts or molecular sieves by studying the motions of cations, zeolitic water and adsorbed species. Dielectric studies have been carried out in different types of zeolites and under several experimental conditions [2-5]. They have been performed mainly for high-symmetry synthetic zeolites and sometimes have revealed controversial and inconsistent results. The latter could be a consequence of the lack of an adequate microscopic theory, which enable us to correlate the experimental results with any particular feature of each zeolitic structure.

In this work we used the Jonscher's screened hopping model [6], for a qualitative interpretation of the dielectric response observed in the natural clinoptilolite from a cuban deposit. The model used was the starting point for the known Universal Dielectric Response (UDR) [7], whose predictions have resulted to be in great accordance with experimental results on materials of most diverse chemical compositions and structures.

EXPERIMENTAL

Sample preparation

Some homoionic forms of natural clinoptilolite from Tasajeras deposit, for the cations Na^+, K^+, Li^+, Ca^{2+}, Mg^{2+} and Zn^{2+} were classified in order to obtain 40 μm powders. The samples were set at room humidity and temperature for 24 h. Later some of them were placed in a stove during 24 h for water extraction, and then they were grouped as in Table I. Samples of groups A and B were treated at 388 and 413 K respectively; samples of group C were not thermally treated; and samples of group D were also placed at 413 K.

Measurements

We used a planar-parallel type cell, with cooper electrodes. The thickness of the samples after being tamped the powder inside the cell was always less than 2.50 mm, and the diameter, 26 mm, was fixed by the cell electrodes. The dielectric data were obtained from a Tesla BM 507 Impedance Meter.

Samples of groups A, B and C were measured at room temperature, and samples of group D were measured for several temperatures in the range from about 473 K down to room temperature.

RESULTS AND DISCUSSION

Water content influence

The dielectric response is presented in terms of the dielectric permittivity in log-log plots for the real and imaginary parts. Figure 1 shows the response of the divalent-cations forms of group A. Two relaxation domains are distinguished: one in the

Table I

Separation of the samples used according to the preparation and measurements

A	B	C	D
Na^+	Na^+	Na^+	Na^+
K^+	K^+	Ca^{2+}	Ca^{2+}
Li^+	Li^+	-	-
Ca^{2+}	Ca^{2+}	-	-
Mg^{2+}	Mg^{2+}	-	-
Zn^{2+}	Zn^{2+}	-	-

low frequency part of our measurement window, which we will refer to as domain I, and another in the high frequency part referred to as domain II.

Domain I is characterized by an increase of both components of the permittivity towards low frequencies. The behaviour of the polarization component, although yet apparent, indicates the incipient formation of the anomalous phenomenon known as Low Frequency Dispersion (LFD), which is characterized by the rapid increase of both components of the permittivity with the same fractional power law with the decrease of frequency, and is proper for systems of charge carriers with some degree of mobility [7]. Domain II is typical from dipolar systems.

An analogous representation for the divalent-cation samples from group B is shown in Figure 2. It can be observed the shift of the loss peaks towards frequencies higher than those of the similar samples in group A. The peak for CaC is not observed within the frequency window. The magnitude of the response is lower for this group than for group A. The number of experimental points only allowed us notice a small independence with cation nature towards lower frequencies in the domain.
The effect of the water content in the response for the divalent-cation samples can clearly be seen in Figure 3. There we show the response for samples CaA, CaB and CaC. The shift of both domains can be noticed. Domain I spreads towards higher frequencies and domain II does to lower frequencies with a higher water content. The displacement of the domain II is in opposite direction to the shift observed in other types of zeolites [1]. Besides, the magnitude of the response increases with the increase of water content. The loss peak gets wider with higher water content. Figure 4 shows the dielectric response for the monovalent-cation samples of groups A and B. In this case only domain I can be observed with a better trend to the anomalous LFD, although it is not completely established. Similarly it is

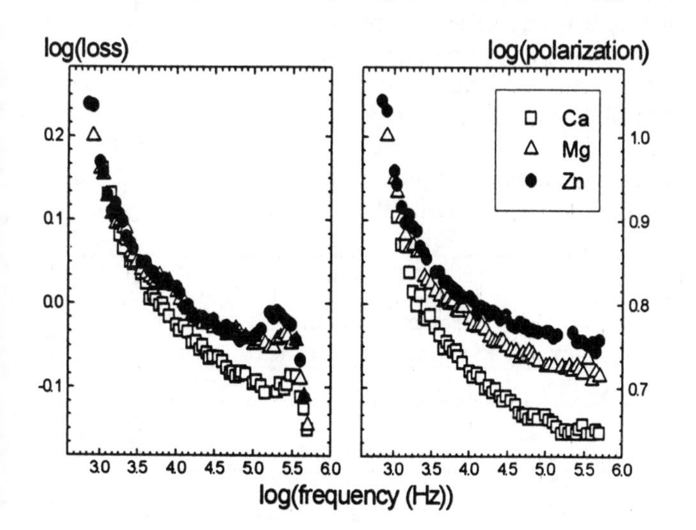

Figure 1. Log-log presentation of the dielectric permittivity of the divalent-cation samples from group A measured at room temperature and treated at 388 K.

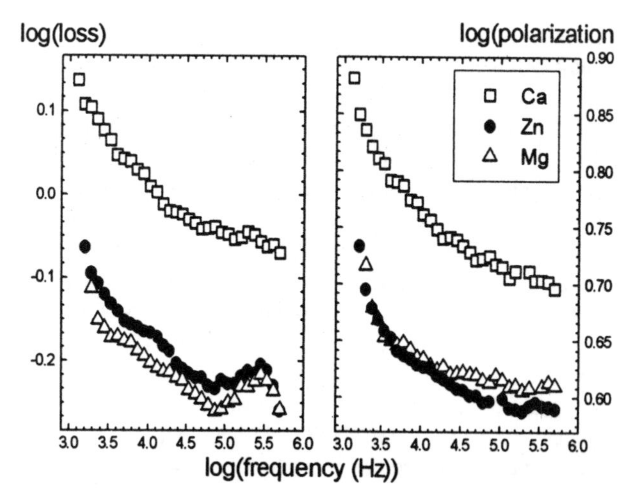

Figure 2. Log-log presentation of the dielectric permittivity of the divalent-cation samples from group B, measured at room temperature and treated at 413 K.

seen that this phenomenon should occur at higher frequencies with increasing water content. The water content influence is similar to that for the divalent-cation samples and is clearly presented for the sodium form in Figure 5. Table II shows the values of the UDR index for the high frequency region of domain I [7]. Notice in all cases that the index is greater with less water content. This parameter is greater for the divalent-cation samples than for the monovalent-cation ones.

Regarding domain I, from the behaviour observed we can assure that the crossover frequency f_c, characteristic for the LFD process, increases with the increasing water content. Large scale transport would occur at lower frequencies outside our work window, which must cause an increase of the losses with a concomitant increase of the polarization. The anomalous LFD has been studied on synthetic zeolites before [5]. For such systems it could not be possible to precise the nature of the transport mechanisms which formed it. If we keep in mind that the clinop-

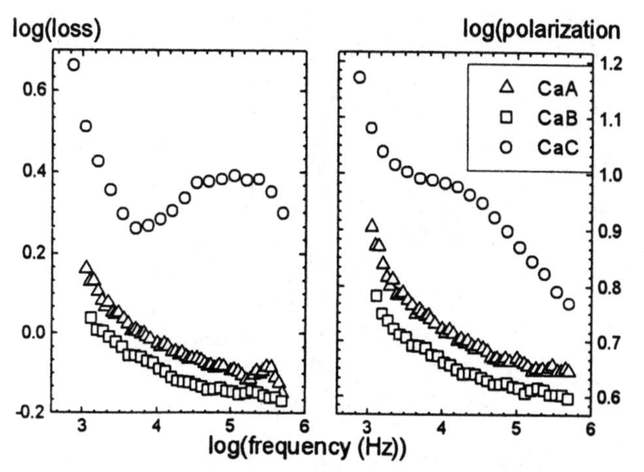

Figure 3. Log-log presentation of the permittivity of the samples CaA, CaB and CaC, measured at room temperature.

tilolite is a zeolite with a high Si/Al ratio and thus a low accessible-site density, and that with an increase of hydration the system geometry strongly hinders the cation motions, then we can suggest the existence of correlated motions between cations and water-induced species, which can be more with greater water contents. This idea gets stronger when we observe the independence from cation nature for the response during the formation of the dispersive phenomenon in Figure 1. This independence is not observed in the response for the monovalent-cation samples whose cations have more freedom in their motions and thus can also participate in the transport better than the divalent ones.

Concerning domain II, we can infer that the cation nature is a determinant factor in this relaxation. Clinoptilolite geometry, as pointed above, should impose a strong hindrance to the cation motions. Divalent cations are more strongly bonded to the structure than the monovalent ones, and moreover coordinate more amount of water, forming a complex of entities with a relaxation that must be slower than that of the complex formed with monovalent cations, which have higher mobility and coordinate a less amount of water, and thus must relax at higher frequencies. This relaxation due to the system cation-water-anion framework is in good agreement with the screened hopping model proposed by Jonscher [6]. During its reorientation, water molecules can interact with the dipole moment formed by the

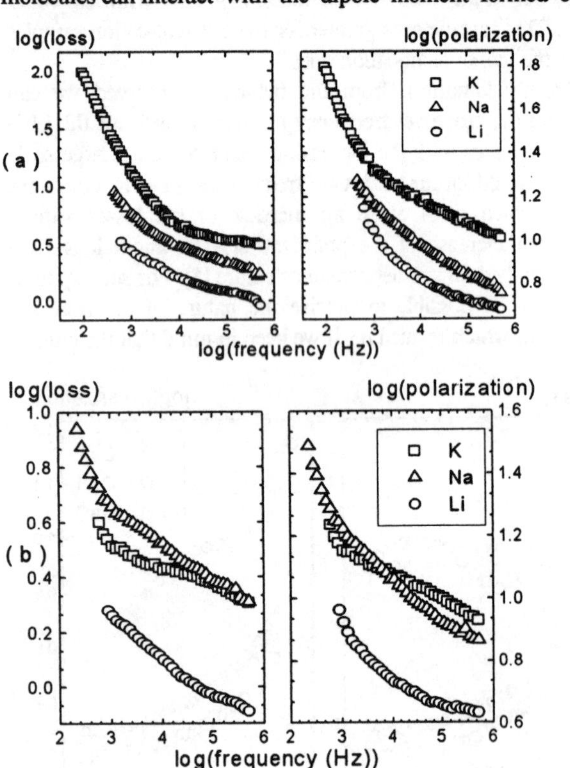

Figure 4. Log-log presentations of the permittivity for the monovalent-cation samples. a) Samples from group A. b) Samples from group B.

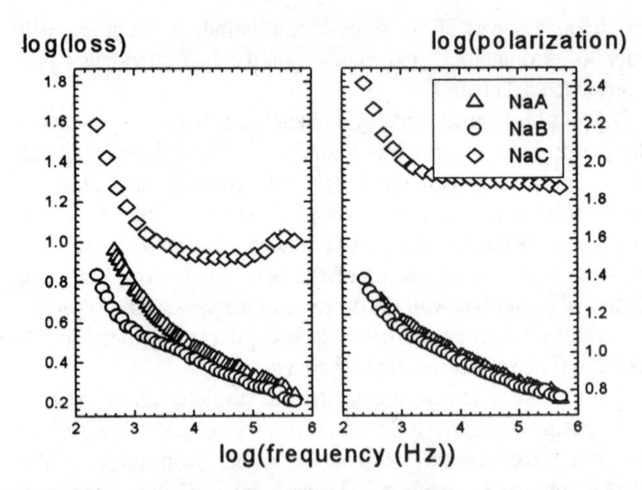

Figure 5. Log-log presentation of the permittivity of the sodium form of clinoptilolite of groups A, B, and C. Notice the appearance of domain II for the most hydrated sample, that of the group C.

subsystem cation-framework, weakening the effective dipole moment of the whole system, with the subsequent increase of the magnitude of the dielectric loss peak, whose widening with the ascendant hydration would be related to an increase of the heterogeneity of the system, as in the case of sample CaC, measured at room humidity. Both ladders of the loss peaks follow fractional power laws as observed by the UDR [7].

Behaviour with temperature

The normalized data of the dielectric response of CaD and NaD samples is presented in Figure 6. Normalization was made in order to study domain I, and the reference isotherm was always the highest temperature one. The reference points are also included, which were equally defined as the first point of the parallel behaviour of both components of the permittivity in the high frequency region of the domain. For sample CaD, a little separation of the isotherms is observed, while for sample NaD normalization resulted almost perfect. For sample CaD we cannot observe the LFD phenomenon while for NaD the LFD is practically established. Only domain I is observed for NaD, and for CaD the separation of the response for the highest fre-

Table II

Values of the parameter *n* of the UDR for the high frequency region of domain I for the samples analyzed

Cation	Group B	Group C
Mg^{2+}	0.94	0.98
Ca^{2+}	0.94	0.95
Zn^{2+}	0.96	0.97
Li^{+}	0.90	0.92
Na^{+}	0.85	0.86
K^{+}	0.89	0.92

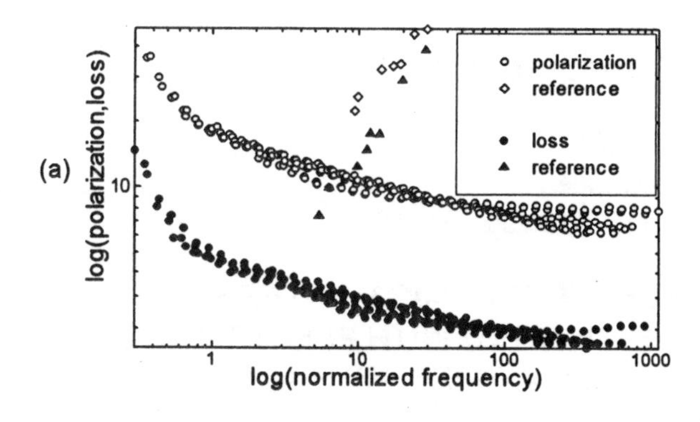

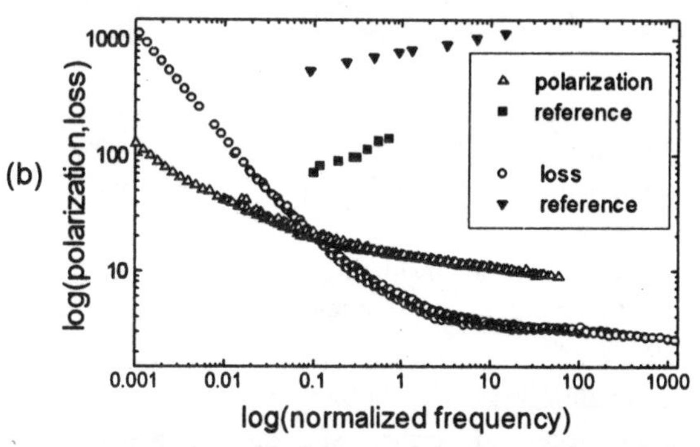

Figure 6. Normalized dielectric permittivity of the samples of group D, measured for various temperatures. a) Sample CaD, reference temperature of 470 K and critical frequency of 1.7 kHz. b) Sample NaD, reference temperature of 472 K and critical frequency of 37 kHz. Both samples were treated at 413 K. The reference points for each isotherm are also shown.

quencies indicates the formation of domain II, which is in nature different from domain I, as pointed above. The characteristic parameters for this domain, namely the UDR index n and the activation energy E_a are presented in Table III.

These results allow us to establish that the dynamic behaviour of these homoionic forms does not vary in the temperature range used. The values of activation energies seem to be controversial with the results pointed out above, but if we suggest that this domain is mainly due to the motions of water-induced species, we can understand the results. Divalent cations are strongly bonded to the structure, and a small displacement due to thermal deformations can lead the structure to collapse. Monovalent cations can move along longer distances. As the system geometry hinders ca-tion movements then it seems likely to think of movements of protonic nature, induced by the residual water. This correlation between water content, cation nature and system geometry is still obscured and will be an underlying theme for future works.

Table III

Parameters of the normalized response for the samples of group D

Sample	n	E_a (kJ/mol)
NaD	0.94	32.8
CaD	0.92	10.5

CONCLUSIONS

We can conclude that two relaxation processes exist in natural clinoptilolite at room temperature in the frequency range from 34 Hz to 500 kHz. One process is characteristic of hopping charge systems and the other one of dipolar systems. Although the frequency window did not allow to distinguish the anomalous LFD, it was proposed this imperfect transport phenomenon to be due to correlated motions between cationic and water-induced species. Water content affects the response with the increase of its magnitude and the spreading of both relaxation domains, being the hopping carrier domain spread towards higher frequencies, and the dipolar systems domain towards lower frequencies with the ascendant water content. The dynamic and structural behaviour of the homoionic forms for the cations Na^+ and Ca^{2+} does not vary in the frequency ranges 34 Hz - 44.3 MHz and 340 Hz - 10 MHz respectively.

ACKNOWLEDGEMENT

This work was partially supported by an Alma Mater Grant 95 of the University of Havana.

REFERENCES

1. Breck, D.R. *Zeolites Molecular Sieves.* Wiley - Interscience, New York, 1974.
2. Barrer, R.M and E.A. Saxon-Napier. "Dielectric Properties of some Ion-Exchanged Analcites". Trans. Faraday Soc. Vol.58. No.156, 1962, p. 4853.
3. Carru,J.C., D. Delafosse and M. Kermarec. "Study of Large Frequency Band Dielectric Relaxation in X, Y and M Type Zeolites Containing Protons". J. Chim. Phys. Vol.86. No.2. 1989, p. 263.
4. Szás, A.; S.S. Al Rahim and J. Liszi. "Dielectric Relaxation in Zeolites I. Study on Dielectric Relaxation of Linde A Zeolite". Acta Chim. Hung., Vol. 125. No. 1, 1988, p. 37.
5. Haidar, A.R. and A.K. Jonscher. "The Dielectric Properties of Zeolites in Variable Temperature and Humidity". J. Chem. Soc. Faraday Trans. 1. Vol.82. No.12, 1986, p. 3535.
6. Jonscher, A.K. "A Many-Body Model of Dielectric Polarisation in Solids II. Phys. Stat. Solidi B. Vol.84. No.1, 1977, p. 159.
7. Jonscher, A.K. *Dielectric Relaxation in Solids.* Chelsea Dielectrics Press, London, 1983.

Conference Record of the 1996 IEEE International Symposium on Electrical Insulation, Montreal, Quebec, Canada, June 16-19, 1996

Electrochemical Nature of Metal-Insulator Interfaces

Ashok K. Vijh, Fellow IEEE
Institut de recherche d'Hydro-Québec
1800 boul. Lionel-Boulet, Varennes, Qué., Canada, J3X 1S1

***Abstract* - At the interface between a metal and a dielectric (e.g., vacuum, sulfur hexafluoride, polymers, oils, detonated nitromethane, etc.) a variety of charge transfer events can take place which, basically, are electrochemical in nature. Such interfacial charge transfers are usually investigated by solid-state physicists and electrical engineers; for the fundamental understanding of the phenomena involved, it is necessary, however, to think of these interfacial charge transfers in the conceptual framework of electrochemical physics.**

Some typical cases of electrochemical processes at metal-dielectric interfaces are described. For example, metal-vacuum-metal sandwiches subjected to high-field dielectric breakdown can be shown to behave as metal-electrolyte-metal electrochemical cells. The field-induced dielectric breakdown of the vacuum creates a low-pressure plasma that acts as a gaseous electrolyte in that it contains ions and electrons that conduct electricity. The situation obtaining at such interfaces is quite similar to that at metal-electrolyte interfaces.

1 INTRODUCTION

The use of electrical insulation materials in power industry is indispensable and covers a wide spectrum of gaseous, liquid, solid or other (*e.g.*, vacuum) dielectrics. Although there are many factors involved in the choice of a dielectric material for a particular application, the most crucial property appears to be the dielectric strength, *i.e.*, the capacity of the insulator to withstand applied electrical stress without undergoing a dielectric breakdown.

Since most insulating materials under bias are held between metallic electrodes, the behaviour of such metal-insulator metal configurations under applied field is of interest. Before an electric field high enough to cause a dielectric breakdown is reached, one has so-called a pre-breakdown regime in which some significant current (called the "leakage" current) can be passed through the metal-insulator-metal "sandwich"; this current (and more so the high current observed in the dielectric breakdown region) can cause physico-chemical events at the metal-insulator interfaces (as well as within the bulk of the dielectric itself). The interfacial events at the metal-dielectric interfaces are generally electrochemical in nature, *i.e.*, they cause chemical changes associated with the passage of electrical charge across the metal-dielectric interfaces. The object of this paper is to point out the electrochemical nature of the phenomena at the metal-dielectric interfaces, a fact not widely appreciated by electrical engineers who usually study these materials for their application in the power industry.

2 ELECTROCHEMICAL INTERFACES ASSOCIATED WITH ELECTRICAL INSULATION AND DIELECTRIC MATERIALS

All dielectrics show characteristics of electrolytes, although high resistance electrolytes, *e.g.*, pure water, insulating polymers, oils etc.

Conceptually, a dielectric "capacitor" is a battery with a high internal resistance; a battery can be viewed as a "super-capacitor" that stores a lot of "charge" but has very low internal resistance.

Dielectrics are used as insulating materials in electrical power equipment and give rise to a number of electrochemical interfaces, as follows:

(1) Metal-Gas Interfaces

These interfaces arise when SF_6, air etc. are used as interrupting media in large power circuit breakers. Also in pressurized transmission cables, coaxial cables, capacitors for 1500 kV systems. The electrochemistry of such interfaces has been explored[1].

(2) Metal-Vacuum Interfaces

Metal-vacuum-metal sandwiches are the basis of high-power vacuum switches, electronic valves, low-loss high frequency capacitors, electrostatic generators etc. Many interfacial problems can arise[2].

(3) Metal-Polymer Interfaces

Metal-polymer interfaces can arise in cables and other places where polymers are used as insulating materials. Such interfaces provide a plethora of detailed data amenable to an electrochemical treatment[3,4].

(4) Metal-Plasma Interfaces

These electrochemical interfaces are formed in arcs[5], circuit breakers[1], "flashing-over" in contaminated insulators[6], and, in controlled nuclear fusion devices[7]. Essentially, a plasma is a gaseous electrolyte in which the ionic species are not solvated, in contrast to the usual liquid (*i.e.*, aqueous and non-aqueous) electrolytes.

(5) Metal-Oil Interfaces

Metal-oil interfaces are formed in transformers, capacitors, etc. Some aspects point to the importance of electrochemical reactions[8].

(6) Metal-Ethylene Glycol Interfaces

The recent defence technologies ("Star Wars") using high pulsed power lasers require enormous capacitors for high-voltage application, formed by large tanks full of ethylene glycol or pure water. Importance of events at the electrode-dielectric interfaces has been realized[9,10].

(7) Metal-Nitromethane Interface

Nitromethane and other liquid dielectrics are used as detonation materials in rocket and other applications, giving rise to electrochemical reactions at the metal-dielectric interface[11,12].

(8) Metal-Salt (*e.g.*, Alkali Halides) Interfaces

These electrochemical interfaces arise, *e.g.*, in flash-over voltages observed on the salt-contaminated SiO_2 plates, *i.e.*, outdoor pollution of insulators employed in electric power industry[6].

(9) Metal-Water-Air or Metal-Air Interfaces

These interfaces arise in corona discharges from cables[13].

(10) Metal-Water (Electrolyte) Interfaces

These interfaces arise in the propagation of electrochemical water trees in polyethylene that eventually destroy the dielectric[14].

(11) Metal/Oxide/Gas/Polymer Interfaces

Charge transfers across such interfaces are involved in electrical tree initiation in polyethylene[15].

(12) Biological Membrane-Electrolyte Interfaces

Recall that biological membranes, lipids, proteins and enzymes etc. are electronically-conducting insulators when dry, but, conduct by protons and other ions when wet. An understanding of electrochemical charge transfers across such interfaces[16] is essential in elucidating the effects of high voltage induced currents in animal and human bodies.

(13) Liquid Insulation-Water (+ Ions) Interfaces

The presence of traces of water (plus the ubiquitous traces of impurity ions dissolved in it) in liquid dielectrics (*e.g.*, transformer oils) can cause, under applied voltage stress, tree growth in a manner similar to that observed for solid polymers[17]. The type of ion present in water determines the extent and rate of tree growth, thus indicating the electrochemical nature of the phenomenon[17].

3 THE METAL-VACUUM INTERFACE

A metal-vacuum-metal sandwich under high voltage bias can conduct electricity due to the dielectric breakdown of the vacuum. A vacuum is conceptually not nothingness but a low-pressure plasma formed by the erosion, evaporation and ionization of the electrode (cathode) surface.

In order to appreciate this subject from the point of view of an electrochemist, the vacuum may be compared to pure water in that both are insulators and require a very high value of applied bias between the electrodes in order to conduct electric current; when the bias value is appreciably high, the "solvent" becomes conducting either by means of the charged species injected from the electrodes into it or by the production of charged species by the solvent itself through its electrical breakdown. In any real case, the presence of trace ionic impurities also contributes to the charged species responsible for the conduction.

The dielectric breakdown of a metal-vacuum-metal sandwich creates a conducting plasma (a gaseous electrolyte) and gives rise to a situation quite analogous to that of a metal-electrolyte-metal electrochemical cell (Table 1). The voltage distance relationship in such a metal-vacuum-metal "electrochemical" cell is depicted in Fig. 1; at the metal-plasma interfaces created within this cell, an interfacial voltage profile very similar to an electrochemical double layer can arise (Fig. 2).

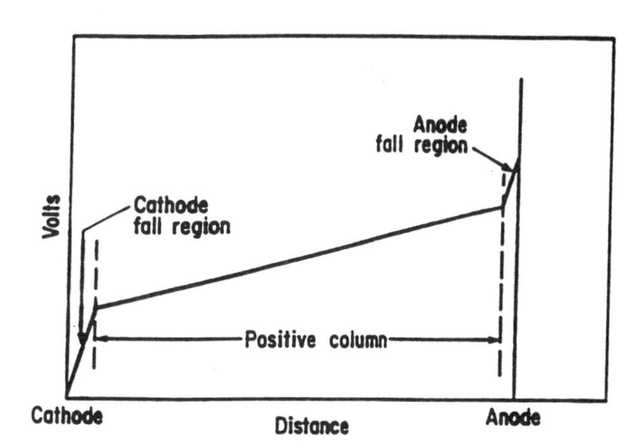

Figure 1. The voltage-distance relation in an arc. The sharp voltage drops near the cathode and anode with a less steep drop in the "bulk" of the arc (*i.e.*, the positive column) may be compared to the analogous case of steep cathode and anode potential drop and the less steep potential profile in the electrolyte, for an electrochemical cell.

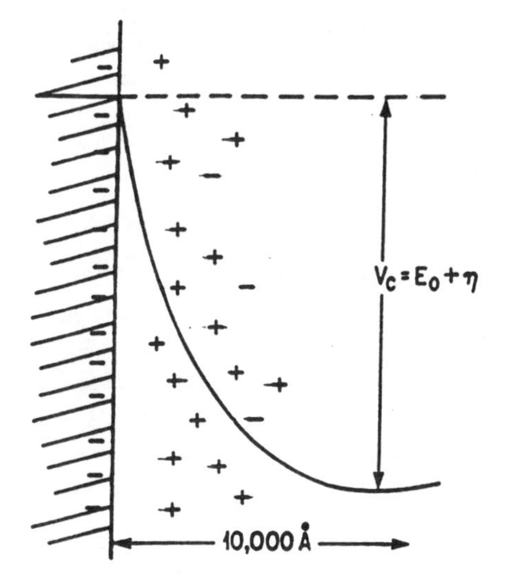

Figure 2. A schematic depiction of the potential profile at a cathode undergoing the reaction $M \rightarrow M^+ + e$ at the metal/plasma interface in an arc discharge.

Table 1
A Qualitative Comparison and Contrast of Electrochemical Features at the Metal-Electrolyte-Metal, and, Metal-Vacuum-Metal Electrolysis Situation

Item/Parameter	Metal-electrolyte-metal situation	Metal-vacuum-metal arc
(1) Conceptual description under appreciable applied bias	Electrolysis cell: metal-conductive fluid-metal configuration	Electrolysis cell: metal-conductive fluid-metal configuration
(2) Role of metal electrodes	Anode and cathode	Anode and cathode
(3) Solvent	Water (*e.g.*, in aqueous electrolysis)	Vacuum (a solvent "analogue")
(4) Solutes	Cations and anions of a salt	Metal ions and electrons produced by erosion of the electrodes (mostly cathode)
(5) Any solvation of ions	Yes	No
(6) Nature of electrode potential drops during electrolysis	Anode and cathode overpotentials (~ 0.1 - 1.0 V)	Anode and cathode falls (several volts)
(7) Approximate Debye length of the space charge region near the electrodes	Usually ≤ 1 μm in the normally used concentrated solutions	Usually several micrometers
(8) Total applied voltage	Cell voltage (= anode overpotential + cathode overpotential + ir drop across the electrolyte)	Arc voltage (= anode fall + cathode fall + column voltage - ohmic - across the arc)
(9) Inert electrolyte - its nature	A number of ionized but nondischarging salts dissolved in the electrolyte	"Inert" gases such as oxygen, nitrogen, air, usually as negative ions due to electron attachment

REFERENCES

1. Vijh, A.K., "The nature of metal electrodes/SF_6 decomposition due to direct current interruption under simulated circuit-breaker conditions", IEEE Trans. on Electr. Ins., EI-11, 157-160 (1976).
2. Vijh, A.K., "The influence of solid state cohesion of conditioned metal electrodes on the electrical strength of the vacuum gap", IEEE Trans. on Electr. Ins., EI-11, 160-162 (1976).
3. Vijh, A.K., "Electrochemical effects as the source of electromotive force in $metal_1$ polymer-$metal_2$ systems", J. App. Phys., 49, 3621-3624 (1978); idem, ibid, 50, 3764-3765 (1979).
4. Crine, J.-P., Vijh, A.K., "The electrochemical origins of open-circuit voltages observed during heating of metal-polymer-metal systems", Mater. Chem. and Phys., 10, 85-98 (1984).
5. Vijh, A.K., "An electrochemical approach to the interfaces in metal-vacuum-metal arcs: comparison and contrast with metal-electrolyte-metal situation", Surface Technology, 25, 335-341 (1985).
6. Vijh, A.K., "Fundamental interpretation of the salt effects on limiting flashover voltages of the salt-contaminated SiO_2 plates in terms of solid state electrochemistry", IEEE Trans. on Electr. Ins., EI-11, 59-62 (1976).
7. Dimoff, K., Vijh, A.K., "The reduction of surface oxides and carbon during discharge cleaning in Tokamaks: some kinetic-mechanistic aspects", Surface Technology, 25, 177-195 (1985).
8. Vijh, A.K., Duval, M., Crine, J.-P., "Open-circuit corrosion as the origin of electrical instability arising from the dissolved metal content in insulating oils in ehv current transformers", Mater. Chem. and Phys., 19, 397-402 (1988).
9. Zahn, M., Ohki, Y., Fenneman, D.G., Gripshover, R.J., and Gehman, V.H., "Dielectric properties of water and water/ethylene glycol mixtures for use in pulsed power system design", Proc. IEEE, 74, 1182-1221 (1986).
10. Jones, H.M., and Kunhardt, E.E., "Pulsed dielectric breakdown of pressurized water and salt solutions", J. App. Phys., 77, 795-805 (1995).
11. Vijh, A.K., "Electrochemical potentials in nitromethane detonation", Acta Astronautica, 3, 167-170 (1976).
12. Vijh, A.K., "The electrochemical nature of the electrical signals at metal electrodes inserted in nitromethane during detonation", J. Chim. Phys., 73, 641-644 (1976).
13. Brandvold, D.K., Martinex, P., and Dogruel, D., "Polarity dependence of N_2O formation from Corona discharge", Atmospheric Environment, 23, 1881-1883 (1989).

14. Xu, J.J., and Boggs, S.A., "The chemical nature of water-treeing: theories and evidence", IEEE Electrical Insulation Magazine, 10 (#5), 29-37 (Sep.-Oct. 1995).

15. Vijh, A.K., and Crine, J.-P., "Influence of metallic electrodes on electrical tree initiation in polyethylene", J. App. Phys., 65, 398-399 (1989).

16. Vijh, A.K., "Conceptually-similar origins of the interfacial events involved in unipolar arcs, corrosion of metals and some biological (*e.g.*, enzymatic) oxidation-reductions", App. Physics Communications, 5, 211-221 (1986).

17. Auckland, D.W., Chandraker, K., Golra, M.A., and Varlow, B.R., "Water treeing in insulating liquids", IEE Proc.-Sci. Meas. Technol., 142, 157-161 (1995).

Conference Record of the 1996 IEEE International Symposium on Electrical Insulation, Montreal, Quebec, Canada, June 16-19, 1996

Organic Electronic Crystals N-phenyl-phthalic imidine and their Properties*

D. YANG, W. DENG, Y. JIANG, S. WANG, Z. WU
Dept. of Materials Science and Engineering, EPRL
University of Electronic Science and Technology of China
Chengdu 610054, China

Abstract: The significance of a new type organic electronic crystals based on substituted N-phenylphthalic imidine have been illustrated in this paper. Two crystal materials: N-(phenyl) phthalic imidine (I, $C_{14}H_9NO_2$) and N-(p-methylphenyl) phthalic imidine (II, $C_{15}H_{11}NO_2$) were prepared and confirmed by IR, NMR, MS and elementary analysis. The crystal and molecular structures were determined by X-ray diffraction. The relative dielectric constants ε_r, piezoelectric parameters d_{33} and refractive index n were measured respectively. The synthesis preparation, molecular structure, dielectric and piezoelectric and optic properties were discussed.

INTRODUCTION

Inorganic crystal materials were studied mainly in the international academic field in the past few decades. Since 1950's, Europe and America began to study organic crystal materials, and analysized the electronic processes of metallic crystal[1], inonic crystal, covalent crystal and molecular crystal. However, since 1980's, the domestic and international studies of organic single crystal concentrated mainly on nonelectronic properties including medicine and biological activities.

Authors advanced a concept of organic electronic crystal (OEC) at 8th International Symposium on Electrest, and reported the dielectric and piezoelectric properties of organic piezoelectric crystals of substituted phenylthiourea. Therefore, in modern electronic science of functional materials, a series new fields of organic electronic crystal high-and new-tech materials, including organic crystal dielectric , organic piezoelectric, pyroelectric, ferroelectric single crystal and organic photoelectric crystal materials, were opened up based on organic electronic crystal amd materials with electronics functional effect. With a great deal of studies and analysis, we have discovered two systems of organic electronic crystal, including substituted phenylthiourea and subtituted phthalic imidine with excellent electronic and optic properties, which have remarkable academic significance and applied value in molecular based electronic (MbE) and molecular electronics (ME).

EXPERIMENTAL

Synthesis Preparation and Analysis Characterization

Two phthalic imidine derivatives N-(phenyl) phthalic imidine (I) and N-(p-methylphenyl) phthalic imidine (II) were synthesized by the reaction of phthalic anhydride and aniline, p-methylaniline respectively. The reaction scheme is schematically shown follow:

Where (I) R=H, (II) R=CH_3.

Two compounds were purified by recrystallization in ethanol solutions and confirmed by IR, NMR, MS and elementary analysis.

Crystal Growth and Structure Determination

Crystals (I) and (II) were grown in acetone solutions of corresponding compounds.

A amount of pure N-(phenyl) phthalic imidine (I) sample was added in a clear beaker, added acetone, with stirring. After the solid sample was resolved completely, the solution was filter to separate microquantity mechanical impurity. Then the solution was removed into both bath apparatus which was ready before, and added acetone again. The temperature of solution is controled, and days later the colourless transparent crystal N-

* Supported by the National Natural Science Foundation of China

(phenyl) phthalic imidine (I) growth. Similarly, colourless transparent crystal (II) growth.

Crystals structures and molecular structures were determined by X-ray four circle diffraction on an Enraf-Nonius CAD_4 diffractometer with PDP11/44 computer and SDP program package[3,4].

Taking some sample (I), (II) to prepare KBr chip, then collect absorbing peak from wavenumber 4000 cm^{-1} to 350 cm^{-1} on a Nicolet-170SX Infra-red Spectrograph respectively. A certainty amount of (I) and (II), resolved with $(CD_3)_2CO$ and DMSO mixture resolver, on a JNM-FX90Q Nuclear Magnetic Resonance Spectrograph, TMS act as reference standard, collect NMR spectrum in 2000Hz frequcency band field at 21.3℃. Similary, on a Finnigan-MAT4510 Mass Spectrometer, molecular ionic peak and fragment ionic peak were collected.

Crystal (I) and (II) have three dimensions of 0.15 × 0.35 × 0, 30mm and 0.15 × 0.20 × 0.25mm respectively.

Dielectric, Piezoelectric Parameter and Refractive Index Measurement

Crystal samples with three dimensional size, capacitances were measured on a HP 4275 MULTI-FREQUENCY LCR METER, then calculated the relative dielectric constant respectively. Crystal (I) and (II) piezoelectric parameter d_{33} were measured with a ZJ－2 quasi-static measuring apparatus after polarization at different electric field, different temperature and different times on a CY 2671 omnipotent breakdown apparatus. Using oil-soaked method, refrqctive index n were obtained on a ORTHPLAN-POL polarizing microscope.

RESULTS AND DISCUSSION

Composition Characterizaton

The composition of synthesis produts, characterized with IR, NMR, MS and elementary analysis, are consistent with chemical formula $C_{14}H_9NO_2$ (I) and $C_{15}H_{11}NO_2$ (II) respectively.

Molecular Structure

Infra-red spectrogram of crystal (I) is shown in Fig. 1

For (I), benzene ring C-H, C-C stretching vibration and bending vibration peak $v(C_6H_4)$ is 3023, 1592 and 1465cm^{-1}, v(CO) is 1705cm^{-1}. For (II), $v(C_6H_4)$ is 3071, 1519 and 1386cm^{-1}, v(CO) is 1718cm^{-1}, $v(CH_3)$ is 2916cm^{-1}.

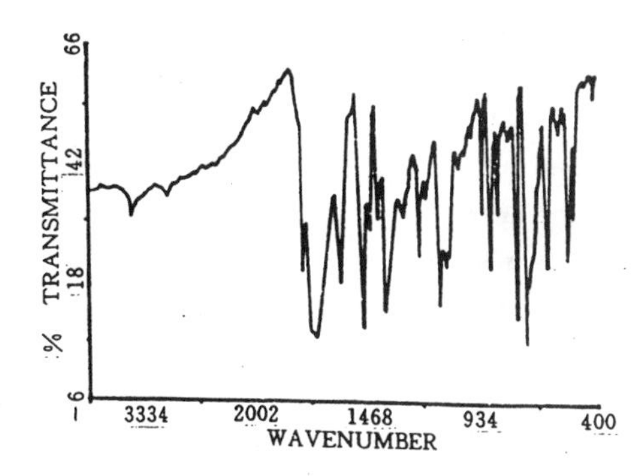

Fig. 1 Infra-red spectrogram of N-(phenyl) phthalic imidine.

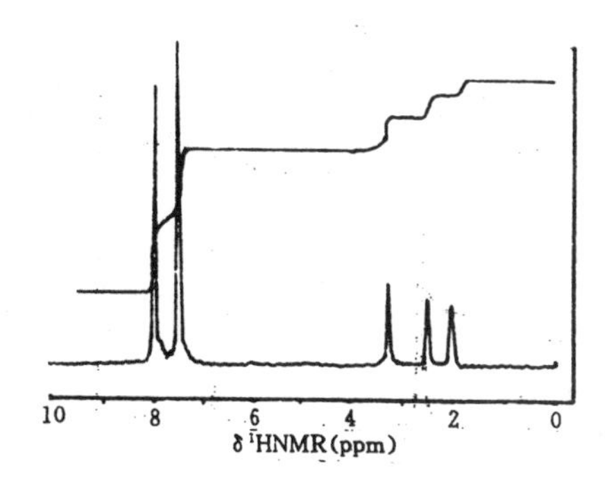

Fig. 2 δ1HNMR(ppm) for N-(phenyl) phthalic imidine.

1H Nuclear magnetic resonance of (I) is shown in Fig. 2. Chemical shift δ1HNMR(ppm) are as follow: for (I), 3.29(1H, m, NH), 7.51(5H, s, C_6H_5), 7.96(4H, s, C_6H_4); for (II), 2.40(3H, s, CH_3), 3.16(1H, m, NH), 7.36 (4H, s, p-C_6H_4), 7.94 (4H, s, o-C_6H_4), in which , s means strong, m means middle.

Mass spectrogram of (I) is shown in FIg. 3 . For (I) and (II), both have stable molecular ionic peak, and (II) molecular ionic peak is based peak, molecular ion break down continually into fragment ion.

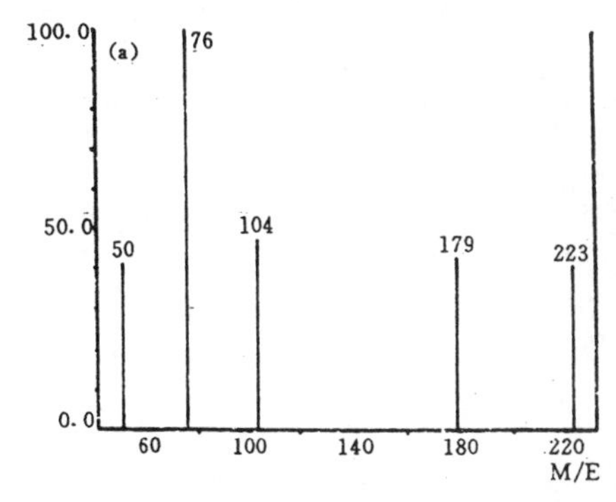

Fig. 3 Mass spectrogram of N-(phenyl) phthalic imidine.

In a word, all analysis results on elementary analysis,

IR, NMR and MS, indicate that synthetic products composition coincide with chemical formula $C_{14}H_9NO_2$ for (Ⅰ) and $C_{15}H_{11}NO_2$ for (Ⅱ) respectively.

The fractional atomic coordinates and equivalent isotropic temperature factors for crystal (Ⅰ) and (Ⅱ) all are omitted. The crystallographic data are shown in Table 1. The perspective view of molecular structure of crystal (Ⅰ) and (Ⅱ)are shown in Fig. 4 and Fig. 5.

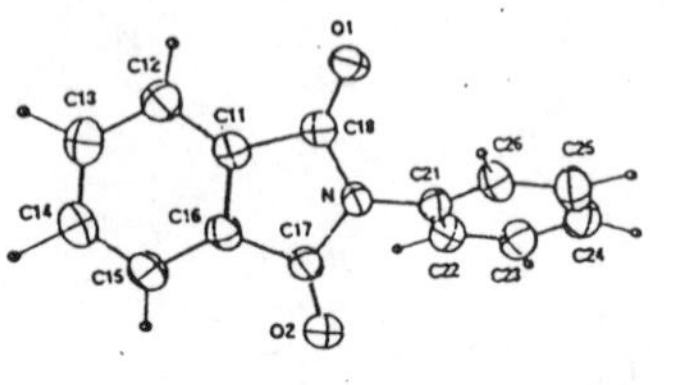

Fig. 4 The perspective view of N-(phenyl) phthalic imidine.

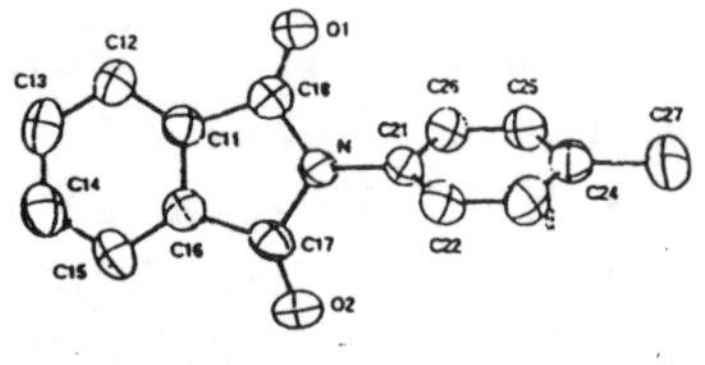

Fig. 5 The perspective view of N-(p-methylphenyl) phthalic imidine.

Table Ⅰ
The crystallographic data for N-phenyl-phthalic imidine

crystal	(Ⅰ)	(Ⅱ)
formula	$C_{14}H_9NO_2$	$C_{15}H_{11}NO_2$
crystal system	orthorhombic	orthorhombic
space group	Pcab	Pbna2
a(Å)	7.649(4)	7.642(2)
b(Å)	11.659(2)	11.237(1)
c(Å)	23.739(3)	13.856(2)
v(Å^3)	2117.0	1187.0
Z	8	4
D_c(g.cm^{-3})	1.401	1.328
μ(cm^{-1})	0.885	2.746

In molecules (Ⅰ) and (Ⅱ), the atoms O(1), O(2), C(17), C(18) and N, are coplanar with the benzo ring, the CH_3 group carbon atom C (27) is coplanar with the substituent benzene ring. The dihedral angles between the two planes are 58.4° and 56.2° respectively. The detailed description and disscussion have been reported in elsewhere[3,4].

Dielectric and Piezoelectric and Refractive Properties

The ε_r data of crystals (Ⅰ), (Ⅱ) are in range 25~55. The d_{33} data of (Ⅰ) and (Ⅱ) are 50 to 250 PC/N. We have investigated the frequency and temperature properties of relative dielectric constant[2,5]. The experimental aresults suggest that substituted phenylth alic imidine crystals have excellent dielectric and piezoelectric properties. Relative dielectric constant ε_r of crystal (Ⅰ) and (Ⅱ) at different frequency are shown in Figure 7. We have observed that ε_r change dramaticly with increasing of frequency. Figure 8 shown the change of d_{33} with different E_p.

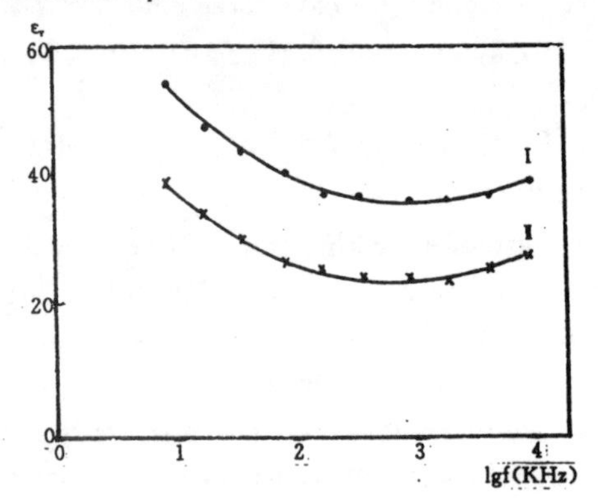

Fig. 6 The frequency propety of ε_r of crystal Ⅰ (upper) and Ⅱ (lower)

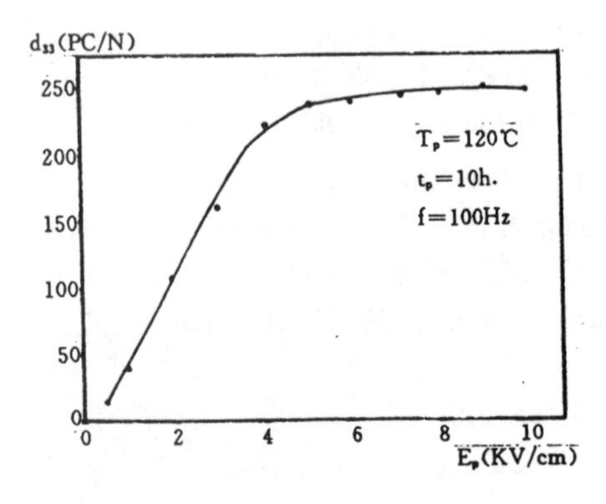

Fig. 7 The d_{33}~E_p curve of crystal Ⅰ

The change of relative dielectric constant for crystal (Ⅰ) and (Ⅱ) is shown as figure. Compare crystal (Ⅰ) with (Ⅱ), former ε_r is lager than latter, for molecular polarity of (Ⅰ) is stonger than (Ⅱ).

The refractive index of crystals at two directions are shown in Figure 8. Only two directions data were measured. Owing to crystals naturally laided on level, to obtain the third dimension data is difficult. Further studies on the relationship between structures and properties of crystals are in progress and will be reported elsewhere.

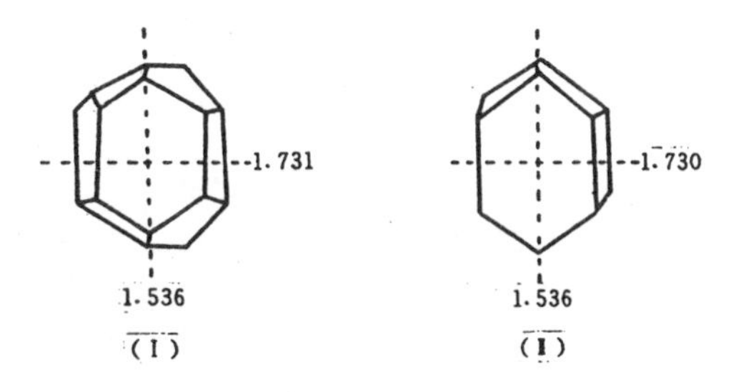

Fig. 8 Refractive index of crystals at two direction

CONCLUSION

Two types of organic electronic crystals, substituted phthalic imidine with excellent dielectric and piezoelectric and optic properties, have been prepared and confirmed, which are highly ordered, oriented and polar crystals.

The study and exploitation of organic electronic cryystal, a great deal of organic crystals with excellent dielectric, piezoelectric , pyroelectric, ferroelectric, electret, nonlinear optics and bioelectric properties, have been opened up for modern electronics science of functional materials. These crystals will have been of widely applied properties in molecular electronics.

REFERENCES

1. M. Pope and C. E. Swenberg, *Electronic Processes in Organic Crystals* , Oxford, 1982.

2. D. B. Yang and W. L. Deng, 8th Int. Symp. Electrets Proc. 1994. p. 690.

3. W. L. Deng and D. B. Yang, Chinese J. Struct. Chem. , 13, 1994. p. 451.

4. Z. H. Mao, Z. H. Zhou, W. L. Deng, Z. Hong and S. X. Shan, Chinese J. Struct. Chem. , 12, 1993. p. 197.

5. Daben Yang, Wenli Deng, AMF-1, 1995. p. 188.

Conference Record of the 1996 IEEE International Symposium on Electrical Insulation, Montreal, Quebec, Canada, June 16-19, 1996

Influence of Distillation Fraction on Breakdown Strength of Organopolysiloxane Gel

Zhang Feng Xu ChuanXiang Li YangPing
Department of Electrical Engineering
Xi'an Jiaotong University, China

Abstract: **The electrical properties of the lightly cured organopolysiloxane gel with excellent viscoelastic properties have been studied. The influences of the distillation fraction on conductivity, initial discharge voltage and the breakdown strength of the gel are discussed. The distillates collected at different temperatures are analyzed as well. The voids formed in the curing process only influence the lnρ-T semi-logarithmic line but also reduce the initial discharge voltage. The integrity of the gel structure is destroyed by the cyclic partial discharge and thus, cause the breakdown strength to decline.**

INTRODUCTION

High molecular weight polymer materials are widely used in electrical insulation because they possess excellent electrical and physical properties. Although the high voltage strength of polymer have been studied for a long time[1], the mechanism of the polymer insulation at high voltage remains unclear, even for the homogeneous polymer.

Considering the uniqueness in engineering practices, the breakdown of materials might be caused by the weak points in the polymer under uniform electrical field or uneven field[2]. The weak voids can be formed from polymer impurity, thin point and the voids formed during the curing process. These points concentrate the electrical strength, and breakdown first. Being irregular, the existence of weak points restricts the polymer breakdown strength. A viscoelastic polymer used as protective material for silicon rectifiers has been investigated in this laboratory[3].

Despite the intensive research performed, litter is known about the breakdown strength of organopolysiloxane gel and the behavior of the distillate. Having distillable fractions, the influence of the distillation fractions on breakdown strength has been studied.

EXPERIMENTAL

The type of electrodes used is shown in Fig. 1. Rod-plant electrodes were made of copper. The gap spacing was 1 mm and the curvature of the rod was 5 mm.

Organopolysiloxane gel (JUS), is made of JUS-A (Vinyl organopolysiloxane with platinum catalyst) with no distillation, JUS-B(methyl hydrogen organopolysiloxane compounds) with distillation, which is cured at 150℃ for 60 minutes.

In order to avoid any flashover, the hole arrangement of electrodes and gel were put into a test cell of polyvinyl fluoride with a fluid of silicone oil.

Material JUS1 was cured with distillate of 8 percentage of volume. Material JUS2 was cured after being distilled at 200℃. The distillates of JUS1 was divided in three parts by different distillate temperature: (I) one below 60℃. (II) one between 60℃-120℃. (III) one between 120℃-200℃. The resultant infrared (IR) spectrum was shown in Fig. 2. As one can see from this figure, part (II) and (III) had few Si-H functional groups which was consumed in the addition reaction of polymerization. The part (I) had OH function group only as shown by the 3200-3500 cm^{-1} band and had no Si-H functional group.

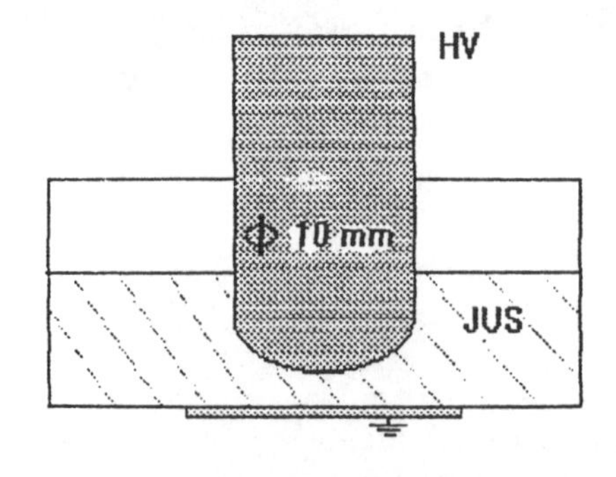

Figure 1. The rod-plant electrode used in practice.

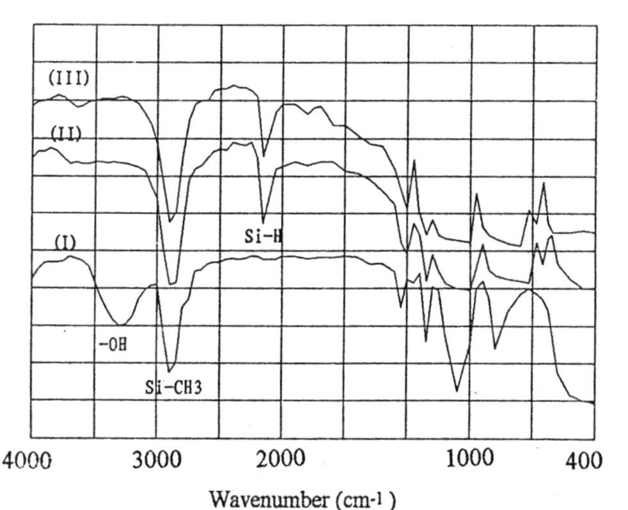

Figure 2. The infrared spectrum of the three parts of distillation fraction with different distillate temperature

Experimental Instrument:

Automatic booster equipment: set up rate 10 kV/min
frequency 50 Hz
highest voltage 200 kV

MODER 5 ac initial discharge measurement equipment.

ZC-36 resistometer

260-50 IR spectrophotometer.

RESULTS

Table 1 shows the value of breakdown voltage and average breakdown strength values(as ABS: kV/mm) of the material JUS1 and JUS2. Fig 3. shows the resistance-temperature curve of the material JUS1 and JUS2.

Table 1. Measurement of breakdown voltage of two kinds of gel.

	1	2	3	4	5	ABS
JUS1 kV	14	16	17	13	14	15 kV/mm
JUS2 kV	38	33	35	37	35	35 kV/mm

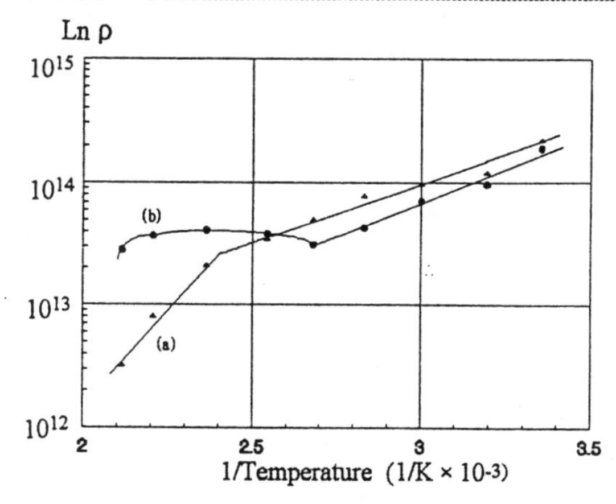

Figure 3. The effect of temperature on the resistance of the gel with (a) no distillation fraction, (b) 8% of weight of distillation fraction.

Material JUS1 starts to discharge when the ac voltage rises more than 2 kV. If the voltage continues to rise up, the sample would breakdown at a voltage of 15 kV. There is a range (6-8 mm diameter),under the rod electrode being damaged with some white and black powder, which is analyzed to be SiO_2 and carbon black. Experiment with Material JUS2, the initial discharge starts at above 15 kV, after breakdown, we can find the damaged point (diameter less 1 mm) below the rod electrode.

To measure the accurate initial discharge of this rod-plant electrode with organopolysiloxane gel, we use the MODER 5 ac initial discharge measurement equipment. Filled with material JUS2, the initial discharge of the sample is 10 kV, which is invisible to the naked eye. The capacitance of the sample is 5 pF.

From the Figure of ln ρ -T there is a unusual range in material JUS1 with distillation fraction, the resistantivity rise up above temperature 100℃. The curve of the material JUS2 with no distillation fraction shows a broken line.

DISCUSSION

Compared the average breakdown strength of the material JUS1 and JUS2, we find that the strength of JUS1 is obviously less than that of JUS2. During curing process at temperature of 150℃ the distillation volatilized, causing the exist of voids in materials . With applied ac high voltage, the voids causes discharge , makes the concentration of the electric field, so decrease its breakdown strength.

To analyze the potential discharge of solid dielectric, generally a series connection equivalent circuit was used. Suppose the voids g is a rounded hole, the equivalent circuit of the material with voids show as Fig. 4. (the dielectric loss $\tan\delta < 5.0 \times 10^{-4}$). Cg is capacitance of voids, Cb is capacitance series with Cg. Ca is the Capacitance parallel with the Cg and Cb. If a ac voltage U is set up across the dielectric.

$$U = U_m \cdot Cos\omega t \qquad (1)$$

then

$$U = \left(1 + \frac{d-g}{\varepsilon \cdot g}\right)Ug \qquad (2)$$

considering that d/g>>1,then

For a kind of polymer, its ln ρ -T curve is approximately linear[4]:

$$\gamma=\frac{1}{\rho}=N_0 q\mu=\frac{A_0 q^2\delta^2 v}{6kT}e^{-\frac{2u_0+u_a}{kT}}=Ae^{-\frac{B}{T}} \quad (7)$$

consideration the influence of the impurity, it may be:

$$\gamma=A_1 e^{-\frac{B_1}{T}}+A_2 e^{-\frac{B_2}{T}} \quad (8)$$

the measurement of JUS2 is a broken line, which is conform to the equivalence of (8).. The unusual of the JUS1 with distillation faction is caused by the voids, which increases the equivalent resistantivity. So it is very easy to distinguish whether there is a great amount voids by measurement the relationship of ln ρ -T curve.

CONCLUSION

In this paper, influence of the distillation fraction on the characteristic of high field breakdown of polymer is studied. The distillation fraction forms voids during the process of material curing at moderate temperature, causes the initial discharge voltage decrease, and, with the increasing of ac voltage, the cycle discharge makes the electric-tree developed, causing the damage of polymer, reduces the breakdown strength. The influence of the voids made from distillation also shows the unusual change of the ln ρ -T semi-logical curve at high temperature. The polymer JUS2 without distillation, especially the distillation with no Si-H bond, express a good electrical and physical properties.

Reference

1. T. Deming, "Space charges of polymers mechanism of electrical breakdown". Journal of Xi'an Jiaotong University (Chinese). Vol. 24, No. 5, 1990 , pp109-122.
2. M. G. Danikas, " On the breakdown strength of Silicon Rubber". IEEE Transaction on Dielectric and Electrical Insulation. Vol. 1, No. 6, 1994, pp1196-1200.
3. F. Zhang. "Research and Produce Organosiloxane Polymers as Protective Materials of Power Electronic Device".(Chinese) The 5th International Power Electronic conference . Chendu; China, 1993, pp198-200.
4. Anderso J. C *Dielectric* Printed in Great Britain by Spottiswoode, Ballantyne & Co. Ltd. 1964 , pp98.

$$U=\left(1+\frac{1}{\varepsilon}\frac{d}{g}\right)Ug \qquad (3)$$

ε is the dielectric constant .

When Ug rise up ,with the increasing of U, to the initial discharge voltage of air, the discharge of cavity happened, it neutralized the changes concentrated in the surface of materials , causing the rapidly decrease of voltage on the cavity, while the Cb being charged by the Cg. As the applied voltage rise this is followed by discharge at other sites, until the whole cavity is affected. With further increase in voltage new discharges occur at already partially discharges sites.

When an alternating voltage is applied discharges will occur during each half-cycle. If we suppose that the discharge occurs just at the crest of the first positive half-cycle, and the area completely discharges, then, as the alternating voltage falls from its positive crest value, a reverse field builds up across the cavity. This would cause a new discharge near the point where the alternating voltage crosses the zero axis. A further discharge would occur then the voltage reaches negative crest value and again as it passes through zero, as illustrated in Fig. 5.

$$Ui=\left(1+\frac{1}{\varepsilon}\frac{d}{g}\right)Us \qquad (4)$$

The initial discharge voltage Ui decreases with the cavity g increase . That is the reason the initial discharge voltage of JUS1 is lower than that of JUS2.

The frequency of ac voltage is 50 Hz, causing the discharge of cavity to be observed continuously in naked eye, but before the breakdown of specimen, the voltage could be rise up.

when discharge occurs, there would be a ac voltage U applied on the material of the thickness of (d-g) mm:

$$U=E_b\times(d-g) \qquad (5)$$

Breakdown strength of air is 0.3 kV/mm. Average breakdown strength of JUS1 with voids is 15 kV, which means the discharge already occur. We calculate the breakdown voltage of JUS2 as material strength

$$E_b=\frac{U}{d}=35kV/mm \qquad (6)$$

Using (5),Thus

g=0.5-0.6 mm

In practice, although there is plenty voids disperse materials, as it is expressed as:

$$g=\sum_i g_i \qquad (6)$$

it is impossible to have that value. Thus, the cycle discharge of voids will damage the material, causing the occurrence of electrical-tree, make a large area of material damaged, decrease the breakdown strength. In our case, the voids is formed during the process of curing with the existing of distillation fraction.

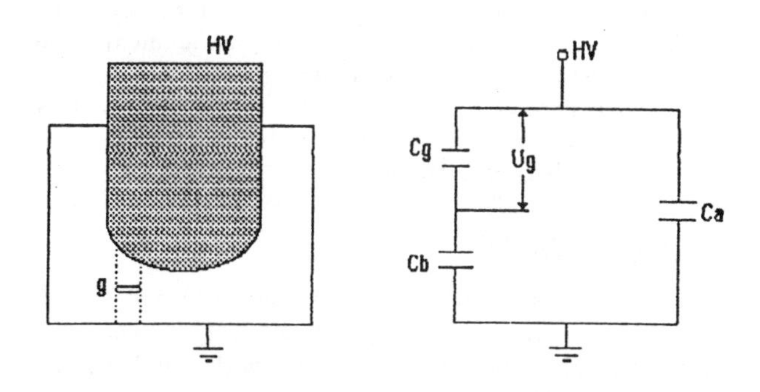

Figure 4. The effective circuit of the material with the voids forming by the distillation fraction.

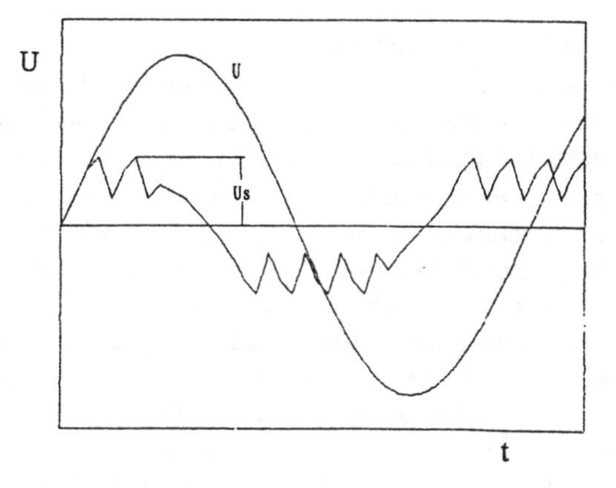

Figure 5. The change of voltage of voids when the particle discharge occurs.

Conference Record of the 1996 IEEE International Symposium on Electrical Insulation, Montreal, Quebec, Canada, June 16-19, 1996

EFFECT OF WATER CONDUCTIVITY ON ITS PULSE ELECTRIC STRENGTH

Piotr Lubicki, James D. Cross, Shesha Jayaram
University of Waterloo
Electrical & Computer Engineering Department
Waterloo, Ontario, N2L 3G1 Canada

Jacek Staron, Boleslaw Mazurek
Technical University of Wroclaw
Institute of Electrotechnics and Electrotechnology
pl. Grunwaldzki 13, Wroclaw 51, Poland

Abstract: The dependence of the impulse electrical strength of water on conductivity was investigated in non-uniform electric field - point-plate electrode system. The voltage pulses, with the rise time t_r=1 μs, and time to half-decay $t_{1/2}$= 30 μs, were generated by a Marx bank. Conductivity was varied by the use of different $CuSO_4$ concentrations in distilled water. The voltage and current during the electrical breakdown were recorded by using a pulse voltage divider and current shunt. The dynamic resistance calculated from current and voltage waveforms during electrical breakdown, is also presented. On the basis of these waveforms, the dependencies of the time to breakdown and breakdown voltage on the conductivity for two polarizations of the point electrode are plotted. Investigations made by means of static photography were carried out. It was found that the electrical strength of water depends strongly on the polarity of point electrode and the conductivity of the water. The electrical strength of water decreases, and time to breakdown increases, as water conductivity is increased.

INTRODUCTION

Water has been used in high voltage resistors and capacitors as a dielectric for pulse applications and as a medium for high voltage commutators with relatively low losses [1-5]. Phenomena accompanying its impulse breakdown have been used in high energy physics, and in technological processes (metal machining, generation of high intensity light, generation of acoustic waves used for hydroacoustic and hydrogeological investigations) since as early as 1950 [2]. Besides, discharges in water have been investigated as a possible means for inactivation of microorganisms in consumable liquids [6-9].

Depending on the purpose, whether breakdown is desirable or not, the high voltage properties of water under pulse application must be known [4,8,9].

The experiments described in this paper are part of the research undertaken to investigate the possibilities of developing a non-thermal sterilization technique using high voltage pulses. Any discharges in liquid (water) must be avoided in a wide range of water conductivity, as the quality of the sterilized liquid is affected by electrolysis, temperature rise, free radicals formation, injection of electrode materials and others [1,2,4,8,9]. The results described in the paper are useful to estimate the range of rise time, peak voltage, and the voltage polarity which could be applied for sterilization of consumable liquids without causing any undesirable electrical breakdowns.

EXPERIMENTAL SETUP

The investigations were carried out in an asymmetrical point-plate electrode system (plate grounded) with electrode spacing, d = 10 mm and the radius of the point electrode, r = 0.2 mm. Such an electrode system gives an electric field ~ 1000 kV/cm.

HV pulses were generated by using a one stage modified Marx bank [4]. The generator setup is shown in Fig. 1. The water capacitor C_2 was used to shape the front of the pulse. A rise time of T_n= 1 μs, and time to half-decay, $T_{1/2}$ = 30 μs was obtained without any sample connected across the generator terminals.

Breakdown current was measured by means of the tube shunt. The dc resistance of the shunt is equal to, R=1.1 Ω. The shunt was connected by BNC socket and 1:10 measuring probe with the KIKUSUI COM7101E oscilloscope with the sampling rate 200 MS/s.

Voltage was recorded with the use of HV pulse divider with a ratio of 2000. The accuracy of the current and voltage recording is 7%. The dynamic resistance r=f(t) was calculated on the basis of current and voltage waveforms recorded during breakdown. The waveforms obtained were also used to estimate breakdown voltage and time to breakdown depending on conductivity. The breakdown voltage is defined as a point at which the resistance rapidly decreases and the current increases to its maximum. This point is referred as the beginning of breakdown. The difference between the time at which the voltage pulse is applied and the beginning of breakdown is defined as the time to breakdown. The measurements were carried out for both polarizations of the applied voltage. Conductivity of the samples was varied by varying the concentration of $CuSO_4$ dissolved in distilled water. The conductivity was measured before and after the breakdown by means of conductometer INCO N5721 with the probe type PS-27. The investigations were carried out for the solution conductivities within the range of γ=5 mS/m to 15 mS/m. This range of conductivity matches the conductivity of consumable liquids and drinking water. Additionally, the static photograph method was applied to investigate the influence of conductivity on the development of impulse breakdown in water. In this case, the Pentacon sixTL camera was used.

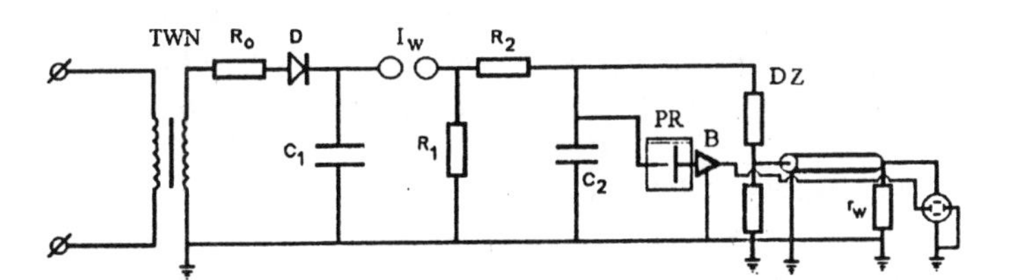

Figure 1. Experimental setup used for measurements of breakdown voltages and currents. TWN - high voltage transformer, I_w - spark gap, PR - sample, B - current shunt, DZ - voltage divider, C_2 - water capacitor C=3.7 nF, r_w - terminating resistor.

RESULTS

Figures 2-4 present the typical waveforms of current and voltage obtained for impulse breakdown of water with different conductivities for positively polarized point electrode. Figure 2a shows the waveforms obtained for water with conductivity $\gamma=2\times10^{-3}$ S/m. Breakdown voltage is approximately equal to U_b=37 kV. Time to breakdown is T_b=0.7 μs with respect to the origin of voltage waveform. The current and voltage waveforms were used to calculate the resistance-time dependence which is presented in Fig. 2b. The resistance rapidly decreases after the time of t=0.8 μs with respect to the origin of voltage waveform.

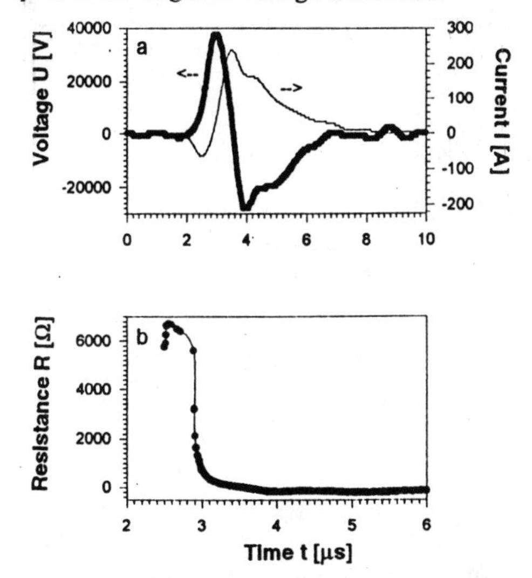

Figure 2. a) Current and voltage waveforms, b) resistance vs. time for impulse breakdown of $CuSO_4$ water solution with conductivity $\gamma=2\times10^{-3}$ S/m, positive polarization of the point electrode.

Fig. 3 shows current and voltage and dynamic changes of resistance during the breakdown of the water solution with conductivity $\gamma=2\times10^{-2}$ S/m. The voltagewaveform allows to estimate the value of breakdown voltage U_b=42 kV. The time to breakdown in this case equals to T_b=0.9 μs.

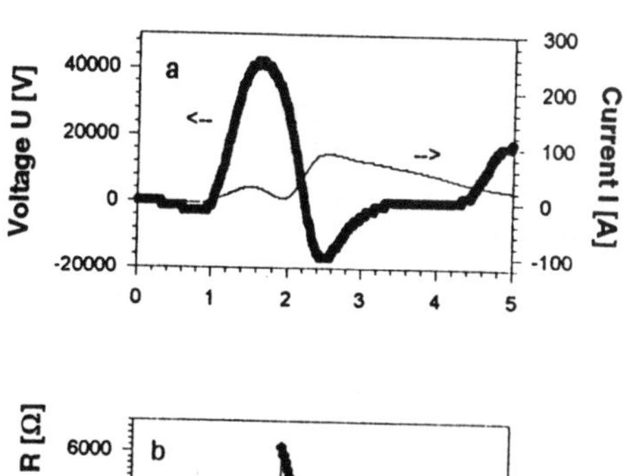

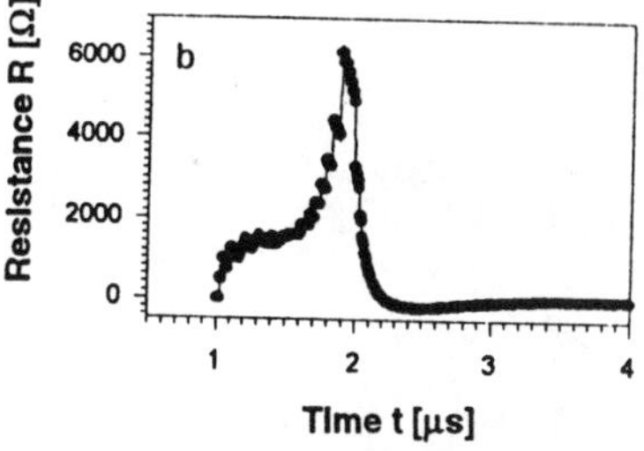

Figure 3. a) Current and voltage waveforms, b) resistance vs. time for impulse breakdown of $CuSO_4$ water solution with conductivity $\gamma=2\times10^{-2}$ S/m, positive polarization of the point electrode.

Fig. 4a shows current and voltage waveforms for positive polarization of point electrode during electrical impulse breakdown of the water solution with conductivity $\gamma=5\times10^{-2}$ S/m.

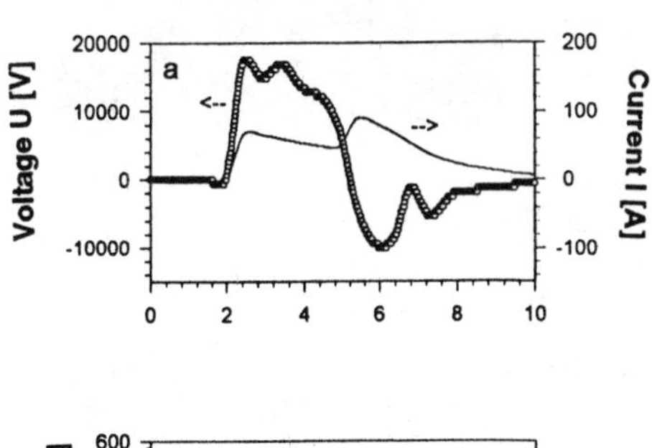

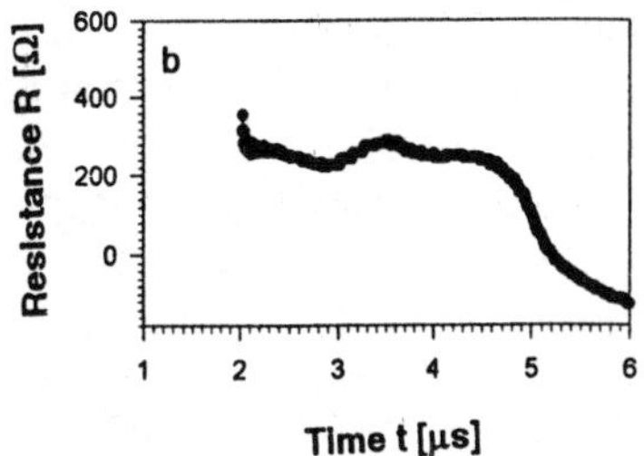

Figure 4. a) Current and voltage waveforms, b) resistance vs. time for impulse breakdown of $CuSO_4$ water solution with conductivity $\gamma=5\times10^{-2}$ S/m, positive polarization of the point electrode.

The breakdown voltage is significantly lower than in previously described cases U_b=7.5 kV, and the time to breakdown is remarkably longer T_b=2.9 μs. The current waveform differs from previous ones. The dependence r=f(t) (Fig. 4b) shows that the time to minimum value of resistance is equal to 4 μs.

Fig. 5 shows the photographs taken during the impulse breakdown of $CuSO_4$ water solutions with conductivity equal to $\gamma=2x10^{-3}$ and $5x10^{-2}$ S/m for positive polarization of the point electrode. Branched and complicated preliminary discharges and few main channels coming to plate electrode are visible (Fig. 5). The discharge channels are thin and long. This is the evidence that in this case the collision ionization processes are very fast.

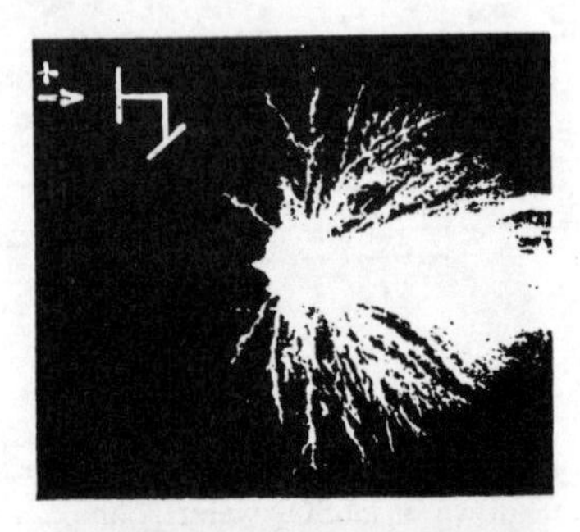

Figure 5. Photographs of electrical breakdown of $CuSO_4$ water solution with conductivity: a) $\gamma=2x10^{-3}$ S/m, b.) $5x10^{-2}$ S/m, for positive polarization of point electrode.

The second series of experiments were carried out for negatively polarized point electrode. The voltage and current waveforms and the dynamic resistance characteristics for impulse breakdown of $CuSO_4$ solutions with conductivity respectively $\gamma=2x10^{-3}$, $9x10^{-3}$, and $2x10^{-2}$ S/m for negative polarization of point electrode are shown in Fig. 6-8. The breakdown voltage decreases with an increase in conductivity. This voltage is equal to U_b=28.6 kV for water with $\gamma=2x10^{-3}$ S/m (Fig. 6a), and it decreases to U_b=3.6 kV for water with $\gamma=2x10^{-2}$ S/m (Fig. 8a). Time to breakdown is much longer than that in the case of positive polarization of point electrode. The breakdown occurs during the decay of the HV pulse. Time to breakdown is T_b=8.4 μs for water with conductivity $\gamma=2x10^{-3}$ S/m, and T_b=17.1 μs for conductivity of the solution $\gamma=2x10^{-2}$ S/m (Fig. 6-8). The dynamic resistance characteristics (Fig. 6b-8b) differ from those obtained for positive polarization of the point electrode.

The photographs of discharges in water solutions with conductivity equal to respectively $\gamma=10^{-3}$ and $5x10^{-3}$ S/m for negative polarization of the point electrode are presented in Fig. 9. The discharge channels are bigger. Branches of those channels are filled with a material displaying a different light deflection coefficient from the water solution sample. It is most likely that this is a gas phase.

DISCUSSION AND CONCLUSIONS

The results presented (Fig. 2-9) illustrate an electrical breakdown process of $CuSO_4$ water solution depending on its conductivity for both polarizations of the HV pulses applied. Because the breakdown takes place under microsecond regime, it should be thought that the bubble mechanism is the cause for the electrical breakdown in both cases of the point electrode polarization [1,2,10-11]. It was observed in works [1] and [2] that the gas bubbles in electrical impulse breakdown of water are formed in the vicinity of point electrode for its positive polarization as well as for negative one.

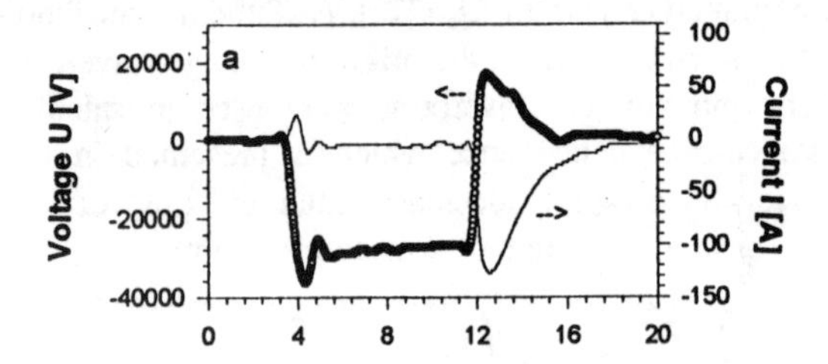

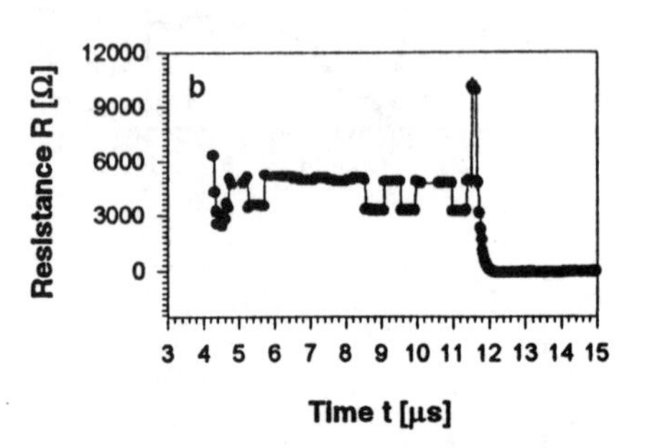

Figure 6. a) Current and voltage waveforms, b) resistance vs. time for impulse breakdown of $CuSO_4$ water solution with conductivity $\gamma=2x10^{-3}$ S/m, negative polarization of the point electrode.

A phase transition liquid-gas must take place for electrical breakdown to occur. Energetic interaction between charged particles (in the case of short duration HV pulses - mainly electrons) and bulk liquid causes energy dissipation, local heating and vaporization. In this way, the gas bubbles are created. The ionization processes well known from the

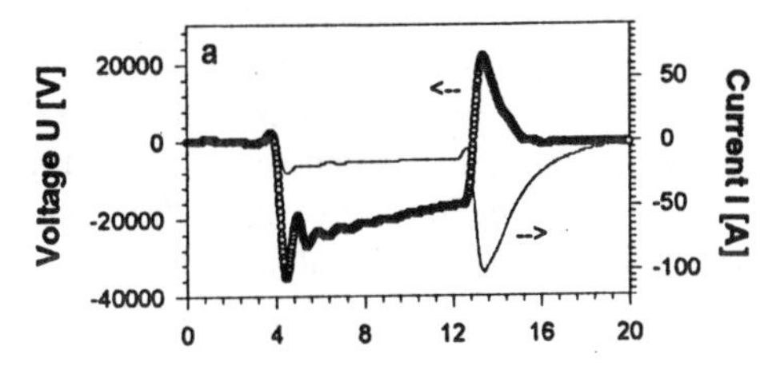

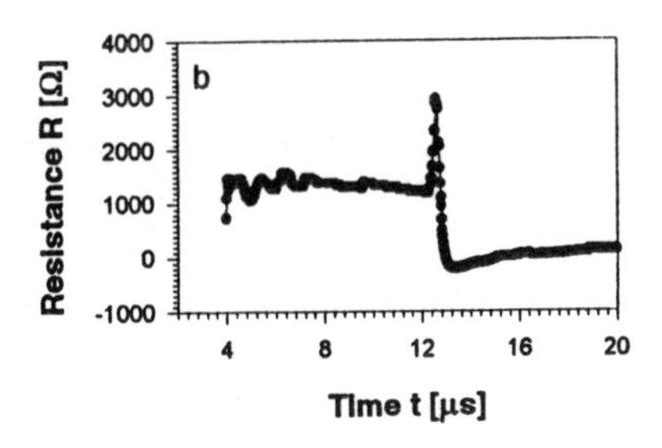

Figure 7. a) Current and voltage waveforms, b) resistance vs. time for impulse breakdown of $CuSO_4$ water solution with conductivity $\gamma=9x10^{-3}$ S/m, negative polarization of the point electrode.

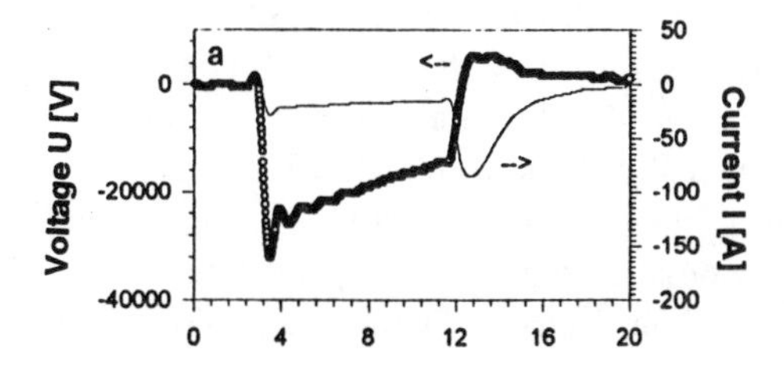

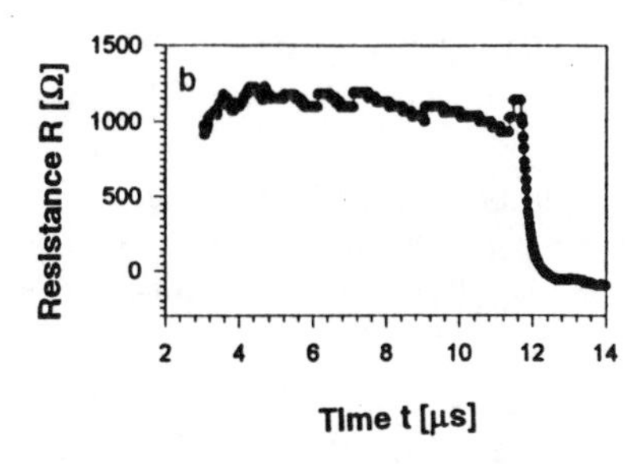

Figure 8. a) Current and voltage waveforms, b) resistance vs. time for impulse breakdown of $CuSO_4$ water solution with conductivity $\gamma=2x10^{-2}$ S/m, negative polarization of the point electrode.

theory of electrical breakdown in gases take place in those gaseous bubbles. Those areas are filled with a space charge polarization which depends on polarization of electrodes. The space charge plays a major role in discharge development, because it creates a new electric field distribution.

For positive polarization of point electrode, electrons (displaying higher mobility than ions) can be found in its vicinity. This causes an increase of electric field in this area. The higher the electric field, the faster the development of the electrical breakdown, and the lower the breakdown voltage. Microchannels surrounding the positive point electrode presented in Fig. 5 are filled with a gas formed by an interaction of electrons accelerated in electric field with insulation medium. The light from main breakdown channel lighting up the area of the discharge is diffracted on liquid-gas interfaces and makes it possible to take a photograph of the breakdown process. An increase of conductivity of the medium is connected with a higher concentration of $CuSO_4$ which dissociates in water into ions of Cu^{2+} and SO_4^{2-}. An increase of the positive ion concentration causes a decrease in

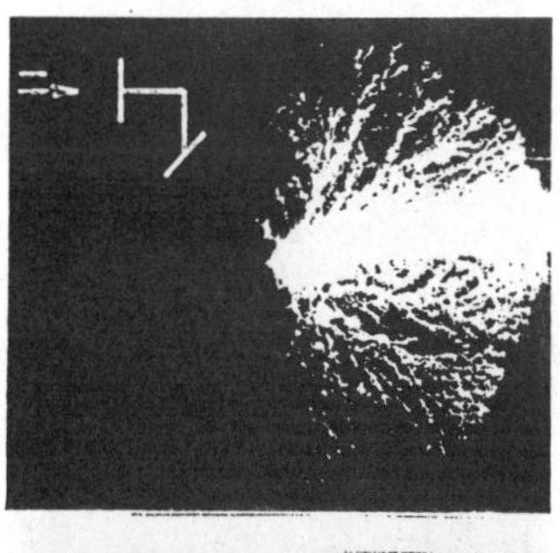

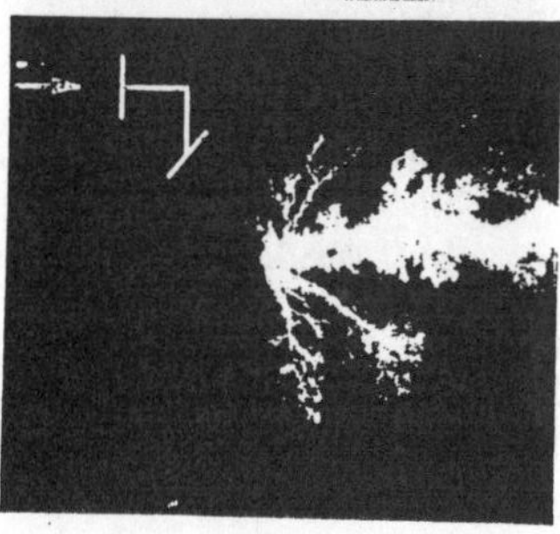

Figure 9. Photographs of electrical breakdown of $CuSO_4$ water solution with conductivity: a) $\gamma=1x10^{-3}$ S/m, b), $5x10^{-3}$ S/m, for positive polarization of point electrode.

a number of free electrons, and that is why the number of the breakdown microchannels decrease and an influence of a space charge on the electric field distribution is lowered with an increase of conductivity of the $CuSO_4$ water solution. At the same time the time to breakdown increases with an increase of conductivity (Fig. 2-4), and a vaporization process of gas bubble formation is intensified by electrolysis processes [1,2] - the number of channels decreases, but their diameters are greater (Fig. 5c).

The breakdown process also begins in the vicinity of point electrode for its negative polarization (Fig. 9), but the movement of electrons takes place towards positive plate electrode. The interaction between electrons and liquid cause an energy dissipation and formation of gas bubbles. For negative polarization of point electrode, the electrical breakdown takes place during the HV pulse decay (Fig. 6-8). This means that the impulse electric strength of water for the pulse duration being less than 4 μs and for negative polarization of point electrode is higher than the impulse electric strength in the case of positive polarization of the point electrode (Fig. 10). At the same time, as it is shown in Fig. 10, the time to breakdown is much longer than that for positive point (5-20 μs).

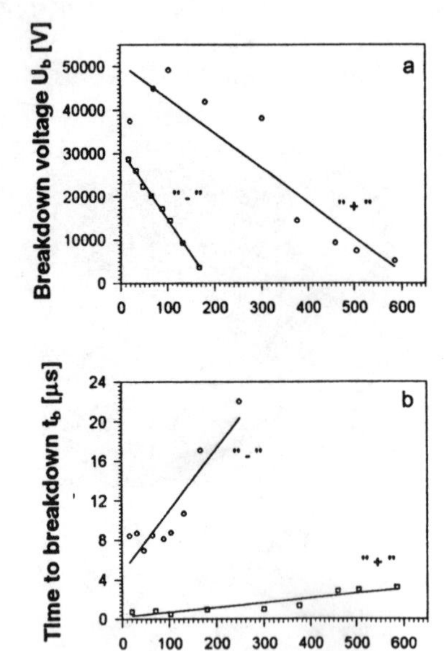

Figure 10. a.) Breakdown voltage, and b.) time to breakdown for $CuSO_4$ water solution vs. its conductivity for both polarizations of point electrode.

The photographs show that the development of breakdown for negative point is accompanied by formation of a disturbance area which can be identified as a concentration of gas microchannels and gas microbubbles (Fig. 9). This area filled with a space negative charge is "screening" the point electrode and causes the more homogeneous electric field distribution [1,2]. This is most likely to be the reason of a more difficult development of electrical breakdown for negative point electrode polarization.

Time to breakdown for positive point is significantly shorter that that for negative point. In the case of negative point, time to breakdown remarkably increases even with a slight increase of conductivity. For both polarizations of the point electrode, the impulse electrical strength of $CuSO_4$ water solution decreases with an increase of its conductivity.

The typical range of conductivity of drinking water sources and consumable liquids is 10^2-10^3 μS/cm [4]. On the basis of the presented results of the impulse electrical strength investigations, it can be concluded that HV pulses used for drinking water disinfection and for consumable liquids sterilization should be negatively polarized, and the duration of the pulse should not exceed 4 μs. Then the breakdown risk is lowered and, as a consequence, the degradation of the quality of a sterilized liquid can be avoided even for a very high electric field.

1. W. Ya. Ushakow, *Impulsnyî elektricheskiî proboî zhidkosteî*, Izdatelstvo Tomskovo Universiteta, Tomsk, 1975.

2. K. A. Naugolnykh, N. A. Roy., *Elektricheskie razryady v vode*, Moscow, Izdatelstvo Nauka, 1971.

3. N. G. Beruchev, E. P. Bol'shakov, V. V. Vecherkovskii, Yu. A. Istomin, F. A. Fedorov, "Iskra 4-MV test bench", Instruments and Experimental Techniques Vol. 33, 5(1990)1111.

4. P. Lubicki, B. Mazurek, Z. Staroniewicz, "Effect of short duration high voltage pulses on Gram-positive, Gram-negative bacteria and yeastlike fungi", IEEE Transactions on Dielectrics and Electrical Insulation 2(1995)418.

5. N. F. Kovsharov, N. A. Ratakhin, V. F. Feduschchak, "2-MV water-insulated commutator", Instruments and Experimental Techniques, Vol. 33, 2(1990)359.

6. Li Zheng-ying, Wang Yan, "Effects of high voltage pulse discharges on microorganisms suspended in liquid", 8th Int. Symp. on High Voltage Engineering, Yokohama, Japan, (1993)551.

7. Y. Matsumoto, N. Shioji, T. Satake., A. Sakuma, "Inactivation of microorganisms by pulsed high voltage application", IEEE Trans. on Industry Applications, 25(1991)652.

8. Jayaram S., Castle G. S. P., Margaritis A., "Influence of liquid conductivity and dc pulse width on high field interaction of *Lactobacillus brevis*", Applied Microbiology and Biotechnology 40(1993)117.

9. Palaniappan S., Sastry S. K., Richter E. R., "Effect of Electricity on Microorganisms: A Review", Journal of Food Processing and Preservation 14(1990)393.

10. Szklarczyk M., "Electrical Breakdown of Liquids", in *Modern Aspects of Electrochemistry*, edited by B. E. Conway, J. O'M. Bockris, R. E. White, Plenum Press, New York, 25(1993)253.

11. Kuskova N. I., "Mechanisms of Electrical Breakdown in Water", Soviet Technical Physics Letters 15(1989)936.

International Symposium on Electrical Insulation, Montreal, Quebec, Canada, June 16-19, 1996

ctrical tree growth in solid insulating materials using cellular automata

dis, A. Thanailakis
y of Thrace
omputer Engineering
GREECE

A. M. Bruning
Lectromechanical Design Co.
101-H Executive Drive, Sterling
VA 20166-9557, USA

Ar.
15/82

Abstra... the breakdown mechanisms of the solid insulating materials are based, among others, on electromagnetic theory, avalanche theory and fractals. In this paper the breakdown of insulating materials is simulated using von Neumann's Cellular Automata (CAs). A algorithm for solid dielectric breakdown simulation based on CAs is presented with a point/plane electrode arrangement. The algorithm is also used to simulate breakdown in a solid dielectric having a spherical void.

INTRODUCTION

Solid dielectric breakdown depends on the experimental arrangement and conditions, the nature and morphology of the insulation under test as well as on the form and type of the applied voltage. Breakdown due to solely electrical causes arises from processes that are not in evidence until very close to breakdown [1]. Some authors have pointed out the importance of space charges at the crystalline-amorphous boundary for breakdown [2]. Yet others proposed a cumulative model of breakdown in solid dielectrics, according to which charge carriers of low energies can extend pre-existing defects and increase their density so that clusters of interacting defects will result. These clusters can grow into macroscopic defects in the direction of the applied field which in turn will form a single channel of high probability of continuous conduction between the electrodes [3].

The aim of the present paper, is to model the electrical tree growth in solid dilectrics with a point/plane electrode arrangement using von Newmann's CAs [8]. The behaviour of a solid dielectric containing a spherical void is also investigated and the possibility of tree initiation from such a void is discussed. A algorithm for dielectric breakdown simulation based on CAs was developed, in the framework of this research work. The algorithm has been used to simulate solid dielectric breakdown with a point/plane electrode arrangement. The algorithm has also been used to simulate breakdown in a solid dielectric having a spherical void. The algorithm is fast and numerically stable. Computation time goes roughly as $O(N^3)$,where N is the number of CA cells in one dimension. The results of the simulation are in good agreement with physical appearance of trees shown by experimental photographs.

TREEING

The phenomenon of treeing is often a necessary prelude to a total breakdown. In the context of the present work, we are interested in electrical trees. The significance of trees for polymeric insulation became evident since the classical studies by Vahlsrom et al., who investigated polyethylene cables of 15 kV and 22 kV which have been in service [4,5]. Void formation plays a role in tree formation in epoxy resin [6]. A tree channel can be initiated due to pure partial breakdown of the polymer. Tree propagation can be induced by internal gas discharge in the tree. The electric potential of the point electrode is transferred to the tip of the tree channel through the conductive plasma of gas discharge [7]. Treeing apparently plays a critical role in the process of breakdown in solid dielectrics. It is thus understood that the simulation of the electrical breakdown in solid dielectrics presupposes the simulation of the phenomenon of treeing.

CELLULAR AUTOMATA

CAs, first introduced by von Neumann [8], have been applied to several physical problems, where local interactions are involved [9-12]. In spite of the simplicity of their structure, CAs exhibit complex dynamical behaviour and can describe many physical systems and processes. CAs are idealisations of physical systems in which space and time are discrete and

interactions are local. A CA consists of a regular uniform n-dimensional lattice (or array).
At each site of the lattice (cell) a physical quantity takes values. This physical quantity is the global state of the CA, and the value of this quantity at each cell is the local state of this cell. Each cell is restricted to local neighbourhood interaction only, and as a result it is incapable of immediate global communication.

	(i-1, j-1)	(i-1, j)	(i-1, j+1)	
	(i, j-1)	(i, j)	(i, j+1)	
	(i+1, j-1)	(i+1, j)	(i+1, j+1)	

Figure 1. The neighbourhood of the (i,j) cell is formed by the same and the eight marked cells.

The neighbourhood of a cell, shown in Figure 1, is taken to be the cell itself and some of (or all) the immediately adjacent cells. The states at each cell are updated simultaneously at discrete time steps, based on the states in their neighbourhood at the preceding time step. The algorithm the cell uses to compute its successor state is referred to as the CA local rule. Usually the same local rule applies to all cells of the CA. The state of a cell at time step t+1 is affected by the states of all eight cells in its neighbourhood at time step t, and by its own state at time step t:

$$C^{t+1}_{i,j} = F(C^{t}_{i-1,j-1}, C^{t}_{i-1,j}, C^{t}_{i-1,j+1}, C^{t}_{i,j-1}, C^{t}_{i,j}, C^{t}_{i,j+1}, C^{t}_{i+1,j-1}, C^{t}_{i+1,j}, C^{t}_{i+1,j+1}) \quad (1)$$

This function is the CA local rule. $C^{t}_{i,j}$ and $C^{t+1}_{i,j}$ are the states of the (i,j) cell at time steps t and t+1, respectively.

SIMULATION

The solid dielectric is divided into a matrix of identical square cells, with side length a, and is represented by a CA, by considering each cell of the dielectric as a CA cell. The algorithm which will be presented has been used to simulate dielectric breakdown with a point/plane electrode arrangement. The voltage applied was taken to be equal to 20 kV and the distance between the tip of the point electrode and the plane electrode was taken equal to 5 mm. The non-homogeneous dielectric constant varied randomly between 2.1 and 2.25.

In order to determine the potential distribution before the onset of tree formation, Laplace equation is solved numerically using the finite differences method [12]. The potential at the point electrode is taken equal to the applied voltage and the potential at the plane electrode is taken equal to zero. Figure 2 shows the initial potential distribution for the point/plane electrode arrangement. The value of the potential at the center of each CA cell is taken to be the potential of this cell.

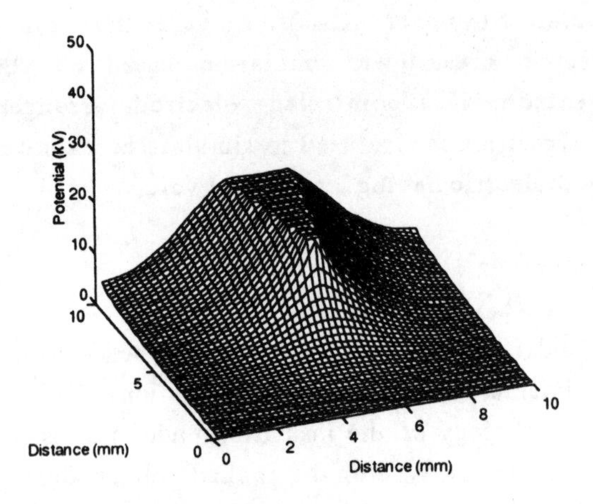

Figure 2 The initial potential distribution.

We assume that at the end of the tip of the point electrode the maximum electric stress E_{max} is given from:

$$E_{max} = \frac{2d\,E_{avg}}{r\left(ln\left(1+\frac{4d}{r}\right)\right)} \quad (2)$$

where, d is the electrode gap spacing, r the radius of the tip of the point electrode and E_{avg} the average electric stress applied to the gap spacing. We further assume that the value of E_{max} is transferred from the tip of the electrode to the end of the treeing tip i.e. the tip forms a conducting extension of the point electrode. This view is based upon the supposition that the electric stress at the end of the tip quite often approaches the intrinsic dielectric strength of the material and that progressive breakdown can occur by destruction of the material. Because of tree advancement, the potential distribution into the insulating material changes with time and should be calculated at each time step. The potential distribution into the dielectric does not vary significantly at two successive time steps (as shown in Figures 3 and 4), and therefore the values of the potential at each CA cell at time step n were used as initial values for the calculation of the potential at time step t+1, resulting in significant reduction of the number of interations. Figures 3 and 4 show the potential distribution at two successive time steps.

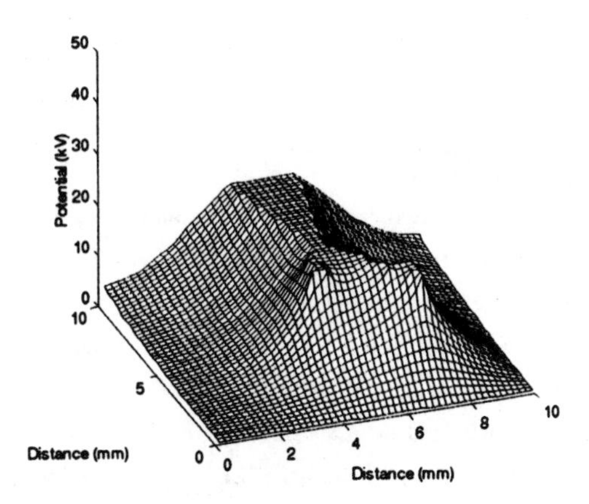

Figure 3 Potential distribution at time step t = n.

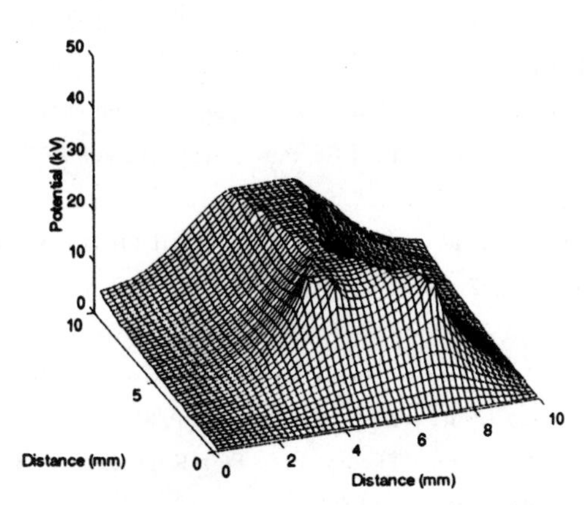

Figure 4 Potential distribution at time step t = n+1.

Every CA cell can be considered as an elementary parallel plane electrode arrangement. The state of a cell which does not belong in the tree structure is defined to be 0, whereas the state of a cell belonging to the tree structure is defined to be 1. The neighbourhood of each cell of the CA ((i,j) cell) is formed by the cells shown in Figure 1.

The simulation presented in this paper is based on the inherently small fluctuation of the dielectric constant of the insulating material under investigation. The dielectric constant of the dielectric between the elementary electrodes can take values from a certain value range. We can thus assume that, because of the inherent variation of the dielectric constant, in an elementary square two dielectric materials exist with slightly different dielectric constants. Remembering some high voltage theory the following equation can apply for the electric field E developed at the tip of the tree (i.e. at (i,j) cell) in the direction towards some other cell in the neighbourhood (for example the (i+1,j+1) cell), at time step t :

$$E^t_{i+1,j+1} = \varepsilon_{i+1,j+1} \frac{V^t_{i+1,j+1}}{a_{i+1,j+1}} \tag{3}$$

where, $\varepsilon_{i+1,j+1}$ is the permittivity of the material at the (i+1,j+1) cell, $V^t_{i+1,j+1}$ is the potential at the center of this cell at time step t and $a_{i+1,j+1}$ is the distance between the centers of the (i+1,j+1) and (i,j) cell. It is also based on the assumption that the electric field at the end of the tip is comparable to that developed at the end of the point electrode. This assumption is in agreement with the other assumption we make, namely, that the insulating material is inherently inhomogeneous. Here is also assumed that the tree will progress taking into account the local dielectric strength. The tree path follows the path where the insulating material is locally weakest..

In order to simulate the tree propagation the following CA rule has been applied :

- If the state of the (i,j) cell is 1 at time step (t), it will not change at time step (t+1).
- If the state of the (i,j) cell is 0 at time step (t) *and* none of its neighbours is at state 1, then the state of the (i,j) cell will not change at time step (t+1).
- If the state of the (i,j) cell is 0 at time step (t) *and* one or more of its neighbours is at state 1, then the state of the (i,j) cell will change to 1 at time step (t+1) only if the local electric field, which is calculated using (3), is greater than a certain value E_C. (E_C represents the local dielectric strength of the dielectric).

The results of the simulation are shown in Figure 5. Figure 6 shows the simulation results of tree formation in the same dielectric used in Figure 5 when it contains a spherical void. As the tip of the tree approaches the void the potential difference between points at the void perimeter increases resulting in electrical discharges which onset a secondary tree as shown in Figure 6.

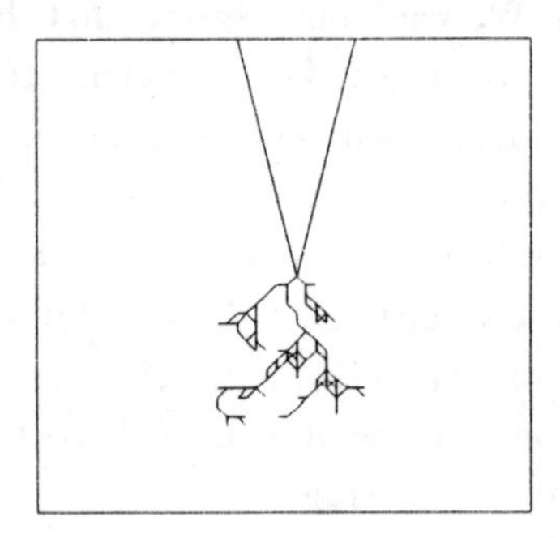

Figure 5. Tree formation in a dielectric.

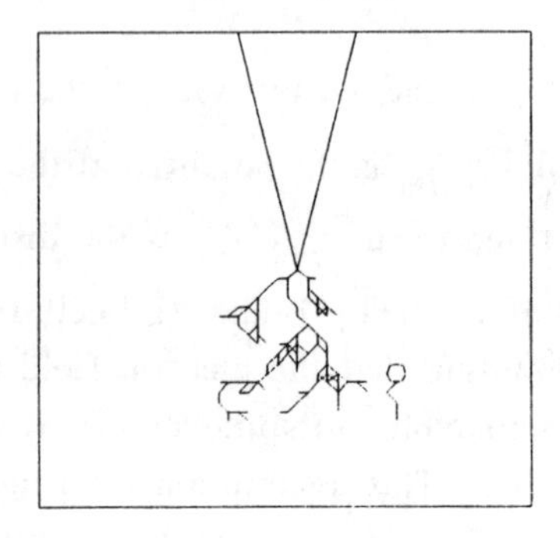

Figure 6. Tree formation in the same dielectric as in Figure 5 but containing a spherical void

The formation of treeing from a spherical void is not surprising because we know that as discharges take place inside a spherical void, a weakly conducting layer is created with time on the wall surfaces. This conducting layer - consisted of the by-products of the discharges - is by no means uniform throughout the wall surfaces of the void. This layer contributes to the temporary diminution or even cessation of the partial discharges. Consequently, sites on the wall surfaces which are not covered by the weakly conducting layer will most likely be sources of further partial discharge activity. The end result will be the initiation of treeing from the walls of the void. This explanation is physically sound and in accordance with the theories developed until now about the behaviour of the discharges in a void.

The algorithm is fast and numerically stable. The presented algorithm run on an i486/66 computer and the computation time was 2 min and 35 sec for a 50X50 CA grid. Computation time was found to go roughly as $O(N^3)$,where N is the number of CA cells in one dimension.

CONCLUSIONS

A simulation of the breakdown of insulating materials using CAs was presented in this paper. Using a very simple rule the tree formation was successfully reproduced in a dielectric with a point/plane electrode arrangement. The change with time of the potential distribution into the insulating material has been taken into account and the potential distribution was calculated at each time step. Tree formation has also been successfully reproduced in the case where a a spherical void is present in the dielectric.

REFERENCES

1 J.J. O'Dwyer, "Breakdown in solid dielectrics", IEEE Trans. Elec.Insul., vol. 17, pp. 484-487, 1982.

2 Y. Inuishi, "Effect of space charge and structure on breakdown of liquids and solids", Ann. Rep. Conf. Elec. Insul. Diel. Phen., Amherst/USA, pp. 328-338, 1982.

3 A.K. Jonscher and R. Lacoste, "On a cumulative model of dielectric breakdown in solids", IEEE Trans. Elec. Insul., vol. 19, pp. 567-577, 1984.

4 W. Vahlstrom, Jr., "Investigation of insulation deterioration in 15 kV and 22 kV polyethylene cables removed from service", IEEE Trans. Power App. Sys., vol. PAS-91, no. 1-3, pp. 1023-1035, 1972.

5 J.H. Lawson and W. Vahlstrom, Jr., "Investigation of insulation deterioration in 15 kV and 22 kV polyethylene cables removed from service - Part II", IEEE Power App. Sys.,vol. PAS-92, no. 1-3, pp. 824-835, 1973.

6 Y. Shibuya, "Void formation and electrical breakdown in epoxy resin", IEEE Trans. Power App. Sys., vol. PAS-96, no. 1, pp. 198-206, 1977.

7 M. Ieda, "Dielectric breakdown process of polymers", IEEE Trans. Elec. Insul., vol. 15, pp. 206-224, 1980.

8 J. von Neumann "Theory of Self-Reproducing Automata" University of Illinois, 1966

9 M. Gerhard, H. Schuster "A Cellular Automaton describing the formation of spatially ordered structures in chemical systems" Physica D, vol 36, pp. 209-221, 1989.

10 M. Gerhard, H. Schuster, J.J. Tyson "A Cellular Automaton model for Excitable Media" Physica D, vol 46 , pp. 392-415, 1990

11 I. Karafyllidis, A. Thanailakis, "Simulation of two-dimenional photoresist etching process in integrated circuit fabrication using cellular automata", Modelling and Simulation in Matterial science and Engineering, vol. 3, pp 629-642, 1995

12 A. M. Bruning, "Design of electrical insulation equipment", Ph. D. Diss., Univ. of Missouri-Columbia, 1984.

Author's Index

I

J

K

L

M

N

O

P

Q

R

S

T

U

V

W

X

Y

Z

Notes

Notes

Notes

Notes